Neurobiology and Clinical Aspects of the Outer Retina

The front cover illustration shows a transverse section of the parafocal area of a monkey's outer retina. The inserts on the back cover are tangential views of the outer nuclear layer. (Courtesy of Dr Ron Douglas, photomicrography by Sinclair Stammers.)

Neurobiology and Clinical Aspects of the Outer Retina

Edited by

M.B.A. Djamgoz

Professor of Neurobiology
Department of Biology
Imperial College of Science, Technology and Medicine
London
UK

S.N. Archer

Senior Research Fellow
International Marine Centre
Oristano
Italy

and

S. Vallerga

Senior Researcher
Institute of Cybernetics and Biophysics
Italian National Research Council
Genoa
Italy

CHAPMAN & HALL
London · Glasgow · Weinheim · New York · Tokyo · Melbourne · Madras

Published by Chapman & Hall, 2–6 Boundary Row, London SE1 8HN, UK

Chapman & Hall, 2–6 Boundary Row, London SE1 8HN, UK

Blackie Academic & Professional, Wester Cleddens Road, Bishopbriggs, Glasgow G64 2NZ, UK

Chapman & Hall GmbH, Pappelallee 3, 69469 Weinheim, Germany

Chapman & Hall USA, 115 Fifth Avenue, New York, NY 10003, USA

Chapman & Hall Japan, ITP-Japan, Kyowa Building, 3F, 2–2–1 Hirakawacho, Chiyoda-ku, Tokyo 102, Japan

Chapman & Hall Australia, 102 Dodds Street, South Melbourne, Victoria 3205, Australia

Chapman & Hall India, R. Seshadri, 32 Second Main Road, CIT East, Madras 600 035, India

First edition 1995

Typeset in 10/12pt Palatino by Photoprint, Torquay, Devon

Printed in Great Britain by The Alden Press, Oxford

ISBN 0 412 60080 3

A catalogue record for this book is available from the British Library

Library of Congress Catalog Card Number: 95–67595

∞ Printed on acid-free text paper, manufactured in accordance with ANSI/NISO Z39.48–1992 (Permanence of Paper).

Contents

Contributors

Andrea Antal
Division of Neurology
University of Nebraska Medical Center
660 S 42nd Street
Omaha
NE 68198–2045
USA

Eckart Apfelstedt-Sylla
Department of Neuro-ophthalmology and Pathophysiology of Vision
University Eye Hospital
Schleichstrasse 12–16
72076 Tübingen
Germany

Simon Archer
International Marine Centre
Lungomare Eleonora d'Arborea 22
09072 Torregrande
Oristano
Italy

Steven Barnes
Neuroscience Research Group and Lions Sight Centre
University of Calgary
Faculty of Medicine
3330 Hospital Drive N.W.
Calgary
Alberta
Canada T2N 4N1

Alan C. Bird
Institute of Ophthalmology
Moorfields Eye Hospital
Department of Clinical Ophthalmology
City Road
London EC1V 2PD
UK

Ivan Bodis-Wollner
Department of Neurology
SUNY Health Science Center at Brooklyn
450 Clarkson Avenue
Brooklyn
NY 11203–2098
USA

Mustafa B.A. Djamgoz
Department of Biology
Neurobiology Group
Imperial College of Science, Technology and Medicine
Prince Consort Road
London SW7 2BB
UK

Mark W. Hankins
Gunnar Svaetichin Laboratory
Imperial College at Silwood Park
Ascot
Berkshire SL5 7PY
UK

P. Michael Iuvone
Emory University School of Medicine
Department of Pharmacology
Atlanta
GA 30322–3090
USA

Satoru Kawamura
Department of Physiology
Keio University, School of Medicine
Shinano-machi 35
Shinjuku-ku
Tokyo 160
Japan

Helga Kolb
John A. Moran Eye Center
University of Utah School of Medicine
Salt Lake City
UT 84132
USA

Ellis R. Loew
Physiology
Cornell University
Ithaca
NY 14853
USA

Robert E. Marc
John Moran Eye Center
University of Utah Health Services Center
50 North Medical Drive
Salt Lake City
UT 84132
USA

Ann H. Milam
Department of Ophthalmology RJ-10
University of Washington
Seattle
Washington
USA

Ralph Nelson
Laboratory of Neurophysiology
National Institute of Neurological Disorders and Stroke
National Institutes of Health
Bethesda
MD 20892
USA

Marco Piccolino
Instituto di Fisiologia Generale
Università di Ferrara
Via L. Borsari 46
44100 Ferrara (Italy)
and
Instituto di Neurofisiologia del CNR
Via S. Zeno 51
56110 Pisa
Italy

Rukmini Rao
University of Pennsylvania
Department of Neuroscience
Philadelphia
PA 19104–6058
USA

Pamela A. Raymond
Department of Anatomy and Cell Biology
University of Michigan Medical School
Ann Arbor
MI 48109–616
USA

Andreas Riechenbach
Carl Ludwig Institute of Physiology
Leipzig University
Liebigstrasse 27
D 04103
Leipzig
Germany

Stephen R. Robinson
Vision, Touch and Hearing Research Centre
The University of Queensland
Brisbane
Australia

Klaus Rüther
Department of Neuro-ophthalmology and Pathophysiology of Vision
University Eye Hospital
Schleichstrasse 12–16
72176 Tübingen
Germany

Richard Shiells
Department of Physiology
University College London
Gower Street
London WC1E 6BT
UK

Robert G. Smith
University of Pennsylvania
Department of Neuroscience
Philadelphia
PA 19104–6058
USA

Peter Sterling
University of Pennsylvania
Department of Neuroscience
Philadelphia
PA 19104–6058
USA

Masao Tachibana
Department of Psychiatry
Faculty of Letters
The University of Tokyo
7–3–1 Hongo
Bunkgo-ku
Tokyo 113
Japan

Noga Vardi
University of Pennsylvania
Department of Neuroscience
Philadelphia
PA 19104–6058
USA

Hans-Joachim Wagner
Eberhard-Karls-Universität Tübingen
Anatomisches Institut Osterberstrasse 3
D 72–74 Tübingen
Germany

Paul Witkovsky
Department of Ophthalmology
New York University Medical Center
550 First Avenue
New York
NY 10016
USA

Stephen Yazulla
Department of Neurobiology and Behavior
University of Stony Brook
Stony Brook
NY 11794–5230
USA

Eberhart Zrenner
Department of Neuro-ophthalmology and Pathophysiology of Vision
University Eye Hospital
Schleichstrasse 12–16
72076 Tübingen
Germany

Preface

This book deals with the cellular biology, biochemistry and physiology of photoreceptors and their interactions with the second-order neurons, bipolar and horizontal cells. The focus is upon the contributions made by these neurons to vision. Thus the basic neurobiology of the outer retina is related to the visual process, and visual defects that could arise from abnormalities in this part of the retina are highlighted in the first 16 chapters. Since all vertebrate retinas have the same basic structure and physiological plan, examples are given from a variety of species, with an emphasis upon mammals, extending to human vision. The last four chapters approach the problem from the other end. This part of the book covers a range of clinical conditions involving visual abnormalities that are due to cellular defects in the outer retina. Although the contents of this book do not represent the proceedings of a conference, the concept arose at an international symposium on 'Recent Advances in Retinal Research' which was held at the International Marine Centre in Oristano, Sardinia. We hope that the book will give a coherent, up to date review of the neurobiology and clinical aspects of the outer retina and encourage further integration of these areas. Retinal neurobiology has been an intense field of investigation for several decades. More recently, it has seen significant advances with the application of modern techniques of cell and molecular biology. We hope that the book will be instrumental in the ultimate exploitation of this knowledge in clinical practice.

In trying to achieve these aims, we depended on numerous people to whom we are grateful. We particularly appreciated the patience and co-operation of all the contributing authors, especially at times when we pressed them to make modifications to their chapters in order to maximize the integration of the neurobiological and clinical aspects of the topics being covered, and to meet deadlines. In choosing a restricted group of authors to realize the envisaged product, we inadvertantly could not involve many other experts in the field. However, we trust that all major findings essential to the philosophy of the book have been sufficiently represented. Finally, we would like to thank Rachel Young and staff at Chapman & Hall for their unfailing advice and support (and patience) in realizing the aims of the project.

M.B.A. Djamgoz, S.N. Archer and S. Vallerga
London and Oristano
1994

1

Development and morphological organization of photoreceptors

PAMELA A. RAYMOND

1.1 INTRODUCTION

A common theme in developmental biology is that orderly structures arise out of initially unorganized assemblages of undifferentiated, proliferating cells, and that cells with specific identities often become organized into highly regular patterns of repeating units. A hallmark of neuronal organization is the partitioning or clustering of neurons into repeating functional modules that are collected into more complex structures. In the vertebrate retina, this type of modular organization is of paramount importance for the proper functioning of the neural circuitry, and regular mosaics are found among all classes of neurons and at all levels (Wässle and Reimann, 1978; Marc, 1986). The general organizational principle of the retina is that major categories of neurons are stratified, like the sheets in a layer cake, and within a given lamina, subtypes of neurons are distributed with a regular spacing, typically abutting but often not overlapping one another, like the tiles on a floor. Photoreceptors, the subject of this chapter, form the outermost layer at the apical surface of the retinal epithelium. In many vertebrate species, and most notably in teleost fish, the cone photoreceptors are arrayed in a two-dimensional crystalline lattice of highly regular and repeating mosaic units each with three or four different types of cones at specific positions in the array.

Photoreceptors in the insect compound eye, which has been studied extensively, are similarly arranged in a regular array, where the repeating unit is an ommatidium containing eight different photoreceptor types (numbered R1 to R8). In developing *Drosophila* ommatidia, genetic analysis has shown convincingly that the commitment of presumptive receptor cells to a specific identity involves a series of inductive events in which the spacing and pattern of the ommatidial units is first established by the early commitment of the R8 photoreceptors. The other photoreceptors then follow in a precise and orderly sequence that depends on inductive signals passed from the committed cells to their uncommitted neighbors. One of the objectives of this chapter is to compare what is known about patterning events that underlie the formation of photoreceptor mosaics in *Drosophila* and in vertebrates, especially fish

Neurobiology and Clinical Aspects of the Outer Retina
Edited by M.B.A. Djamgoz, S.N. Archer and S. Vallerga
Published in 1995 by Chapman & Hall, London
ISBN 0 412 60080 3

and mammals, and to examine the possibility that these superficially similar developmental processes might in fact involve common cellular and molecular mechanisms.

The mechanisms of cell determination in the vertebrate nervous system are not well understood, although lateral inductive interactions among differentiating cells are thought to be of primary importance in determining the phenotypic identity of individual neurons (McConnell, 1991). There is growing evidence that the photoreceptor mosaic in vertebrates might arise from an initial 'prepattern' laid down by the early commitment and differentiation of a specific subtype of cone photoreceptor, much like the situation in the insect compound eye. This chapter deals with the development and morphological organization of vertebrate photoreceptors and presents a hypothetical model for pattern formation in the developing teleost cone mosaic.

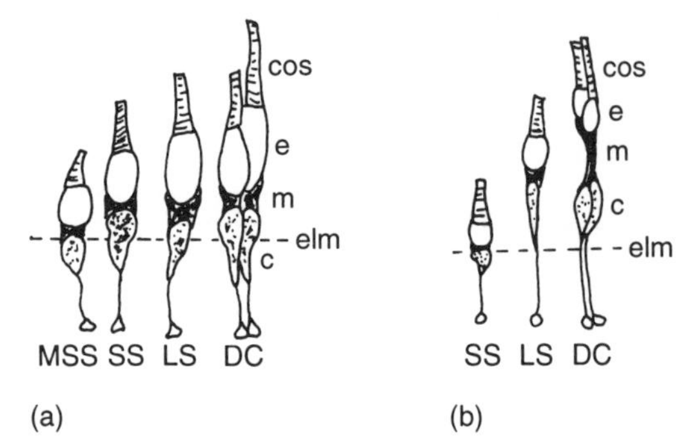

Figure 1.1 Morphological subtypes of cones in goldfish and zebrafish. (a) Goldfish. Cone types include double cones (DC), long single (LS), short single (SS) and miniature short single (MSS) cones. (b) Zebrafish cones include DC, LS and SS cones. Based on microspectrophotometric measurements of pigments and *in situ* hybridization of cone opsin probes, the double cones in goldfish and zebrafish are homologous, with the principal (longer) member containing red and the accessory member containing a green pigment. The blue cones are the SS cones in goldfish and the zebrafish LS cones. The goldfish MSS and zebrafish SS cones are ultraviolet receptors. Cone outer segment, cos; ellipsoid, e; myoid, m; cone nucleus, c; external limiting membrane, elm. (Reprinted with permission from Raymond *et al.*, 1993.)

1.2 PHOTORECEPTORS AND CONE MOSAICS

1.2.1 CLASSICAL MORPHOLOGICAL DESCRIPTIONS AND IDENTIFICATION OF SPECTRAL SUBCLASSES

Historically, vertebrate photoreceptors have been classified into two categories, rods and cones, on the basis of morphological, biochemical and functional criteria (Cohen, 1972). In most species, the retina contains a single type of rod and variable numbers of cone types containing visual pigments with different spectral absorption maxima. It is the multiplicity of cone types that underlies the ability to distinguish different wavelengths of light and therefore mediates color vision (Walls, 1967).

The cone photoreceptors in teleost fish are arrayed with geometric precision in a two-dimensional lattice (Lyall, 1957a; Engström, 1960, 1963; Ahlbert, 1968). Most fish have several morphologically distinct cone types, including single and double cones (Figure 1.1). Double cones, a consistent feature of teleost retinas, typically consist of two individual cones fused along their inner segments (Lyall, 1957a; Engström, 1960, 1963; Walls, 1967; Ahlbert, 1968). Although the functional significance of double cones is uncertain, it has been suggested that they mediate sensitivity to polarized light (Cameron and Pugh, 1991).

In general, the single and double cones are segregated into repeating patterns, which vary somewhat among different species and during ontogeny of a given species. The 'primitive' plan consists of single and double cones arranged in alternating centripetal rows* with the two kinds of single cones alternating along the rows (Figure 1.2a). These rows are orthogonal to the retinal margin, *i.e.* they intersect the circumferential boundary of the retina at a right angle (Lyall, 1957a; Larison and BreMiller, 1990). The prototype of the teleost cone mosaic consists of a repeating, square or rhomboid mosaic pattern with four pairs of double cones arranged orthogonally around a central single cone; additional single cones of a different morphological type are at each corner (Figure 1.2b). This square pattern arises out of the row pattern by a gradual shifting and rotation of the elements (see below). The degree to which the mosaic in a given retina consists of the row pattern versus the square pattern varies with both the maturity of the photoreceptors (the row pattern being the immature form) and with the species (Müller, 1952; Lyall, 1957b; Engström, 1960, 1963; Ahlbert, 1968). In more visual teleost species, including many marine fish (especially reef fishes) as well as certain freshwater groups such as percids (sunfish, perch), cichlids (gouramis, betas) and poecilids (guppies, platys and swordtails), the square mosaic pattern predominates (Engström, 1963; Ahlbert, 1968). At the periphery of the adult retina in goldfish (*Carassius auratus*), where new cones are being added continuously (Johns, 1977), and in larval fish, the row pattern is found, whereas the square arrangement occurs in central (more mature) regions of the adult goldfish retina. In contrast, in a related cyprinid, the zebrafish (*Brachydanio rerio*), the row pattern is retained throughout the adult retina (Engström, 1960; Larison and BreMiller, 1990). A number of

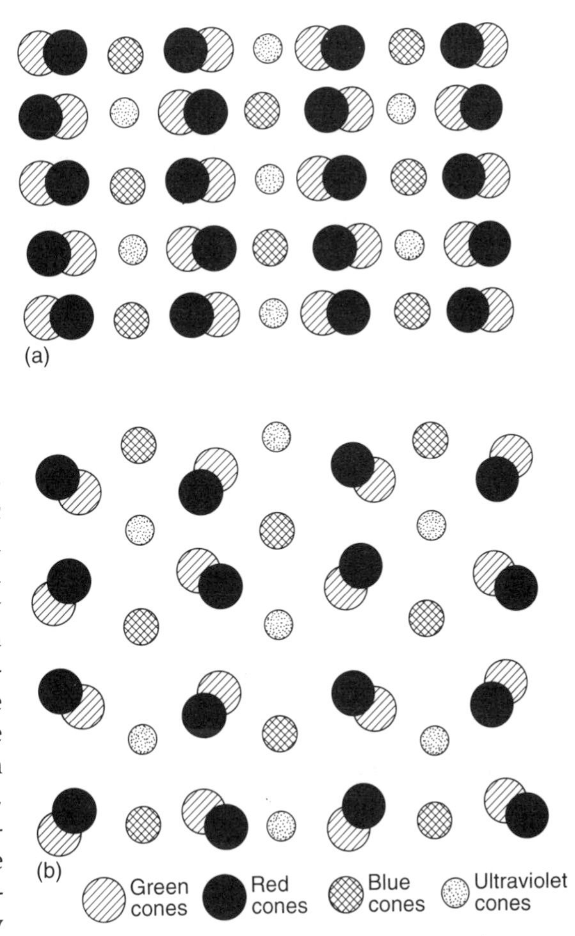

Figure 1.2 Idealized teleost cone mosaic patterns. (a) Row pattern. (b) Square pattern. The retinal margin is parallel to the top of the page; alternating rows of double and single cones run orthogonal to the margin in (a). Note that along a row of single cones, blue and ultraviolet cones alternate; in the double cone rows, the red and green members are alternately to the left and right. In the goldfish retina, the square pattern predominates, although LS cones are distributed randomly throughout and elements of the mosaic unit are sometimes missing, degrading the regularity of the pattern.

modifications in the basic square or rhomboid pattern are found in other teleost species; for example, in some species the double cones

* The direction 'centripetal' refers to the dimension from the retinal margin toward the center, as in the spokes on a wheel (as if the retina were a flattened disk). The term 'radial' is reserved for the dimension orthogonal to the retinal laminae, along radii eminating from the center of the globe. 'Circumferential' refers to the circular retinal margin, the location of the germinal zone.

are identical (in morphology and spectral properties), and are therefore termed 'twin cones' (Engström, 1960).

The morphological cone subtypes in teleost fish correspond to spectral classes (Stell and Hárosi, 1975; Marc and Sperling, 1976). In many species of birds, which also have highly developed color vision and four or five different cone pigments, morphological subtypes also correspond to spectral classes, but the cones are not arranged in an obvious mosaic pattern. Statistically, it can be demonstrated that the different subtypes of cones are spaced regularly in the chicken retina, but this is not obvious on examination (Morris, 1970). In humans and Old World primates (such as macaques) there are three cone pigments, but morphological subtypes are not easily recognized (de Monasterio *et al.*, 1981; Ahnelt *et al.*, 1987). By the use of antibodies specific for visual pigments, red/green cones† can be distinguished from blue, although the high degree of homology between red and green opsins has thus far prevented the development of antibodies that distinguish red from green. In macaque monkeys red/green cones outnumber blue cones by about 10 : 1 (Wikler and Rakic, 1990), but the distribution of red/green cones is less regular than blue cones, which form a remarkably precise, hexagonal lattice in many primate retinas (Marc and Sperling, 1977, de Monasterio *et al.*, 1981; Wikler and Rakic, 1990).

The spectral characteristics of identified morphological classes have been determined with microspectrophotometry in a very large number of teleost species (Levine and MacNichol, 1979), including goldfish (Stell and Hárosi, 1975; Marc and Sperling, 1976) and zebrafish (Nawrocki *et al.*, 1985; Robinson, *et al.*, 1993). With this technique the absorption of individual, isolated photoreceptors is measured. In goldfish the principal member of the double cone is sensitive to long wavelengths (red), the accessory member to medium wavelengths (green), the long single cones are either red or green, and the short single cones are sensitive to short wavelengths (blue). The short single blue cone is at the center of the square mosaic pattern; the cones at the corners are miniature singles (Marc and Sperling, 1976). Missing and ectopic photoreceptors often disrupt the mosaic in the goldfish retina, but the precise hexagonal array of short single cones is typically the most regular feature of the pattern (Marc and Sperling, 1976). More recently, an ultraviolet pigment was found by microspectrophotometry in miniature single cones in several cyprinids, including roach, *Rutilus rutilus* (Avery *et al.*, 1983), goldfish (Bowmaker *et al.*, 1991), carp, *Cypripus carpio* (Hawryshyn and Hárosi, 1991) and Japanese dace, *Tribolodon hakonensis* (Hashimoto, *et al.*, 1988). In addition, roach and goldfish demonstrate behavioral sensitivity to ultraviolet light (Douglas, 1986; Hawryshyn, 1991; Neumeyer, 1992). In some teleost species, such as trout (*Salmo trutta*), the corner cones disappear in adult fish (Lyall, 1957b; Bowmaker and Kunz, 1987), and concomitantly, the animals lose ultraviolet sensitivity (Bowmaker and Kunz, 1987), consistent with the suggestion that the corner cones are the ultraviolet-sensitive elements.

In zebrafish, in which the row pattern is retained in the mature retina, the double

† Traditionally, the human cone pigment sensitive to long wavelengths has been called 'red-sensitive', and the cone containing this pigment the 'red cone' even though the wavelength of maximal absorbance (λ_{max}) of the pigment, at 552 nm to 557 nm, is in the yellow part of the spectrum (Dartnell *et al.*, 1983; Merbs and Nathans, 1992). The human 'green' and 'blue' cone pigments absorb maximally at the color for which they are named. The short, middle and long wavelength-sensitive pigments in different species of teleost fish cover a broad spectral range, with λ_{max} from 410 nm to 630 nm for the traditional pigments (Levine and MacNichol, 1979), and down to 360 nm for the ultraviolet pigment (Robinson *et al.*, 1993). Despite these inconsistencies between absorption maxima and color names, the cone pigments are here referred to as 'red', 'green', 'blue' or 'ultraviolet' for ease of description and comparison across species.

cone pairs are similar to those in goldfish: red cones are the principal members and green cones are the accessory members, long single cones are blue and short single cones contain an ultraviolet-sensitive pigment (Robinson *et al.*, 1993). These new microspectrophotometric data revise the assignment of spectral types to morphological subclasses that was presented in an earlier study of zebrafish cones (Nawrocki *et al.*, 1985).

1.2.2 STUDIES OF MOSAIC FORMATION WITH OPSIN ANTIBODIES AND cRNA PROBES

Since the mosaic constitutes a precise arrangement of spectral classes of cones, the study of pattern formation requires a suitable marker for the visual pigments. Relying strictly on morphological criteria is inadequate, since at early stages the cones are not yet differentiated enough to allow subtypes to be distinguished (Branchek and BreMiller, 1984; Raymond, 1985; Larison and BreMiller, 1990). The classical technique for studying spectral properties of individual photoreceptors, microspectrophotometry, is unsuitable for developmental studies as the cones must be dissociated and they must also be relatively large for the technique to be practical. Specific antibodies for the different cone pigments have been somewhat useful in this regard, especially for developing mammalian retinas (Wikler and Rakic, 1991; Szél *et al.*, 1993). Unfortunately, however, many of the available opsin or other photoreceptor-specific antibodies label both rods and cones in teleost fish (Raymond *et al.*, 1993). Although the biochemistry of rod pigments (rhodopsins) has been extensively studied, cone pigments are notoriously difficult to extract from retinas, and biochemical isolation of cone pigments has only been achieved in chickens (Wald *et al.*, 1955; Okano *et al.*, 1992), which have strongly cone-dominated retinas.

In contrast, substantial progress in this area has been achieved with the tools of molecular biology, and over the past ten years the genes encoding over 20 different opsins (the protein component of visual pigments) have been identified (see Chapter 4). Opsins constitute a family of proteins that have been highly conserved over hundreds of millions of years of evolution, so that opsins from arthropods (*Drosophila*), cephalopod molluscs (octopus), and vertebrates all share a common lineage (Goldsmith, 1990). The gene encoding bovine rhodopsin was the first vertebrate opsin to be cloned, and shortly thereafter the genes for human rhodopsin and the three human cone pigments were identified by homology (Falk and Applebury, 1987). Genes coding for red and green cone opsins in the cavefish, *Astyanax mexicanus* (Yokoyama and Yokoyama, 1990a,b), and the red cone opsin in chicken (Kuwata *et al.*, 1990; Tokunaga *et al.*, 1990), were then identified by homology with the human cone pigments. In humans and cavefish, the red and green cone opsins are nearly identical (96% at the amino acid level), and represent independent gene duplication events (Yokoyama and Yokoyama, 1990b). The human blue cone opsin is as similar to rhodopsin as it is to the red/green opsins (Nathans *et al.*, 1986; Goldsmith, 1990), suggesting that the bipartite categorization of photoreceptors as rods or cones may not be reflected in the molecular genetics of the pigments. In support of this somewhat heretical idea, an unusual rhodopsin-like chicken pigment gene has been described; the predicted amino acid sequence is 80% identical to chicken rhodopsin but only 40–50% identical to the human, chicken and cave-fish cone opsins (Okano *et al.*, 1992; Wang *et al.*, 1992). Araki *et al.* (1984) had previously noted rhodopsin-like immunoreactivity in single cones of chicken retinas with a polyclonal antibody prepared against bovine rhodopsin, and they speculated that these might be the green cones. They further suggested that the green cone pigment might be immunologically similar to chicken rhodopsin since both have λ_{max} around 500 nm. Evidence from recent work in fish (described below) supports this suggestion.

Several laboratories (Johnson *et al.*, 1993; Raymond *et al.*, 1993; Hisatomi *et al.*, 1994) have independently isolated partial or complete cDNA clones for goldfish opsins. Nakanishi's group has found two rhodopsin-like genes that are highly homologous to the chicken rhodopsin-like gene. These two rhodopsin-like clones are only 90% identical at the nucleotide level and are therefore believed to represent two different genes (Johnson *et al.*, 1993). Because goldfish are tetraploid (Risinger and Larhammar, 1993) the other opsin genes are also likely to be duplicated. The evolutionary origin of the rhodopsin-like green cone opsin gene in chickens and goldfish is not clear, but it almost certainly represents an example of convergent evolution.

The goldfish opsin cDNA clones have been used to prepare sense and antisense cRNA probes tagged with digoxigenin for colorimetric *in situ* hybridization (Raymond *et al.*, 1993). The results demonstrated that both of the rhodopsin-like pigments are expressed in the accessory green cone member of the double cone pair in goldfish. Some of the long single cones also expressed rhodopsin-like pigments, consistent with the earlier spectral studies (Stell and Hárosi, 1975; Marc and Sperling, 1976). It is not yet certain, however, whether the two rhodopsin-like genes are expressed in different cones or simultaneously in the same cones; expression of multiple visual pigments in a single photoreceptor has been described in other teleosts (Chapter 2, and Archer and Lythgoe, 1990). The suggestion that green cones contain a rhodopsin-like pigment is further supported by the observation that two different monoclonal antibodies prepared against mammalian rhodopsins – RET-P1 (from C. Barnstable) and rho4D2 (from D. Hicks and R. Molday) – label both rods and green cones in goldfish (Raymond *et al.*, 1993). It is curious that the green cone pigment in goldfish apparently differs molecularly from that in *Astyanax* since both species are in the cyprinid (carp, minnow) family. However, since the putative red and green *Astyanax* pigment genes have not yet been cloned into a cell expression system, and therefore the spectral properties of the proteins they code for have not been evaluated, it is possible that the color assignments, which were based solely on sequence homology with human visual pigments, are incorrect, and that the two *Astyanax* genes actually represent polymorphic variants of a red pigment. Polymorphism of red opsins has been described previously in teleosts (Archer and Lythgoe, 1990).

The *in situ* hybridization studies in goldfish retina further showed that the blue cone opsin is expressed in short single cones (Figure 1.3), and the red cone opsin is expressed in the principal member of the double cone pair and in some single cones (Raymond *et al.*, 1993). These results again match the earlier morphological identification of the three spectral classes in goldfish (Stell and Hárosi, 1975; Marc and Sperling, 1976). The goldfish rod and cone opsin probes also hybridize specifically to appropriate pigments in other teleost species, including zebrafish (Raymond *et al.*, 1993) and green sunfish, *Leposmis cyanellus* (L. Barthel, D. Cameron and P. Raymond, unpublished observations).

Recently, a partial cDNA clone coding for a putative ultraviolet pigment has been identified in goldfish based on sequence homology with chicken violet (Hisatomi *et al.*, 1994). Riboprobes transcribed from this putative ultraviolet clone hybridize to miniature single cones located at the corners of the mosaic in the goldfish retina (P. Raymond and L. Barthel, unpublished observations). This localization is consistent with the identification

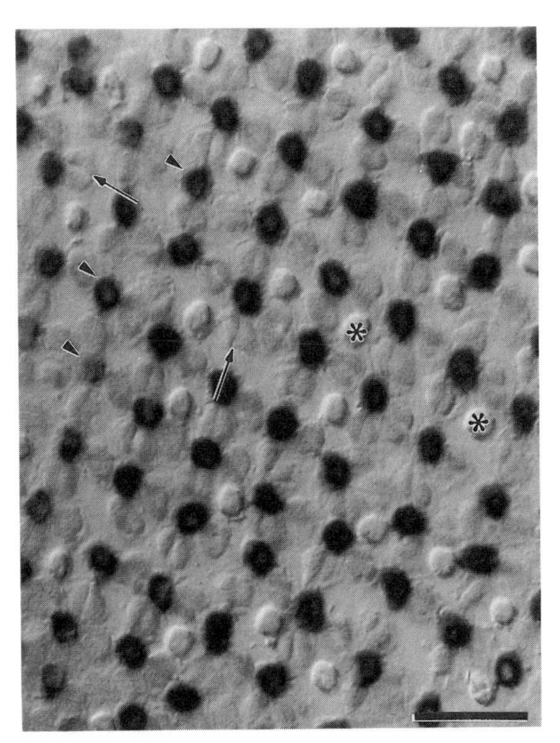

Figure 1.3 *In situ* hybridization with the blue cone opsin probe applied to goldfish retina; whole mount preparation. The focus is at the level of the myoids of the short single cones, which contain the hybridization signal (arrowheads). For double cones (arrows) the focus is also within the myoid, but miniature short single cones are focused at the level of the refractile ellipsoid (asterisks). Calibration bar = 20 μm. (Reprinted with permission from Raymond *et al.*, 1993.)

of miniature single cones as ultraviolet receptors based on earlier work cited above. The possible molecular similarity of the zebrafish ultraviolet pigment, recently cloned and sequenced by Robinson and colleagues (Robinson *et al.*, 1993), and the goldfish pigment cannot be determined until the full length goldfish sequence is available. Curiously, however, the zebrafish ultraviolet pigment appears to be more homologous to lamprey rhodopsin than to chicken violet, on the basis of the predicted amino acid sequence. In both goldfish and zebrafish, the ultraviolet pigment is expressed in the smallest of the photoreceptor subtypes: the miniature short single cones of goldfish and the short single cones of zebrafish (Figure 1.1). Interestingly, the ultraviolet receptors in zebrafish and goldfish share some unique features: their nuclei do not penetrate through the outer limiting membrane of the retina, as do the nuclei of the other cone types, and the ultraviolet photoreceptors do not show photomechanical movements in response to light and dark adaptation as do the other cones. The short single cones in the guppy (*Poecilia reticulatus*), which have a λ_{max} of 410 nm and are located at the corners of the square mosaic unit, have similar features. Photomechanical movements involve light-evoked contraction of the myoid process, which brings the ellipsoid/outer segment of the cone closer to the nucleus, and corresponding elongation of the myoid process during the dark (Walls, 1967). The significance of the lack of photomechanical movements in the ultraviolet receptors is unknown.

The availability of teleost cone opsin probes and the development of a colorimetric *in situ* hybridization method with subcellular resolution provides a unique opportunity to investigate pattern formation in the teleost cone mosaic.

1.3 PHOTORECEPTOR DEVELOPMENT

Many fish continue to grow as adults, and in those species which do so, neurogenesis continues in the retina, such that annuli of new cells are continually added from a circumferential germinal zone (Müller, 1952; Johns, 1977). Thus, in the adult teleost retina there is a circular gradient of cellular maturation, from mitotic precursors at the margin to mature neurons in the center; this spatiotemporal gradient of development allows all

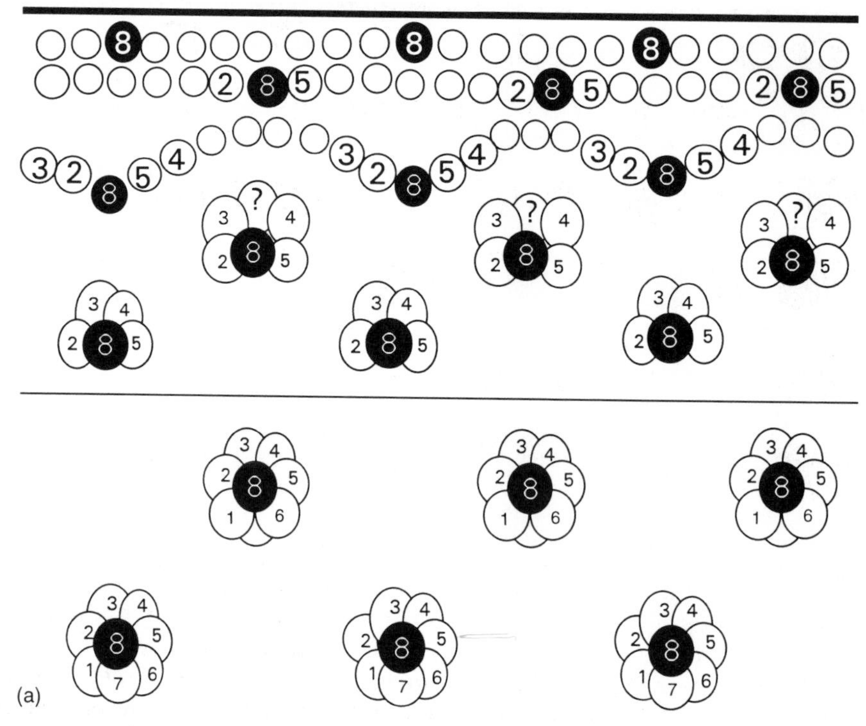

Figure 1.4 (a) Model for photoreceptor determination in *Drosophila*. The horizontal line at the top represents the morphogenetic furrow. Successive generations of increasingly more mature ommatidial units are arrayed below the furrow, and mature ommatidia are along the bottom. Each mature ommatidial unit is composed of eight photoreceptors, numbered 1 to 8. In the first row of undifferentiated, presumptive photoreceptors immediately below the morphogenetic furrow, R8 photoreceptors begin to differentiate at regular intervals. The R8 photoreceptors induce their immediate neighbors to differentiate as R2 or R5 (identical except for their position to the left or right of R8). The adjoining R3/4 pairs are then induced, and the five photoreceptors (plus a variable number of mystery cells, so called because they are only transiently associated with the ommatidial cluster and their ultimate fate is unknown) pinch off to form preclusters. After another round of mitotic division, indicated by the thinner horizontal line, the R1/6 pair joins each cluster, followed in the end by R7. (Redrawn from Banerjee and Zipursky, 1990.)

steps in the formation of the cone mosaic pattern to be examined in a single retina from an adult animal, thus providing a powerful experimental tool.

Larison and BreMiller (1990) made a fundamental observation which implies that the classical view of the row mosaic pattern – alternating parallel rows of single and double cones radiating outward like spokes of a wheel and intersecting the retinal margin at right angles – may be misleading. When they examined the adult zebrafish retina with light and electron microscopy it was clear that the closest appositions are between adjacent single and double cone photoreceptors in the adjoining centripetal rows of double and single cones. This is an important observation because it suggests that commitment of presumptive cone photoreceptors to a specific identity and the resultant precise spacing

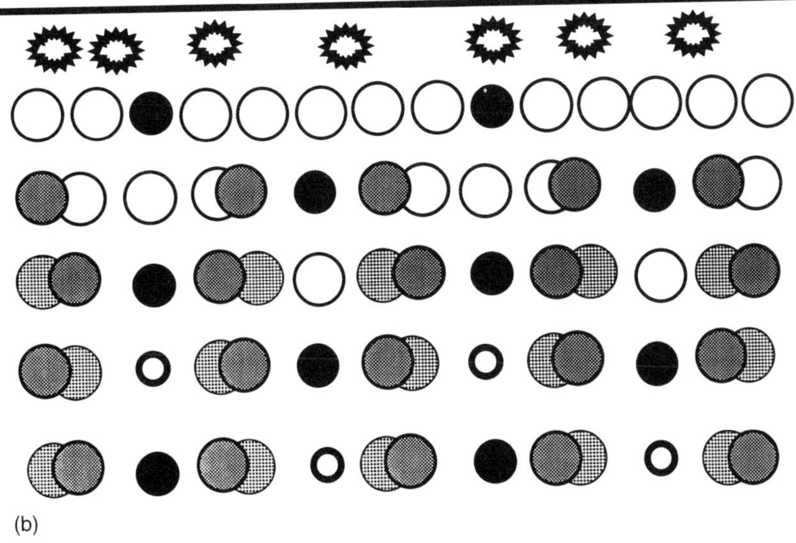

Figure 1.4 (*cont.*) (b) Model for photoreceptor determination in teleost fish. The horizontal line at the top represents the germinal zone, and starbursts are mitotic neuroepithelial cells. (The germinal zone is actually a large annulus at the retinal margin, but it can be approximated as a line segment over this short interval.) Successive generations of increasingly more mature photoreceptors are arrayed below. Assume that the first photoreceptors to differentiate are the blue cones (solid dots) located in the first (youngest) row adjacent to the germinal zone. The blue cones then induce their immediate neighbors to become red cones (heavy stippling), which are the principal members of the double cone pair. The red cones in turn induce the adjacent cells to become accessory green cone (light stippling) members of the double cone. Finally, four rows away from the germinal zone, the green cones induce the remaining undifferentiated cells to become ultraviolet receptors (heavy outline), and the row mosaic pattern is thereby established.

of cone types in a mosaic pattern might be accomplished by lateral inductive influences between neighboring cone types all differentiating at approximately the same time. This appears similar to the inductive events that have been shown to regulate photoreceptor commitment in *Drosophila*, as outlined in the next section.

1.3.1 PATTERN FORMATION AND CONSTRUCTION OF A MODEL

An elegant series of genetic mosaic analyses (Ready *et al.*, 1976, 1986) showed that the photoreceptor cells in each ommatidium of the *Drosophila* compound eye do not share an obligatory cell lineage relationship, but instead inductive cell–cell interactions are responsible for directing the choice of fate of presumptive photoreceptor cells. Banerjee and Zipursky (1990) proposed a model in which determination of photoreceptors is accomplished by a cascade of inductive events that take place just after the precursor cells withdraw from the mitotic cycle and while they are aligned in a linear array at the posterior edge of the morphogenetic furrow (Figure 1.4a). The 'founder' cell is the R8 photoreceptor, and a stochastic process underlies the original determinative event. Lateral inhibitory interactions perhaps mediated by a diffusible agent prevent neigh-

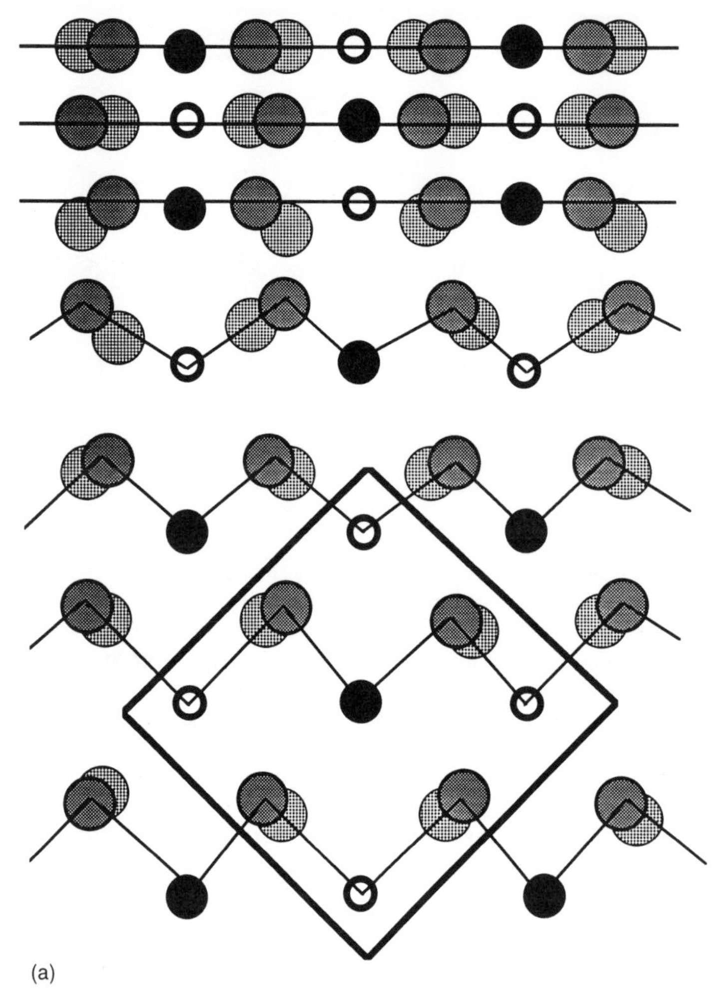

Figure 1.5 Transformation of the row pattern to the square mosaic unit. The germinal zone is toward the top of both panels, and lines connect photoreceptors of a single generation (i.e., generated approximately simultaneously at the germinal zone). (a) One might first try rotating the double cone pairs inward toward the blue cones (solid dots), and shifting the single cones downward. However, this results in an incorrect pattern for the square mosaic unit. Note that the square unit here is bilaterally symmetric, whereas the actual square unit is not (compare Figure 1.2b.)

boring cells within a certain distance from adopting the R8 fate, and this leads to an evenly spaced array of R8 cells. R8 induces its immediate neighbors on either side to become R2/R5; R3/4 are in turn induced by R2/5. These five cells (along with two more mystery cells that later disappear) pinch off to form a precluster. After a second round of mitosis, three cells join the developing cluster; two of these are induced by R2/5 to become R1/6, perhaps by the same signal used to induce R3/4. Finally, R8 induces R7.

In Figure 1.4b a hypothetical model is presented which describes cone photorecep-

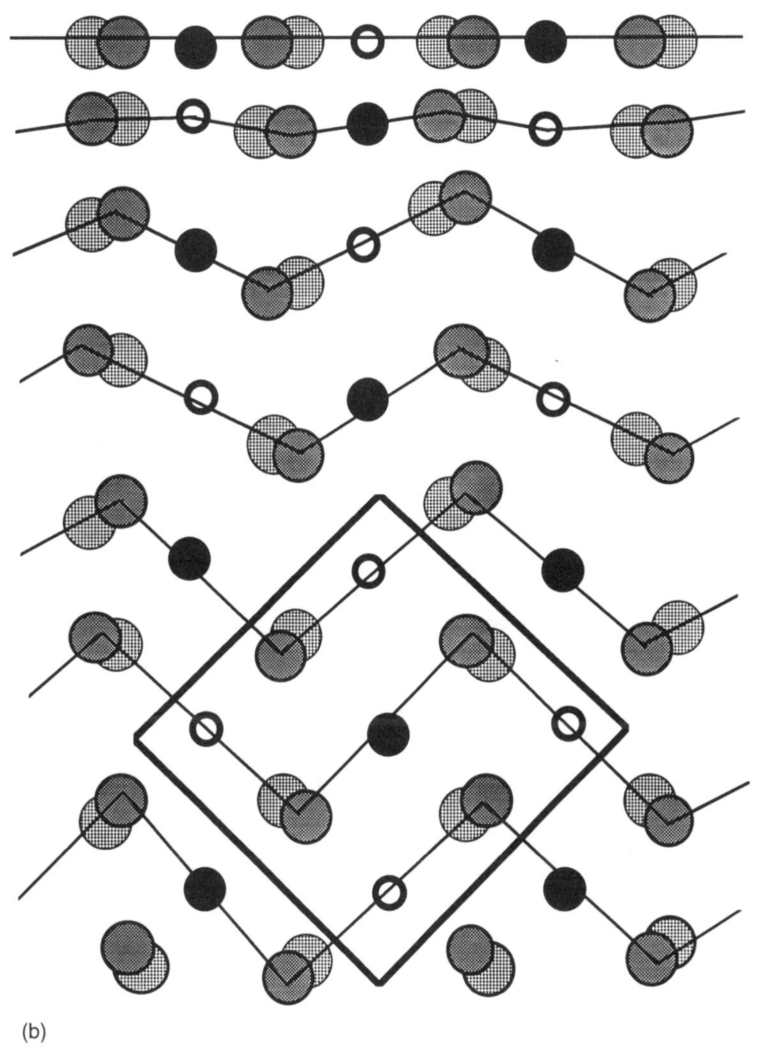

Figure 1.5 (*cont.*) (b) The simplest topological rearrangement that leads to the correct square unit pattern involves synchronous rotation of double cone pairs within a generation, but from one generation to the next, the rotation is alternately clockwise and counterclockwise. Slippage of alternate rows of double cones centripetally and centrifugally then generates a square lattice pattern.

tor commitment in the teleost retina based on similar principles. Instead of a morphogenetic furrow, teleost retinas are surrounded by a circumferential germinal zone (Johns, 1977). Cone photoreceptors, along with other retinal neurons, are therefore generated along a circular, two-dimensional front of mitotic activity. Each successive generation represents a circumferential row of putative cone photoreceptors, all of approximately equivalent age and all as yet uncommitted to a specific phenotype. The model proposes that the R8 equivalent, the cone type that differentiates first and establishes the spacing of the mosaic unit, might be one of the two short wavelength-sensitive single cone types. The reason for postulating that blue or ultraviolet cones might adopt this primary role is that in many species of teleost fish (Engström, 1960) and in primates (Marc and Sperling, 1977; Wikler and Rakic, 1990), the spacing of short wavelength-sensitive, i.e. blue and/or ultraviolet, cones is more regular than the other cone types. Furthermore, the double cones in the teleost retina are mirror-symmetrically arranged with respect to the single cones, and therefore it would follow that one of the single cone types marks the center and the other the boundary of the repeating mosaic units (Figure 1.2). For the sake of argument then, assume the blue cone is the founder, and establishes the prototype of the mosaic pattern, inhibiting nearby cells from adopting a similar fate. The founder blue cones then induce the neighboring presumptive photoreceptor cells to become double cones, with the red member of the pair nearest and the green member toward the outside. Last to differentiate in this sequence would be the ultraviolet cones. The ultraviolet cones are flanked by the green members of the double cone pairs, as the pattern repeats. This model makes several specific predictions which are experimentally testable, as discussed in the next section.

In those fish in which the primitive row pattern of repeating, mirror-symmetric units is further refined into the square mosaic shown in Figure 1.2(b) (e.g. goldfish), a second set of patterning events transforms the mosaic organization. It has been suggested that this transformation involves movement of individual cones (Müller, 1952; Lyall, 1957b), although the details of the required movements have not been previously considered. Constraints on the predicted cell movements are imposed by the final orientation of the double cone members in the square pattern: the red and green cones alternate on adjacent sides, much like the four partners are positioned in a square dance. A correctly oriented square pattern cannot be generated by the synchronous 45° rotation of all the double cones in given birth cohort, i.e., those along a single circumferential row, as this would produce square mosaics with bilaterally symmetric double cones (Figure 1.5a). The simplest topological transformation that achieves the desired pattern is shown in Figure 1.5(b). If double cone pairs in alternate generations rotated 45° in opposite directions, and then moved centripetally or centrifugally with respect to the adjacent single cones as shown in Figure 1.5(b), the proper orientation of double cone pairs in the square mosaic could be achieved. The required shearing movements along rows of double cones are alternately centrifugal and centripetal, as indicated by the zigzag lines which connect members of a given generation. Note that this topological transformation would imply that the members of a given mosaic unit (enclosed by a rhombus in Figure 1.5b) actually belong to three successive generations, or 'birth cohorts' produced by the germinal zone. Again, these proposed patterning events are experimentally testable, as described in the next section.

1.3.2 SEARCH FOR CELLULAR MECHANISMS UNDERLYING PATTERNING EVENTS

The model in Figure 1.4 makes several predictions about how inductive interactions could influence determination of cone spectral types and how the final organization of the mosaic units might be achieved. The first prediction of the model is that within a given generation, that is, within a circumferential row of cells adjacent to the germinal zone that became postmitotic at roughly equivalent times, different spectral cone types ought to differentiate in a specific order determined by nearest neighbor relationships. Accordingly, the order should be either blue then red then green then ultraviolet or the reverse. An orderly sequence of cone differentiation should be reflected in a sequential pattern of expression as measured with the different cone opsin cRNA probes, and the spacing of the earliest opsin-expressing cones ought to reflect the repeat distance of the mosaic unit.

These predictions can be tested experimentally by using *in situ* hybridization with cone opsin cRNA probes on whole mounted adult goldfish retinas, and examining opsin expression in the immature retina adjacent to the circumferential germinal zone, as outlined in Figures 1.6 and 1.7. In the first experiment (Figure 1.6), a cocktail containing a mixture of cRNA opsin probes (red, green, blue, ultraviolet) is hybridized to the retina, and if the model is incorrect, and cone types differentiate simultaneously, then the newest circumferential row of cones should show a pattern of continuous hybridization. If the model (Figure 1.6a) is correct, then in the youngest (least mature) areas labeled cones should be separated by several intervening unlabeled (and undifferentiated) cells as shown in Figure 1.6(b). Results from preliminary experiments are consistent with the model (D. Stenkamp and P. Raymond, unpublished observations).

A second test of the model is to estimate the order of differentiation by examining the distance from the germinal zone at which various cone subtypes begin expressing opsin (Figure 1.7). The mitotic cells in the germinal zone can be labeled with bromodeoxyuridine (BrdU), and since *in situ* hybridization is compatible with BrdU immunocytochemistry, the tissue can be processed for both (Biffo *et al.*, 1992). Double-label preparations can be made separately for each of the cone opsin cRNA probes, and the distance between the BrdU-labeled cells in the germinal zone and the nearest hybridizing cones can be measured. Because distance

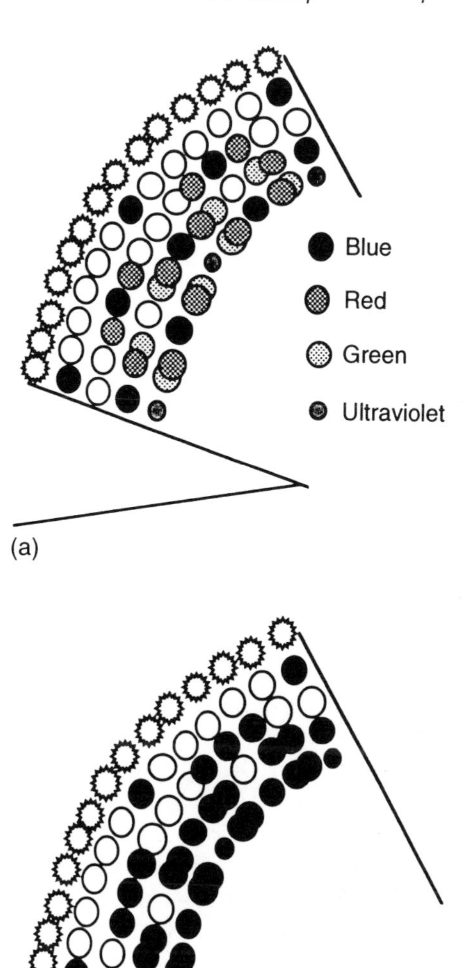

Figure 1.6 Cartoons illustrating a fragment of a retinal whole mount taken adjacent to the margin. Proliferating cells in the germinal zone are indicated by starbursts. (a) If the model in Figure 1.4(b) is correct, then the pattern of differentiation of the four cone types should be as indicated. Mature cones in central regions (more than five rows in from the germinal zone) are omitted for clarity. (b) Results predicted by the model following *in situ* hybridization with a 'cocktail' probe containing a mixture of all four cone opsin cRNA probes.

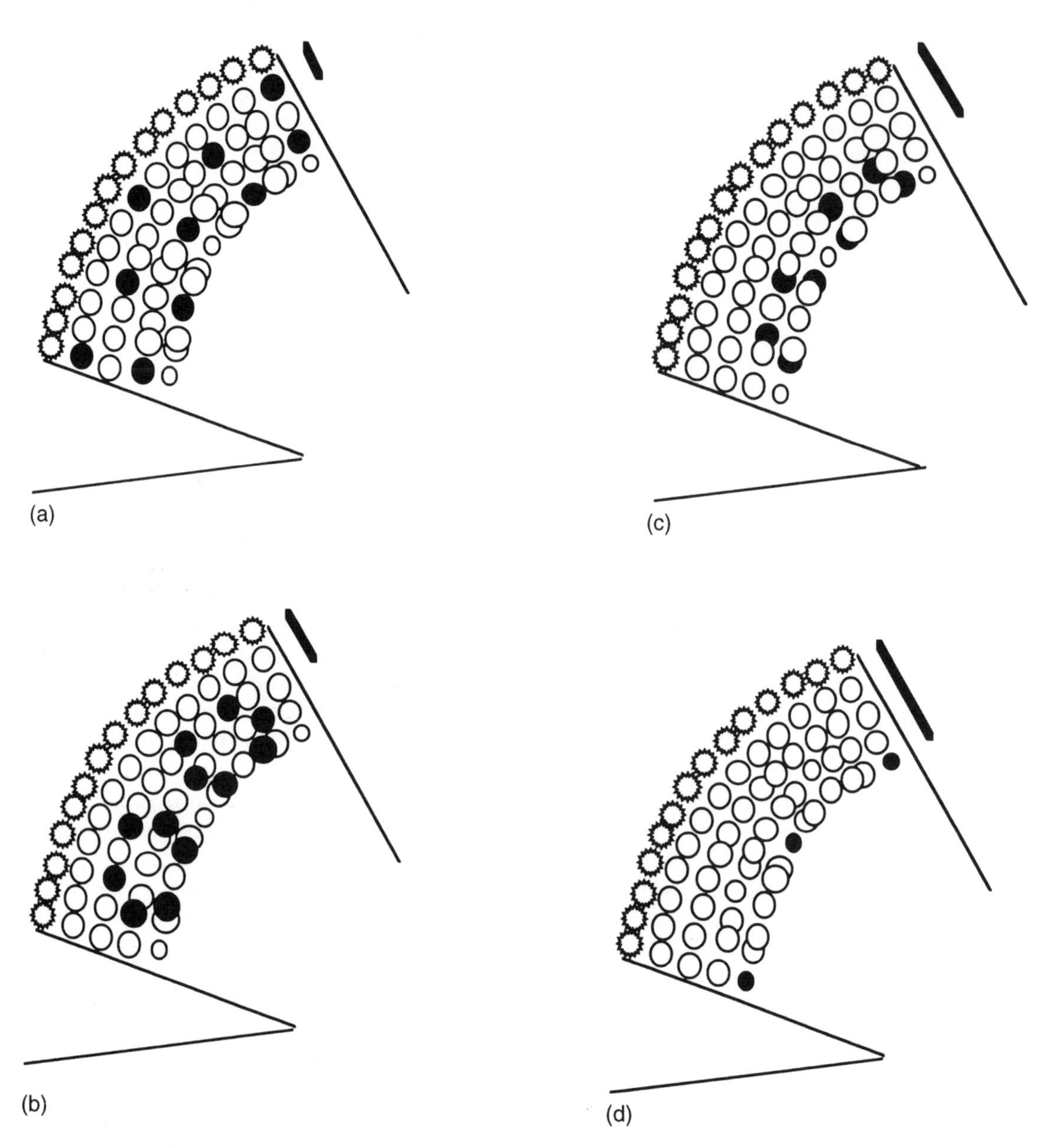

Figure 1.7 Cartoons illustrating results predicted by the model for *in situ* hybridization of retinal whole mounts with the (a) blue, (b) red, (c) green and (d) ultraviolet cRNA probes. The respective distances from the germinal zone, indicated by starbursts (BrdU-labeled proliferating cells) to the first row containing hybridization signals are indicated by line segments. Note that the lengths of the line segments vary, with blue < red < green < ultraviolet.

from the germinal zone is an index of cell birthdate, as shown previously (Johns, 1977), if labeling of different cone subtypes commences at different distances from the germinal zone, then cones must be generated sequentially. The ones labeled closest to the germinal zone would be the earliest to differentiate, and so on. The spacing between cones expressing the same opsin can be measured directly on whole mounts, allowing a test of the prediction that early-differentiating cones establish the repeat distance for the mosaic unit. Finally, double *in situ* hybridization with two different cone pigment probes could be used to allow a direct pairwise comparison of birth order. If there are areas near the germinal zone in which cones express one but not the other probe, then the former must be generated before the latter. Double-label *in situ* hybridization can be done if one probe is prepared with digoxigenin and the other with biotin; the probes are then visualized by immunofluorescence with antibodies conjugated to different tags for visualization (*e.g.*, different enzymes or different fluorochromes).

Similar questions can also be examined in developing embryonic retinas. Several histological descriptions of the development of the cone mosaic in a variety of teleost fish have been published (Müller, 1952; Lyall, 1957a; Ahlbert, 1968). With a single exception (Larison and BreMiller, 1990), however, this previous work relied exclusively on morphological criteria, which are not easy to interpret given the extreme diversity of cone types in teleost fish and the problems inherent in identifying morphologically immature cells. With the use of cone-specific monoclonal antibodies, Larison and BreMiller (1990) provided some evidence that the onset of patterning of cone photoreceptors in zebrafish is an early event. The early stages of retinal development in zebrafish are illustrated in Figure 1.8(a). The first cones become postmitotic at about 48 h after fertilization (Nawrocki, 1985). Also at 48 h double cones immunoreactive with a cell-specific monoclonal antibody (FRet 43) appear, and by 54 h they are organized into rows (Larison and BreMiller, 1990). Expression of opsin message and protein has also been observed as early as 48 h with monoclonal antibodies RET-P1 and rho4D2 and by *in situ* hybridization with opsin riboprobes (Raymond *et al.*, 1994). The first photoreceptors are located in a small patch in the ventral retina (Nawrocki, 1985; Kljavin, 1987), and within the next 10 h the patch enlarges to fill the embryonic retina. At this early stage of development, the cone photoreceptors are morphologically indistinguishable, but they have become committed to a particular subtype as evidenced by their expression of specific opsins (Figure 1.8b). The model predicts that the different spectral classes will differentiate in a specific sequence; this is currently being examined by using *in situ* hybridization with probes specific for the various cone opsins.

Finally, there is additional evidence consistent with a model of sequential differentiation of cone photoreceptor subtypes in the mammalian retina, where several groups are examining the earliest stages in the organization of a cone mosaic pattern. Wikler and Rakic (1991) showed that in primates (*Macaca mulata*) a subset of cones precociously express opsin-immunoreactivity, and these early differentiating cones are positioned in a precise hexagonal array with a spacing similar to that of the blue cones in the adult retina. Their results are somewhat difficult to interpret, however, since the antibody that labeled the precocious cones was shown in adult retinas to label red/green cones. The authors suggested that blue cones might initially express red or green opsin, and then later switch to the blue, or alternatively that the precocious cones may represent a subset of the red/green population. Recently, however, with different opsin antibodies (monoclonal antibodies OS-2 and COS-1), it has been shown that in

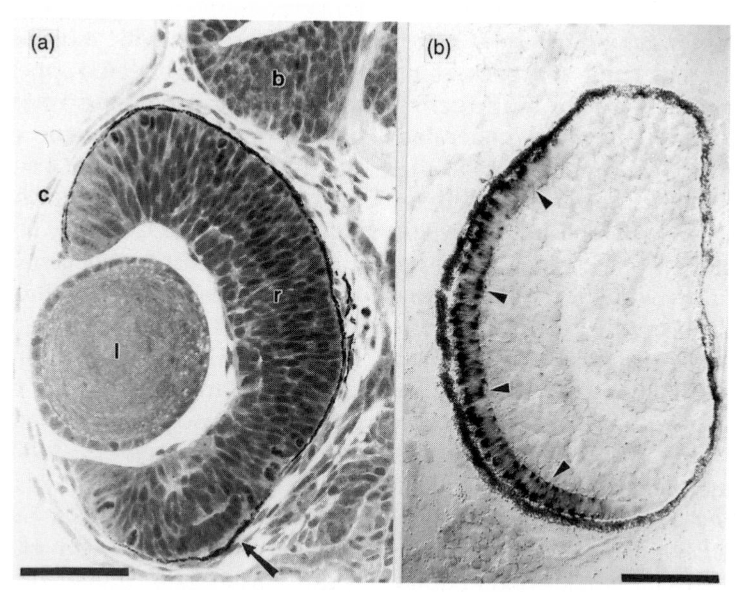

Figure 1.8 (a) Histological section of a zebrafish embryo at 2 days postfertilization (methacrylate-embedded, 3 μm thickness). Dorsal is up. Note the small patch of laminated retina in the ventral retina (arrow). The remainder of the presumptive retina (r) consists of proliferating neuroepithelial cells. Cornea, c; lens, 1; brain, b. (b) *In situ* hybridization with the red opsin probe on a cryosection of a newly hatched zebrafish, at 3.5 days postfertilization. Note that some of the immature cones, which as yet lack outer segments, have abundant opsin message (arrowheads). Bar = 50 μm.

mouse (Szél *et al.*, 1993) and *Macaca* monkeys (Bumsted *et al.*, 1994), the earliest cones to differentiate, as evidenced by opsin immunoreactivity, are the blue cones. However, these are preliminary results and more work is needed before definitive conclusions can be reached.

The next question to be addressed is: do inductive interactions play a role in determination of cone photoreceptor identity? If the model in Figure 1.4 is correct, and cone photoreceptors commit to a specific identity and organize into a mosaic pattern under the influence of lateral, inductive interactions between neighboring cones, then the cone mosaic should have self-assembling properties. The experimental strategy used by Ready and colleagues to demonstrate that photoreceptors in the *Drosophila* ommatidium have autonomous, self-organizing capabilities (Lebovitz and Ready, 1986) was to culture fragments of the presumptive retina (the eye imaginal disk). It was then asked whether the information required to generate the organized cone mosaic is an intrinsic property of the cells that are becoming organized, and whether new photoreceptors are recruited into the pattern, similar to the way a crystal grows, by using the organized structure of earlier generations as a template. The same strategy can be applied to the teleost retina by using whole organ culture of embryonic zebrafish eyes at a stage prior to the appearance of the first cone photoreceptors. Eyes can be removed to culture before 48 h, when the first cones become postmitotic, so that differentiation of cones, if it occurs, must have occurred in culture. Differentiation of photoreceptors *in vitro* from mitotic precursor cells, has been demonstrated by many laboratories (Adler, 1986; Reh, 1991; Altshuler *et al.*, 1991), but the generation of the cone mosaic pattern has not yet been investigated *in vitro*. To demonstrate directly that cones develop in culture, BrdU can be added to the culture medium at the beginning of the experiment, and the eyes processed for both BrdU immunocytochemistry and *in situ* hybridization (as above). In these preparations double-labeled cells must have derived from precursor cells that underwent their final mitotic division in culture. Fragments of eyes including or excluding the ventral patch of early differentiating cells can also be cultured separately. If fragments from dorsal retina, which lack the ventral region where the early-differentiating patch of photoreceptors is located (Figure 1.8a), are capable of generating a cone mosaic pattern, this result would argue against the hypothesis that pattern formation depends on the presence of a 'template', represented in the early embryonic retina by the specialized ventral patch of early-differentiating photoreceptors (Kljavin, 1987) or, in the postembryonic retina, by the most recently formed annulus of cones (Johns, 1977).

Finally, the topology of transformation from row to square mosaic pattern as illustrated in Figure 1.5(b) requires a carefully orchestrated ballet of movements involving rotation and displacement of double cones. The proposed choreography that yields the correct spatial pattern predicts that successive generations of cones will initially comprise curvilinear arrays which are subsequently distorted into a zig-zag pattern with a phase equal to the repeat distance of the square mosaic pattern. This prediction could be tested by a careful analysis of patterns of cones labeled by short pulses of BrdU.

The mechanisms that might generate the predicted rotational and shearing forces remain a mystery, although there are reports in the literature that twin cones in some species can rotate around their long axis, essentially converting a row pattern into a square pattern as a function of changes in ambient illumination (Kunz, 1980; Wahl, 1994). In the walleye (*Stizostedion vitreum*) these twisting movements occur only in an annular region of central retina, which represents a transitional zone between the area around the optic disc which has a permanent square mosaic and the periphery which has the row pattern characteristic of the immature retina (Wahl, 1994). In the guppy, the cone mosaic shifts from squares during the day to rows at night (Kunz, 1980). As to the shearing force, lateral slippage of cone photoreceptors independent of other retinal cells, and independent of movements of the overlying pigmented epithelium, have been documented during foveal development in the primate (Packer *et al.*, 1990). During the first year after birth in *Macaca* monkeys, differentiated cones move centripetally inward toward the fovea, producing a fivefold increase in foveal cone density; the axons of the migrating cones remain attached to perifoveal postsynaptic partners, producing the long, oblique fibers of Henle on the banks of the foveal pit. The movements of teleost cones predicted by the model in Figure 1.5b are fundamentally similar to those observed in the primate, including their unexpectedly late developmental occurrence in cells already morphologically differentiated.

1.4 REGENERATION OF PHOTORECEPTORS

It has been known for some time that following surgical or chemical lesions, the adult goldfish retina will regenerate (reviewed in Raymond, 1991; Hitchcock and Raymond, 1992). The regenerated neurons derive from a source of proliferating neuroepithelial cells intrinsic to the retina. Neurogenesis in the teleost fish retina continues throughout life, and neuroepithelial cells with pluripotent, unrestricted developmental potential are retained in these retinas. As adult fish grow larger, the proportion of rods to cones increases, and these new rods derive from proliferating rod precursors. Rod precursors originate from residual neuroepithelial cells that are initially sequestered in the inner nuclear layer (INL) associated with radial Müller glia, along which they migrate to colonize the outer nuclear layer (ONL), which at first consists of a single row of postmitotic cones (reviewed in Raymond, 1991). The neuroepithelial cells in the ONL are self-renewing stem cells in that some of their progeny (rod precursors) continue to proliferate and others normally differentiate into rods, but apparently also have the capacity to generate other types of retinal neurons.

Although cones are restored to normal or greater than normal density in goldfish retinas following their ablation (Raymond *et al.*, 1988; Hitchcock *et al.*, 1992), it is unclear whether the cone mosaic pattern is reconstructed. Preliminary evidence suggests that following laser lesions of goldfish retina that selectively destroy photoreceptors (Braisted *et al.*, 1994), or surgical removal of patches of sunfish retina (D. Cameron and S. Easter, personal communication), cones regenerate but the pattern is irregular. Nothing is known about the phenotypic identity of the regenerated cones. Other neurons are also arranged in a regular array in the goldfish retina, for example, the dopaminergic interplexiform cells, but after regeneration, the pattern is less orderly than normal (Braisted and Raymond, 1992; Hitchcock and Vanderyt, 1994). These observations suggest that formation of mosaic patterns may involve mechanisms that are active during development, but fail to be reactivated, or are only partially functional, during regeneration. It is believed that creating mosaic arrays of neurons provides a means to ensure equal coverage of all retinal loci with the appropriate cellular elements and to bring into correct spatial register neurons in adjacent retinal laminae that are synaptically interconnected. If the mosaic patterns of the regenerated retina are degraded, it is possible that synaptic connectivity patterns are also less precise, and if so, functional recovery may be compromised. As yet, there are only a few reports which attempt to assess the functional status of the regenerated goldfish retina, and these have used fairly crude measures such as the electroretinogram (Mensinger and Powers, 1993). These preliminary data suggest that the A-wave of the electroretinogram, which is thought to reflect photoreceptor activity, returns to normal in regenerated retinas after several months, suggesting that at least photoreceptors have recovered normal function.

Regeneration of retinal neurons has also been reported in adult newts, larval frog tadpoles, and chicks, but in these cases the regenerative process involves transdifferentiation of a non-neural tissue, the retinal pigmented epithelium (RPE) (reviewed in Park and Hollenberg, 1993). The retina and RPE both derive embryologically from the optic vesicle, and the ability to switch between these two cellular phenotypes is a property of embryonic stages in all vertebrates examined, including mammals, but is lost with maturation in most species, including teleost fish (Knight and Raymond, 1995). The only exceptions are the urodele amphibians, which spontaneously regenerate a neural retina from RPE. In chicks and frogs, the ability to transdifferentiate apparently depends on access to a source of peptide growth factors such as basic fibroblast growth factor (bFGF). The source of bFGF may be endogenous or exogenous, and it has been implicated as a signalling agent that can induce RPE cells to proliferate and thence differentiate into retinal neurons (Park and

Hollenberg, 1993). Interestingly, bFGF and other cytokines and peptide growth factors have been implicated as survival factors for photoreceptors in rats (LaVail *et al.*, 1992), although the relationship between these trophic effects and regeneration via RPE transdifferentiation is unknown.

1.5 CONCLUSIONS AND FUTURE PERSPECTIVES

The long-term goal of the studies discussed in this chapter is to understand the cellular and molecular mechanisms of vertebrate photoreceptor determination. One strategy that has been suggested by many is to take advantage of the explosion of information about the genes involved in photoreceptor development and pattern formation in *Drosophila* ommatidia (Ready *et al.*, 1986; Banerjee and Zipursky, 1990). The unexpected discovery that many of the genes involved in early development and pattern formation in *Drosophila*, such as homeobox genes, segmentation genes, and the neurogenic genes, have homologues in vertebrates and indeed appear to be involved in early developmental events has led to a new appreciation of the conservation of genetic mechanisms and the universality of developmental regulatory events in organisms separated by vast evolutionary distances. Despite the very different evolutionary origins of vertebrate and arthropod eyes, it is not unreasonable to suggest that the control of photoreceptor determination might share common genetic mechanisms. Relevant genes that regulate development of photoreceptors in the compound eye include the neurogenic gene *Notch*, thought to play a permissive rather than an instructive role in cell commitment events; genes coding for growth factor receptors (such as a member of the epidermal growth factor (EGF) receptor family, *Ellipse*); genes coding for protein tyrosine kinase receptors (*sevenless*) and their ligands (*boss*); genes coding for proteins involved in signal transduction (*Sos*); transcription factors (*glass, rough*); and proto-oncogenes (especially c-ras homologues). Homologues of *Notch* have been cloned in several vertebrate species, including human, rat, mouse and *Xenopus* (Coffman *et al.*, 1990; Weinmaster *et al.*, 1991). We have recently isolated two partial cDNA clones for the goldfish homologue of *Notch* (Sullivan *et al.*, 1992), and we have shown by RNase protection and *in situ* hybridization that *Notch* is expressed in the embryonic and growing adult goldfish retina (S. Sullivan, B. Largent, P. Raymond, unpublished observations). Future experiments will assess the possibility that *Notch* might play a role in the development of the cone mosaic.

There is also a strong rationale to examine similar questions in multiple species of teleost fish, being cognizant of the unique advantages of each system. The goldfish retina, for example, is a classic model system for investigations at all levels (anatomical, biochemical, physiological, behavioral). Furthermore, the regenerative capabilities of the goldfish retina (Raymond *et al.*, 1988; Hitchcock and Raymond, 1992) and the development of cell culture systems for *in vitro* studies with teleost fish (Powers, 1989) provide useful experimental tools. Another teleost species that is becoming the model of choice in many laboratories is the zebrafish. This species is well suited to genetic and embryonic studies due to the ease with which they can be bred and the rapid (a few months) generation time. With the awakening of interest in zebrafish as a genetic model, and the growing list of laboratories beginning to screen for zebrafish mutants, the future holds promise that detailed molecular genetic investigations on zebrafish retina will soon be possible. The transparent fish embryo has unique advantages for studies dealing with pattern formation in a whole organ, like the eye. For all of these reasons, the teleost retina is an ideal system in which to study inductive interactions important in neuronal determination.

REFERENCES

Adler, R. (1986) Trophic interactions in retinal development and in retinal degenerations: *in vivo* and *in vitro* studies, in *The Retina: a Model for Cell Biology Studies* (eds R. Adler and D. Farber), Academic Press, New York, pp.111–50.

Ahlbert, I.-B. (1968). The organization of the cone cells in the retinae of four teleosts with different feeding habits (*Perca fluviatilis, Lucioperca lucioperca, Acerina cernua,* and *Coregonus albula). Arkiv for Zoologi*, **22**, 445–81.

Ahnelt, P.K., Kolb, H. and Pflug, R. (1987) Identification of a subtype of cone photoreceptor, likely to be blue sensitive, in the human retina. *Journal of Comparative Neurology*, **255**, 18–34.

Altshuler, D.M., Turner, D.L. and Cepko, C.L. (1991) Specification of cell type in the vertebrate retina, in *Development of the Visual System* (eds D.M.-K. Lam and C.J. Shatz), MIT Press; Cambridge; Retina Res. Fnd. Symp. Vol. 3, pp.37–58.

Araki, M., Watanabe, K. and Yasuda, K. (1984) Immunocytochemical localization of rhodopsin-like immunoreactivity in the outer segments of the rods and single cones of chick retina. *Cell Structure and Function*, **9**, 1–12.

Archer, S.N. and Lythgoe, J.N. (1990) The visual pigment basis for cone polymorphism in the guppy, *Poecilia reticulata. Vision Research*, **30**, 225–33.

Avery, J.A., Bowmaker, J.K., Djamgoz, M.B.A. and Downing, J.E.G. (1983) Ultra-violet receptors in a freshwater fish. *Journal of Physiology*, **334**, 23–4.

Banerjee, U. and Zipursky, S.L. (1990) The role of cell–cell interaction in the development of the *Drosophila* visual system. *Neuron*, **4**, 177–87.

Biffo, S., di Cantogno, L.V. and Fasolo, A. (1992) Double labeling with non-isotopic in situ hybridization and BrdU immunocytochemistry: calmodulin (CaM) mRNA expression in post-mitotic neurons of the olfactory system. *Journal of Histochemistry and Cytochemistry*, **40**, 535–40.

Bowmaker, J.K. and Kunz, Y.W. (1987) Ultraviolet receptors, tetrachromatic colour vision and retinal mosaics in the brown trout (*Salmo trutta*): age-dependent changes. *Vision Research*, **27**, 2101–8.

Bowmaker, J.K., Thorpe, A. and Douglas, R.H. (1991) Ultraviolet-sensitive cones in the goldfish. *Vision Research*, **31**, 349–52.

Braisted, J.E. and Raymond, P.A. (1992) Regeneration of dopaminergic neurons in goldfish retina. *Development*, **114**, 913–19.

Braisted, J.E., Essman, T.F. and Raymond, P.A. (1994) Selective regeneration of photoreceptors in goldfish retina. *Development*, **120**, 2409–19.

Branchek, T. and BreMiller, R. (1984) The development of photoreceptors in the zebrafish, *Brachydanio rerio*. I. Structure. *Journal of Comparative Neurology*, **224**, 107–15.

Bumsted, K., Lerea, C., Szél, A. *et al.* (1994) Appearance of cone and rod opsins in the developing primate retina. *Investigative Ophthalmology and Visual Science Supplement*, **35**, 1728.

Cameron, D.A. and Pugh, E.N. (1991) Double cones as a basis for a new type of polarization vision in vertebrates. *Nature*, **353**, 161–4.

Coffman, C., Harris, W. and Kintner, C. (1990) Xotch, the Xenopus homolog of *Drosophila* Notch. *Science*, **249**, 1438–41.

Cohen, A.I. (1972) Rods and cones, in *Handbook of Sensory Physiology* (ed. M.G.F. Fourtes), Springer-Verlag, New York, Vol. VII/2, pp.63–110.

Dartnall, H.J., Bowmaker, J.K. and Mollon, J.D. (1983) Human visual pigments: microspectrophotometric results from the eyes of seven persons. *Proceedings of the Royal Society of London Series B: Biological Sciences*, **220**, 115–30.

de Monasterio, F.M., Schein, S.J. and McCrane, E.P. (1981) Staining of blue-sensitive cones of the macaque retina by a fluorescent dye. *Science*, **213**, 1278–81.

Douglas, R.H. (1986) Photopic spectral sensitivity of a teleost fish, the roach (*Rutilus rutilus*), with special reference to its ultraviolet sensitivity. *Journal of Comparative Physiology A*, **159**, 415–21.

Engström, K. (1960) Cone types and cone arrangement in the retina of some cyprinids. *Acta Zoologica*, **41**, 277–95.

Engström, K. (1963) Cone types and cone arrangements in teleost retinae. *Acta Zoologica*, **44**, 179–243.

Falk, J.D. and Applebury, M.L. (1987) The molecular genetics of photoreceptor cells. *Progress in Retinal Research*, **7**, 89–112.

Goldsmith, T.H. (1990) Optimization, constraint, and history in the evolution of eyes. *Quarterly Review of Biology*, **65**, 281–322.

Hashimoto, Y., Hárosi, F.I., Ueki, K. and Fukurotani, K. (1988) Ultra-violet-sensitive cones in the color-coding systems of cyprinid retinas. *Neuroscience Research Supplement*, **8**, 81–96.

Hawryshyn, C.W. (1991) Light-adaptation properties of the ultraviolet-sensitive cone mechanism in comparison to the other receptor mechanisms of goldfish. *Visual Neuroscience*, **6**, 293–301.

Hawryshyn, C.W. and Hárosi, F.I. (1991) Ultraviolet photoreception in carp: microspectrophotometry and behaviorally determined action spectra. *Vision Research*, **31**, 567–76.

Hisatomi, O., Kayada, S., Aoki, Y. *et al.* (1994) Phylogenetic relationships among vertebrate visual pigments. *Vision Research* **34**, 3097–3102.

Hitchcock, P.F. and Raymond, P.A. (1992) Retinal regeneration. *Trends in Neuroscience*, **15**, 103–8.

Hitchcock, P.F., and Vanderyt, J.T. (1994) Regeneration of dopamine-cell mosaic in the retina of the goldfish. *Visual Neuroscience*, **11**, 209–17.

Hitchcock, P.F., Lindsey Myhr, K.J., Easter, S.S., *et al.* (1992) Local regeneration in the retina of the goldfish. *Journal of Neurobiology*, **23**, 187–203.

Johns, P.A. Raymond (1977) Growth of the adult goldfish eye. III. Source of the new retinal cells. *Journal of Comparative Neurology*, **176**, 343–58.

Johnson, R.L., Grant, K.B., Zankel, T.C. *et al.* (1993) Cloning and expression of goldfish opsin sequences. *Biochemistry*, **32**, 208–14.

Kljavin, I.J. (1987) Early development of photoreceptors in the ventral retina of the zebrafish embryo. *Journal of Comparative Neurology*, **260**, 461–71.

Knight, J.K. and Raymond, P.A. (1995) Retinal pigmented epithelial (RPE) cells do not transdifferentiate in adult goldfish. *Journal of Neurobiology*, in press.

Kunz, Y.W. (1980) Cone mosaics in a teleost retina: changes during light and dark adaptation. *Experientia*, **36**, 1371–4.

Kuwata, O., Imamoto, Y., Okano, T. *et al.* (1990) The primary structure of iodopsin, a chicken red-sensitive cone pigment. *FEBS Letters*, **272**, 128–32.

Larison, K.D. and BreMiller, R. (1990) Early onset of phenotype and cell patterning in the embryonic zebrafish retina. *Development*, **109**, 567–76.

LaVail, M.M., Unoki, K., Yasumura, D. *et al.* (1992) Multiple growth factors, cytokines, and neurotrophins rescue photoreceptors from the damaging effects of constant light. *Proceedings of the National Academy of Sciences, USA*, **89**, 11249–53.

Lebovitz, R.M. and Ready, D.F. (1986) Ommatidial development in *Drosophila* eye fragments. *Developmental Biology*, **117**, 663–71.

Levine, J.S. and MacNichol, E.F. (1979) Visual pigments in teleost fishes: effects of habitat, microhabitat, and behavior on visual system evolution. *Sensory Proceedings*, **3**, 95–131.

Lyall, A.H. (1957a) Cone arrangements in teleost retinae. *Quarterly Journal of Microscopical Science*, **98**, 189–201.

Lyall, A.H. (1957b) The growth of the trout retina. *Quarterly Journal of Microscopical Science*, **98**, 101–10.

Marc, R.E. (1986) The development of retinal networks, in *The Retina: a Model for Cell Biology Studies* (eds R. Adler and D. Farber), Academic Press, New York, pp.17–65.

Marc, R.E. and Sperling, H.G. (1976) The chromatic organization of the goldfish cone mosaic. *Vision Research*, **16**, 1211–24.

Marc, R.E and, Sperling, H.G. (1977) Chromatic organization of primate cones. *Science*, **196**, 454–6.

McConnell, S.K. (1991) The generation of neuronal diversity in the central nervous system. *Annual Review of Neuroscience*, **14**, 269–300.

Mensinger, A.F. and Powers, M.K. (1993) Visual function following surgical removal of retinal tissue. *Investigative Ophthalmology and Visual Science Supplement*, **34**, 1176.

Merbs, S.L. and Nathans, J. (1992) Absorption spectra of human cone pigments. *Nature*, **356**, 433–5.

Morris, V.B. (1970) Symmetry in a receptor mosaic demonstrated in the chick from the frequencies, spacing and arrangement of the types of retinal receptor. *Journal of Comparative Neurology*, **140**, 359–98.

Müller, H. (1952) Bau und Wachstum der Netzhaut des Guppy (*Lebistes reticulatus*). *Zoologische Jahrbucher Abteilung für Allegmeine Zoologie und Physiologie der Tiere*, **63**, 275–324.

Nathans, J., Thomas, D. and Hogness, D. (1986) Molecular genetics of human color vision: the genes encoding blue, green and red pigments. *Science*, **232**, 193–232.

Nawrocki, L.W. (1985) Development of the neural retina in the zebrafish, *Brachydanio rerio*. PhD Thesis, University of Oregon.

Nawrocki, L., BreMiller, R., Streisinger, G. and Kaplan, M. (1985) Larval and adult visual pigments of the zebrafish, *Brachydanio rerio*. *Vision Research*, **25**, 1569–76.

Neumeyer, C. (1992) Tetrachromatic color vision in goldfish: evidence from color mixture experiments. *Journal of Comparative Physiology [A]*, **171**, 639–49.

Okano, T., Kojima, D., Kukada, Y. *et al.* (1992) Primary structures of chicken cone visual pigments: vertebrate rhodopsins have evolved out of cone visual pigments. *Proceedings of the National Academy of Science, USA*, **89**, 5932–6.

Packer, O., Hendrickson, A.E. and Curcio, C.A. (1990) Developmental redistribution of photoreceptors across the *Macaca nemestrina* (pigtail macaque) retina. *Journal of Comparative Neurology*, **298**, 472–93.

Park, C.M. and Hollenberg, M.J. (1993) Growth factor-induced retinal regeneration *in vivo*. *International Review of Cytology*, **146**, 49–74.

Powers, D.A. (1989) Fish as model systems. *Science*, **246**, 352–8.

Raymond, P.A. (1985) Cytodifferentiation of photoreceptors in larval goldfish: delayed maturation of rods. *Journal of Comparative Neurology*, **236**, 90–105.

Raymond, P.A. (1991) Retinal regeneration in teleost fish, in *Regeneration of Vertebrate Sensory Receptor Cells*, (ed.) E. Rubel, *Ciba Foundation Symposium vol. 160*, Wiley, Chichester, pp.171–91.

Raymond, P.A., Reifler, M.J. and Rivlin, P.K. (1988) Regeneration of goldfish retina: rod precursors are a likely source of regenerated cells. *Journal of Neurobiology*, **19**, 431–63.

Raymond, P.A., Barthel, L.K., Rounsifer, M.E. *et al* (1993). Expression of rod and cone visual pigments in goldfish and zebrafish: a rhodopsin-like gene is expressed in cones. *Neuron*, **10**, 1161–74.

Raymond, P.A., Barthel, L.K. and Curran, G.A. (1994) Recruitment of photoreceptors in zebrafish embryos. *Molecular Biology of the Cell*, **4**, 374a.

Ready, D.F., Hanson, T.E. and Benzer, S. (1976) Development of the *Drosophila* retina, a neurocrystalline lattice. *Developmental Biology*, **53**, 217–40.

Ready, D.F., Tomlinson, A. and Lebovitz, R.M. (1986) Building an ommatidium: geometry and genes, in *Cell and Developmental Biology of the Eye* (eds S.R. Hilfer and J.B. Sheffield), Springer-Verlag, New York, pp.97–125.

Reh, T.A. (1991) Determination of cell fate during retinal histogenesis: intrinsic and extrinsic mechanisms, in *Development of the Visual System* (eds D.M. Lam and C.J. Shatz), MIT Press, Cambridge, *Retina Research Foundation Symposium, Vol. 3, pp.79–94.*

Risinger, C. and Larhammar, D. (1993) Multiple loci for synapse protein SNAP-25 in the tetraploid goldfish. Proceedings of the National Academy of Science, USA, **90**, 10598–602.

Robinson, J., Schmitt, E.A., Hárosi, F.I. *et al.* (1993) Zebrafish ultraviolet visual pigment: absorption spectrum, sequence, and localization. *Proceedings of the National Academy of Science, USA*, **90**, 6009–12.

Stell, W.K. and Hárosi, F.I. (1975) Cone structure and visual pigment content in the retina of the goldfish. *Vision Research*, **16**, 647–57.

Sullivan, S.A., Largent, B.L., Goldman, D.J. and Raymond, P.A. (1992) Isolation of a goldfish homologue to *Drosophila* Notch. *Investigative Ophthalmology and Visual Science Supplement*, **33**, 1062.

Szél, A., Roehlich, P., Mieziewska, K. *et al.* (1993) Spatial and temporal differences between the expression of short- and middle-wave sensitive cone pigments in the mouse retina: a developmental study. *Journal of Comparative Neurology*, **331**, 564–77.

Tokunaga, F., Iwasa, T., Miyagishi, M. and Kayada, S. (1990) Cloning of cDNA and amino acid sequence of one of the chicken cone visual pigments. *Biochemical and Biophysical Research Communication*, **173**, 1212–17.

Wahl, C.M. (1994) Periodic cone cell twists in the walleye, *Stizostendion vitreum;* a new type of retinomotor activity, *Vision Research*, **34**, 11–18.

Wald, G., Brown, P.K. and Smith, P.H. (1955)

Iodopsin. *Journal of General Physiology*, **38**, 623–81.

Walls, G.L. (1967) *The Vertebrate Eye and its Adaptive Radiation*. Hafner, New York, 787pp.

Wang, S.-Z., Adler, R. and Nathans, J. (1992) A visual pigment from chicken that resembles rhodopsin: amino acid sequence, gene structure, and functional expression. *Biochemistry*, **31**, 3309–15.

Wässle, H. and Reimann, H.J. (1978) The mosaic of nerve cells in the mammalian retina. *Proceedings of the Royal Society of London Series B: Biological Sciences*, **200**, 441–61.

Weinmaster, G., Roberts, V.J. and Lemke, G. (1991) A homolog of *Drosophila* Notch expressed during mammalian development. *Development*, **113**, 199–205.

Wikler, K.C. and Rakic, P. (1990) Distribution of photoreceptor subtypes in the retina of diurnal and nocturnal primates. *Journal of Neuroscience*, **10**, 3390–401.

Wikler, K.C. and Rakic, P. (1991) Relation of an array of early-differentiating cones to the photoreceptor mosaic in the primate retina. *Nature*, **351**, 397–400.

Yokoyama, R. and Yokoyama, S. (1990a) Isolation, DNA sequence and evolution of a color visual pigment gene of the blind cave fish *Astyanax fasciatus*. *Vision Research*, **6**, 807–16.

Yokoyama, R. and Yokoyama, S. (1990b) Convergent evolution of the red- and green-like visual pigment genes in fish, *Astyanax fasciatus*, and human. *Proceedings of the National Academy of Sciences USA*, **87**, 9315–18.

2

Cell biology and metabolic activity of photoreceptor cells: light-evoked and circadian regulation

P. MICHAEL IUVONE

2.1 INTRODUCTION

Retinal photoreceptor cells are specialized sensory neurons that convert light energy into electrical and neurochemical signals for transmission to other neurons. The two main photoreceptor subtypes, rods and cones, function over several log units of light intensity to provide information about luminance, contrast, movement, and color in the visual environment. Rod photoreceptor cells are more sensitive to light than cones, and are the predominant photoreceptor mediating scotopic or night vision. Cones respond to higher, photopic intensities of light, and mediate color vision. Rods and cones hyperpolarize in response to absorption of photons, with a consequent decrease in neurotransmitter release.

Photoreceptors are highly polarized cells (Figure 2.1). They consist of an outer segment, an inner segment, and a synaptic terminal. The outer segment is the light-sensitive portion of the cell. It contains a stack of membranous discs, rich in the visual pigment rhodopsin and other components of the visual transduction cascade (Chapters 4 and 5). Rod discs are closed and not contiguous with the plasma membrane, whereas cone discs are continuous with each other and exposed to the extracellular space. The outer segment is connected to the inner segment by a relatively short, non-motile cilium. This connecting cilium appears to provide a conduit for transport of soluble components between the inner and outer segments, and is the site of outer segment disc formation. The inner segment consists of the myoid, ellipsoid and nuclear regions. The ellipsoid is densely packed with mitochondria, and is generally thought to be the major source of high energy phosphates for both inner and outer segments. The myoid contains most of the cell's rough endoplasmic reticulum and Golgi apparatus, and the majority of protein synthesized there ultimately becomes associated with the outer segment (Young and Droz, 1968). The nuclear region, which contains the genetic machinery of the cell, is connected to the

Neurobiology and Clinical Aspects of the Outer Retina
Edited by M.B.A. Djamgoz, S.N. Archer and S. Vallerga
Published in 1995 by Chapman & Hall, London
ISBN 0 412 60080 3

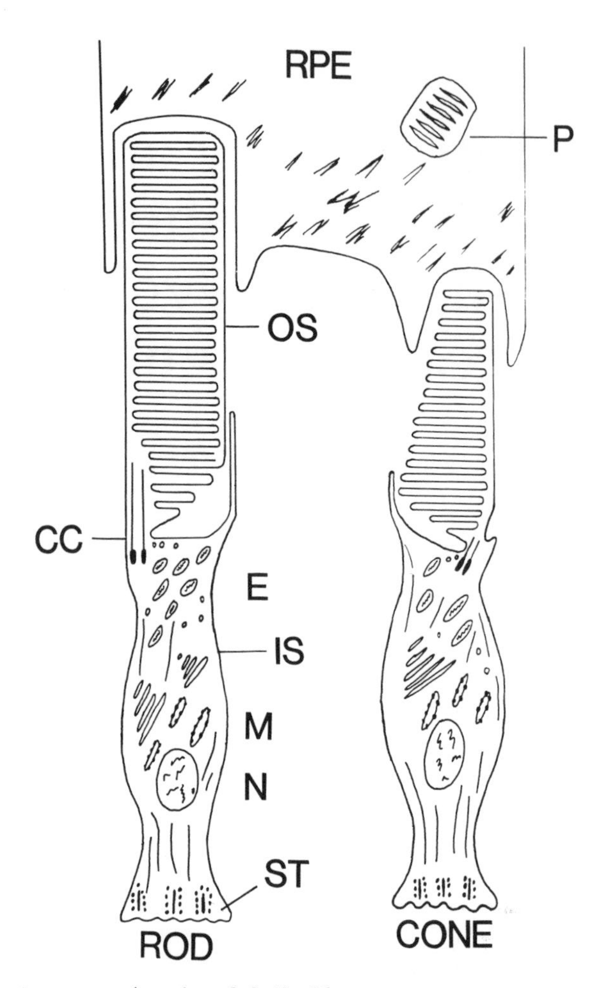

Figure 2.1 Photoreceptor cells and retinal pigment epithelium. Photoreceptors are highly polarized cells that are classified on morphological and physiological criteria as rods and cones. Rod and cone photoreceptors have three primary regions: an outer segment (OS), inner segment (IS) and synaptic terminal (ST). OS membranes of rods and cones contain the visual pigment proteins that impart light-sensitivity to the photoreceptor cells. The OS of rod photoreceptors contain stacks of visual pigment-containing membranous discs that are isolated from one another and the plasma membrane. The photosensitive discs of cone outer segments are continuous with each other and the plasma membrane. The discs and plasma membrane of photoreceptor cells are renewed by evagination of the membrane of the connecting cilium (CC), which links the IS and OS. The distal tips of photoreceptor cells are ensheathed by apical processes of the retinal pigment epithelial (RPE) cells. Distal tips of the OS are shed and phagocytosed by the RPE, which degrade the disc-containing phagosomes (P). The IS is divided into ellipsoid (E), myoid (M), and nuclear (N) regions. The IS is connected to the ST by a short fiber. The ST contains synaptic ribbons with their associated neurotransmitter-containing vesicles.

synaptic terminal, the site of neurotransmitter release, by a short axon or fiber.

The outer segment of rods and cones is surrounded by the apical processes of the retinal pigment epithelial (RPE) cells, which interact with the photoreceptor cells in many important ways. For example, the RPE is required for normal outer segment development (e.g., Hollyfield and Witkovsky, 1974), regeneration of the visual pigment chromophore following its isomerization by light (Bernstein *et al.*, 1987), expression of the opsin gene (Stiemke *et al.*, 1994), and phagocytosis and digestion of shed outer segment discs in the terminal step of outer segment turnover (section 2.2.1). Photoreceptor outer segments degenerate when detached from the RPE (Anderson *et al.*, 1983).

Light- and dark-adaptive mechanisms allow the retina to process visual stimuli over a large range of light intensities. Much of this adaptation occurs on a biochemical level within the outer segments, and occurs rapidly in response to changes in luminance. Other aspects of adaptation occur more slowly, frequently as diurnal or circadian rhythms. These latter aspects of adaptation involve inner retinal neurons as well as photoreceptor cells, and may occur as changes in synaptic strength, coupling and uncoupling of neurons through gap junctions, and release of neurotransmitters or neuromodulators that influence metabolism in the photoreceptor–pigment epithelial complex (Besharse *et al.*, 1988a, Dowling, 1989; and Chapter 7).

Several topics related to photoreceptor cell biology are explored in detail in other chapters of this book, including development (Chapter 1), visual transduction (Chapter 5), and synaptic neurotransmission (Chapters 6 and 7). The goal of the present chapter is to review select aspects of photoreceptor cell biology and metabolism, with an emphasis on those processes that are regulated by light, circadian clocks, and neurotransmitters.

2.2 LIGHT-EVOKED AND CIRCADIAN METABOLISM IN PHOTORECEPTOR CELLS

Many important aspects of photoreceptor cell biology and metabolism are controlled by light, photoperiod or circadian clocks. Some processes are acutely affected by light exposure. Others may be temporally regulated as rhythmic events, controlled as a function of the photoperiod. These rhythms are diurnal or circadian in nature. Diurnal photoreceptor rhythms are day–night rhythms that require the presence of a light–dark cycle for expression. In contrast, circadian rhythms are daily rhythms controlled by endogenous oscillators or clocks that are entrained by the light–dark cycle but that persist in constant darkness. Although the functions of diurnal and circadian rhythms may be similar, circadian mechanisms have the unique capacity to anticipate the transition between light and darkness, and can *prepare* a cell, tissue, or organism for the next phase of the photoperiod. Circadian clocks have been localized in both invertebrate and vertebrate retinas (Eskin, 1971; Besharse and Iuvone, 1983; Underwood *et al.*, 1988, 1990; Remé *et al.*, 1991; Cahill *et al.*, 1991; Cahill and Besharse, 1991, 1993). While multiple circadian oscillators may be involved in the regulation of retinal physiology, at least one of these circadian clocks appears to be in vertebrate photoreceptor cells (Zawilska and Iuvone, 1992; Cahill and Besharse, 1993; Pierce *et al.*, 1993; Thomas *et al.*, 1993). This section reviews several processes related to photoreceptor metabolism that are regulated by light and circadian oscillators.

2.2.1 ASSEMBLY AND TURNOVER OF PHOTORECEPTOR OUTER SEGMENT: DISC SHEDDING

Photoreceptor cells, like other neurons, are terminally differentiated cells and, once formed, do not undergo replication by cell division. With few exceptions, the photoreceptor cells formed during development must last the lifetime of the individual. Photoreceptor cells appear to remain viable through a process of renewal, in which most of the cells' contents, except their nuclear DNA, are dynamically replaced by synthesis and turnover (reviewed by Young, 1976). This renewal process is most striking for the outer segment of rod photoreceptor cells (Figure 2.2). Following the transport of newly synthesized proteins over time by autoradiography, Young and colleagues established that proteins synthesized in the inner segment are transported to the distal inner segment membrane near the connecting cilium and then assembled into membranous discs at the base of the outer segment (Young, 1967; Young and Droz, 1968; Young and Bok, 1969). These discs are gradually displaced distally by other newly forming discs. They ultimately become detached or shed from the outer segment and are phagocytosed by RPE cells. Young interpreted these observations as the mechanism whereby the entire outer segment, including the

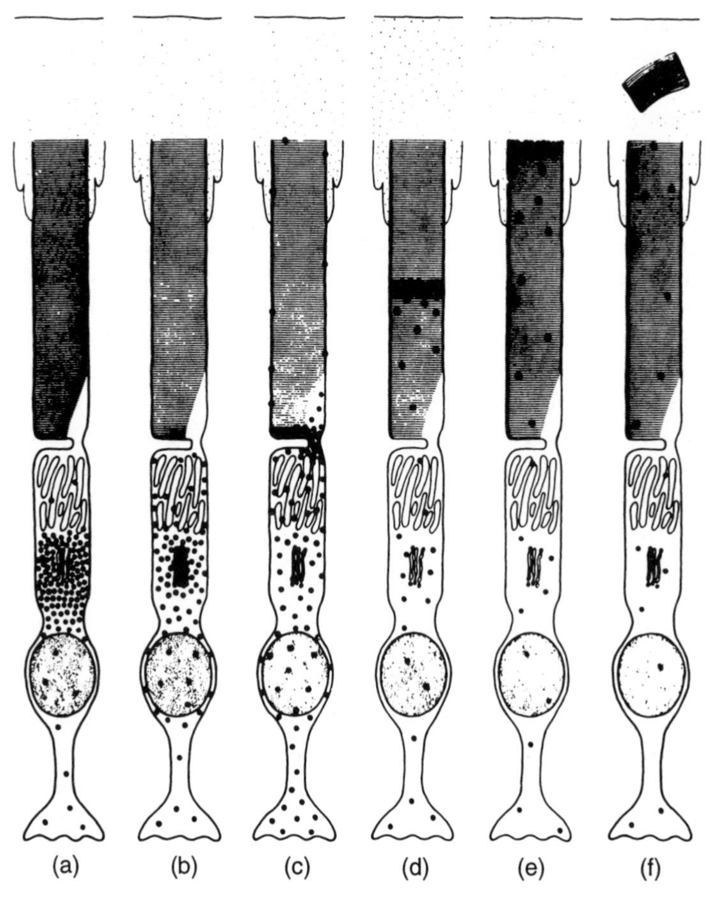

Figure 2.2 Diagram illustrating the renewal of protein in rod visual cells as revealed in the autoradiographic studies by Young and colleagues. Following administration of radioactive amino acids, newly synthesized protein is first concentrated in the myoid region of the inner segment (a). Labeled proteins then begin to be transported throughout the cell, with many proteins passing through the Golgi complex (b). Much of the radioactive protein subsequently moves to the periciliary region of the inner segment and is incorporated into newly forming disc membranes at the base of the outer segment, resulting in a band of radioactive discs that lie perpendicular to the long axis of the outer segment (c). Repeated formation of additional, non-radioactive discs displaces the labeled discs distally along the outer segment (d). Ultimately, the labeled band of discs reaches the tip of the outer segment (e), becomes detached from the photoreceptor cell, and is phagocytosed and degraded by the RPE cells (f). Reproduced from Young (1976) with the permission of the C.V. Mosby Co., St Louis, MO.

disc membranes and associated opsin, are renewed.

Approximately 90% of the intrinsic disc-membrane protein is opsin (Hall *et al.*, 1969; Papermaster and Dreyer, 1974). Thus, opsin was the major radioactive protein observed in the autoradiographic studies of outer segment renewal (Young and Droz, 1968). Radioactive amino acids initially incorporated in the rough endoplasmic reticulum of the inner segment were observed to move through the Golgi complex and subsequently through the cytoplasm to the periciliary region. Opsin is co-translationally inserted into lipid bilayers, and remains membrane-associated as it is transferred through the inner segment to the newly forming discs of the outer segment (e.g., Papermaster *et al.*, 1975; reviewed in Besharse, 1986).

Disc membranes are apparently formed at the base of the outer segment by evagination of the membrane of the inner face of the connecting cilium. In the model of disc morphogenesis proposed by Steinberg *et al.*, (1980), the ciliary membrane is specified into alternating regions of disc *surface* formation and disc *rim* assembly. The two sides of the evaginating ciliary membrane form the surfaces of adjacent discs. At this time, the disc membranes are continuous with the plasma membrane (Figure 2.1) and increase in surface volume as they are displaced distally along the connecting cilium. The disc rim originates from regions of the ciliary membrane between adjacent disc surface evaginations and grows around the circumference of the expanding disc surfaces, joining the two surfaces into a closed disc and sealing off the plasma membrane from the disc. This model provides an explanation for the distinct contents of the disc surface, which is rich in opsin, and the disc rim, which contains a high-molecular-weight protein and peripherin but apparently lacks opsin (Papermaster *et al.*, 1978; Molday *et al.*, 1987). It also establishes a possible mechanism for the renewal of the outer segment plasma membrane. More extensive information on opsin synthesis and transport, and disc assembly can be found in reviews by Bok (1985), Besharse (1986), Besharse and Horst (1990) and Williams (1991).

Disc morphogenesis may be regulated by an imposed light/dark cycle. Diurnal expression of opsin mRNA levels has been observed in mouse, fish and toad retinas (section 2.2.3). Disc addition to rod outer segments in *Xenopus* retina is regulated as a diurnal rhythm (Figure 2.3). Based on the number of open discs (partially assembled discs) at the base of the outer segment and the rate of radioactive disc band displacement following pulse-labeling of the photoreceptor with [^{3}H]leucine, it is apparent that most disc assembly occurs during the first half of the light phase of the light/dark cycle (Besharse *et al.*, 1977). Similar data are not available for cone disc assembly because cone discs remain open through the length of the outer segment (Figure 2.1) and radioactive proteins diffuse throughout the outer segments and do not form discrete bands like those in rods.

In the terminal process of outer segment turnover, disc shedding, stacks of discs and their associated plasma membrane at the distal tip of the outer segment become detached from the photoreceptor and are phagocytosed and digested by the adjacent RPE (reviewed by Besharse, 1982, 1986; Bok, 1985, 1993). Disc shedding is thought to involve active processes by both photoreceptor cells and RPE cells, and does not appear to occur in the absence of the RPE. Photoreceptor outer segment tips are surrounded by RPE processes that participate in the mechanisms of detachment and phagocytosis. Disc vesiculation may occur at the point of detachment from the outer segment. Pseudopods from the RPE processes then engulf the fragment of outer segment to be shed. The outer segment fragment detaches into the cytoplasm of the RPE cell as a phagosome, and is transported to the RPE cell body where it fuses with lysosomes that digest it.

The process of shedding and phagocytosis may involve specific receptor–ligand interactions on the RPE cell surface (e.g., Hall and Abrams, 1987), and as such, second messengers may participate in this phase of outer segment turnover. This hypothesis is consistent with observations that phagocytosis of outer segment fragments is accompanied by changes in protein phosphorylation in RPE cells (Heth and Schmit, 1991). Several second

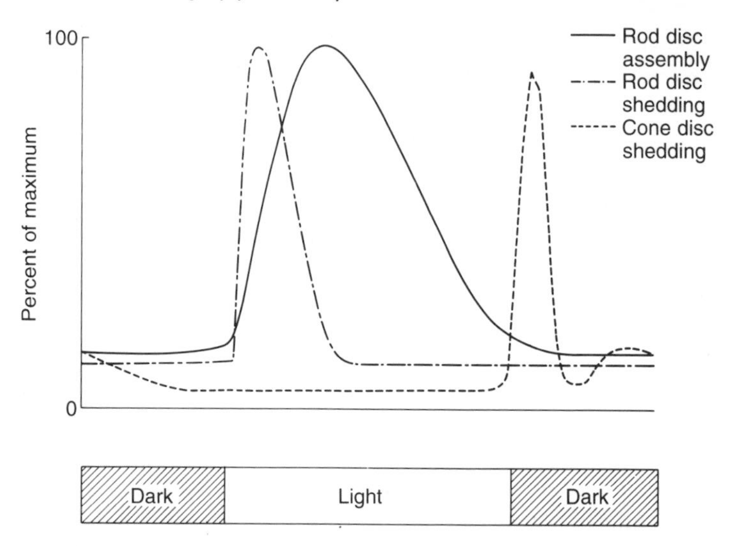

Figure 2.3 Schematic illustration of the temporal relationship of outer segment disc assembly and disc shedding to the 12 h light: 12 h dark cycle. The majority of rod disc assembly occurs during the first half of the light phase of the light/dark cycle (Besharse *et al.*, 1977). Rod disc shedding occurs during a short interval just after light offset, requires a preceding period of darkness, and is suppressed later in the light phase. Cone disc shedding occurs primarily at the beginning of the dark phase, although some cone disc shedding has been observed later in the dark phase (e.g., Young, 1978). Although conclusive evidence is lacking, outer segment membrane renewal in cones, like disc shedding, may be out of phase with that in rods by approximately 12 h (Young, 1978).

messengers have been implicated in the regulation of disc shedding and phagocytosis. Cyclic AMP analogues and phosphodiesterase inhibitors inhibit disc shedding in *Xenopus* eye cups (Besharse *et al.*, 1982). Cyclic AMP also inhibits phagocytosis of outer segment fragments by cultured RPE cells (Edwards and Bakshian, 1980), indicating that at least part of the effect of cyclic AMP on disc shedding is mediated within the epithelial cell. Low concentrations of Ca^{2+} also inhibit disc shedding (Greenberger and Besharse, 1983), but the mechanism of this effect is unknown. In cultured RPE cells, phagocytosis of outer segment fragments has been associated with activation of phospholipase C (Heth and Marescalchi, 1994), which generates two second messengers: the protein kinase C activator diacylglycerol, and the intracellular Ca^{2+} mobilizing agent inositol trisphosphate (Nishizuka, 1992).

Disc shedding and phagocytosis are temporally regulated processes, and their relation to the photoperiod differs for rods and cones (Figure 2.3). Rod disc shedding is restricted to a period just after light onset, whereas cone disc shedding generally occurs at night, just after light offset. At least for rods, the period of maximal disc shedding immediately precedes that of maximal assembly. Young (1978) suggested that both rods and cones renew their outer segment membranes in a manner that follows daily rhythms, and the rhythms for rods and cones are out of phase by approximately 12 h. The role of light and circadian oscillators in rod disc shedding has been more thoroughly studied than that in cones, and differs between species. In frogs such as *Rana pipiens*, rod disc shedding appears to be exclusively light driven (Basinger *et al.*, 1976). Disc shedding in the African clawed frog, *Xenopus laevis*, is regulated by both light onset and by circadian oscillators (Besharse *et al.*, 1977), whereas that in rats is primarily circadian and persists in constant darkness (LaVail, 1976; Goldman *et al.*, 1980; Tierstein *et al.*, 1980).

The temporal regulation of rod disc shedding has been extensively studied *in vitro* using frog eye cups (reviewed by Besharse *et al.*, 1988a). Rod outer segment disc shedding occurs during a period of 1–1.5 h after light onset. It requires a preceding period of darkness, as shedding is dramatically suppressed in constant light. In animals maintained in constant light for three days, a brief period of darkness (30–60 min) is sufficient to initiate shedding. These observations have led to a model for the temporal regulation of disc shedding that involves processes of 'dark priming', 'light triggering' and 'light inhibition'. Light *per se* is not required for the mechanical processes of shedding or phagocytosis, but serves as an initiating stimulus. This was clearly demonstrated by exposing eye cups to a brief pulse of light followed by a return to darkness (Besharse *et al.*, 1986). The light pulse initiated a cascade of events resulting in disc detachment and phagocytosis in darkness.

In rats that have been entrained to a light/dark cycle and then kept in constant darkness, the morning shedding response persists and is similar in magnitude to that seen in cyclic light and dark (LaVail, 1976; Goldman *et al.*, 1980; Tierstein *et al.*, 1980). The circadian rhythm of disc shedding in rats is suppressed in constant light. Under these conditions, exposure to darkness for 2 h will initiate shedding, but does so most effectively near the usual time of light onset (Goldman *et al.*, 1980). Similarly, sensitivity of shedding to light exposure appears to be regulated by a circadian clock, as light pulses administered to rats exposed to constant darkness are most effective in stimulating disc shedding near the subjective dawn (Remé *et al.*, 1986). Like other circadian rhythms, that of disc shedding can be phase shifted by changing the light/dark cycle to which rats are exposed (Tierstein *et al.*, 1980).

Circadian disc shedding is locally controlled by oscillators within the eye. Circadian shedding in rat retina persists following removal of pituitary and pineal glands (LaVail and Ward, 1978; Tamai *et al.*, 1978), cutting the optic nerves (Tierstein *et al.*, 1980), or lesioning the suprachiasmatic nucleus (SCN) of the hypothalamus (Terman *et al.*, 1993), the site of the circadian clock that regulates most behavioral and endocrine rhythms in mammals. Furthermore, the circadian rhythms of disc shedding in the two eyes of individual animals can be dissociated by occluding one eye and re-entraining the shedding rhythm of the open eye to a shifted light/dark cycle (Tierstein *et al.*, 1980). Lastly, evidence has been presented for continued rhythmicity of shedding in isolated eyes of *Xenopus laevis* (Flannery and Fisher, 1984).

2.2.2 REGULATION OF RETINOMOTOR MOVEMENTS BY THE LIGHT/DARK CYCLE

Photoreceptor cells of lower vertebrates have the capacity to elongate or contract via actin-based and microtubule-based cytoskeletal systems in the inner segment region of the cell (reviewed by Burnside and Dearry, 1986). In animals kept on a daily light/dark cycle, cones contract in the light and elongate in darkness. In some animals (e.g., fish), the response of rods is opposite to that in cones, with elongation in light and contraction in darkness. However, other animals demonstrating cone retinomotor movements, such as *Xenopus laevis*, show little or no rod movement. The role of these movements is still unclear, but it has been suggested that they represent morphological components of light- and dark-adaptation (Burnside and

Nagle, 1983), and may serve to the position the cone outer segments in the plane of focus during the day and out of the plane of focus at night, when rod vision predominates. In some species, pigment migration in the RPE is also regulated by the light/dark cycle. In light, pigment granules disperse into the villous processes of the RPE cells that extend into the space between outer segments, whereas in darkness the pigment granules migrate out of the processes and aggregate in the cell body.

Cone movement is regulated by circadian clocks. Welsh and Osborne (1937) noted that the day/night cycle of contraction and elongation in a teleost fish persisted for several days in constant darkness. Furthermore, contraction of cones may precede the onset of light during the light/dark cycle, presumably preparing the retina for dawn (Burnside and Ackland, 1984; Kohler *et al.*, 1990; McCormack and Burnside, 1991).

Retinomotor movements involve cyclic AMP and Ca^{2+}-dependent processes (Burnside and Dearry, 1986). Cyclic AMP promotes dark-adaptive rod and cone movements and RPE pigment aggregation. Cyclic AMP and Ca^{2+} appear to have antagonistic roles in regulating cone position and pigment granule migration in RPE. Cyclic AMP activates a microtubule-based motor that promotes cone elongation, and interferes with Ca^{2+}-calmodulin-dependent contraction, which is an actin-based process. Dark-adaptive pigment aggregation is also promoted by cyclic AMP and inhibited by high extracellular concentrations of Ca^{2+}. In rods, cyclic AMP and Ca^{2+} may act together to promote dark-adaptive contraction. The observation that cyclic AMP promotes the dark-adaptive response in all three cell types suggests that it may be the key mediator of darkness throughout the photoreceptor–RPE complex (Besharse *et al.*, 1988a). This hypothesis is consistent with observations on cyclic AMP levels in photoreceptor cells (section 2.2.5).

2.2.3. EXPRESSION OF VISUAL TRANSDUCTION GENES

The visual transduction cascade is reviewed in detail in Chapter 5. Briefly, photoreceptor cells are depolarized in darkness due to an inward cation current through cyclic GMP-gated channels in the outer segment plasma membrane. Light decreases this conductance by an action on rhodopsin. Absorption of a photon activates rhodopsin, which in turn activates a heteromeric guanyl nucleotide binding protein, transducin. Activated transducin dissociates into free α and $\beta\gamma$ subunits. Transducin-α binds to cyclic GMP phosphodiesterase, resulting in a dramatic stimulation of cyclic GMP hydrolysis and the consequent closing of the cyclic GMP-gated channel. The cascade is terminated in part by the phosphorylation of rhodopsin and the binding of arrestin to the phosphoprotein, which prevents its further interaction with transducin.

Expression of the mRNAs for several of the proteins in the visual transduction cascade is regulated as a function of the light/dark cycle (Figure 2.4). Diurnal rhythms of opsin (rhodopsin apoprotein) mRNA abundance have been observed in mouse, fish and toad (Bowes *et al.*, 1988; Korenbrot and Fernald, 1989; McGinnis *et al.*, 1992). Opsin mRNA level peaks in early morning, near the time of light onset, and decreases thereafter to reach a minimum in the early to mid-dark phase. Toad retinal opsin mRNA levels may be regulated by both a circadian oscillator and light. Korenbrot and Fernald (1989) reported that the morning increase in mRNA levels occurs in toads kept in darkness, and that a 30 min period of light exposure increases opsin mRNA levels at night.

Transcription of the iodopsin gene in chicken retina is regulated by a circadian clock (Pierce *et al.*, 1993). Iodopsin is a cone visual pigment of avian retinas. Iodopsin mRNA levels increase late in the light phase of the light/dark cycle, reaching a peak near

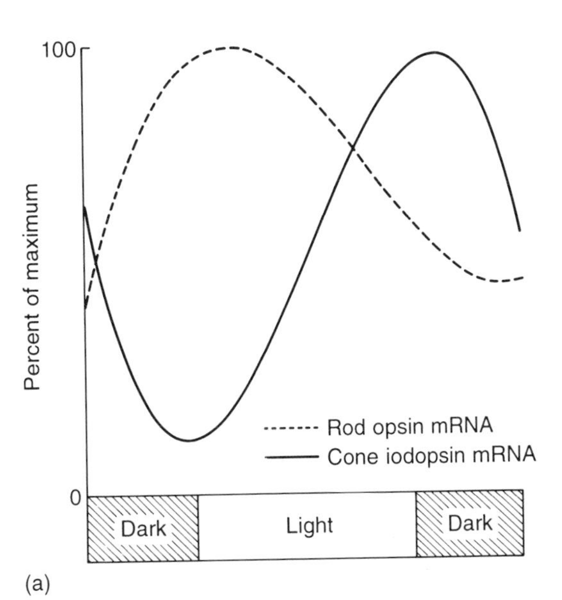

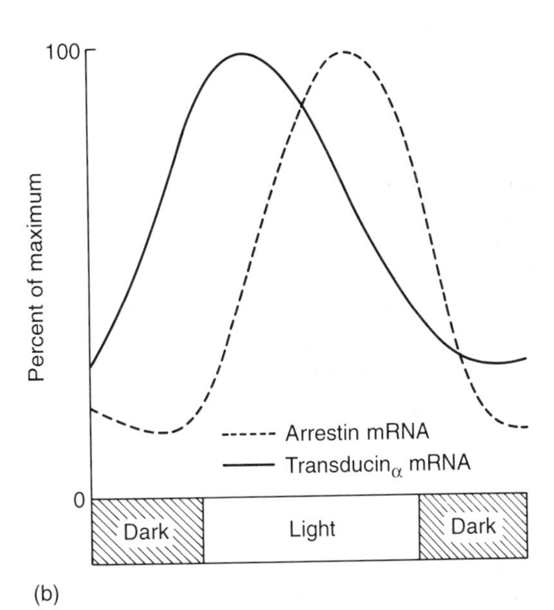

Figure 2.4 Schematic representation of the rhythms in mRNA levels of outer segment proteins. (a) Visual pigments. Rod opsin mRNA fluctuates rhythmically, with a peak near the beginning of the light phase of the light/dark cycle (Bowes *et al.*, 1988; Korenbrot and Fernald, 1989; McGinnis *et al.*, 1992), whereas the mRNA levels for cone iodopsin peak near the end of the light phase or beginning of the dark phase (Pierce *et al.*, 1993). The latter rhythm has been shown to be regulated by a circadian clock and to represent a rhythm in mRNA transcription rather than stability. The times of maximal mRNA levels for rod and cone visual pigments correspond to times of outer segment disc shedding in the two types of photoreceptors. (b) Other visual transduction proteins. The mRNA for the α-subunit of transducin in rods fluctuates with a rhythm similar to that of opsin, with a peak near the time of light onset (Brann and Cohen, 1987; Bowes *et al.*, 1988). Arrestin mRNA in rods also fluctuates rhythmically, but the peak in arrestin mRNA levels comes later in the light phase (Bowes *et al.*, 1988; Craft *et al.*, 1990; McGinnis *et al.*, 1992). The immunocytochemical localization of the protein products of these mRNA also differs in light and dark; in dark-adapted retinas, immunoreactivity for transducin is high in the outer segment but low in the inner segment, whereas that of arrestin shows the opposite distribution. The relative distribution of immunoreactivity for these two proteins reverses rapidly upon light exposure (Philp *et al.*, 1987; Brann and Cohen, 1987; Whelan and McGinnis, 1988).

the time of light offset (Figure 2.4). The rhythm of iodopsin mRNA abundance persists in constant darkness. The fluctuations in mRNA abundance reflect changes in iodopsin gene transcription. Rhythmic iodopsin gene expression occurs *in vivo* and in dissociated retinal cell cultures from embryonic chick.

The mRNA of transducin-α, which couples rhodopsin activation to cyclic GMP metabolism, is also regulated as a function of the light/dark cycle (Brann and Cohen, 1987; Bowes *et al.*, 1988). Transducin-α mRNA levels in mouse and rat retinas fluctuate in a diurnal fashion that resembles the rhythm of opsin mRNA levels (Figure 2.4). It is unclear if the rhythm of the G-protein mRNA levels is circadian or light driven, and if it reflects transcriptional activation or a change in mRNA stability. Arrestin (S-antigen, 48 kDa protein) mRNA levels in mouse and rat retinas display robust diurnal regulation with peak expression in the middle or late light phase (Bowes *et al.*, 1988; Craft *et al.*, 1990; McGinnis *et al.*, 1992). As with transducin, the role of circadian oscillators in the regulation of arrestin mRNA levels is unclear. Relatively short (0.25–1 h) light pulses delivered 4 h into the dark phase increase arrestin mRNA levels, demonstrating a role for light in arrestin mRNA expression (Craft *et al.*, 1990). Based on kinetic arguments related to the rate of change and half-life of arrestin mRNA, McGinnis *et al.* (1992) have concluded that the diurnal fluctuation in mRNA levels are due to changes in transcriptional activity rather than changes in mRNA stability. The increase in arrestin mRNA levels during light phase is accompanied by an increased rate of arrestin biosynthesis (Agarwal *et al.*, 1994).

The biological significance of the different temporal patterns of mRNA levels for outer segment gene products is not known. It is of interest that the peaks of maximal expression of the photopigment protein mRNAs for rods and cones, which are out of phase by approximately 12 h, correspond to times just prior to disc shedding in rods and cones, which are similarly out of phase with one another. It has been suggested that the expression of opsin mRNA is anticipatory to rod disc shedding and functions to provide new opsin molecules for insertion into newly forming discs at the base of the outer segment (McGinnis *et al.*, 1992). A similar function can be proposed for the temporal regulation of iodopsin mRNA in cones (Pierce *et al.*, 1993). However, arrestin mRNA levels in rods is also out of phase with rod opsin mRNA expression. Thus, the mRNAs for different outer segment proteins appear to be differentially regulated by photoperiod. Perhaps some indication of the functions of these different rhythms can be gained from an understanding of the distribution and transport of the various gene products within the photoreceptor cells.

2.2.4 CELLULAR TRAFFICKING OF SIGNAL TRANSDUCTION PROTEINS

Immunohistochemical localization of a variety of proteins within the rod photoreceptor cell shows dramatic light/dark differences, which may participate in adaptive mechanisms. Transducin-α and -β and calpain II, a Ca^{2+}-activated protease, are localized primarily in the outer segment in dark-adapted retinas. Upon light exposure, immunoreactivity for these proteins rapidly redistributes to the inner segment (Philp *et al.*, 1987; Brann and Cohen, 1987; Whelan and McGinnis, 1988; Azarian *et al.*, 1993). Arrestin shows the opposite light-induced change in immunolocalization, and cyclic GMP phosphodiesterase immunoreactivity displays no differential distribution in light and dark (Philp *et al.*, 1987; Whelan and McGinnis, 1988). The effects of light on outer segment content of transducin, arrestin, and phosphodiesterase have been confirmed by immunoblots and Coomassie staining of SDS-polyacrylamide gels of purified outer segments (Philp *et al.*, 1987; Whelan and McGinnis, 1988). The molecular motors that regulate the transport of proteins between the inner and outer segments are not known. Whelan and McGinnis (1988) have made the analogy to microtubule-based bidirectional axoplasmic flow in neurons, which involves distinct translocator proteins for movement in each direction (Vale *et al.*, 1985).

The effect of light on the ultrastructural localization of arrestin-like immunoreactivity has been examined in mouse photoreceptor cells (Nir and Ransom, 1993). In darkness, the majority of labeling was observed in the connecting cilium, inner segments, nuclei and synaptic terminals. The role of arrestin in these areas is not known. Within the outer segment, labeling was low in discs, but areas of cytoplasm, most notably on the ciliary side of the outer segment and in caps above the apical disc, were densely labeled. In light, labeling was low in the inner segment, whereas within the outer segment labeling was densely associated with disc membranes, and not in cytoplasm. These latter observations suggest that light may cause an intrasegmental redistribution of arrestin, and is consistent with the effect of light to promote rhodopsin phosphorylation and the binding of arrestin to the phosphorylated visual pigment protein in disc membranes. Nir and Ransom (1993) have also questioned the concept that arrestin is transported bidirectionally between the inner and outer segment. They note that arrestin synthesis in the inner segment is maximal in light, yet arrestin immunoreactivity in the inner segment is very low. Thus the transport of newly synthesized arrestin into the outer segment must be very rapid in light-exposed retinas. The apparent transport from the outer to the inner segment in darkness could be due to inhibition of outward transport and accumulation of newly synthesized arrestin. Arrestin synthesis might be expected to continue for some time after the onset of darkness, as the half-life of arrestin mRNA is approximately 3.5 h (McGinnis *et al.*, 1992). It is also important to recognize the possibility that apparent redistribution, based on immunocytochemical localization, could be due to masking of epitopes on the proteins rather than to protein transport.

Although no functional relationship can yet be ascribed to the correlation, it is interesting that two soluble transduction proteins, transducin and arrestin, show opposite cellular redistributions in response to light, and their mRNA levels have opposite photoperiod phase relationships.

2.2.5 SECOND MESSENGERS

A variety of second messenger cascades in photoreceptor cells appear to be regulated by light. The prominent role of light-evoked cyclic GMP metabolism in the visual transduction is reviewed in Chapter 5. Another important cyclic nucleotide, cyclic AMP, is also regulated by light, which decreases cyclic AMP levels (e.g. Orr *et al.*, 1976; DeVries *et al.*, 1978; Farber *et al.*, 1981; Burnside *et al.*, 1982). Microdissection studies (Orr *et al.*, 1976; DeVries *et al.*, 1979) suggest that the light-evoked decreases of cyclic AMP levels occur primarily in the inner segment, nuclear, and synaptic regions of the cell, rather than in outer segments.

The mechanism whereby light reduces cyclic AMP levels is not fully understood. In the rod dominant mouse retina, Cohen (1982) found that the inhibition of cyclic AMP levels by light was evident in the presence of cyclic nucleotide phosphodiesterase inhibition, indicative of a reduction of adenylate cyclase activity. In contrast, Denton *et al.* (1992) reported light-evoked increases in both adenylate cyclase activity and cyclic AMP phosphodiesterase activity in isolated *Anolis* cones. The effect of light on cyclic AMP levels in the cell body/ synaptic terminal compartments of the cell is presumably indirect, and may be regulated by changes in membrane potential. This possibility was explored in photoreceptor-enriched cell cultures prepared from embryonic chick retina (Iuvone *et al.*, 1991a). The photoreceptor cells in these cultures have morphologically distinct inner segments and cell bodies and a single short axon, but no organized outer segment (Adler *et al.*, 1984). Depolarizing concentrations of extracellular K^+ increased cyclic AMP accumulation in the presence or absence of phosphodiesterase inhibitors, demonstrating a stimulation of cyclic AMP formation (Iuvone *et al.*, 1991b). The effect of K^+ involved Ca^{2+} influx through L-type voltage-gated channels, which are present in the inner segment and terminal regions of photoreceptor cells (Fain *et al.*, 1980; Corey *et al.*, 1984; Barnes and Hille, 1989; Gleason *et al.*, 1992). These observations are consistent with a model in which cyclic AMP synthesis in photoreceptor cells is stimulated in depolarized, dark-adapted photoreceptor cells via a mechanism that involves elevated intracellular Ca^{2+}. According to the model, light reduces cyclic AMP formation by hyperpolarizing the photoreceptor and decreasing intracellular free Ca^{2+}.

Cyclic AMP appears to be involved in regulating several aspects of photoreceptor metabolism. In addition to playing a key role in the regulation of melatonin biosynthesis (section 2.2.7), the cyclic nucleotide promotes dark-adaptive cone elongation and rod contraction (section 2.2.2). Phosducin, which may play multiple roles in photoreceptor physiology (Lee *et al.*, 1987; Abe *et al.*, 1990; Pagh-Roehl *et al.*, 1993), is phosphorylated in darkness by cyclic AMP-dependent protein kinase (Lee *et al.*, 1984, 1990). Cyclic AMP analogues inhibit rod outer segment disc shedding (Besharse *et al.*, 1982), but it is unclear if the primary target of the analogs is the photoreceptor or the retinal pigment epithelium.

Membrane phospholipid-derived second messengers are also involved in photoreceptor metabolism. Phosphatidylinositol 4,5-bisphosphate (PIP2) is a minor membrane phospholipid that is metabolized by phospholipase C into two second messenger molecules: inositol trisphosphate, which mobilizes intracellular stores of calcium, and diacylglycerol, which activates protein kinase C (reviewed by Nishizuka, 1992; Berridge, 1993). Light activates PIP2-specific phospholipase C in whole retina of frogs and rats (Ghalayini and Anderson, 1984; Jung *et al.*, 1993) and in outer segments of frog, rat and chick (Hayashi and Amakawa, 1985; Millar *et al.*, 1988). This observation is consistent with observations that protein kinase C may participate in the phosphorylation and desensitization of rhodopsin (Newton and Williams, 1993a,b). Ghalayini and Anderson (1992) have demonstrated that arrestin activates soluble phospholipase C, and proposed a model for light-activated PIP2 hydrolysis in which a soluble phospholipase C–arrestin complex binds to bleached and phosphorylated rhodopsin enabling the phospholipase to hydrolyze PIP2 in disc membranes. Phospholipase C may be regulated differently in the cell body/synaptic terminal regions of the cell. K^+-evoked depolarization of photoreceptor-enriched cell cultures stimulates phospholipase C activity and inositol phosphate accumulation as a consequence of calcium influx (Gan and Iuvone, 1993). The stimulation of inositol phosphate accumulation appears localized to photoreceptor cells, as it is greater in photoreceptor-enriched cell cultures than in cultures where multipolar neurons are the dominant cell type, and is not diminished by pretreatment with neurotoxins that destroy multipolar neurons but not photoreceptor cells. Inositol phosphate receptors, which regulate Ca^{2+} mobilization, are localized in photoreceptor cells to the presynaptic terminals (Peng *et al.*, 1991). These findings suggest a role for phospholipase C and inositol trisphosphate in modulating photoreceptor neurotransmitter release in darkness.

Phospholipase A_2 metabolizes a variety a membrane phospholipids, liberating free arachidonic acid that is subsequently converted into multiple first and second messengers (reviewed by Axelrod *et al.*, 1988; Shimizu and Wolfe, 1990). Light or GTPγS activate phospholipase A_2 in rod outer segment preparations, apparently by a mechanism that involves activated transducin (Jelsema and Axelrod, 1987). The roles of arachidonate metabolites in photoreceptor function have not been established. Prostaglandins, which are generated from arachidonic acid via a cyclo-oxygenase pathway, have been shown to stimulate cone and RPE retinomotor movements (Cavallaro and Burnside, 1988). However, the effect of prostaglandins may be indirect, as they are not observed in isolated photoreceptor cells. Furthermore, prostaglandins stimulate dark-adaptive cone elongation, suggesting that the relevant prostaglandins are not produced by light-activated phospholipase A_2.

2.2.6 IMMEDIATE EARLY GENES

Genes that are expressed in neurons in response to synaptic input or membrane depolarization generally fall into two categories: (1) those that are expressed rapidly and transiently within minutes of neuronal stimulation, referred to as cellular immediate early genes; and (2) those that are expressed or repressed with a time frame of hours, called late response genes (reviewed by Sheng and Greenberg, 1990). Immediate early genes may function as transcription factors, regulating the expression of several late response genes. One of the more extensively studied immediate early genes is c-*fos*, the protein product of which combines with other immediate early gene products to form transcriptional activators or repressors.

Using *in situ* hybridization to localize c-*fos* mRNA in rat retina, Yoshida *et al.* (1993) described a differential pattern of c-*fos* expression in the inner and outer retina during the light/dark cycle. In the inner nuclear and ganglion cell layers, labeling was transiently increased in a subpopulation of cells shortly after light onset. In contrast, c-*fos* expression in outer nuclear layer cells, presumably photoreceptor cells, increased rapidly early in the dark phase of the light/dark cycle, remained elevated throughout the night, and was low during the light phase (Figure 2.5). This rhythm persisted in constant darkness, indicating that it was controlled by a circadian oscillator. The expression of c-*fos* mRNA at night was rapidly suppressed to daytime levels by exposure to light for 30 min. A similar pattern of c-*fos* expression was observed by Northern blot analysis of mouse retina, but attempts to demonstrate circadian control in mouse retina were not successful (Nir and Agarwal, 1993). The role of c-*fos* expression in photoreceptor metabolism is currently unknown, but the c-*fos* gene product is a candidate transcriptional regulator for one or more of the genes in photoreceptor cells that are expressed in a diurnal or circadian fashion.

The expression of mRNA for NGF1-A, another immediate early gene, in mouse retina has also recently been shown to fluctuate in a diurnal fashion, with high expression at night (Agarwal, 1994). However, the cellular localization of this rhythm has not yet been established.

2.2.7 MELATONIN BIOSYNTHESIS AND RELEASE

Melatonin is a putative neuromodulator that influences many aspects of dark-adaptive and rhythmic retinal physiology (reviewed by Besharse *et al.*, 1988a). Photoreceptor cells are the primary, if not exclusive, retinal cell type that synthesizes melatonin (e.g., Redburn and-Mitchell, 1989; Iuvone *et al.*, 1990; Cahill and Besharse, 1993; Thomas *et al.*, 1993; Wiechmann and Craft, 1993). Melatonin biosynthesis in retina occurs as a diurnal or circadian rhythm (reviewed by Iuvone, 1986a; Cahill *et al.*, 1991; Cahill and Besharse, 1991; Zawilska and Nowak, 1992). Synthesis and release of melatonin are high at night and are acutely suppressed by light exposure (Figure 2.5). The circadian regulation of melatonin biosynthesis has been extensively examined in retinas of chickens, quail, and African clawed frogs (e.g., Hamm and Menaker, 1980; Besharse and Iuvone, 1983; Iuvone and Besharse, 1983; Underwood *et al.*, 1988). Diurnal

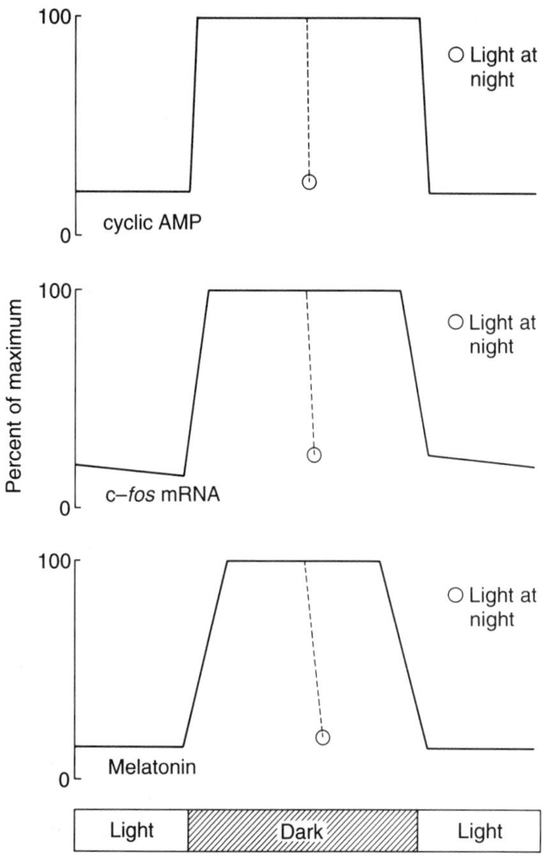

Figure 2.5 Schematic illustration of the similar rhythms of cyclic AMP levels, c-*fos* mRNA abundance, and melatonin biosynthesis in photoreceptor cells. Cyclic AMP induces the expression of c-*fos* and the activity of serotonin *N*-acetyltransferase, a key regulatory enzyme in the biosynthetic pathway for melatonin (see text for references). Cyclic AMP levels, c-*fos* expression and melatonin synthesis in photoreceptors are low in light (daytime), and increase in darkness (at night). Exposure to light at night decreases cyclic AMP to daytime levels within 15 min (e.g., DeVries *et al.*, 1978). Light exposure decreases c-*fos* mRNA abundance and *N*-acetyltransferase activity to daytime levels within 30 and 60 min, respectively (e.g., Hamm *et al.*, 1983; Yoshida *et al.*, 1993).

rhythms of melatonin levels and the activity of serotonin *N*-acetyltransferase, a key regulatory enzyme in melatonin biosynthesis, have been observed in a variety of other animals, including several mammalian species, but it is unknown if the rhythms in these animals are driven by the light/dark cycle or by circadian oscillators.

In retinas of chicks, quail and African clawed frogs, the circadian rhythm of melatonin biosynthesis is generated by parallel rhythms in the activities of two regulatory enzymes in the biosynthetic pathway, tryptophan hydroxylase and serotonin *N*-acetyltransferase (Hamm and Menaker, 1980; Iuvone and Besharse, 1983; Cahill and Besharse, 1990; Thomas and Iuvone, 1991; H. Underwood and P.M. Iuvone, unpublished observations). The circadian oscillator that regulates these rhythms is in the retina and appears to be contained within the photoreceptor cells (Besharse and Iuvone, 1983; Underwood *et al.*, 1988, 1990; Zawilska and Iuvone, 1992; Cahill and Besharse, 1993; Thomas *et al.*, 1993). The nocturnal expression of both these enzymes is sensitive to protein synthesis inhibitors, suggesting involvement of *de novo* protein synthesis (Iuvone and Besharse, 1983; Iuvone *et al.*, 1990; Thomas and Iuvone, 1991). The expression of *N*-acetyltransferase activity is also blocked by the RNA synthesis inhibitor actinomycin D (Iuvone *et al.*, 1990). Recently, Green and colleagues (Green and Besharse, 1994; Green *et al.*, 1994) demonstrated that tryptophan hydroxylase mRNA in frog retina is expressed primarily in photoreceptor cells and that the rate of transcription of the tryptophan hydroxylase gene is regulated by the circadian clock.

Light has two distinct effects on the synthesis of melatonin: (1) light exposure at night acutely and rapidly reduces *N*-acetyltransferase activity and melatonin synthesis and release (Hamm *et al.*, 1983; Iuvone and Besharse, 1983; Cahill and Besharse, 1991); and (2) light entrains the circadian clock that regulates melatonin biosynthesis (e.g., Besharse and Iuvone, 1983; Cahill and Beshare, 1991; Thomas and Iuvone, 1991).

The induction of serotonin *N*-acetyltransferase activity in chick, *Xenopus* and rat retinas, and in cultured chick photoreceptor cells and human retinoblastoma cells is stimulated by cyclic AMP (Iuvone and Besharse, 1983, 1986c; Iuvone, 1990; Iuvone *et al.*, 1990; Nowak, 1990; Wiechmann *et al.*, 1990; Zawilska *et al.*, 1991; Janavs *et al.*, 1991). Melatonin production and release from retinoblastoma cells is also stimulated by cyclic AMP (Pierce *et al.*, 1989; Wiechmann *et al.*, 1990; Deng *et al.*, 1991; Janavs *et al.*, 1991). The dark-dependent, nocturnal induction of *N*-acetyltransferase activity in *Xenopus* and chicken retinas is dependent on extracellular Ca^{2+} and is blocked by inhibitors of L-type voltage-gated Ca^{2+} channels (Iuvone and Besharse, 1986b; Zawilska and Nowak, 1990). Similarly, depolarization-evoked induction of *N*-acetyltransferase activity in photoreceptor-enriched chick retinal cell cultures requires Ca^{2+} influx through L-type channels (Avendano *et al.*, 1990) and involves a Ca^{2+} dependent activation of cyclic AMP formation (Iuvone *et al.*, 1991 a,b). These observations are consistent with a working model in which melatonin biosynthesis at night is stimulated in darkness via a cascade involving depolarization of the photoreceptor cell, Ca^{2+} influx through voltage-gated channels and a consequent accumulation of intracellular cyclic AMP, and activation of cyclic AMP-dependent protein kinase. The kinase phosphorylates proteins that regulate gene expression and protein synthesis. In light, the photoreceptor cells hyperpolarize, closing the Ca^{2+} channels and decreasing cyclic AMP formation, resulting in a decrease in *N*-acetyltransferase activity and melatonin production. The decrease in cyclic AMP may lower *N*-acetyltransferase activity in multiple ways, including decreasing the protein synthesis-dependent expression of enzyme activity and enhancing the rate of *N*-acetyltransferase inactivation (A. Alonso Gomez and P.M. Iuvone, unpublished observations).

The temporal pattern of c-*fos* expression in photoreceptor cells closely resembles that of melatonin production (Figure 2.5). In other systems, c-*fos* expression has been shown to be regulated by Ca^{2+} and cyclic AMP-dependent mechanisms (Sheng and Greenberg, 1990). These correlations suggest that the c-*fos* gene product in photoreceptor cells may play a role in the transcriptional regulation of the serotonin *N*-acetyltransferase and tryptophan hydroxylase genes.

2.2.8 IS THE OCULAR CIRCADIAN CLOCK IN PHOTORECEPTOR CELLS?

The circadian clock that regulates rhythms of melatonin production is in the eye, and it is likely that this clock regulates other photoreceptor rhythms as well. Recent evidence suggests that the clock may be a component of the photoreceptor cell. Circadian rhythms of serotonin *N*-acetyltransferase activity and tryptophan hydroxylase activity in chick retina persist following retinal lesions with kainic acid, a neurotoxin that destroys most inner retinal neurons but spares photoreceptor cells (Zawilska and Iuvone, 1992; Thomas *et al.*, 1993). Furthermore, circadian release of melatonin from isolated, cultured *Xenopus* photoreceptor layers has been demonstrated (Cahill and Besharse, 1993). Although these observations do not exclude the presence of circadian clocks in other retinal cell types, they strongly support the hypothesis that photoreceptor cells contain the clock that regulates melatonin biosynthesis. The circadian clock that regulates the transcription of the iodopsin gene also appears to reside within the photoreceptor, as the rhythm was observed in dissociated cell cultures from embryonic chick retina (Pierce *et al.*, 1993).

2.3 INFLUENCE OF NEUROTRANSMITTERS ON METABOLISM IN PHOTORECEPTOR CELLS

Although the primary input to photoreceptor cells is light and the control of many of the functions of photoreceptor cells are self-contained within the visual cell or the photoreceptor–pigment epithelial complex, aspects of photoreceptor metabolism are subject to modulation by neurotransmitters or neuromodulators. This section reviews studies demonstrating effects of dopamine, melatonin, adenosine and glutamate on photoreceptor cell biology and metabolism.

2.3.1 DOPAMINE

Dopamine is a neuromodulator synthesized and released from amacrine and/or interplexiform cells of the inner retina (reviewed by Iuvone, 1986a; Dowling, 1989; Djangoz and Wagner, 1992). These dopamine neurons are activated by steady or flickering light (e.g., Kramer, 1971; Dowling and Watling, 1981; Iuvone *et al.*, 1978; Godley and Wurtman, 1988; Boatright *et al.*, 1989; Kirsch and Wagner, 1989; Dearry and Burnside, 1989; Kolbinger *et al.*, 1990). An extensive literature now documents effects of dopamine on many aspects of photoreceptor metabolism. Photoreceptor cells of mammalian and non-mammalian vertebrates contain dopamine receptors. Pharmacologically and immunologically, these dopamine receptors appear to be of the D2 and/or D4 subtypes (e.g., Zarbin *et al.*, 1986; Brann and Young, 1986; Vuvan *et al.*, 1993; Muresan and Besharse, 1993; Wagner and Behrens, 1993), and mRNA for these receptors has been localized to photoreceptor cells by *in situ* hybridization (Dearry *et al.*, 1991; Cohen *et al.*, 1992). Activation of dopamine receptors on mouse photoreceptor cells inhibits cyclic AMP formation, presumably in the same pool of cyclic nucleotide that is reduced by light exposure (Cohen and Blazynski, 1990). Stimulation of D2/D4 receptors also appears to reduce cyclic AMP formation in *Xenopus* and chick photoreceptor cells (Iuvone, 1986b; Iuvone *et al.*, 1990). In contrast, no evidence of D2 receptor-mediated inhibition of adenylate cyclase was observed in bovine rod outer segments (Sitaramayya *et al.*, 1993). This could be due to species differences or to selective localization of D2/D4 receptors to cell body and synaptic terminal areas, where light-evoked inhibition of cyclic AMP levels is most apparent (Orr *et al.*, 1976; DeVries *et al.*, 1979).

Dopamine regulates photoreceptor retinomotor movements in teleost fish and *Xenopus* retinas. Application of dopamine or D2 dopamine receptor agonists to retinas or isolated cone inner/outer segments in darkness elicits light-adaptive cone contraction (Pierce and Besharse, 1985; Dearry and Burnside, 1986; Burnside *et al.*, 1993). Dopamine also mimics the effect of light on photomechanical movement of cultured chick photoreceptors (Stenkamp *et al.*, 1994). D2 receptors in photoreceptor cells are negatively coupled to cyclic AMP formation (see above), and cyclic AMP induces dark-adaptive cone elongation and inhibits cone contraction (Besharse *et al.*, 1982; Burnside *et al.*, 1982). Thus, dopamine can influence retinomotor movements by decreasing cyclic AMP formation. D2 dopamine receptor antagonists inhibit light-evoked cone contraction and promote cone elongation in retinas prepared and incubated in light. Furthermore, intraocular injection of D2 antagonists in fish inhibits cone contraction at expected dawn (Douglas *et al.*, 1992; McCormack and Burnside, 1992). These observations suggest a functional role for dopamine in light-evoked and circadian retinomotor movements. However, 6-hydroxydopamine-induced lesions of dopamine neurons in fish retina reduce, but do not eliminate, circadian cone movements, indicating the presence of other regulatory mechanisms for controlling circadian cone retinomotor movements (Douglas *et al.*, 1992).

Dopamine and D2 dopamine receptor agonists inhibit rod outer segment disc shedding in *Rana pipiens* eye cups *in vitro* (Pierce and Besharse, 1986; Besharse *et al.*, 1988a). Application of the agonists during the dark-priming period, but not during the light-triggering transition from darkness to light, dramatically decreased shedding and phagocytosis. Dopamine receptor agonists have no apparent effect on the processes of disc detachment or phagocytosis. Thus, dopamine may be a mediator of daytime or light-mediated suppression of shedding. Consistent with these observations, clorgyline and methamphetamine, drugs that increase dopamine levels and stimulate release, respectively, reduced the rhythm of disc shedding in rat retina (Remé *et al.*, 1984; 1991).

Dopamine and D2 dopamine receptor agonists mimic the effect of acute light exposure on *N*-acetyltransferase activity and melatonin biosynthesis (Iuvone, 1986b; Iuvone and Besharse, 1986a; Zawilska and Iuvone, 1989; Cahill and Besharse, 1991). Activation of dopamine receptors at night also shifts the phase of the circadian clock that regulates melatonin biosynthesis, with a phase-response relationship similar to that produced by pulses of light (Cahill and Besharse, 1991). Responses to dopamine persist in reduced retinas in which most of the inner retina is destroyed by neurotoxins or detergent lysis (Zawilska and Iuvone, 1992; Cahill and Besharse, 1992), in low-density photoreceptor-enriched cell cultures (Iuvone *et al.*, 1990), and in isolated photoreceptor cells (Cahill and Besharse, 1993). Thus, these effects appear to be direct, receptor-mediated responses in photoreceptor cells. The inhibitory effects of dopamine appear to be mediated by D2- and/or D4-like dopamine receptors, and involve inhibition of cyclic AMP formation (Iuvone, 1986b; Zawilska and Iuvone, 1989; Cahill and Besharse, 1991; Iuvone *et al.*, 1991b; Zawilska and Nowak., 1994).

The physiological role of dopamine in regulating melatonin biosynthesis is as yet unclear. In cultured *Xenopus* eye cups, dopamine receptor antagonists do not block the effects of light on *N*-acetyltransferase activity, melatonin release, or the phase of the circadian rhythm of melatonin release (Iuvone *et al.*, 1987; Cahill and Besharse, 1991). However, dopamine antagonists in combination with cyclic nucleotide phosphodiesterase inhibitors do attenuate the acute inhibitory effects of light (Iuvone *et al.*, 1987). *In vivo*, administration of dopamine receptor antagonists alone antagonize the acute inhibition of *N*-acetyltransferase activity by light in chick retina (Zawilska *et al.*, 1991; Kazula *et al.*, 1993). The ability of dopamine receptor antagonists to reverse the effects of light varies as a function of light intensity, being most effective in very dim light (Iuvone *et al.*, 1992). This latter observation suggests that dopamine may function as an amplifying signal for regulating melatonin biosynthesis under conditions of dim light, such as that which occurs at dawn.

2.3.2 MELATONIN

Melatonin, which is synthesized in photoreceptor cells in darkness, regulates photoreceptor metabolism. However, the effects of melatonin may be indirect. Radioligand binding studies using the high affinity ligand 2-[^{125}I]iodomelatonin indicate that melatonin receptors are localized primarily to the inner nuclear and plexiform layers of the neural retina and to the retinal pigment epithelium (Laitinen and Saavedra, 1990; Blazynski and Dubocovich, 1991; Chong and Sugden, 1991; Wiechmann Wirsig–Wiechmann, 1991), although some apparent specific binding over photoreceptor cells has also been observed.

Melatonin promotes dark-adaptive cone

elongation in light-exposed *Xenopus* retinas (Pierce and Besharse, 1985, 1987) and mimics the effect of darkness on photoreceptor length in chick retinal cell cultures (Stenkamp *et al.*, 1994). The effect of melatonin on cone retinomotor movements may be indirect and related to the regulation of dopamine release. Exogenous dopamine or D2 dopamine receptor agonists prevent melatonin-induced cone elongation in light, and inhibit dark-induced cone elongation (Pierce and Besharse, 1985). In contrast, dopamine receptor antagonists and melatonin have the same effect on cone position. Melatonin inhibits light-evoked dopamine release in *Xenopus* retina (Boatright and Iuvone, 1989; Boatright *et al.*, 1994). Inhibition of dopamine release by melatonin has also been demonstrated in chick and rabbit retinas (Dubocovich, 1983; Dubocovich and Takahashi, 1987; Nowak, 1988). Thus, it seems likely that melatonin regulates cone retinomotor movement indirectly, by inhibiting dopamine release. This hypothesis is consistent with the observed localization of melatonin receptors to the inner nuclear and plexiform layers, where the cell bodies and processes of dopaminergic neurons are found (reviewed by Ehinger, 1977).

Melatonin also influences rod outer segment disc shedding. Melatonin activates shedding in *Xenopus* eye cups cultured under conditions that do not support a normal shedding response (Besharse and Dunis, 1983; Besharse *et al.*, 1984; Pierce and Besharse, 1986). In these experiments, melatonin had to be applied in darkness, before light onset, for shedding to occur. In eye cups prepared from constant light-treated frogs, brief exposure to either darkness or melatonin increases disc shedding. Thus, melatonin may be a component of the dark-priming mechanism for regulation of outer segment turnover. The site of action for activation of disc shedding by melatonin is unknown. As dopamine inhibits dark priming (section 2.3.1), melatonin's effect on disc shedding may be partially mediated by suppression of dopamine release. Melatonin may also influence disc shedding by an action on the pigment epithelium. Cyclic AMP suppresses disc shedding (section 2.2.1), at least in part by an action on the pigment epithelium (Edwards and Bakshian, 1980). In many cell types, including cultured retinal cells, melatonin receptor activation inhibits cyclic AMP formation (Iuvone and Gan, 1994; reviewed in Morgan *et al.*, 1994). Thus, melatonin may activate shedding by reducing cyclic AMP levels in pigment epithelial cells.

2.3.3 ADENOSINE

Adenosine functions as a neuromodulator in the brain and retina (reviewed by Fredholm *et al.*, 1993; Blazynski and Perez, 1991). Adenosine receptors have been localized throughout the retina, with different subtypes found in the inner and outer retina. Adenosine receptors are broadly divided into A1 and A2 subtypes (reviewed by Collis and Hourani, 1993). A1 adenosine receptors have been localized primarily to the inner nuclear, inner plexiform and ganglion cell layers (Braas *et al.*, 1987; Blazynski, 1990). In these areas, adenosine may function to regulate neurotransmitter release (Blazynski *et al.*, 1992) and to modulate neurotransmitter receptor–effector coupling (Paes de Carvalho and deMello, 1985). In contrast, A2 adenosine receptors are localized primarily on photoreceptor cells and pigment epithelial cells (Friedman *et al.*, 1989; Blazynski, 1990; McIntosh and Blazynski, 1994). A2 adenosine receptors in retina and in pigment epithelium stimulate adenylate cyclase (Friedman *et al.*, 1989; Blazynski and McIntosh, 1993).

Adenosine receptors are involved in the regulation of melatonin biosynthesis in photoreceptor cells. Intravitreal administration of adenosine receptor agonists stimulates serotonin *N*-acetyltransferase activity in light-exposed chick retinas (Gan *et al.*, 1994).

Adenosine receptor agonists also stimulated the induction of *N*-acetyltransferase activity in low-density, photoreceptor-enriched retinal cell cultures, suggesting a direct effect on photoreceptor cells. The stimulation of *N*-acetyltransferase activity is receptor-mediated and blocked by adenosine receptor antagonists. Preliminary pharmacological characterization of the response indicates the involvement of an A2-like adenosine receptor (Gan *et al.*, 1994).

Adenosine is also a potential modulator of photoreceptor disc shedding. Activation of A2 adenosine receptors on cultured pigment epithelial cells inhibited phagocytosis of outer segment fragments (Gregory *et al.*, 1994). This may be related to the stimulation of cyclic AMP accumulation in RPE cells caused by adenosine receptor activation.

In chick retina, adenosine-like immunoreactivity has been observed in photoreceptor cells (Paes de Carvalho *et al.*, 1992), but in most mammalian species, immunoreactivity has been observed primarily in ganglion cells and inner nuclear cells, presumably amacrine cells (Braas *et al.*, 1987). However, A2 adenosine receptors are located on photoreceptor and pigment epithelial cells of a variety of mammalian species, including human (Braas *et al.*, 1987; Blazynski, 1990; Friedman *et al.*, 1989). This observation raises the question of the source of endogenous adenosine that occupies these receptors in the outer retina. Adenosine may diffuse from the inner retina, as has been proposed for dopamine (reviewed in Besharse *et al.*, 1988a; Witkovsky and Dearry, 1991; Besharse and Iuvone, 1992). Adenosine may be produced in outer retina as a function of photoperiod, perhaps at night. Thus, adenosine-like immunoreactivity would not be observed in retinas fixed during the day. Alternatively, extracellular adenosine may be derived from adenine nucleotides via the action of ectonucleotidases, such as that found in synaptic clefts of photoreceptor synapses (Kreutzberg and Hussain, 1984). Another possibility is that adenosine receptors in the outer retina are only occupied under certain pathophysiological conditions, such as ischemia, which are known to stimulate the production and release of adenosine in the nervous system (Fredholm *et al.*, 1993).

2.3.4 GLUTAMATE

The putative photoreceptor neurotransmitter glutamate, as well as aspartate, massively stimulate rod photoreceptor disc shedding in a light-independent manner in amphibian eye cups (Greenberger and Besharse, 1985). The shedding response to glutamate is 3–5 times greater than the usual light-evoked shedding response, and is accompanied by the formation of large pseudopods from the apical surface of the RPE that ensheathe the distal tips of the outer segments (Matsumoto *et al.*, 1987). Glutamate also produced an apparent increase in adhesion between the RPE and photoreceptor cells (Matsumoto *et al.*, 1987), but this appears to be unrelated to the stimulation of disc shedding (Defoe *et al.*, 1989).

Although the mechanism involved in the shedding response to excitatory amino acids is unknown, it appears to be receptor-mediated. The effect of glutamate on disc shedding is mimicked by micromolar concentrations of kainic acid, an agonist of a subtype of ionotrophic glutamate receptors (Greenberger and Besharse, 1985; Besharse and Spratt, 1988). Shedding is not stimulated by *N*-methyl-D-aspartate (NMDA), an agonist of a different subtype of ionotropic glutamate receptor, and only weakly stimulated by quisqualic acid, an agonist of G-protein-coupled glutamate receptors. Kainate-induced disc shedding is inhibited by the glutamate receptor antagonists kynurenic acid, D-*O*-phosphoserine, and *cis*-piperidine dicarboxylic acid (*cis*-PDA) (Besharse *et al.*, 1988b). Remarkably, kynurenate and D-*O*-

phosphoserine, but not *cis*-PDA, also blocked light-evoked disc shedding, suggesting that activation of kynurenate-sensitive glutamate receptors may be involved in the 'light-triggering' mechanism.

The location of the glutamate receptors involved in the regulation of disc shedding is unknown. Glutamate depolarizes photoreceptor cells, but this response may be related to electrogenic uptake of glutamate into photoreceptor cells rather than an action on cell surface glutamate receptors (Chapter 6). The receptors may be located in the inner retina, where excitatory amino acids cause profound neurotoxicity. In addition to blocking kainate-induced disc shedding, kynurenate and other glutamate receptor antagonists block the neurotoxic effect of kainic acid (Besharse *et al.*, 1988b). However, the neurotoxicity may be unrelated to disc shedding, because NMDA caused a similar neurotoxic effect but failed to stimulate disc shedding, and other amino acids, such as glycine and taurine, stimulated disc shedding while having little or no neurotoxic effects (Greenberger and Besharse, 1985).

Glutamate may stimulate disc shedding by a direct action on the RPE. Glutamate binds to high affinity, saturable binding sites in membranes prepared from primary cultures of chick RPE (López–Colomó *et al.*, 1993). Interestingly, specific binding of glutamate to these putative receptors is enhanced by glycine and taurine.

2.4 SPECULATIONS ON THE ROLE OF RHYTHMIC METABOLISM IN PHOTORECEPTOR PATHOLOGY

Having discovered the dynamics of outer segment turnover, it became obvious that a balance must be achieved between disc assembly and disc shedding for the maintenance of normal outer segment length and function (Young, 1976). The importance of disc shedding is perhaps best illustrated with the Royal College of Surgeons (RCS) rat, whose RPE fail to phagocytize outer segment discs (Herron *et al.*, 1969; Bok and Hall, 1971). Initially, photoreceptor cells of RCS rats appear to develop normally, with usual morphology and electroretinographic (ERG) function (Dowling and Sidman, 1962). This is followed later in development by an accumulation of rhodopsin, a disorganized lamellar network of outer segment membranes between the photoreceptor cells and RPE cells, and a depression of the a-wave (photoreceptor component) of the ERG. Ultimately, the photoreceptor cells degenerate and die. The site of the genetic mutation is the pigment epithelium (Mullen and LaVail, 1976).

Thus, disruption of circadian disc shedding may be a cause of photoreceptor cell death. Disruption of other aspects of rhythmic photoreceptor metabolism may also influence photoreceptor survival. For example, *rds* mice, which have a mutation in the peripherin gene that ultimately leads to photoreceptor cell death (Travis, 1991), fail to show the normal diurnal expression of the arrestin gene (Agarwal *et al.*, 1994). Arrestin gene expression appears to be locked in the daytime, light-exposed pattern. Exposing the retina to continuous light, which disrupts circadian retinal physiology, causes profound photoreceptor degeneration (Noell, 1980). Recently, Fain and Lisman (1993) proposed the 'equivalent light hypothesis' to explain the photoreceptor cell death associated with vitamin A deficiency and retinitis pigmentosa caused by rhodopsin mutations. They suggested that in these disorders constitutively active opsin may continuously activate the visual transduction cascade, producing the equivalent of continuous light. This equivalent light might disrupt retinal circadian rhythms, such as those of disc shedding, protein synthesis or neuromodulator production, leading to photoreceptor degeneration.

Neuromodulators that are produced in a

rhythmic fashion may also influence retinal degeneration. Administration of melatonin enhances light-induced photoreceptor degeneration in rats (Bubenik and Purtill, 1980; Leino *et al.*, 1984; Wiechmann and O'Steen, 1992). In contrast, dopamine has been suggested to have a survival-promoting effect on photoreceptor cells (Bubenik and Purtill, 1980). Bromocriptine, a D2 dopamine receptor agonist, apparently reduces light-induced photoreceptor degeneration and the degeneration in RCS rat retina (Bubenik and Purtill, 1980). In *rds* mice, the diurnal regulation of dopaminergic activity is absent or greatly reduced, and abnormally low dopamine synthesis is present prior to the onset of photoreceptor cell death (Nir and Iuvone, 1994). Thus, if dopamine indeed has a survival-promoting, trophic action on photoreceptor cells, the low level of dopaminergic activity may contribute to the photoreceptor cell death associated with the *rds*/peripherin mutation.

2.5 CONCLUSIONS AND FUTURE PERSPECTIVES

It is now clear that many aspects of photoreceptor cell biology are influenced by light, circadian clocks and neuromodulators. At least some of the cellular processes regulated by these stimuli are essential for the proper functioning and even survival of the cell. It is, as yet, unclear why several processes regulated by photoperiod show different patterns of expression, and why the rhythms of similar processes differ in rods and cones. The molecular basis of the circadian clock that regulates photoreceptor metabolism is unknown. In many cases it is also unclear which rhythmic processes in photoreceptor cells are light-driven and which are coupled to a circadian clock. Although the role of a photoreceptor-based clock has been emphasized in this chapter, it is important to recognize that networks of oscillators, within and outside the eye, may exist and function in a coordinate manner to regulate photoreceptor metabolism *in vivo* (e.g., Tierstein *et al.*, 1980). In this regard, it is noteworthy that retinal dopamine neurons, which influence many aspects of photoreceptor metabolism, are targets of the centrifugal neurons that project from the brain to the retina in some species (Zucker and Dowling, 1987; Behrens *et al.*, 1993). Continued research into rhythmic photoreceptor metabolism holds promise for gaining new insights into visual cell function, and possibly into the processes that lead to disease-related photoreceptor cell death.

Much remains to be learned about rhythmic retinal physiology. Although psychophysical and electroretinographic studies indicate that circadian processes influence visual sensitivity and adaptation, little or no electrophysiological data are available on the effects of clock output at the cellular level. Such an approach will be essential in fully understanding the ways in which the photoreceptor circadian clock regulates the adaptive state of the retina. Another important future direction is to understand the molecular mechanism of the circadian clock. Having recently identified genes in photoreceptor cells whose expression is regulated in a circadian fashion, it should now be possible to determine the regulatory elements of those genes that are specifically controlled by clock output. A particularly exciting prospect is the possibility that pharmacological approaches could be used to manipulate rhythmic photoreceptor metabolism in retinal disease. If the equivalent light hypothesis of Fain and Lisman (1993) is correct and dopamine receptor agonists entrain the photoreceptor circadian clock in humans as it does in *Xenopus* (Cahill and Besharse, 1991), it may be possible to prevent or delay the progression of photoreceptor cell death in some forms of retinitis pigmentosa using drugs to manipulate the clock or mimic its output.

ACKNOWLEDGEMENTS

The author wishes to acknowledge the many colleagues who spent time discussing issues related to this chapter, who sent preprints and reprints, and brought key references to my attention, especially Cheryl Craft, Jeff Boatright, Joe Besharse, Charlotte Remé and David Williams. Special thanks to Jeff Boatright for his helpful comments on the manuscript, to Bonnie Johnson for her help in preparing figures, to Joe Besharse who greatly influenced my thinking on the subject of this chapter, and to the members of my laboratory, Jiwei Gan, Angel Alonso, Chad Friel, Leon Goe and B. Johnson for their accomplishments and patience while my attentions were directed to this chapter. Retinal research in the author's lab is supported by NIH grants RO1-EY04864 and RO1-EY09737.

REFERENCES

Abe, T., Nakabayashi, H., Tamada, H. *et al.* (1990) Analysis of the human, bovine and rat 33-kDa proteins and cDNA in retina and pineal gland. *Gene*, **91**, 209–15.

Adler, R., Lindsey, J.D. and Elsner, C.L. (1984) Expression of cone-like properties by chick embryo retina cells in glial-free monolayer cultures. *Journal of Cell Biology*, **99**, 1173–8.

Agarwal, N. (1994) Diurnal expression of NGF1-A mRNA in retinal degeneration slow (*rds*) mutant mouse retina. *FEBS Letters*, **339**, 253–7.

Agarwal, N., Nir, I. and Papermaster, D.S. (1994) Loss of diurnal arrestin gene expression in *rds* mutant mouse retinas. *Experimental Eye Research*, **58**, 1–8.

Anderson, D.H., Stern, W.H., Fisher, S.K. *et al.* (1983) Retinal detachment in the cat: the pigment epithelial-photoreceptor interface. *Investigative Ophthalmology and Visual Science*, **24**, 906–26.

Avendano, G., Butler, B.J. and Iuvone, P.M. (1990) K^+-Evoked depolarization increases serotonin *N*-acetyltransferase activity in photoreceptor-enriched retinal cell cultures. Involvement of calcium influx through L-type calcium channels. *Neurochemisry International*, **17**, 117–26.

Axelrod, J., Burch, R.M. and Jelsema, C.L. (1988) Receptor-mediated activation of phospholipase A_2 via GTP-binding proteins: arachidonic acid and its metabolites as second messengers. *Trends in Neuroscience*, **11**, 117–23.

Azarian, S.M., Schlamp, C.L. and Williams, D.S. (1993) Characterization of calpain II in the retina and photoreceptor outer segments. *Journal of Cell Science*, **105**, 787–98.

Barnes, S. and Hille, B. (1989) Ionic channels of the inner segment of tiger salamander cone photoreceptors. *Journal of General Physiology*, **94**, 719–43.

Basinger, S., Hoffman, R. and Matthes, M. (1976) Photoreceptor shedding is initiated by light in the frog retina. *Science*, **194**, 1074–6.

Behrens, U.D., Douglas, R.H. and Wagner, H.-J. (1993) Gonadotropin-releasing hormone, a neuropeptide of efferent projections to the teleost retina induces light-adaptive spinule formation on horizontal cell dendrites in dark-adapted preparations kept in vitro. *Neuroscience Letters*, **164**, 59–62.

Bernstein, P.S., Law, W.C. and Rando, R. (1987) Isomerization of all-*trans* retinoids to 11-*cis* retinoids in vitro. *Proceedings of the National Academy of Sciences, USA*, **84**, 1849–53.

Berridge, M.J. (1993) Inositol trisphosphate and calcium signalling. *Nature*, **361**, 315–25.

Besharse, J.C. (1982) The daily light–dark cycle and rhythmic metobolism in the photoreceptor–pigment epithelial complex, in *Progress in Retinal Research* (eds N. Osborne and G. Chader), Pergamon, Oxford pp.81–124.

Besharse, J.C. (1986) Photosensitive membrane turnover: differentiated membrane domains and cell–cell interaction, in *The Retina: A Model for Cell Biology Studies*, Part I (eds R. Adler and D. Faber), Academic Press, Orlando, pp.297–352.

Besharse, J.C. and Dunis, D.A. (1983) Methoxyindoles and photoreceptor metabolism: activation of rod shedding. *Science*, **219**, 1341–3.

Besharse, J.C. and Horst, C.J. (1990) The photoreceptor connecting cilium: a model for the transition zone, in *Ciliary and Flagellar Membranes* (ed. R.A. Bloodgood), Plenum, New York, pp.389–417.

Besharse, J.C. and Iuvone, P.M. (1983) Circadian clock in Xenopus eye controlling retinal serotonin *N*-acetyltransferase. *Nature*, **305**, 133–5.

Besharse, J.C. and Iuvone, P.M. (1992) Critique: is dopamine a light-adaptive or a dark-adaptive modulator in retina. *Neurochemistry International*, **20**, 193–9.

Besharse, J.C. and Spratt, G. (1988) Excitatory amino acids and rod photoreceptor disc shedding: analysis using specific agonists. *Experimental Eye Research*, **47**, 609–20.

Besharse, J.C., Hollyfield, J.G. and Rayborn, M.E. (1977) Turnover of rod photoreceptor outer segments: II. Membrane addition and loss in relationship to light. *Journal of Cell Biology*, **75**, 507–27.

Besharse, J.C., Dunis, D.A. and Burnside, B. (1982) Effects of cyclic adenosine 3′, 5′-monophosphate on photoreceptor disc shedding and retinomotor movement. Inhibition of shedding and stimulation of cone elongation. *Journal of General Physiology*, **79**, 775–90.

Besharse, J.C., Dunis, D.A. and Iuvone, P.M. (1984) Regulation and possible role of serotonin *N*-acetyltransferase in the retina. *Federation Proceedings*, **43**, 2704–8.

Besharse, J.C., Spratt, G. and Forestner, D.M. (1986) Light-evoked and kainic acid-induced disc shedding by rod photoreceptors: differential sensitvity to extracellular calcium. *Journal of Comparative Neurology*, **251**, 185–97.

Besharse, J.C., Iuvone, P.M. and Pierce, M.E. (1988) Regulation of rhythmic photoreceptor metabolism: a role for post-receptoral neurons, in *Progress in Retinal Research*, 7th edn (eds N. Osborne and G.J. Chader), Perganon Press, Oxford, pp.21–61.

Besharse, J.C., Spratt, G. and Reif-Lehrer, L. (1988) Effects of kynurenate and other excitatory amino acid antagonists as blockers of light- and kainate-induced retinal rod photoreceptor disc shedding. *Journal of Comparative Neurology*, **274**, 295–303.

Blazynski, C. (1990) Discrete distributions of adenosine receptors in mammalian retina. *Journal of Neurochemistry*, **54**, 648–55.

Blazynski, C. and Dubocovich, M.L. (1991) Localization of 2-[^{125}I]iodomelatonin binding sites in mammalian retina. *Journal of Neurochemistry*, **56**, 1873–80.

Blazynski, C. and McIntosh, H. (1993) Characterization of adenosine A_2 receptors in bovine retinal membranes. *Experimental Eye Research*, **56**, 585–93.

Blazynski, C. and Perez, M.-T.R. (1991) Adenosine in vertebrate retina: localization, receptor characterization, and function. *Cellular and Molecular Neurobiology*, **11**, 463–84.

Blazynski, C., Woods, C. and Mathews, G.C. (1992) Evidence for the action of endogenous adenosine in the rabbit retina: modulation of the light-evoked release of acetylcholine. *Journal of Neurochemistry*, **58**, 761–7.

Boatright, J.H. and Iuvone, P.M. (1989) Melatonin suppress the light-evoked release of endogenous dopamine from retinas of frogs (*Xenopus laevis*). *Society for Neuroscience Abstracts*, **15**, 1395.

Boatright, J.H., Hoel, M.J. and Iuvone, P.M. (1989) Stimulation of endogenous dopamine release and metabolism in amphibian retina by light and K^+-evoked depolarization. *Brain Research*, **482**, 164–8.

Boatright, J.H., Rubim, N.M. and Iuvone, P.M. (1994) Regulation of endogenous dopamine release in amphibian retina by melatonin: the role of GABA. *Visual Neuroscience*, **11**, 1013–18.

Bok, D. (1985) Retinal photoreceptor–pigment epithelium interactions. *Investigative Ophthalmology and Visual Science*, **26**, 1659–94.

Bok, D. (1993) The retinal pigment epithelium: a versatile partner in vision. *Journal of Cell Science*, **106** Suppl. 17, 189–95.

Bok, D. and Hall, M.O. (1971) The role of the pigment epithelium in the etiology of inherited retinal dystrophy in the rat. *Journal of Cell Biology*, **49**, 664–82.

Bowes, C., VanVeen, T. and Farber, D.B. (1988) Opsin, G-protein and 48-kDa protein in normal and *rd* mouse retinas: developmental expression of mRNAs and proteins and light/dark cycling of mRNAs. *Experimental Eye Research*, **47**, 369–90.

Braas, K.M., Zarbin, M.A. and Snyder, S.H. (1987) Endogenous adenosine and adenosine receptors localized to ganglion cells of the retina. *Proceedings of the National Academy of Sciences, USA*, **84**, 3906–10.

Brann, M.R. and Cohen, L.V. (1987) Diurnal expression of transducin mRNA and translocation of transducin in rods of rat retina. *Science*, **235**, 585–7.

Brann, M.R. and Young, W.S.I. (1986) Dopamine receptors are located on rods in bovine retina. *Neuroscience Letters*, **69**, 221–6.

Bubenik, G.A. and Purtill, R.A. (1980) The role of melatonin and dopamine in retinal physiology. *Canadian Journal of Physiology and Pharmacology*, **58**, 1457–62.

Burnside, B. and Ackland, N. (1984) Effects of circadian rhythm and cAMP on retinomotor movements in the green sunfish, *Lepomis cyanellus*. *Investigative Ophthalmology and Visual Science*, **25**, 539–45.

Burnside, B. and Dearry, A. (1986) Cell motility in the retina, in *The Retina: A Model for Cell Biology Studies, Part I* (eds R. Adler and D. Farber), Academie Press, Orlando, pp.151–206.

Burnside, B. and Nagle, B. (1983) Retinomotor movements of photoreceptors and retinal pigment epithelium: mechanisms and regulation. *Progress in Retinal Research*, **2**, 67–109.

Burnside, B., Evans, M., Fletcher, R.T. and Chader, G.J. (1982) Induction of dark-adaptive retinomotor movement (cell elongation) in teleost retinal cones by cyclic adenosine 3′, 5′-monophosphate. *Journal of General Physiology*, **79**, 759–74.

Burnside, B., Wang, E., Pagh-Roehl, K. and Rey, H. (1993) Retinomotor movements in isolated teleost retinal cone inner-outer segment preparations (CIS-COS): effects of light, dark and dopamine. *Experimental Eye Research*, **57**, 709–22.

Cahill, G.M. and Besharse, J.C. (1990) Circadian regulation of melatonin in the retina of *Xenopus laevis*: limitation by serotonin availability. *Journal of Neurochemistry*, **54**, 716–19.

Cahill, G.M. and Besharse, J.C. (1991) Resetting the circadian clock in cultured *Xenopus* eyecups: regulation of retinal melatonin rhythms by light and D_2 dopamine receptors. *Journal of Neuroscience*, **11**, 2959–71.

Cahill, G.M. and Besharse, J.C. (1992) Light-sensitive melatonin synthesis by *Xenopus* photoreceptors after destruction of the inner retina. *Visual Neuroscience*, **8**, 487–90.

Cahill, G.M. and Besharse, J.C. (1993) Circadian clock functions localized in Xenopus retinal photoreceptors. *Neuron*, **10**, 573–7.

Cahill, G.M., Grace, M.S. and Besharse, J.C. (1991) Rhythmic regulation of retinal melatonin: metabolic pathways, neurochemical mechanisms, and the ocular circadian clock. *Cellular and Molecular Neurobiology*, **11**, 529–60.

Cavallaro, B. and Burnside, B. (1988) Prostaglandins E_1, E_2 and D_2 induce dark-adaptive retinomotor movements in teleost retinal cones and RPE. *Investigative Ophthalmology and Visual Science*, **29**, 882–91.

Chong, N.W.S. and Sugden, D. (1991) Guanine nucleotides regulate 2-[^{25}I]iodomelatonin binding sites in chick retinal pigment epithelium but not in neuronal retina. *Journal of Neurochemistry*, **57**, 685–9.

Cohen, A.I. (1982) Increased levels of 3′, 5′-cyclic adenosine monophosphate induced by cobaltous ion or 3-isobutylmethylxanthine in the incubated mouse retina: evidence concerning location and response to ions and light. *Journal of Neurochemistry*, **38**, 78–96.

Cohen, A.I. and Blazynski, C. (1990) Dopamine and its agonists reduce a light-sensitive pool of cyclic AMP in mouse photoreceptors. *Visual Neuroscience*, **4**, 43–52.

Cohen, A.I., Todd, R.D., Harmon, S. and O'Malley, K.L. (1992) Photoreceptors of mouse retinas possess D_4 receptors coupled to adenylate cyclase. *Proceedings of the National Academy of Sciences, USA*, **89**, 12093–7.

Collis, M.G. and Hourani, S.M.O. (1993) Adenosine receptor subtypes. *Trends in Pharmacological Science*, **14**, 360–6.

Corey, D.P., Dubinsky, J.M. and Schwartz, E.A. (1984) The calcium current in inner segments of rods from the salamander (*Ambystoma tigrinum*) retina. *Journal of Physiology (London)*, **354**, 557–75.

Craft, C.M., Whitmore, D.H. and Donoso, L.A. (1990) Differential expression of mRNA and protein encoding retinal and pineal S-antigen during the light/dark cycle. *Journal of Neurochemistry*, **55**, 1461–73.

Dearry, A. and Burnside, B. (1986) Dopaminergic regulation of cone retinomotor movement in isolated teleost retinas. I. Induction of cone contraction is mediated by D2 receptors. *Journal of Neurochemistry*, **46**, 1006–21.

Dearry, A. and Burnside, B. (1989) Light-induced dopamine release from teleost retinas acts as a light-adaptive signal to the retinal pigment epithelium. *Journal of Neurochemistry*, **53**, 870–8.

Dearry, A., Falardeau, P., Shores, C. and Caron, M.G. (1991) D_2 dopamine receptors in the human retina: cloning of cDNA and localization of mRNA. *Cellular and Molecular Neurobiology*, **11**, 437–53.

Defoe, D.M., Matsumoto, B. and Besharse, J.C. (1989) Cytochalasin D inhibits L-glutamate-induced disc shedding without altering L-glutamate-induced increase in adhesiveness. *Experimental Eye Research*, **48**, 641–52.

Deng, M.H., Lopez G.-Coviella, I., Lynch, H.J. and Wurtman, R.J. (1991) Melatonin and its precursors in Y79 human retinoblastoma cells: effect of sodium butyrate. *Brain Research*, **561**, 274–8.

Denton, T.L., Yamashita, C.K. and Farber, D.B. (1992) The effects of light on cyclic nucleotide metabolism of isolated cone photoreceptors. *Experimental Eye Research*, **54**, 229–237.

DeVries, G.W., Cohen, A.I., Hall, I.A. and Ferrendelli, J.A. (1978) Cyclic nucleotide levels in normal and biologically fractionated mouse retina: effects of light and dark adaptation. *Journal of Neurochemistry*, **31**, 1345–51.

DeVries, G.W., Cohen, A.I., Lowry, O.H. and Ferrendelli, J.A. (1979) Cyclic nucleotides in the cone-dominant ground squirrel retina. *Experimental Eye Research*, **29**, 315–21.

Djamgoz, M.B.A. and Wagner, H.-J. (1992) Localization and function of dopamine in the adult vertebrate retina. *Neurochemistry International*, **20**, 139–91.

Douglas, R.H., Wagner, H.-J., Zaunreiter, M. *et al.* (1992) The effect of dopamine depletion on light-evoked and circadian retinomotor movements in the teleost retina. *Visual Neuroscience*, **9**, 335–43.

Dowling, J.E. (1989) Neuromodulation in the retina: the role of dopamine. *Seminars in the Neurosciences*, **1**, 35–43.

Dowling, J.E. and Sidman, R.L. (1962) Inherited retinal dystrophy in the rat. *Journal of Cell Biology*, **14**, 73–109.

Dowling, J.E. and Watling, K.J. (1981) Dopaminergic mechanisms in the teleost retina. II. Factors affecting the accumulation of cyclic AMP in pieces of intact carp retina. *Journal of Neurochemistry*, **36**, 569–79.

Dubocovich, M.L. (1983) Melatonin is a potent modulator of dopamine release in the retina. *Nature*, **306**, 782–4.

Dubocovich, M.L. and Takahashi, J.S. (1987) Use of 2-[^{125}I]iodomelatonin to characterize melatonin binding sites in chicken retina. *Proceedings of the National Academy of Sciences, USA*, **84** 3916–20.

Edwards, R.B. and Bakshian, S. (1980) Phagocytosis of outer segments by cultured retinal pigment epithelium: reduction by cyclic AMP and phosphodiesterase inhibitors. *Investigative Ophthalmology*, **19**, 1184–8.

Ehinger, B. (1977) Synaptic connections of the dopaminergic retinal neurons. *Advances in Biochemical Psychopharmacology*, **16**, 299–306.

Eskin, A. (1971) Properties of the *Aplysia* visual system: in vitro entrainment of the circadian rhythm and centrifugal regulation of the eye. *Zeitschrift fur Vergleichende Physiologie*, **74**, 353–71.

Fain, G.L. and Lisman, J.E. (1993) Photoreceptor degeneration in vitamin A deprivation and retinitic pigmentosa: the equivalent light hypothesis. *Experimental Eye Research*, **57**, 335–40.

Fain, G.L., Gerschenfeld, H.M. and Quandt F.N. (1980) Calcium spikes in toad rods. *Journal of Physiology (London)*, **303**, 495–513.

Farber, D.B., Souza, D.W., Chase, D.G. and Lolley, R.N. (1981) Cyclic nucleotides of cone-dominant retinas. Reduction of cyclic AMP levels by light and cone degeneration *Investigative Ophthalmology and Visual Science*, **20**, 24–31.

Flannery, J.G. and Fisher, S.K. (1984) Circadian disc shedding in Xenopus retina in vitro. *Investigative Ophthalmology and Visual Science*, **25**, 229–32.

Fredholm, B.B., Johansson, B., Van der Ploeg, I. *et al.* (1993) Neuromodulatory roles of purines. *Drug Development Research*, **28**, 349–53.

Friedman, Z., Hackett, S.F., Linden, J. and Campochiaro, P.A. (1989) Human retinal pigment epithelial cells in culture possess A_2-adenosine receptors. *Brain Research*, **492**, 29–35.

Gan, J. and Iuvone, P.M. (1993) Calcium influx through dihydropyridine sensitive channels stimulates inositol phosphate accumulation in photoreceptor-enriched retinal cell cultures.

Investigative Ophthalmology and Visual Science, **34**, 1326(Abstract).

Gan, J., Rosenbaum, S., Mowrey, J.O. *et al.* (1994) Melatonin biosynthesis in chick retinal photoreceptor cells: stimulation of serotonin *N*-acetyltransferase activity by adenosine receptor agonists. *Investigative Ophthalmology and Visual Science*, **35**, 1583 (Abstract).

Ghalayini, A. and Anderson, R.E. (1984) Phosphatidylinositol 4, 5-bisphosphate: light-mediated breakdown in the vertebrate retina. *Biochemical and Biophysical Research Communications*, **124**, 503–6.

Ghalayini, A.J. and Anderson, R.E. (1992) Activation of bovine rod outer segment phospholipase C by arrestin. *Journal of Biological Chemistry*, **267**, 17977–82.

Gleason, E., Mobbs, P., Nuccitelli, R. and Wilson, M. (1992) Development of functional calcium channels in cultured avian photoreceptors. *Visual Neuroscience*, **8**, 315–27.

Godley, B.F. and Wurtman, R.J. (1988) Release of endogenous dopamine from the superfused rabbit retina in vitro: effect of light stimulation. *Brain Research*, **452**, 393–5.

Goldman, A.I., Tierstein, P.S. and O'Brien, P.J. (1980) The role of ambient lighting in circadian disc shedding in the rod outer segment in the rat retina. *Investigative Ophthalmology and Visual Science*, **19**, 1257–67.

Green, C.B. and Besharse, J.C. (1994) Tryptophan hydroxylase expression is regulated by a circadian clock in *Xenopus laevis* retina. *Journal of Neurochemistry*, **62**, 2420–8.

Green, C.B., Cahill, G.M. and Besharse, J.C. (1994) Tryptophan hydroxylase in *Xenopus laevis* retina: localization to photoreceptors and circadian rhythm *in vitro*. *Investigative Ophthalmology and Visual Science*, **35**, 1701 (Abstract).

Greenberger, L.M. and Besharse, J.C. (1983) Photoreceptor disc shedding in eye cups: inhibition by deletion of extracellular divalent cations. *Investigative Ophthalmology and Visual Science*, **24**, 1456–64.

Greenberger, L.M. and Besharse, J.C. (1985) Stimulation of photoreceptor disc shedding and pigment epithelial phagocytosis by glutamate, aspartate, and other amino acids. *Journal of Comparative Neurology*, **239**, 361–72.

Gregory, C.Y., Abrams, T.A. and Hall, M.O. (1994) Stimulation of A_2 adenosine receptors inhibits the ingestion of photoreceptor outer segments by retinal pigment epithelium. *Investigative Ophthalmology and Visual Science*, **35**, 819–25.

Hall, M.O. and Abrams, T. (1987) Kinetic studies of rod outer segment binding and ingestion by cultured rat RPE cells. *Experimental Eye Research*, **45**, 907–22.

Hall, M.O., Bok, D. and Bacharach, A.D.E. (1969) Biosynthesis and assembly of the rod outer segment membrane system. Formation and fate of visual pigment in the frog retina. *Journal of Molecular Biology*, **45**, 397–406.

Hamm, H.E. and Menaker, M. (1980) Retinal rhythms in chicks – circadian variation in melatonin and serotonin *N*-acetyltransferase. *Proceedings of the National Academy of Sciences, USA*, **77**, 4998–5002.

Hamm, H.E., Takahashi, J.S. and Menaker, M. (1983) Light-induced decrease of serotonin *N*-acetyltransferase activity and melatonin in the chicken pineal gland and retina. *Brain Research*, **266**, 287–93.

Hayashi, F. and Amakawa, T. (1985) Light-mediated breakdown of phosphatidylinositol-4,5-bisphosphate in isolated rod outer segments of frog photoreceptor. *Biochemical and Biophysical Research Communications* **128**, 954–9.

Herron, W.L., Reigel, B.W., Myers, O.E. and Rubin, M.L. (1969) Retina dystrophy in the rat – a pigment epithelial disease. *Investigative Ophthalmology*, **8**, 595–604.

Heth, C.A. and Marescalchi, P.A. (1994) Inositol triphosphate generation in cultured rat retinal pigment epithelium. *Investigative Ophthalmology and Visual Science*, **35**, 409–16.

Heth, C.A. and Schmidt, S.Y. (1991) Phagocytic challenge induces changes in phosphorylation of retinal pigment epithelium proteins. *Current Eye Research*, **10**, 1049–57.

Hollyfield, J.G. and Witkovsky, P. (1974) Pigmented retinal epithelium involvement in photoreceptor development and function. *Journal of Experimental Zoology*, **189**, 357–8.

Iuvone, P.M. (1986a) Evidence for a D2 dopamine receptor in frog retina that decreases cyclic AMP accumulation and serotonin *N*-acetyltransferase activity. *Life Sciences*, **38**, 331–42.

Iuvone, P.M. (1986b) Neurotransmitters and neuromodulators in the retina: regulation, interactions and cellular effects, in *The Retina, a Model for Cell Biology Studies,* Part II (eds R. Adler and D. Farber), Academic Press, Orlando, pp. 1–72.

Iuvone, P.M. (1990) Development of melatonin biosynthesis in chicken retina: regulation of serotonin *N*-acetyltransferase activity by light, circadian oscillators, and cyclic AMP. *Journal of Neurochemistry,* **54**, 1562–8.

Iuvone, P.M. and Besharse, J.C. (1983) Regulation of indoleamine *N*-acetyltransferase activity in the retina: effects of light and dark, protein synthesis inhibitors and cyclic nucleotides. *Brain Research,* **273**, 111–19.

Iuvone, P.M. and Besharse, J.C. (1986a) Dopamine receptor-mediated inhibition of serotonin *N*-acetyltransferase activity in retina. *Brain Research,* **369**, 168–76.

Iuvone, P.M. and Besharse, J.C. (1986b) Involvement of calcium in the regulation of serotonin *N*-acetyltransferase in retina. *Journal of Neurochemistry,* **46**, 82–8.

Iuvone, P.M. and Besharse, J.C. (1986c) Cyclic AMP stimulates serotonin *N*-acetyltransferase activity in Xenopus retina in vitro. *Journal of Neurochemistry,* **46**, 33–9.

Iuvone, P.M. and Gan, J. (1994) Melatonin receptor-mediated inhibition of cyclic AMP accumulation in chick retinal cell cultures. *Journal of Neurochemistry,* **63**, 118–24.

Iuvone, P.M., Galli, C.L., Garrison-Gund, C.K. and Neff, N.H. (1978) Light stimulates tyrosine hydroxylase activity and dopamine synthesis in retinal amacrine neurons. *Science,* **202**, 901–2.

Iuvone, P.M., Boatright, J.H. and Bloom, M.M. (1987) Dopamine mediates the light-evoked suppression of serotonin *N*-acetyltransferase activity in retina. *Brain Research,* **418**, 314–24.

Iuvone, P.M., Avendano, G., Butler, B.J. and Adler, R. (1990) Cyclic AMP-dependent induction of serotonin *N*-acetyltransferase activity in photoreceptor-enriched chick retinal cell cultures: characterization and inhibition by dopamine. *Journal of Neurochemistry,* **55**, 673–82.

Iuvone, P.M., Gan, J. and Avendano, G. (1991a) K^+-evoked depolarization stimulates cyclic AMP accumulation in photoreceptor-enriched retinal cell cultures: role of calcium influx through hydropyridine-sensitive calcium channels. *Journal of Neurochemistry,* **57**, 615–21.

Iuvone, P.M., Gan, J., Avendano, G. and Butler, B.J. (1991b) Role of cyclic AMP in K^+-evoked, Ca^{2+}-dependent induction of serotonin *N*-acetyltransferase activity in photoreceptor-enriched retinal cell cultures. *Journal of Neurochemistry,* **57**, S67D(Abstract).

Iuvone, P.M., Taylor, M., Tigges, M. *et al.* (1992) Melatonin biosynthesis in chick retinal photoreceptor cells: evidence for a rod pathway to dopamine to cone regulatory mechanism. *Experimental Eye Research,* **35** (Suppl. 1), 5182(Abstract).

Janavs, J.L., Pierce, M.E. and Takahashi, J.S. (1991) *N*-acetyltransferase and protein synthesis modulate melatonin production by Y79 human retinoblastoma cells. *Brain Research,* **540**, 138–44.

Jelsema, C.L. and Axelrod, J. (1987) Stimulation of phospholipase A2 activity in bovine rod outer segments by βγ subunits of transducin and its inhibition by the α subunit. *Proceedings of the National Academy of Sciences, USA,* **84**, 3623–7.

Jung, H.H., Remé, C.E. and Pfeilschifter, J. (1993) Light evoked inositol trisphosphate release in the rat retina *in vitro. Current Eye Research,* **12**, 727–32.

Kazula, A., Nowak, J.Z. and Iuvone, P.M. (1993) Regulation of melatonin and dopamine biosynthesis in chick retina: The role of GABA. *Visual Neuroscience,* **10**, 621–9.

Kirsch, M. and Wagner, H.-J. (1989) Release pattern of endogenous dopamine in teleost retinae during light adaptation and pharmacological stimulation. *Vision Research,* **29**, 147–54.

Kohler, K., Kolbinger, W., Kurz-Isler, G. and Weiler, R. (1990) Endogenous dopamine and cyclic events in the fish retina, II: Correlation of retinomotor movement, spinule formation, and connexon density of gap junctions with dopamine activity during light/dark cycles. *Visual Neuroscience,* **5**, 417–28.

Kolbinger, W., Kohler, K., Oetting, H. and Weiler, R. (1990) Endogenous dopamine and cyclic events in the fish retina, I: HPLC assay of total content release, and metabolic turnover during different light/dark cycles. *Visual Neuroscience,* **5**, 143–9.

Korenbrot, J.I. and Fernald, R.D. (1989) Circadian rhythm and light regulate opsin mRNA in rod photoreceptors. *Nature*, **337**, 454–7.

Kramer, S.G. (1971) Dopamine: a retinal neurotransmitter. I. Retinal uptake, storage, and light stimulated release of H3-dopamine in vivo. *Investigative Ophthalmology*, **10**, 438–52.

Kreutzberg, G.W. and Hussain, S.T. (1984) Cytochemical localization of 5′-nucleotidase activity in retinal photoreceptor cells. *Neuroscience*, **11**, 857–66.

Laitinen, J.T. and Saavedra, J.M. (1990) The chick retinal melatonin receptor revisited: localization and modulation of agonist binding with guanine nucleotides. *Brain Research*, **528**, 349–52.

LaVail, M.M. (1976) Rod outer segment disc shedding in rat retina: relationship to cyclic lighting. *Science*, **194**, 1071–4.

LaVail, M.M. and Ward, P.A. (1978) Studies on the hormonal control of circadian outer segment disc shedding in the rat retina. *Investigative Ophthalmology and Visual Science*, **17**, 1189–93.

Lee, R.H., Brown, B.M. and Lolley, R.N. (1984) Light-induced dephosphorylation of a 33K protein in rod outer segments of rat retina. *Biochemistry*, **23**, 1972–7.

Lee, R.H., Brown, B.M. and Lolley, R.N. (1990) Protein kinase A phosphorylates retinal phosducin on serine 73 *in situ*. *Journal of Biological Chemistry*, **265**, 15860–6.

Lee, R.H., Lieberman, B.S. and Lolley, R.N. (1987) A novel complex from bovine visual cells of a 33 000-dalton phosphoprotein the β- and γ-transducin: purification and subunit structure. *Biochemistry*, **26**, 3983–90.

Leino, M., Aho, I.-M., Kari, E. *et al.* (1984) Effects of melatonin and 6-methoxy-tetrahydro-β-carboline in light induced retinal damage: a computerized morphometric method. *Life Sciences*, **35**, 1997–2001.

López-Colomé, A.M., Salceda, R. and Fragoso, G. (1993) Specific interaction of glutamate with membranes from cultured retinal pigment epithelium. *Journal of Neuroscience Research*, **34**, 454–61.

Matsumoto, B., Defoe, D.M. and Besharse, J.C. (1987) Membrane turnover in rod photoreceptors: ensheathment and phagocytosis of outer segment distal tips by pseudopodia of the retinal pigment epithelium. *Proceedings of the Royal Society of London B*, **230**, 339–54.

McCormack, C.A. and Burnside, B. (1991) Effects of circadian phase on cone retinomotor movements in the Midas cichlid. *Experimental Eye Research*, **52**, 431–8.

McCormack, C.A. and Burnside, B. (1992) A role for endogenous dopamine in circadian regulation of retinal cone movement. *Experimental Eye Research*, **55**, 511–20.

McGinnis, J.F., Whelan, J.P. and Donoso, L.A. (1992) Transient, cyclic changes in mouse visual cell gene products during the light-dark cycle. *Journal of Neuroscience Research*, **31**, 584–90.

McIntosh, H.H. and Blazynski, C. (1994) Characterization and localization of adenosine A_2 receptors in bovine rod outer segments. *Journal of Neurochemistry*, **62**, 992–7.

Millar, F.A., Fisher, S.C., Muir, C.A. *et al.* (1988) Phosphoinositide hydrolysis in response to light stimulation of rat and chick retina and retinal rod outer segments. *Biochimica et Biophysica Acta*, **970**, 205–11.

Molday, R.S., Hicks, D. and Molday, L.L. (1987) Peripherin: a rim-specific protein of rod outer segment discs. *Investigative Ophthalmology and Visual Science*, **28**, 50–61.

Morgan, P.J., Barrett, P., Howell, H.E. and Helliwell, R. (1994) Melatonin receptors: localization, molecular pharmacology and physiological significance. *Neurochemistry International*, **24**, 101–46.

Mullen, R.J. and LaVail, M.M. (1976) Inherited retinal dystrophy: primary defect in pigment epithelium determined by experimental chimeras. *Science*, **192**, 799–801.

Muresan, Z. and Besharse, J.C. (1993) D2-like dopamine receptors in amphibian retina: localization with fluorescent ligands. *Journal of Comparative Neurology*, **331**, 149–60.

Newton, A.C. and Williams, D.S. (1993a) Does protein kinase C play a role in rhodopsin densensitization? *Trends in Biochemical Science*, **18**, 275–277.

Newton, A.C. and Williams, D.S. (1993b) Rhodopsin is the major *in situ* substrate of protein kinase C in rod outer segments of photoreceptors. *Journal of Biological Chemistry*, **268**, 18181–6.

Nir, I. and Agarwal, N. (1993) Diurnal expression

of *c-fos* in the mouse retina. *Molecular Brain Research*, **19**, 47–54.

Nir, I. and Iuvone, P.M. (1994) Alterations in light-evoked dopamine metabolism in dystrophic retinas of mutant *rds* mice. *Brain Research*, **649**, 85–94.

Nir, I. and Ransom, N. (1993) Ultrastructural analysis of arrestin distribution in mouse photoreceptors during dark/light cycle. *Experimental Eye Research*, **57**, 307–18.

Nishizuka, Y. (1992) Intracellular signaling by hydrolysis of phospholipids and activation of protein kinase C. *Science*, **258**, 607–14.

Noell, W.K. (1980) Possible mechanism of photoreceptor damage by light in mammalian eyes. *Vision Research*, **20**, 1163–71.

Nowak, J.Z. (1988) Melatonin inhibits [^{3}H]-dopamine release from the rabbit retina evoked by light, potassium and electrical stimulation. *Medical Science Research*, **16**, 1073–5.

Nowak, J.Z. (1990) Control of melatonin formation in vertebrate retina. *Advances in Pineal Research*, **4**, 81–90.

Orr, H.T., Lowry, O.H., Cohen, A.I. and Ferrendelli, J.A. (1976) Distribution of 3′,5′-cyclic AMP and 3′,5′-cyclic GMP in rabbit retina in vivo: selective effects of dark and light adaptation and ischemia. *Proceedings of the National Academy of Sciences, USA*, **73**, 4442–5.

Paes de Carvalho, R., Braas, K.M., Adler, R. and Snyder, S.H. (1992) Developmental regulation of adenosine A_1 receptors, uptake sites and endogenous adenosine in the chick retina. *Developmental Brain Research*, **70**, 87–95.

Paes de Carvalho, R. and deMello, F.G. (1985) Expression of A_1 adenosine receptors modulating dopamine-dependent cyclic AMP accumulation in the chick embryo retina. *Journal of Neurochemistry*, **44**, 845–51.

Pagh–Roehl, K., Han, E. and Burnside, B. (1993) Identification of cyclic nucleotide-regulated phosphoproteins, including phosducin, in motile rod inner-outer segments of teleosts. *Experimental Eye Research*, **57**, 679–91.

Papermaster, D.S. and Dreyer, W.J. (1974) Rhodopsin content in the outer segment membranes of bovine and frog retinal rods. *Biochemistry*, **13**, 2438–44.

Papermaster, D.S., Converse, C.A. and Siu, J. (1975) Membrane biosynthesis in the retina: opsin transport in the photoreceptor cell. *Biochemistry*, **14**, 1343–52.

Papermaster, D.S., Schneider, B.G., Zorn, M.A. and Kraehenbuhl, J.P. (1978) Immunocytochemical localization of a large intrinsic membrane protein to the incisures and margins of frog rod outer segment disks. *Journal of Cell Biology*, **78**, 415–25.

Peng, Y.-W., Sharp, A.H., Snyder, S.H. and Yau, K.-W. (1991) Localization of the inositol 1,4,5-trisphosphate receptor in synaptic terminals in the vertebrate retina. *Neuron*, **6**, 525–31.

Philp, N.J., Chang, W. and Long, K. (1987) Light-stimulated protein movement in rod photoreceptor cells of the rat retina. *FEBS Letters*, **225**, 127–32.

Pierce, M.E. and Besharse, J.C. (1985) Circadian regulation of retinomotor movements. I. Interaction of melatonin and dopamine in the control of cone length. *Journal of General Physiology*, **86**, 671–89.

Pierce, M.E. and Besharse, J.C. (1986) Melatonin and dopamine interactions in the regulation of rhythmic photoreceptor metabolism, in *Pineal and Retinal Relationships* (eds P.J. O'Brien and D.C. Klein), Academic Press, Orlando, pp.219–37.

Pierce, M.E. and Besharse, J.C. (1987) Melatonin and rhythmic photoreceptor metabolism: melatonin-induced cone elongation is blocked by high light intensity. *Brain Research*, **405**, 400–4.

Pierce, M.E., Barker, D., Harrington, J. and Takahashi, J.S. (1989) Cyclic AMP-dependent melatonin production in Y79 human retinoblastoma cells. *Journal of Neurochemistry*, **53**, 307–10.

Pierce, M.E., Sheshberadaran, H., Zhang, Z. *et al.* (1993) Circadian regulation of iodopsin gene expression in embryonic photoreceptors in retinal cell culture. *Neuron*, **10**, 579–84.

Redburn, D.A. and Mitchell, C.K. (1989) Darkness stimulates rapid synthesis and release of melatonin in rat retina. *Visual Neuroscience*, **3**, 391–403.

Remé, C., Wirz–Justice, A., Aeberhard, B. and Rhyner, A. (1984) Chronic clogyline dampens rat retinal rhythms. *Brain Research*, **298**, 99–106.

Remé, C., Wirz–Justice, A., Rhyner, A. and Hoffman, S. (1986) Circadian rhythm in the light response of rat retinal disk shedding and autophagy. *Brain Research*, **369**, 356–60.

Remé, C.E., Wirz-Justice, A. and Terman, M. (1991) The visual input stage of the mammalian circadian pacemaking system: I. Is there a clock in the mammalian eye. *Journal of Biological Rhythms*, **6**, 5–29.

Sheng, M. and Greenberg, M.E. (1990) The regulation and function of c-*fos* and other immediate early genes in the nervous system. *Neuron*, **4**, 477–85.

Shimizu, T. and Wolfe, L.S. (1990) Arachidonic acid cascade and signal transduction. *Journal of Neurochemistry* **55**, 1–15.

Sitaramayya, A., Lombardi, L. and Margulis, A. (1993) Influence of dopamine on cyclic nucleotide enzymes in bovine retinal membrane fractions. *Visual Neuroscience*, **10**, 991–6.

Steinberg, R.H., Fisher, S.K. and Anderson, D.H. (1980) Disc morphogenesis in vertebrate photoreceptors. *Journal of Comparative Neurology*, **190**, 501–18.

Stenkamp, D.L., Iuvone, P.M. and Adler, R. (1994) Photomechanical movements of cultured embryonic photoreceptors: regulation by exogenous neuromodulators and by a regulable source of endogenous dopamine. *Journal of Neuroscience*, **14**, 3083–96.

Stiemke, M.M., Landers, R.A., Al-Ubaidi, M.R., Rayborn, M.E. and Hollyfield, J.G. (1994) Photoreceptor outer segment development in *Xenopus laevis*: influence of the pigment epithelium. *Developmental Biology*, **162**, 169–80.

Tamai, M., Tierstein, P., Goldman, A. *et al.* (1978) The pineal gland does not control rod outer segment shedding and phagocytosis in the rat retina and pigment epithelium. *Investigative Ophthalmology and Visual Science*, **17**, 558–62.

Terman, J.S., Remé, C.E. and Terman, M. (1993) Rod outer segment disk shedding in rats with lesions of the suprachiasmatic nucleus. *Brain Research*, **605**, 256–64.

Thomas, K.B. and Iuvone, P.M. (1991) Circadian rhythm of tryptophan hydroxylase activity in chicken retina. *Cellular and Molecular Neurobiology*, **11**, 511–27.

Thomas, K.B., Tigges, M. and Iuvone, P.M. (1993) Melatonin synthesis and circadian tryptophan hydroxylase activity in chicken retina following destruction of serotonin immunoreactive amacrine and bipolar cells by kainic acid. *Brain Research*, **601**, 303–7.

Tierstein, P.S., Goldman, A.I. and O'Brien, P.J. (1980) Evidence for both local and central regulation of rat rod outer segment disc shedding. *Investigative Ophthalmology and Visual Science*, **19**, 1268–73.

Travis, G.H. (1991) Molecular characterization of the retinal degeneration slow (*rds*) mutation in mouse. *Progress in Clinical and Biological Research*, **362**, 87–114.

Underwood, H., Siopes, T. and Barrett, R.K. (1988) Does a biological clock reside in the eye of quail? *Journal of Biological Rhythms*, **3**, 323–31.

Underwood, H., Barrett, R.K. and Siopes, T. (1990) The quail's eye: a biological clock. *Journal of Biological Rhythms*, **5**, 257–65.

Vale, R.D., Schnapp, B.J., Mitchison, T. *et al.*, (1985) Different axoplasmic proteins generate movement in opposite directions along microtubules in vitro. *Cell*, **43**, 623–32.

Vuvan, T., Geffard, M., Denis, P. *et al.* (1993) Radioimmunoligand characterization and immunohistochemical localization of dopamine D_2 receptors on rods in the rat retina. *Brain Research*, **614**, 57–64.

Wagner, H.-J. and Behrens, U.D. (1993) Microanatomy of the dopaminergic system in the rainbow trout retina. *Vision Research*, **33**, 1345–58.

Welsh, J.H. and Osborne, C.M. (1937) Diurnal changes in the retina of the catfish, *Ameiurus nebulosus*. *Journal of Comparative Neurology*, **66**, 349–60.

Whelan, J.P. and McGinnis, J.F. (1988) Light-dependent subcellular movement of photoreceptor proteins. *Journal of Neuroscience Research*, **20**, 263–70.

Wiechmann, A.F. and Craft, C.M. (1993) Localization of mRNA encoding the indolamine synthesizing enzyme, hydroxyindole-*O*-methyltransferase, in chicken pineal gland and retina by in situ hybridization. *Neuroscience Letters*, **150**, 207–11.

Wiechmann, A.F. and O'Steen, W.K. (1992) Melatonin increases photoreceptor susceptibility to light-induced damage. *Investigative Ophthalmology and Visual Science*, **33**, 1894–902.

Wiechmann, A.F. and Wirsig-Wiechmann, C.R. (1991) Localization and quantification of high-affinity melatonin binding sites in *Rana pipiens* retina. *Journal of Pineal Research*, **10**, 174–9.

Wiechmann, A.F., Kyritsis, A.P., Fletcher, R.T. and Chader, G.J. (1990) Cyclic AMP and butyrate modulate melatonin synthesis in Y79 human retinoblastoma cells. *Journal of Neurochemistry*, **55**, 208–14.

Williams, D.S. (1991) Actin filaments and photoreceptor membrane turnover. *BioEssays*, **13**, 171–8.

Witkovsky, P. and Dearry, A. (1991) Functional roles of dopamine in the retina, in *Progress in Retinal Research*, Vol. 11 (eds N.N. Osborne and G.J. Chader), Pergamon, Oxford, pp. 247–92.

Yoshida, K., Kawamura, K. and Imaki, J. (1993) Differential expression of c-*fos* mRNA in rat retinal cells: Regulation by light/dark cycle. *Neuron*, **10**, 1049–54.

Young, R.W. (1967) The renewal of photoreceptor cell outer segments. *Journal of Cell Biology*, **33**, 61–72.

Young, R.W. (1976) Visual cells and the concept of renewal. *Investigative Ophthalmology*, **15**, 700–25.

Young, R.W. (1978) The daily rhythm of shedding and degradation of rod and cone outer segment membranes in the chick retina. *Investigative Ophthalmology*, **17**, 105–16.

Young, R.W. and Bok, D. (1969) Participation of the retinal pigment epithelium in the rod outer segment renewal process. *Journal of Cell Biology*, **42**, 392–403.

Young, R.W. and Droz, B. (1968) The renewal of protein in retinal rods and cones. *Journal of Cell Biology*, **39**, 169–84.

Zarbin, M.A., Wamsley, J.K., Palacios, J.M. and Kuhar, M.J. (1986) Autoradiographic localization of high affinity GABA, benzodiazepine, dopaminergic, adrenergic and muscarinic receptors in the rat, monkey, and human retina. *Brain Research*, **374**, 75–92.

Zawilska, J. and Iuvone, P.M. (1989) Catecholamine receptors regulating serotonin *N*-acetyltransferase activity and melatonin content of chicken retina and pineal gland: D_2-dopamine receptors in retina and *alpha*-2 adrenergic receptors in pineal gland. *Journal of Pharmacology and Experimental Therapeutics*, **250**, 86–92.

Zawilska, J.B. and Iuvone, P.M. (1992) Melatonin synthesis in chicken retina: effect of kainic acid-induced lesions on the diurnal rhythm and D_2-dopamine receptor-mediated regulation of serotonin *N*-acetyltransferase activity. *Neuroscience Letters*, **135**, 71–4.

Zawilska, J.B. and Nowak, J.Z. (1990) Calcium influx through voltage-sensitive calcium channels regulates in vivo serotonin *N*-acetyltransferase (NAT) activity in hen retina and pineal gland. *Neuroscience Letters*, **118**, 17–20.

Zawilska, J.B. and Nowak, J.Z. (1992) Regulatory mechanisms in melatonin biosynthesis in retina. *Neurochemistry International*, **20**, 23–36.

Zawilska, J.B. and Nowak, J.Z. (1994) Does D_2 dopamine receptor mediate the inhibitory effect of light on melatonin biosynthesis in chick retina. *Neuroscience Letters*, **166**, 203–6.

Zawilska, J.B., Kazula, A., Zurawska, E. and Nowak, J.Z. (1991) Serotonin *N*-acetyltransferase activity in chicken retina: in vivo effects of phosphodiesterase inhibitors, forskolin, and drugs affecting dopamine receptors. *Journal of Pineal Research*, **11**, 116–22.

Zucker, C.L. and Dowling, J.E. (1987) Centrifugal fibers synapse on dopaminergic interplexiform cells in the teleost retina. *Nature*, **330**, 166–8.

3

Determinants of visual pigment spectral location and photoreceptor cell spectral sensitivity

ELLIS R. LOEW

3.1 INTRODUCTION

What factors determine where the visual pigments of an animal will be spectrally located? This question remains a stimulus for scientific speculations and investigations today as it has for the last 120 years (see historical reviews in Crescitelli, 1991a,b). In the early 1970s, the umbrella term 'Visual Ecology' was coined to define more formally a specific approach to this and similar questions (e.g. why do animals have the number of visual pigments they do? or why do photoreceptor cells look the way they do?), which always take into account the visual tasks of an animal within a given photic environment.

A subdiscipline of visual ecology concerned specifically with visual pigment spectral location was born from studies that attempted to correlate the spectral location of extractable visual pigments (almost exclusively from rods) with the characteristics of the photic environment (see review in Munz and McFarland, 1975, 1977). With the development of microspectrophotometry (MSP) this approach was extended to cones with some success (Loew and Lythgoe, 1978, 1985; Levine and MacNichol, 1979; Lythgoe, 1991; Lythgoe and Partridge, 1989, 1991; Partridge, 1990; see Bowmaker, 1991). Almost all of these survey data are for fish, although some data both for visual pigments and oil droplets in birds (Partridge, 1989; Varela *et al.*, 1993) and reptiles (Crescitelli, 1991a) are available. A number of excellent reviews covering the evolution and ecology of visual pigments, particularly as regards colour vision, from the molecular to the environmental have been published over the past four years (Appendix 1). The aims of this chapter are instead to review the photochemistry of visual pigments as it relates to photoreceptor cell spectral sensitivity, outline the biophysical/biochemical mechanisms providing the variation upon which selection can act to shift and locate visual pigments and photoreceptor cell spectral sensitivity, and present some of the as yet unexplained aspects of visual pigment spectral sensitivity placement.

Neurobiology and Clinical Aspects of the Outer Retina
Edited by M.B.A. Djamgoz, S.N. Archer and S. Vallerga
Published in 1995 by Chapman & Hall, London
ISBN 0 412 60080 3

3.2 PHOTOCHEMICAL CONSIDERATIONS

The most fundamental law of vision, and photobiology in general, is that only those photons absorbed are effective in stimulating a photic process. This means that for retinal photoreceptor cells, the photochemical properties of the contained visual pigment set the fundamental limit on photon capture probability and, therefore, visual perception. The only way to characterize a visual pigment in this context is to measure its light-absorbing properties. Ideally, this measurement should be made on single molecules, but this is obviously impossible. Rather, the properties of single molecules must be deduced from measurements of numerous molecules in solution or *in situ*. The methods of extraction- and microspectrophotometry (MSP) have been covered in depth by Dartnall (1957) and Knowles and Dartnall (1977) along with the photochemical principles needed for proper data analysis and presentation. For this discussion, the following terminology of Knowles and Dartnall (1977) has been adopted. Absorption is the algebraic difference between the number of quanta incident on a solution or photoreceptor cell and the number exiting along a parallel path. The fractional absorption (the absorption divided by the number of incident quanta) is the absorptance. Absorbance (sometimes called optical density, density or extinction) is the $\log_{10}$ of the number of quanta incident divided by the number of quanta exiting. Variations of these with wavelength yield absorptance or absorbance spectra. The shape of an absorbance spectrum normalized to its peak is concentration independent and thus is representative of single molecules.

A visual pigment is characterized not only by the shape of its normalized absorbance spectrum but also by its photosensitivity, where photosensitivity is the product of the extinction coefficient and the quantum efficiency of bleaching (Dartnall, 1957; Knowles and Dartnall, 1977). Thus, photosensitivity is a measure of photochemical bleaching under defined conditions and is directly related to the probability that a photon will be absorbed (the effective reaction cross-sectional area). Rhodopsin solutions follow the Beer–Lambert laws, and the quantum efficiency for bleaching is essentially independent of wavelength (above 430 nm), concentration, temperature and pH, being approximately 0.67 for rhodopsins and 0.64 for porphyropsins (Dartnall, 1968). Comparisons between photosensitivity measurements made at different wavelengths and the absorbance spectra for rhodopsin are coincident for wavelengths above 430 nm (Dartnall, 1957), meaning that the extinction coefficient varies exactly as the absorbance spectrum. Thus, the probability that a photon of a given wavelength will be absorbed can be obtained directly from the absorbance spectrum, and photons having the same probability of absorption will have the same photochemical effect (i.e. produce the same photoreceptor cell response) regardless of their wavelength. This is known as the Principle of Univariance.

As mentioned, the probability that any single molecule of visual pigment, whether in solution or in a photoreceptor cell, will absorb a photon at a particular wavelength falling within its capture cross-section is predicted by the normalized absorbance spectrum and is concentration and path-length independent. However, the probability that a photon will encounter a visual pigment molecule on its passage through a solution or photoreceptor cell is predicted by the absorptance spectrum. The term 'axial density' is often used for photoreceptor cells and is the absorptance for a photoreceptor of known length along its optic axis. The more concentrated the pigment and the longer the pathlength, the greater the probability of absorption.

One consequence of this concentration- and pathlength-dependence is that as either increases, the total fraction of photons absorbed increases and the absorption spectrum becomes broader. In the limit of a single molecule, the absorbance and absorptance spectra are identical. For an infinitely concentrated or long photoreceptor, all photons would be absorbed and the cell would be perfectly black over its spectral absorbance range.

Vision proceeds directly from the absorption of photons that produce chromophore isomerization leading to activation of the visual pigment and transductional cascade. Thus, it is the absorptance spectrum that should predict the spectral sensitivity of a photoreceptor cell, not the absorbance spectrum of its visual pigment. The more optically dense the cell, the greater will be its luminous sensitivity and the broader its spectral sensitivity. This was first validated by Dartnall and Goodeve (1937) who found that the calculated absorptance spectrum for rhodopsin in the frog retina agreed with Granit's electrophysiologically derived frog scotopic spectral sensitivity data much better than the narrower rhodopsin absorbance spectrum (see also Dartnall, 1953).

It is rarely possible to measure the actual absorption of visual photoreceptors *in situ*. Rather, the axial density is calculated from the specific absorbance of the pigment (optical density per unit path in micrometers), and measurements of outer segment length. In innumerable cases, the absorption spectra calculated this way has been fitted to presumed photoreceptor spectral sensitivities or visual system action spectra quite well. However, this has not always been the case. In fitting human psychophysical data to visual pigment absorption curves for rods and cones, calculated from MSP absorbance data modified by the contributions of pre-receptoral filtering, it was found that the CIE (Commission Internationale de l'Eclairage) scotopic (rod) sensitivity was narrower than the absorptance spectrum (Bowmaker and Dartnall, 1980). Instead, this CIE curve was best fitted by the absorptance spectrum calculated for zero pathlength, that is, the absorbance spectrum. Since the absorbance and absorptance spectra are essentially identical for densities less than 0.1 (Dartnall, 1957), this sets a limit on the 'effective' length of human rods at less than 6.6 μm, assuming a specific density of 0.015 (Bowmaker and Dartnall, 1980; based their absorption calculations on a peak axial density of 0.475).

The explanation for this discrepancy remains obscure, particularly as this matching process worked well for human cones. The suggestion by Bowmaker and Dartnall (1980) that only the portion of the rod outer segment close to the inner segment is visually effective helps, but is not sufficient. Neither is the idea that only a small proportion of the visual pigment present in a human rod is capable of initiating the events leading to transduction. It is better to suppose in this, and perhaps other cases, that the effective axial density is not strictly predicted by transverse specific density estimates and photoreceptor cell length. Structural and waveguiding properties, to be discussed below, may also influence photon absorption and, therefore, the magnitude of the axial density.

Theoretically, photoreceptor quantum catch could be increased by evolving visual pigments having greater photosensitivity, increasing visual pigment concentration per unit distance, or increasing outer segment length. Of these three, the latter seems to be the most prevalent solution. However, increasing absorptance to increase sensitivity is not without other consequences. First, the 'dark' noise produced by spontaneous thermal isomerizations of visual pigment will increase in proportion to the concentration of pigment thereby degrading signal detection capabilities (Barlow, 1964, see Lythgoe, 1988). Second, the absorptance spectrum will flatten out which will decrease hue discrimination if the photoreceptor is part of a colour vision mechanism. In fact, it may be necessary to keep the axial density of cones low, and their outer segments short, to preserve this function.

All of the above considerations have been directed at the main, or the α-absorbance band. However, visual pigments show substantial absorbance (never less than about 20%) on the short-wavelength side of the α-band peak well into the ultraviolet (UV) (Abrahamson and Japar, 1972; Abrahamson and Wiesenfeld, 1972; Morton, 1972). There is a large, visually ineffective protein absorbance due to aromatic amino acids at 278 nm, and a β-band in the near-UV due to the chromophore. For rhodopsin, the β-band peaks between 350 nm and 370 nm, and represents a higher energy state of the visual pigment. Its maximum absorbance is about 30% of the α-peak, due probably to a smaller extinction coefficient (Abrahamson and Wiesenfeld, 1972). Since the α-band electron transitions overlap those responsible for the β-band (Abrahamson and Japar, 1972, Abrahamson and Wiesenfeld, 1972), it could be that it is only the α-band tail absorbance that has the most visual relevance. Nevertheless, photons absorbed in the β-band region bleach visual pigment in the usual way and can presumably be visually effective (Dartnall, 1957; Knowles and Dartnall, 1977). Unfortunately, the presence of photoproducts and other 'contaminants' make matching of spectral sensitivity curves and absorption spectra unreliable in this spectral region. Consequently, the question of β-band efficacy remains open.

3.3 SPECTRAL TUNING OF PHOTORECEPTOR CELL SPECTRAL SENSITIVITY

Photoreceptor cell types can be defined by their spectral sensitivity, the mechanisms responsible for this sensitivity and their biochemical/biophysical characteristics. Any mechanism altering spectral sensitivity can act as a selective focus. In an evolutionary sense it is not necessarily important how this spectral sensitivity is achieved. As with all adaptive processes, there must be some phenotypic polymorphism within the population as a whole upon which selection can act. The main determinant for selection will be the conferring of a fitness advantage to some variant(s).

A number of mechanisms are available by which animals can alter or tune the spectral sensitivity of their visual photoreceptors. These are not only the mechanisms responsible for 'classical' spectral differences seen between, say, red- and blue-sensitive cone classes, but also those producing small differences within the spectral types that lead to class polymorphism within an animal or a group. Phenotypic variation of visual photoreceptor cell spectral sensitivity may result from the following.

1. Opsin differences which originally derive from random assortment or gene mutation (i.e. amino acid sequence alterations) and possible differences in post-translational modification of the expressed opsins.
2. Variation in the expression and/or activity of the enzymes involved in chromophore synthesis and/or cycling.
3. Variation in pathways responsible for regulating the ionic environment in and around the photoreceptors.
4. Variation in the synthetic and/or regulatory pathways responsible for inner segment absorptive or scattering filter production and location.

5. Variation in the structural genes responsible for photoreceptor cell size and shape.

The phenotype of an individual may be determined at conception and show little variation throughout life, or there may be much variation which is either random among retinal cells or shows definite life-cycle correlations. Since a given animal may have demonstrable variation of more than one of these factors, the potential for tuning is great. How 'fit' a given individual will be and how much variation may exist in a population will, of course, be determined by the selective pressures acting on spectral sensitivity.

3.3.1 OPSIN TUNING

Primary structure (i.e. amino acid sequence) differences between opsins are the classic explanation for visual pigment absorbance differences among photoreceptor cell classes (see historical treatment in Knowles and Dartnall, 1977). Whereas the protonated Schiff base linkage between the opsin and the chromophore (assumed here to be retinaldehyde) produces a spectral shift out of the ultraviolet (360 nm peak) and into the blue (430 nm peak), conferring 'color' to the visual pigment, it is the 3-dimensional energy state around this bond and within the hydrophobic chromophore 'pocket' that is responsible for further red-shifting. Amino acid substitutions at sites far from this pocket may influence spectral position through long-range interactions, and certainly counter-ion position will also have effects. In some cases, it appears that a very small set of amino acids at defined positions may be responsible for most, if not all, spectral tuning (Bowmaker, 1991; Tovee, 1994). It is not necessarily the case that each spectral state of an opsin is achieved by a single amino acid sequence – different sequences may produce isochromic visual pigments. Thus, rods and cones may have the same absorbance spectrum, but differ substantially in the primary sequence of their opsins in the region of the binding pocket.

Before the introduction of quantitative MSP and modern molecular biological techniques, it was widely believed that there was little, if any, variation in the wavelength-determining domain of an opsin expressed by a given class of photoreceptor within a species. Any variation seen was put to experimental error, A_1/A_2 chromophore mixing, photoproduct contamination or the presence of other retinal isomers (see discussion in Knowles and Dartnall, 1977). The implication from this belief in invariability was that the spectral position of a visual pigment was so critical, that any variation was maladaptive and quickly selected against. However, if there was no variability, what did selection act upon to shift λ_{max} when it was adaptive to do so? The first paper to suggest that there was, indeed, opsin variability capable of producing small spectral shifts within a single class of photoreceptor came from *in situ* and MSP measurements on frogs (Bowmaker *et al.*, 1975). It had already been shown that the λ_{max} of extracted and *in situ* rhodopsin could differ by as much as 3 nm in a given animal (Bowmaker, 1973), but variations between animals of up to 8 nm were totally unexpected; such variation was never seen within the rods of a single animal.

The idea of intraspecific spectral differences in separate photoreceptor cell classes was tested again with the behavioral and MSP data from primates (see review of early work in Mollon *et al.*, 1984; also see Tovee, 1994). For example, Jacobs *et al.* (1981) found a bimodal distribution of long-wavelength sensitivity in squirrel monkeys that correlated with polymorphic visual pigments in the red-sensitive receptor mechanism with λ_{max} values varying by about 20 nm; the cluster wavelengths were at 552 nm and 568 nm. It is now well established that there are many amino acid substitutions that can produce spectral shifts in λ_{max} while still meeting the requirements for a functional visual pigment. This area is covered extensively in Chapter 4.

One last area for discussion here is the concept of visual pigment 'clustering'. Dartnall and Lythgoe (1965) considered the question of whether visual pigments were continuously tunable, or whether there were constraints on the λ_{max} position that a visual pigment could attain. This could have obvious evolutionary consequences. A histogram of visual pigment λ_{max} values for the rhodopsins or porphyropsins from all vertebrates examined to that date indicated that there were certain preferred spectral locations or clusters. The distance between clusters was approximately 6 nm for rhodopsins and 12 nm for porphyropsins. The model proposed to explain that this phenomenon involved discrete charge interactions along the length of the chromophore separate from those at the Schiff's base – that is, opsin tuning (Dartnall and Lythgoe, 1965). More recent examination of data from deep sea fish using frequency domain analysis, and mammalian data also support spectral clustering (Partridge *et al.*, 1989; Jacobs and Neitz, 1985; Jacobs, 1993). Although this type of interspecific analysis is suggestive, the fact that there are still λ_{max} values falling between cluster points leaves the question of continuous tunability for opsins within a given species or individual unanswered. Restating the side-chain interaction model of Dartnall and Lythgoe (1965) in more contemporary terms, one can ask whether the amino acid substitutions known to be involved in spectral tuning and intraspecific polymorphism interact to shift λ_{max} in discrete steps. The answer would appear to be 'yes'. In attempting to rationalize the spectral positions of primate red- and green-sensitive cones in terms of specific amino acid substitutions, Neitz *et al.* (1991) came to the conclusion that spectral shifts occurred in discrete steps. Their model involved amino acid substitutions (hydroxyl-bearing for non-polar residues) at three positions (180, 277 and 285) which produced spectral shifts of 5.33, 9.5 and 15.5 nm. The actual λ_{max} of a pigment produced by different opsins is determined by simple summation of the sequence differences at these three locations. In this model, the minimum shift possible by a single amino acid substitution would be 5–6 nm, remarkably similar to the distance between interspecific clusters! The 5 nm minimum also seems to hold for invertebrates like octopus (Morris *et al.*, 1993). Whether this 'rule of five' holds for all visual pigment tuning at the primary structure level remains to be seen.

3.3.2 CHROMOPHORE TUNING

The λ_{max} of a visual pigment depends not only on the structure of the opsin, but also on the form of vitamin A – aldehyde used as the chromophore. The four naturally occurring retinaldehydes used for visual pigment synthesis are retinaldehyde (A_1), 3,4-didehydroretinaldehyde (A_2), 3-hydroxyretinaldehyde and 4-hydroxyretinaldehyde. For the same opsin, substitution of the chromophore results in a spectral shift. The 3-hydroxyretinaldehyde chromophore is found only in insects where it is always associated with short-wavelength-sensitive pigments (Smith and Goldsmith, 1990; see Goldsmith, 1990). The 4-hydroxy form has only been found in the firefly squid, *Watasenia scintillans*, and produces a blue-shift when substituted for retinaldehyde in bovine rhodopsin (Kito *et al.*, 1992).

In vertebrates, it is $A_1 - A_2$ substitution that is most widely used for spectral sensitivity tuning. The A_1 chromophore produces rhodopsin and iodopsin, whereas A_2 produces porphyropsin and cyanopsin (see Crescitelli, 1972 for historical treatment). Substitution of A_2 for A_1 produces a red-shift in the absorbance spectrum of the visual pigment. The amount of this shift depends on the spectral location of the A_1 parent pigment – the longer the wavelength of the parent pigment, the greater the shift with chromophore substitution. For 562 nm iodopsin, A_2 substitution produced cyanopsin at 620 nm (Wald *et al.*, 1953). Interestingly, it was not until 12 years later that MSP data from goldfish proved the existence of cyanopsin (Marks, 1965). The quantitative relationship between A_1/A_2 pigment pairs was first published by Dartnall and Lythgoe (1965) and later refined by Whitmore and Bowmaker (1989). A_2 substitution carries with it consequences other than a spectral shift. The extinction coefficient is less, the absorbance spectrum broader and the visual pigment less thermally stable (Bridges, 1972).

It was realized early on that the chromophore used by vertebrates could be associated with its environment. Freshwater species or developmental stages utilized A_2 whereas marine and terrestrial forms used A_1 (see reviews by Bridges, 1972, and Beatty, 1984). The consequences were obvious – freshwater forms could be more inherently red-sensitive than marine/terrestrial forms. As with any general rule, however, it is the exceptions that often prove more interesting. It is now known that deep-sea marine species can synthesize A_2-based pigments (Bowmaker *et al.*, 1988; see Partridge, 1990) as can at least one fully terrestrial vertebrate, the common chameleon *Anolis carolinensis* (Provencio *et al.*, 1992).

The utilization of A_1 or A_2 chromophore need not be exclusive within a cell or across a retina. Many species are known to incorporate mixtures of the two chromophores within a single photoreceptor. In these cases the mixture ratio will determine the spectral sensitivity of the cell which can be truly continuous between the pure chromophore extremes. In some cases, such as certain wrasses, the mixture ratio appears to be fairly constant (Lythgoe, 1979). However, the more usual situation is for the ratio to vary depending on the developmental stage, life history or age. During downstream migration, anadromous or catadromous species switch from A_2 to A_1 presumably as an adaptation to the 'bluer' marine environment. The switch is reversed during upstream migration (see extensive review in Crescitelli, 1972). In some species, mixture variation can be seasonal (Beatty, 1984). Lastly, there seems to be a tendency for older, post-reproductive salmonids and cyprinids to 'peg out' in the pure A_2 condition (E. Loew and H.J.A. Dartnall, unpublished observations; Bridges, 1972). The adaptive significance of chromophore shifting remains conjectural. The simple idea that animals in 'redder' environments utilize A_2 to maximize sensitivity does not hold up to scrutiny. Neither is there any real evidence to show that in seasonally variable species the time of increased retinal A_2 correlates with a 'redder' environment or 'red' visual tasks.

The use of the A_1 or A_2 chromophore by a photoreceptor appears to be non-selective in all retinas except those of a few deep sea species (Bowmaker *et al.*, 1988; Partridge, 1990). It is the composition of the local chromophore pool established by the retinal pigment epithelium (RPE) that will determine the chromophore ratio of contiguous photoreceptors. This goes for both rods and cones as they both draw from the same local chromophore pool (Loew and Dartnall, 1976). Within the RPE there must be a specific dehydrogenase for synthesizing the A_2 chromophore from the A_1 form. The final ratio will be established by the activity of this enzyme as well as by a probable 3,4-hydrogenase operating in the reverse direction (Bridges, 1972; Provencio *et al.*, 1992).

It is likely that the selective pressures favoring A_2 are applied at the photopic (cone) rather than the scotopic (rod) level. In particular, it appears to be the red-sensitive photoreceptor mechanism that is targeted. First, A_2 substitution for the long-wavelength cone moves the animal's spectral sensitivity range into previously unavailable regions. For example, a 625 nm A_2-based visual pigment can utilize the reflective edge of chlorophyll above 730 nm at photopic levels for contrast enhancement (J. Lythgoe and E. Loew, unpublished calculations). A similar argument for spectral sensitivity extension may also explain the use of A_2 by deep sea fish (Bowmaker *et al.*, 1988). It is noteworthy that in the chameleon the sensitivity of the 'red' cone is shifted to 625 nm, whereas the opsins of the other cones have been selected so that their A_2 forms have λ_{max} values similar to those found for pure A_1 lizards (Provencio *et al.*, 1992; E. Loew, unpublished observations). Second, small variations in chromophore ratio can produce large variations in sensitivity for the longer-wavelength cones. Lastly, the action spectrum for the chromophore shift produced by the action of light in rainbow trout and the common shiner does not match the rhodopsin spectrum, but peaks above 550 nm suggesting photopic control (Munz and McFarland, 1977).

It should be pointed out that just because an animal does not use the A_2 chromophore for visual pigment production does not mean that this potential is genetically excluded. Humans do not use A_2 and, in fact, appear to have a powerful hydrogenase in the RPE that converts any A_2 present into A_1 (see discussion in Yoshikami *et al.*, 1969). However, vitamin A_2 is found in great abundance in skin showing that the enzymatic pathways for its production are in the human genome (Torma and Vahlquist, 1990).

3.3.3 IONIC TUNING

Any change in the energy state of the opsin/chromophore interaction may be expected to affect the absorbance spectrum. Such changes need not come about solely by amino acid substitution. The presence of ions bound at or near the binding site or pocket may also affect the energy state and produce spectral shifts.

Crescitelli (1977) was the first to demonstrate an ionochromic effect on visual pigments. He found that extraction of the gecko 521 nm pigment in NaCl-free solution gave instead a visual pigment at 483 nm. When NaCl was added to this extract, the λ_{max} shifted to 505 nm. The degree of red-shift was concentration dependent and reversible. Crescitelli (1977) demonstrated that the Cl^- was responsible for the effect and not the cation. The 'chloride effect' could not be demonstrated for rhodopsins that have very hydrophobic binding pockets. However, chicken iodopsin was found to be Cl^- sensitive (Knowles, 1976). These findings clearly support the presence of an anion-specific binding site in the region of the protonated Schiff base.

It now appears that the long-wavelength pigments from a number of other species are also Cl^- sensitive *in situ* (Novitskii *et al.*, 1989; Kleinschmidt and Hárosi, 1992). This also holds for closely related pigments such as the green/red pair in primates and may, in fact, apply to all long wavelength-sensitive pigments (Kleinschmidt and Hárosi, 1992). Spectral shifts of up to 55 nm can be seen in some cases. Other anions (e.g. NO_3^-, SO_4^{2-}, ClO_4^-) can also shift λ_{max} in either direction but it appears that only Cl^- and Br^- produce native pigment λ_{max} values when bound to opsin. On the assumption that the Cl^- binding site on the long-wavelength opsin is always saturated, which seems reasonable given that the K_D is one to two orders of magnitude lower than the intracellular Cl^- concentration (Kleinschmidt and Hárosi, 1992), the above suggests that Cl^- binding has been selected for as a tuning mechanism in this class of pigments. In fact, it may be that the only way to attain λ_{max} values for retinaldehyde-based pigments above 540 nm is via an ionochromic mechanism. This raises some interesting questions concerning the origins of the different cone classes. There is also the possibility that anion concentration can be used for rapid and gradual shifts in λ_{max}.

3.3.4 ABSORPTIVE TUNING

This results from the presence of inner segment filter elements which can significantly modify the spectral distribution of photons reaching the outer segment. The best example is the use of highly colored oil droplets by reptiles and birds (see review in Bowmaker, 1991). The filtering properties of oil droplets are determined both by the nature of the filtering molecule, almost always a carotenoid, and its concentration. Although it is not possible to obtain really accurate absorbance spectra for oil droplets *in situ* due to their often high optical density (O.D. >20) and refractive problems, comparisons within species show little obvious variation among normal individuals (Partridge, 1989; Bowmaker 1991; E. Loew, unpublished observations). However, carotenoid-stressed individuals may show saturation decreases that can progress to a colorless state (Meyer *et al.*, 1971; Bowmaker *et al.*, 1993). How this affects fitness due to alterations in visual capabilities is unknown, and in any case would be difficult to assess as most carotenoid-deficient animals suffer from other significant health problems.

After examining the phylogenetic distribution of oil droplets and relating this to ecotype, Walls (1967) proposed that oil droplets appeared early in vertebrate evolution along with hue discrimination. As a species moved toward nocturnality, the density of pigment in the oil droplets decreased and eventually the oil droplet was lost. Once this occurred, any descendants that moved toward diurnality could not regain the droplets.

The function of oil droplets is presumed to be improvement in hue discrimination by spectral sharpening (Govardovskii, 1983; Partridge, 1989; Bowmaker, 1991). This is a pure photopic function linked directly to the adaptive value of color vision. The positioning of the oil droplet absorptive edges would be selected so that, together with the post-droplet visual pigments, hue discrimination in ecologically important spectral regions would be enhanced. For example, subtle differences in hue along the red and magenta axes might signal fruit ripeness or carbohydrate content to birds (Lythgoe, 1979).

Oil droplets are not the only inner segment filters possible. In reptiles, one class of cone contains a pigment, presumably a carotenoid, distributed diffusely throughout the inner segment. Its density is much lower than in an oil droplet and refractive effects are minimal, but its function may be the same. Mitochondrial cytochromes can also act as filters if in high concentration along the light path. This has been found in guppies where at least some of the cone spectral sensitivities are narrowed by cytochrome *c* in the ellipsosome (MacNichol *et al.*, 1978).

3.3.5 STRUCTURAL TUNING

Since the refractive index of photoreceptors is different from that of the surrounding medium, these cells can be expected to act as optical waveguides. The fact that waveguide modal patterns can be viewed at the tips of the outer segments is direct evidence for this process (Enoch and Tobey, 1981). This waveguiding shows wavelength selectivity so that light reaching the outer segments has already passed through a 'structural filter' (Enoch, 1963). The effect of such filtering is to shift the photoreceptor's absorptance and, presumably its spectral sensitivity. The magnitude of this filtering depends on many factors including the refractive indices of all optical compartments, visual pigment density and cell geometry. Calculations by Horowitz (1981), assuming a 500 nm rhodopsin in several model cells, yield 'blue-shifts' of as much as 20 nm! Even without resorting to variations in refractive indices, structural differences in photoreceptor size, particularly radius, should theoretically affect spectral sensitivity – the smaller the radius, the greater the blue-shift. It should be noted that along with this shift is a decrease in absolute absorptance and a broadening of the spectrum. There is tremendous diversity in the size and shape of retinal photoreceptors, not only between species and individuals, but even within the same retina. Although outer segment diameter, length and shape can be rationalized in terms of adaptations for quantum catch, the same cannot be said for inner segment metrics. The only realistic quantitative model incorporating different inner segment shapes is that of Rowe *et al.* (1994) done within the context of birefringence due to waveguiding by fish double cones (see also Hárosi, 1981). However, there is no unifying theory as to why photoreceptors are shaped as they are. And there is certainly no direct evidence that there is significant spectral tuning using cell shape or size changes alone. It is even possible that any spectral tuning that would arise by changes in geometry are offset by refractive index changes so that any potential structural filtering is negated. This whole area remains virtually unexplored and certainly demands more attention.

3.4 ENVIRONMENTAL AND ECOLOGICAL CONSIDERATIONS

3.4.1 THE SENSITIVITY HYPOTHESIS

This was the first, and is the most fundamental, of the hypotheses attempting to explain the spectral location of visual pigments in an environmental context. It is also the one for which there is the most supporting evidence. Since it deals with sensitivity, it is almost always applied to the visual pigments of rods, although it can also be applied to cones with less certainty (Lythgoe and Partridge, 1989, 1991). It states that scotopic visual pigments will be spectrally located so as to maximize quantal catch **in the spectral environment for which maximal sensitivity and the visual tasks subserved by this sensitivity are of greatest survival value** (the author's emphasis). Although not particularly obvious for broadband environments, one would expect this hypothesis to be testable in spectrally narrow, dim ones, such as blue-shifted oceanic waters at depth. This was proposed by Clarke (1936) and Bayliss *et al.* (1936), although the latter authors failed to find the expected blue-shift in their extract data. Since that time, numerous studies have confirmed the general idea that for oceanic species, there is a blue-shift in the λ_{max} of the scotopic pigments correlated with habitat depth (Figure 3.1 (Bowmaker, 1990, 1991; Crescitelli *et al.*, 1985; Denton, 1990)). The apparent contradictory findings of long-wavelength pigments in the rods of deep-sea fishes can be resolved by taking into account ambient bioluminescence or vertical migratory behaviors (Crescitelli *et al.*, 1985; Denton, 1990).

Are all scotopic pigments located for maximal quantum catch? If so, what does this say about the tight clustering of the rhodopsins around 500 nm? The sensitivity hypothesis would predict that this position would maximize quantal catch under night-time illumination coming from the moon and stars. However, spectroradiometric measurements of the night-time sky show that there are relatively more long-wavelength photons present than in daylight. This means that in switching between photopic and scotopic systems (the 'Purkinje shift'), the spectral sensitivity of many vertebrates goes in the wrong direction (Munz and McFarland, 1977). The sensitivity hypothesis does, however, apply if the twilight sky is considered to be the selective illuminant. During this

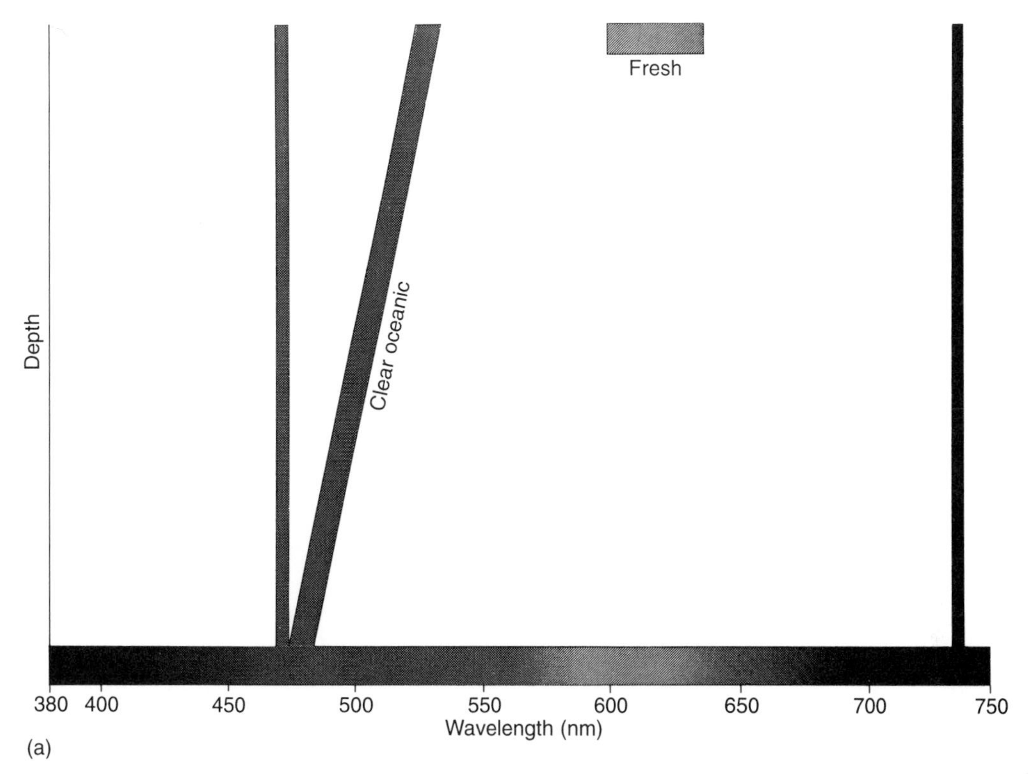

Figure 3.1 Relationship between λ_{max} of the best match visual pigment for fish living at different depths in various waters, and the wavelength of maximal transmission (i.e. penetration) for the particular water. (a) Clear oceanic water has maximal transmission between 470 and 480 nm (vertical line). With depth, the λ_{max} of the best match pigment decreases from about 530 nm to 480 nm. This is predicted by the sensitivity hypothesis. The maximal transmission of the very red fresh water of Rancocas Creek, New Jersey is shown. The long-wave pigments of fish from this water are in the region of 600–640 nm. The lack of match is probably due to biophysical constraints on λ_{max}.

period there is substantial long-wavelength absorption which shifts the quantal distribution towards the blue. The match between the position of this absorptance and the long-wavelength limb of 500 nm rhodopsin is quite good. The 'twilight hypothesis' of Munz and McFarland (1977) eliminates the enigma of the Purkinje shift, at least qualitatively.

The sensitivity hypothesis applied to cones is more tentative, but still operative in a broad sense. For detection of objects at a distance in water at either scotopic or photopic light levels, a visual pigment matched to the background spacelight is most adaptive given the filtering characteristics of water (see historical review of this idea in Munz and McFarland, 1977). This holds for silhouette detection against downwelling light at depth, as well as shallow horizontal detection at a distance. The cone extraction data of Munz and McFarland (1975) and almost all MSP surveys (Bowmaker, 1991) show that there is always at least one cone type fairly well matched, at least qualitatively, to the dominant background spacelight. This also holds for freshwater species where the long-wavelength cone tends to track the 'redness' of the water (Lythgoe, 1979, 1991; Lythgoe

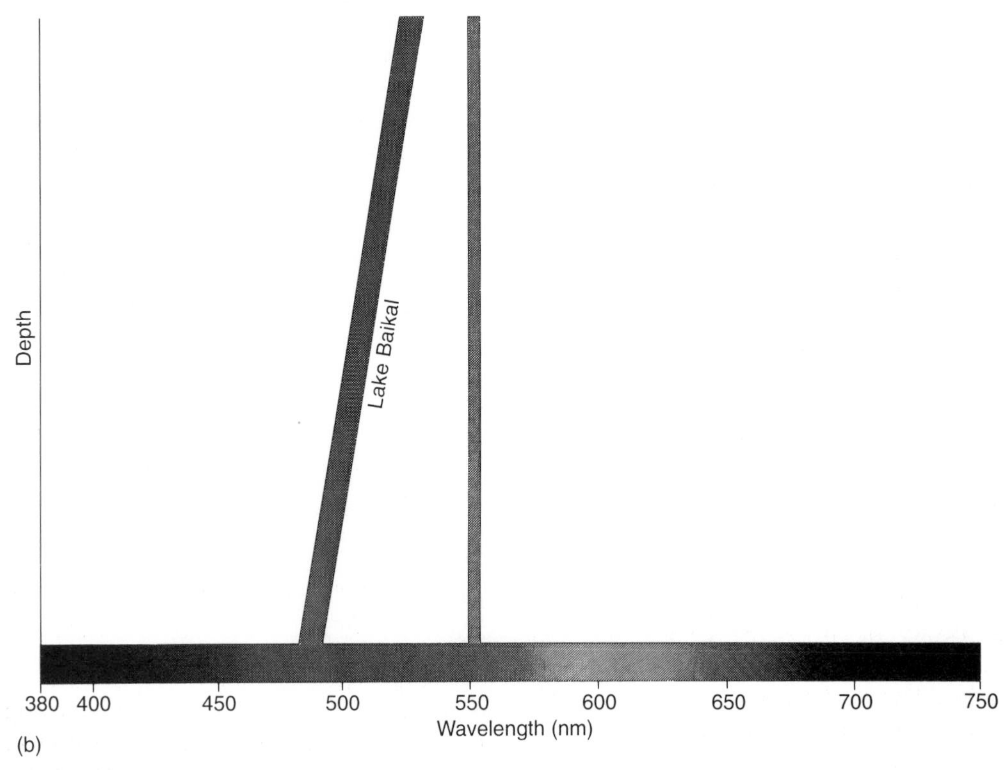

Figure 3.1 (*cont.*) (b) For Lake Baikal, the visual pigment λ_{max} shows a shift toward shorter wavelengths with depth, as seen for oceanic species, but the estimated transmission maximum is around 550 nm. This situation is in apparent violation of the sensitivity hypothesis.

and Partridge, 1989, 1991, E. Loew and J. Lythgoe, unpublished observations).

An apparent contradiction to the above 'rules' occurs for the cottoid fish of Lake Baikal (Bowmaker *et al.*, 1994). Both the rod and cone pigments show a clear blue-shift in λ_{max} with depth as seen for oceanic species (Figure 3.1b). However, the downwelling light becomes greener, not bluer, with depth. These freshwater species also use vitamin A_1 as their visual pigment chromophore. The sensitivity hypothesis can still be applied by supposing that the 'greening' of Baikal is a geologically recent event and the deep-water fish are showing adaptations for the earlier, 'bluer' condition (J. Bowmaker, personal communication). This suggests that the deep-water species are somehow maladapted in their present environment and could be excellent subjects for future fitness studies.

There is no doubt that the sensitivity hypothesis is qualitatively correct. However, the exact λ_{max} of those pigments adapted for sensitivity are subject to other pressures and restrictions besides just quantum catch. Our lack of knowledge about what is important visually to most species, the spectral characteristics of relevant targets, and the actual spectral sensitivity of the visual photoreceptors containing the visual pigment in question limits our ability to employ the sensitivity hypothesis in quantitative models.

3.4.2 THE CONTRAST HYPOTHESIS

Maximizing quantal catch in dim, spectrally narrow environments by adjusting photoreceptor absorption probability is expected of any photoreceptor, be it plant or animal. However, visual systems are concerned not just with detecting the presence or absence of environmental photons, but with two-dimensional target detection. For this task there must not only be sufficient photons absorbed to produce a perceptual signal above the background noise, but there must also be adequate contrast between the target and the background against which it is being viewed. This applies for scotopic, single pigment target detection as well as multi-pigment detection or color vision systems. For detection of an absorptive, dark or silhouetted target against a spectrally filtered background, a pigment matched to the brighter background radiance will maximize both detection and contrast. The same can be assumed for a broadband reflective target viewed at moderate distance in dim light where detection is limiting (see discussion in Lythgoe, 1968, 1972; and Munz and McFarland, 1977). This is the sensitivity hypothesis applied to target detection problems. However, in brighter, photopic conditions where quantal flux is not limiting, detection is less of a problem than contrast. For a broadband reflective (i.e. bright) target viewed against a spectrally narrower background, contrast may be improved by offsetting the visual pigment λ_{max} from the spectral peak of the background. This will have the effect of making the background appear darker and, thus improve the contrast for the near-field, broadband reflector. It is possible to calculate the optimal spectral position for a visual pigment that will maximize contrast under these conditions as long as all necessary spectral data are available (Lythgoe, 1968; Munz and McFarland, 1977). These include the spectral radiance of the target and background, the optical characteristics of the medium between the target and detector, and the contributions from any pre-photoreceptor filters. Rarely are all these parameters available, but using reasonable assumptions, Munz and McFarland (1977) were able to calculate optimal offsetting for pigments under different conditions, and compare their results to the positions of putative cone pigments for a number of coral reef fish. Their results certainly support the foundations of the hypothesis. For the dichromatic species in their sample, one visual pigment was always matched to the background spacelight optimizing that photoreceptor class for sensitivity and dark target detection. The other visual pigment was offset to a degree in rough agreement with predictions of the model applied to the particular species, and apparently maximizes bright target contrast at the expense of absolute detection.

The spectral positions of cone visual pigments subserving color vision involve adjustments presumed to maximize just noticeable differences between relevant targets differing in spectral reflectance against a range of backgrounds and under various illuminants. The necessity of dealing with both contrast and signal-to-noise parameters for a system that must handle both luminance and chromatic information can place conflicting requirements on cone spectral position (Osorio and Bossomaier, 1992) and make modeling particularly difficult. However, mathematical implementations of the contrast hypothesis suitable for assessing the ability of various pigment pairs to discriminate objects on the basis of hue alone have been formulated and applied (Lythgoe and Partridge, 1989, 1991; Lythgoe *et al.*, 1994). An example of this type of modeling is seen in Figure 3.2 for the case of fish living in the highly red-stained water of Rancocas Creek in the New Jersey Pine Barrens (E. Loew and J. Lythgoe, unpublished observations). This water has a maximum spectral transmission in the region of

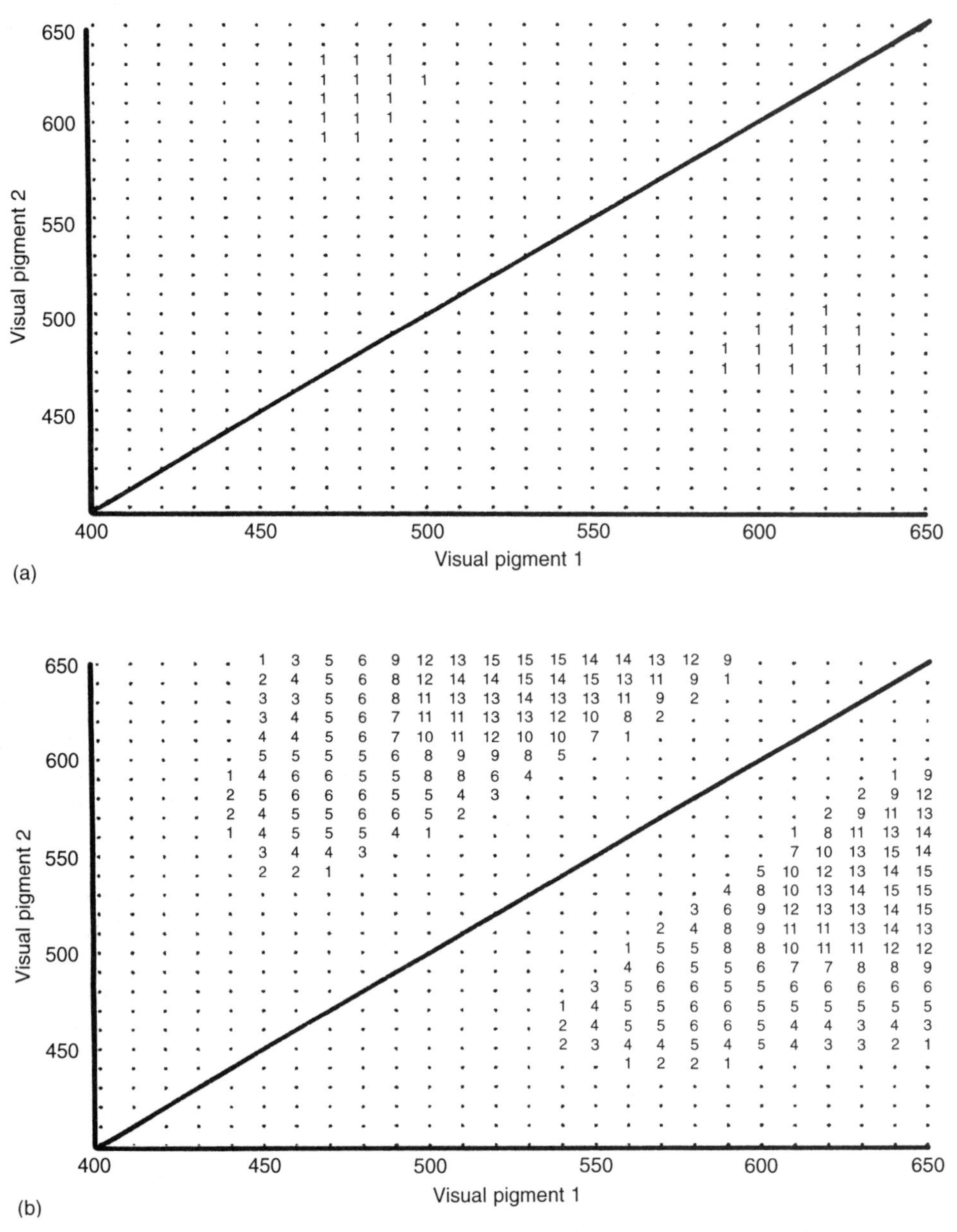

Figure 3.2 Results from applying the computational model of Lythgoe and Partridge (1991) to the visual situation in the deep red water of Rancocas Creek, NJ. The number in each cell is the number of objects that can be discriminated by the combination of two visual pigment λ_{max} values intersecting at that cell. (a) For a gray object, the two 'best' visual pigments have λ_{max} around 620 nm (matching) and 470 nm (offset). (b) With spectral reflectance data for 30 'brown' objects input into the model, and using a discrimination threshold of 50%, the two best pigments peak in the region of 630 nm and 530 nm (see text).

740 nm. The question addressed by the model is: what visual pigment pairs are best able to distinguish between a number of colored objects in a given photic environment based only on their spectral reflectance? A threshold of 50% discrimination is set, in this case 15 out of 30 objects. The model uses the spectral reflectances of the objects and the radiance of the spacelight. Figure 3.2(a) shows the case of the model applied to a single 'gray' object. The model predicts a pigment with λ_{max} matching the spacelight and one offset into the blue (a limit of 630 nm was put on the available visual pigment λ_{max}). This is expected from the predictions of the Contrast Hypothesis. When the model is applied to 30 'brown' objects collected from Rancocas Creek, the graph in Figure 3.2(b) is obtained. A matching pigment pushed as far out into the red as possible, and a second pigment having λ_{max} in the region of 520–540 nm is predicted. This is not far from what is found in fish endemic to the Pine Barrens, except that the λ_{max} of the second, 'green' pigment is somewhat longer than is actually measured (E. Loew and J. Lythgoe, unpublished observations). However when both green and brown objects are used in the model, a second pigment closer to 500 nm is predicted. It is hard to assess the success of these models since the number of assumptions needed due to lack of actual data is great. However, they do yield results that are qualitatively predictive of visual pigment location in a number of dichromatic and some trichromatic fish species (Lythgoe and Partridge, 1991; Lythgoe, *et al.*, 1994).

Models attempting to rationalize the spectral position of the three cone pigments in humans and other trichromats have also been put forward (Goldsmith, 1990; Mollon, 1991; Osorio and Bossomaier, 1992). These require more simplifying assumptions concerning chromatic variability of relevant targets and backgrounds than needed for dichromatic models. However, both the dichromatic model of Lythgoe and Partridge (1989, 1991) for fish and the trichromatic one of Osorio and Bossomaier (1992) for humans give great evolutionary significance to the need for target discrimination against chlorophyll (both positive and negative contrast), and simplify the rest of the world into 'broadband grey' and 'red–grey/brown–grey' backgrounds.

Given the number of assumptions that must be made when applying these models, it is not surprising that they have generally failed when used to try and predict which visual pigments a given animal will actually have under a given set of visual circumstances. The same can be said for techniques adopted from remote sensing image analysis which use principal components to determine which set of optical filter functions will maximize target discrimination. True, we lack the specific knowledge for any species needed to test the models predictively, but there could be something more fundamentally wrong. Perhaps the error is not in the specifics used or assumptions made, but in the belief that visual pigments measured today will be optimally located in a given species. Maybe the measured spectral sensitivities represent adaptations to past visual tasks and environments and are actually not optimized at present, as suggested above for the Baikal fish. Assuming that the present environment remains stable in Baikal, could we expect selection to 'push' the Baikal fish to the calculated 'optimal' positions? Thus, in the absence of a true measure of fitness conferred on a given animal by its spectral sensitivities, predictive failure of a model does not necessarily mean that the assumptions and data are insufficient or that the model is inherently wrong.

3.4.3 THE NOISE HYPOTHESIS

Fundamental to all detection systems is the necessity of separating the signal from noise. Some of the noise sources are external, such as temporal and spatial uncertainties in photon flux due to quantum statistics (Barlow, 1964; Falk and Fatt, 1972; Lythgoe, 1988; Laming, 1991; Dusenbery, 1992). Others are internal, such as membrane and synaptic noise. However, for most animals studied, the detection limit is set by photoreceptor cell noise, in particular the thermal activation of visual pigment (Barlow, 1988; Barlow *et al.*, 1993). This 'dark noise' or 'dark light' (Barlow, 1964) would be indistinguishable from the normal photochemical activation of visual pigment. Obviously, this limit must be taken into account in any quantitative model based on the sensitivity or contrast hypotheses.

One might expect that there would be much selective pressure to minimize noise of this type by favoring more thermally stable molecular configurations. This could take the form of selecting for particular opsins, or minimizing the formation of thermally unstable visual pigment configurations resulting from ionic or membrane lipid interactions affecting the chromophore binding site (Barlow *et al.*, 1993). In either case, these attempts to minimize noise might affect the spectral location of the pigments via one of the mechanisms outlined above (section 3.3). That is, there might be a relationship between spectral position of a visual pigment and its thermal stability.

Although this idea is certainly not new, and has been incorporated in explanations for spectral clustering (Dartnall and Lythgoe, 1965) and noise minimization in vertebrate rods associated with the use of 500 nm rhodopsin (Falk and Fatt, 1972), formalization of this idea into a 'noise hypothesis' for spectral selection has only recently been done by Govardovskii and his collaborators (Govardovskii, 1972; Donner *et al.*, 1990; Firsov and Govardovskii, 1990; and V. Govardovskii, personal communications). According to this hypothesis, the spectral position for scotopic pigments is not an adaptation for spectral matching as predicted by the sensitivity hypothesis (section 3.4.1), but rather an adaptation for noise reduction. Thus, the prevalence of 500 nm rhodopsins in dim, broadband spectral environments, and the blue-shift seen in aquatic species with depth is determined by noise minimization rather than quantal catch maximization. The only hard evidence for this is the finding by Firsov and Govardovskii (1990) that there is a minimum dark-noise for visual pigments located around 500 nm, with increasing noise as one moves in either spectral direction. The noise hypothesis can explain the apparent violation of the sensitivity hypothesis mentioned above for Lake Baikal fish.

Porphyropsins are inherently more 'noisy' than rhodopsins (Donner *et al.*, 1990) and it seems logical to suppose that in their evolutionary quest for noise reduction, all scotopic or 'dim light' photoreceptors would use the vitamin A_1 chromophore. However, the spectral characteristics of the environment may necessitate the use of porphyropsins. Firsov and Govardovskii (1990) have calculated that the signal-to-noise ratio in yellow water is maximal for rods containing pigments with λ_{max} between 530 nm and 540 nm. It would appear that the only way to achieve these λ_{max} values in rods is through the use of the A_2 chromophore.

Even though the number of photons is not limiting, and the photochemical bleaching rate is usually far greater than that resulting from thermal bleaching, photopic systems must still cope with noise as this will certainly limit luminous and color contrast. Here, the noise hypothesis predicts that whereas qualitative spectral positioning is determined by the simultaneous requirements of maximizing both contrast and sensitivity for particular visual situation, exact spectral location would be selected for minimal thermal noise. This is a restatement of the cluster hypothesis. Only experimental measurements of receptor noise for a wide range of photoreceptor cell spectral classes under physiological normal conditions of pH, anionic concentration and temperature can adequately test this hypothesis.

3.5 CONCLUSIONS AND FUTURE PERSPECTIVES

It is clear that there are many available mechanisms upon which selection can act to spectrally tune visual pigments, and many hypotheses attempting to rationalize how environmental and biophysical considerations conspire to 'push' visual pigments to particular spectral locations. Most of these factors have been incorporated into predictive models that have been qualitatively successful in explaining why the visual pigments of some species are spectrally located where they are. However, there is no case where it can be said with certainty why the visual pigments are spectrally located exactly where they are. Our level of ignorance about visual pigment spectral constraints, the relative importance of all the potential tuning factors, and, finally, what visual tasks are most important to an animal in determining fitness is too great at present.

Where should future studies be directed to alleviate these deficiencies? Our ignorance as to the exact molecular mechanisms underlying spectral tuning is certainly temporary given the almost exponential increase in the number of reports appearing on this subject each month. What is not clear is the extent to which polymorphism is allowable within a selective context. Certainly, red/green polymorphism in primates is evolutionarily tolerated, but is the same true for other species? Or put another way, how wide is the selectively neutral spectral band for a given photoreceptor cell class. As previously mentioned, this speaks to the question of fitness, and only studies on populations over many generations can address these questions. Of course, any question of fitness must be asked within the context of relevant visual tasks. Here is where field ecologists and animal behaviorists must concentrate their attention as regards vision. Lastly, mathematical models for fitness assessment must be refined and made more easily manipulated. This will require the application of newer spectroradiometric technologies along with state-of-the-art visualization routines. In the end, we are sure to understand why visual pigments are spectrally located where they are, at least for a few 'simple' systems like larval or deep-sea fish. However, adequate modeling of species, like man, with very complex visual systems and behavioral repertoires is likely to be a much more difficult task.

APPENDIX: SELECTED REVIEWS SINCE 1990

1. Signal detection

Laming, D. (1991) On the limits of visual detection, in *Limits of Vision* (eds J.J. Kulikowski, V. Walsh and I.J. Murry), CRC Press, Boca Raton, FL, pp.6–14.

Dusenbery, D.B. (1992) *Sensory Ecology*, W.H. Freeman, New York.

2. General

Goldsmith, T.H. (1990) Optimization, constraint and history in the evolution of eyes. *Quarterly Review of Biology*, **65**, 281–322.

3. Environmental light and target characteristics

Denton, E.J. (1990) Light and vision at depths greater than 200 metres, in *Light and Life in the Sea* (eds P.J. Herring, A.K. Campbell, M. Whitfield and L. Maddock), Cambridge University Press, Cambridge, pp.127–48.

Endler, J.A. (1990) On the measurement and classification of color in studies of animal color patterns. *Biological Journal of the Linnean Society*, **41**, 315–52.

Endler, J.A. (1993) The color of light in forests and its implications. *Ecological Monographs*, **63**, 1–27.

Loew, E.R. and McFarland, W.N. (1990) The underwater visual environment, in *The Visual System of Fish* (eds R.H. Douglas and M.B.A. Djamgoz), Chapman and Hall, London, pp.1–44.

Lythgoe, J.N. (1991) Evolution of visual behavior, in *Evolution of the Eye and Visual System* (eds J.R. Cronly-Dillon and R.L. Gregory), CRC Press, Boca Raton, FL, pp.3–14.

McFall–Ngai, M.J. (1993) Crypsis in the pelagic environment. *American Zoologist*, **30**, 175–88.

McFarland, W.N. (1990) Light in the sea: the optical world of elasmobranchs. *Journal of Experimental Zoology, Supplement*, **5**, 3–12.

McFarland, W.N. (1991) The visual world of coral reef fishes, in *The Ecology of Fishes on Coral Reefs* (ed. P.F. Sale), Academic Press, New York, pp.45–58.

Muntz, W.R.A. (1990) Stimulus, environment and vision in fishes, in *The Visual System of Fish* (eds R.H. Douglas and M.B.A. Djamgoz), Chapman and Hall, London, pp.491–512.

4. Visual pigments: molecular

Nathans, J. (1992) Rhodopsin: structure, function and genetics. *Biochemistry*, **31**, 4923–31.

Tovee, M.J. (1994) The molecular genetics and evolution of primate color vision. *Trends in Neuroscience*, **17**, 30–7.

5. Visual pigments: evolution

Bowmaker, J.K. (1990) Visual pigments of fish, in *The Visual System of Fish* (eds R.H. Douglas and M.B.A. Djamgoz), Chapman and Hall, London, pp.81–107.

Bowmaker, J.K. (1991) The evolution of vertebrate visual pigments and photoreceptors, in *Evolution of the Eye and Visual System* (eds J.R. Cronly-Dillon and R.L. Gregory), CRC Press, Boca Raton, FL, pp.63–81.

Crescitelli, F. (1990) Adaptations of visual pigments to the photic environment of the deep sea. *Journal of Experimental Zoology, Supplement*, **5**, 66–75.

Crescitelli, F. (1991) The natural history of visual pigments: 1990, in *Progress in Retinal Research*, vol. 11 (eds N. Osborne and G. Chader), Pergamon, Oxford, pp.2–31.

6. Color vision

Jacobs, G.H. (1992) Ultraviolet vision in vertebrates. *American Zoologist*, **32**, 544–54.

Jacobs, G.H. (1993) The distribution and nature of color vision among the mammals. *Biological Reviews*, **68**, 413–71.

Mollon, J.D. (1992) Uses and evolutionary origins of primate color vision, in *Evolution of the Eye and Visual System* (eds J.R. Cronly-Dillon and R.L. Gregory), CRC Press, Boca Raton, FL, pp.306–19.

Neumeyer, C. (1991) Evolution of color vision, in *Evolution of the Eye and Visual System* (eds J.R. Cronly-Dillon and R.L. Gregory), CRC Press, Boca Raton, FL, pp.284–305.

Partridge, J.C. (1990) The color sensitivity and vision of fishes, in *Light and Life in the Sea* (eds P.J. Herring, A.K. Campbell, M. Whitfield and L. Maddock), Cambridge University Press, Cambridge, pp.167–84.

Thompson, E., Palacios, A. and Varela, F.J. (1992) Ways of coloring: comparative color vision as a case study for cognitive science. *Behavioral and Brain Science*, **15**, 1–74.

Varela, F.J., Palacios, A.G. and Goldsmith, T.H. (1993) Color vision in birds, in *Vision, Brain, and Behavior in Birds* (eds H.P. Zeigler and H-J. Bischof), Bradford Press, Cambridge, pp.77–98.

REFERENCES

Abrahamson, E.W. and Japar, S.M. (1972) Principles of the interaction of light and matter, in *Handbook of Sensory Physiology* VII/1 (ed H.J.A. Dartnall), Springer-Verlag, Berlin, pp.1–32.

Abrahamson, E.W. and Wiesenfeld, J.R. (1972) The structure, spectra and reactivity of visual pigments, in *Handbook of Sensory Physiology* VII/1 (ed H.J.A. Dartnall), Springer-Verlag, Berlin, pp.69–121.

Barlow, H.B. (1964) The physical limits of visual discrimination, in *Photophysiology* (ed. A.C. Giese), Academic Press, New York, pp.163–202.

Barlow, H.B. (1988) Thermal limits of seeing. *Nature (London)*, **334**, 296–7.

Barlow, R.B., Birge, R.R., Kaplan, E. and Tallent, J.R. (1993) On the molecular origin of photoreceptor noise. *Nature (London)*, **366**, 64–6.

Bayliss, L.E., Lythgoe, R.J. and Tansley, K. (1936) Some new forms of visual purple found in sea fishes with a note on the visual cells of origin. *Proceedings of the Royal Society of London B*, **120**, 95–114.

Beatty, D.D. (1984) Visual pigments and the labile scotopic visual system of fish. *Vision Research*, **24**, 1563–74.

Bowmaker, J.K. (1973) Spectral sensitivity and

visual pigment absorbance. *Vision Research*, **13**, 783–92.

Bowmaker, J.K. (1990) Visual pigments of fish, in *The Visual System of Fish* (eds R.H. Douglas and M.B.A. Djamgoz), Chapman and Hall, London, pp.81–107.

Bowmaker, J.K. (1991) The evolution of vertebrate visual pigments and photoreceptors, in *Evolution of the Eye and Visual System* (eds J.R. Cronly-Dillon and R.L. Gregory), CRC Press, Boca Raton, FL, pp.63–81.

Bowmaker, J.K. and Dartnall, H.J.A. (1980) Visual pigments of rods and cones in a human retina. *Journal of Physiology (London)*, **298**, 501–11.

Bowmaker, J.K., Loew, E.R. and Liebman, P.A. (1975) Variation in the λ_{max} of rhodopisn from individual frogs. *Vision Research*, **15**, 997–1003.

Bowmaker, J.K., Dartnall, H.J.A. and Herring, P.J. (1988) Longwave-sensitive visual pigments in some deep-sea fishes: segregation of 'paired' rhodopsins and porphyropsins. *Journal of Comparative Physiology A*, **163**, 685–98.

Bowmaker, J.K., Kovach, J.K., Whitmore, A.V. and Loew, E.R. (1993) Visual pigments and oil droplets in genetically manipulated and carotenoid deprived quail: A microspectrophotometric study. *Vision Research*, **33**, 571–78.

Bowmaker, J.K., Govardovskii, V.I., Shukolyukov, S.A. *et al.* (1994) Visual pigments and the photic environment: the cottoid fish of Lake Baikal. *Vision Research*, **34**, 591–605.

Bridges, C.D.B. (1972) The rhodopsin–porphyropsin visual system, in *Handbook of Sensory Physiology* VII/1 (ed. H.J.A. Dartnall), Springer-Verlag, Berlin, pp.417–80.

Clarke, G.L (1936) On the depth at which fish can see. *Ecology*, **17**, 452–6.

Crescitelli, F. (1972) The visual cells and visual pigments of the vertebrate eye, in *Handbook of Sensory Physiology* VII/1 (ed. H.J.A. Dartnall), Springer-Verlag, Berlin, pp.245–363.

Crescitelli, F. (1977) The visual pigments of geckos and other vertebrates: an essay in comparative biology, in *Handbook of Sensory Physiology* VII/5 (ed F. Crescitelli), Springer-Verlag, Berlin, pp.391–450.

Crescitelli, F. (1991a) The natural history of visual pigments: 1990, in *Progress in Retinal Research*, Vol. 11 (eds N. Osborne and G. Chader), Pergamon, Oxford, pp.2–31.

Crescitelli, F. (1991b) The scotopic photoreceptors and their visual pigments of fishes: functions and adaptations. *Vision Research*, **31**, 339–48.

Crescitelli, F., McFall–Ngai, M. and Horowitz, J. (1985) The visual pigment sensitivity hypothesis: further evidence from fishes of varying habitats. *Journal of Comparative Physiology A*, **157**, 323–33.

Dartnall, H.J.A. (1953) The interpretation of spectral sensitivity curves. *British Medical Bulletin*, **9**, 24–30.

Dartnall, H.J.A. (1957) *The Visual Pigments*, Methuen, London.

Dartnall, H.J.A. (1968) The photosensitivities of visual pigment in the presence of hydroxylamine. *Vision Research*, **8**, 339–58.

Dartnall, H.J.A. and Goodeve, C.F. (1937) Scotopic luminosity curve amd the absorption spectrum of visual purple. *Nature (London)*, **139**, 409–11.

Dartnall, H.J.A. and Lythgoe, J.N. (1965) The spectral clustering of visual pigments. *Vision Research*, **5**, 81–100.

Denton, E.J. (1990) Light and vision at depths greater than 200 metres, in *Light and Life in the Sea* (eds P.J. Herring, A.K. Campbell, M. Whitfield and L. Maddock), Cambridge University Press, Cambridge, pp.127–48.

Donner, K., Firsov, M.L. and Govardovskii, V.I. (1990) The frequency of isomerization-like 'dark' events in rhodopsin and porphyropsin rods of the bull-frog retina. *Journal of Physiology (London)*, **428**, 673–92.

Dusenbery, D.B. (1992) *Sensory Ecology*, W.H. Freeman, New York.

Enoch, J.M. (1963) Optical properties of the retinal photoreceptors. *Journal of the Optical Society of America*, **53**, 71–85.

Enoch, J.M. and Tobey, F.L. Jr (1981) Waveguide properties of retinal receptors: techniques and observations, in *Vertebrate Photoreceptor Optics* (eds J.M. Enoch and F.L. Tobey, Jr), Springer-Verlag, Berlin, pp.169–218.

Falk, P. and Fatt, G. (1972) Physical changes induced by light in the rod outer segment of vertebrates, in *Handbook of Sensory Physiology* VII/1 (ed. H.J.A. Dartnall), Springer-Verlag, Berlin, pp.200–44.

Firsov, M.L. and Govardovskii, V.I. (1990) Dark

noise of visual pigments with different absorption maxima. *Sensory Systems*, **4**, 25–34.

Goldsmith, T.H. (1990) Optimization, constraint and history in the evolution of eyes. *Quarterly Review of Biology*, **65**, 281–322.

Govardovskii, V.I. (1972) Possible adaptive significance of the position of the visual pigment absorption maxima. *Journal of Evolutionary Physiology and Biochemistry*, **8**, 8–17.

Govardovskii, V.I. (1983) On the role of oil drops in the color vision. Vision Research, **23**, 1739–40.

Hárosi, F.I. (1981) Microspectrophotometry and optical phenomena: birefringence, dichroism and anomalous dispersion, in *Vertebrate Photoreceptor Optics* (eds J.M. Enoch and F.L. Tobey, Jr), Springer-Verlag, Berlin, pp.337–400.

Horowitz, B.R. (1981) Theoretical considerations of the retinal photoreceptor as a waveguide, in *Vertebrate Photoreceptor Optics* (eds J.M. Enoch and F.L. Tobey, Jr), Springer-Verlag, Berlin, pp.219–300.

Jacobs, G.H. (1993) The distribution and nature of color vision among the mammals. *Biological Reviews*, **68**, 413–71.

Jacobs, G.H., Bowmaker, J.K. and Mollon, J.D. (1981) Behavioural and microspectrophotometric measurements of color vision in monkeys. *Nature (London)*, **292**, 541–3.

Jacobs, G.H. and Neitz, J. (1985) Spectral positioning of mammalian cone pigments. *Journal of the Optical Society of America*, A, **2**, P23.

Kito, Y., Partridge, J.C., Seidou, M. *et al.* (1992) The absorbance spectrum annd photosensitivity of a new synthetic 'visual pigment' based on 4-hydroxyretinal. *Vision Research*, **32**, 3–10.

Kleinschmidt, J. and Hárosi, F.I. (1992) Anion sensitivity and spectral tuning of cone visual pigments *in situ*. *Proceedings of the National Academy of Sciences USA*, **89**, 9181–5.

Knowles, A. (1976) The effects of chloride ions upon chicken visual pigments. *Biochemical and Biophysical Research Communications*, **73**, 56–62.

Knowles, A. and Dartnall, H.J.A. (1977) The photobiology of vision, in *The Eye*, vol 2B, 2nd edn (ed. H. Davson), Academic Press, London.

Laming, D. (1991) On the limits of visual detection, in *Limits of Vision* (eds. J.J. Kulikowski, V. Walsh and I.J. Murry), CRC Press, Boca Raton, FL, pp.6–14.

Levine, J.S. and MacNichol, E.F. Jr (1979) Visual pigments in teleost fishes: Effects of habitat, microhabitat, and behavior on visual system evolution. *Sensory Processes*, **3**, 95–131.

Loew, E.R. and Dartnall, H.J.A. (1976) Vitamin A_1/A_2-based visual pigment mixtures in cones of the rudd. *Vision Research*, **16**, 891–6.

Loew, E.R. and Lythgoe, J.N. (1978) The ecology of cone pigments in teleost fishes. *Vision Research*, **18**, 715–22.

Loew, E.R. and Lythgoe, J.N. (1985) The ecology of colour vision. *Endeavour*, **9**, 170–4.

Lythgoe, J.N. (1968) Visual pigments and visual range underwater. *Vision Research*, **8**, 997–1012.

Lythgoe, J.N. (1972) The adaptation of visual pigments to the photic environment, in *Handbook of Sensory Physiology* VII/1 (ed. H.J.A. Dartnall), Springer-Verlag, Berlin, pp.566–403.

Lythgoe, J.N. (1979) *The Ecology of Vision*, Clarendon Press, Oxford.

Lythgoe, J.N. (1988) Light and vision in the aquatic environment, in *Sensory Biology of Aquatic Animals* (eds J. Atema, R.R. Fay, A.N. Popper and W.N. Tavolga), Springer-Verlag, Berlin, pp.57–82.

Lythgoe, J.N. (1991) Evolution of visual behaviour, in *Evolution of the Eye and Visual System* (eds J.R. Cronly-Dillon and R.L. Gregory), CRC Press, Boca Raton, FL, pp.3–14.

Lythgoe, J.N. and Partridge, J.C. (1989) Visual pigments and the modelling of visual information. *Journal of Experimental Biology*, **146**, 1–20.

Lythgoe, J.N. and Partridge, J.C. (1991) The modelling of optimal visual pigments of dichromatic teleosts in green coastal waters. *Vision Research*, **31**, 361–71.

Lythgoe, J.N., Muntz, W.R.A., Partridge, J.C. *et al.* (1994) The ecology of the visual pigments of snappers (Lutjanidae) on the Great Barrier Reef. *Journal of Comparative Physiology* A, **174**, 461–67.

MacNichol, E.F. Jr, Kunz, Y.W., Levine, J.S. *et al.* (1978). Ellipsosomes: organelles containing a cytochrome-like pigment in the retinal cones of certain fishes. *Science*, **200**, 549–51.

MacNichol, E.F. Jr, Levine, J.S., Mansfield, R.J.W. *et al.* (1983) Microspectrophotometry of visual pigments in primate photoreceptors, in *Colour Vision* (eds J.D. Mollon and L.T. Sharpe), Academic Press, London, pp.13–38.

Marks, W.B. (1965) Visual pigments of single goldfish cones. *Journal of Physiology (London)* **178**, 14–32.

Meyer, D.B., Stuckey, S.R. and Hudson, R.A. (1971) Oil droplet carotenoids of avian cones – I. Dietary exclusion. Models for biochemical and physiological studies. *Comparative Biochemistry and Physiology*, **40B**, 61–70.

Mollon, J.D. (1991) Uses and evolutionary origins of primate colour vision, in *Evolution of the Eye and Visual System* (eds J.R. Cronly-Dillon and R.L. Gregory), CRC Press, Boca Raton, FL, pp.306–19.

Mollon, J.D., Bowmaker, J.K. and Jacobs, G.H. (1984) Variations of colour vision in a New World primate can be explained by a polymorphism of retinal photopigments. *Proceedings of the Royal Society of London B*, **222**, 373–99.

Morris, A., Bowmaker, J.K. and Hunt, D.M. (1993) The molecular basis of a spectral shift in the rhodopsins of two species of squid from different photic environments. *Proceedings of the Royal Society of London B*, **254**, 233–40.

Morton, R.A. (1972) The chemistry of the visual pigments, in *Handbook of Sensory Physiology* VII/1 (ed. H.J.A. Dartnall), Springer-Verlag, Berlin, pp.33–68.

Munz, F.W. and McFarland, W.N. (1977) Evolutionary adaptations of fishes to the photic environment, in *Handbook of Sensory Physiology* VII/5 (ed. F. Crescitelli), Springer-Verlag, Berlin, pp.193–274.

Munz, F.W. and McFarland, W.N. (1975) Part I: Presumptive cone pigments extracted from tropical marine fishes. *Vision Research*, **15**, 1045–62.

Neitz, M., Neitz, J. and Jacobs, G.H. (1991) Spectral tuning of pigments underlying red-green color vision. *Science*, **252**, 971–4.

Novitskii, I.Y., Zak, P.P. and Ostrovskii, M.A. (1989) The effect of anions on absorption spectrum of the long wavelength retinal-containing pigment iodopsin in native frog cones: a microspectrophotometric study. *Bioorganicheskaya Khimiya*, **15**, 1037–43.

Osorio, D. and Bossomaier, T.R.J. (1992) Human cone-pigment spectral sensitivities and the reflectances of natural surfaces. *Biological Cybernetics*, **67**, 217–22.

Partridge, J.C. (1989) The visual ecology of avian oil droplets. *Journal of Comparative Physiology, A*, **165**, 415–26.

Partridge, J.C. (1990) The colour sensitivity and vision of fishes, in *Light and Life in the Sea* (eds P.J. Herring, A.K. Campbell, M. Whitfield and L. Maddock), Cambridge University Press, Cambridge, pp.167–84.

Partridge, J.C., Shand, J., Archer, S.N. *et al.* (1989) Interspecific variation in the visual pigments of deep sea fishes. *Journal of Comparative Physiology A*, **164**, 513–30.

Provencio, I., Loew, E.R. and Foster, R.G. (1992) Vitamin A_2-based visual pigments in fully terrestrial vertebrates. *Vision Research*, **32**, 2201–8.

Rowe, M.P., Engheta, N. and Easter, S.S. (1994) Graded-index model of a fish double cone exhibits differential polarization sensitivity. *Journal of the Optical Society of America A*, **11**, 55–65.

Smith, W.C. and Goldsmith, T.H. (1990) Phyletic aspects of the distribution of 3-hydroxretinal in the class Insecta. *Journal of Molecular Evolution*, **30**, 72–84.

Torma, H. and Vahlquist, A. (1990) Vitamin A esterification in human epidermis in relation to keratinocyte differentiation. *Journal of Investigative Dermatology*, **94**, 132–8.

Tovee, M.J. (1994) The molecular genetics and evolution of primate colour vision. *Trends in Neuroscience*, **17**, 30–7.

Varela, F.J., Palacios, A.G. and Goldsmith, T.H. (1993) Color vision of birds, in *Vision, Brain, and Behavior in Birds* (eds H.P. Zeigler and H.-J. Bischof), Bradford Press, Cambridge, pp.77–98.

Wald, G. Brown, P.K. and Smith, P.H. (1953) Cyanopsin, a new pigment of cone vision. *Science*, **119**, 505–8.

Walls, G.L. (1967) *The Vertebrate Eye and Its Adaptive Radiation*, Hafner, New York.

Whitmore, A.V. and Bowmaker, J.K. (1989) Seasonal variation in cone sensitivity and short-wave absorbing visual pigments in the rudd, *Scardinius erythrophthalmus*. *Journal of Comparative Physiology A*, **166**, 103–15.

Yoshikami, S., Pearlman, J.T. and Crescitelli, F. (1969) Visual pigments of the vitamin A-deficient rat following vitamin A_2 administration. *Vision Research*, **9**, 633–46.

4

Molecular biology of visual pigments

SIMON ARCHER

4.1 INTRODUCTION

The vertebrate retina is a tissue that has become adapted through evolution to act as an efficient collector of photons of light. The retina acts as a two-dimensional array detector that samples light from different points in space within set time intervals, integrating the signal and counting the frequency of photon arrival. Cells in the neural retina perform some initial processing and enhancement on the captured image before it is relayed to higher order processing areas in the brain. One of the most astonishing features of the retina is that, in performing this function of converting a complex visual scene into a neural representation, all of the information initially passes via one type of cellular membrane 'receptor', the so-called 'visual pigments'. When visual pigments were first extracted from the retina two important characteristics were noted. One formed the basis of the now largely abandoned visual pigment naming system and concerned the fact that the visual pigments were colored (i.e. rhodopsin, porphyropsin, cyanopsin etc.). The other was that this coloration soon bleached away in the light and demonstrated that the visual pigments were the photosensitive component of the retina.

The fact that visual pigments absorb light selectively underlines another very important function of the retina which is the color coding of the image. Human beings have been aware for some time that light can be split up into its spectral components and our varying sensitivity to different regions of the spectrum forms the basis of our color vision. We now know that our sensitivity to different wavelengths is based on the existence of three different classes of cone photoreceptor in our retinas and that each of these contains a visual pigment that samples light from a different region of the spectrum. The distribution of these different cone types in the retina ensures that the retinal array, or at least part of it, not only detects luminance but also distinguishes between wavelengths.

In the time since the first visual pigments were extracted there has been great advance in spectrophotometric techniques that now allow the measurement of visual pigment absorbance within single photoreceptors. The ability to characterize the spectral absorbance of visual pigments in single cells has enabled the investigation of how visual sensitivity in animals has become adapted to

Neurobiology and Clinical Aspects of the Outer Retina
Edited by M.B.A. Djamgoz, S.N. Archer and S. Vallerga
Published in 1995 by Chapman & Hall, London
ISBN 0 412 60080 3

different environments, which has led to an appreciation of the ecology of vision (Chapter 3). Very recently, we have begun to understand how the structure of the visual pigment molecule is responsible for absorbing light of different wavelengths, and much progress has been made from studying the molecular biology of the human visual pigments where there is a desire to explain both normal and abnormal functions. There are also many other vertebrates (and invertebrates) with advanced color vision capabilities and many of these possess visual pigments that have sensitivities which are different from those of humans. Molecular studies of visual pigments from other vertebrates could explain how the adaptation of visual sensitivity is achieved. It is the purpose of this chapter to explain the features of the photoreceptor visual pigment molecule that are important for converting photons of light into a neural signal and to discuss how visual pigment membrane receptors may have evolved to become sensitive to many different light wavelengths thus conferring visual adaptability on the vertebrates that possess them.

4.2 THE GENERAL MOLECULAR STRUCTURE OF VISUAL PIGMENTS

The photoreceptors in the vertebrate retina can generally be divided into rods that provide for scotopic, dim-light vision and cones that are responsible for photopic, bright-light vision. Both types of photoreceptor are made up of an outer segment, inner segment, nucleus and synaptic body and the two types of retinal photoreceptor have been called collectively 'rods' and 'cones' because the outer segments of each resemble these forms. Apart from this shape difference and the manner in which the outer segment membrane folds, the outer segments of rods and cones are very similar and are highly adapted for the absorption of light photons. Whereas the inner segment contains numerous mitochondria and other cellular machinery, the outer segment is composed almost entirely of folded membrane stacked along the length of the outer segment with the plane of the membrane perpendicular to the axis of the cell (Figure 4.1a). This membrane stack provides a large surface area through which photons of light pass as they travel through the photoreceptor. It is here that the light-sensitive visual pigment molecules are positioned, embedded in the outer segment membrane in high density (Figure 4.1b). A frog rod outer segment can contain around 10^9 visual pigment molecules (Saibil, 1986), and one photon of light is sufficient to isomerize one molecule which can, via amplification, generate a signal (Khorana, 1992).

Visual pigment molecules consist of a protein, called 'opsin', coupled with a chromophore. In vertebrates the chromophore is a derivative of either vitamin A_1 (retinal) or vitamin A_2 (3-dehydroretinal) which are primarily manufactured from beta-carotene ingested with food. The opsin protein and the chromophore are bound by a Schiff's base at a specific lysine in the peptide. Various analyses of the opsin protein structure (e.g. proteolysis, antibody and lectin binding, X-ray diffraction) have revealed that about 50% of the protein is composed of regions that are buried within the membrane connected by regions that are outside the membrane (Hargrave *et al.*, 1983; Chabre and Deterre, 1989; Hargrave and McDowell, 1992). Recent cloning of opsin genes has also demonstrated that there are seven hydrophobic amino acid domains of roughly equal length separated by shorter, more hydrophilic sections (Figure 4.5). This evidence strongly suggests that opsin is a transmembrane protein with seven membrane spanning, helical sections and that on average the protein is 348 amino acids in length with an overall mass of 40 kDa. Bacteriorhodopsin is a proton pump found in the membranes of some bacteria and is structurally very similar to opsin. X-ray

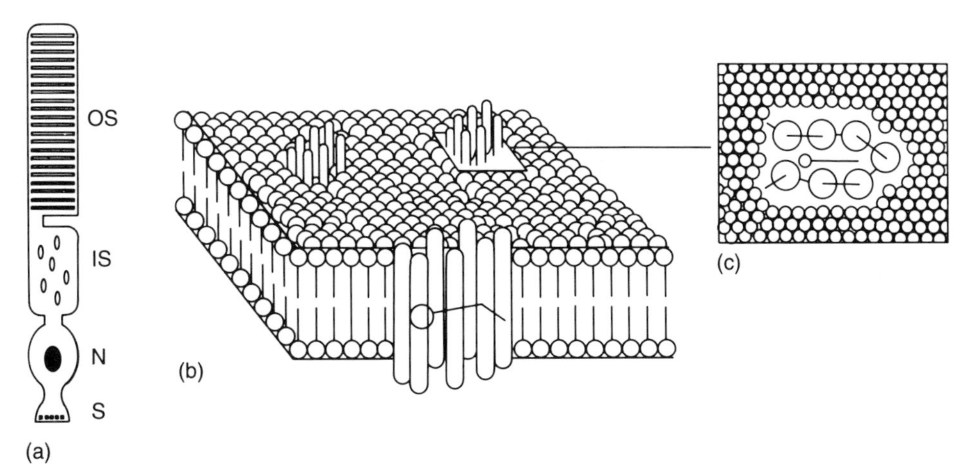

Figure 4.1 Schematic representation of the visual pigment molecule *in situ*. (a) The rod photoreceptor cell is divided into outer segment (OS), inner segment (IS), nucleus (N), and synapse (S). The outer segment contains numerous membrane disks stacked along its length. Light enters from the bottom of the cell and travels vertically through it. (b) and (c) (cytoplasmic view from above). The outer segment disk membrane contains many visual pigment molecules. The seven transmembrane bundles of the opsin component of the molecule are embedded in the membrane and connected by short protein loops. The retinal chromophore (represented here as a simple bent line) lies in the plane of the membrane, surrounded by the opsin transmembrane bundles.

diffraction measurements made with crystals of that protein showed that the seven transmembrane domains surround the chromophore which is buried in a central pocket (Henderson and Schertler, 1990) and it is likely that visual pigments have a similar spatial structure in the membrane (Figures 4.1b, c and 4.2).

Visual pigments are members of the superfamily of G-protein coupled receptors that includes the beta-adrenergic, muscarinic acetylcholine and dopamine receptors (Hall, 1987 and Trumpp-Kallmeyer *et al.*, 1992 give discussion and model of group). The rhodopsin receptor protein has several distinct domains: an N-terminal tail in the intradiscal space, a C-terminal in the cytoplasmic space, and seven transmembrane helices separated by cytoplasmic and intradiscal loops (Figure 4.2). Specific functions have been assigned to each of these domains and these will be discussed in section 4.3. The current model for the visual pigment molecule is one where the seven helices of the opsin protein surround the chromophore which lies horizontally within the membrane (Figure 4.1c). The chromophore is bound at a lysine residue in the seventh helix but is believed to have functionally important interactions with the other helices (section 4.3). Upon absorption of a photon of light, the chromophore isomerizes from an 11-*cis* form to an all-*trans* form. This isomerization causes some conformational change in the molecule which allows the binding of the G-protein transducin to opsin which, in turn, initiates a cascade of reactions that eventually leads to the hyperpolarization of the cell membrane (Chapter 5).

Visual pigments do not absorb light of only one wavelength but instead have bell-shaped absorbance curves with a broad wavelength bandwidth (Figures 4.3 and 4.4). The wavelength of maximum absorbance (λ_{max}) serves

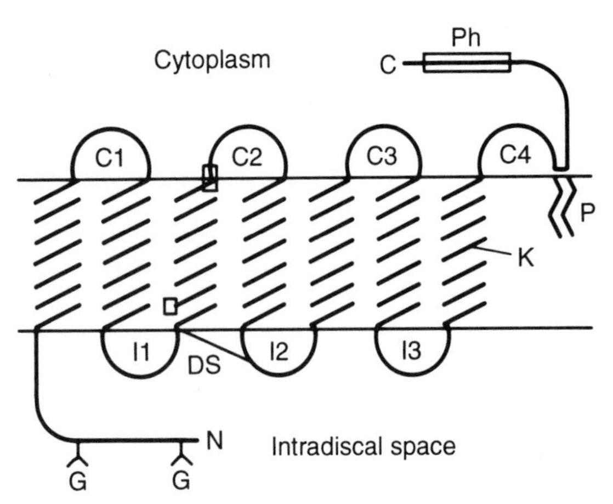

Figure 4.2 Plan view of the rhodopsin protein in the membrane. There are seven hydrophobic transmembrane segments (groups of angled lines) and retinal is bound by the Schiff base at a lysine in the seventh segment (labeled K). The counterion (Glu 113) for the Schiff base is found in segment three (box at base of segment). In the cytoplasm there are four amino acid loops (labeled C1–C4) connecting the transmembrane segments, the fourth being formed by palmityolation at cysteines 322 and 323. A Glu/Arg residue pair important for G protein binding is located at the boundary of loop C2 (boxed at top of segment three). The main site of phosphorylation in the carboxyl terminus (labeled C) is boxed (labeled Ph). The three intradiscal loops are labeled I1–I3. The structurally important disulfide bond between cysteines 110 and 187 is shown by a line connecting loop I2 and the base of transmembrane segment three. The two sites of glycosylation in the amino terminus (labeled N) are labeled G.

as a useful label for distinguishing visual pigments with different wavelength sensitivities. The human cone visual pigments are commonly described as absorbing in the blue, green and red regions of the spectrum but can be more accurately described as having λ_{max} values of around 420, 530 and 560 nm (but see section 4.4 and Figure 4.3 for more detail). Color vision requires at least two visual pigments with overlapping absorbance spectra so that the output of one photoreceptor can be compared to the other thus coding the wavelength of light illuminating both. Rod visual pigments in many vertebrates have a λ_{max} around 500 nm (Lythgoe, 1972, 1979; Lythgoe and Partridge, 1989).

Not all vertebrates have visual pigments with the same λ_{max} values as humans and this is especially true for aquatic vertebrates that have been well studied and shown to be able to adapt the sensitivity of their visual pigments to the different types of colored water they inhabit (Lythgoe and Partridge, 1989; and Chapter 3). The λ_{max} of a visual pigment can be altered by either substituting one chromophore for another or by changing the amino acid sequence of the opsin protein (sections 4.3 and 4.4). Vertebrate visual pigments are usually referred to as rhodopsins if they incorporate an A_1 chromophore, or porphyropsins if they incorporate an A_2 chromophore. The majority of land and marine vertebrates examined possess rhodopsins in their photoreceptors whereas many freshwater fish possess porphyropsins (Lythgoe, 1979). Some fish that move between marine and fresh water swap from rhodopsin to porphyropsin (or have a changing mixture of both) in their retinas (Beatty, 1984). The switch from rhodopsin to porphyropsin is achieved by substituting one chromophore with the other within the visual pigment molecule (as can be demonstrated by HPLC) and causes a long-wave directed jump of around 25 nm for a rhodopsin of λ_{max} 500 nm (Wood *et al.*, 1992; Figure 4.4 and Chapter 3). This change in λ_{max} is predictable if the starting λ_{max} of the visual pigment is known and a linear numerical relationship exists that can predict this translation for any wavelength (Lythgoe, 1979). Interestingly, there appears to be a point around 465 nm where no shift in λ_{max} occurs and below this point the shift occurs in the opposite direction.

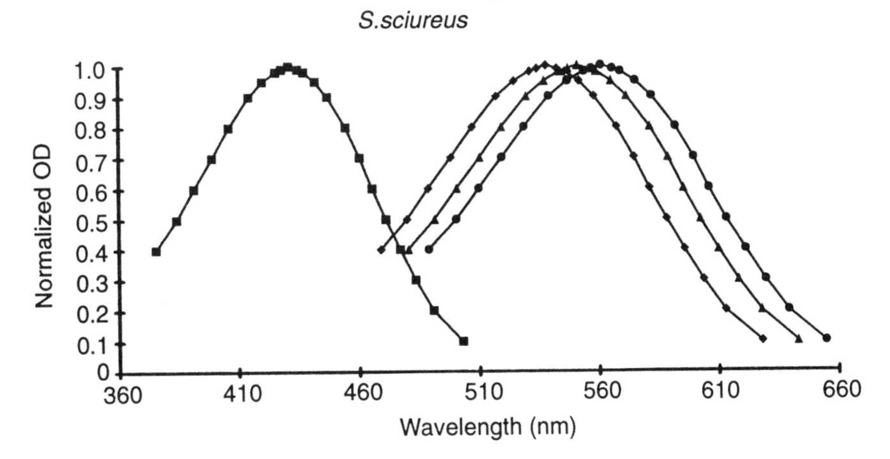

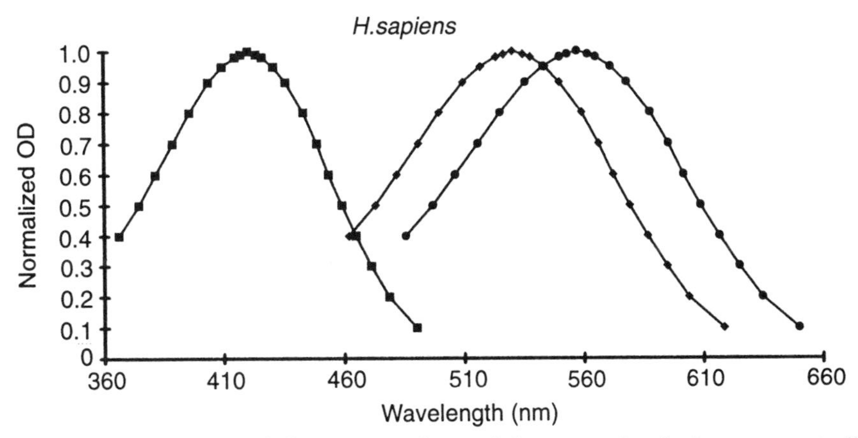

Figure 4.3 The normalized optical density templates of the cone visual pigments typically found in New World primates (this example, *Saimiri sciureus*, squirrel monkey) and Old World primates (this example, *Homo sapiens*, human). Both species have a cone pigment with a similar blue sensitivity (approximate λ_{max} values of 431 nm, *S. sciureus*, and 420 nm, *H. sapiens*). In the long-wavelength region of the spectrum the squirrel monkey has three cone pigments with λ_{max} values around 538, 551 and 561 nm, whereas *H. sapiens* has two cone pigments with λ_{max} values around 530 and 557 nm. The two systems are genetically very different. In the squirrel monkey there is one X-linked, long-wave opsin gene locus and three alleles that give rise to the three observed visual pigments. That system produces three male and three female dichromat phenotypes (the blue pigment plus any one of the long-wave pigments) and, because of X chromosome inactivation, three female trichromat phenotypes (the blue pigment plus any pair of the long-wave pigments). In humans there are two X-linked, long-wave opsin genes (positioned next to each other) which means that males and females are normally trichromats. The situation in humans is complicated by the fact that the X-linkage and misaligned recombination events between the long-wave genes gives rise to a proportion of abnormal male long-wave phenotypes. These include dichromats which lack one or other of the long-wave pigments and anomalous trichromats that posses one or other of the long-wave pigments plus an anomalous pigment (not shown in figure) that has a λ_{max} intermediate to those of the normal long-wave pigments. In addition the more red-sensitive human long-wave pigment is polymorphic and can have a λ_{max} of 552 or 557 nm. See main text for more detail and references.

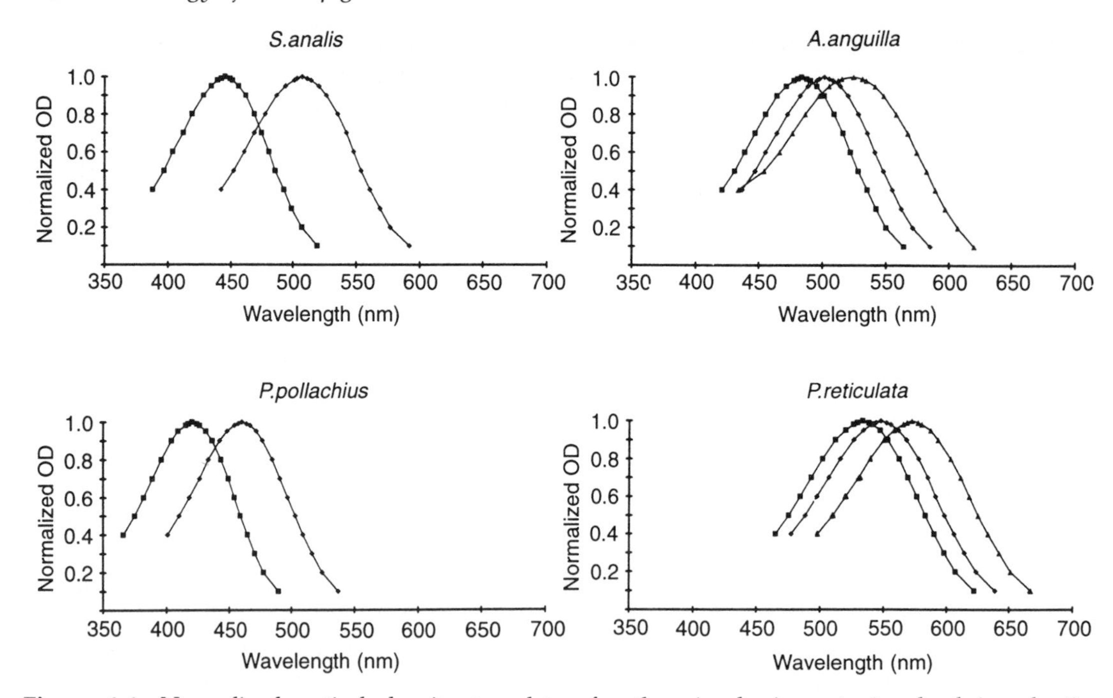

Figure 4.4 Normalized optical density templates for the visual pigments involved in adaptive mechanisms in four species of fish. In *Scoperlarchus analis* there may be a swap from a juvenile pigment (λ_{max} 505 nm (A_1), line with diamond symbols) to an adult pigment (λ_{max} 444 nm (A_1), line with square symbols) and this probably involves a change in opsin gene expression (Partridge *et al.*, 1992). In the eel, *Anguilla anguilla*, several visual pigment changes can occur. Eel rods can contain a variable mixture of two visual pigments (λ_{max} 501 and 523 nm, lines with diamond and triangular symbols respectively) that have the same opsin but differ in whether they contain an A_1 (former) or an A_2 (latter) chromophore. This mixture is replaced by a single A_1 pigment based on a new opsin when eels migrate back to the deep sea (λ_{max} 482 nm, line with squares) (Wood and Partridge, 1993). When larval pollack (*Pollachius pollachius*) move from shallow sea water to deeper water when they mature they change one of their cone visual pigments from a violet-sensitive pigment (λ_{max} 420 nm (A_1), line with squares) to a blue-sensitive one (λ_{max} 460 nm (A_2), line with diamonds). This change appears to be a gradual one involving the substitution of one opsin with another (Shand *et al.*, 1988). The long-wavelength cone pigment classes in the guppy, *Poecilia reticulata*, have similar spectral sensitivities to the primate long-wavelength sensitive pigments (λ_{max} 533 nm, line with squares, 543 nm, line with diamonds, and 572 nm, line with triangles, all A_1). Like the primates, guppies are also polymorphic for these pigments and individual fish can posses any combination of the three and it is possible that the polymorphism in the guppy is based on the differential expression of several different opsin genes (Archer and Lythgoe, 1990). See main text for more details.

The possibilities open to an animal for changing the spectral position of its λ_{max} value by swapping chromophores are limited. Another option is to somehow modify the interaction between the protein and the chromophore so that different wavelengths of light are more or less effective at producing the necessary isomerization of the chromophore and the conformational change in the molecule. It is now clear that this is the mechanism that can explain the wide variety of visual pigment λ_{max} values that are found in vertebrates and the next section describes what is presently known about how the opsin amino acid sequence can determine visual pigment sensitivity.

4.3 VISUAL PIGMENT STRUCTURE AND FUNCTION

Progress in determining the structure and function of rhodopsin has been aided by the high density of visual pigment present in the retina. Much of what has been learnt from the structure of rhodopsin now forms a model for other members of the GTP-binding protein receptors and rhodopsin was the first member of this family to have its gene sequence (and hence protein sequence) determined (Nathans and Hogness, 1983). The rhodopsin molecule is composed of three main domains: the membrane domain, the cytoplasmic domain and the intradiscal domain. In the following sections these domains and their functions are discussed (refer also to Figure 4.2).

4.3.1 THE MEMBRANE DOMAIN

The portion of the visual pigment that resides in the membrane is almost certainly the most important for determining the spectral sensitivity function of the molecule. The seven transmembrane sections of the opsin protein form a pocket in which the chromophore lies and interacts. It is the nature of this interaction that determines the visual pigment sensitivity. Free, unprotonated retinal in a methanol solution has an absorbance spectrum with a bell shape that resembles the absorbance of the visual pigment. The λ_{max} of the absorbance is 380 nm (Nathans, 1990a) and the interaction between the chromophore and the opsin protein must modify this absorbance and shift the λ_{max} into the absorbance range of the ultraviolet to far-red visual pigments that are known to exist. One of the main factors responsible for shifting the λ_{max} of retinal is the Schiff base that binds the chromophore to the opsin (Nathans, 1990a).

The chromophore is bound by the Schiff base at a particular lysine residue at roughly the midpoint in the seventh transmembrane helix, and this residue is conserved in all the visual pigments that have so far been examined (Figure 4.5; Yokoyama, 1994). The Schiff base is stabilized by a counterion which has been shown to be a conserved glutamate residue (Glu 113 in bovine rhodopsin) in the third transmembrane segment (Sakmar *et al.*, 1989; Nathans, 1990a). Substitution of this glutamate for glutamine shifts the normal λ_{max} of bovine rhodopsin from around 500 nm down to 380 nm. The λ_{max} of the mutated form can be shifted back up to around 500 nm with the addition of protonating halides, confirming the stabilizing role of the conserved glutamate. In order for the negatively charged side chain of the glutamate residue to interact electrostatically with the lysine Schiff base, transmembrane helices three and seven must be in close proximity (Khorana, 1992) and this arrangement has been demonstrated for the equivalent residues in bacteriorhodopsin (Henderson and Schertler, 1990; Baldwin, 1993).

Protonation of free retinal in solution moves the peak absorbance from 380 to around 440 nm (Nathans, 1990a). The interaction between the chromophore and the opsin protein produces the 'opsin shift' which moves the λ_{max} to other regions of the spectrum. During transduction, the positive charge at the Schiff base is delocalized and the opsin shift in different visual pigments is probably attributable to combinations of factors that make delocalization more or less easy. These factors must perturb the chromophore/opsin interaction, altering the energy difference between ground and excited states and leading to a change in the absorbance spectrum (Nathans, 1992). Changes in interaction between the chromophore and the opsin in the binding pocket could be achieved in several ways. Increasing the distance between the Schiff base and the counterion would decrease their interaction and produce a red shift. A second negative charge, in addition to the counterion, placed somewhere along the polyene chain of the retinal (the point charge model) could encourage delocalization and cause a red shift in absorbance. For mutants of bovine rhodopsin the latter has been shown not to be effective (Zhukovsky and Oprian, 1989; Nathans, 1990a, b) but its importance in other visual pigments cannot be ruled out. Related to this is the possibility that the UV sensitivity in the zebrafish may be achieved by the presence of a basic lysine amino acid in helix three that could negate the effect of the counterion and shift the absorbance back to around 360 nm (Robinson *et al.*, 1993). The presence of polar or polarizable amino acids (in positions that can interact with the polyene chain) could also affect delocalization, and hydoxyl bearing amino acid side chains appear to be important in this respect (section 4.4). Visual pigments that absorb maximally in the blue region of the spectrum close to 440 nm may have very little opsin shift due to chromophore perturbation, whereas those absorbing in the green or red regions of the spectrum will show increasingly more. Examples of the latter type of λ_{max} tuning can be found in the primates, including humans and these will be discussed specifically in section 4.4.

As described, the absorbance of the visual pigment is sensitive to the charge environment of the Schiff base counterion complex. Recently it has been demonstrated that anions in solution can also affect visual pigment spectral tuning by interacting with this charge complex (Kleinschmidt and Harosi, 1992; Harosi and Kleinschmidt, 1993; and Chapter 3). It appears that short wavelength-sensitive (UV, violet and blue sensitive cones) and medium wavelength-sensitive (rods and some green-sensitive cones) visual pigments are insensitive to anion, whereas long wavelength-sensitive (green and red-sensitive cones) visual pigments are anion sensitive. The anion sensitive cone pigments can have their λ_{max} shifted by up to 55 nm depending on the anion applied. These cone pigments may have a complex Schiff base counterion like that found in bacteriorhodopsin where the complex is made up of three charged residues (Marti *et al.*, 1992). Wild-type bacteriorhodopsin is anion insensitive but if any of the three counterion residues is mutated to an uncharged residue, then the molecule becomes sensitive to anions which can compensate for the charge loss (Marti *et al.*, 1992). Kleinschmidt and Harosi (1992) proposed that anions can interact with the counterion complex in anion sensitive cones and that native tuning of these pigments is achieved by the binding of Cl^- at the complex. Cl^- gives the maximum λ_{max}, but progressive blue shifts in λ_{max} are observed when Cl^- is removed or replaced with other anions (e.g. nitrate). These observations are obviously closely linked to the results obtained with bovine rhodopsin where Cl^- applied in solution can restore almost normal spectral sensitivity in a mutant where the counterion Glu 113 has been replaced with Gln (Nathans, 1990a). Even though it is not known how physiologically significant this type of tuning is in general, it is evident that the modification caused by anions can play an important role in the λ_{max} shift found in some long-wavelength cone pigments.

4.3.2 THE CYTOPLASMIC DOMAIN

Once the visual pigment has absorbed light and retinal has isomerized, the molecule must somehow transfer the signal to the other components of the transduction cascade, and when this has occurred activation must be terminated (Chapter 5 gives more details). The cytoplasmic domain plays an important role in both of these activities.

When rhodopsin is bleached it passes

through several distinct intermediates. Upon absorption of a photon, isomerization of retinal occurs in a matter of femtoseconds. After several milliseconds, the intermediate metarhodopsin II is formed and it is in that form that the molecule transmits its signal to the cytoplasmic cascade that culminates in the hyperpolarization of the cell. Metarhodopsin II is conformationally different from rhodopsin and provides cytoplasmic binding sites with altered configurations that allow the binding of other proteins involved in transduction (for review see Hargrave and McDowell, 1992; and Chapter 5).

The first protein to bind and provide the link between metarhodopsin II and the cytoplasm is the opsin-coupled G-protein, transducin. Transducin is a heterotrimeric protein made up of α-, β- and γ-subunits. When transducin binds to metarhodopsin II it becomes activated. The α-subunit binds GTP and dissociates from the βγ-component. The activated α-subunit with bound GTP then activates cGMP-phosphodiesterase and the hydrolysis of cGMP that follows leads to the closure of cGMP-gated channels in the membrane and the associated hyperpolarization (for more details see Chapter 5).

Rhodopsin has three cytoplasmic loops (labelled C1, C2 and C3 in Figure 4.2) that connect membrane spanning helices and one cytoplasmic loop (labelled C4 in Figure 4.2) that connects the seventh helix with two palmityolated cysteines attached to the membrane (bovine Cys322 and Cys323). Loop C4 has been implicated in transducin binding (Konig *et al.*, 1989) but the true functional significances of the loop and the palmityolation sites are unclear, especially because cone opsins do not appear to have them. Site-directed mutagenesis has now demonstrated that loops C2 and C3 are important for transducin binding and activation (Franke *et al.*, 1992; Khorana, 1992). Large deletions in loop C3 and smaller deletions in C2 do not affect retinal binding. A large deletion in C2 does affect retinal binding but normal binding can be restored by replacing the deleted amino acids with unrelated amino acids. These results indicate as expected, that, the absolute size of C2 is critical for correct opsin configuration but that the size of C3 is not so important. The latter is not expected and may mean that helices 5 and 6 are very close together, or that the C3 loop is bigger than thought. Deletions to C2 do not allow transducin activation, and opsin with amino acid replacements in this loop also fails to activate transducin suggesting that this loop and its amino acid sequence are important for transducin activation. Similarly, all types of deletion to C3 reduce or abolish transducin activation depending on the exact position in the loop. Transducin activity is more sensitive to mutations in regions of C3 closer to the C terminus so this part of the loop is supposed to be more interactive. Single amino acid substitutions have also revealed that a Glu/Arg charge pair, located in helix 3 close to the membrane boundary with loop C2, is very important in transducin interaction and is required for transducin activation (Franke *et al.*, 1992). The exact positioning of this pair relative to the membrane is somewhat unclear but it is worth noting that this pair (or Asp/Arg and in particular the Arg residue) is found in almost all G-protein coupled receptors so far examined. It has been found that mutating the Glu residue (bovine Glu134) of this pair to Gln increases the activity of opsin in the absence of a bound chromophore and this indicates that the negative charge at the Glu residue is important for stabilizing the inactive state of the opsin (Cohen *et al.*, 1993).

Once the visual signal has been transduced via the G-protein, the activation is terminated. Activation of the metarhodopsin II–transducin complex involves phosphorylation by rhodopsin kinase and subsequent binding of the protein 'arrestin'. There are several threonine and serine residues in the carboxyl terminus that become phosphorylated (McDowell *et al.*, 1993). There are also two threonine residues and a serine residue in cytoplasmic loop C3 that also become phosphorylated (Franke *et al.*, 1992) and cleavage of this loop reduces phosphorylation indicating that the C3 loop is also important in rhodopsin kinase binding. When metarhodopsin II has become phosphorylated then arrestin can bind to the cytoplasmic domain and exclude transducin thus terminating activation. Phosphorylation of rhodopsin is important for the binding of arrestin and there are carboxyl-terminal domains in arrestin that bind specifically to the phosphorylated form (Gurevich and Benovic, 1992).

4.3.3 THE INTRADISCAL DOMAIN

Much of the correct functioning of rhodopsin must rely on the opsin protein forming the correct tertiary structure. A specific configuration of opsin must be necessary to align the seven helices in the correct positions to form the interactive retinal binding pocket. The membrane and cytoplasmic domains obviously contribute to forming tertiary structure. For example, the sections of hydrophobic residues that traverse the membrane are important in determining how the protein is inserted into the membrane (Hargrave and McDowell, 1992) and the size of the cytoplasmic loops must have some influence on the spacing of the helices. From the bacteriorhodopsin model (Henderson and Schertler, 1990; Baldwin, 1993) it is likely that a precise rhodopsin structure is required for specific interaction with the chromophore (although this interaction must be adaptable to provide different wavelength sensitivity). Whereas the membrane and cytoplasmic domains have other active functions, the intradiscal domain appears to have a primary role in maintaining correct structure (Doi *et al.*, 1990).

The intradiscal domain of rhodopsin has two glycosylation sites (Asp2 and Asp15 in bovine) in the N terminal region and these sites are conserved in many opsins (Figure 4.5). There are also two cysteines (Cys110 and Cys187 in bovine) that are highly conserved among the opsins and in other G-protein linked receptors. These two cysteines have been shown to be essential for forming a disulfide bond that is necessary for the formation of a functional rhodopsin and it is likely that the disulfide bond is important for configuring the helices in the membrane domain and the loops in the cytoplasmic domain (Karnik and Khorana, 1990). Extensive mutation studies on the residues of the intradiscal domain have shown that almost all deletions to the intradiscal loops (labeled I1–I3 in Figure 4.2) gave rise to defective phenotypes (Doi *et al.*, 1990). A small number of single deletions in loop I2 had no effect on phenotype and these correspond to a small section of residues in this loop that appear to be completely unconserved among the opsins and may have no functional or structural value other than spacing (in Figure 4.5 this section is obvious in the rod opsins and corresponds to bovine residues 194–198). All other mutants regenerate the rhodopsin chromophore poorly or not at all and show aberrant glycosylation. When these mutants are expressed in tissue culture most of them are retained in the endoplasmic reticulum confirming that they are structurally defective (Khorana, 1992).

4.4 THE GENETIC BASIS FOR PRIMATE COLOR VISION AND COLOR VISION VARIATION

Not surprisingly, the primate opsin genes have been the most studied to date. These studies have revealed mechanisms that give rise to wavelength tuning and have proposed models that account for the observed wavelength sensitivity variation. The situation, however, is far from resolved with the results of some studies contradicting the findings of others. One result seems clear, however: the variation in visual sensitivity in humans may be far more subtle than previously imagined and more investigation is required before we can completely appreciate the mechanisms concerned.

4.4.1 COLOR VISION IN PRIMATES

As expected, visual pigment sensitivities among the primates that have been studied are very similar. Both Old World (catarrhines) and New World (platyrrhines) primates, in common with many land vertebrates, have a rod pigment with a λ_{max} close to 500 nm and a blue cone pigment with a λ_{max} in the 420–436 nm range (Tovee, 1994). The groups differ, however, in the long-wavelength cone pigments that they possess (Figure 4.3). Old World primates (including humans) normally possess a green and a red sensitive cone pigment with λ_{max} values close to 533 and 566 nm whereas New World primates have the equivalent green and red pigments and an additional long-wavelength cone pigment with an intermediate λ_{max} (Bowmaker *et al.*, 1991; Tovee, 1994). Microspectrophotometric and behavioral studies have shown that the platyrrhines are sexually dimorphic for long-wavelength visual pigment sensitivity (Bowmaker *et al.*, 1987; Hunt *et al.*, 1993; Tovee, 1994). Generally, females are either trichromats (possessing a blue absorbing pigment and the green and red pigment) or dichromats (lacking one of the long-wavelength pigments) whereas males are always dichromats (with only one long-wavelength pigment plus the short-wavelength pigment). Males can have any one of the three green/red pigments and females can have any two (Figure 4.3). The model that has been proposed to explain these visual pigment phenotypes is that there are two gene loci, one for the blue cone pigment and one for the green/red cone pigments. The long-wavelength pigment locus is linked to the X chromosome and has three alleles that give rise to the three green/red pigments. Because of X chromosome inactivation, females can express one or two long-wavelength alleles and be trichromats or dichromats, whereas males only ever express one allele and are dichromats. The catarrhines, on the other hand, are normally trichromatic (Bowmaker *et al.*, 1991; Tovee, 1994). In this case, there are two long-wavelength cone pigment loci that are also linked to the X chromosome and individuals have blue, green and red sensitive cones (Figure 4.3). Because the green and red cone pigment genes are X linked, some sex dependent polymorphism does occur and this is discussed for humans in section 4.4.2.

The genes coding for the opsins of the human rod visual pigment and cone visual pigments have been cloned (Nathans and Hogness, 1984; Nathans *et al.*, 1986a, b). The amino acid sequences of the green and red opsins are very similar (Figure 4.5) and their genes are arranged in a tandem array on the distal tip of the X chromosome (Nathans *et al.*, 1986a; Vollrath *et al.*, 1988). The green and red opsin amino acid sequences are 96% identical compared to around 43% identity for all other human opsin comparisons. Consequently, there are very few candidate amino acid substitutions in the membrane domain (only 13) that could be responsible for the approximately 25 nm shift in λ_{max} from the green cone pigment to the red cone pigment. Of the 13 amino acid differences only seven are considered likely to be important for spectral tuning because they involve non-conservative substitutions of amino acids with or without hydroxyl groups and these groups are likely to influence tuning via interaction with the chromophore (Nathans *et al.*, 1992). Tamarins and squirrel monkeys have a long-wavelength red sensitive cone pigment in common with humans (λ_{max} around 560 nm determined by electroretinography) and three cone pigments with sensitivities between the human red and green pigments (λ_{max} = 547 nm in the squirrel monkey and 556 and 541 nm in the tamarin) (Neitz *et al.*, 1991). Neitz *et al.* (1991) obtained opsin sequence for these cone pigments from the platyrrhines and compared the seven amino acid sites that were suspected to be important in green/red wavelength tuning in the human opsins. They found that different combinations of substitutions at three of the seven sites were responsible for the observed λ_{max} differences and that these three substitutions are present in the human cone pigments where they probably combine to produce the observed 30 nm shift in λ_{max}. These sites and their amino acids in green or red cone opsins (respectively) are: 180 Ala or Ser, 277 Phe or Tyr and 285 Ala or Thr. Subsequently, it was also found that a fourth site is probably involved in long-wavelength tuning in catarrhines (Ibbotson *et al.*, 1992). The results from six species of catarrhines and also humans indicate that there is a substitution at site 233 between all the green and red absorbing cone pigments and the substitution is also between amino acids with and without hydroxyl groups. This difference also occurs in two of the squirrel monkey long-wavelength sensitive pigments (Neitz *et al.*, 1991) and also in the long-wavelength pigments that have been isolated from the blind cavefish (Yokoyama and Yokoyama, 1990a,b). These amino acid sites have also been identified as underlining long-wavelength tuning in New World marmosets where the key sites are thought to be 180, 233 and 285 (Williams *et al.*, 1992; Hunt *et al.*, 1993). This small number of amino acid substitutions is undoubtedly important for long-wavelength cone pigment sensitivity tuning in humans with normal color vision. The human population is polymorphic for color vision, however, and models have been proposed that provide a genetic explanation for the observed polymorphic phenotypes.

4.4.2 LONG-WAVELENGTH COLOR VISION POLYMORPHISM

Because the human green and red cone pigment opsin genes are so similar and lie next to each other on the X chromosome (Vollrath *et al.*, 1988), it is assumed that they have arisen from a gene duplication event and that the Old World primate trichromatic vision evolved from a dichromatic form in ancestral mammals. Some 8% of caucasian males are defined as being color-blind because they demonstrate abnormal visual phenotypes in the long-wavelength region of the spectrum and it seems certain that this polymorphism arises from the close proximity of the green and red opsin genes. Because the two genes and their intervening sequences are very similar and they lie next to each other it is possible that they can misalign during recombination and cross over unequally (Nathans *et al.*, 1986b). Intragenic and intergenic recombination could give rise to hybrid red/green opsin genes and multiple or deleted green opsin genes. Multiple or deleted red opsin genes are not thought to occur because the red opsin gene lies at the 5′ end of the array and sequence upstream from the red gene and that upstream from the green gene are too divergent to allow pairing (Nathans *et al.*, 1986a, b). Behavioral and also microspectrophotometric (e.g. Dartnall *et al.*, 1983) analysis has demonstrated several male 'color-blind' phenotypes and these are grouped into dichromats that lack red or green sensitivity, and anomalous trichromats where either the normal green or red sensitivity is replaced by an intermediate sensitivity. In the model first proposed by Nathans *et al.* (1986a, b) a red-sensitive dichromat would either lack the green opsin gene, possess a non-functional red/green hybrid gene in place of the normal green gene, or possess a red/green hybrid gene that produced the red cone pigment. Green-sensitive dichromats would have a non-functional or green coding red/green hybrid gene in place of the normal red gene. The situation for anomalous trichromats is a little more complicated but rests on the fact that red/green hybrid genes can form in recombination. It is proposed that the sensitivity of the anomalous hybrid opsin will depend upon exactly where the cross over occurs and the proportion of the red and green genes that make up the hybrid gene (assuming that the hybrid is functional). The mechanism seems to involve different combinations of the wavelength tuning amino acid sites described in the previous section and it has been shown that hybrid human red and green opsin gene constructs (when expressed and measured spectrophotometrically *in vitro*) give rise to the anomalous visual pigments predicted by the model (Merbs and Nathans, 1992a).

In addition to the main model described above, it has been noticed that the long wavelength- (red-) sensitive mechanism shows a bimodal sensitivity between individuals classified as 'color normal' (Dartnall *et al.*, 1983; Neitz and Jacobs, 1990). This bimodality has been shown to be due to a substitution of Ser for Ala at site 180 in the red opsin gene. A study of 50 'color normal' males showed that 62% possessed Ser at the site and 38% possessed Ala (Winderickx *et al.*, 1992). An independent study found that the spectral absorbance phenotypes of these two opsin forms had λ_{max} values of 552 and 557 nm when recombined with chromophore (Merbs and Nathans, 1992b). The results from both these studies describe the bimodality recorded very well (Dartnall *et al.*, 1983). It is also possible that there are two forms of the green cone opsin gene showing the same substitution difference at site 180 (Neitz *et al.*, 1993). This would explain the bimodality also noted by microspectrophotometry for the green cone pigments (Dartnall *et al.*, 1983) and is in agreement with the earlier proposition (Neitz and Jacobs, 1990) that there may be two variants of each long-wavelength cone opsin gene. This possibility has important implications. First, it means that there may be more subtly different color vision phenotypes than originally thought. Second, because females have two X chromosomes they have two loci for the green and red opsin genes. If these loci are occupied by a mixture of different red and green gene variants or red/green hybrids then they could be tetrachromatic or pentachromatic (for discussion, see Tovee, 1994). That would imply that females could make more subtle color discriminations compared with males. Another potential problem is that the model proposed does not rule out the possibility that a green opsin gene and a red opsin gene could be expressed in the same photoreceptor. If there are upstream sequences on the green and red genes that regulate the expression of the genes in the 'correct' cone photoreceptor then the model could give rise to a normal gene and a hybrid gene that both instruct the expression of different genes in the same cell, perhaps giving rise to an intermediate or anomalous sensitivity.

4.5 THE ADAPTATION OF VISUAL SENSITIVITY AND THE EVOLUTION OF OPSIN GENES

The absorbance curves of many visual pigments have been measured spectrophotometrically (mainly from fish) and this has demonstrated a clear relationship between environmental light conditions and visual pigment sensitivity (Lythgoe and Partridge, 1989). Research into this relationship has now developed into a separate subject in its own right which has been called the 'ecology of vision' (Lythgoe, 1979; and Chapter 3). The ecology of vision is concerned mainly with understanding the visual system of an animal in terms of the visual tasks it must perform in its natural light environment. Research in this field has shown how the visual sensitivity of animals is adapted to the habitat and how in some animals sensitivity can be changed as conditions or visual tasks change. Much of that sensitivity adaptation will most probably be due to visual pigments with different opsins and changes in visual pigment sensitivity could be achieved by differential opsin gene expression. Investigation of opsin genes and their expression in animals that demonstrate this type of visual adaptation could now provide much information on how genotypes can fine tune visual phenotypes. This section expands upon the relevance of combining the ecology of vision with the molecular biology of vision.

4.5.1 THE SPECTRAL DIVERSITY AND CLUSTERING OF VISUAL PIGMENTS

Vertebrate visual pigments have been found that have λ_{max} values ranging from the UV to the far red (Lythgoe and Partridge, 1989). Fish that live in very shallow water can show sensitivity to the UV light that penetrates the surface water and fish that live in freshwater which transmits predominantly red light can have visual pigments that incorporate an A_2 chromophore and enable absorbance in the far-red. Within this range, there are other visual pigments with λ_{max} values at many different spectral locations. Most of this information has come from fish which make excellent models for the study of visual sensitivity because they inhabit a wide variety of different colored waters. Freshwater can be stained green or brown due to dissolved chlorophyll and tannin, shallow coastal marine water can be spectrally very clear, moderate depth coastal water can be green in color whereas deep oceanic water transmits predominantly blue light until around 1000 m where no more light penetrates. Studies with fish from such diverse environments has demonstrated that the sensitivity of the visual pigments of fish are shifted to suit the band of light available for vision in different conditions (Lythgoe and Partridge, 1989; Partridge, 1990). Within these bands it has been noticed that the λ_{max} values of the visual pigments tend to cluster at discrete wavelength positions. This was first pointed out by Dartnall and Lythgoe (1965) but has recently been confirmed in mammals (Jacobs, 1993) and in deep-sea fish (Partridge *et al.*, 1989; and Chapter 3). Dartnall and Lythgoe (1965) found that the cluster points were evenly spaced in the spectrum and they proposed that this was not the product of precise sensitivity adaptation to the environment but reflected restrictions placed upon the sensitivity of the molecule by the configuration of the chromophore. In their model, each cluster point would be produced by a certain active configuration of the chromophore that was stabilized by a particular opsin sequence, and this theory has similarities with those that are presently proposed to account for opsin wavelength tuning (section 4.3). From this two themes can be established: (1) that alterations to the opsin sequence must account for much of the wide variety of visual pigment wavelength sensitivities that exist, and (2) that the possible alterations to the opsin are finite and may give rise to a restricted family of visual pigments that are more or less evenly distributed throughout the spectrum. Future work with opsin sequences should focus on these themes by comparing visual pigments from different cluster points to determine how opsin sequence affects sensitivity between clusters and maintains the same sensitivity within clusters.

4.5.2 THE ECOLOGICAL AND EVOLUTIONARY IMPORTANCE OF SPECTRAL SENSITIVITY TUNING AND ADAPTATION

Spectral sensitivity tuning to available light in fish is a widespread phenomenon and obviously confers biological advantages that can be selected through evolution. As we have seen, freshwater fish tend to use the A_2 chromophore and have visual pigments with a longer λ_{max} than marine fish that use the A_1 chromophore. The relationship is so pronounced that some fish (e.g. the eel) that migrate between marine and freshwater swap from one chromophore to another or even mix the chromophores in a ratio that can be altered (Figure 4.4; Wood *et al.*, 1992). Evidence has been found in some freshwater fish that this ratio can be altered in response to changing environmental factors such as temperature, light and season (Beatty, 1984). Apart from chromophore exchange, fish tend to adapt the spectral sensitivity of their visual pigments to the type of light environment that they inhabit so that species that live in clear, shallow rockpool water have a broader spectral sensitivity than fish that live in green coastal water which will lack long-wavelength sensitivity (Lythgoe and Partridge, 1989). A good example of visual pigment tuning is provided by many species of deep-sea fish. The majority of these fish have a single visual pigment with a λ_{max} around 485 nm which matches the narrow waveband of light that penetrates the deep oceanic water (Partridge *et al.*, 1988, 1989, 1992). Many deep-sea fish visual pigments have been examined and they also demonstrate the clustering hypothesis with discrete groups of λ_{max} values placed around a main cluster between 470 and 490 nm (Partridge *et al.*, 1989). Within these examples of wavelength tuning, there must also exist species-specific fine tuning and this could explain much of the smaller λ_{max} variation seen among animals from the same type of light environment.

Several species of fish are now known to change their visual pigments during development or migration. The juvenile trout has a class of UV receptive cones in its retinal mosaic that disappear when the fish matures and starts to feed at greater depth where UV does not penetrate (Bowmaker and Kunz, 1987). Similarly, juvenile pollack possess a violet sensitive cone with a λ_{max} at 420 nm that shifts its λ_{max} up to 460 nm when it moves to deeper water (Shand *et al.*, 1988 Figure 4.4). Winter flounder begin life in surface waters where they possess only a cone-like photoreceptor with λ_{max} of 519 nm but when they metamorphose, changing from a symmetrical form to an asymmetrical flattened form, they become bottom dwellers and possess cone pigments with λ_{max} values of 457, 531 and 547 nm and a rod pigment of λ_{max} 506 nm (Evans *et al.*, 1993). Larval goatfish live in surface waters where they have a long-wave sensitive cone with a λ_{max} of 580 nm. Upon metamorphosis and settlement in reef beds this cone class changes its λ_{max} from 580 to 530 nm (Shand, 1993). Eels spend much of their life in inland waters but migrate back to deep-sea breeding grounds when sexually mature. When they do this, they change the λ_{max} of their rod pigment from an A_1/A_2 mixture between 501 and 523 nm to a virtually pure A_1 pigment with λ_{max} of 501 nm and then to a pigment with a λ_{max} at 482 nm which is more adapted to photoreception in deep-sea water (Wood and Partridge, 1993; Figure 4.4).

It is highly likely that all of the above examples of visual pigment adaptation during development or migration have their basis in differential opsin gene expression. In the trout, an opsin gene for the UV pigment may have its expression terminated with maturation. In the winter flounder, an opsin gene may be turned off and four alternative genes turned on during metamorphosis. In the pollack and the goatfish, there appears to be a switch from one opsin to another in the same cone class. Also in the eel, there appears to be a switch from one opsin to another in the rod cells and there is evidence that this switch in the eel is indeed due to differential opsin expression (Archer, Hope and Partridge, in preparation). During the switch in the eel, the λ_{max} of the rod pigment shifts gradually between 501 and 482 nm (Wood and Partridge, 1993). This is probably caused by the insertion of new opsin at the outer segment base which then gradually rises up the outer segment as old opsin is removed by disk shedding at the top. Interestingly, this mechanism may already have been observed in the large rods of the deep-sea fish *Scoperlarchus analis* which were found (by microspectrophotometry) to have a 445 nm pigment in the proximal half of the outer segment and a 507 nm pigment in the distal half (Partridge *et al.*, 1992). The fish examined in this study may have been changing from a shallow larval habitat to a deeper adult habitat and it is possible that an opsin switch was replacing the 507 nm pigment in the rods with a shorter absorbing 445 nm pigment (Figure 4.4).

Visual pigment polymorphisms where individuals of the same species have different visual pigments (as in the primates) are not so common. Variable A_1/A_2 ratios have been shown to produce this type of polymorphism (Bridges, 1964) but this is not directly relevant to variable opsin expression. There is one species of fish where visual pigment polymorphism does exist that may be due to variable opsin expression. The long-wavelength visual pigment polymorphism in the guppy is similar in some respects to the variation found in the primates. There appears to be three classes of cone pigment, one of which may be composed of a mixture of the other two in the same photoreceptor (Archer *et al.*, 1987; Archer and Lythgoe, 1990; Figure 4.4). Individual fish can have any combination of these three pigments and the variation is not sex linked. In the wild, the fish live in different types of freshwater ranging from very clear springs to highly stained river basins. Whereas fish from two areas of clear water could possess any of the three pigments, one group of fish that inhabited highly green stained water did not possess the most red-absorbing long-wavelength pigment (Archer, 1988). So, this polymorphism could be a complicated example of the differential expression of at least two opsin genes which can possibly be expressed together in the same cell to produce a third pigment phenotype. If the ratio of the mixed opsin expression could be altered, then presumably the λ_{max} of the mixed pigment could be moved as well, as was observed in the λ_{max} variation of the data from this class (Archer and Lythgoe, 1990). The opsin gene for the red-absorbing pigment may be switched off in individuals in green-stained water or evolution could select for those fish that do not express the gene. Initial results show that fish moved from one type of artificial light environment to another can change their long-wavelength sensitivity very quickly and that this change is reversible (S.N. Archer, unpublished observations). This would suggest that if this variation is opsin based, then the opsin gene expression can be regulated very quickly in individuals. The expression of two opsin genes in one photoreceptor is an unusual proposition but is one that has now also been suggested in the goldfish to account for the finding of two very similar opsin genes that, from *in situ* hybridization results, appear to be expressed in the same cone type (Johnson *et al.*, 1993; Raymond *et al.*, 1993; and Chapter 1). In addition, the two opsins have λ_{max} values of 505 and 511 nm when combined with A_1 but the λ_{max} of the cone type that they are supposedly expressed in is 509 nm. That value is midway between the other values which would be expected from a mixture of two opsins. The goldfish and chicken (Wang *et al.*, 1992) green opsin sequences are also unusual in that they have similarities with both rod and cone opsins and that may reflect the proximity of their λ_{max} values to most vertebrate rod pigments. It is also worth remembering that the present model used to explain polymorphic long-wavelength visual phenotypes in humans (section 4.4.2) does not rule out the possibility of two long-wavelength opsin genes being expressed in the same photoreceptor.

The vertebrate opsin sequences obtained so far (with some exceptions) are presented together in Figure 4.5. Even though there are not too many representatives from different visual pigment clusters, some common features between fish and primate long-wavelength absorbing pigments have already been established (Yokoyama and Yokoyama, 1990 a, b). Blue absorbing visual pigments are very common in fish and represent a strong cluster. Two blue pigment opsins from fish have been isolated (Johnson *et al.*, 1993; Yokoyama and Yokoyama, 1993) but more should be examined. These two blue opsins, together with the chicken blue opsin, show obvious similarities but differences from the human blue and the chicken violet which are also similar to each other. In fact, from the order of the sequences in Figure 4.5 several distinct groups can be seen. These consist of: (1) the chicken violet and human blue, (2) the goldfish, chicken and cavefish blues, (3) the rod opsins, and (4) the green and red opsins. The zebrafish UV opsin is unusual in that it shows a high similarity to the rod opsins, and the gecko blue opsin shows some similarities to the blue opsins and some to the rod opsins. In addition, the chicken green and the goldfish green (GFgrl and GFgr2) opsins are more similar to the rod opsins in places. The other green and red opsins (apart from the human) also have well-conserved amino terminal sequence that is lacking in the other opsins. The consensus sequence identifies some of the more important residues (section 4.3) and one area of strong conservation (GWSR in intradiscal loop I2). The function of the latter region is unclear but it must be specifically important to opsin because it is not conserved in dopamine receptors, for example (Van Tol *et al.*, 1991). The sequence has been placed on the border with helix four in intradiscal loop I2. As discussed in section 4.3, this loop has been shown to be important for producing correct tertiary structure and other residues in this loop are also conserved.

As the list of opsin sequences grows, we should be able to question the exact ecological requirements for fine sensitivity tuning. The fine tuning could enhance sensitivity or contrast or reduce noise by positioning the λ_{max} of the molecule in a more thermally stable spectral position (Aho *et al.*, 1988; Barlow, 1988; and Chapter 3). The latter hypothesis may explain why some fish that have variable A_1/A_2 ratios switch to more A_1 in the summer, when it is warmer, because A_1 is more thermally stable than A_2 (Allen and McFarland, 1973). It is also possible to begin to understand how opsin wavelength sensitivity tuning has been driven by evolution. An opsin-like protein seems to have appeared very early in evolution being present in several unicellular organisms (e.g. halobacteria, Henderson and Schertler, 1990). Natural selection has, therefore, had a long time to work on the structure of opsin to give rise to the numerous spectral sensitivity phenotypes that we see around us. Visual phenotypes that conferred advantage to the individual must have occurred for the selective process to work. In some cases evolution may have given rise to individuals that could exploit new light environment niches, a new food source, avoid detection by predators or be able to break the camouflage of prey. There are many examples of the possible results of such evolution. Two species of squid (*A. subulata* and *L. forbesi*) that live at different depths have different visual pigments and this has been shown to be due to opsin sequence differences (Morris *et al.*, 1993). Also, whereas most deep-sea fish only have sensitivity to blue light, there are some deep-sea fish that emit green and red bioluminescence and possess visual pigments with the long-wavelength sensitivity necessary to detect it (O'Day and Fernandez, 1974; Partridge *et al.*, 1989). This must represent a break away from the standard blue sensitivity to provide a species specific communication channel. Presumably, it also requires the bioluminescent emitter and detector to have evolved together. Another potential example of evolutionary visual adaptation can be seen in the guppy (Archer *et al.*, 1987; Archer, 1988; Endler, 1991). The long-wavelength polymorphism in these fish gives rise to individuals with several different phenotypes which could enable detection of different food sources and promote resource partitioning. In addition, guppies are sexually dimorphic and females choose males on the basis of skin coloration and pattern and consequently male skin coloration is highly polymorphic. Females prefer brightly colored males but in areas of high predation less conspicuous males could be selected for. Genetic variation in female mate choice is required to promote this type of sexual selection and it is possible that the long-wavelength visual pigment polymorphism has provided the variation and become integral in the sexual selection process. In the wild, guppies have inhabited many different types of interconnected river environments, and long-wavelength visual pigment polymorphism could also enable adaptation to the different light conditions encountered (e.g. loss of red sensitivity in fish living in green-colored water). This complicated example of the selective evolution of opsin gene expression is surely not alone and it will be very interesting to investigate the opsin genetics behind this and other examples.

Figure 4.5 (Pages 97–99) Most of the currently known vertebrate opsin sequences. Because the primate green and red opsins are so similar, only those for human are shown (see references in text for other sequences). Only partial alignment in the terminal regions has been attempted. The sequences are arranged in order of increasing maximum absorbance (λ_{max}). The λ_{max} values have been obtained from the references that list the sequence and where appropriate are the values for the expressed clones. In all other cases the values used have been taken from Lythgoe (1972). λ_{max} values are not given for the blind cavefish. A consensus line indicates the positions where all opsins have the same amino acid. The transmembrane segments are indicated by overlying bars (based on bovine model). The numbering system adopted is that used for bovine rhodopsin and the numbering for all other sequences starts at the equivalent residue aligned with the bovine start site. Also for convenience the two spaces inserted at the beginning to accommodate the goldfish, chicken and cavefish blue opsins have been ignored in the numbering. The abbreviations for the sequences and the references are: Zfuv (zebrafish UV, Robinson *et al.*, 1993), Chviol (chicken violet, Okano *et al.*, 1992), Humblu (human blue, Nathans *et al.*, 1986a), Gfblu (goldfish blue, Johnson *et al.*, 1993), Chblu (chicken blue, Okano *et al.*, 1992), Cfblu (cavefish blue, Yokoyama and Yokoyama, 1993), Gekblu (gecko blue, Kojima *et al.*, 1992), Gfrod (goldfish rod, Johnson *et al.*, 1993), Chrod (chicken rod, Takao *et al.*, 1988), Lamrod (lamprey rod, Hisatomi *et al.*, 1991), Humrod (human rod, Nathans and Hogness, 1984), Bovrod (bovine rod, Nathans and Hogness, 1983), Mserod (mouse rod, Baehr *et al.*, 1988), Sgrod (sand goby rod, Archer *et al.*, 1992), Hamrod (hamster rod, Gale *et al.*, 1992), Rprod (frog rod, Pittler *et al.*, 1992), Gfgre2 (goldfish green GFgr-2, Johnson *et al.*, 1993), Chgre (chicken green, Okano *et al.*, 1992), Gfgrel (goldfish green GFgr-1, Johnson *et al.*, 1993), Gekgre (gecko green, Kojima *et al.*, 1992), Cfgrel (cavefish G101af, Yokoyama and Yokoyama, 1990a), Cfgre2 (cavefish G103af, Yokoyama and Yokoyama, 1990b), Gfred (goldfish red, Johnson *et al.*, 1993), Humgre (human green, Nathans *et al.*, 1986a), Humred (human red, Nathans *et al.*, 1986a), Chred (chicken red, Kuwata *et al.*, 1990), Cfred (cavefish red, Yokoyama and Yokoyama 1990a), Anolred (chameleon red, Kawamura and Yokoyama, 1993).

Species & Opsin	λmax	Sequence					
			1	20			40
Zfuv	360		MNGTEGTAFY	VPMSNATGVV	..RSPYEYPQ	YYLVAPWAYG	FVAAYMFFLI
Chviol	415		MSSDDDFY	L.FTNGS.VP	...GPWDGPQ	YHIAPPWAFY	LQTAFMGIVF
Humblu	420		MRKMSEEEFY	L.FKNISSV.	...GPWDGPQ	YHIAPVWAFY	LQAAFMGTVF
Gfblu	441	MK	QVPEFHEDFY	IPIPLDINNL	SAYSPFLVPQ	DHLGNQGIFM	AMSVFMFFIF
Chblu	455	MHPPR	PTTDLPEDFY	IPMALDAPNI	TALSPFLVPQ	THLGSPGLFR	AMAAFMFLLI
Cfblu	---	MKS	RPQEFQENFY	IPIPLDTNNI	TALSPFLVPQ	DHLGGSGIFM	IMTVFMFFLF
Gekblu	467		MNGTEGINFY	VPLSNKTGLV	..RSPFEYPQ	YYLADPWKFK	VLSFYMFFLI
Gfrod	492		MNGTEGDMFY	VPMSNATGIV	..RSPYDYPQ	YYLVAPWAYA	CLAAYMFFLI
Chrod	495		MNGTEGQDFY	VPMSNKTGVV	..RSPFEYPQ	YYLAEPWKFS	ALAAYMFMLI
Lamrod	497		MNGTEGDNFY	VPFSNKTGLA	..RSPEVYPQ	YYLAEPWKYS	ALAAYMFFLI
Humrod	497		MNGTEGPNFY	VPFSNATGVV	..RSPFEYPQ	YYLAEPWQFS	MLAAYMFLLI
Bovrod	498		MNGTEGPNFY	VPFSNKTGVV	..RSPFEAPQ	YYLAEPWQFS	MLAAYMFLLI
Mserod	498		MNGTEGPNFY	VPFSNVTGVG	..RSPFEQPQ	YYLAEPWQFS	MLAAYMFLLI
Sgrod	501		MNGTEGPFFY	IPMVNTTGIV	..RSPYEYPQ	YYLVNPAAYA	ALGAYMFFLI
Hamrod	502		MNGTEGPNFY	VPFSNATGVV	..RSPFEYPQ	YYLAEPWQFS	MLAAYMFLLI
Rprod	502		MNGTEGPNFY	IPMSNKTGVV	..RSPFDYPQ	YYLAEPWKYS	VLAAYMFLLI
Gfgre2	505		MNGTEGNNFY	VPLSNRTGLV	..RSPFEYPQ	YYLAEPWQFK	LLAVYMFFLI
Chgre	508		MNGTEGINFY	VPMSNKTGVV	..RSPFEYPQ	YYLAEPWKYR	LVCCYIFFLI
Gfgre1	511		MNGTEGKNFY	VPMSNRTGLV	..RSPFEYPQ	YYLAEPWQFK	ILALYLFFLM
Gekgre	521	MTEAWNVAVFAARRSRDD	DDTTRGSVFT	YTNTNNT...	..RGPFEGPN	YHIAPRWVYN	LVSFFMIIVV
Cfgre1	---	MAAHEPVFAARRHNE.	.DTTRESAFV	YTNANNT...	..RDPFEGPN	YHIAPRWVYN	VSSLWMIFVV
Cfgre2	---	MAAHADEPVFAARRYNE.	.ETTRESAFV	YTNANNT...	..RDPFEGPN	YHIAPRWVYN	LASLWMIIVV
Gfred	525	MAEQWGDAIFAARRRGD.	.ETTRESMFV	YTNSNNT...	..RDPFEGPN	YHIAPRWVYN	LATVWMFFVV
Humgre	530	MAQQWSLQRLAGRHPQDSY	EDSTQSSIFT	YTNSNST...	..RGPFEGPN	YHIAPRWVYH	LTSVWMIFVV
Humred	560	MAQQWSLQRLAGRHPQDSY	EDSTQSSIFT	YTNSNST...	..RGPFEGPN	YHIAPRWVYH	LTSVWMIFVV
Chred	571	MAAWEAAFAARRRHEE	EDTTRDSVFT	YTNSNNT...	..RGPFEGPN	YHIAPRWVYN	LTSLWMIFVV
Cfred	---	MGDQWGDAVFAARRRGD.	.DTTREAAFT	YTNSNNT...	..KDPFEGPN	YHIAPRWVYN	LATCWMFFVV
Anolred	625	MAGTVTEAWDVAVFAARRRND.	EDTTRDSLFT	YTNSNNT...	..RGPFEGPN	YHIAPRWVYN	ITSVWMIFVV
Cons			F		P P		

Species & Opsin	λmax						
			60		80		100
Zfuv	360	ITGFPVNFLT	LYVTIEHKKL	RTPLNYILLN	LAIADLFMVF	GGFTTTMYTS	LHGYFVFGRL
Chviol	415	AVGTPLNAVV	LWVTVRYKRL	RQPLNYILVN	ISASGFVSCV	LSVFVVFVAS	ARGYFVFGKR
Humblu	420	LIGFPLNAMV	LVATLRYKKL	RQPLNYILVN	VSFGGFLLCI	FSVFPVFVAS	CNGYFVFGRH
Gfblu	441	IGGASINILT	ILCTIQFKKL	RSHLNYILVN	LSIANLFVAI	FGSPLSFYSF	FNRYFIFGAT
Chblu	455	ALGVPINTLT	IFCTARFRKL	RSHLNYILVN	LALANLLVIL	VGSTTACYSF	SQMYFALGPT
Cfblu	---	IGGTSINVLT	IVCTVQYKKL	RSHLNYILVN	LAISNLLVST	VGSFTAFVSF	LNRYFIFGPT
Gekblu	467	AAGMPLNGLT	LFVTFQHKKL	RQPLNYILVN	LAAANLVTVC	CGFTVTFYAS	WYAYFVFGPI
Gfrod	492	ITGFPVNFLT	LYVTIEHKKL	RTPLNYILLN	LAISDLFMVF	GGFTTTMYTS	LHGYFVFGRV
Chrod	495	LLGFPVNFLT	LYVTIQHKKL	RTPLNYILLN	LVVAHLFMVF	GGFTTTMYTS	MNGYFVFGVT
Lamrod	497	LYGFPYNFLT	LFVTVQHKKL	RTPLNYILLN	LAMANLFMVL	FGFTVTMYTS	MNGYFVFGPT
Humrod	497	VLGFPINFLT	LYVTVQHKKL	RTPLNYILLN	LAVADLFMVL	GGFTSTLYTS	LHGYFVFGPT
Bovrod	498	MLGFPINFLT	LYVTVQHKKL	RTPLNYILLN	LAVADLFMVF	GGFTTTLYTS	LHGYFVFGPT
Mserod	498	VLGFPINFLT	LYVTVQHKKL	RTPLNYILLN	LAVADLFMVF	GGFTTTLYTS	LHGYFVFGPT
Sgrod	501	LTGFPINFLT	LYVTLEHKKL	RTALNLILLN	LAVADLFMVF	GGFTTTMYTS	MHGYFVLGRL
Hamrod	502	VLGFPINFLT	LYVTVQHKKL	RTPLNYILLN	LAVADLFNVF	GGFTTTLYTS	LHGYFVFGPT
Rprod	502	LLGLPINFMT	LYVTIQHKKL	RTPLNYILLN	LGVCNHFMVL	CGFTITMYTS	LHGYFVFGQT
Gfgre2	505	CLGLPINGLT	LICTAQHKKL	RQPLNFILVN	LAVAGAIMVC	FGFTVTFYTA	INGYFALGPT
Chgre	508	STGLPINLLT	LLVTFKHKKL	RQPLNYILVN	LAVADLFMAC	FGFTVTFYTA	WNGYFVFGPV
Gfgre1	511	SMGLPINGLT	LVVTAQHKKL	RQPLNFILVN	LAVAGTIMVC	FGFTVTFYTA	INGYFVLGPT
Gekgre	521	IASCFTNGLV	LVATAKFKKL	RPHLNWILVN	LAFVDLVETL	VASTISVFNQ	IFGYFILGHP
Cfgre1	---	IASVFTNGLV	IVATAKFKKL	QHPLNWILVN	LAIADLGETV	LASTISVINQ	IFGYFILGHP
Cfgre2	---	IASIFTNSLV	IVATAKFKKL	RHPLNWILVN	LAIADLGETV	LASTISVFNQ	VFGYFVLGHP
Gfred	525	VASTFTNGLV	LVATAKFKKL	RHPLNWILVN	LAVADLAETL	LASTISVTNQ	FFGYFILGHP
Humgre	530	IASVFTNGLV	LAATMKFKKL	RHPLNWILVN	LAVADLAETV	IASTISVVNQ	VYGYFVLGHP
Humred	560	TASVFTNGLV	LAATMKFKKL	RHPLNWILVN	LAVADLAETV	IASTISIVNQ	VSGYFVLGHP
Chred	571	AASVFTNGLV	LVATWKFKKL	RHPLNWILVN	LAVADLGETV	IASTISVINQ	ISGYFILGHP
Cfred	---	VASTVTNGLV	LVASAKFKKL	RHPLNWILVN	LAIADLLETL	LASTISVCNQ	FFGYFILGHP
Anolred	625	IASIFTNGLV	LVATAKFKKL	RHPLNWILVN	LAIADLGETV	IASTISVINQ	ISGTFILGHP
Cons		N	L	LN IL N			F G

Species & Opsin	λmax	Sequence					
			120		140		160
Zfuv	360	GCNLEGFFAT	LGGEMGLKSL	VVLAIERWMV	VCKPVSNFRF	GENHAIMGVA	FTWVMACSCA
Chviol	415	VCELEAFVGT	HGGLVTGWSL	AFLAFERYIV	ICKPFGNFRF	SSRHALLVVV	ATWLIGVGVG
Humblu	420	VCALEGFLGT	VAGLVTGWSL	AFLAFERYIV	ICKPFGNFRF	SSKHALTVVL	ATWTIGIGVS
Gfblu	441	ACKIEGFLAT	LGGMVGLWSL	AVVAFERWLV	ICKPLGNFTF	KTPHAIAGCI	LPWISALAAS
Chblu	455	ACKIEGFAAT	LGGMVSLWSL	AVVAFERFLV	ICKPLGNFTF	RGSHAVLGCV	ATWVLGFVAS
Cfblu	---	ACKIEGFVAT	LGGMVSLWSL	SVVAFERWLV	ICKPVGNFSF	KGTHAIIGCA	LTWFFALLAS
Gekblu	467	GCAIEGFFAT	IGGQVALWSL	VVLAIERYIV	ICKPMGNFRF	SATHAIMGIA	FTWFMALACA
Gfrod	492	GCNPEGFFAT	LGGEMGLWSL	VVLAFERWMV	VCKPVSNFRF	GENHAIMGVV	FTWFMACTCA
Chrod	495	MCYIEGFFAT	LGGEIALWSL	VVLAVERYVV	VCKPMSNFRF	GENHAIMGVA	FSWIMAMACA
Lamrod	497	GCSIEGFFAT	LGGEVALWSL	VVLAIERYIV	ICKPMGNFRF	GNTHAIMGVA	FTWIMALACA
Humrod	497	GCNLEGFFAT	LGGEIALWSL	VVLAIERYVV	VCKPMSNFRF	GENHAIMGVA	FTWVMALACA
Bovrod	498	GCNLEGFFAT	LGGEIALWSL	VVLAIERYVV	VCKPMSNFRF	GENHAIMGVA	FTWVMALACA
Mserod	498	GCNLEGFFAT	LGGEIALWSL	VVLAIERYVV	VCKPMSNFRF	GENHAIMGVV	FTWIMALACA
Sgrod	501	GCNVEGFFAT	LGGEIALWSL	VVLAVERWVV	VCKPISNFRF	TENHAIMGVA	FSWIMAATCA
Hamrod	502	GCNLEGFFAT	LGGEIALWSL	VVLAIERYVV	ICKPMSNFRF	GENHAIMGVV	FTWIMALACA
Rprod	502	GCYFEGFFAT	LGGEIALWSL	VVLAIERYIV	VCKPMSNFRF	GENHAMMGVA	FTWIMALACA
Gfgre2	505	GCAVEGFMAT	LGGEVALWSL	VVLAIERYIV	VCKPMGSFKF	SSTHASAGIA	FTWVMAMACA
Chgre	508	GCAVEGFFAT	LGGQVALWSL	VVLAIERYIV	VCKPMGNFRF	SATHAMMGIA	FTWVMAFSCA
Gfgre1	511	GCAVEGFMAT	LGGEVALWSL	VVLAIERYIV	VCKPMGSFKF	SSSHAFAGIA	FTWVMALACA
Gekgre	521	LCVIEGYVVS	SCGITGLWSL	AIISWERWFV	VCKPFGNIKF	DSKLAIIGIV	FSWVWAWGWS
Cfgre1	---	MCVFEGWTVS	VCGITALWSL	TIISWERWVV	VCKPFGNVKF	DGKWAAGGII	FSWVWAIIWC
Cfgre2	---	MCIFEGWTVS	VCGITALWSL	TIISWERWVV	VCKPFGNVKF	DGKWAAGGII	FAWTWAIIWC
Gfred	525	MCIFEGFTVS	VCGIAGLWSL	TVISWERWVV	VCKPFGNVKF	DAKWASAGII	FSWVWSAIWC
Humgre	530	MCVLEGYTVS	LCGITGLWSL	AIISWERWMV	VCKPFGNVRF	DAKLAIVGIA	FSWIWAAVWT
Humred	560	MCVLEGYTVS	LCGITGLWSL	AIISWERWLV	VCKPFGNVRF	DAKLAIVGIA	FSWIWSAVWT
Chred	571	MCVVEGYTVS	ACGITALWSL	AIISWERWFV	VCKPFGNIKF	DGKLAVAGIL	FSWLWSCAWT
Cfred	---	MCVFEGFTVA	TCGIAGLWSL	TVISWERWVV	VCKPFGNVKF	DGKMATAGIV	FTWVWSAVWC
Anolred	625	MCVLEGYTVS	TCGISALWSL	AVISWERWVV	VCKPFGNVKF	DAKLAVAGIV	FSWVWSAVWT
Cons		C E	G SL	ER V	CKP F	A	W

Species & Opsin	λmax						
			180		200		220
Zfuv	360	VPPLVGWSRY	IPEGMQCSCG	VDYYTRTPGV	NNESFVIYMF	IVHFFIPLIV	IFFCYGRLVC
Chviol	415	LPPFFGWSRY	MPEGLQGSCG	PDWYTVGTKY	RSEYYTWFLF	IFCFIVPLSL	IIFSYSQLLS
Humblu	420	IPPFFGWSRF	IPEGLQCSCG	PDWYTVGTKY	RSESYTWFLF	IFCFIVPLSL	ICFSY.QLLR
Gfblu	441	LPPLFGWSRY	IPEGLQCSCG	PDWYTTNNKY	NNESYVMFLF	CFCFAVPFGT	IVFCYGQLLI
Chblu	455	APPLFGWSRY	IPEGLQCSCG	PDWYTTDNKW	HNESYVLFLF	TFCFGVPLAI	IVFSYGRLLI
Cfblu	---	TPPLFGWSRY	IPEGLQCSCG	PDWYTTENKY	NNESYVMFLF	CFCFGFPFTV	ILFCYGQLLF
Gekblu	467	GPPLFGWSRF	IPEGMQCSCG	PDYYTLNPDF	HNESYVIYMF	IVHFTVPMVV	IFFSYGRLVC
Gfrod	492	VPPLVGWSRY	IPEGMQCSCG	VDYYTRPQAY	NNESFVIYMF	IVHFIIPLIV	IFFCYGRLVC
Chrod	495	APPLFGWSRY	IPEGMQCSCG	IDYYTLKPEI	NNESFVIYMF	VVHFMIPLAV	IFFCYGNLVC
Lamrod	497	APPLVGWSRY	IPEGMQCSCG	PDYYTLNPNF	NNESVVYYMF	VVHFLVPVI.	IFFCYGRLLC
Humrod	497	APPLAGWSRY	IPEGLQCSCG	IDYYTLKPEV	NNESFVIYMF	VVHFTIPMII	IFFCYGQLVF
Bovrod	498	APPLVGWSRY	IPEGMQCSCG	IDYYTPHEET	NNESFVIYMF	VVHFIIPLIV	IFFCYGQLVF
Mserod	498	APPLVGWSRY	IPEGMQCSCG	IDYYTLKPEV	NNESFVIYMF	VVHFTIPMIV	IFFCYGQLVF
Sgrod	501	VPPLVGWSRY	IPEGMQCSCG	VDYYTRAEGF	NNESFVIYMF	IVHFLAPLIV	IFFCYGRLLC
Hamrod	502	APPLVGWSRY	IPEGMQCSCG	VDYYTLKPEV	NNESFVIYMF	VVHFTIPLIV	IFFCYGQLVF
Rprod	502	VPPLFGWSRY	IPEGMQCSCG	VDYYTLKPEV	NNESFVIYMF	VVHFLIPLII	ISFCYGRLVC
Gfgre2	505	APPLVGWSRY	IPEGIQCSCG	PDYYTLNPEY	NNESYVLYMF	ICHFILPVTI	IFFTYGRLVC
Chgre	508	APPLFGWSRY	MPEGMQCSCG	PDYYTHNPDY	HNESYVLYMF	VIHFIIPVVV	IFFSYGRLIC
Gfgre1	511	APPLFGWSRY	IPEGMQCSCG	PDYYTLNPDY	NNESYVIYMF	VCHFILPVAV	IFFTYGRLVC
Gekgre	521	APPIFGWSRY	WPHGLKTSCG	PDVSSGSVEL	GCQSFMLTLM	ITCCFLPLFI	IIVCYLQVWM
Cfgre1	---	TPPIFGWSRY	WPHGLKTSCG	PDVFSGSEDP	GVASYMITLM	LTCCIPLLSI	IIICYIFVWS
Cfgre2	---	TPPIFGWSRY	WPHGLKTSCG	PDVFSGSEDP	GVASYMVTLL	LTCCILPLSV	IIICYIFVWN
Gfred	525	APPIFGWSRF	WPHGLKTSCG	PDVFSGSEDP	GVQSYMIVLM	ITCCIIPLAI	IILCYIAVWL
Humgre	530	APPIFGWSRY	WPHGLKTSCG	PDVFSGSSYP	GVQSYMIVLM	VTCCITPLSI	IVLCYLQVWL
Humred	560	APPIFGWSRY	WPHGLKTSCG	PDVFSGSSYP	GVQSYMIVLM	VTCCIIPLAI	IMLCYLQVWL
Chred	571	APPIFGWSRY	WPHGLKTSCG	PDVFSGSSDP	GVQSYMVVLM	VTCCFFPLAI	IILCYLQVWL
Cfred	---	APPIFGWSRY	WPHGLKTSCG	PDVFSGSEDP	GVQSYMIVLM	ITCCFIPLGI	IILCYIAVWW
Anolred	625	APPVFGWSRY	WPHGLKTSCG	PDVFSGSDDP	GVLSYMIVLM	ITCCFIPLAV	ILLCYLQVWL
Cons		PP GWSR	P G SCG	D			I Y

```
Species &   λmax                        Sequence
Opsin

                                 240                  260                   280
Zfuv        360     TVKEAARQQQ ESETTQRAER EVTRMVIIMV IAFLICWLPY AGVAWYIFTH QGSEFGPVFM
Chviol      415     ALRAVAAQQQ ESATTQKAER EVSRMVVVMV GSFCLCYDPY AALAMYMVNN RDHGLDLRLV
Humblu      420     ALKAVAAQQQ ESATTQKAER EVSRMVVVMV GSLCVCYVPY AAFAMYMVNN RNHGLDLRLV
Gfblu       441     TLKLAAKAQA DSASTQKAER EVTKMVVVMV LGFLVCWAPY ASFSLWIVSH RGEEFDLRMA
Chblu       455     TLRAVARQQE QSATTQKADR EVTKMVVVMV LGFLVCWAPY TAFALWVVTH RGRSFEVGLA
Cfblu       ---     TLKSAAKAQA DSASTQKAER EVTKMVVVMV MGFLVCWLPY ASFALWVVFN RGQSFDLRLG
Gekblu      467     KVREAAAQQQ ESATTQKAEK EVTRMVILMV LGFLLAWTPY AATAIWIFTN RGAAFSVTFM
Gfrod       492     TVKEAAAQHE ESETTQRAER EVTRMVVIMV IGFLICWIPY ASVAWYIFTH QGSEFGPVFM
Chrod       495     TVKEAAAQQQ ESATTQKAEK EVTRMVIIMV IAFLICWVPY ASVAFYIFTN QGSDFGPIFM
Lamrod      497     TVKEAAAAQQ ESASTQKAEK EVTRMVVLMV ISFLVCWVPY ASVAFYIFTH QGSDFGAIFM
Humrod      497     TVKEAAAQQQ ESATTQKAEK EVTRMVIIMV IAFLICWVPY ASVAFYIFTH QGSNFGPIFM
Bovrod      498     TVKEAAAQQQ ESATTQKAEK EVTRMVIIMV IAFLICWLPY AGVAFYIFTH QGSDFGPIFM
Mserod      498     TVKEAAAQQQ ESATTQKAEK EVTRMVIIMV IFFLICWLPY ASVAFYIFTH QGSNFGPIFM
Sgrod       501     AVKEAAAAQQ ESETTQRAER EVTRMVIIMV IGFLTSWLPY ASVAWYIFTH QGTEFGPLFM
Hamrod      502     TVKEAAAQQQ ESATTQKAEK EVTRMVILMV VFFLICWFPY AGVAFYIFTH QGSNFGPIFM
Rprod       502     TVKEAAAQQQ ESATTQKAEK EVTRMVIIMV IFFLICWVPY AYVAFYIFTH QGSEFGPIFM
Gfgre2      505     TVKAAAAQQQ DSASTQKAER EVTKMVILMV LGFLVAWTPY ATVAAWIFFN KGAAFSAQFM
Chgre       508     KVREAAAQQQ ESATTQKAEK EVTRMVILMV LGFMLAWTPY AVVAFWIFTN KGADFTATLM
Gfgre1      511     TVKAAAAQQQ DSASTQKAER EVTKMVILMV FGFLIAWTPY ATVAAWIFFN KGADFSAKFM
Gekgre      521     AIRAVAAQQK ESESTQKAER EVSRMVVVMI VAFCICWGPY ASFVSFAAAN PGYAFHPLAA
Cfgre1      ---     AIHQVAQQQK DSESTQKAEK EVSRMVVVMI LAFIVCWGPY ASFATFSAVN PGYAWHPLAA
Cfgre2      ---     AIHQVAQQQK DSESTQKAEK EVSRMVVVMI LAFILCWGPY ASFATFSALN PGYAWHPLAA
Gfred       525     AIRTVAQQQK DSESTQKAEK EVSRMVVVMI FAYCFCWGPY TFCACFAAAN PGYAFHPLAA
Humgre      530     AIRAVAKQQK ESESTQKAEK EVTRMVVVMV LAFCFCWGPY AFFACFAAAN PGYPFHPLMA
Humred      560     AIRAVAKQQK ESESTQKAEK EVTRMVVVMI FAYCVCWGPY TFFACFAAAN PGYASHPLMA
Chred       571     AIRAVAAQQK ESESTQKAEK EVSRMVVVMI VAYCFCWGPY TFFACFAAAN PGYAFHPLAA
Cfred       ---     AIRTVAQQQK DSESTQKAEK EVSRMVVVMI MAYCFCWGPY TFFACFAAAN PGYAFHPLAA
Anolred     625     AIRAVAAQQK ESESTQKAEK EVSRMVVVMI IAYCFCWGPY TVFACFAAAN PGYAFHPLAA
Cons                     A      S  TQ A   EV  MV  M          PY


                                300                   320                     340
Zfuv        360     TLPAFFAKTS AVYNPCIYIC MNKQFRHCMI TTLCCGKNPF EEEE.GASTT ASKTEASSVS SSSVSPA
Chviol      415     TIPAFFSKSA CVYNPIIYCF MNKQFRACIM E.TVCGK.PL TDD.SDASTS AQRTEVSSVS SSQVGPT
Humblu      420     TIPSFFSKSA CIYNPIIYCF MNKQFQACIM K.MVCGKAMT ..DESDTCSS .QKTEVSTVS STQVGPN
Gfblu       441     TIPSCLSKAS TVYNPVIYVL MNKQFRSCMM K.MVCGKN.I EEDE..ASTS SQVTQVSS.. ...VAPEK
Chblu       455     SIPSVFSKSS TVYNPVIYVL MNKQFRSCML KLLFCGRSPF GDDE.DVSGS SQATQVSSVS SSHVAPA
Cfblu       ---     TIPSCFSKAS TVYNPVIYVF MNKQFRSCMM KLIFCGKSPF GDDEE.ASSS SQVTQVSS . ...VGPEK
Gekblu      467     TIPAFFSKSS SIYNPIIYVL LNKQFRNCMV TTICCGKNPF GDEDVSSSVS QSKTEVSSVS SSQVAPA
Gfrod       492     TLPAFFAKTA AVYNPCIYIC MNKQFRHCMI TTLCCGKNPF EEEE.GASTT ASKTEASSVS SS.VSPA
Chrod       495     TIPAFFAKSS AIYNPVIYIV MNKQFRNCMI TTLCCGKNPL GDED....TS AGKTETSSVS TSQVSPA
Lamrod      497     TLPAFFAKSS ALYNPVIYIL MNKQFRNCMI TTLCCGKNPL GDDESGAST. .SKTEVSSVS TSPVSPA
Humrod      497     TIPAFFAKSA AIYNPVIYIM MNKQFRNCML TTICCGKNPL GDDE..ASAT VSKTET.... .SQVAPA
Bovrod      498     TIPAFFAKTS AVYNPVIYIM MNKQFRNCMV TTLCCGKNPL GDDE..ASTT VSKTET.... .SQVAPA
Mserod      498     TLPAFFAKSS SIYNPVIYIM LNKQFRNCML TTLCCGKNPL GDDD..ASAT ASKTET.... .SQVAPA
Sgrod       501     TIPAFFAKSS ALYNPMIYIC MNKQFRHCMI TTLCCGKNPF EEEE.GASTT ..KTEASSVS SSSVSPA
Hamrod      502     TLPAFFAKSS SIYNPVIYIM MNKQFRNCML TTLCCGKNIL GDDE..ASAT ASKTET.... .SQVAPA
Rprod       502     TVPAFFAKSS AIYNPVIYIM LNKQFRHCMI TTLCCGKNPF GDDD.ASSAA TSKTEATSVS TSQVSPA
Gfgre2      505     AIPAFFSKTS ALYNPVIYVL LNKQFRSCML TTLFCGKNPL GDEE.SSTVS TSKTEVSS.. ...VSPA
Chgre       508     AVPAFFSKSS SLYNPIIYVL MNKQFRNCMI TTICCGKNPF GDEDVSSTVS QSKTEVSSVS SSQVSPA
Gfgre1      511     AIPAFFSKSS ALYNPVIYVL LNKQFRNCML TTIFCGKNPL GDDE.SSTVS TSKTEVSS.. ...VSPA
Gekgre      521     ALPAYFAKSA TIYNPVIYVF MNRQFRNCIM Q.LF.GKKV. ..DDGSEAST TSRTEVSSVS NSSVAPA
Cfgre1      ---     AMPAYFAKSA TIYNPIIYVF MNRQFRSCIM Q.LF.GKKV. ..ED.ASEVS GSTTEVSTAS
Cfgre2      ---     ALPAYFAKSA TIYNPIIYVF MNRQFRSCIM Q.LF.GKKV. ..ED.ASEVS GSTTEVSTAS
Gfred       525     AMPAYFAKSA TIYNPIIYVF MNRQFRVCIM Q.LF.GKKV. ..DDGSEVST .SKTEVSS.. ...VAPA
Humgre      530     ALPAFFAKSA TIYNPVIYVF MNRQFRNCIL Q.LF.GKKV. ..DDGSELSS ASKTEVSSVS S..VSPA
Humred      560     ALPAYFAKSA TIYNPVIYVF MNRQFRNCIL Q.LF.GKKV. ..DDGSELSS ASKTEVSSVS S..VSPA
Chred       571     ALPAYFAKSA TIYNPIIYVF MNRQFRNCIL Q.LF.GKKV. ..DDGSEVST .SRTEVSSVS NSSVSPA
Cfred       ---     AMPAYFAKSA TIYNPVIYVF MNRQFRVCIM Q.LF.GKKV. ..DDGSEVST .SKTEVSS.. ...VAPA
Anolred     625     ALPAYFAKSA TIYNPIIYVF MNRQFRNCIM Q.LF.GKKV. ..DDGSELSS TSRTEVSSVS NSSVSPA
Cons                  P    K     YNP IY    N QF  C        G                   T          V P
```

4.6 CONCLUSIONS AND FUTURE PERSPECTIVES

Our understanding of the molecular mechanisms underlying photoreception has increased rapidly in the last decade. Just over ten years ago, the first opsin gene sequence was obtained by Nathans and Hogness (1983). Determining that sequence represented an important turning point because it provided a handle with which to grasp other opsin sequences. This led rapidly to the isolation of the human visual pigment genes and the first chance to make correlations between genotype and spectral sensitivity phenotypes in humans. Subsequent work on the molecular biology of human opsin genes has now provided an important understanding of how our color vision and its variants are determined genetically and this represents a great achievement. During the same period, the opsin sequence for bovine rhodopsin became the model that researchers adopted and tested to determine the structural and functional relationships of opsin. Based on this model, we now have a very good idea of how the rhodopsin molecule functions and are in a position to make some very confident predictions about the three-dimensional structure of the molecule in the membrane. No doubt crystal structure data will soon be available to confirm these predictions. Also, as we understand more about how the components of the opsin protein are important for the correct function of the visual pigment opsin molecule, it is becoming possible to highlight relationships between inherited visual disease and mutations in different regions of the opsin gene.

For good color vision across the spectrum we use three cone visual pigments with different wavelength sensitivities and these visual pigments are constructed with three different opsin proteins. This should remind us of one of the fundamental characteristics of visual pigments that is so important for vision, which is that they have variable wavelength sensitivity. In nature, this potential for variability has been well exploited so that we know of visual pigment sensitivities that range from the UV to the far-red, with many examples in between. The potential is obviously something that has been molded by evolution and we can see this in animals living in extreme environments that have driven their visual pigment sensitivities to particular spectral bands. We now appreciate that much of this wavelength sensitivity variation must have genetic foundations in opsin sequence variation. This high level of sensitivity variation makes the visual pigments one of the most interesting of the G-protein linked membrane receptors to study and at the same time should prepare us to expect similar variable function in other members of the family.

The recent advances that have been made in understanding the molecular structure and function of opsin have been met by a greater awareness of the subject of visual ecology. Research in the latter is now uncovering many examples of how the visual system is adapted to functioning in different light environments and how it can also become adapted to different conditions during the lifetime of an animal. The challenge for the future will be to combine the two areas of research (which has already started) to investigate how different visual pigments in the wild are determined by opsin amino acid sequence and coded for genetically. As more examples are found of changes in visual pigment sensitivity during development, we should also be looking carefully at the control of opsin gene expression. Finally we should remember that our own view of the world is specifically colored and we should keep an open mind about the way color is perceived by other animals and even by other humans.

REFERENCES

Aho, A.C., Donner, K., Hyden, C. *et al.* (1988) Low retinal noise in animals with low body temperature allows high visual sensitivity. *Nature, London*, **334**, 348–50.

Allen, D.M. and McFarland, W.N. (1973) The effect of temperature on rhodopsin–porphyropsin ratios in a fish. *Vision Research*, **13**, 1303–9.

Archer, S.N. (1988) A microspectrophotometric study of visual pigment polymorphism in the guppy, *Poecilia reticulata*. PhD Thesis, University of Bristol, Bristol, U.K.

Archer, S.N. and Lythgoe, J.N. (1990) The visual pigment basis for cone polymorphism in the guppy, *Poecilia reticulata. Vision Research*, **30**, 225–33.

Archer, S.N., Endler, J.A., Lythgoe, J.N. and Partridge, J.C. (1987) Visual pigment polymorphism in the guppy, *Poecilia reticulata. Vision Research*, **27**, 1243–52.

Archer, S.N., Lythgoe, J.N. and Hall, L. (1992) Rod opsin cDNA sequence from the sand goby (*Pomatoschistus minutus*) compared with those of other vertebrates. *Proceedings of the Royal Society of London B*, **248**, 19–25.

Baehr W.W., Falk, J.D., Bugra, K. *et al.* (1988) Isolation and analysis of the mouse opsin gene. *Federation of European Biochemical Societies Letters*, **238**, 253–6.

Baldwin, J. (1993) The probable arrangement of the helices in G protein-coupled receptors. *EMBO Journal*, **12**, 1693–703.

Barlow, H.B. (1988) The thermal limit to seeing. *Nature, London*, **334**, 296–7.

Beatty, D.D. (1984) Visual pigments and the labile scotopic visual system of fish. *Vision Research*, **24**, 1563–73.

Bowmaker, J.K. and Kunz, Y.W. (1987) Ultraviolet receptors, tetrachromatic colour vision and retinal mosaics in the brown trout (*Salmo trutta*): age-dependent changes. *Vision Research*, **27**, 2101–8.

Bowmaker, J.K., Jacobs, G.H. and Mollon, J.D. (1987) Polymorphism of photopigments in the squirrel monkey: a sixth phenotype. *Proceedings of the Royal Society of London B*, **231**, 383–90.

Bowmaker, J.K., Astell, S., Hunt, D.M. and Mollon, J.D. (1991) Photosensitive and photostable pigments in the retinae of Old World monkeys. *Journal of Experimental Biology*, **156**, 1–19.

Bridges, C.D.B. (1964) Variation of visual pigment amongst individuals of an American minnow, *Notemigonu crysoleycas boscii. Vision Research*, **4**, 233–9.

Chabre, M. and Deterre, P. (1989) Molecular mechanisms of visual transduction. *European Journal of Biochemistry*, **179**, 255–66.

Cohen, G.B., Yang, T., Robinson, P.R. and Oprian, D.D. (1993) Constitutive activation of opsin: influence of charge at position 134 and size at position 296. *Biochemistry*, **32**, 6111–15.

Dartnall, H.J.A. and Lythgoe, J.N. (1965) The clustering of fish visual pigments around discrete spectral positions, and its bearing on chemical structure, in *Ciba Foundation Symposium on Physiolosy and Experimental Psychology of Colour Vision* (eds G.E.W. Wolstenholme and J. Knight, J. & A. Churchill, London, pp.3–21.

Dartnall, H.J.A., Bowmaker, J.K. and Mollon, J.D. (1983) Human visual pigments: microspectrophotometric results from the eyes of seven persons. *Proceedings of the Royal Society of London B*, **220**, 115–30.

Doi, T., Molday, R.S and Khorana, H.G. (1990) Role of the intradiscal domain in rhodopsin assembly and function. *Proceedings of the National Academy of Sciences USA*, **87**, 4991–5.

Endler, J.A. (1991) Variation in the appearance of guppy color patterns to guppies and their predators under different visual conditions. *Vision Research*, **31**, 587–608.

Evans, B.I., Harosi, F.I. and Fernald, R.D. (1993) Photoreceptor spectral absorbance in larval and adult winter flounder (*Pseudopleuronectes americanus*). *Visual Neuroscience*, **10**, 1065–71.

Franke, R.R., Sakmar, T.P., Graham, R.M. and Khorana, H.G. (1992) Structure and function in rhodopsin. Studies of the interaction between the rhodopsin cytoplasmic domain and transducin. *Journal of Biological Chemistry*, **267**, 14767–74.

Gale, J.M., Tobey, R.A. and D'Anna, A. (1992) Localization and DNA sequence of a replication origin in the rhodopsin gene locus of Chinese hamster cells. *Journal of Molecular Biology*, **224**, 343–58.

Gurevich, V.V. and Benovic, J.L. (1992) Cell-free expression of visual arrestin. *Journal of Biological Chemistry*, **267**, 21919–23.

Hall, Z.W. (1987) Three of a kind: the β-adrenergic receptor, the muscarinic acetylcholine receptor, and rhodopsin. *Trends in Neuroscience*, **10**, 99–101.

Hargrave, P.A. and McDowell, J.H. (1992) Rhodopsin and phototransduction. *International Review of Cytology*, **137 B**, 49–97.

Hargrave, P.A., McDowell, J.H., Curtis, D.R. *et al.* (1983) The structure of bovine rhodopsin. *Biophysics of Structure and Mechanism*, **9**, 235–44.

Harosi, F.I. and Kleinschmidt, J. (1993) Visual pigments in the sea lamprey, *Petromyzon marinus*. *Visual Neuroscience*, **10**, 711–15.

Henderson, R. and Schertler, G.F.X. (1990) The structure of bacteriorhodopsin and its relevance to the visual opsins and other seven-helix, G-protein coupled receptors. *Philosophical Transactions of the Royal Society, London, B*, **326**, 379–89.

Hisatomi, O., Iwasa, O.T., Tokunaga, F. and Yasui, A. (1991) Isolation and characterization of lamprey rhodopsin cDNA. *Biochemical and Biophysical Research Communications*, **174**, 1125–32.

Hunt, D.M., Williams, A.J., Bowmaker, J.K. and Mollon, J.D. (1993) Structure and evolution of the polymorphic photopigment gene of the marmoset. *Vision Research*, **33**, 147–54.

Ibbotson, R.E., Hunt, D.M., Bowmaker, J.K. and Mollon, J.D. (1992) Sequence divergence and copy number of the middle- and long-wave photopigment genes in Old World monkeys. *Proceedings of the Royal Society of London B*, **247**, 145–54.

Jacobs, G.H. (1993) The distribution and nature of colour vision among the mammals. *Biological Reviews*, **68**, 413–71.

Johnson, R.L., Grant, K.B., Zankel, T.C. *et al.* (1993) Cloning and expression of goldfish opsin sequences. *Biochemistry*, **32**, 208–14.

Karnik, S. and Khorana, H.G. (1990) Assembly of functional rhodopsin requires a disulphide bond between cysteine residues 110 and 187. *Journal of Biological Chemistry*, **265**, 17520–4.

Kawamura, S. and Yokoyama, S. (1993) Molecular characterization of the red visual pigment gene of the American chameleon (*Anolis carolinensis*). *Federation of European Biochemical Societies Letters*, **323**, 247–51.

Khorana, H.G. (1992) Rhodopsin, photoreceptor of the rod cell. *Journal of Biological Chemistry*, **267**, 1–4.

Kleinschmidt, J. and Harosi, F.I. (1992) Anion sensitivity and spectral tuning of cone visual pigments *in situ*. *Proceedings of the National Academy of Science USA*, **89**, 9181–5.

Kojima, D., Okano, T., Fukada, Y. *et al.* (1992) Cone visual pigments are present in rod cells. *Proceedings of the National Academy of Sciences USA*, **89**, 6841–5.

Konig, B., Arendt, A., McDowell, J.H. *et al.* (1989) Three cytoplasmic loops of rhodopsin interact with transducin. *Proceedings of the National Academy of Sciences USA*, **86**, 6878–82.

Kuwata, O., Imamoto, Y., Okano, T. *et al.* (1990) The primary structure of iodopsin, a chicken red-sensitive cone pigment. *Federation of European Biochemical Societies Letters*, **272**, 128–32.

Lythgoe, J.N. (1972) List of vertebrate visual pigments, in *Handbook of Sensory Physiology*, vol VII/1 (ed. H.J.A. Dartnall), Springer Verlag, New York, pp. 604–24.

Lythgoe, J.N. (1979) *The Ecology of Vision*. Clarendon Press, Oxford.

Lythgoe, J.N. and Partridge, J.C. (1989) Visual pigments and the acquisition of visual information. *Journal of Experimental Biology*, **146**, 1–20.

Marti, T., Otto, H., Rosselet, S.J. *et al.* (1992) Anion binding to the Schiff base of the bacteriorhodopsin mutants Asp-85→Asn/Asp-212→Asn and Arg-82→Gln/Asp-85→Asn/Asp-212→Asn. *Journal of Biological Chemistry*, **267**, 16922–7.

McDowell, J.H., Nawrocki, J.P. and Hargrave, P.A. (1993) Phosphorylation sites in bovine rhodopsin. *Biochemistry*, **32**, 4968–74.

Merbs, S.L. and Nathans, J. (1992a) Absorption spectra of the hybrid pigments responsible for anomalous color vision. *Science*, **258**, 464–6.

Merbs, S.L. and Nathans, J. (1992b) Absorption spectra of human cone pigments. *Nature*, **356**, 433–5.

Morris, A., Bowmaker, J.K. and Hunt, D.M. (1993) The molecular basis of a spectral shift in the rhodopsins of two species of squid from different photic environments. *Proceedings of the Royal Society of London B*, **254**, 233–40.

Nathans, J. (1990a) Determinants of visual pigment absorbance: identification of the retinylidene Schiff's base counterion in bovine rhodopsin. *Biochemistry*, **29**, 9746–52.

Nathans, J. (1990b) Determinants of visual pigment absorbance: role of charged amino acids in the putative transmembrane segments. *Biochemistry*, **29**, 937–42.

Nathans, J. (1992) Rhodopsin: structure, function and genetics. *Biochemistry*, **31**, 4923–31.

Nathans, J. and Hogness, D.S. (1983) Isolation, sequence analysis, and intron–exon arrangement of the gene encoding bovine rhodopsin. *Cell*, **34**, 807–14.

Nathans, J. and Hogness, D.S. (1984) Isolation and nucleotide sequence of the gene encoding human rhodopsin. *Proceedings of the National Academy of Sciences USA*, **81**, 4851–5.

Nathans, J., Thomas, D. and Hogness, D.S. (1986a) Molecular genetics of human color vision; the genes encoding blue, green and red pigments. *Science*, **232**, 193–202.

Nathans, J., Piantanida, T.P., Eddy, R.L. *et al.* (1986b) Molecular genetics of inherited variation in human color vision. *Science*, **232**, 203–10.

Nathans, J., Merbs, S.L., Sung, C. *et al.* (1992) Molecular genetics of human visual pigments. *Annual Review of Genetics*, **26**, 403–24.

Neitz, J. and Jacobs, G.H. (1990) Polymorphism in normal human color vision and its mechanism. *Vision Research*, **30**, 621–36.

Neitz, M., Neitz, J. and Jacobs, G.H. (1991) Spectral tuning of pigments underlying red–green colour vision. *Science*, **252**, 971–4.

Neitz, J., Neitz, M. and Jacobs, G.H. (1993) More than three different cone pigments among people with normal colour vision. *Vision Research*, **33**, 117–22.

O'Day, W.T. and Fernandez, H.R. (1974) *Aristostomias scintillans* (Malacosteidae): a deep-sea fish with visual pigments apparently adapted to its own bioluminescence. *Vision Research*, **14**, 545–50.

Okano, T., Kojima, D., Fukada, Y. *et al.* (1992) Primary structures of chicken cone visual pigments: vertebrate rhodopsins have evolved out of cone visual pigments. *Proceedings of the National Academy of Sciences USA*, **89**, 5932–6.

Partridge, J.C. (1990) The colour sensitivity and vision of fishes, in *Light and Life in the Sea* (eds P.J. Herring, A.K. Cambell, M. Whitfield and L. Maddock), Cambridge University Press, Cambridge.

Partridge, J.C., Archer, S.N. and Lythgoe, J.N. (1988) Visual pigments in the individual rods of deep sea fishes. *Journal of Comparative Physiology A*, **162**, 543–50.

Partridge, J.C., Shand, J., Archer, S.N. *et al.* (1989) Interspecific variation in the visual pigments of deep-sea fishes. *Journal of Comparative Physiology A*, **164**, 513–29.

Partridge, J.C., Archer, S.N. and van Oostrum, J. (1992) Single and multiple visual pigments in deep-sea fishes. *Journal of the Marine Biological Association UK*, **72**, 113–30.

Pittler, S.J., Fliesler, S.J. and Baehr, W. (1992) Primary structure of frog rhodopsin. *Federation of European Biochemical Societies Letters*, **313**, 103–8.

Raymond, P.A., Barthel, L.K., Rounsifer, M.E. *et al.* (1993) Expression of rod and cone visual pigments in goldfish and zebrafish: a rhodopsin-like gene is expressed in cones. *Neuron*, **10**, 1161–74.

Robinson, J., Schmitt, E.A., Harosi, F.I. *et al.* (1993) Zebrafish ultraviolet visual pigment: absorption spectrum, sequence, and localization. *Proceedings of the National Academy of Sciences USA*, **90**, 6009–12.

Saibil, H.R. (1986) From photon to receptor potential: the biochemistry of vision. *News in Physiological Sciences*, **1**, 122–5.

Sakmar, T.P., Franke, R.R. and Khorana, H.G. (1989) Glutamic acid-113 serves as the retinylidene Schiff's base counterion in bovine rhodopsin. *Proceedings of the National Academy of Science USA*, **86**, 8309–13.

Shand, J. (1993) Changes in the spectral absorption of cone visual pigments during the settlement of the goatfish *Upeneus tragula*: the loss of red sensitivity as a benthic existence begins. *Journal of Comparative Physiology A*, **173**, 115–21.

Shand, J., Partridge, J.C., Archer, S.N. *et al.* (1988) Spectral absorbance changes in the violet/blue sensitive cones of the juvenile pollack, *Pollachius pollachius*. *Journal of Comparative Physiology A*, **163**, 699–703.

Sibley, D.R., Benovic, J.L., Caron, M.G. and Lefkowitz, R.J. (1987) Regulation of transmembrane signalling by receptor phosphorylation. *Cell*, **48**, 913–22.

Takao, M., Yasui, A. and Tokunaga, F. (1988) Isolation and sequence determination of the chicken rhodopsin gene. *Vision Research*, **28**, 471–80.

Tovee, M.J. (1994) The molecular genetics and evolution of primate colour vision. *Trends in Neuroscience*, **17**, 30–7.

Trumpp-Kallmeyer, S., Hoflack, J., Bruinvels, A. and Hibert, M. (1992) Modelling of G-protein-coupled receptors: application to dopamine, adrenalin, serotonin, acetylcholine, and mammalian opsin receptors. *Journal of Medical Chemistry*, **35**, 3448–62.

Van Tol, H.H.M., Bunzow, J.R., Guan, H.-G. *et al.* (1991) Cloning the gene for a human dopamine D4 receptor with high affinity for the antipsychotic clozapine. *Nature*, **350**, 610–14.

Vollrath, D., Nathans, J. and Davis, R.W. (1988) Tandem array of human visual pigment genes at Xq28. *Science*, **240**, 1669–72.

Wang, S., Adler, R. and Nathans, J. (1992) A visual pigment from chicken that resembles rhodopsin: amino acid sequence, gene structure, and functional expression. *Biochemistry*, **31**, 3309–15.

Williams, A.C., Hunt, D.M., Bowmaker, J.K. and Mollon, J.D. (1992) The polymorphic photopigments of the marmoset: spectral tuning and genetic basis. *EMBO Journal*, **11**, 2039–45.

Winderickx, J., Lindsey, D.T., Sanocki, E. *et al.* (1992) Polymorphism in red photopigment underlies variation in colour matching. *Nature*, **356**, 431–3.

Wood, P. and Partridge, J.C. (1993) Opsin substitution induced in retinal rods of the eel (*Anguilla anguilla* (L.)): a model for G-protein-linked receptors. *Proceedings of the Royal Society, London, B*, **254**, 227–32.

Wood, P., Partridge, J.C. and De Grip, W.J. (1992) Rod visual pigment changes in the elver of the eel *Anguilla anguilla* L. measured by microspectrophotometry. *Journal of Fish Biology*, **41**, 601–11.

Yokoyama, S. (1994) Gene duplications and evolution of the short wavelength-sensitive visual pigments in vertebrates. *Molecular Biology and Evolution*, **11**, 32–9.

Yokoyama, R. and Yokoyama, S. (1990a) Isolation, DNA sequence and evolution of a color visual pigment gene of the blind cave fish, *Astyanax fasciatus*. *Vision Research*, **30**, 807–16.

Yokoyama, R. and Yokoyama, S. (1990b) Convergent evolution of the red- and green-like visual pigment genes in fish, *Astyanax fasciatus*, and human. *Proceedings of the National Academy of Science USA*, **87**, 9315–18.

Yokoyama, R. and Yokoyama, S. (1993) Molecular characterization of a blue visual pigment gene in the fish *Astyanax fasciatus*. *Federation of European Biochemical Societies Letters*, **334**, 27–31.

Zhukovsky, E.A. and Oprian, D.D. (1989) Effect of carboxylic acid side chains on the absorption maximum of visual pigments. *Science*, **246**, 928–30.

5

Phototransduction, excitation and adaptation

SATORU KAWAMURA

5.1 INTRODUCTION

Vertebrate photoreceptors respond to light and transmit electrical signals to secondary neurons. In the dark, a steady current flows into the outer segment. The current is blocked in the light so that the cell elicits a hyperpolarizing response which is the primary phototransduction signal. The mechanisms generating this photoresponse are now well characterized, especially in the case of rod photoreceptors. The phototransduction process consists of an enzyme cascade including; a visual pigment (rhodopsin), a GTP-binding protein (transducin) and an enzyme hydrolyzing cGMP (cGMP-phosphodiesterase).

A photoreceptor cell is not simply a detector of light; it can also actively adapt to environmental light. Following light absorption, rhodopsin changes its color from red to pale yellow, i.e. it is 'bleached'. When a light flash is given in the dark, a rod photoresponse is half-saturated by bleaching of 30–50 rhodopsin molecules per outer segment which contains approximately 3×10^9 rhodopsin molecules in the case of amphibians. The response is practically fully saturated by bleaching of 10^3 rhodopsin molecules per outer segment. However, when a rod is exposed to a continuous light for many minutes, the response gradually recovers towards the dark level and then the cell becomes responsive to a light flash that would evoke a fully saturated response under dark-adapted conditions. This desensitizing process is called 'light-adaptation' which accompanies acceleration of the time course of the photoresponse.

Light-adaptation is classified into two types depending on light intensity, or amount of visual pigment bleached: one is background adaptation and the other is bleaching adaptation. Background adaptation takes place under weak light when a negligible amount of rhodopsin ($\sim 10^{-8}$ of total) is bleached. In the case of background adaptation, the light-sensitivity of the cell fully recovers, even in isolated rods, when the light stimulus ceases. A light-induced decrease in the cytoplasmic $[Ca^{2+}]$ is responsible for this adaptation. Background adaptation is attained by modifying the phototransduction cascade. Bleaching adaptation takes place when the light intensity is high and a substantial amount of rhodopsin ($\sim 10^{-1}$ of total) is bleached. In contrast to background adaptation, the light-sensitivity of isolated rods does not recover to

Neurobiology and Clinical Aspects of the Outer Retina
Edited by M.B.A. Djamgoz, S.N. Archer and S. Vallerga
Published in 1995 by Chapman & Hall, London
ISBN 0 412 60080 3

the dark level even after the light stimulation, i.e. the light-sensitivity decreases irreversibly. However, the sensitivity recovers fully in the dark when rods remain attached to the pigment epithelium such that rhodopsin can be regenerated. The recovery of sensitivity from either background or bleaching adaptation is generally called 'dark adaptation.' The mechanisms of phototransduction and background adaptation but not bleaching adaptation are known in detail. Much of the space of this chapter has, therefore, been devoted to the mechanisms of phototransduction and background adaptation. Furthermore, most of the work has been done on rod photoreceptors.

Since as indicated above, background adaptation is attained by modification of phototransduction, the mechanisms responsible for turning the cascades on and off are reviewed first and then adaptation is covered. Recent other review articles also deal with some of these areas (McNaughton, 1990; Stryer, 1991; Kaupp and Koch, 1992; Detwiler and Gray-Keller, 1992; Lagnado and Baylor, 1992; Kawamura, 1993a; Koutalos and Yau, 1993).

5.2 PHOTOTRANSDUCTION CASCADE

5.2.1 TURN-ON MECHANISMS

A summary scheme of the basic mechanisms of photoreceptor excitation and adaptation is shown in Figure 5.1. The characteristics of the photoreceptor proteins involved in these mechanisms are listed in Table 5.1.

In the rod outer segment, cGMP binds to a cation channel (cGMP-gated channel) in the dark and keeps it open (Figure 5.1, a). On light absorption, rhodopsin (R) is activated. Activated rhodopsin (R*) interacts with a GTP-binding protein, transducin (T), and catalyzes the exchange of bound-GDP for GTP on the transducin molecule. The GTP-bound form of transducin (T_{GTP}) is active and interacts with a cGMP hydrolyzing enzyme, cGMP-phosphodiesterase (PDE) and stimulates it (Figure 5.1, PDE*). As a consequence, cGMP concentration decreases and leads to the dissociation of cGMP from the channel and the suppression of the inward current in the light (Figure 5.1, b). Since rhodopsin, transducin and PDE are all associated with the disk membrane, this is where these reactions take place. The following section reviews these molecular events in greater detail.

(a) Photoreception by rhodopsin

Rhodopsin (39 kDa) absorbs green light at around 500 nm maximal and therefore, its color is red when we see it with a light flash in the dark. Rhodopsin is a member of a family of G-protein-coupled receptors which have seven transmembrane helices. This family also includes β-adrenergic receptors and muscarinic acetylcholine receptors. Rhodopsin consists of a protein moiety (opsin) and a chromophore (11-*cis* retinal) that is covalently linked to opsin through a protonated Schiff base at the Lys-296 residue of bovine rhodopsin (Chapters 3 and 4). Due to the protonation, the absorption wavelength of the chromophore shifts from the UV to the visible region of the spectrum. The positive charge at the protonated Schiff base is stabilized by a carboxyl counter ion at Glu-113 (Zhukovsky and Oprian, 1989; Sakmar *et al.*, 1989; Nathans, 1990).

Upon light absorption, the 11-*cis* retinal isomerizes to an all-*trans* form, which initiates a series of conformational changes in the rhodopsin molecule that results in bleaching which is the photoreception signal of rhodopsin. During the course of bleaching, several intermediates can be identified by spectrophotometry. Among them, metarhodopsin II (meta II) is an important link to the subsequent phototransduction cascade. After a light flash, meta II is formed within 10 ms

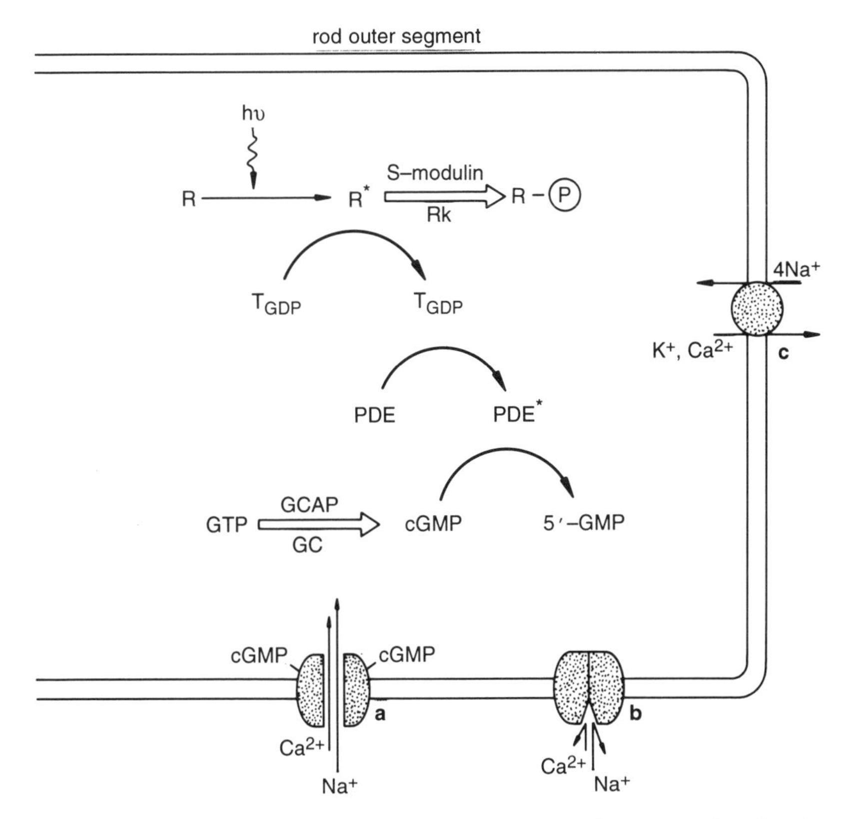

Figure 5.1 Schematic diagram of photoreceptor excitation and adaptation. For details, see text. R, inactive rhodopsin; R* activated rhodopsin; Rk, rhodopsin kinase; T_{GDP}, inactive transducin; T_{GTP}, active transducin; PDE, cGMP phosphodiesterase; PDE*, active cGMP phosphodiesterase; GC, guanylate cyclase; GCAP, guanylate cyclase activating protein. **a** cGMP-gated channel in the open state. **b** cGMP-gated channel in the closed state. **c** Na^+/Ca^{2+}, K^+ exchanger. Open arrow indicates the acceleration of the reaction at low $[Ca^{2+}]$, i.e. during light-adaptation.

which is comparable to the time range of the latency of a photoreceptor potential. It has been shown that meta II activates transducin (Fukada and Yoshizawa, 1981). However, meta II becomes phosphorylated (section 5.2.2) and this form of meta II does not have the activating capacity. Since phosphorylated and non-phosphorylated meta II cannot be distinguished spectrophotometrically, the term 'activated rhodopsin' (R*; Figure 5.1) is used to specify the non-phosphorylated, active form of meta II throughout this chapter.

In meta II, the Schiff base is unprotonated (Doukas *et al.*, 1978), which raises the possibility that deprotonation is a necessary step for the formation of R*. However, a rhodopsin mutant having an unprotonated Schiff base can activate transducin in a light-dependent manner (Fahmy and Sakmar, 1993b). Moreover, the mutation of either Lys-296 or Glu-113 results in a constitutive activation of transducin. From these results, deprotonation does not seem to be required for the R* formation. Instead, a salt bridge between Lys-296 and Glu-113 has been suggested to help to constrain opsin in an inactive conformation (Robinson *et al.*, 1992; Cohen *et al.*,

Table 5.1 Photoreceptor proteins involved in phototransduction and adaptation

	Molecular weight (kDa)	Molar ratio to rhodopsin	Concentration (μM) or density* (μm^{-2})	Reference
Rhodopsin	39	1000	4000	Harosi (1975), Hamm and Bownds (1986)
Transducin		130	520	Hamm and Bownds (1986)
α	40			
β	35			
γ	5			
Phosphodiesterase		5	20	Hamm and Bownds (1986)
α	95			
β	94			
γ	11			Hurley and Stryer (1982)
Guanylate cyclase	112	10	40	Koch (1991), Hayashi and Yamazaki (1991)
Rhodopsin kinase	65	3	12	Sitaramayya (1986)
Arrestin	48	30	120	Hamm and Bownds (1986)
Phosducin	33	6	24	Estimated from Lee *et al.* (1987)
S-modulin Recoverin	26	10	40	Kawamura and Murakami (1991), Dizhoor *et al.* (1991)
GCAP	20			Gorczyca *et al.* (1994)
Calmodulin	16	1	4	Kohnken *et al.* (1981), Nagao *et al.* (1987)
Component I	13	2	8	Polans *et al.* (1979)
II	12	2	8	Polans *et al.* (1979)
$Na^+/Ca^{2+}, K^+$ exchanger	220	1	4 450–1400*	Cook and Kaupp (1988), Nicoll and Applebury (1989), Reid *et al.* (1990)
cGMP-gated channel		2	8 400–3000*	Haynes *et al.* (1986), Zimmerman and Baylor (1986)
α	63			Cook *et al.* (1987)
β	71			Chen *et al.* (1993)

1993). Isomerization of the chromophore would induce breakage of this salt bridge and expose the sites that are necessary for activation of transducin. Site-directed mutagenesis studies showed that the second, third and fourth cytoplasmic loops of rhodopsin are required for this activation (König *et al.*, 1989; Franke *et al.*, 1992; Weitz and Nathans, 1992; Fahmy and Sakmar, 1993a). Highly conserved amino acid residues are found in these regions of other G-protein-coupled receptors, indicating that these loops may be involved in the activation of G-proteins generally.

(b) Signaling from rhodopsin to transducin

Transducin transmits the photoreception signal from rhodopsin to cGMP-phosphodiesterase. Transducin is a heteromeric trimer composed of alpha (T_α; 40 kDa), beta (T_β; 35 kDa) and gamma (T_γ; 5 kDa) subunits. The T_α subunit possesses the site for GTP binding. Activated rhodopsin (R*) catalyzes the exchange of GDP for GTP on the transducin molecule and, with this exchange, transducin converts into the active form (T_{GTP} in Figure 5.1; Fung and Stryer, 1980). The T_β and T_γ subunits are essential for the exchange of GDP for GTP on T_α (Fung, 1983). Transducin has intrinsic GTPase activity and hydrolyzes GTP to GDP, and for this reason, transducin binds GDP in the dark. The GDP-bound form of transducin (T_{GDP}) is inactive.

The photoreception signal generated by a rhodopsin molecule is amplified during activation of transducin. One R* molecule catalyzes the exchange of GDP for GTP in 500 (Fung and Stryer, 1980), and even tens of thousands (Gray-Keller *et al.*, 1990) of transducin molecules, which results in the formation of an enormous number of active transducin. This is the first step of the amplification of the photoreception signal.

The binding of T_α to R* follows a sigmoidal relationship with a Hill coefficient near 2, which means that the probability of a second T_α binding is increased by 40 times after the first T_α binding (Wessling-Resnick and Johnson, 1987). K_d values are about 50 nM and 2 nM for the first and second T_α bindings, respectively (Willardson *et al.*, 1993).

Recently, a 2.2 Å crystal structure of T_α was obtained (Noel *et al.*, 1993); T_α was found to consist of a domain structurally homologous to small GTPases (for review, see Boguski and McCormick, 1993) and a helical domain unique to heteromeric G-proteins. Since the GTP-binding site is buried deep in a cleft formed between the GTPase and the helical domains, R* should open the cleft for the exchange of GDP for GTP. T_α possesses a significantly higher level of basal GTPase activity than small GTPases, and the helical domain may serve as a GTPase activating protein (GAP) which is present in the small GTPase systems. The binding site to R* is near the carboxyl terminal (Fung and Nash, 1983; Hamm *et al.*, 1988; Dratz *et al.*, 1993; for review, see Pfister *et al.*, 1993).

T_α and T_β are post-translationally modified. T_γ is farnesylated and methylated (Fukada *et al.*, 1990; Lai *et al.*, 1990); T_α is heterogeneously fatty acylated (Kokame *et al.*, 1992; Neubert *et al.*, 1992). These modifications are essential for the function of transducin.

T_α is also known to be ADP-ribosylated by both cholera and pertussis toxins. The reactions affected by the toxins are known to be different from each other (Bornancin *et al.*, 1992). With modification by cholera toxin, GTP hydrolysis is inhibited so that PDE activation is prolonged. With pertussis toxin, interaction with R*, and, therefore, GDP/GTP exchange is inhibited so that activation of transducin is inhibited. In rod outer segments, endogenous ADP-ribosylase is present, and this enzyme is activated by nitric oxide (NO), a recently proposed messenger for adaptation in brain (Ehret-Hilberer *et al.*, 1992; Pozdnyakov *et al.*, 1993). The enzyme synthesizing NO, NO synthase, is also present in rod outer segments (Venturini *et al.*, 1991). The reaction affected by the endogenous ATP-ribosylase and its functional role are not known yet.

(c) Signaling from transducin to cGMP-phosphodiesterase

The GTP-bound form of transducin activates cGMP-phosphodiesterase (PDE; Figure 5.1). Bovine rod PDE is thought to be a heterotetrameric molecule consisting of one P_α (95 kDa), one P_β (94 kDa) and two P_γ (11 kDa) subunits. P_γ is a inhibitory subunit and P_α is a catalytic subunit. P_β has a similar amino acid sequence as P_α, and recent work using expressed P_α and P_β showed that both have catalytic activity (Piriev *et al.*, 1993). Each of the two P_γ subunits can bind to either P_α or P_β (Fung *et al.*, 1990; Clerc *et al.*, 1992).

The GTP-bound form of transducin forms a complex with P_γ to remove its inhibitory effect on $P_{\alpha\beta}$. The exact mechanism of this has been a subject of argument; one suggestion was that T_α bound with GTP carries P_γ to dissociate from the catalytic subunits; another was that the T_α/P_γ complex remains associated with $P_\alpha P_\beta$ but liberates the inhibitory constraint of P_γ. Recent studies suggest, however, that this may depend on the experimental conditions used (Clerc and Bennet, 1992; Catty *et al.*, 1992; Otto–Bruc *et al.*, 1993). At low protein concentrations, T_α/P_γ physically dissociates from $P_{\alpha\beta}$, but under high (physiological) concentrations of protein, the T_α/P_γ complex remains associated with membranes. These studies were done in bovine photoreceptors. In amphibians, however, cGMP determines the fate of P_γ (section 5.5.1).

As described above, photoreception is amplified during the transmission of signal from R* to transducin. However, the step from transducin to PDE, seems to involve one-to-one activation, the GTP-bound form of T_α binds to one P_γ and activates one P_α or one P_β. One activated PDE molecule hydrolyzes several hundred cGMP molecules per second (Sitaramayya *et al.*, 1986), and, therefore, one R* molecule leads to hydrolysis of more than 10^5 molecules of cGMP as a total amplification from rhodopsin to cGMP (Yee and Liebman, 1978).

It has been shown that the carboxyl terminal region of P_α (Swanson and Applebury, 1983; Ong *et al.*, 1989) and P_β (Catty and Deterre, 1991) are also modified by methylation and, possibly, lipidation. Furthermore, P_α and P_β appear to be differentially prenylated; P_α is modified with a farnesyl group and P_β with a geranylgeranyl group (Anant *et al.*, 1992). These modifications are thought to be important for correct positioning of the molecules in the disk membrane (Qin *et al.*, 1992).

The central region of P_γ contains many cations and interacts with T_α (Lipkin *et al.*, 1988; Morrison *et al.*, 1989; Brown, 1992; Artemyev *et al.*, 1993) and $P_{\alpha\beta}$ (Lipkin *et al.*, 1988; Artemyev and Hamm, 1992). The site on P_γ, essential for inhibition of P_α or P_β, is at the carboxyl terminal (Lipkin *et al.*, 1988; Brown and Stryer, 1989; Artemyev and Hamm, 1992) and elimination of the last five residues totally abolishes its inhibitory effect (Brown, 1992). The corresponding site on T_α for binding to P_γ is near the carboxyl terminal, next to the R* binding site (Rarick *et al.*, 1992; Artemyev *et al.*, 1992); the site on $P_{\alpha\beta}$ for binding to P_γ is near the amino acid terminal of P_α or P_β (Oppert *et al.*, 1991; Oppert and Takemoto, 1991).

(d) cGMP-gated channel

The cGMP-gated channel was characterized by Fesenko *et al.* (1985) after a long dispute about the identity of the intracellular messenger in phototransduction. This channel is a cation channel situated exclusively in the photoreceptor plasma membrane (Bauer, 1988; Cook *et al.*, 1989; Molday *et al.*, 1991). Most cations can permeate the channel (Hodgkin *et al.*, 1985); the permeability ratios for Ca^{2+} and Mg^{2+} to Na^+ are 12.5 and 2.5, respectively (Nakatani and Yau, 1988b; for review, see Yau and Baylor, 1989). The gating of the channel is co-operatively regulated by cGMP and the Hill coefficient is 2–3. Half saturation is observed at 10–50 μM cGMP. In the presence of divalent cations, this channel has a low conductance (~ 100 fS), whereas in their absence, this rises to around 25 pS (Haynes *et al.*, 1986; Zimmerman and Baylor, 1986). The proportion of the channels in their open state is only 1–2% under *in situ* conditions (Nakatani and Yau, 1988c), suggesting that the intracellular cGMP concentration in intact cells is several micromolar.

The phototransduction channel protein (63 kDa) has been purified (Cook *et al.*, 1987); and the corresponding DNA cloned and functionally expressed (Kaupp *et al.*, 1989).

However, the expressed channel differed from the native channel in some aspects. The purified channel did not show sensitivity to l-*cis* diltiazem which blocks the cGMP-gated channel in intact cells; also the channel open times were long and stable compared to the flickering activity seen in intact cells. Chen *et al.* (1993) described a new subunit of the cGMP-gated channel, called hRCN2a (71 kDa), which by itself did not form functional channels. However, when hRCN2a was co-expressed with the previously obtained 63 KDa protein, the resulting channel activity much more resembled the properties of the native channel. These results suggest that the cGMP-gated channel is a hetero-oligomer. Recently it has been shown that the cGMP-gated channel is a member of voltage-gated channels and the highest conservation in the amino acid sequence is observed in the putative pore-forming region (Guy *et al.*, 1991; Goulding *et al.*, 1993).

5.2.2 TURN-OFF MECHANISMS

(a) Phototransduction turn-off

All of the activated components in the phototransduction cascade are inactivated after termination of the light stimulus. R* is inactivated by phosphorylation of as many as seven amino acid residues in the carboxyl terminal by rhodopsin kinase (Figure 5.1). When the phosphorylation site is removed by proteolysis, inactivation of the phototransduction cascade becomes slow (Sitaramayya and Liebman, 1983a; Miller and Dratz, 1984; Palczewski *et al.*, 1991). Only R* is the substrate of rhodopsin kinase (Weller *et al.*, 1975; McDowell and Kühn, 1977), although phosphorylation of unbleached rhodopsin has also been reported (Binder *et al.*, 1990; Dean and Akhtar 1993). Overall, it is thought that the phosphorylation site of rhodopsin is exposed to the cytoplasmic surface after bleaching so that the kinase can access the phosphorylation site. However, exposure of the site is not sufficient because the carboxyl terminal peptide including the phosphorylation site is phosphorylated only in the presence of R* (Fowles *et al.*, 1988; Palczewski *et al.*, 1991). Cleavage of the V–VI cytoplasmic loop of rhodopsin reduces the phosphorylation. This loop seems to be the binding site of the kinase and responsible for the activation of this enzyme. After weak light stimulation ($\sim 10^{-5}$ fractional bleach), the phosphorylation is complete in less than 2 s, a physiologically significant time range (Sitaramayya and Liebman, 1983b).

The amino acids involved in the phosphorylation of rhodopsin have been identified. In bovine rhodopsin, the major initial phosphorylation site is Ser-339; subsequent phosphorylation takes place either at Ser-343 or Thr-336 (Ohguro *et al.*, 1993; McDowell *et al.*, 1993; Papac *et al.*, 1993). It has been suggested that the rate of incorporation of the first phosphate is slow but the initial phosphorylation facilitates the subsequent incorporation of phosphates (Adamus *et al.*, 1993a). However, the reason(s) for multiple phosphorylation of rhodopsin is not known yet.

Rhodopsin kinase (65 kDa) is autophosphorylated (Lee *et al.*, 1982; Kelleher and Johnson, 1990), and this seems to be the mechanism of dissociation of this enzyme from R* (Buczylko *et al.*, 1991). The major autophosphorylation sites are Ser-488 and Thr-489 (Palczewski *et al.*, 1992b). The kinase is also farnesylated at its carboxyl terminal (Inglese *et al.*, 1992a; Anant and Fung, 1992) and this is essential for the activity and the membrane-binding of this enzyme (Inglese *et al.*, 1992b).

Arrestin (48 kDa), also called 'S-antigen' or 48K protein, is another factor involved in inactivating R*. This protein binds to phosphorylated R* and thereby prevents the binding of T_α to R* (Wilden *et al.*, 1986). For the binding of arrestin, both rhodopsin activation and phosphorylation of rhodopsin are necessary. Binding to unbleached/phosphorylated rhodopsin, or bleached/unphosphorylated rhodopsin, is weak (Gurevich and Benovic, 1993). Arrestin has been suggested to move from outer to inner segment during dark-adaptation, and from inner to outer segment during light-adaptation; it has also been suggested to be a major Ca^{2+}-binding protein and to have ATPase activity. However, these characteristics have only tentatively been established (Palczewski and Hargrave, 1991; Nir and Ranson, 1993).

The role of arrestin was investigated electrophysiologically in internally dialyzed, detached rod outer segments. Phytic acid, an inhibitor of arrestin binding to phosphorylated R*, inhibited and prolonged the recovery of a photoresponse, which indicated that arrestin inactivates the phototransduction cascade and facilitates the recovery process of the photoresponse (Palczewski *et al.*, 1992a). Arrestin binds to bleached rhodopsin as long as all-*trans* retinal is associated with opsin (see below).

R* (meta II) decays into opsin and all-*trans* retinal within a time range of minutes. Rhodopsin regenerates from opsin by re-association with 11-*cis* retinal. During these processes, arrestin and the phosphate groups are removed from the protein moiety. Arrestin dissociates from the phosphorylated photoproduct of rhodopsin once all-*trans* retinal is reduced to all-*trans* retinol by retinol dehydrogenase (Hofmann *et al.*, 1992). The phosphates are removed from the protein moiety by a type-2A phosphatase (Palczewski *et al.*, 1989; Fowles *et al.*, 1989).

Inactivation of transducin is attained by its intrinsic GTPase activity. Earlier biochemical work done under fully bleached conditions (Wheeler and Bitensky, 1977) showed that GTP hydrolysis is complete within a time range of minutes. According to the present scheme of the phototransduction mechanisms, PDE activation persists as long as GTP stays on T_α. Therefore, GTP hydrolysis occurring over minutes is too slow to explain the physiological time course of a photoresponse terminating in seconds. Recent work has shown, however, that under physiological conditions, GTP hydrolysis is rapid. Using frog material, Arshavsky and Bownds (1992) showed that GTP hydrolysis accelerated as the concentration of P_γ was increased. This indicated that P_γ acts like a GTPase-activating protein (GAP) which was originally found in a small GTP-binding protein system (for review, see Boguski and Cormick, 1993). Using microcalorimetry, Vuong and Chabre (1991) also detected rapid GTP hydrolysis in a more intact situation. In one single rod, P_γ is present at a concentration of about 10% of transducin (Table 5.1). Therefore, activated transducin exceeding this ratio cannot bind to P_γ and thus cannot hydrolyze GTP at a fast rate, and this explains the slow GTP hydrolysis observed biochemically under fully bleached conditions. The same explanation can be applied to the observation that PDE inactivation becomes fast as the intensity of the stimulus decreases (Kawamura, 1983).

PDE is activated by removal of its inhibitory subunit P_γ from the catalytic subunits $P_{\alpha\beta}$ (section 5.2.1); therefore, inactivation of PDE requires reassociation of P_γ with $P_{\alpha\beta}$. If the T_α/P_γ complex physically dissociates from $P_{\alpha\beta}$, it will not be easy for P_γ liberated from T_α to reassociate with $P_{\alpha\beta}$. The mechanism of this reassociation is not yet known.

According to the widely accepted scheme, PDE inactivation becomes possible after hydrolysis of GTP on T_α. However, Erickson *et al.* (1992) observed PDE inactivation in the presence of GTP–γS, a non-hydrolyzable GTP analog, and proposed a mechanism which does not require GTP hydrolysis for termination of PDE activation. One possibility is that free P_γ acts as an inhibitor of $P_{\alpha\beta}$ independently of GTP hydrolysis.

(b) Recovery of cGMP concentration

After termination of a light stimulus, the phototransduction cascade turns-off and cGMP concentration recovers to the original dark level with supplement of cGMP synthesis by guanylate cyclase. Guanylate cyclases are classified into two groups: one is soluble or cytosolic and the other is membrane-associated. Only the membrane-associated form is found in photoreceptors (Troyer *et al.*, 1978; Yoshikawa *et al.*, 1982), where it is present in the outer segment (Kawamura and Murakami, 1989a). However, sodium nitroprusside, an NO donor, altered dark voltage and photoresponses of isolated frog rods. Since NO is known to activate soluble guanylate cyclase, the result suggests also the presence of this form of the enzyme (Schmidt *et al.*, 1992).

The membrane-associated guanylate cyclase in rods has been purified (Koch, 1991; Hayashi and Yamazaki, 1991) and the corresponding cDNA cloned (Shyjan *et al.*, 1992). In frogs, the protein is a doublet of 110 and 115 kDa polypeptides, whereas the bovine enzyme is a single 110–112 kDa polypeptide. Both frog and bovine cyclases are separated into 4–5 variants on the basis of isoelectric focusing (Hayashi and Yamazaki, 1991). The significance of these variations is not known.

5.3 MECHANISM OF BACKGROUND ADAPTATION

Under dark-adapted conditions, rod photoreceptors can respond to a range of light intensities over 3 log units. The relation between light intensity and rod photoresponse can roughly be expressed with a Michaelis–Menten equation. When a photoreceptor is light-adapted, the working range shifts to a higher light intensity by approximately 3 log units and the cell can still respond to light over the range of 3 log units. Thus, with light-adaptation, rods can cover light intensities over a range of $\sim$ 6 log units. Enzyme reactions usually saturate with substrate concentrations over a range of only 3 log units. Recent work has shown that the light-induced decrease in cytoplasmic $[Ca^{2+}]_i$ in the outer segment has an important role for widening the range of the light intensities covered. As shown below, the $[Ca^{2+}]$ decrease is the cause of background adaptation which is fully reversible. Due to the $[Ca^{2+}]_i$ decrease, light-sensitivity of PDE is reduced, guanylate cyclase is activated and sensitivity of the cGMP-gated channel to cGMP is increased. The molecular mechanisms of these processes are reviewed in the following sections.

5.3.1 $[Ca^{2+}]_i$ DECREASE IN THE LIGHT

Sillman *et al.* (1969) originally found that Na^+ is the major charge carrier of the steady current that flows into the outer segment in the dark. Subsequently, Ca^{2+} was also found to permeate the channel (for a review, see Yau and Baylor, 1989). The entering Ca^{2+} is pumped out by a Na^+/Ca^{2+}, K^+ exchanger situated in the plasma membrane of the outer segment (Figure 5.1, c). The stoichiometry of the exchange is such that the efflux of $1Ca^{2+}$ requires the efflux of $4Na^+$ and the efflux of $1K^+$ (Figure 5.1) (Cervetto *et al.*, 1989; Schnetkamp *et al.*, 1989).

The exchanger operates continuously, irrespectively of light/dark conditions. The light-induced decrease in the rod cytoplasmic $[Ca^{2+}]$ could be explained in the following way. In the dark, the cGMP-gated channel is open and Ca^{2+} enters the cell (Figure 5.1, a). The cytoplasmic $[Ca^{2+}]_i$ is maintained at the level where the entry and extrusion of Ca^{2+} are balanced. When the channel is closed in the light, Ca^{2+} entry is blocked (Figure 5.1, b) whereas the extrusion continues (Figure 5.1, c). As a result, cytoplasmic $[Ca^{2+}]_i$ decreases in the light (Yau and Nakatani, 1985a). According to measurements with fura-2 or quin-2, $[Ca^{2+}]_i$ in the dark is less than 220–270 μM (Ratto *et al.*, 1988; Korenbrot and Miller, 1989) and decreases in the light to nearly 140 nM (Ratto *et al.*, 1988). However, the Ca^{2+} decrease may be more complex because of the presence of Ca^{2+} buffers in the outer segments; the decrease in $[Ca^{2+}]_i$ may be more rapid in the areas adjacent to the plasma membranes, compared with the center of the rod outer segment (McCarthy *et al.*, 1993).

The Na^+/Ca^{2+}, K^+ exchanger protein has been purified (Cook and Kaupp, 1988; Nicoll and Applebury, 1989; Reid *et al.*, 1990); and localized exclusively to the plasma membrane and not in the disk membrane (Haase *et al.*, 1990; Reid *et al.*, 1990). Recently, the cDNA for the exchanger has been obtained from the bovine retina (Reiländer *et al.*, 1992). According to the deduced amino acid sequence, the calculated molecular weight was ~130 kDa; expression of the cDNA produced functional proteins.

A single exchanger molecule can exchange 30 Ca^{2+}/s (Cook and Kaupp, 1988). Because of this high exchange rate, $[Ca^{2+}]_i$ decreases with a time constant of about 0.5 s (McNaughton *et al.*, 1986; Nakatani and Yau, 1988b). The exact mechanism of ion transport by the exchanger is not known. However, Perry and McNaughton (1993) suggested a consecutive model in which unbinding of Na^+ on either the inner or the outer membrane surface is followed by binding of intracellular Ca^{2+} and then K^+.

5.3.2 ROLE OF $[Ca^{2+}]_i$ DECREASE IN BACKGROUND ADAPTATION

The role of the $[Ca^{2+}]_i$ decrease in background adaptation was examined electrophysiologically by suppressing the $[Ca^{2+}]_i$ decrease (Matthews *et al.*, 1988; Nakatani and Yau, 1988a; Fain *et al.*, 1989). In these studies, the inner segment of lower vertebrates was sucked into a macropipet electrode and the outer segment was exposed first to Ringer solution and then to a test solution in which Na^+ was replaced with lithium or guanidium. These ions permeate the cGMP-gated channel and therefore maintained the steady inward current in the dark. However, since there was no external Na^+, cytoplasmic Ca^{2+} could not be extruded by the exchanger. Moreover, in this test solution, external $[Ca^{2+}]$ was kept low so that the Ca^{2+} entry was minimal. The outer segment was exposed to the test solution just before a light stimulus. Since there was no entry or extrusion of Ca^{2+}, cytoplasmic $[Ca^{2+}]_i$ would be kept at the high value of the dark even in the light. With the $[Ca^{2+}]_i$ decrease suppressed, the response elicited by a steady light became larger and peaked later, and the recovery was much slower than normal. The result, therefore, indicated that both low light-sensitivity and facilitated photoresponse time course during background adaptation are due to the decrease in the cytoplasmic $[Ca^{2+}]_i$. A similar result was obtained in mammalian rods (Matthews, 1991; Nakatani *et al.*, 1991; Tamura *et al.*, 1991).

Another noticeable point of the above experiments was that the intensity–response curve obtained in the absence of the $[Ca^{2+}]_i$ decrease deviated from the Michaelis–Menten relation (seen with normal Ringer solution) and instead followed an exponential function, i.e. a steeper relation than the Michaelis–Menten equation. Since the probability of one photon falling on a disk membrane is an exponential function of light intensity, the above finding would indicate that the Michaelis–Menten-type relation is the result of the light-induced $[Ca^{2+}]_i$ decrease. This idea is also supported by the finding that the amplitude of the photoresponse is an exponential function of the light intensity at the very beginning of the light stimulus, presumably due to the negligible decrease in $[Ca^{2+}]_i$ (Lamb *et al.*, 1981).

The above result suggested that the $[Ca^{2+}]_i$ decrease leads to both the decrease in the efficiency of the light-induced cGMP hydrolysis and the facilitation of the recovery of cGMP level. There are at least two possible sites where $[Ca^{2+}]_i$ could have a direct or indirect effect on (a) PDE and (b) guanylate cyclase. Previous biochemical work suggested that both enzymes are, in fact, regulated by $[Ca^{2+}]_i$. Light-induced activation of PDE was reported to be affected by Ca^{2+} (Robinson *et al.*, 1980; Kawamura and Bownds, 1981; Del Priore and Lewis, 1983; Barkdoll *et al.*, 1989). PDE activation was lower and the recovery of its activity faster in low $[Ca^{2+}]_i$ (Kawamura and Bownds, 1981). On the other hand, guanylate cyclase activity was found to be higher at low $[Ca^{2+}]$ (Krishnan *et al.*, 1978; Troyer *et al.*, 1978; Fleischman and Denisevich, 1979). The Ca^{2+}-dependent regulation of these enzymes could explain both the suppression of cGMP concentration decrease and its accelerated recovery under light-adapted conditions where $[Ca^{2+}]_i$ is low. However, in these studies, the Ca^{2+} effects were studied at millimolar levels (but see sections 5.3.3 and 5.3.4).

Previous electrophysiological work also suggested Ca^{2+}-dependent regulation of guanylate cyclase and/or PDE. Intracellular injection of a Ca^{2+} chelator, BAPTA, into isolated rods slowed the recovery of photoresponse (Torre *et al.*, 1986). In the presence of BAPTA, $[Ca^{2+}]_i$ would be maintained at a high, 'dark' level, and therefore, the effect would be due to the blockage of the $[Ca^{2+}]_i$ decrease. This Ca^{2+} effect was attributed to the prolonged activation of PDE. Hodgkin and Nunn (1988) and Cobbs (1991) found little effect of Ca^{2+} on the kinetics of PDE activation. Based on these findings, it was suggested that guanylate cyclase is activated during background adaptation, and hence at low $[Ca^{2+}]_i$.

5.3.3 REGULATION OF GUANYLATE CYCLASE

Direct biochemical evidence that guanylate cyclase activity is regulated at physiological levels of $[Ca^{2+}]_i$ was demonstrated by Koch and Stryer (1988). They measured the cyclase activity at different $[Ca^{2+}]_i$ and found that 50–200 nM Ca^{2+} inhibits the cyclase with half-maximum inhibition occurring at around 100 nM. The Ca^{2+}-dependent regulation on the cyclase required a soluble protein(s) since the Ca^{2+} effect disappeared, reversibly, by washing the disk membranes with a low ionic strength-buffer solution (also see below).

Using a preparation of 'truncated rod outer segments' (Yau and Nakatani, 1985b) or more informatively 'inside-out rod outer segments' (I/O ROS), we also showed electrophysiologically that the cyclase is activated at low/physiological levels of $[Ca^{2+}]_i$ (Kawamura and Murakami, 1989a). In this preparation, an open-ended rod outer segment was held in an electrode measuring the membrane current and, the cytoplasmic side perfused from the open end with a solution containing known chemicals. The I/O ROS preparation retains both disk membranes and plasma membranes, and therefore, contains most of the components necessary for phototransduction. Application of cGMP to I/O ROS induced a membrane current which was suppressed by light, presumably due to hydrolysis of cGMP caused by light-activated PDE (Yau and Nakatani, 1985b). Application of GTP instead of cGMP, generated a light-suppressible current which was indistinguishable from the cGMP-induced current and indicated that cGMP was synthesized from GTP by guanylate cyclase in I/O ROS. The light-induced current amplitude (i.e. cGMP concentration) increased when $[Ca^{2+}]_i$ was lowered. The apparent increase in the cGMP concentration was mainly due to the activation of guanylate cyclase and partly because of the decrease in the PDE activity. The Ca^{2+} effect was minimum at 1 μM $[Ca^{2+}]_i$ and saturated at 10 nM (with half maximum at around 100 nM), in agreement with the biochemical study done by Koch and Stryer (1988). The Ca^{2+} effect was observed in less than one second after the reduction of $[Ca^{2+}]_i$ (Kawamura and Murakami, 1989b).

(a) Recoverin and GCAP

As mentioned above, regulation of guanylate cyclase by Ca^{2+} requires soluble protein(s). Using bovine retina, Dizhoor *et al.* (1991) and Lambrecht and Koch (1991a) reported the isolation of a 26 kDa soluble protein, named 'recoverin', that activated the cyclase at low $[Ca^{2+}]_i$. Slightly earlier and using frog, we found a soluble 26 kDa protein, named 'sensitivity-modulating protein' (S-modulin) that increased the light-sensitivity of PDE at high $[Ca^{2+}]_i$ (section 5.3.4). Because of the similarities in molecular weight and Ca^{2+}-dependency, it was suggested that S-modulin and recoverin are the same protein having two functions, one on the cyclase at low $[Ca^{2+}]_i$ and the other on PDE through the phototransduction cascade at high $[Ca^{2+}]_i$ (Stryer, 1991). However, other studies showed that recoverin acts on the phototransduction cascade (Kawamura *et al.*, 1993) but not on the cyclase (Gorodovikova and Philippov, 1993; Gray-Keller *et al.*, 1993; Hurley *et al.*, 1993). Recently, it has been shown that the cyclase is activated at low $[Ca^{2+}]$ by a 20 kDa Ca^{2+}-binding protein named 'guanylate cyclase activating protein' (GCAP). Intracellular application of this protein to isolated rod outer segments induced facilitation of photoresponse recovery, which was expected from activation of the cyclase by GCAP at low $[Ca^{2+}]_i$ (Gorczyca *et al.*, 1994).

5.3.4 REGULATION OF PHOSPHODIESTERASE

As described above (section 5.3.2), previous results suggested that PDE is activated at high, but rather non-physiological levels of $[Ca^{2+}]_i$ (Robinson *et al.*, 1980; Kawamura and Bownds, 1981; Del Priore and Lewis, 1983; Barkdoll *et al.*, 1989). However, our recent studies showed that physiological concentrations of Ca^{2+} also affect both PDE activation and inactivation via S-modulin (Kawamura and Murakami, 1991; Kawamura, 1993b).

(a) S-modulin

Using the frog I/O ROS preparation, we applied cGMP at different levels of $[Ca^{2+}]_i$ and measured the light-suppressible current. The peak amplitude of the photoresponse became smaller and the recovery faster when $[Ca^{2+}]_i$ was lowered from 1 μM to 30 nM. Since the GTP concentration was kept low, the contribution of the cyclase could be ignored and the observed effects of Ca^{2+} were attributed to PDE (Kawamura and Murakami, 1991). However, the Ca^{2+} effect was lost irreversibly when the inside of the I/O ROS was perfused with a low-$[Ca^{2+}]_i$ solution, which raised the possibility that a protein responsible for the Ca^{2+}-dependent regulation on PDE was washed out. A 26 kDa protein bound to disk membranes was found to fit the criteria; as it increased the light-sensitivity of PDE at high $[Ca^{2+}]_i$ it was called 'sensitivity-modulating' protein (S-modulin) (Kawamura and Murakami, 1991; Kawamura, 1993b). Since the maximum activity of PDE elicited by a saturating light was not affected, S-modulin appeared to modulate the efficiency of the signaling at a certain stage(s) of the phototransduction cascade. Lagnado and Baylor (1994) confirmed that PDE activation is regulated by both $[Ca^{2+}]_i$ and a diffusible substance(s).

Kawamura (1993b) showed unequivocally

that S-modulin inhibits rhodopsin phosphorylation at high $[Ca^{2+}]_i$ (Figure 5.1). The effect was observed at 30 nM–1 μM $[Ca^{2+}]_i$ with a half-maximum effect at around 100 nM. Since rhodopsin phosphorylation is the mechanism for turning off R* (section 5.2.2), the inhibition of this reaction would result in an increase in the lifetime of R* so that the light-sensitivity of PDE would increase and the time course of PDE inactivation slowed down. Hermolin *et al.* (1982) had observed the inhibition of rhodopsin phosphorylation by Ca^{2+} but the involvement of soluble proteins like S-modulin was not reported. Wagner *et al.* (1989) also suggested the control of rhodopsin phosphorylation by Ca^{2+} by measuring the change of light scattering.

The structure of S-modulin is characteristic of the calmodulin family proteins, a group of well-known Ca^{2+}-binding proteins (Kawamura *et al.*, 1993). Ca^{2+}/calmodulin complex in other tissues is known to activate protein kinases, whereas Ca^{2+}/S-modulin inhibits the kinase reaction. In this regard, the rhodopsin/rhodopsin kinase/S-modulin system is unique.

Though the ultimate effect of reaction of S-modulin has been identified as rhodopsin phosphorylation, the target molecule(s) is not known yet. Both rhodopsin kinase and rhodopsin are candidates (Kawamura and Murakami, 1991). In addition to the target protein, the Ca^{2+}/S-modulin complex has been shown also to bind to disk membranes at high $[Ca^{2+}]_i$ (Kawamura and Murakami, 1991; Kawamura, 1992).

Recoverin, a bovine homolog of S-modulin, has been crystallized and subject to detailed structural studies (Ray *et al.*, 1992; Flaherty *et al.*, 1993). A major conformational change was found to occur following binding of Ca^{2+} (Kataoka *et al.*, 1993). Recoverin has been shown to be acylated near its amino terminal (Dizhoor *et al.*, 1992); and this was found to be necessary for the membrane binding of the Ca^{2+}/recoverin complex (Zozulya and Stryer, 1992; Dizhoor *et al.*, 1993). However, its role in inhibiting rhodopsin phosphorylation is not known. As shown above, the inhibition of rhodopsin phosphorylation is observed at <1 μM $[Ca^{2+}]_i$ (Kawamura, 1993b), whereas the membrane binding is observed at >1 μM $[Ca^{2+}]_i$ (Kawamura *et al.*, 1992). In terms of $[Ca^{2+}]_i$, therefore, there is no simple correlation between the binding of S-modulin to disk membranes and its inhibitory effect on the rhodopsin phosphorylation.

Recoverin has been shown to be phosphorylated (Lambrecht and Koch, 1991b). S-modulin and recombinant recoverin have been suggested to act directly on both PDE and the cGMP-gated channel (Nikonov *et al.*, 1993). However, details of those mechanisms are not known.

It has been reported that recoverin is the antigen of an autoimmune disease, cancer-associated retinopathy (Polans *et al.*, 1991; Adamus *et al.*, 1993b) (Chapter 19). In patients of this disease, photoreceptors degenerated due to the presence of an autoantibody to recoverin produced in cancer cells outside the retina. This finding raised the possibility that recoverin-like proteins and similar regulatory mechanism(s) are also present in other tissues. In fact, several S-modulin-like proteins have been found in brain (Lenz *et al.*, 1992; Okazaki *et al.*, 1992; Kobayashi *et al.*, 1992; Kajimoto *et al.*, 1993).

5.3.5 REGULATION OF THE cGMP-GATED CHANNEL

It has been known for some time that calmodulin is present in rods (Kohnken *et al.*, 1981; Nagao *et al.*, 1987). In other cells, Ca^{2+}/calmodulin complexes activates phosphodiesterases or kinases. However, these effects are not found in rods. Instead, recent work has indicated that Ca^{2+}/calmodulin binds to the cGMP-gated channel via a 240 kDa protein and decreases the sensitivity of the channel to cGMP at physiological levels of

$[Ca^{2+}]_i$ (50–300 nM) (Hsu and Molday, 1993). During background adaptation, both $[Ca^{2+}]_i$ and cGMP concentrations are reduced; however, since the cGMP-gated channel is sensitized by low $[Ca^{2+}]_i$, it can still respond to small light-induced changes in the cGMP concentration.

5.3.6 SUMMARY OF THE MECHANISM OF BACKGROUND ADAPTATION

The mechanism of background adaptation can be summarized as follows (Figure 5.1). When a photoreceptor absorbs light, cGMP concentration decreases resulting in a closure of the cGMP-gated channel. Consequently, the cytoplasmic $[Ca^{2+}]_i$ decreases, which leads to (a) desensitization of PDE mediated by S-modulin, (b) activation of guanylate cyclase mediated by GCAP, and (c) sensitization of the cGMP-gated channel mediated by calmodulin.

When a continuous light is presented to rods, the photoresponse may initially be saturated because of massive hydrolysis of cGMP. When $[Ca^{2+}]_i$ decreases in the light, however, the cGMP level recovers to some extent owing to the desensitization of PDE and activation of the cyclase. The recovery of the cGMP level may not be perfect, due to continental presence of light. However, the cGMP-gated channel may re-open to a similar extent as in the dark due to the sensitization to cGMP. As a result, the inward current could recover close to the dark level.

When a light flash is superimposed on the continuous background desensitization the incremental decrease in cGMP concentration should be smaller than that normally caused by the same light flash given due to desensitization of the PDE. This decreased flash effect on the decrease in the cGMP concentration is counteracted by the sensitization of the channel to cGMP. Even though the relative contribution of the PDE desensitization and the channel sensitization is not known, the observed behavior of photoreceptors suggests that the PDE desensitization overcomes the channel sensitization. As a result, the effect of the light flash on channel closure is smaller in the light than in the dark. In this way, photoreceptors are desensitized.

Thus, background adaptation appears to be entirely dependent on the $[Ca^{2+}]_i$ decrease. Unfortunately, local volumes of $[Ca^{2+}]_i$ near the plasma membrane (for the channel) or the disk membrane (for PDE and cyclase) are not known. Furthermore, our knowledge about time-dependent changes in $[Ca^{2+}]_i$ is very limited. For a quantitative understanding, we need to know spatial and temporal information about $[Ca^{2+}]_i$ changes in rods.

5.4 MECHANISM OF BLEACHING ADAPTATION

When the light stimulus is very bright and a substantial portion of (~ 10%) rhodopsin is bleached, the cells are irreversibly desensitized; even in the dark, the desensitization persists for a period at least some hours. The level of desensitization exceeds that expected from the loss of rhodopsin alone. If the components of the phototransduction cascade are all inactivated, the sensitivity must return to the level which is determined by the loss of rhodopsin. However, meta II, transducin and PDE have been shown to be inactivated over a time scale of seconds to minutes. Therefore, some mechanism(s) for this irreversible desensitization must be present. One can predict that a diffusible substance (cf. Ca^{2+}) is not required for bleaching adaptation. This conclusion was drawn from an experiment where a rod was illuminated transversely with a slit of light. The space constant (the distance where the effect reduces to e^{-1} along the long axis of the rod) for bleaching adaptation was 2.5 μm (Cornwall *et al.*, 1990); this is significantly shorter than that for background adaptation (6 μm; Lamb *et al.*, 1981). The bleaching product(s) of rhodopsin must be closely related to bleaching adaptation, since restoration of rhodopsin conformation leads immediately to the recovery of the photoreceptor light-sensitivity (section 5.6).

5.5 OTHER PROPOSED REGULATORY MECHANISMS FOR LIGHT ADAPTATION

Some other possible mechanisms in addition to Ca^{2+} have been proposed for photoreceptor light adaptation, as follows. However, the significance and plausibility of these mechanisms are yet to be determined fully.

5.5.1 REGULATION OF GTPase ACTIVITY BY cGMP

It is known that $P_{\alpha\beta}$ has non-catalytic cGMP-binding sites (Yamazaki *et al.*, 1980; Gillespie and Beavo, 1989; Cote and Brunnock, 1993). Using frog material, Arshavsky *et al.* (1991) and Arshavsky and Bownds (1992) showed that cGMP inhibits, and P_γ increases GTPase activity of transducin. It was also shown that even though P_γ dissociates from disk membranes in the presence of GTP during light adaptation, it remains associated with the membranes if cGMP is present (Arshavsky *et al.*, 1992). Based on these findings, the following scheme was proposed. First, in the dark, cGMP binds to the non-catalytic cGMP binding site. Second, the formation of the T_α/P_γ complex in the light removes inhibition of the catalytic sites of PDE to result in activation of PDE. However, assuming that the cGMP concentration is not too low, cGMP binds to the non-catalytic site and the T_α/P_γ complex still remains bound to $P_{\alpha\beta}$; consequently, GAP activity of P_γ is reduced. Third, when light stimulation continues and the cGMP concentration decreases to a very low level, cGMP dissociates from the non-catalytic site. As a result, T_α/P_γ dissociates from $P_{\alpha\beta}$ physically and GAP activity is fully restored to result in rapid GTP hydrolysis.

According to this hypothesis, during light-adaptation when the cGMP concentration could be very low, GTP hydrolysis would be rapid if T_α is formed by superimposed stimulus. As a result, the efficiency of PDE activation is low and PDE activity recovers rapidly to the pre-flash level. This could explain the low light-sensitivity but facilitated time course of the photoresponse observed during light-adaptation. However, reversible binding of cGMP to the non-catalytic site has been observed so far only in amphibians. In bovine PDE, cGMP binds firmly to the non-catalytic binding sites and practically does not dissociate under *in vitro* conditions (Gillespie and Beavo, 1989). In addition, GAP activity was found in bovine PDE but not in P_γ (Pages *et al.*, 1992; Antonny *et al.*, 1993; Angleson and Wensel, 1993).

5.5.2 REGULATION OF TRANSDUCIN BY PHOSDUCIN (MEKA PROTEIN)

Phosducin (Lee *et al.*, 1987), or MEKA protein (Kuo *et al.*, 1989), is a soluble 33 kDa phosphoprotein. It binds tightly to $T_{\beta\gamma}$, and thereby competes with T_α for $T_{\beta\gamma}$. It is phosphorylated by protein kinase A and dephosphorylated by type-2A phosphatase. In the dark, because of the high level of cGMP present, phosducin is phosphorylated, but in the light, it is rapidly dephosphorylated. The dephosphorylated form seems to bind to $T_{\beta\gamma}$ more tightly than the phosphorylated form. At the first step of activation of transducin, the GTP-bound form of T_α dissociates from $T_{\beta\gamma}$. The reassociation of T_α with $T_{\beta\gamma}$ could be inhibited by dephosphorylated phosducin (Lee *et al.*, 1992). With this mechanism, the efficiency of light activation of T_α is expected to decrease, which could explain the low light-sensitivity of photoreceptors during light adaptation.

5.5.3 PHOSPHORYLATION OF OTHER PHOTORECEPTOR PROTEINS

Rhodopsin has been shown to be phosphorylated not only by rhodopsin kinase but also by protein kinase C (PKC) (Kelleher and Johnson, 1986). Immunoreactivity against PKC has been detected in the rod outer segments of the rat (Wood *et al.*, 1988); the enzyme was also purified from bovine rod outer segments (Kelleher and Johnson, 1986). Pre-treatment with a phorbol ester, an activator of PKC, increased the light-dependent phosphorylation of rhodopsin. This kinase phosphorylates both bleached and unbleached rhodopsin after a light stimulus whereas it does not phosphorylate them in the dark (Newton and Williams, 1991). Whether the site(s) phosphorylated by PKC is the same as for rhodopsin kinase is yet to be examined (Newton and Williams, 1993). In addition to rhodopsin, a recent report showed that P_α and P_γ are also phosphorylated by PKC (Udovichenko *et al.*, 1993).

In frog rod outer segments, two membrane-associated, 12–13 kDa proteins (components I and II) are phosphorylated in the dark in the presence of cGMP (Polans *et al.*, 1979). These proteins form a complex with $T_{\beta\gamma}$ (Hamm, 1990) as in the case of phosducin (see above). It is interesting that components I and II cannot be found in mammalian retinas whereas phosducin cannot be found in frogs. An intriguing possibility is that components I and II and phosducin are functional homologs in amphibians and mammalians (Hamm, 1990). In addition to $T_{\beta\gamma}$, components I and II have been reported to form a complex with arrestin (Krapivinsky *et al.*, 1992).

5.5.4 LIGHT-INDUCED ACTIVATION OF GUANYLATE CYCLASE

Using the I/O ROS preparation to observe guanylate cyclase activity, we created a set of experimental conditions whereby $[Ca^{2+}]_i$ would increase as a result of a light flash (Kawamura and Murakami, 1989a). In the presence of GTP, a train of light flashes induced light adaptation determined by the amplitudes of the photoresponses becoming smaller while the time course was accelerated. The effect was reversible since the response amplitude recovered to the dark-adapted level when the preparation was kept in the dark. An elevation of cGMP concentration during the recovery to the dark level was observed. From these results, we suggested the presence of a Ca^{2+}-independent, light-dependent mechanism activating guanylate cyclase.

5.6 MECHANISM OF DARK ADAPTATION

As shown above, background adaptation is dependent on the $[Ca^{2+}]_i$ decrease. After termination of the light stimulus, the phototransduction cascade is inactivated so that $[Ca^{2+}]_i$ returns to the dark level. In parallel with the restoration of $[Ca^{2+}]_i$ the light-sensitivity of the rod is expected to recover. Dark adaptation following from background light adaptation, therefore, is dependent on the restoration of $[Ca^{2+}]_i$ and this leads to recovery of light-sensitivity.

In contrast to background adaptation, recovery from the condition of bleaching adaptation is entirely dependent on the regeneration of rhodopsin (Pepperberg *et al.*, 1978). Under *in situ* conditions, the re-isomerized chromophore 11-*cis* retinal is supplied from the pigment epithelium (Bridges, 1977; Bernstein *et al.*, 1987) by a retinal binding protein (Okajima *et al.*, 1990). For this reason, rhodopsin cannot regenerate in rods detached from the pigment epithelium, and the light-sensitivity of isolated rods does not recover completely from the level attained during bleaching adaptation. In retina attached to the pigment epithelium, therefore, the sensitivity recovers without addition of exogenous 11-*cis* retinal and the recovery time course follows the formation of rhodopsin (Dowling, 1960; Rushton, 1961).

Not only 11-*cis* retinal but also 9-*cis* retinal (Pepperberg *et al.*, 1978) and 11-cis-locked analogs of retinal (Corson *et al.*, 1990) are able to restore light-sensitivity. Moreover, β-ionone, a ring portion of retinal able to make contact with the binding site of 11-*cis* retinal (Matsumoto and Yoshizawa, 1975), can also restore the light-sensitivity (Jin *et al.*, 1993). The results available so far support the idea that bleaching product(s) of rhodopsin is responsible for bleaching adaptation. Addition of hydroxylamine, which facilitates decomposition of the photoproducts to opsin and all-*trans* retinal-oxime, did not affect the recovery of sensitivity (Brin and Ripps, 1977); this suggests that opsin is the photoproduct responsible for bleaching adaptation. Since rhodopsin is an integral disk membrane protein, longitudinal spread of bleaching adaptation is restricted (Cornwall *et al.*, 1990; section 5.4).

5.7 CONCLUSIONS AND FUTURE PERSPECTIVES

This chapter has reviewed phototransduction and adaptation mechanisms in vertebrate photoreceptors. Vertebrate photoreceptors, especially rods, can be isolated in large quantities as a single cellular species very easily. This is a great advantage for biochemical studies of photoreceptors, since, contamination from other cells which would often introduce complications or erroneous results, can be neglected in the study of photoreceptors. Owing to this advantage, phototransduction and adaptation have become the best characterized mechanisms in the study of signal transduction in general. It may, therefore, be possible to use vertebrate photoreception as a model for understanding cellular signal transduction mechanisms in general.

However, as reviewed in the text, phototransduction and adaptation mechanisms are not so simple. It should be emphasized that these mechanisms seem to be designed on the basis of multiple regulation. For example, quenching of R* is granted by both rhodopsin phosphorylation and arrestin binding (section 5.2.2); PDE activity is regulated not only by Ca^{2+} (section 5.3.4) but also by feedback regulation of transducin by cGMP (section 5.5.1); photoreceptor light-sensitivity is controlled by Ca^{2+}-dependent regulations of guanylate cyclase (section 5.3.3), PDE (section 5.3.4) and cGMP-gated channel (section 5.3.5). Other mechanisms have been proposed to explain the physiology of photoreceptors (see section 5.5). Taken together, it is very likely that the biochemical or electrophysiological processes in the outer segment are designed to accomplish the photoreceptor function in a concerted manner. The multiple regulations probably serve as fail-safe mechanism in phototransduction and generally may be applicable to other biological functions.

There had been a long dispute about the second messenger in vertebrate photoreceptors. The work done by Fesenko *et al.* (1985) finally identified that cGMP is the messenger. At that time, many scientists probably thought that not many things were left to study phototransduction. However, since then, remarkable progress has been made in the detailed understanding of biochemical and molecular mechanisms of photoreceptor functions. These studies have brought about the most advanced knowledge about sensory transduction and adaptation among signal transduction mechanisms in general. What will come in the next ten years? That nobody knows.

One ultimate goal of these studies would be to use the information gained for the maintenance or improvement of vision. For example, it has been known that mutation in human rhodopsin causes retinal diseases such as retinitis pigmentosa (Nathans *et al.*, 1992; Oprian, 1992; and Chapters 17 and 18). Also, mutation in P_β causes retinal degeneration in the *rd* mouse (Bowes *et al.*, 1990; Pittler and Baehr, 1991), whereas cancer-associated retinopathy is an autoimmune disease involving recoverin (Polans *et al.*, 1991; Adamus *et al.*, 1993b). There are probably other retinal diseases related to photoreceptor proteins. Currently, we are able to determine the causes of these diseases at the molecular level, and can expect in the future also to develop 'tools' to treat photoreceptor dysfunction.

ACKNOWLEDGEMENT

This work was supported by the Ministry of Education, Science and Culture of Japan (05454635) and the Tokyo Biochemical Research Foundation.

REFERENCES

Adamus, G., Arendt, A., Hargrave, P.A. *et al.* (1993a) The kinetics of multiphosphorylation of rhodopsin. *Archives of Biochemistry and Biophysics*, **304**, 443–7.

Adamus, G., Guy, J., Schmied, J.L. *et al.* (1993b) Role of anti-recoverin autoantibodies in cancer-associated retinopathy. *Investigative Ophthalmology and Visual Science*, **34**, 2626–33.

Anant, J.S. and Fung, B.K.-K. (1992) *In vivo* farnesylation of rat rhodopsin kinase. *Biochemical and Biophysical Research Communications*, **183**, 468–73.

Anant, J.S., Ong, O.C., Xie, H. *et al.* (1992) *In vitro* differential prenylation of retinal cyclic GMP phosphodiesterase catalytic subunits. *Journal of Biological Chemistry*, **267**, 687–90.

Angleson, J.K. and Wensel, T.G. (1993) A GTPase accelerating factor for transducin, distinct from its effector cGMP phosphodiesterase, in rod outer segment membranes. *Neuron*, **11**, 939–49.

Antonny, B., Otto-Bruc, A., Chabre, M. and Vuong, T.M. (1993) GTP hydrolysis by purified α-subunit of transducin and its complex with the cyclic GMP phosphodiesterase inhibitor. *Biochemistry*, **32**, 8646–53.

Arshavsky, V.Y. and Bownds, M.D. (1992) Regulation of deactivation of photoreceptor G proteins by its target enzyme and cGMP. *Nature*, **357**, 416–17.

Arshavsky, V.Y., Gray-Keller, M.P. and Bownds, M.D. (1991) cGMP suppresses GTPase activity of a portion of transducin equimolar to phosphodiesterase in frog rod outer segments. *Journal of Biological Chemistry*, **266**, 18530–7.

Arshavsky, V.Y., Dumke, C.L. and Bownds, M.D. (1992) Noncatalytic cGMP-binding sites of amphibian rod cGMP phosphodiesterase control interaction with its inhibitory γ-subunits. *Journal of Biological Chemistry*, **267**, 24501–7.

Artemyev, N.O. and Hamm, H.E. (1992) Two-site high affinity interaction between inhibitory and catalytic subunits of rod cyclic GMP phosphodiesterase. *Biochemical Journal*, **283**, 273–9.

Artemyev, N.O., Rarick, H.M., Mills, J.S. *et al.* (1992) Sites of interaction between rod G-protein α-subunit and cGMP-phosphodiesterase γ-subunit. *Journal of Biological Chemistry*, **267**, 25067–72.

Artemyev, N.O., Mills, J.S., Thornburg, K.R. *et al.* (1993) A site on transducin α-subunit of interaction with the polycationic region of cGMP phosphodiesterase inhibitory subunit. *Journal of Biological Chemistry*, **268**, 23611–15.

Barkdoll III, A.E., Pugh Jr E.N. and Sitaramayya, A. (1989) Calcium dependence of the activation and inactivation kinetics of the light-activated phosphodiesterase of retinal rods. *Journal of General Physiology*, **93**, 1091–108.

Bauer, P.J. (1988) Evidence for two functionally different fractions in bovine retinal rod outer segments. *Journal of Physiology*, **401**, 309–27.

Bernstein, P.S., Law, W.C. and Rando, R.R. (1987) Isomerization of all-*trans*-retinoids to 11-*cis*-retinoids *in vitro*. *Proceedings of the National Academy of Sciences, USA*, **84**, 1849–53.

Binder, B.M., Biernbaum, M.S. and Bownds, M.D. (1990) Light activation of one rhodopsin molecule causes the phosphorylation of hundreds of others. *Journal of Biological Chemistry*, **265**, 15333–40.

Boguski, M.S. and McCormick, F. (1993) Proteins regulating Ras and its relatives. *Nature*, **366**, 643–54.

Bornancin, F., Franco, M., Bigay, J. and Chabre, M. (1992) Functional modifications of transducin induced by cholera or pertussis-toxin-catalyzed ADP-ribosylation. *European Journal of Biochemistry*, **210**, 33–44.

Bowes, C., Li, T., Danciger, M. *et al.* (1990) Retinal degeneration in the *rd* mouse is caused by a defect in the β subunit of rod cGMP-phosphodiesterase. *Nature*, **347**, 677–80.

Bridges, C.D.B. (1977) Rhodopsin regeneration in rod outer segments: utilization of 11-*cis* retinal and retinol. *Experimental Eye Research*, **24**, 571–80.

Brin, K.P., and Ripps, H. (1977) Rhodopsin photoproducts and rod sensitivity in the skate retina. *Journal of General Physiology*, **69**, 97–120.

Brown, R.L. (1992) Functional regions of the inhibitory subunit of retinal rod cGMP phosphodiesterase identified by site-specific mutagenesis and fluorescence spectroscopy. *Biochemistry*, **31**, 5918–25.

Brown, R.L. and Stryer, L. (1989) Expression in bacteria of functional inhibitory subunit of retinal rod cGMP phosphodiesterase. *Proceedings of the National Academy of Sciences, USA*, **86**, 4922–6.

Buczylko, J., Gutmann, C. and Palczewski, K. (1991) Regulation of rhodopsin kinase by autophosphorylation. *Proceedings of the National Academy of Sciences, USA*, **88**, 2568–72.

Catty, P. and Deterre, P. (1991) Activation and solubilization of the retinal cGMP-specific phosphodiesterase by limited proteolysis. *European Journal of Biochemistry*, **199**, 263–9.

Catty, P., Pfister, C., Bruckert, F. and Deterre, P. (1992) The cGMP phosphodiesterase–transducin complex of retinal rods. *Journal of Biological Chemistry*, **267**, 19489–93.

Cervetto, L., Lagnado, L., Perry, R.J. *et al.* (1989) Extrusion of calcium from rod outer segments is driven by both sodium and potassium gradients. *Nature*, **337**, 740–3.

Chen, T.-Y., Peng, Y.-W., Dhallan, R.S. *et al.* (1993) A new subunit of the cyclic nucleotide-gated cation channel in retinal rods. *Nature*, **362**, 764–7.

Clerc, A. and Bennet, N. (1992) Activated cGMP phosphodiesterase of retinal rods. *Journal of Biological Chemistry*, **267**, 6620–7.

Clerc, A., Catty, P. and Bennett, N. (1992) Interaction between cGMP-phosphodiesterase and transducin α-subunit in retinal rods. *Journal of Biological Chemistry*, **267**, 19948–53.

Cobbs, W.H. (1991) Light and dark active phosphodiesterase regulation in salamander rods. *Journal of General Physiology*, **98**, 575–614.

Cohen, G.B., Yang, T., Robinson, P.R. and Oprian, D.D. (1993) Constitutive active opsin: influence of charge at position 134 and size at position 296. *Biochemistry*, **32**, 6111–15.

Cook, N.J. and Kaupp, U.B. (1988) Solubilization, purification, and reconstitution of the sodium–calcium exchanger from bovine retinal ros outer segments. *Journal of Biological Chemistry*, **263**, 11382–88.

Cook, N.J., Hanke, W. and Kaupp, U.B. (1987) Identification, purification, and functional reconstitution of the cyclic GMP-dependent channel from rod photoreceptors. *Proceedings of the National Academy of Sciences, USA*, **84**, 585–9.

Cook, N.J., Molday, L.L., Reid, D. *et al.* (1989) The cGMP-gated channel of bovine rod photoreceptors is localized exclusively in the plasma membrane. *Journal of Biological Chemistry*, **264**, 6996–9.

Cornwall, M.C., Fein, A. and MacNichol Jr, E.F. (1990) Cellular mechanisms that underlie bleaching and background adaptation. *Journal of General Physiology*, **96**, 345–72.

Corson, D.W., Cornwall, M.C., MacNichol Jr, E.F. *et al.* (1990) Sensitization of bleached rod photoreceptors by 11-*cis*-locked analogues of retinal. *Proceedings of the National Academy of Sciences, USA*, **87**, 6823–7.

Cote, R.H. and Brunnock, M.A. (1993) Intracellular cGMP concentration in rod photoreceptors is regulated by binding to high and moderate affinity cGMP binding sites. *Journal of Biological Chemistry*, **268**, 17190–98.

Dean, K.R. and Akhtar, M. (1993) Phosphorylation of solubilized dark-adapted rhodopsin. *European Journal of Biochemistry*, **213**, 881–90.

Del Priore, L.V. and Lewis, A. (1983) Calcium-dependent activation and deactivation of rod outer segment phosphodiesterase is calmodulin-independent. *Biochemical and Biophysical Research Communications*, **113**, 317–24.

Detwiler, P.B. and Gray-Keller, M.P. (1992) Some unsolved issues in the physiology and biochemistry of phototransduction. *Current Opinion in Neurobiology*, **2**, 433–8.

Dizhoor, A.M., Ray, S., Kumar, S. *et al.* (1991) Recoverin: a calcium sensitive activator of retinal rod guanylate cyclase. *Science*, **251**, 915–18.

Dizhoor, A.M., Ericsson, L.H., Johnson, R.S. *et al.* (1992) The NH_2 terminus of retinal recoverin is acylated by a small family of fatty acids. *Journal of Biological Chemistry*, **267**, 16033–6.

Dizhoor, A.M., Chen, C.-K., Olshevskaya, E. *et al.* (1993) Role of the acylated amino terminus of recoverin in Ca^{2+}-dependent membrane interaction. *Science*, **259**, 829–32.

Doukas, A.G., Aton, B., Callender, R.H. and Ebrey, T.G. (1978) Resonance Raman studies of bovine metarhodopsin I and metarhodopsin II. *Biochemistry*, **17**, 2430–5.

Dowling, J.E. (1960) Chemistry of visual adaptation in the rat. *Nature*, **188**, 114–18.

Dratz, E.A., Furstenau, J.E., Lambert, C.G. *et al.* (1993) NMR structure of a receptor-bound G-protein peptide. *Nature*, **363**, 276–81.

Ehret–Hilberer, S., Nullans, G., Aunis, D. and Virmaux, N. (1992) Mono ADP-ribosylation of transducin catalyzed by rod outer segment extract. *FEBS Letters*, **309**, 394–8.

Erickson, M.A., Robinson, P.R. and Lisman, J. (1992) Deactivation of visual transduction without guanosine triphosphate hydrolysis by G protein. *Science*, **257**, 1255–8.

Fahmy, K. and Sakmar, T.P. (1993a) Regulation of the rhodopsin–transducin interaction by a highly conserved carboxylic acid group. *Biochemistry*, **32**, 7229–36.

Fahmy, K. and Sakmar, T.P. (1993b) Light-dependent transducin activation by an ultraviolet-absorbing rhodopsin mutant. *Biochemistry*, **32**, 9165–71.

Fain, G.L., Lamb, T.D., Matthews, H.R. and Murphy, R.L.W. (1989) Cytoplasmic calcium as the messenger for light adaptation in salamander rods. *Journal of Physiology*, **416**, 215–43.

Fesenko, E.E., Kolesnikov, S.S. and Lyubarsky, A.L. (1985) Induction by cyclic GMP of cationic conductance in plasma membrane of retinal rod outer segment. *Nature*, **313**, 310–13.

Flaherty, K.M., Zozulya, S., Stryer, L. and McKay, D.B. (1993) Three-dimensional structure of recoverin, a calcium sensor in vision. *Cell*, **75**, 709–16.

Fleischman, D. and Denisevich, M. (1979) Guanylate cyclase of isolated bovine retinal rod axonemes. *Biochemistry*, **18**, 5060–6.

Fowles, C., Sharma, R. and Akhtar, M. (1988) Mechanistic studies on the phosphorylation of photoexcited rhodopsin. *FEBS Letters*, **238**, 56–60.

Fowles, C., Akhtar, M. and Cohen, P. (1989) Interplay of phosphorylation and dephosphorylation in vision: protein phosphatases of bovine rod outer segments. *Biochemistry*, **28**, 9385–91.

Franke, R.R., Sakmar, T.P., Graham, R.M. and Khorana, H.G. (1992) Structure and function of rhodopsin. *Journal of Biological Chemistry*, **267**, 14767–74.

Fukada, Y. and Yoshizawa, T. (1981) Activation of phosphodiesterase in frog rod outer segment by an intermediate of rhodopsin photolysis. II. *Biochimica et Biophysica Acta*, **675**, 195–200.

Fukada, Y., Takao, T., Ohguro, H. *et al.* (1990) Farnesylated γ-subunit of photoreceptor G protein indispensable for GTP-binding. *Nature*, **346**, 658–60.

Fung, B.K.-K. (1983) Characterization of transducin from bovine retinal rod outer segments. *Journal of Biological Chemistry*, **258**, 10495–502.

Fung, B.K.-K. and Nash, C.R. (1983) Characterization of transducin from bovine retinal rod outer segments. *Journal of Biological Chemistry*, **258**, 10502–10.

Fung, B.K.-K. and Stryer, L. (1980) Photolyzed rhodopsin catalyzes the exchange of GTP for bound GDP in retinal rod outer segments. *Proceedings of the National Academy of Sciences, USA*, **77**, 2500–4.

Fung, B.K.-K., Young, J.H., Yamane, H.K. and Griswold–Prenner, I. (1990) Subunit stoichiometry of retinal rod cGMP phosphodiesterase. *Biochemistry*, **29**, 2657–64.

Gillespie, P.G. and Beavo, J.A. (1989) cGMP is tightly bound to bovine retinal rod phosphodiesterase. *Proceedings of the National Academy of Sciences, USA*, **86**, 4311–15.

Gorczyca, W.A. Gray–Keller, M.P., Detwiller, P.B. and Palczewski, K. (1994) Purification and physiological evaluation of a guanylate cyclase activating protein from retinal rods. *Proceedings of the National Academy of Sciences, USA*, **91**, 4014–18.

Gorodovikova, E.N. and Philippov, P.P. (1993) The presence of a calcium-sensitive p26-containing complex in bovine retina rod cells. *FEBS Letters*, **335**, 277–9.

Goulding, E.H., Tibbs, G.R., Liu, D. and Slegelbaum, S.A. (1993) Role of H5 domain in determining pore diameter and ion permeation through cyclic nucleotide-gated channel. *Nature*, **364**, 61–4.

Gray–Keller, M.P., Biernbaum, M.S. and Bownds, M.D. (1990) Transducin activation in electropermeabilized frog rod outer segments is highly amplified, and a portion equivalent to phosphodiesterase remains membrane-bound. *Journal of Biological Chemistry*, **265**, 15323–32.

Gray–Keller, M.P., Polans, A.S., Palczewski, K. and Detwiler, P. (1993). The effect of recoverin-like calcium-binding proteins on the photoresponse of retinal rods. *Neuron*, **10**, 523–31.

Gurevich, V.V. and Benovic, J.L. (1993) Visual arrestin interaction with rhodopsin. *Journal of Biological Chemistry*, **268**, 11628–38.

Guy, H.R., Durell, S.R., Warmke, J. *et al.* (1991). Similarities in amino acid sequence of *Drosophila* ear and cyclic nucleotide-gated channels. *Science*, **254**, 730.

Haase, W., Friese, W., Gordon, R.D. *et al.* (1990) Immunological characterization and localization of the Na^+/Ca^{2+}-exchanger in bovine retina. *Journal of Neuroscience*, **10**, 1486–94.

Hamm, H. (1990) Regulation by light of cyclic nucleotide-dependent protein kinases and their substrates in frog rod outer segments. *Journal of General Physiology*, **95**, 545–67.

Hamm, H.E. and Bownds, M.D. (1986) Protein complement of rod outer segments of frog retina. *Biochemistry*, **25**, 4512–23.

Hamm, H.E., Deretic, D., Arendt, A. *et al.* (1988) Site of G protein binding to rhodopsin mapped with synthetic peptides from the α subunit. *Science*, **241**, 832–5.

Harosi, F.I. (1975) Absorption spectra and linear dichroism of some amphibian photoreceptors. *Journal of General Physiology*, **66**, 357–82.

Hayashi, F. and Yamazaki, A. (1991) Polymorphism in purified guanylate cyclase from vertebrate rod photoreceptors. *Proceedings of the National Academy of Sciences, USA*, **88**, 4746–50.

Haynes, L.W., Kay, A.R. and Yau, K.-W. (1986) Single cyclic GMP-activated channel activity in excised patches of rod outer segment membrane. *Nature*, **321**, 66–70.

Hermolin, H.M., Karell, M.A., Hamm, H.E. and Bownds, M.D. (1982) Calcium and cyclic GMP regulation of light-sensitive protein phosphorylation in frog photoreceptor membranes. *Journal of General Physiology*, **79**, 633–55.

Hodgkin, A.L. and Nunn, B.J. (1988) Control of light-sensitive current in salamander rods. *Journal of Physiology*, **403**, 439–71.

Hodgkin, A.L., McNaughton, P.A. and Nunn, B.J. (1985) The ionic selectivity and calcium dependence of the light-sensitive pathway in toad rods. *Journal of Physiology*, **358**, 447–468.

Hofmann, K.P., Pulvermüller, A., Buczylko, J. *et al.* (1992) The role of arrestin and retinoids in the regeneration pathway of rhodopsin. *Journal of Biological Chemistry*, **267**, 15701–6.

Hsu, Y.-T. and Molday, R.S. (1993) Modulation of the cGMP-gated channel of rod photoreceptor cells by calmodulin. *Nature*, **361**, 76–9.

Hurley, J.B. and Stryer, L. (1982) Purification and characterization of the regulatory subunit of the cyclic GMP phosphodiesterase from retinal rod outer segments. *Journal of Biological Chemistry*, **257**, 11094–9.

Hurley, J.B. Dizhoor, A.M., Ray, S. and Stryer, L. (1993) Recoverin's role: conclusion withdrawn. *Science*, **260**, 740.

Inglese, J., Glickman, J.F., Lorenz, W. *et al.* (1992a) Isoprenylation of a protein kinase. *Journal of Biological Chemistry*, **267**, 1422–5.

Inglese, J., Koch, W.J., Caron, M.G. and Lefkowitz, R.J. (1992b) Isoprenylation in regulation of signal transduction by G-protein-coupled receptor kinases. *Nature*, **359**, 147–50.

Jin, J., Crouch R.K., Corson. D.W. *et al.* (1993) Noncovalent occupancy of the retinal-binding pocket of opsin diminishes bleaching adaptation of retinal cones. *Neuron*, **11**, 513–22.

Kajimoto, Y., Shirai, Y., Mukai, H. *et al.* (1993) Molecular cloning of two additional members of the neural visinin-like Ca^{2+}-binding protein gene family. *Journal of Neurochemistry*, **61**, 1091–6.

Kataoka, M., Mihara, K. and Tokunaga, F. (1993)

Recoverin alters its surface properties depending on both calcium-binding and N-terminal myristoylation. *Journal of Biochemistry, 114*, 535–40.

Kaupp, U.B. and Koch, K.-W. (1992) Role of cGMP and Ca^{2+} in vertebrate photoreceptor excitation and adaptation. *Annual Review of Physiology*, **54**, 153–75.

Kaupp, U.B., Niidome, T., Tanabe, T. *et al.* (1989) Primary structure and functional expression from complementary DNA of the rod photoreceptor cyclic GMP-gated channel. *Nature*, **342**, 762–6.

Kawamura, S. (1983) Involvement of ATP in activation and inactivation sequence of phosphodiesterase in frog rod outer segments. *Biochimica et Biophysica Acta*, **732**, 276–81.

Kawamura, S. (1992) Light-sensitivity modulating protein in frog rods. *Photochemistry and Photobiology*, **56**, 1173–80.

Kawamura, S. (1993a) Molecular aspects of photoreceptor adaptation in vertebrate retina. *International Review of Neurobiology*, **35**, 43-86.

Kawamura, S. (1993b) Rhodopsin phosphorylation as a mechanism of cyclic GMP phosphodiesterase regulation by S-modulin. *Nature*, **362**, 855–7.

Kawamura, S. and Bownds, M.D. (1981) Light adaptation of the cyclic GMP phosphodiesterase of frog photoreceptor membranes mediated by ATP and calcium ions. *Journal of General Physiology*, **77**, 571–91.

Kawamura, S. and Murakami, M. (1989a) Regulation of cGMP levels by guanylate cyclase in truncated frog rod outer segments. *Journal of General Physiology*, **94**, 649–68.

Kawamura, S. and Murakami, M. (1989b) Control of cGMP concentration by calcium ions in frog rods. *Neuroscience Research, Suppl.* **10**, S15–S22.

Kawamura, S. and Murakami, M. (1991) Calcium-dependent regulation of cyclic GMP phosphodiesterase by a protein from frog retinal rods. *Nature*, **349**, 420–3.

Kawamura, S., Takamatsu, K. and Kitamura, K. (1992) Purification and characterization of S-modulin, a calcium-dependent regulator on cGMP phosphodiesterase in frog rod photoreceptors. *Biochemical and Biophysical Research Communications*, **186**, 411–17.

Kawamura, S., Hisatomi, O., Kayada, S. *et al.* (1993) Recoverin has S-modulin activity in frog rods. *Journal of Biological Chemistry*, **268**, 14579–82.

Kelleher, D.J. and Johnson, G.L. (1986) Phosphorylation of rhodopsin by protein kinase C *in vitro*. *Journal of Biological Chemistry*, **261**, 4749–57.

Kelleher, D.J. and Johnson, G.L. (1990) Characterization of rhodopsin kinase purified from bovine rod outer segments. *Journal of Biological Chemistry*, **265**, 2632–9.

Kobayashi, M., Takamatsu, K., Saitoh, S. *et al.* (1992) Molecular cloning of hippocalcin, a novel calcium-binding protein of the recoverin family exclusively expressed in hippocampus. *Biochemical and Biophysical Research Communications*, **189**, 511–17.

Koch, K.-W. (1991) Purification and identification of photoreceptor guanylate cyclase. *Journal of Biological Chemistry*, **266**, 8634–7.

Koch, K.-W. and Stryer, L. (1988) Highly cooperative feedback control of retinal rod guanylate cyclase by calcium ions. *Nature*, **334**, 64–6.

Kohnken, R.E., Chafouleas, J.G., Eadie, D.M. *et al.* (1981) Calmodulin in bovine rod outer segments. *Journal of Biological Chemistry*, **256**, 12517–22.

Kokame, K., Fukada, Y., Yoshizawa, T. *et al.* (1992) Lipid modification at the N terminus of photoreceptor G-protein α-subunit. *Nature*, **359**, 749–52.

König, B., Arendt, A., McDowell, J.H. *et al.* (1989) Three cytoplasmic loops of rhodopsin interact with transducin. *Proceedings of the National Academy of Sciences, USA*, **86**, 6878–82.

Korenbrot, J.I. and Miller, D.L. (1989) Cytoplasmic free calcium concentration in dark-adapted retinal rod outer segments. *Vision Research*, **29**, 939–48.

Koutalos, Y. and Yau, K.-W. (1993) A rich complexity emerges in phototransduction. *Current Opinion in Neurobiology*, **3**, 513–19.

Krapivinsky, G.B., Malenyov, A.L., Zaikina, I.V. and Fesenko, E.E. (1992) Low molecular mass phosphoproteins from the frog rod outer segments form a complex with 48 kDa protein. *Cellular Signalling*, **4**, 583–93.

Krishnan, N., Fletcher, R.T., Chader, G.J. and Krishna, G. (1978) Characterization of guanylate cyclase of rod outer segments of the bovine retina. *Biochimica et Biophysica Acta*, **523**, 506–15.

Kuo, C.-H., Akiyama, M. and Miki, N. (1989) Isolation of a novel retina-specific clone (MEKA cDNA) encoding a photoreceptor soluble protein. *Molecular Brain Research*, **6**, 1–10.

Lagnado, L. and Baylor, D.A. (1992) Signal flow in visual transduction. *Neuron*, **8**, 995–1002.

Lagnado, L. and Baylor, D.A. (1994) Calcium controls light-triggered formation of catalytically active rhodopsin. *Nature*, **367**, 273–7.

Lai, R.K., Pérez–Sala, D., Canãda, F.J. and Rando, R.R. (1990) The γ subunit of transducin is farnesylated. *Proceedings of the National Academy of Sciences, USA*, **87**, 7673–7.

Lamb, T.D., McNaughton, P.A. and Yau, K.W. (1981) Spatial spread of activation and background desensitization in toad rod outer segments. *Journal of Physiology*, **319**, 463–96.

Lambrecht, H.-G. and Koch, K.-W. (1991a) A 26 kd calcium binding protein from bovine rod outer segments as modulator of photoreceptor guanylate cyclase. *EMBO Journal*, **10**, 793–8.

Lambrecht, H.-G. and Koch, K.-W. (1991b) Phosphorylation of recoverin, the calcium-sensitive activator of photoreceptor guanylyl cyclase. *FEBS Letters*, **294**, 207–9.

Lee, R.H., Brown, B.M. and Lolley, R.N. (1982) Autophosphorylation of rhodopsin kinase from retinal rod outer segments. *Biochemistry*, **21**, 3303–7.

Lee, R.H., Lieberman, B.S. and Lolley, R.N. (1987) A novel complex from bovine visual cells of a 33000-dalton phosphoprotein with β- and γ-transducin: purification and subunit structure. *Biochemistry*, **26**, 3983–90.

Lee, R.H., Ting, T.D., Lieberman, B.S. *et al.* (1992) Regulation of retinal cGMP cascade by phosducin in bovine rod photoreceptor cells. *Journal of Biological Chemistry*, **267**, 25104–12.

Lenz, S.E., Henschel, Y., Zopf, D. *et al.* (1992) VILIP, a cognate protein of the retinal calcium binding proteins visinin and recoverin, is expressed in the developing chicken brain. *Molecular Brain Research*, **15**, 133–140.

Lipkin, V.M., Dumler, I.L., Muradov, K.G. *et al.* (1988) Active sites of the cyclic GMP phosphodiesterase γ-subunit of retinal rod outer segments. *FEBS Letters*, **234**, 287–90.

Matsumoto, H. and Yoshizawa, T. (1975) Existence of a β-ionone ring-binding site in the rhodopsin molecule. *Nature*, **258**, 523–6.

Matthews, H.R. (1991) Incorporation of chelator into guinea-pig rods shows that calcium mediates mammalian photoreceptor light adaptation. *Journal of Physiology*, **436**, 93–105.

Matthews, H.R., Murphy, R.L.W., Fain, G. and Lamb, T.D. (1988) Photoreceptor light adaptation is mediated by cytoplasmic calcium concentration. *Nature*, **334**, 67–9.

McCarthy, S.T., Younger, J.P. and Owen, W.G. (1993). Disc structure, buffering and kinetics of cytosolic free calcium in bullfrog rod outer segments. *ARVO abstracts* 1327.

McDowell, J.H. and Kühn, H. (1977) Light-induced phosphorylation of rhodopsin in cattle photoreceptor membranes: substrate activation and inactivation. *Biochemistry*, **16**, 4054–60.

McDowell, J.H., Nawrocki, J.P. and Hargrave, P.A (1993) Phosphorylation sites in bovine rhodopsin. *Biochemistry*, **32**, 4968–74.

McNaughton, P.A. (1990) Light response of vertebrate photoreceptors. *Physiological Reviews*, **70**, 847–83.

McNaughton, P.A., Cervetto, L. and Nunn, B.J. (1986) Measurement of the intracellular free calcium concentration in salamander rods. *Nature*, **322**, 261–3.

Miller, J.L. and Dratz, E.A. (1984) Phosphorylation at sites near rhodopsin's carboxyl-terminus regulates light initiated cGMP hydrolysis. *Vision Research*, **24**, 1509–21.

Molday, R.S., Molday, L.L., Dosé, A. *et al.* (1991) The cGMP-gated channel of the rod photoreceptor cell characterization and orientation of the amino terminus. *Journal of Biological Chemistry*, **266**, 21917–22.

Morrison, D.F., Cunnick, J.M., Oppert, B. and Takemoto, D.J. (1989) Interaction of the γ-subunit of retinal rod outer segment phosphodiesterase with transducin. *Journal of Biological Chemistry*, **264**, 11671–81.

Nagao, S., Yamazaki, A. and Bitensky, M.W. (1987) Calmodulin and calmodulin binding proteins in amphibian rod outer segments. *Biochemistry*, **26**, 1659–65.

Nakatani, K. and Yau, K.-W. (1988a) Calcium and light adaptation in retinal rods and cones. *Nature*, **334**, 69–71.

Nakatani, K. and Yau, K.-W. (1988b) Calcium and

magnesium fluxes across the plasma membrane of the toad rod outer segment. *Journal of Physiology*, **395**, 695–729.

Nakatani, K. and Yau, K.-W. (1988c) Guanosine 3′,5′-cyclic monophosphate-activated conductance studied in a truncated rod outer segment. *Journal of Physiology*, **395**, 731–53.

Nakatani, K., Tamura, T. and Yau, K.-W. (1991) Light adaptation in retinal rods of the rabbit and two other nonprimate mammals. *Journal of General Physiology*, **97**, 413–35.

Nathans, J. (1990) Rhodopsin: structure, function, and genetics. *Biochemistry*, **31**, 4923–31.

Nathans, J., Merbs, S.L., Sung, C.-H. *et al.* (1992) Molecular genetics of human visual pigments. *Annual Review of Genetics*, **26**, 403–24.

Neubert, T.A., Johnson, R.S., Hurley, J.B. and Walsh, K.A. (1992) The rod transducin α subunit amino terminus is heterogeneously fatty acylated. *Journal of Biological Chemistry*, **267**, 18274–7.

Newton, A.C. and Williams, D.S. (1991) Involvement of protein kinase C in the phosphorylation of rhodopsin. *Journal of Biological Chemistry*, **267**, 17725–8.

Newton, A.C. and Williams, D.S. (1993) Rhodopsin is the major *in situ* substrate of protein kinase C in rod outer segments of photoreceptors. *Journal of Biological Chemistry*, **268**, 18181–6.

Nicoll, A. and Applebury, M.L. (1989) Purification of the bovine rod outer segment Na^+/Ca^{2+} exchanger. *Journal of Biological Chemistry*, **264**, 16207–13.

Nikonov, S.S., Filatov, G.N. and Fesenko, E.E. (1993) On the activation of phosphodiesterase by a 26 kDa protein. *FEBS Letters*, **316**, 34–6.

Nir, I. and Ranson, N. (1993) Ultrastructural analysis of arrestin distribution in mouse photoreceptor during dark/light cycle. *Experimental Eye Research*, **57**, 307–18.

Noel, J.P., Hamm, H.E. and Sigler, P.B. (1993) The 2.2 Å crystal structure of transducin-α complexed with GTPγS. *Nature*, **366**, 654–63.

Ohguro, H., Palczewski, K., Ericsson, L.H. *et al.* (1993) Sequential phosphorylation of rhodopsin at multiple sites. *Biochemistry*, **32**, 5718–24.

Okajima, T.L., Pepperberg, D.R., Ripps, H. *et al.* (1990) Interphotoreceptor retinoid-binding protein promotes rhodopsin regeneration in toad photoreceptors. *Proceedings of the National Academy of Sciences, USA*, **87**, 6907–11.

Okazaki, K., Watanabe, M., Ando, Y. *et al.* (1992) Full sequence of neurocalcin, a novel calcium-binding protein abundant in central nervous system. *Biochemical and Biophysical Research Communications*, **185**, 147–53.

Ong, O.C., Ota, I.M., Clarke, S. and Fung, B.K.-K. (1989) The membrane binding domain of rod cGMP phosphodiesterase is posttranslationally modified by methyl esterification at a C-terminal cysteine. *Proceedings of the National Academy of Sciences, USA*, **86**, 9238–42.

Oppert, B. and Takemoto, D.J. (1991) Identification of the γ-subunit interaction sites on the retinal cyclic GMP phosphodiesterase β-subunit. *Biochemical and Biophysical Research Communications*, **178**, 474–9.

Oppert, B., Cunnick, J.M., Hurt, D. and Takemoto, D.J. (1991) Identification of the retinal cyclic GMP phosphodiesterase inhibitory γ-subunit interaction sites on the catalytic α-subunit. *Journal of Biological Chemistry*, **266**, 16607–13.

Oprian, D.D. (1992) Molecular determinants of spectral properties and signal transduction in the visual pigments. *Current Opinion in Neurobiology*, **2**, 428–32.

Otto-Bruc, A., Antonny, B., Vuong, T.M. *et al.* (1993) Interaction between the retinal cyclic GMP phosphodiesterase inhibitor and transducin. Kinetics and affinity studies. *Biochemistry*, **32**, 8636–45.

Pagès, F., Deterre, P. and Pfister, C. (1992) Enhanced GTPase activity of transducin when bound to cGMP phosphodiesterase in bovine retinal rods. *Journal of Biological Chemistry*, **267**, 22018–21.

Palczewski, K. and Hargrave, P.A. (1991) Studies of ligand binding to arrestin. *Journal of Biological Chemistry*, **266**, 4201–6.

Palczewski, K., Hargrave, P.A., McDowell, J.H. and Ingebritsen, T.S. (1989) The catalytic subunit of phosphatase 2A dephosphorylates rhodopsin. *Biochemistry*, **28**, 415–19.

Palczewski, K., Buczylko, J., Kaplan, M.W. *et al.* (1991) Mechanism of rhodopsin kinase activation. *Journal of Biological Chemistry*, **266**, 12949–55.

Palezewski, K., Rispoli, G. and Detwiler, P.B.

(1992a) The influence of arresting (48K protein) and rhodopsin kinase on visual transduction. *Neuron*, **8**, 117–26.

Palczewski, K., Buczylko, J., Van Hooser, P. *et al.* (1992b) Identification of the autophosphorylation sites in rhodopsin kinase. *Journal of Biological Chemistry*, **267**, 18991–8.

Papac, D.I., Oatis, J.E., Crouch, R.K. and Knapp, D.R. (1993) Mass spectrometric identification of phosphorylation sites in bleached bovine rhodopsin. *Biochemistry*, **32**, 5930–4.

Pepperberg, D.R., Brown, P.K., Lurie, M. and Dowling, J.E. (1978) Visual pigment and photoreceptor sensitivity in the isolated state retina. *Journal of General Physiology*, **71**, 369–96.

Perry, R.J. and McNaughton, P.A. (1993) The mechanism of ion transport by the Na^+-Ca^{2+}, K^+ exchange in rods isolated from the salamander retina. *Journal of Physiology*, **466**, 443–80.

Pfister, C., Bennet, N., Bruckert, F. *et al.* (1993) Interactions of a G-protein with its effector: transducin and cGMP phosphodiesterase in retinal rods. *Cellular Signalling*, **5**, 235–51.

Piriev, N.I., Yamashita, C., Samuel, G. and Farber, D.B. (1993) Rod photoreceptor cGMP-phosphodiesterase: analysis of α and β subunits expressed in human kidney cells. *Proceedings of the National Academy of Sciences, USA*, **90**, 9340–4.

Pittler, S.J. and Baehr, W. (1991) Identification of a nonsense mutation in the rod photoreceptor cGMP phosphodiesterse β-subunit gene of the *rd* mouse. *Proceedings of the National Academy of Sciences, USA*, **88**, 8322–6.

Polans, A.S., Hermolin, J. and Bownds, M.D. (1979) Light-induced dephosphorylation of two proteins in frog rod outer segments. *Journal of General Physiology*, **74**, 595–613.

Polans, A.S., Buczylko, J., Crabb, J. and Palczewski, K. (1991) A photoreceptor calcium binding protein is recognized by autoantibodies obtained from patients with cancer-associated retinopathy. *Journal of Cell Biology*, **112**, 981–9.

Pozdnyakov, N., Lloyd, A., Reddy, V.N. and Sitaramayya, A. (1993) Nitric oxide-regulated endogenous ADP-ribosylation of rod outer segment proteins. *Biochemical and Biophysical Research Communications*, **192**, 610–15.

Qin, N., Pittler, S.J. and Baehr, W. (1992) *In vitro* isoprenylation and membrane association of mouse rod photoreceptor cGMP phosphodiesterase α and β subunits expressed in bacteria. *Journal of Biological Chemistry*, **267**, 8458–63.

Rarick, H.M., Artemyev, N.O. and Hamm, H.E. (1992) A site on rod G protein α subunit that mediates effector activation. *Science*, **256**, 1031–3.

Ratto, G.M., Payne, R., Owen, W.G. and Tsien, R.Y. (1988) The concentration of cytosolic free calcium in vertebrate rod outer segments measured with fura-2. *Journal of Neuroscience*, **8**, 3240–6.

Ray, S., Zozulya, S., Niemi, G.A. *et al.* (1992) Cloning, expression, and crystallization of recoverin, a calcium sensor in vision. *Proceedings of the National Academy of Sciences, USA*, **89**, 5705–9.

Reid, D.M., Friedel, U., Molday, R.S. and Cook, N.J. (1990) Identification of the sodium–calcium exchanger as the major ricin-binding glycoprotein of bovine rod outer segments and its localization to the plasma membrane. *Biochemistry*, **29**, 1601–17.

Reiländer, H., Achilles, A., Friedel, U. *et al.* (1992) Primary structure and functional expression of the Na/Ca, K-exchanger from bovine rod photoreceptors. *EMBO Journal*, **11**, 1689–95.

Robinson, P.R., Kawamura, S., Abramson, B. and Bownds, M.D. (1980) Control of the cyclic GMP phosphodiesterase of frog photoreceptor membranes. *Journal of General Physiology*, **76**, 631–45.

Robinson, P.R., Cohen, G.B., Zhukovsky, E.A. and Oprian, D.D. (1992) Constitutively active mutants of rhodopsin. *Neuron*, **9**, 719–25.

Rushton, W.A.H. (1961) Rhodopsin measurement and dark-adaptation in a subject deficient in cone vision. *Journal of Physiology*, **156**, 193–205.

Sakmar, T.P., Franke, R.R. and Khorana, H.G. (1989) Glutamic acid-113 serves as the retinylidene Schiff base counterion in bovine rhodopsin. *Proceedings of the National Academy of Sciences, USA*, **86**, 8309–13.

Schmidt, K.-F., Nöll, G.N. and Yamamoto, Y. (1992) Sodium nitroprusside alters dark voltage and light responses in isolated retinal rods during whole-cell recording. *Visual Neuroscience*, **9**, 205-9.

Schnetkamp, P.P.M., Basu, D.K. and Szerensei, R.T. (1989) Na–Ca exchange in the outer segments of bovine rod photoreceptors requires and transports potassium. *American Journal of Physiology*, **257**, C153–7.

Shyjan, A.W., de Sauvage, F., Gillett, N.A. *et al.* (1992) Molecular cloning of a retina–specific membrane guanylyl cyclase. *Neuron*, **9**, 727–37.

Sillman, A.J., Ito, H. and Tomita, T. (1969) Studies on the mass receptor potential of the isolated frog retina. II. On the basis of the ionic mechanism. *Vision Research*, **9**, 1443–51.

Sitaramayya, A. (1986) Rhodopsin kinase prepared from bovine rod disk membranes quenches light activation of cGMP phosphodiesterase in a reconstituted system. *Biochemistry*, **25**, 5460–8.

Sitaramayya, A. and Liebman, P.A. (1983a) Mechanism of ATP quench of phosphodiesterase activation in rod disc membranes. *Journal of Biological Chemistry*, **258**, 1205–9.

Sitaramayya, A. and Liebman, P.A. (1983b) Phosphorylation of rhodopsin and quenching of cyclic GMP phosphodiesterase activity by ATP at weak bleaches. *Journal of Biological Chemistry*, **258**, 12106–9.

Sitaramayya, A., Harkness, J., Parkes, J.H. *et al.* (1986) Kinetic studies suggest that light-activated cyclic GMP phosphodiesterase is a complex with G-protein subunits. *Biochemistry*, **25**, 65–6.

Stryer, L. (1991) Visual excitation and recovery. *Journal of Biological Chemistry*, **266**, 10711–4.

Swanson, R.J. and Applebury, M.L. (1983) Methylation of proteins in photoreceptor rod outer segments. *Journal of Biological Chemistry*, **258**, 10599–605.

Tamura, T., Nakatani, K. and Yau, K.-W. (1991) Calcium feedback and sensitivity regulation in primate rods. *Journal of General Physiology*, **98**, 95–130.

Torre, V., Matthews, H.R. and Lamb, T.D. (1986) Role of calcium in regulating the cyclic GMP cascade of phototransduction in retinal rods. *Proceedings of the National Academy of Sciences, USA*, **83**, 7109–13.

Troyer, E.W., Hall, I.A. and Ferrendelli, J.A. (1978) Guanylate cyclase in CNS: enzymatic characteristics of soluble and particulate enzymes from mouse cerebellum and retina. *Journal of Neurochemistry*, **31**, 825–33.

Udovichenko, I.P., Cunnick, J., Gonzales, K. and Takemoto, D.J. (1993) Phosphorylation of bovine rod photoreceptor cyclic GMP phosphodiesterase. *Biochemical Journal*, **295**, 49–55.

Venturini, C.M., Knowels, R.G., Palmer, R.M.J. and Moncada, S. (1991) Synthesis of nitric oxide in the bovine retina. *Biochemical and Biophysical Research Communications*, **180**, 920–5.

Vuong, T.M. and Chabre, M. (1991) Deactivation kinetics of the transduction cascade of vision. *Proceedings of the National Academy of Sciences, USA*, **88**, 9813–17.

Wagner, R., Ryba, N. and Uhl, R. (1989) Calcium regulates the rate of rhodopsin disactivation and the primary amplification step in visual transduction. *FEBS Letters*, **242**, 249–54.

Weitz, C.J. and Nathans, J. (1992) Histidine residues regulate the transition of photoexcited rhodopsin to its active conformation, metarhodopsin II. *Neuron*, **8**, 465–72.

Weller, M., Virmaux, N. and Mandel, P. (1975) Light-stimulated phosphorylation of rhodopsin in the retina: the presence of a protein kinase that is specific for photobleached rhodopsin. *Proceedings of the National Academy of Sciences, USA*, **72**, 381–5.

Wessling-Resnick, M. and Johnson, G.L. (1987) Allosteric behavior in transducin activation mediated by rhodopsin. *Journal of Biological Chemistry*, **262**, 3697–705.

Wheeler, G.L. and Bitensky, M.W. (1977) A light-activated GTPase in vertebrate photoreceptors: regulation of light-activated cyclic GMP phosphodiesterase. *Proceedings of the National Academy of Sciences, USA*, **74**, 4238–42.

Wilden, U., Hall, S.W. and Kühn, H. (1986) Light-dependent phosphorylation of rhodopsin: number of phosphorylation sites. *Proceedings of the National Academy of Sciences, USA*, **83**, 1174–8.

Willardson, B.M., Pou, B., Yoshida, T. and Bitensky, M.W. (1993) Cooperative binding of the retinal rod G-protein, transducin, to light activated rhodopsin. *Journal of Biological Chemistry*, **268**, 6371–82.

Wood, J.G., Hart, C.E., Mazzei, G.J. *et al.* (1988) Distribution of protein kinase C immunoreactivity in rat retina. *Histochemical Journal*, **20**, 63–8.

Yamazaki, A., Sen, I. and Bitensky, M.W. (1980) Cyclic GMP-specific, high affinity, noncatalytic binding sites on light-activated phosphodiesterase. *Journal of Biological Chemistry*, **255**, 11619–24.

Yau, K.-W. and Nakatani, K. (1985a) Light-induced reduction of cytoplasmic free calcium in retinal rod outer segment. *Nature*, **313**, 579–87.

Yau, K.-W. and Nakatani, K. (1985b) Light-suppressible, cyclic GMP-sensitive conductance in the plasma membrane of a truncated rod outer segment. *Nature*, **317**, 252–5.

Yau, K.-W. and Baylor, D.A. (1989) Cyclic GMP-activated conductance of retinal photoreceptor cells. *Annual Review of Neuroscience*, **12**, 289–327.

Yee, R. and Liebman, P.A. (1978) Light-activated phosphodiesterase of the rod outer segment. *Journal of Biological Chemistry*, **253**, 8902–9.

Yoshikawa, K., Nishimura, C. and Kuriyama, K. (1982) Characterization of guanylate cyclase in frog retina using nitrosoguanidine and superoxide dismutase. *Neurochemistry International*, **4**, 129–33.

Zhukovsky, E.A. and Oprian, D.D. (1989) Effect of carboxylic amino acid side chain on the absorption maximum of visual pigments. *Science*, **246**, 928–30.

Zimmerman, A.L. and Baylor, D.A. (1986) Cyclic GMP-sensitive conductance of retinal rods consists of aqueous pores. *Nature*, **321**, 70–2.

Zozulya, S. and Stryer, L. (1992) Calcium-myristoyl protein switch. *Proceedings of the National Academy of Sciences, USA*, **89**, 11569–73.

6

Photoreceptor synaptic output: neurotransmitter release and photoreceptor coupling

STEVEN BARNES

6.1 INTRODUCTION

A distinctive feature of retinal anatomy is the stratification formed by sheets of neuronal somata interspersed by layers of neurites and synaptic terminals. This organization is responsible for the center-surround interactions that dominate the receptive fields of many retinal neurons. Even at the level of photoreceptors, the cells are coupled together to form functional networks. Direct synaptic interactions occur laterally between the photoreceptors, before signals propagate 'vertically' to bipolar and horizontal cells. Coupling endows the network of photoreceptors with important, specific properties. The focus of this chapter is on the structural and functional properties of photoreceptors that mediate lateral interactions and synaptic output. First, to describe the pathways mediating information flow between photoreceptors, features of interphotoreceptor synaptic coupling are summarized. Then, to account for transmission of the photoreceptor signal to second-order cells, the presynaptic mechanisms responsible for photoreceptor output are described.

6.2 LATERAL INTERACTIONS OF PHOTORECEPTORS IN THE RETINA

Early descriptions of cone responses to light showed that these cells are not only sensitive to light falling on their own outer segments, but they respond also to light falling on their neighbors. Light-induced hyperpolarization measured in cones from turtle (*Pseudemys scripta elegans*) increased as the diameter of the light stimulus increased up to about 50 μm (Baylor *et al.*, 1971). Similar phenomena were also reported for the rod network in toad (Fain, 1975, 1976) and tiger salamander (Werblin, 1978). It was clear that photoreceptors are coupled and interact directly with one another in a summative manner. In turtle, a negative feedback mechanism was

Neurobiology and Clinical Aspects of the Outer Retina
Edited by M.B.A. Djamgoz, S.N. Archer and S. Vallerga
Published in 1995 by Chapman & Hall, London
ISBN 0 412 60080 3

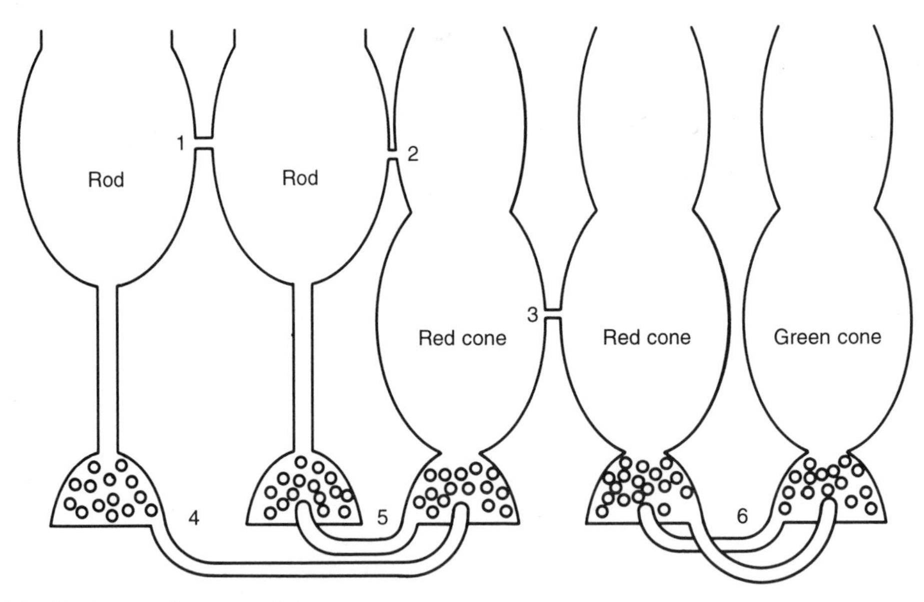

Figure 6.1 Summary diagram of photoreceptor interconnections identified anatomically or indicated electrophysiologically. Represented are (1) rod-to-rod gap junctions (Witkovsky *et al.*, 1974), (2) rod-to-cone gap junctions (Gold and Dowling, 1979), (3) cone-to- (spectrally similar) cone gap junctions (Baylor *et al.*, 1971), (4) rod-to-cone telodendria (Lasansky, 1973), (5) cone-to-rod telodendria (Lasansky, 1973), and (6) cone-to- (spectrally different) cone telodendria (Scholes, 1975). Red-sensitive and green-sensitive cone interactions are shown but telodendria couple other cone types. The telodendria provide interphotoreceptor connections at ribbon synapses, at basal junctions and at unspecified contacts. Additional references to original material describing these contacts are given in the text.

demonstrated with large (600 μm) spots. Inhibitory synaptic interactions between photoreceptors such as these are mediated by second-order interneurons and are described more completely in Chapter 9.

Photoreceptor coupling serves as a substrate for important lateral interactions that affect quantal sensitivity, chromatic sensitivity, temporal response and spatial frequency response (Attwell, 1990). Coupling may be seen in the simplest terms as a mechanism for noise reduction, which requires that a balance be struck between sensitivity at low light intensity (optimization of signal to noise ratio) and visual acuity (response to high spatial frequencies).

6.3 STRUCTURAL BASES FOR PHOTORECEPTOR COUPLING

Two types of lateral processes extend from photoreceptors in all vertebrate retinas (Figure 6.1). Short-range connections between photoreceptors are provided by processes called 'fins'. Cajal (1892) noted also that fine neurites emanate from the synaptic terminals of Golgi-stained photoreceptors. Called 'telodendria', these neurites mediate different types of photoreceptor interconnections, the pattern of which is species specific. Overall, the lateral processes have most thoroughly been characterized in fishes and reptiles, but the same organization appears to exist throughout the vertebrates.

6.3.1 ANATOMICAL EVIDENCE FOR ROD–ROD COUPLING

Gold and Dowling (1979) described fins, not telodendria, in rods of a toad (*Bufo marinus*). These are lateral processes that extend from the inner segment and interdigitate with their counterparts on immediate neighbors. Here, electron microscopy revealed close appositions resembling gap junctions seen elsewhere in the nervous system (Figure 6.3, right). Gap junctions serve as electrical synapses, allowing passive flow of current between cells. Calculations based on the electrical recordings that demonstrated coupling between 'red' rods of toad, suggest that the conductance and presumed physical size of the gap junction channels in photoreceptors are smaller by a factor of ten than those found coupling other types of cells (Gold, 1979). This may explain why dye coupling, which often occurs via gap junctions in other settings, is not typically observed between photoreceptors. Similar types of contacts probably exist between rods in other lower vertebrates and are a likely substrate of short-range summative coupling.

Rod coupling in the amphibian retina appears to be 'color'-specific. Electrical recordings indicate that 'green' rods are not coupled to 'red' rods in toad, although the 'red' rods are strongly coupled among themselves (Gold and Dowling, 1979). However, no gap junctions have been identified between rods in retinas of rabbit, or monkey (Raviola and Gilula, 1975), or in cat (Kolb, 1977), where rods lack the fins described in toad.

In most species examined, with the notable exception of mammals (Kolb, 1977; Dowling, 1987), rods also elaborate telodendria that provide pathways for longer range lateral interactions, modifying the response and the receptive fields of the cells. Electron micrographs show that these rod telodendria contact processes of neighboring rods freely in the outer plexiform layer of turtle retina (Lasansky, 1971). Rod telodendria also extend to cones in salamander retina (Lasansky, 1973). Typically, these are seen to run laterally up to about 40 μm, with occasional *en passant* swellings. Electron microscopic examination has suggested that telodendria take part in both chemical and electrical junctions, but interpretation of these contacts is not unambiguous.

6.3.2 ANATOMICAL EVIDENCE FOR CONE–CONE COUPLING

Cone telodendria contact the synaptic endings, called 'pedicles', of neighboring cones in turtle retina (Lasansky, 1971; Ohtsuka and Kawamata, 1990). In axolotl and tiger salamander retinas, these telodendria terminate within the outer plexiform layer, but the nature of their contacts has not been defined (Custer, 1973; Lasansky, 1973). In turtle, the cone telodendria invaginate to terminate at basal junctions of neighboring cones with dyads, not triads (Ohtsuka and Kawamata, 1990). In spite of the presence of the cone telodendria in salamander retina, there is no physiological evidence for direct electrical coupling between cones (Attwell *et al.*, 1984), so it is unlikely that telodendria mediate electrical coupling.

Failure to conform to univariance was observed in red-sensitive cones from turtle, and the ultrastructural foundation for this was assumed to be the telodendria that connect red- and green-sensitive cones (Normann *et al.*, 1984; Kolb and Jones, 1985). The connecting telodendria terminated commonly as central elements in invaginations and less often at basal junctions. Scholes (1975) showed that the telodendria of teleost (rudd) cones specifically invaginate spectrally different cone neighbors, according to the following scheme: green-sensitive cones contact red-sensitive cones, red-sensitive cones contact green-sensitive cones, and blue-sensitive cones contact green-sensitive cones. In goldfish, the pattern of telodendrial specificity is similar, but with red- and green-sensitive cones contacting blue-sensitive cones in addition (Stell, 1980). Here, the terminations of the lateral processes also approach the ribbon synapses in neighboring cell invaginations. Gap junctions also appear to occur between cone pairs in carp (Witkovsky *et al.*, 1974) and goldfish (Stell, 1980). However, in carp the close match between physiologically determined spectral sensitivity of cones and microspectrophotometric analysis of cone pigments suggests that chromatic tuning of the cone response by direct lateral interactions is of little importance (Tomita *et al.*, 1967).

The efficiency with which signals arising at the telodendria are propagated to the cell body, and vice versa, has been studied via computer simulations assuming observed properties of cone photoreceptor morphology and membrane ion channels. The extent to which synaptic inputs from other cells arising distally in the telodendria are electrotonically decremented at the cell body would limit the degree of lateral interaction. Models based on walleye and salamander cone morphology have indicated that bidirectional coupling between cone cell bodies and the telodendria is good (Kraft and Burkhardt, 1986; Merchant and Barnes, 1991). These simulation studies suggest that telodendria should provide effective lateral transfer of graded potentials by virtue of their short electrotonic length. In addition, direct measurement of coupling between the cell body and pedicle of turtle cones shows this pathway to be effectively coupled (Lasater *et al.*, 1989).

6.3.3 THE SPECIAL CASES OF TWIN AND DOUBLE CONES

Cones, occurring in pairs as double cones or twin cones, are also observed in the vertebrate retina. The design of this intimate physical relation between the two separate cells invites suggestions for function, but there is no compelling hypothesis for their presence in this form (see also Chapter 3). It has been suggested that the double cone structure in the sunfish (*Lepomis cyanellus*) serves in fact as a polarized light-sensitive waveguide (Cameron and Pugh, 1991). In teleost fishes, twin cones are formed by the close apposition of identical member cones. In the walleye (*Stizostedion vitreum*), there is no dye coupling between the two members, which have the same photopigment, nor is there anatomical evidence for gap junction coupling of the apposed inner segments (Burkhardt, 1986). On the other hand, a majority of twin cones send telodendria to the pedicles of immediately neighboring cones and secondary neighbors, where both anatomical and physiological evidence for synaptic junctions do exist (Burkhardt, 1986). In striped bass (*Morone saxitilis*), electrophysiological evidence obtained from isolated twin cones, which have lost their telodendria during isolation, suggests that the two photoreceptors have the same photopigment and that they are tightly electrically coupled (Miller and Korenbrot, 1993).

In other vertebrate retinas, double cones are formed by a principal (larger) member that is closely associated with an accessory (smaller) member. In accordance with the description above for twin cones, in some species (such as salamander, *Ambystoma tigrinum*) the two members of the double cone contain the same visual pigment and do not appear under electrophysiological testing to be electrotonically coupled (Attwell *et al.*, 1984). In other species such as turtle (*Chelydra serpentina*), each member of the pair produces different chromatic responses (Ohtsuka, 1985). Most double cones also appear to interact with other photoreceptors via telodendria. In an analysis of the chromatic interactions between the cones of turtle (*Geoclemys reeve-*

sii), horse radish peroxidase (HRP)-injected, chromatically identified members of double cones were shown to send telodendria to neighboring cones of selective, but different chromatic type (Ohtsuka and Kawamata, 1990). As in the case of teleost cone coupling described above, green-sensitive cells contacted primarily red- but also blue-sensitive cones, red-sensitive cells contacted green- and blue-sensitive cones, and blue-sensitive cones sent telodendria to the terminals of red- and green-sensitive cones. In the cases of both twin and double cones, their telodendria terminated at the pedicles of other photoreceptors with specializations suggesting either chemical or electrical contact.

A novel substrate for coupling between the two members of double cones in the tench retina has been proposed (Baldridge *et al.*, 1987). Electrical coupling has been reported in this species (Marchiafava, 1985; Marchiafava *et al.*, 1985), but ultrastructural analysis has failed to identify a conventional substrate for this interaction. However, a system of subsurface cisternae exists in close apposition with the plasma membrane where the two member cones make contact. This specialized structure could subserve coupling between the double cone members via an unknown mechanism (Baldridge *et al.*, 1987).

6.3.4 ANATOMICAL EVIDENCE FOR ROD–CONE COUPLING

Rods interact with cones electrically in many vertebrate retinas but generally such coupling is weak. As a substrate for these interactions in toad, 'red' rods have been shown to make gap junctions with cones (Gold and Dowling, 1979). However, no anatomical evidence for rod–cone coupling has been found in goldfish (Stell and Lightfoot, 1975; Stell *et al.*, 1982) or turtle (Ohtsuka and Kawamata, 1990). On the other hand, structural evidence for cone–rod coupling has been provided in cat (Kolb and Famiglietti, 1976; Kolb, 1977; Sterling, 1983) and monkey (Raviola and Gilula, 1975).

6.4 FUNCTIONAL SIGNIFICANCE OF PHOTORECEPTOR COUPLING

Psychophysical experiments by Hecht *et al.* (1942) investigating human visual threshold established that the photoreceptor cells of the retina are capable of signaling the absorption of a single quantum of light. Simple calculations permitted by their data suggest that a minimum of about ten photons must impinge on an area of approximately 500 photoreceptors within a few tens of milliseconds of one another in order that they be perceived. This leads to the fundamental conclusion that information from nearby rods is pooled at some point in the visual system, and that in order for it to be perceived, activity in several rods must be temporally coincident.

6.4.1 NOISE AND COINCIDENCE IN THE ROD NETWORK

The suggestion that some form of coincidence detection is necessary for perception of dim stimuli is further borne out by noise analysis experiments (Hagins *et al.*, 1970). This early work showed that the voltage response of a rat rod to a single photon is about 6 μV, three times the thermal noise in the photoreceptor. However, analysis of synaptic noise at the rod-to-bipolar cell synapse suggested that even if a single rod gave its saturated response (tens of millivolts) the noise from other rod synapses on to that bipolar cell would be too great to allow reliable detection of the signal (Falk and Fatt, 1977). Thus, for a signal to be transmitted to the bipolar cells, many photoreceptors must hyperpolarize. This could be accomplished if the rod responses were coupled so that the signal in a single photoreceptor spread to adjacent rods.

A thorough characterization of rod–rod coupling was made by Owen and Copenhagen (1977). Their experiments consisted of injecting current into a rod in dark-adapted turtle retina to polarize the membrane, while recording responses from a neighboring rod. Hyperpolarizing and depolarizing voltage steps were transmitted between neighboring rods reliably, in both directions and in all rod pairs of up to 150 μm separation. Coupling of both negative-going and positive-going voltage signals suggests that gap junctions, not chemical synapses, mediate the interaction. Other work established that the response in a single rod increased as the diameter of a spot of uniform flux was increased, up to a maximum diameter of 300 μm (Copenhagen and Owen, 1976). With 25 μm diameter spots the calculated quantal response was about 28 μV, whereas for a 300 μm spot the quantal response was 720 μV. Similar values were estimated in toad (Fain, 1977). The latency of responses also decreased as spot diameter increased. These observations may be accounted for by a network of rods extensively interconnected by electrotonic gap junctions.

Rod photoreceptor coupling was originally invoked as a mechanism for increasing the signal to bipolar cells (Falk and Fatt, 1972). It was argued that if rods were coupled in an excitatory manner, phenomena such as those observed by Hecht *et al.* (1942) could be explained. Excitatory coupling would provide an improved signal-to-noise ratio by amplifying rod input to bipolar cells. It is now well established that the photoreceptor network is coupled together in a dissipative network, not an excitatory one, and as a result, the response of a single rod to single quanta is reduced by the leakage of hyperpolarizing current into neighboring rods. Although coupling apparently reduces the signal-to-noise ratio of an individual rod, it allows the network of rods to average quantal responses, smoothing fluctuations before transmitting a signal to second order cells (Lamb and Simon, 1976; Baylor, 1978). By equalizing membrane potential between adjacent cells, the adaptational state of photoreceptors is also made uniform (Copenhagen and Green, 1987; Itzhaki and Perlman, 1987).

The increased input to bipolar cells afforded by receptor coupling can be considered to be a coincidence detection mechanism. The bipolar cell serves as the actual coincidence detector, coincidence being the necessary criterion for weak stimuli to rise above the noise of synaptic input to bipolar cells. The effect of coupling is to spread the single photoreceptor's signal over many receptors so that the bipolar cell sees input from many receptors simultaneously. Since the synaptic transfer function between photoreceptors and second-order cells has highest gain for small hyperpolarization, the spread of a quantal response among many photoreceptors may provide for a more detectable postsynaptic response. That is, a larger postsynaptic response may be generated by small signals transmitted with high gain at numerous synaptic sites rather than a larger response generated in few (or one) photoreceptor(s) and transmitted with lower gain (Norman and Perlman, 1979; Attwell *et al.*, 1987).

6.4.2 PROPAGATION OF TRANSIENT SIGNALS IN THE ROD NETWORK

Attwell and Wilson (1980) showed that the lateral spread of signals was affected by voltage-dependent membrane currents in individual rods. Their work indicated that gating of two ionic currents, later described as I_h and I_{Kx} (section 6.5), resists hyperpolarization of the photoreceptor membrane with a slight delay, sharpening the transient hyperpolarization of the response (Bader *et al.*, 1982; Beech and Barnes, 1989). High-

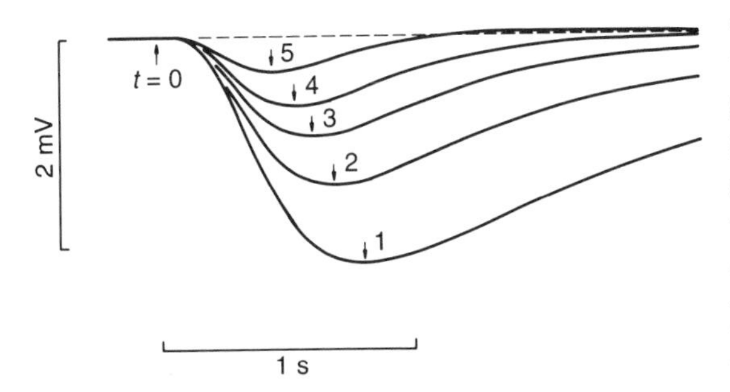

Figure 6.2 The surprising negative propagation velocity of the photoreceptor hyperpolarization crest, analysed in the turtle rod network by Detwiler *et al.* (1978). Compared to the rod which received the light stimulus directly (trace 1), the time-to-peak of the responses in each more distant rod occurs earlier (traces 2, 3, 4 and 5 were calculated for rods 20, 40, 60 and 100 μm away). Passive filtering of the voltage signal progressively slows and diminishes the signal in distant rods. As a result, membrane hyperpolarization is opposed more effectively by the slowly activating, voltage-gated currents, and the peak response occurs earlier. (Modified from Detwiler *et al.* (1978).)

pass temporal filtering results, so that a step of voltage produced in a single rod is transformed into on and off transients as it spreads laterally in the network (Owen and Torre, 1983). Thus a signal originating in a rod would spread laterally, but low temporal frequency components would be attenuated, resulting in a relative enhancement of transient signals propagating through the network.

Detwiler *et al.* (1978) observed that the spread of signals through the turtle rod network occurs with a negative propagation velocity, in as much as the peak response of more distant rods preceded that in the rod in which the signal was generated (Figure 6.2). This surprising result is a consequence of two factors. First, the leading edge of the hyperpolarization is reduced and slowed down in distant rods due to the passive electrotonic decay of membrane potential as it spreads in the network. Second, as described above, gating of I_h and I_{Kx} in the rod inner segments resists hyperpolarization so that each rod in sequence sharpens the response. The peak of the response is reached more rapidly in the distant rods because a smaller, slower hyperpolarization can be opposed by the voltage-gated currents more quickly.

The functional significance of a negative propagation velocity is unclear. It is plausible that the negative propagation focuses attention of higher order neurons upon the centre of the stimulus. Distant rods would respond quickly and briefly to a small stimulus. The bipolar cells would receive a broad initial pulse of input, which would then narrow to the centre of the stimulus where the rod response was longer lasting.

6.4.3 SPATIAL FREQUENCY FILTERING IN THE PHOTORECEPTOR NETWORK

One apparently negative effect of photoreceptor coupling is to reduce the resolving capability of the retina. As the signal spreads among receptors, its precise localization is lost. This system seems to be counterproductive if one considers the purpose of the retina to be simply the point-by-point localization and signaling of light stimuli. On the other hand, components of the visual system, such as the rod network, can be viewed as spatial frequency filters. Another way to describe the effect of receptor coupling, then, is to say that high spatial frequencies are filtered out or that the receptor network has the function of a low pass spatial filter.

If the optical properties of the eye filtered high frequencies to the same extent as the photoreceptor network, the coupling of receptors could serve as an adequate device for matching the modulation transfer functions of the ocular optics and the retina. Any high frequency signals arising neurally would be attributable to noise and they would be filtered out. An improved signal-to-noise ratio would result. In the turtle, Owen and Copenhagen (1977) described rod–rod interactions operating over a 150 μm range, whereas cone–cone interactions were more limited at 60 μm. If the optical modulation transfer function of the turtle eye were matched to a neural modulation transfer function based on 150 μm resolution, the 60 μm resolution available from the cone system would be wasted. However, it is unlikely that this would be the case; rather, the optical high frequency cut-off should more closely match the high frequency cut-off of the cone system. However, since a match then would not exist for the rod system, no improvement in the spatial frequency signal-to-noise ratio could be made for the rod network signals.

At scotopic levels of illumination, spatial frequency sensitivity is much lower than that at photopic levels. Information content is minimal at low light levels and the rod system must devote itself to retrieving what information is available. Analysis of high spatial frequency information occurs in the cone system which operates when light levels are higher and when there is a surplus of information. Thus, the vertebrate retina optimizes visual acuity and sensitivity by utilizing two systems of photoreceptors which operate over different spatial frequency and absolute sensitivity domains. With reduced spatial frequency content at scotopic levels, the task of rods is to reliably detect objects, whose form is generally represented adequately in low spatial frequencies. Rods perform this task by summing dim stimuli over space and time; in order to do this they need a large receptive area. At the same time, the retina cannot be totally devoted to low frequency analysis, because when high frequency information is available it must be exploited. The cones need to occur at high density in order to achieve this. One simple way to satisfy these conflicting requirements is to couple rods around neighboring cones.

6.4.4 CHROMATIC TUNING BY ROD–CONE COUPLING AND CONE–CONE COUPLING

Coupling between photoreceptors of the same spectral sensitivity, which has been observed extensively, does nothing to the spectral response properties of the cells. On the other hand, coupling between photoreceptors with different spectral sensitivities could provide a basis for chromatic tuning of the cell's response, depending on the nature of the coupling.

Coupling between rods and cones of different spectral sensitivities could lead to some chromatic tuning. In comparing rod–rod coupling with rod–cone coupling in the tiger salamander (*Ambystoma tigrinum*), twin electrode impalements showed that cones receive from immediately neighboring rods about one-fourth of the voltage signal that rods in the same proximity receive (Attwell and Wilson, 1980; Attwell *et al.*, 1984). Although second neighbor rods are coupled quite strongly, there is almost no detectable coupling between rods and cones separated by more than one rod. Thus, rod–cone coupling would have an effective range of about 10 μm, i.e. a rod diameter. This would imply that the spatial extent of coupling in the network would be minimal and that the chromatic ramifications arising from the local coupling would be more important. Current injection experiments indicate that coupling is weak between rods and cones in toad and turtle as well (Fain, 1976; Copenhagen and Owen, 1976).

Rod–cone interactions nevertheless do contribute to the spectral responses of photoreceptors in several vertebrate retinas

(Schwartz, 1975; Nelson, 1977; Wu and Yang 1988; Hare and Owen, 1992). Wu and Yang (1988) described rods in salamander retina that receive significant red-sensitive cone input. As described in section 6.4.5, the strength of these important interactions is modulated by light. When one considers that the vertebrate retina is composed of two rather independent light-sensing systems (rods and cones) which operate over different ranges of intensity, wavelength and temporal and spatial frequency, it is surprising that rods and cones are coupled as strongly as the evidence indicates.

One might expect cones to be coupled to each other, at least more than to rods, in spite of the loss of acuity expected from such connections. The anatomical studies described in sections 6.3.2 and 6.3.4 certainly provide support for chromatic interactions between spectrally different cones which have been described electrophysiologically (Normann *et al.*, 1985). Other electrophysiological studies also indicate that complex neural interactions contribute to the chromatic sensitivity of, for example, blue-sensitive cones in turtle, where it is suggested that red-sensitive cones provide an excitatory input (Itzhaki *et al.*, 1992). The apparent lack of direct electrical coupling between spectrally different cones and the fact that cone spectral sensitivities indicate signal sharing taken together would suggest that the interaction occurs via some form of chemical synaptic signaling rather than via gap junctions.

Some cones are nonetheless coupled electrotonically, presumably by gap junctions. Baylor and Hodgkin (1973) found direct electrical coupling between turtle cones, but this was restricted to cones with the same chromatic sensitivity. Consistent with the weaker nature of coupling, a negative propagation velocity for signals in the cone network was not observed, although this network property also bears upon the ion channel content of the cone membrane (Detwiler *et al.*, 1978).

6.4.5 MODULATION OF PHOTORECEPTOR COUPLING

The degree of coupling of the photoreceptor networks appears to be modulated in response to ambient levels of illumination. The spread of transient signals within the rod network depends on background illumination, but this effect has been attributed to changes in membrane conductance activated by hyperpolarization in rods, and not the conductance between rods (Attwell *et al.*, 1985). The strength of coupling between rod and cone photoreceptors is modulated by background light (Wu and Yang, 1988; Hare and Owen, 1992; Witkovsky *et al.*, 1994). Some rods in salamander retina receive significant red-sensitive cone input (Wu and Yang, 1988). When adapted with green light (the rod stimulus), the peak of the spectral sensitivity curve shifted from the normal rod peak near 520 nm to the red-sensitive cone peak near 620 nm. Thus, rod–cone coupling is stronger in light-adapted retinas than in dark-adapted retinas. In the study by Hare and Owen (1992), only a fraction of the rods tested exhibited this Purkinje shift. It was suggested that rods and cones in this retina are gap junction-coupled to varying degrees, and that the strength of this was related to the adaptational state of the retina. It seems a reasonable argument that this adaptability would allow the rod and cone systems to share their synaptic output machinery for most effectively transmitting signals to second order cells under light-adapted conditions (Wu, 1994). The mechanisms by which rod–cone coupling is modulated are not yet known. However, a description of horizontal cell gap junction coupling, and its modulation by dopamine (Chapter 8) might be considered as a model for modulation of coupling between cones.

6.5 THE PHOTORECEPTOR OUTPUT SYNAPSE

The origins of the light response lie in the transduction mechanisms of the photoreceptor outer segment and have been described in Chapter 5. Changes in the current flowing across the outer segment membrane lead to a voltage response that is a graded hyperpolarization from a dark resting potential near −40 mV (Bader *et al.*, 1979; Attwell *et al.*, 1982; Baylor *et al.*, 1984; McNaughton, 1990). Dim light stimuli hyperpolarize the rod membrane by one or two millivolts with a time-to-peak near 1 s. Bright light rapidly hyperpolarizes these cells to −60 or −70 mV, but the membrane quickly depolarizes back to about −50 mV due to activation of a voltage-gated current. Rods may remain hyperpolarized for seconds after a very bright flash, whereas cones, whose transduction process is a hundred times less sensitive and much faster, produce a voltage waveform more closely matching the time course of the stimulus. The photoreceptor responses bear limited resemblance to the transduction current because of voltage response shaping (signal processing) by the ensemble of ion channels in the inner segment (reviewed in Barnes, 1994). The inner segment currents that most strongly affect the voltage waveform include I_{Kx}, a non-inactivating potassium current activated by depolarization that sets resting potential and speeds the time-to-peak of dim light responses (Attwell and Wilson, 1980; Baylor *et al.*, 1984; Beech and Barnes, 1989), and I_h a non-selective monovalent-cation current activated by hyperpolarization (Attwell and Wilson, 1980; Bader *et al.*, 1982; Attwell *et al.*, 1982; Barnes and Hille, 1989; Hestrin, 1987; Maricq & Korenbrot, 1990) that generates the pronounced 'sag' to less hyperpolarized levels during responses to bright light (Fain *et al.*, 1978; Torre, 1983).

Photoreceptor voltage waveforms also bear limited relation to the postsynaptic signal. By means of simultaneous pre- and postsynaptic voltage recordings in salamander retinal slices, Attwell *et al.* (1987) described the synaptic transfer from rods to horizontal cells with a simple exponential function. They showed that rod output was graded over a narrow range of presynaptic voltages (between −40 and −45 mV). Signals of greater hyperpolarization were clipped (Belgum and Copenhagen, 1988). Clipping has been explained within the context of a Ca^{2+} current-triggered release mechanism, using the known gating parameters of photoreceptor Ca^{2+}-channels and cooperativity of Ca^{2+} inferred from a third power relation (Barnes *et al.*, 1993). Plasticity is an important feature of this synapse, as a modifiable balance exists between the relative strength of, for example in salamander, the synaptic inputs of red-versus green-sensitive cones, and red-sensitive cones versus 'green' rods to second-order cells (Hare and Owen, 1992; Hensley *et al.*, 1993; Yang *et al.*, 1994). The following sections describe the substrates and mechanisms of synaptic output from photoreceptors along a vertical pathway to horizontal and bipolar cells.

6.5.1 SPECIALIZED STRUCTURES OF THE PHOTORECEPTOR PRESYNAPTIC TERMINAL

Two anatomically different presynaptic membrane specializations serve as substrates for the synaptic output of photoreceptors (Rodieck, 1973), and are summarized, together with the typical structure of a gap junction, in Figure 6.3. The presynaptic membrane specializations of one resemble those of a typical chemical synapse when examined with the electron microscope (section 7.2.2). At this type, which is called a basal junction (Figure 6.3, left), the apposed pre- and postsynaptic membranes are smoothly parallel within a circular or elliptical

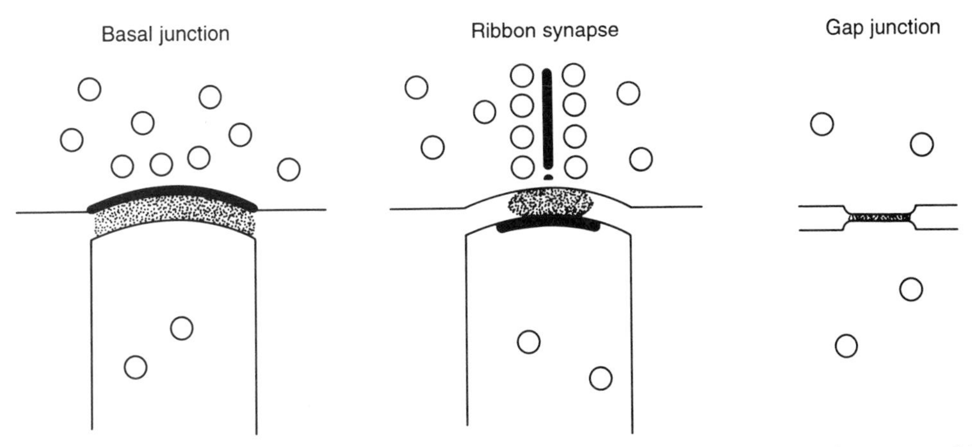

Figure 6.3 Three photoreceptor synaptic structures. Basal junctions (left) are characterized by the apposition of specialized pre- and postsynaptic membranes, with some material, revealed by electron microscopy, evident between the cells, not unlike conventional neuronal chemical synapses. Ribbon synapses (center) are characterized as sites where invaginating postsynaptic processes meet a specialized presynaptic structure, termed the ribbon for the reason that, in section, it resembles a curved ribbon, along the sides of which are organized synaptic vesicles. Gap junction synapses (right) are characterized as sites where the plasma membranes of two neighboring cells align tightly, allowing specialized membrane proteins to form large pores spanning both cells' membranes, making their cytoplasm continuous. (Based on Lasansky (1973) and Dowling (1987).)

contact patch (Lasansky, 1969, 1973). This stands in contrast to the other type of synaptic specialization, the synaptic ribbon, which is associated with the deep invaginations made by horizontal cell dendrites at the terminal regions of rod and cone photoreceptors. A highly organized array of vesicles clustered at the release site is associated only with the ribbon synapses, further differentiating the two synapse types.

Perhaps the most intriguing structure in the presynaptic photoreceptor terminal is the synaptic ribbon, which was first described by Sjöstrand (1953). Figure 6.3 (center), summarizes the general structural features of the ribbon synapse which contains electron-dense lamellae or sheets (the ribbons) which are anchored to the presynaptic membrane via an arciform density and are surrounded by layers of synaptic vesicles (Sjöstrand, 1953). Although it appears that the ribbon has a role in directing and regulating the binding of transmitter-containing vesicles to the presynaptic membrane, both the chemical composition and the functional mechanism of the ribbon remain unclear. It is likely that the ribbon represents a specialized form of vesicle docking structure, as it is found in other non-spiking, graded potential neurons such as retinal bipolar cells, pineal photoreceptors, cochlear hair cells, electroreceptor cells, teleost lateral line cells and vestibular organ cells that occur elsewhere in the nervous system (Mandell *et al.*, 1990).

A variety of specialized molecules have been described at conventional synapses. What molecular components contribute to the photoreceptor presynaptic structures? Intrinsic membrane proteins of small synaptic vesicles, such as SV2, synaptophysin and p65, described commonly at other synapses, are present diffusely within rod and cone photoreceptor synaptic terminals of rat, salamander and monkey (Mandell *et al.*, 1990).

Synapsins, a family of synaptic vesicle-associated proteins that play a key role in regulating neurotransmitter release, are present at nearly all other synapses of the central nervous system, but uniquely absent from the ribbon synapse (Mandell *et al.*, 1990).

An 88 kDa protein of unknown identity, designated as B16 antigen, has been localized specifically to synaptic ribbons of rods and cones at the outer plexiform layer and to bipolar cells of the inner plexiform layer (Balkema, 1991; Bachman and Balkema, 1993). B16 antigen is also associated with ribbon synapses in other brain regions (Mandell *et al.*, 1990), but it is present in the presynaptic densities of conventional chemical synapses. B16 antigen is a component of the photoreceptor ribbon in many vertebrates, including monkey, rat, mouse, rabbit, cat, salamander, lizard, frog, goldfish and chicken (Balkema, 1991). The molecular identity of B16 antigen remains for now undetermined, but it is not clathrin, protein kinase C, synaptophysin, spectrin, ankyrin or brain 4.1 (Balkema, 1991). An important step in understanding the coupling of intracellular free Ca^{2+} concentration with transmitter release at the photoreceptor ribbon synapse will be the structural characterization of B16 antigen and other molecules, including Ca^{2+}-binding proteins.

6.5.2 THE RELEASE OF GLUTAMATE AT THE PHOTORECEPTOR OUTPUT SYNAPSE

Photoreceptors release transmitter maximally in the dark, and reduce release when stimulated by light. Since photoreceptors hyperpolarize in response to light, and since it has been established that neurons typically release transmitter in response to depolarization, this scheme seems reasonable. Consistent with this view, experimental manipulations that reduce transmitter release, either by hyperpolarizing the presynaptic terminal or by blocking calcium influx, mimic the action of light on the activity of horizontal cells (Byzov and Trifonov, 1968; Dowling and Ripps, 1973; Marshall and Werblin, 1978, and section 7.3).

What transmitter substance is released by photoreceptors? The first clue to this question came from testing the responses of postsynaptic cells to putative transmitters. The overall view, originating four decades ago (Furukawa and Hanawa, 1955), is that the effective transmitter released by both rods and cones is the amino acid glutamate. Thus, tonic release of glutamate by rod and cone photoreceptors in darkness maintains horizontal cells and OFF-center bipolar cells in a depolarized state via several types of non-NMDA type glutamate receptors, and ON-center bipolar cells in a hyperpolarized state via APB-type glutamate receptors (Yang and Wu, 1991; Nawy and Jahr, 1990), reflecting the diversity of postsynaptic receptors and responses.

Support for the glutamate hypothesis also comes from demonstrations that glutamate is taken up by the photoreceptor presynaptic terminals (reviewed in Daw *et al.*, 1989). Glutamate is taken up with high affinity by rods and red- and green-sensitive cones in goldfish, although uptake of glutamate by blue-sensitive cones occurs with lower affinity (Marc and Lam, 1981;). In human retina, rods and (more weakly) cones are immunoreactive for glutamate (Davanger *et al.*, 1991; Crooks and Kolb, 1992). *In situ* hybridization indicates the presence of a variety of glutamate receptor subunits in the dendrites of cells postsynaptic to photoreceptors in rat (Hughes *et al.*, 1992; Peng *et al.*, 1992). The GluR4 subunit has been found in postsynaptic membranes apposing all cone types in goldfish (W.K. Stell, L. Barton and K. Schultz, in preparation).

Substances released by photoreceptors in response to membrane depolarization have also been investigated. Consistent with the glutamate hypothesis, it has been shown that toad photoreceptors release glutamate, and indeed other substances that include aspartate (which, like glutamate, is also considered an excitatory amino acid neurotransmitter elsewhere in the nervous system) (Miller and Schwartz, 1983). Depolarization stimulates this release which may have Ca^{2+}-dependent and Ca^{2+}-independent components (Schwartz, 1986), suggested to be associated with activity at ribbon synapses and basal junctions, respectively. Using an optical, biochemical reaction assay, glutamate has been identified as a substance released from reptile cones (Ayoub *et al.*, 1989).

The advent of patch clamping has allowed for very imaginative experiments that exploit novel bioassays for testing transmitter release. An electrode, with a membrane patch containing glutamate-sensitive ion channels harvested from a hippocampal neuron, was brought close to a photoreceptor synaptic terminal (Copenhagen and Jahr, 1989). Subsequent ion channel activity in the test patch indicated that an excitatory amino acid was indeed released by the terminal when the photoreceptor was depolarized or treated with lanthanum, an agent known to induce vesicular release.

6.5.3 CALCIUM CHANNELS CONTROLLING TRANSMITTER RELEASE

Reference has already been made to reports that transmission from photoreceptors to second-order cells is blocked by agents, such as cobalt or cadmium, that block Ca^{2+} channels. What kind of calcium channel has been identified in photoreceptors? Biophysical studies indicate that photoreceptor Ca^{2+} channels share many properties with other high-voltage-activated Ca^{2+} channels. The channels activate near −40 mV, they do not inactivate, Ba^{2+} permeates them better than Ca^{2+}, and they are highly sensitive to block by cadmium (Corey *et al.*, 1984; Barnes and Hille, 1989; MacLeish and O'Brien, 1992). Pharmacologically, the Ca^{2+} channels in photoreceptors are sensitive to dihydropyridines (Maricq and Korenbrot, 1988; Barnes and Hille, 1989; Lasater and Witkovsky, 1991), a class of Ca^{2+} channel drug that defines the channel as 'L-type'. However, with the advent of several new toxins for probing Ca^{2+} channel classifications, evidence has been obtained that the Ca^{2+} current in cone photoreceptors may be composed of current flowing in several types of biophysically similar but pharmacologically differentiable Ca^{2+} channels. The voltage-clamp records of Figure 6.4 indicate, for example, that L- and N-type Ca^{2+} channel currents are present in tiger salamander photoreceptors.

Changes in Ca^{2+} current, arising from sources other than membrane potential, should affect synaptic transmission. Byzov elaborated a model of electrical feedback that would affect transmitter release via the voltage-dependence of transmitter release (Byzov and Shura–Bura, 1986). It proposed that when current flow at postsynaptic sites is of sufficient density, the spatial constraints at the synaptic invaginations, e.g., long and narrow extracellular pathways for current flow, could lead to development of extracellular potentials. If the extracellular potential changed to a non-zero value, the electric field sensed by and gating presynaptic Ca^{2+} channels would change.

Ca^{2+} influx may also be modulated by other mechanisms. Sensitivity of the horizontal cell light responses to experimentally altered external pH has been reported in lower vertebrates (Negishi and Sugawara, 1973; Negishi and Svaetichin, 1966; Kleinschmidt, 1991; Barnes *et al.*, 1993; Harsanyi and Mangel, 1993). These postsynaptic changes have been attributed to the actions of protons on presynaptic Ca^{2+} channels in photoreceptors (Barnes and Bui, 1991; Barnes *et al.*, 1993) and

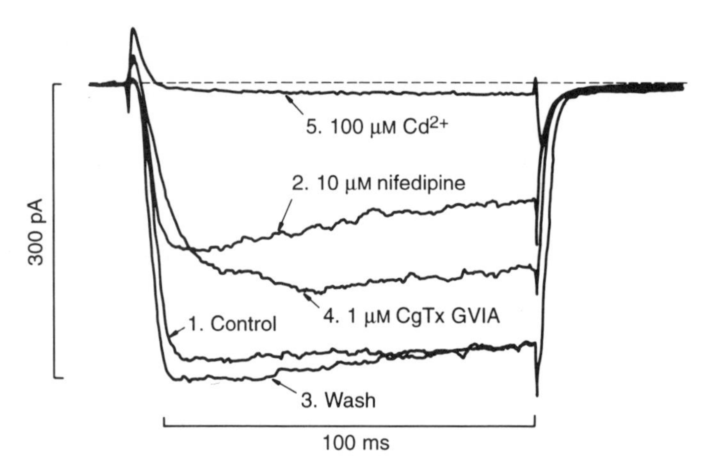

Figure 6.4 Pharmacological separation of L- and N-type calcium channel subtypes in salamander photoreceptors. Isolated cones were voltage clamped with the permeabilized patch technique. Ba^{2+} currents elicited in response to command steps to −25 mV applied from a holding potential of −60 mV are shown in control (trace 1), 10 μM nifedipine (L-type calcium channel blocker, trace 2), wash after nifedipine (trace 3), μm ω-conotoxin (CgTx) GVIA (N-type calcium channel blocker, trace 4) and 100 μM Cd^{+} (non-specific calcium channel blocker, trace 5). (Unpublished work of M.F. Wilkinson and S. Barnes.)

become physiologically relevant when considered together with the 0.02–0.2 pH unit changes that occur in the outer retina during long light exposures (Oakley and Wen, 1989; Borgula *et al.*, 1989; Yamamoto *et al.*, 1992). Additionally, the presence of the nitric oxide synthesizing enzyme has been suggested in the outer retina of rat, cat, fish, toad and tiger salamander by NADPH-diaphorase histochemistry and immunocytochemistry (Vaccaro *et al.*, 1991; Miyachi *et al.*, 1991; Vincent and Kimura, 1992; Yamamoto *et al.*, 1993; Liepe *et al.*, 1993; Kurenny *et al.*, 1994), and modulation of Ca^{2+} channels by nitric oxide has been reported in photoreceptors (Kurenny *et al.*, 1994). Figure 6.5 shows the synaptic transfer function for transmission from rods to horizontal cells in the salamander retina (Attwell *et al.*, 1987). The solid line describes the exponential dependence of postsynaptic potential (measured in horizontal cells) on presynaptic potential (measured in rods) in the dark-adapted salamander retina. The curve drawn with a dashed line is the same transfer function, but shifted by one millivolt to the left to reflect the increased probability of Ca^{2+} channel opening when, for example, external pH becomes 0.1 pH unit more alkaline. This estimate suggests that modest changes in presynaptic Ca^{2+} channel gating may have pronounced affects on postsynaptic response (Barnes *et al.*, 1993).

6.5.4 GLUTAMATE AUTORECEPTION BY PHOTORECEPTORS

Photoreceptors are electrically sensitive to their released neurotransmitter, glutamate. There are no indications that conventional postsynaptic glutamate receptors are responsible for this in photoreceptors, but cones apparently do take up glutamate in a voltage-dependent manner via electrogenic carrier proteins (Eliasof and Werblin, 1993; Tachibana and Kaneko, 1988). Exact agreement is lacking with regard to the details of this

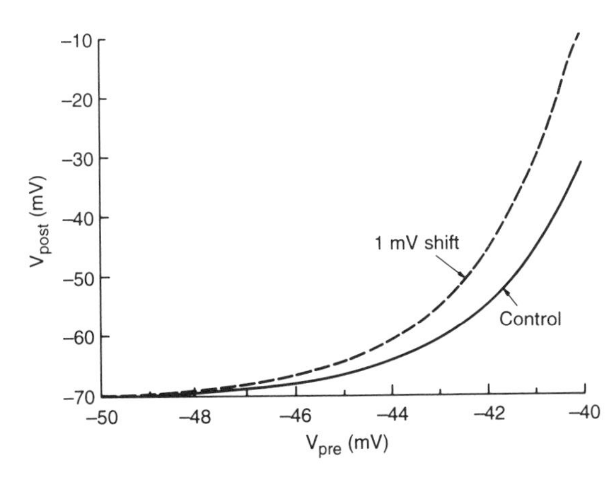

Figure 6.5 Postsynaptic ramifications of a presynaptic calcium channel activation shift. The solid line shows the synaptic transfer function for transmission from salamander rods (membrane potentials plotted as V_{pre}) to horizontal cells (membrane potentials plotted as V_{post}) as measured by Attwell *et al.* (1987). The dashed line describes the synaptic transfer function after it has been shifted 1 mV in the negative direction. The gain of synaptic transfer is increased for small voltage changes in the dark, resulting in increased depolarization in the postsynaptic horizontal cell. This estimate could apply for any treatment that alters calcium channel gating: proton channel interaction (external alkalization of 0.1 pH unit), Byzov's electric feedback model and proposed nitric oxide action. (Modified from Barnes *et al.* (1993).)

mechanism (Sarantis *et al.*, 1988), but the evidence suggests that glutamate is co-transported with three Na^+ and one Cl^- in a reaction that produces a net inward current during the uptake cycle. A recent report suggests that the uptake process is associated with Cl^- channel activity (Picaud *et al.*, 1993).

This form of glutamate autoreception may provide a positive feedback signal in photoreceptors. Since glutamate reception gives rise to a depolarizing signal in photoreceptors, release of glutamate in the dark would reinforce the depolarized state associated with darkness. Furthermore, owing to the voltage-sensitivity of the glutamate co-transporter, uptake could occur in reverse under some conditions, giving rise to release of glutamate. As discussed above, Ca^{2+}-independent release of neurotransmitters from photoreceptors has been proposed (Miller and Schwartz, 1983; Schwartz, 1986), so this mechanism for release and uptake could mediate photoreceptor communication with other cells, including other photoreceptors.

6.6 CONCLUSIONS AND FUTURE PERSPECTIVES

Photoreceptors interact with one another directly via laterally directed projections termed fins and telodendria. Anatomical and electrophysiological evidence indicates that rods are coupled with rods, rods with cones and cones with other cones to varying degrees. Chromatic interactions also vary by species, where sometimes the coupled photoreceptors are of the same spectral sensitivity, and sometimes (most often in the case of cone photoreceptors) cells of different spectral sensitivity are coupled in specific combinations. Gap junctions appear to mediate the immediate neighbor coupling in the rod and cone networks, whereas telodendria carry signals over greater distances and utilize some form of chemical signaling. The interactions provide a host of functional consequences, affecting the quantal and chromatic sensitivity of individual cells and the temporal and spatial frequency response of the coupled photoreceptor networks. Synaptic output from photoreceptors occurs electrotonically via gap junctions and chemically via glutamate release in response to depolarization. The microscopic structures supporting photoreceptor synaptic output include ribbon synapses and basal junctions at contacts with horizontal and bipolar cells and, via telodendria, other photoreceptors.

Glutamate autoreception, possibly mediated by glutamate uptake proteins, may also contribute a component to the photoreceptor response. Characterization of the agents, conditions and mechanisms underlying modulation of photoreceptor synaptic interactions should be expected. It is probable that future investigation will identify the molecular components responsible for photoreceptor synaptic interactions, both gap junctional and chemical, in the lower vertebrate models and in newly accessible mammalian preparations.

ACKNOWLEDGEMENTS

The author acknowledges helpful discussions with Drs W.K. Stell and E.V. Famiglietti. S.B. is supported as a Scholar of the Medical Research Council of Canada and the Alberta Heritage Foundation for Medical Research.

REFERENCES

Attwell, D. (1990) The photoreceptor output synapse. *Progress in Retinal Research*, **9**, 337–62.

Attwell, D., and Wilson, M. (1980) Behaviour of the rod network in tiger salamander retina mediated by properties of individual rods. *Journal of Physiology (London)*, **309**, 287–315.

Attwell, D., Werblin F.S. and Wilson M. (1982) The properties of single cones isolated from tiger salamander retina. *Journal of Physiology (London)*, **328**, 259–83.

Attwell, D., Wilson, M. and Wu, S. (1984) A quantitative analysis of interactions between photoreceptors in the salamander (*Ambystoma*) retina. *Journal of Physiology (London)*, **352**, 703–37.

Attwell, D., Wilson, M. and Wu, S. (1985) The effect of light on the spread of signals through the rod network of the salamander retina. *Brain Research*, **343**, 79–88.

Attwell, D., Borges, S., Wu S.M. and Wilson M. (1987) Signal clipping by the rod output synapse. *Nature*, **328**, 522–4.

Ayoub, G.S., Korenbrot, J.I. and Copenhagen, D.R. (1989) Release of endogenous glutamate from isolated cone photoreceptors of the lizard. *Neuroscience Research Supplement*, **10**, S47–S56.

Bachman, K.M. and Balkema, G.W. (1993) Developmental expression of a synaptic ribbon antigen (B16) in mouse retina. *Journal of Comparative Neurology*, **333**, 109–17.

Bader, C.R., MacLeish, P.R. and Schwartz, E.A. (1979) A voltage-clamp study of the light response in solitary rods of the salamander retina. *Journal of Physiology (London)*, **296**, 1–26.

Bader, C.R., Bertrand, D. and Schwartz, E.A. (1982) Voltage-activated and calcium-activated currents studied in solitary rod inner segments from the salamander retina. *Journal of Physiology (London)*, **331**, 253–84.

Baldridge, W.H., Marchiafava, P.L., Miller, R.G. and Strettoi, E. (1987) Coupling in the absence of gap junctions in the fish double cone: a dye diffusion and freeze fracture study. *Journal of Submicroscopic Cytology*, **19**, 545–54.

Balkema, G.W. (1991) A synaptic antigen (B16) is localized in retinal synaptic ribbons. *Journal of Comparative Neurology*, **312**, 573–83.

Barnes, S. (1994) After transduction: response shaping and control of transmission by ion channels of the photoreceptor inner segment. *Neuroscience*, **58**, 447–59.

Barnes, S. and Bui Q. (1991) Modulation of calcium-activated chloride current via pH-induced changes of calcium channel properties in cone photoreceptors. *Journal of Neuroscience*, **11**, 4015–23.

Barnes, S. and Hille, B. (1989) Ionic channels of the inner segment of tiger salamander cone photoreceptors. *Journal of General Physiology*, **94**, 719–43.

Barnes, S., Merchant, V., and Mahmud, F. (1993) Modulation of transmission gain by protons at the photoreceptor output synapse. *Proceedings of the National Academy of Sciences, USA*, **90**, 10081–5.

Baylor, D.A. (1978) in *Theoretical Approaches in Neurobiology* (eds W.E. Reichhardt and T. Poggio), MIT Press, Cambridge, pp. 18–27.

Baylor, D.A., Fuortes, M.G.F. and O'Bryan, P.M. (1971) Receptive fields of cones in the retina of the turtle. *Journal of Physiology (London)*, **214**, 265–94.

Baylor, D.A. and Hodgkin, A.L. (1973) Detection and resolution of visual stimuli by turtle photoreceptors. *Journal of Physiology (London)*, **234**, 163–198.

Baylor, D.A., Matthews, G. and Nunn, B.J. (1984) Location and function of voltage-sensitive conductances in retinal rods of the salamander, *Ambystoma tigrinum*. *Journal of Physiology (London)*, **354**, 203–23.

Beech, D.J. and Barnes, S. (1989) Characterization of a voltage-gated K^+ channel that accelerates the rod response to dim light. *Neuron*, **3**, 573–81.

Belgum, J.H. and Copenhagen, D.R. (1988) Synaptic transfer of rod signals to horizontal and bipolar cells in the retina of the toad (*Bufo marinus*). *Journal of Physiology (London)*, **396**, 225–46.

Borgula, G.A., Karwoski, C.J. and Steinberg, R.H. (1989) Light-evoked changes in extracellular pH in frog retina. *Vision Research*, **29**, 1069–77.

Burkhardt, D. (1986) Functional properties of twin and double cones. *Neuroscience Research, Supplement*, **4**, S54–S558.

Byzov, A.L. and Shura–Bura, T.M. (1986) Electrical feedback mechanism in the processing of signals in the outer plexiform layer of the retina. *Brain Research*, **26**, 33–44.

Byzov, A.L. and Trifonov, J.A. (1968) The response to electric stimulation of horizontal cells in the carp retina. *Vision Research*, **8**, 817–22.

Cajal, S.R.Y. (1892) La rétine des vertébrés. *La Cellule*, **9**, 17–257.

Cameron, D.A. and Pugh, E.N. Jr (1991) Double cones as a basis for a new type of polarization vision in vertebrates. *Nature*, **353**, 161–4.

Copenhagen, D.R. and Green, D.G. (1987) Spatial spread of adaptation within the cone network of turtle retina. *Journal of Physiology (London)*, **393**, 763–76.

Copenhagen, D.R. and Jahr, C.E. (1989) Release of endogenous excitatory amino acids from turtle photoreceptors. *Nature*, **341**, 536–9.

Copenhagen, D.R. and Owen, W.G. (1976) Functional characteristics of lateral interactions between rods in the retina of the snapping turtle. *Journal of Physiology (London)*, **259**, 251–82.

Corey, D.P., Dubinsky, J.M. and Schwartz, E.A. (1984) The calcium current in inner segments of rods from the salamander (*Ambystoma tigrinum*) retina. *Journal of Physiology (London)*, **354**, 557–75.

Crooks, J. and Kolb, H. (1992) Localization of GABA, glycine, glutamate and tyrosine hydroxylase in the human retina. *Journal of Comparative Neurology*, **315**, 287–302.

Custer, N.V. (1973) Structurally specialized contacts between the photoreceptors of the retina of the axototl. *Journal of Comparative Neurology*, **151**, 35–56.

Davanger, S., Ottersen, O.P. and Storm-Mathisen, J. (1991) Glutamate, GABA and glycine in the human retina: an immunocytochemical investigation. *Journal of Comparative Neurology*, **311**, 483–94.

Daw, N.W., Brunken, W.J. and Parkinson, D. (1989) The function of synaptic transmitters in the retina. *Annual Review of Neuroscience*, **12**, 205–12.

Detwiler, P.B., Hodgkin, A.L. and McNaughton, P.A. (1978) A surprising property of electrical spread in the network of rods in the turtle's retina. *Nature*, **274**, 562–5.

Dowling, J.E. (1987) *The Retina: An Approachable Part of the Brain*, Belknap Press, Cambridge, MA, pp. 57.

Dowling, J.E. and Ripps, H. (1973) Neurotransmission in the distal retina: the effect of magnesium on horizontal cell activity. *Nature*, **242**, 101–3.

Eliasof, S. and Werblin, F. (1993) Characterization of the glutamate transporter in retinal cones of the tiger salamander. *Journal of Neuroscience*, **13**, 402–11.

Fain, G.L. (1975) Quantum sensitivity of rods in the toad retina. *Science*, **187**, 838–41.

Fain, G.L. (1976) Sensitivity of toad rods: dependence on wavelength and background illumination. *Journal of Physiology (London)*, **261**, 71–101.

Fain, G.L. (1977) The threshold signal of photoreceptors, in *Vertebrate Photoreception* (eds H.B. Barlow and P. Fatt), Academic Press, London, pp. 305–23.

Fain, G.L., Quandt, F.N., Bastian, B.L. and Gerschenfeld, H.M. (1978) Contribution of a caesium-sensitive conductance increase to the rod photoresponse. *Nature*, **272**, 467–9.

Falk, G. and Fatt, P. (1972) Physical changes induced by lighting the rod outer segment of vertebrates, in *Handbook of Sensory Physiology*, vol. 7, part 1 (ed. H.J.A. Dartnell), Springer-Verlag, Heidelberg, pp. 200–44.

Furukawa, T. and Hanawa, I. (1955) Effects of some common cations on electroretinogram of the toad. *Journal of Japanese Physiology*, **5**, 289–300.

Gold, G.H. (1979) Photoreceptor coupling in retina of the toad, *Bufo marinus*. I. Physiology. *Journal of Neurophysiology*, **42**, 311–28.

Gold, G.H. and Dowling, J.E. (1979) Photoreceptor coupling in retina of the toad, *Bufo marinus*. I. Anatomy. *Journal of Neurophysiology*, **42**, 292–310.

Hagins, W.A., Penn, R.D. and Yoshikami, S. (1970) Dark current and photocurrent in retinal rods. *Biophysical Journal*, **10**, 380–412.

Hare, W.A. and Owen, W.G. (1992) Effects of 2-amino-4-phosphonobutyric acid on cells in the distal layers of the tiger salamander retina. *Journal of Physiology (London)*, **445**, 741–757.

Harsanyi, K. and Mangel, S.C. (1993) Modulation of cone to horizontal cell transmission by calcium and pH in the fish retina. *Visual Neuroscience*, **10**, 81–91.

Hecht, S., Schlaer, S. and Pirenne, M.M. (1942) Energy, quanta, and vision. *Journal of General Physiology*, **25**, 819–40.

Hensley, S.H., Yang, X.-L. and Wu, S.M. (1993) Relative contribution of rod and cone inputs to bipolar cells and ganglion cells in the tiger salamander retina. *Journal of Neurophysiology*, **69**, 2086–98.

Hestrin, S. (1987) The properties and function of inward rectification in rod photoreceptors of the tiger salamander retina. *Journal of Physiology (London)*, **390**, 319–33.

Hughes, T.E., Hermans-Borgmeyer, I. and Heinemann, S. (1992) Differential expression of glutamate receptor genes (GluR1–5) in the rat retina. *Visual Neuroscience*, **8**, 49–55.

Itzhaki, A. and Perlman, I. (1987) Light adaptation of red cones and L-horizontal cells in the turtle retina: effect of the background spatial pattern. *Vision Research*, **27**, 685–96.

Itzhaki, A., Malik, S. and Perlman, I. (1992) Spectral properties of short-wavelength (blue) cones in the turtle retina. *Visual Neuroscience*, **9**, 235–41.

Kleinschmidt, J. (1991) Signal transmission at the photoreceptor synapse. Role of calcium ions and protons. *Annals of the New York Academy of Sciences*, **635**, 468–70.

Kolb, H. (1977) The organization of the outer plexiform layer in the retina of the cat: electron microscopic observations. *Journal of Neurocytology*, **6**, 131–53.

Kolb, H. and Famiglietti, E.V. (1976) Rod and cone pathways in the retina of the cat. *Investigative Ophthalmology*, **15**, 935–46.

Kolb, H. and Jones, J. (1985) Electron microscopy of Golgi-impregnated photoreceptors reveals connections between red and green cones in the turtle retina. *Journal of Neurophysiology*, **54**, 304–17.

Kraft, T.W. and Burkhardt, D.A. (1986) Telodendrites of cone photoreceptors: structure and probable function. *Journal of Comparative Neurology*, **249**, 13–27.

Kurenny, D.E., Moroz, L.L., Turner, R.W. *et al.* (1994) Modulation of ion channels in rod photoreceptors by nitric oxide. *Neuron*, **13**, 315–24.

Lamb, T.D. and Simon, E.J. (1976) The relation between intercellular coupling and electrical noise in turtle photoreceptors. *Journal of Physiology (London)*, **263**, 257–86.

Lasansky, A. (1969) Basal junctions at synaptic endings of turtle visual cells. *Journal of Cell Biology*, **40**, 577–81.

Lasansky (1971) Synaptic organization of cone cells in the turtle retina. *Philosophical Transactions of the Royal Society of London, B, Biological Sciences*, **262**, 365–81.

Lasansky, A. (1973) Organization of the outer synaptic layer in the retina of the larval tiger salamander. *Philosophical Transactions of the Royal Society of London, B, Biological Sciences*, **265**, 471–89.

Lasater, E.M. and Witkovsky, P. (1991) The calcium current of turtle cone photoreceptor axon terminals. *Neuroscience Research, Supplement*, **15**, S165–S173.

Lasater, E.M., Normann, R.A. and Kolb, H. (1989) Signal integration at the pedicle of turtle cone photoreceptors: an anatomical and electrophysical study. *Visual Neuroscience*, **2**, 553–64.

Liepe, B.A., Stone, C. and Copenhagen, D.R. (1993) Localization of NADPH-diaphorase positive cells in fish and salamander retinas.

Investigative Ophthalmology and Visual Science, **34**, 752.

MacLeish, P.R. and O'Brien, E.V. (1992) Fura-2 imaging of calcium in solitary monkey cones. *Investigative Ophthalmology and Visual Science*, **33**, 753.

Mandell, J.W., Townes–Anderson, E., Czernik, A.J. *et al.* (1990) Synapsins in the vertebrate retina: absence from ribbon synapses and heterogeneous distribution among conventional synapses. *Neuron*, **5**, 19–33.

Marc, R.E. and Lam, D.M.K. (1981) Uptake of aspartric and glutamic acid by photoreceptors in goldfish retina. *Proceedings of the National Academy of Sciences, USA*, **78**, 7185–9.

Marchiafava, P.L. (1985) Cell coupling in double cones of the fish retina. *Proceedings of the Royal Society of London, B*, **226**, 211–15.

Marchiafava, P.L., Strettoi, E. and Alpigiani, V. (1985) Intracellular recording from single and double cone cells isolated from the fish retina, *Tinca tinca*. *Experimental Biology*, **44**, 173–80.

Maricq, A.V. and Korenbrot, J.I. (1988) Calcium and calcium-dependent chloride currents generate action potentials in solitary cone photoreceptors. *Neuron*, **1**, 503–15.

Maricq, A.V. and Korenbrot, J.I. (1990) Inward rectification in the inner segment of single retinal cone photoreceptors. *Journal of Neurophysiology*, **64**, 1917–28.

Marshall, L.M. and Werblin, F.S. (1978) Synaptic transmission to the horizontal cells in the retina of the larval tiger salamander, *Journal of Physiology (London)*, **279**, 321–46.

McNaughton, P.A. (1990) Light response of vertebrate photoreceptors. *Physiological Reviews*, **70**, 847–83.

Merchant, V. and Barnes, S. (1991) Lateral signal propagation via cone telodendrites in tiger salamander retina. *Society for Neuroscience Abstracts*, **17**, 298.

Miller, A.M. and Schwartz, E.A. (1983) Evidence for the identification of synaptic transmitters released by photoreceptors of the toad retina. *Journal of Physiology (London)*, **334**, 325–49.

Miller, J.L. and Korenbrot, J.I. (1993) Phototransduction and adaptation in rods, single cones and twin cones of the striped bass retina: a comparative study. *Visual Neuroscience*, **10**, 653–67.

Miyachi, E.-I., Miyakawa, A. and Murakami, M. (1991) Modulation of electrical coupling between retinal horizontal cells by intracellular messengers. *Neuroscience Research*, **15**, S41–S49.

Nawy, S. and Jahr, C.E. (1990) Suppression by glutamate of cGMP-activated conductance in retinal bipolar cells. *Nature*, **346**, 269–71.

Negishi, K. and Sugawara, K. (1973) Evidence for the anoxia sensitivity of the synaptic region at the outer plexiform layer in the fish retina. *Vision Research*, **13**, 983–7.

Negishi, K. and Svaetichin, G. (1966) Effects of anoxia, CO_2 and NH_3 on S-potential producing cells and on neurons. *Pflugers Archiv*, **292**, 177–205.

Nelson, R. (1977) Cat cones have rod input: a comparison of the response properties of cones and horizontal cell bodies in the retina of the cat. *Journal of Comparative Neurology*, **172**, 109–36.

Normann, R.A. and Perlman, I. (1979) Signal transmission from red cones to horizontal cells in the turtle retina. *Journal of Physiology (London)*, **286**, 509–24.

Normann, R.A., Perlman, I., Kolb, H. *et al.* (1984) Direct excitatory interactions between cones of different spectral types in the turtle retina. *Science*, **224**, 625–7.

Normann, R.A., Perlman, I. and Daly, S.J. (1985) Mixing of color signals by turtle cone photoreceptors. *Journal of Neurophysiology*, **54**, 293–303.

Oakley II, B. and Wen, R. (1989) Extracellular pH in the isolated retina of the toad in darkness and during illumination. *Journal of Physiology (London)*, **419**, 353–78.

Ohtsuka, T. (1985) Spectral sensitivities of seven morphological types of photoreceptors in the retina of the turtle, *Geoclemys reevesii*. *Journal of Comparative Neurology*, **237**, 145–54.

Ohtsuka, T. and Kawamata, K. (1990) Telodendrial contact of HRP-filled photoreceptors in the turtle retina: pathways of photoreceptor coupling. *Journal of Comparative Neurology*, **292**, 599–613.

Owen, W.G. and Copenhagen, D.R. (1977) Characteristics of the electrical coupling between rods in the turtle retina, in *Vertebrate Photorecep-*

tion (eds H.B. Barlow and P. Fatt) Academic Press, London.

Owen, W.G. and Torre, V. (1983) High-pass filtering of small signals by retinal rods. *Biophysical Journal*, **41**, 325–39.

Peng, Y.-W., Blackstone, C.D., Huganir, R.L. and Yau, K.-W. (1992) Distribution of glutamate receptor subtypes in rat retina. *Investigative Ophthalmology and Visual Science*, **33**, 1174.

Picaud, S., Grant, G.B., Larsson, H.P. *et al.* (1993) The glutamate transporter is a chloride channel in salamander cone photoreceptors. *Society for Neuroscience Abstracts*, **19**, 230.

Raviola, E. and Gilula, N.B. (1975) Intramembrane organization of specialized contacts in the outer plexiform layer of the retina. A freeze-fracture study in monkeys and rabbits. *Journal of Cell Biology*, **65**, 192.

Rodieck, R.W. (1973) *The Vertebrate Retina: Principles of Structure and Function*. W.H. Freeman, San Francisco.

Sarantis, M., Everett, K. and Attwell, D. (1988) A presynaptic action of glutamate at the cone output synapse. *Nature*, **332**, 451–3.

Scholes, J.H. (1975) Colour receptors and their synaptic connexions in the retina of a cyprinid fish. *Philosophical Transactions of the Royal Society of London*, B, *Biological Sciences*, **270**, 61–118.

Schwartz, E.A. (1975) Cones excite rods in the retina of the turtle. *Journal of Physiology (London)*, **246**, 639–51.

Schwartz, E.A. (1986) Synaptic transmission in amphibian retinae during conditions unfavorable for calcium entry into presynaptic terminals. *Journal of Physiology (London)*, **376**, 411–28.

Sjöstrand, F.S. (1953) The ultrastructure of the inner segments of the retinal rods of the guinea pig eye as revealed by electron microscopy. *Journal of Cellular and Comparative Physiology*, **42**, 45–50.

Stell, W.K. (1980) Photoreceptor-specific synaptic pathways in goldfish retina: a world of colour, a wealth of connections, in *Colour Vision Deficiencies* V (ed. G. Verriest), Adam Hilger, Bristol.

Stell, W.K. and Lightfoot, D.O. (1975) Color-specific interconnections of cones and horizontal cells in the retina of the goldfish. *Journal of Comparative Neurology*, **159**, 473–502.

Stell, W.K., Kretz, R. and Lightfoot, D.O. (1982) Horizontal cell connectivity in goldfish. in *The S-Potential* (eds B.D. Drujan and M. Laufer), Alan R. Liss, New York, pp.51–75.

Sterling, P. (1983) Microcircuitry of the cat retina. *Annual Reviews of Neuroscience*, **6**, 149–85.

Tachibana, M., and Kaneko, A. (1988) L-Glutamate-induced depolarization in solitary photoreceptors: a process that may contribute to the interaction between photoreceptors in situ. *Proceedings of the National Academy of Science, USA*, **85**, 5315–19.

Tomita, T., Kaneko. A. Murakami, M. and Pautler, E.L. (1967) Spectral sensitivity curves of single cones in carp. *Vision Research*, **7**, 519–31.

Torre. V. (1983) The contribution of the electrogenic sodium–potassium pump to the electrical activity of toad rods. *Journal of Physiology (London)*, **333**, 315–41.

Vaccaro, T.M., Cobcroft, M.D., Provis, J.M., and Mitrofanis, J. (1991), NADPH-diaphorase reactivity in adult and developing cat retinae. *Cell Tissue Research*, **265**, 371–9.

Vincent, S.R. and Kimura, H. (1992) Histochemical mapping of nitric oxide synthase in the rat brain. *Neuroscience*, **46**, 755–84.

Werblin, F.S. (1978) Transmission along and between rods in the tiger salamander retina. *Journal of Physiology (London)*, **280**, 449–70.

Witkovsky, P., Akopian, A., Krizaj, D. and Zhang, J. (1994) A cellular basis for light and dark adaptation in the *Xenopus* retina. *Great Basin Visual Science Symposium*, **1**, 49–57.

Witkovsky, P., Shakib, M. and Ripps, H. (1974), Interreceptoral junctions in the teleost retina. *Investigative Ophthalmology and Visual Science*, **13**, 996–1009.

Wu, S.M. (1994) Synaptic transmission in the outer retina. *Annual Review of Physiology*, **56**, 141–68.

Wu, S.M. and Yang, X.-L. (1988) Electrical coupling between rods and cones in the tiger salamnder retina. *Proceedings of the National Academy of Science* USA, **85**, 275–8.

Yamamoto, F., Borgula, G.A. and Steinberg, R.H. (1992) Effects of light and darkness on pH outside rod photoreceptors in the cat retina. *Experimental Eye Research*, **54**, 689–97.

Yamamoto, R., Bredt, D.S., Snyder, S.H. and Stone, R.A. (1993) The localization of nitric oxide synthase in the rat eye and related cranial ganglia. *Neuroscience*, **54**, 189–200.

Yang, X.-L. and Wu, S.M. (1991) Coexistence and function of glutamate receptor subtypes in the horizontal cells of the tiger salamander retina. *Visual Neuroscience*, **7**, 377–82.

Yang, X.-L., Fan, T.-X. and Shen, W. (1994) Effects of prolonged darkness on light responsiveness and spectral sensitivity of cone horizontal cells in carp retina *in vivo*. *Journal of Neuroscience*, **14**, 326–34.

7

Photoreceptor–horizontal cell connectivity, synaptic transmission and neuromodulation

MUSTAFA B.A. DJAMGOZ,
HANS-JOACHIM WAGNER and
PAUL WITKOVSKY

7.1 INTRODUCTION

Horizontal cells (HCs) are second-order neurons receiving their main synaptic drive from photoreceptors. A minor input may be derived from bipolar cells in the amphibian retina (Lasanky, 1979), whereas in many vertebrates a major modulatory influence is exerted by the interplexiform cells (Chapter 15). HCs occur in a number of morphological forms even in the retina of a given species. They may be axonless or axon-bearing, and the latter may have an axon terminal region which may or may not contact photoreceptors. Thus, in retinas of different species rod and cone photoreceptor inputs can be arranged in a variety of different forms.

The best understood role of HCs in vision is to provide a signal representing the mean luminance, averaged over a broad retinal area. This signal is subtracted from a spatially localized input of photoreceptors to bipolar cells to yield a signal coding for local contrast. The ability of HCs to integrate information from widely separated photoreceptors depends on the permeability of large gap junctions which join neighboring HCs. Another, less well understood, function of HCs is in color vision. In lower vertebrates, a variety of so-called 'chromaticity' HCs are found, characterized by a dependence of response polarity on the wavelength of the stimulating light.

This chapter deals with three main aspects of photoreceptor input to HCs: (a) connectivity, (b) synaptic transmission and (c) neuromodulation. These topics were reviewed previously with varying emphasis by Dowling (1987), Kaneko (1987), Attwell (1986, 1990), Falk (1989) and Wu (1994). HC–photoreceptor feedback is covered in Chapter 9.

7.2 HORIZONTAL CELL–PHOTORECEPTOR CONNECTIVITY

7.2.1 LIGHT MICROSCOPIC CHARACTERIZATION

Horizontal cells form the vitread border of the first synaptic layer of the retina. As their

Neurobiology and Clinical Aspects of the Outer Retina
Edited by M.B.A. Djamgoz, S.N. Archer and S. Vallerga
Published in 1995 by Chapman & Hall, London
ISBN 0 412 60080 3

name implies, these cells are orientated mainly perpendicularly to the incoming light and to the mainstream of neural information leaving the photoreceptor layer for the inner retina and the brain. Indeed, in numerous lower vertebrates, HCs are arranged in prominent tiers which were initially taken for separate membranes (Krause, 1868), fenestrated only to accommodate the radial processes of bipolar and Müller cells. The other conspicuous feature of HCs is the frequent occurrence of laterally oriented, axon-like processes (Figures 7.1–7.4). The latter can extend for several hundred micrometers and terminate as unbranched, cylindrical 'nematode'-like (B.B. Boycott, personal communication) expansions, or as conspicuous, profusely branched telodendria (Boycott *et al.*, 1978).

The light microscopic differentiation of HCs may take into account the following criteria: (a) the number of cell layers and the position of the perikaryon within them; (b) the specific shape of the perikaryon; (c) the presence or absence of a (short) axon; and (d) the degree of specificity with which a HC and/or its axon terminal is synaptically connected to a pattern of photoreceptors.

In the literature, these morphological characteristics have been used in various combinations, often supplemented with light-evoked physiological response characteristics, and this has resulted in a potentially confusing nomenclature. In the following, we describe the classification of HCs based primarily on the last two criteria. Following the suggestion of Gallego (1985), we distinguish two broad groups according to whether a short axon is present or not, i.e. axonless HCs (ALHCs) and short-axon HCs (SAHCs). In each of these two groups, several subtypes may be found which are characterized by their photoreceptor connectivity. For the SAHC, it is further important to take into account a differential pattern of connectivity of perikaryal dendrites as opposed to the axonal telodendron. The degree of specificity with which photoreceptors are contacted concerns not only the segregation of rod and cone channels, but is also of special importance as regards the processing of chromatic information.

The competent evaluation of these questions requires not only a reliable classification of HCs but, at least as importantly, the unequivocal identification of the chromatic type of cone. Immunocytochemical labeling of a specific opsin can be used to identify a spectral type of cone and can be applied in combination with the selective marking (e.g. by intracellular dye injection) of a single HC. However, this approach has not been used to date. Instead, mostly non-analytical, descriptive morphological features have been relied upon. In some reptiles and birds, cone types may be identified by observing the color of oil droplets located at the apical parts of the inner segments (see Table 7.1 for references). In many fishes, chromatic types of cone not only have different shapes and sizes, but often are arranged in crystal-like mosaic patterns, and can also be distinguished by specific ultrastructural features associated

Figure 7.1 Horizontal cells of the primate retina. s, soma; t, at, axon terminals. (a) Polyak's (1941) drawings of Golgi-stained HCs in a wholemount view of the monkey retina (from Gallego, 1985). Small, unlabeled structures denote rod spherules and cone pedicles. (b) Type 2 SAHC. (c) Type 1 SAHC of the parafoveal region. (d) Type 1 SAHC of the peripheral retina. (e) Schematic drawing of the connectivities of the three types of SAHCs according to Ahnelt and Kolb (1994). R, long wavelength-sensitive cone; G, middle wavelength-sensitive cone; B, short wavelength-sensitive cone; unlabeled terminals represent rod spherules.

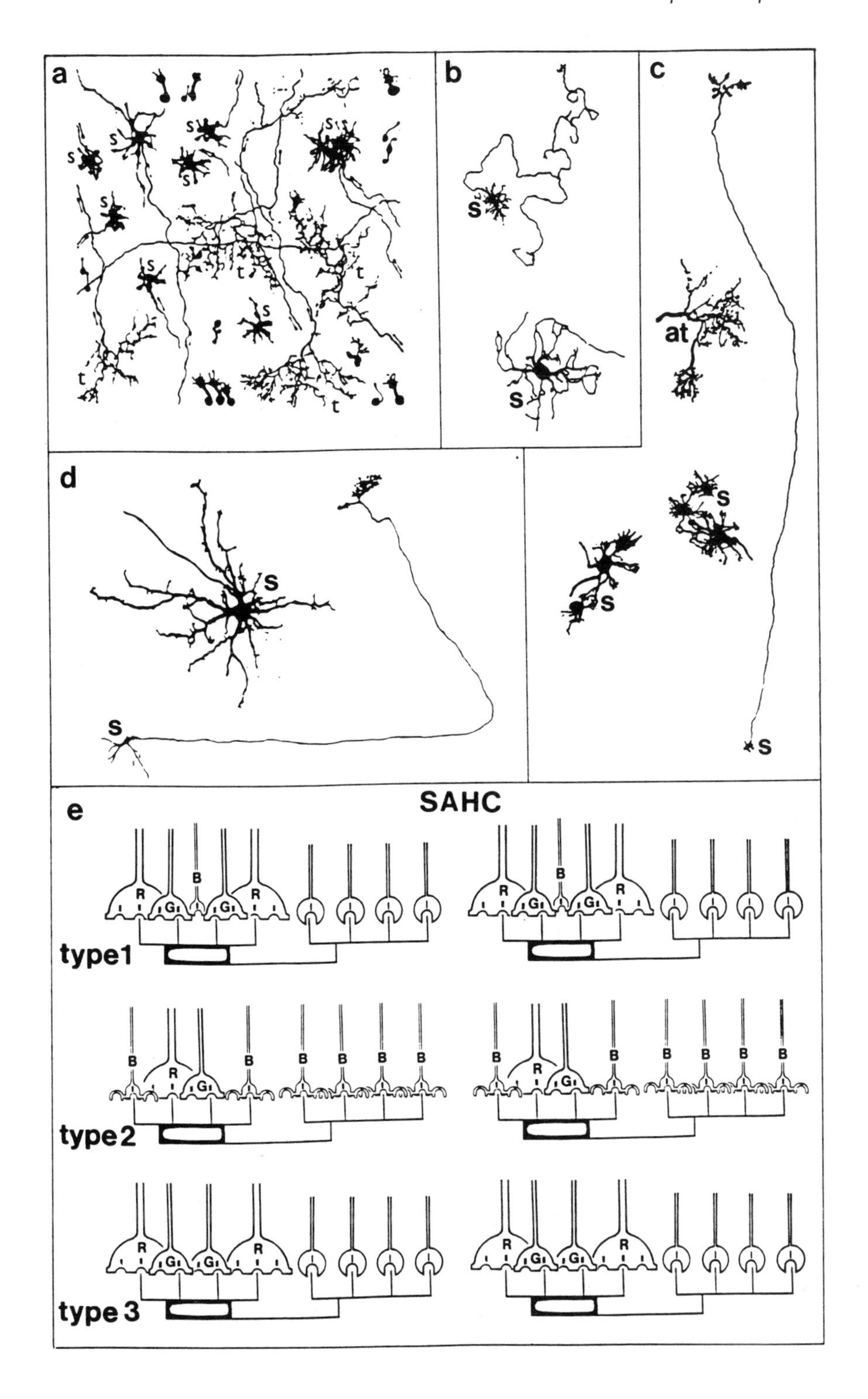
a
b
c
d
e
SAHC
type1
type2
type3
at
R
G
B

with their synaptic terminal (Rodieck, 1988, Wagner, 1990; Wässle and Boycott, 1991). The problem of chromatic identity is most difficult to solve in mammalian, and especially primate and human cones, since morphological differences are not readily apparent and a special arrangement is not obvious either. In the most recent approach, Ahnelt and Kolb (1994) have relied on the specific arrangement and morphology of short wavelength-sensitive (B-) cones in order to analyze the pattern of HC connectivity. Interestingly, their findings indicate that it is important to take into account the specific type of contact made as well as the relative 'weighted' input of the respective cone types.

In order to facilitate an overview of the patterns of connectivity between specific photoreceptors and types of HCs, the available data have been assembled in Table 7.1 where the relevant references are also given. Most vertebrate retinas contain at least one type of ALHC and one type of SAHC (Figure 7.1–7.4). Exceptions are found in primates, rodents and elasmobranchs. In humans and primates, all HCs have short axons (Figure 7.1 a–d). On the other hand, rats and gerbils have only ALHCs but guinea pigs have one type of each. Elasmobranchs (such as sharks), rats and ratfish also have ALHCs as the only kinds of HC.

In the majority of teleost fishes studied, ALHC contact rods exclusively (Figure 7.4a, b). Only in one case of a marine species, *Eugerres*, a type of ALHC has been found with chromaticity responses indicating that this cell receives input from cones. In reptiles and birds, on the other hand, ALHCs contact exclusively cones (Figure 7.3d). In the chick retina, but especially in the turtle, a high degree of specificity has been observed in the chromatic selectivity of cone contacts. In most non-primate mammals, the type-A ALHCs contact cones in an unselective way; only in the gray squirrel are rods also connected to this type of HC.

As for the SAHCs, it was assumed for a long time in mammals, that the perikaryal dendrites receive input exclusively from cones whereas the telodendrial arborizations are synaptically linked to rods. However, recent evidence suggests that the situation may be more complex. In the gray squirrel retina, for instance, the axon terminals are linked to cones and not rods. In humans, three different types of SAHCs have been distinguished and these show clear signs of selective connectivity with the chromatic cone types (Ahnelt and Kolb, 1994). Type 1 HCs conform most closely to the classical concept, since the perikaryal dendrites are linked to all kinds of cones and the axon terminal to rods; however, a detailed analysis of connectivity indicates that B-cones are underrepresented in its connectivity. The type 2 SAHC receives preferential input from B-cones in its perikaryal dendrites; the axon terminal of this type is connected to a pure population of B-cones. The cell body of the type-3 HC avoids input from B-cones altogether. In rabbits, there may be a similar

Figure 7.2 Horizontal cells of non-primate mammalian retina. (a) Cajal's (1893) drawings assembled from Gallego (1985) and relabeled. Radial views of an 'external' (1), and two 'internal' (2,3) SAHCs of the dog retina; (4–6) radial views of 'internal' SAHCs of the bovine retina; (7) 'external' SAHC of the bovine retina in wholemount view. Axon terminals in wholemount view of bovine (8) rabbit (9) and canine (10) retinas. (11) Axon terminal of a bovine SAHC in radial view. (b) ALHC and SAHC of the cat retina (from Fisher and Boycott, 1974). (c) Schematic drawing of the connectivities of the two types of HC in generalized non-primate mammal (e.g. cat) retina. For definitions of the abbreviations, see Figure 7.1 legend.

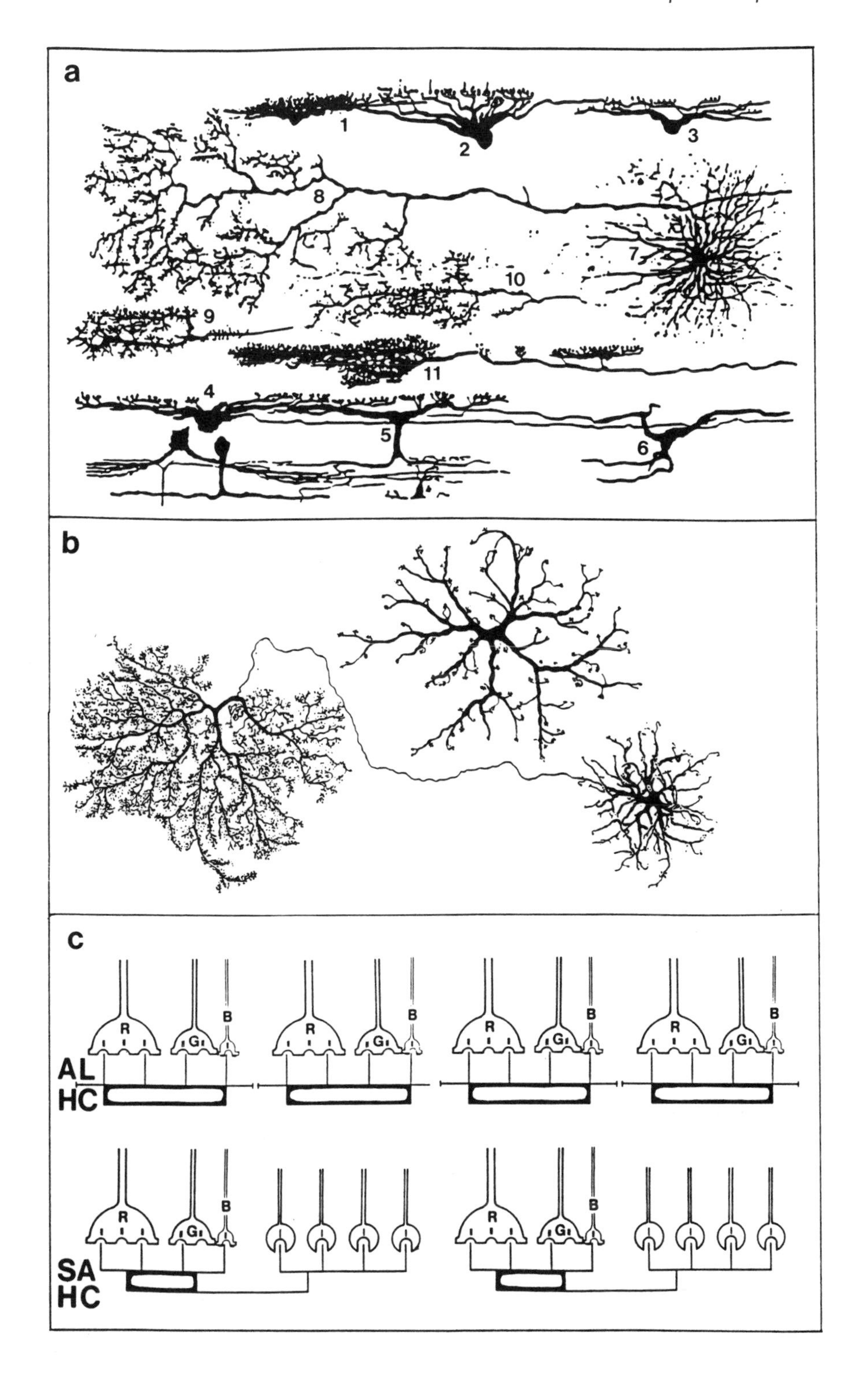
a
1
2
3
8
7
10
9
11
4
5
6
b
c
R
G
B
AL
HC
SA
HC

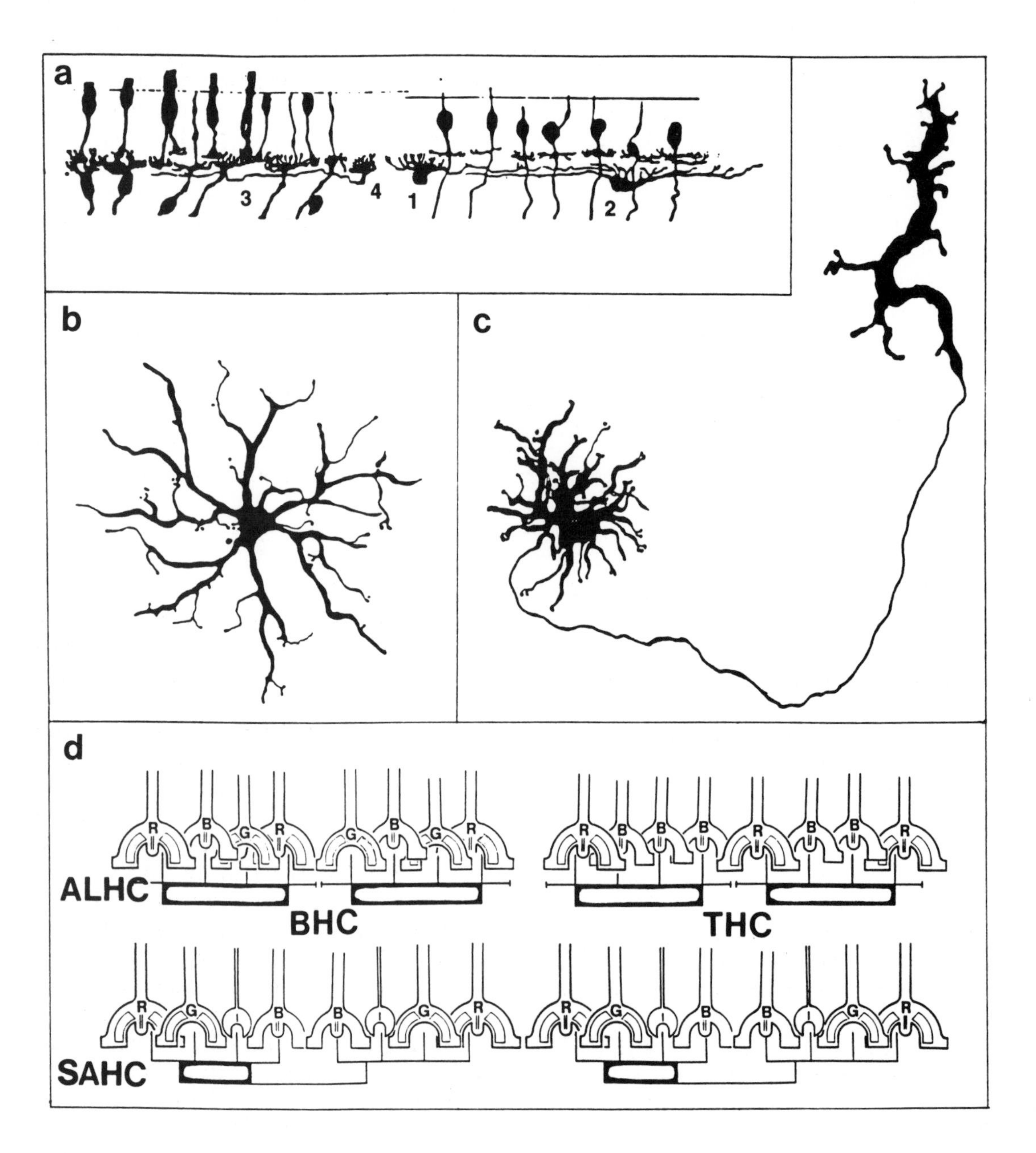

Figure 7.3 Horizontal cells of reptile retina. (a) Cajal's (1893) drawings of Golgi-stained HCs in radial view of the lizard retina assembled from Gallego (1985) and relabeled. (1) 'brush-shaped, inner' SAHC; (2) 'stellate, inner' SAHCs; (3) axons; (4) axon terminals. (b and c) ALHC and SAHC of the turtle retina (from Gallego, 1985). (d) Schematic drawing of the connectivities of HC types in the turtle (based on findings by Ohtsuka and Kouyama, 1986). For abbreviations, see Figure 7.1 legend.

situation with respect to the type-2 cell, since the dendrites of this cell also show a preferential contact pattern with B-cones (Famiglietti, 1990).

In birds, most SAHCs agree with the 'classical' pattern of cone input to the soma and rod input to the axon terminals (Gallego *et al.*, 1975). Only in the chick, a single cone type, in addition to rods, has been found to contact three axon terminals.

A different situation has been found in reptiles and amphibians. In most cases, cell bodies and axon terminals receive input from both rods and cones, with no spectral selectivity. Only long wavelength-sensitive (R-) cones seem to play a special role, sometimes, impinging preferentially on to the axon terminal in some turtle retinas, as well as to the perikarya of SAHCs in *Xenopus*.

In many teleost fish with highly developed color vision, at least three different types of SAHCs are present (Figure 7.4b). In these animals, the axon terminals do not contact photoreceptors but engage in chemical synapses with amacrine cell perikarya. The dendritic tufts originating from the cell bodies are in contact with specific combinations of cone types suggesting a precise and selective pattern of chromatic connectivity. The cyprinid family of teleosts, with goldfish, carp and roach as representative species, has been studied best in this respect. As a general pattern, it has emerged that H1 cells with a luminosity response are linked to all spectral cone types. H2 cells with a biphasic chromaticity response do not contact R-cones, but only medium wavelength-sensitive (G-) and B-cones. Finally, H3-type SAHCs generate a triphasic chromaticity response; they extend their dendrites exclusively into B-cone pedicles (Stell *et al.*, 1982; Djamgoz and Downing, 1988; Downing and Djamgoz, 1989; Greenstreet and Djamgoz, 1994).

In conclusion, this overview shows that there is no clear and consistent pattern in the morphological differentiation of HCs throughout vertebrate retinas. The two broad groups of ALHCs and SAHCs can have vastly overlapping patterns of connectivity with rods and cones, including chromatic selectivity with the latter. Therefore, it is difficult to propose a specific and uniform functional role for each of the two main HC types. However, there appears to be a trend in species with highly developed color vision for specific types of HCs to engage in selective contacts with chromatic cone types. This has been most clearly demonstrated in fish and turtles. In primates, this finding seems less unequivocal. However, the recent findings about a special role of B-cones in HC connectivity in the human retina indicates that, here too, this trend can be recognized.

7.2.2 ULTRASTRUCTURAL CHARACTERISTICS

Photoreceptor–HC synapses are characteristic of an array of multiple complexes which constitute the functional bases of feed-forward and feedback signal transmission, i.e. a special kind of reciprocal synapse (for reviews see Stell, 1972; Dowling, 1987). Typically, the photoreceptor terminals form bulbous swellings and give rise to a number of short telodendria. In most species, the rod terminal has a globular shape, and is therefore often referred to as a 'spherule', whereas in cones, the vitread base is generally flattened, and is called a 'pedicle' (Figure 7.5). The vast majority of the dendrites of secondary neurons contact the photoreceptor terminals from the vitread side. Whereas some bipolar cell dendrites may form 'flat' contacts at the pedicle base, others form invaginating junctions. By contrast, HC processes invariably form conspicuous invaginating contacts, typically terminating as a pair and flanking the plate-like synaptic ribbon. In most mammals, triadic arrays of two HC processes and one bipolar dendrite in the central position form the classical ribbon synapse (Figure 7.5a). However, in lower vertebrates, especially fish, HC processes have also been found in the central position underneath the ribbon (Figure 7.5b). Another notable difference between higher and lower vertebrates concerns the general pattern of invaginating processes into cone pedicles. In mammals, typically there are numerous triads each forming an individual invagination and ribbon synapse at the pedicle base, whereas in teleosts, there is a single cavity within the bell-shaped cone pedicle which may harbor more than 50 bipolar and HC processes. These are arranged in dyads or triads forming secondary invaginations associated with more than a dozen ribbon synapses. On the other hand, in rods, the morphology of the spherules is more uniform across vertebrates. Generally there is a single cavity occupied by one or two ribbon synapses in the classical triad configuration. In conclusion, the high degree of complexity in the organization of the photoreceptor synapses bears little resemblance to a conventional (Gray type I or II) synapse and would suggest that in addition to the classical mechanisms of signal transmission associated with a chemical synapse, special mechanisms may be active involving non-vesicular transmission based on the activity of intramembraneous transporter molecules (Dong *et al.*, 1994) and/or ephaptic effects (Byzov and Shura–Bura, 1986).

Electron microscopy is the most widely used approach in the attempt to identify the sites as well as the direction of signal transmission between photoreceptors and HCs. In the photoreceptor terminals, the most striking features are the multitude of clear vesicles which all but completely fill the cytoplasm, as well as the plate-like synaptic ribbons which consist of 3–5 alternating light and dark sublayers and, at high resolution, electron-dense, globular subunits. Although the arrangement of the synaptic vesicles is mostly random, they show a more regular alignment in the vicinity of the synaptic ribbons. It has therefore been speculated that synaptic ribbons play a special role in delivering vesicles to their site of release, which is of special importance in synapses with tonic release situation (De Vries and Baylor, 1993). Fine structural analysis has provided conclusive evidence that the synaptic ridge corresponds to the active zone of the presynapse where exocytosis of synaptic vesicles occurs and transmitter is released. Closely associated are large intramembraneous particles which are likely candidates for the entry of extracelluar Ca^{2+} into the pedicle. Coated vesicles indicative of membrane retrieval are found close to the base of the synaptic ridge (Schaeffer *et al.*, 1982). The cytoplasm of the invaginating HC dendrites contains only very few vesicles which are mostly irregular in shape. Opposite the flank or apex of the synaptic ridge, their membrane shows a conspicuous thickening. Freeze-fracture preparations show aggregates of P-face particles associated with an array of tiny, 6 nm protrusions arranged in 20 nm diameter

Figure 7.4 Horizontal cells of the fish retina. (a) Cajal's (1893) drawings of Golgi-stained HCs in radial view of the perch retina assembled from Gallego (1985) and relabeled; (1) 'brush-shaped, external' SAHCs; (2) 'stellate, inner' SAHCs; (3) axons; (4) axon terminals. (b) Wholemount view of HCs of the goldfish (modified from Stell, 1975). RH, rod horizontal cell (ALHC); H1–H3, cone-connected SAHCs. (c) Schematic drawing of the connectivities of horizontal cell types in the goldfish (based on Stell *et al.*, 1982). Note that the axons of the cone-connected SAHCs are directed towards the amacrine cell layer. For other abbreviations, see Figure 7.1 legend.

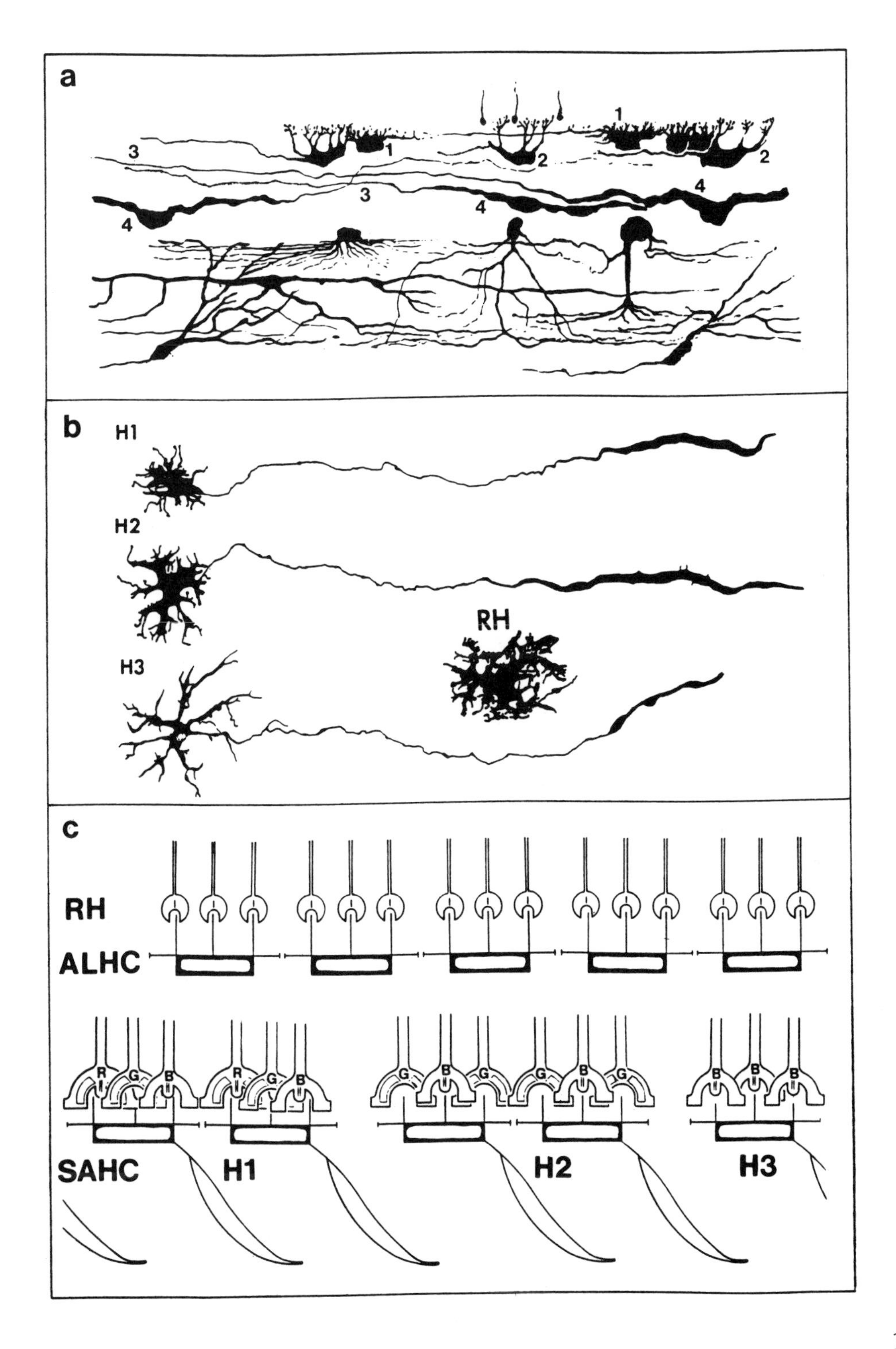
a
1
3
1
2
2
3
4
4
4
b
H1
H2
RH
H3
c
RH
ALHC
R
G
B
R
G
B
G
B
G
G
B
G
B
B
B
SAHC
H1
H2
H3

Table 7.1 Synopsis of horizontal cell types and their connectivity with photoreceptor types in retinas of different classes of vertebrate. Photoreceptor connectivity of axonless (ALHC) and short-axon bearing (SAHC) horizontal cells is given.

	ALHC			SAHC						
	Type 1, 'A'	Type 2	Type 3	Type 1, 'B' soma	Type 1, 'B' axon terminal	Type 2, 'C' soma	Type 2, 'C' axon terminal	Type 3 soma	Type 3 axon terminal	Reference(s)
Mammals										
Humans	–	–	–	Cones only (R,G):B÷9:1	Rods only	Cones only (R;G);B pref.	Only B cones	Cones only R, G; no B	Rods only	Ahnelt and Kolb (1994)
Primates										
Macaque	–	–	–	Cones only all kinds	Rods only	Cones only all kinds	Rods only	–	–	Boycott *et al.* (1987); Röhrenbeck *et al.* (1989)
Monkeys, diverse species	–	–	–	Cones only all kinds	Rods only	Cones only all kinds	Rods only	–	–	Boycott and Kolb (1973); Gallego (1985); Kolb *et al.* (1980)
Non-primates										
Cat	Cones only	–	–	Cones only	Rods only					Kolb (1974); Fisher and Boycott (1974)
Rabbit	Cones only	–	–	Cones only, all kinds	Rods only	Select. cones, possibly B	Bipolar cell dendrites	–	–	Famiglietti (1990); Raviola and Dacheux (1990); Dacheux and Raviola (1982)
Dog	Cones only	–	–	Cones	Rods only	–	–	–	–	Gallego (1985)
Tree shrew	Cones only, unselective	–	–	Cones only all kinds	Rods only					Müller and Peichl (1993)
Gray squirrel	(H2) rods and cones	–	–	(H1) cones only	(H1) cones only	–	–	–	–	West (1978)
Guinea pig	(A-type)	–	–	(B-type)	(B-type)	–	–	–	–	Peichl and Gonzales-Soriano (1993)
Rat	–	–	–	(B-type)	(B-type)	–	–	–	–	Peichl and Gonzales-Soriano (1993)
Gerbil	–	–	–	(B-type)	(B-type)	–	–	–	–	Peichl and Gonzales-Soriano (1993)
Birds										
Chick	Only double cones	Only R-cones		Three types of cones	Rods and 1 cone type					Gallego (1985)

Pigeon	Cones only	–	–	Cones only	Rods and cones	?	?	?	?	Mariani and Leure-Dupree (1987)
Owl	Cones only	–	–	All cones	Rods and (cones)	–	–	–	–	Tarres *et al.* (1986)
Owlet	Cones only	–	–	All cones	Rods and (cones)	–	–	–	–	Tarres *et al.* (1986)
Reptiles										
Turtle (*Geoclemys*)	(BHC): No rods all R cones most G-C all B cones	(THC) No rods some RC no G-C all B-C	–	All rods All R-C All G-C All B-C	All rods All R-C All G-C All B-C	–	–	–	–	Ohtsuka and Kouyama (1986)
Turtles (*Pseudemys*)	'H2' green SC	'H3' only blue SC	'H4' only green DC	'H1' G and R DC	Rods only R DC	–	–	–	–	Leeper (1978 a,b)
(*Chelydra*)	blue SC			R SC; G SC	only R SC					
Amphibians										
Rana pipiens	Small OHC (probably AL) all classes of photo-receptors	Giant OHC (probably AL) only B and R photo-receptors	–	Rods and cones no preference	Rods and cones no preference	–	–	–	–	Ogden *et al.* (1984, 1985)
Xenopus laevis	–	C-cell only B and R photo-receptors	–	Rods and cones prefer-entially from R cones	Rods and cones	–	–	–	–	Stone *et al.* (1990); Witkovsky and Powell (1981); Witkovsky *et al.* (1988)
Mudpuppy	–	biphasic C-cell with pref. B input	–	L-type rod and cone input	–	–	–	–	–	Kim and Miller (1991)

Table 7.1 (*cont.*)

	ALHC			SAHC						
	Type 1, 'A'	Type 2	Type 3	Type 1, 'B' soma	Type 1, 'B' axon terminal	Type 2, 'C' soma	Type 2, 'C' axon terminal	Type 3 soma	Type 3 axon terminal	Reference(s)
Fishes										
Chondrichthyes										
Smooth dogfish	Ext. row: rods only	Middle row rods only	Int. row cones	–	–	–	–	–	–	Stell and Witkovsky (1973)
Skate	Ext. row; axon? rods	Int. row; axon? rods	–	–	–	–	–	–	–	Malchow *et al.* (1990)
Ratfish	Ext. row rods only	Int. row rods only	–	–	–	–	–	–	–	Kretz *et al.* (1982)
Osteichthyes										
Catfish	Rod HC: rods only	–	–	Cone HC cones only (single type)	–	–	–	–	–	Sakai and Naka (1986); Hidaka *et al.* (1986)
Pikeperch	(rod HC rods only)	–	–	'H1': twin cones, occ. sing. cones	–	'H2': twin cones, one type of single cone	–	'H3': single cones only	–	Witkovsky *et al.* (1979)
Eugerres plumieri	H_m (luminosity)	H_i (chromaticity)	H_r rods only	H_c (luminosity)	H_c (luminosity)	–	–	–	–	Negishi *et al.* (1988)
Callionymus lyra	Rods only	–	–	'CH1': all cone types	–	'CH2': pale double cones partners only	–	'CH3': single cones only	–	van Haesendonck and Missoten (1979)
Cyprinids: carp, goldfish, roach	Rod HC rods only	–	–	CH1/ LHC: all cones (R, G, B)	–	$CH2/C_b$ HC: G and B cones only	–	$CH3/C_t$ HC: B cones only	–	Stell *et al.* (1975, 1982); Stell and Lightfoot (1975); Downing and Djamjoz (1989); Djamjoz and Downing (1988); Greenstreet and Djamjoz (1994) Tsukamoto *et al.* (1987)

SC, single cone; DC, double cone; TC, twin cone; B, short wavelength-sensitive cone; G, middle wavelength-sensitive cone; R, long wavelength-sensitive cone. C-cell chromaticity cell; L-type cell luminosity cell; –, not found. Other abbreviations or cell type designations have been adopted from the original references but arranged according to the general classification of horizontal cells into ALHCs and SAHCs

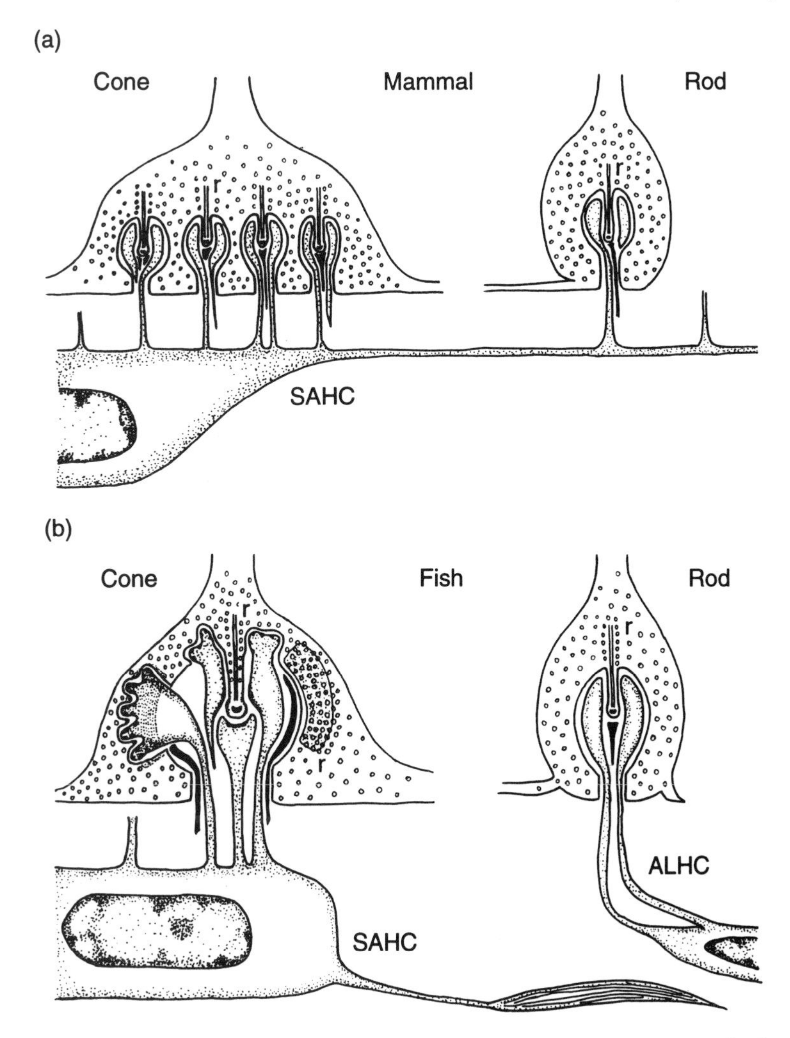

Figure 7.5 Schematic diagram of the basic organization and connectivity of photoreceptors and horizontal cells in a generalized mammal (a) and a generalized fish (b). Cone pedicles are on the left and rod spherules on the right. In a typical subprimate mammal (a), an SAHC contacts cones with its perikaryal dendrites and rods with its axon telodendria. Horizontal cell processes invaginate the bases of the photoreceptor terminals and form the lateral elements of ribbon synapses (bipolar processes are shown in black). In fish (b), by contrast, SAHCs contact only cones, and ALHCs contact rods. In cones, there is a single, large cavity invaginated by all horizontal and bipolar processes destined for ribbon synapses. Horizontal cells processes form both lateral and central elements of ribbon synapses in teleosts; three ribbons are shown in the pedicle with the two on the sides appearing as plate-like structures.

rings. No signs of vesicle exocytosis were observed in HC membranes opposite the synaptic ridge. Staining with ethanolic phosphotungstic acid (EPTA) has been widely used to highlight synaptic structures in the electron microscope. In the case of the photoreceptor–HC junctions, this method gives positive reactions at the synaptic ribbons, the arciform density, the photoreceptor membrane of the synaptic ridge, as well as the HC membrane neighboring the ridge (Wagner, 1980). In summary, these observations are compatible with the concept of feed-forward transmission from photoreceptors to HCs at the apex and the slopes of the synaptic ridge which may be governed by the general principles of a chemical synapse.

Where, then, might the feedback transmission, demonstrated in physiological experiments, take place? In human rods, it has taken serial section analysis and complete reconstruction of the invaginating HC processes to reveal one or two punctate synapses from the axon terminal processes of a type-I SAHC on to the rod spherule. These junctions mostly face away from the ribbon synapse and consist of a small cluster of clear synaptic vesicles, a presynaptic membrane densification on the HC side and no obvious specialization on the presumed postsynaptic rod membrane (Linberg and Fisher, 1988). A similar finding resembling a conventional chemical synapse as the morphological correlate for the feedback transmission within the photoreceptor terminal has not been made in other cases. The only other morphological evidence of a chemical synapse in a position to provide feedback activity comes from the catfish retina, where ALHC dendrites have been observed to synapse with the telodendria of cone photoreceptors (Sakai and Naka, 1986). Given the relative scarcity of these chemical feedback synapses, the question remains whether other conventional chemical transmission sites have been missed due to methodical reasons, or, whether alternative candidates for feedback transmission may be taken into account considering the complex and unique configuration of the synaptic complex in cone pedicles and rod spherules (see also Burkhardt, 1993).

In primates, amphibians (Ogden *et al.*, 1982) and teleosts (Djamgoz *et al.*, 1987; Downing and Djamgoz, 1989) doughnut-shaped HC processes have been observed as lateral elements of the ribbon synapse, which are themselves invaginated by extensions of the photoreceptor terminal. Such reciprocal innervations occur in both rods and cones. They increase the contact area between photoreceptors and HCs and might serve as a low resistance coupling between the two cells. Accordingly, it has been speculated that these morphological reciprocal invaginations might also be involved in reciprocal signal transmission.

In the teleost retina, combination of a number of morphological and electrophysiological observations has led to the development of a concept of feed-forward and feedback transmission between cones and ALHCs (Stell *et al.*, 1975). Based on the finding that HC processes were localized not only in a lateral position with respect to the synaptic ridge but also centrally, i.e. directly beneath its apex, a functional polarization was proposed, in which the central processes would mediate feed-forward and the lateral elements, the feedback transmission. This basic concept has been amended by two further morphological observations. First, no structural differences are discernible in the HC membranes directly adjacent to the synaptic ridge irrespective of whether they are part of the central or the lateral process. Therefore, it is hard to see why, in this region, transmission could differ from the tip of the synaptic ridge. Second, there are conspicuous, finger-like protrusions originating from the lateral processes and directed away from the ribbons; these have been called 'spinules' (Raynauld *et al.*, 1979; Wagner, 1980). Spinules are about 0.3–1.5 μm long and 0.1 μm in diameter. Like the remainder of the HC cytoplasm, they do not contain synaptic vesicles; however, their tips are quite conspicuous due to prominent patches of submembranous material which reacts strongly with the synaptic marker EPTA. The structure of these membrane densities as well as the freeze-fracture appearance of the spinule plasma membranes clearly differs from the feed-forward transmission sites next to the synaptic ribbons. The conclusion based on these observations that spinules may be sites of feedback transmission is supported by electrophysiological results showing that sign-inverting feedback activity in HCs strongly correlates with the presence of these spinules (Weiler and Wagner, 1984; Djamgoz *et al.*, 1988; Downing and Djamgoz, 1989; Kirsch *et al.*, 1990, 1991; Wagner and Djamgoz, 1993). Taking into account these findings, a new version of the 'Stell model' would consider the lateral HC processes as reciprocal synapses with postsynaptic sites opposite the synaptic ridge and the spinules to be presynaptic elements (Chapter 9 has a detailed discussion of HC–photoreceptor feedback).

The molecular organization of photoreceptor terminals and HC processes has been studied using immunocytochemical markers. These indicate the presence of some mechanisms also found in conventional synapses. Among these is a positive immunostaining against synaptophysin, associated with the numerous clear vesicles which fill the photoreceptor synaptic terminals (Kivela *et al.*, 1989; Schmitz and Drenckhahn, 1993a). This protein is part of the vesicle membrane and is involved in the fusion process with the presynaptic membrane during exocytosis. On the other hand, synapsins, serving as links between synaptic vesicles and cytoskeletal actin filaments in conventional synapses are absent from ribbon synapses (Mandell *et al.*, 1990). This indicates that ribbon synapses are endowed with special, as yet unresolved, mechanisms for transporting the synaptic vesicles along the ribbon to the site of exocytosis at the apex of the synaptic ridge. Furthermore, the molecular identity of the constituents of the synaptic ribbons is also still unknown. However, there are clear signs that these ribbons are organized differently in rods and cones, at least in teleosts (Wagner, 1973; Schmitz and Drenckhahn, 1993b). As for the postsynaptic side, actin has been localized both in horizontal and bipolar cell dendrites. It is associated with dystrophin, known to link the cytoskeletal actin microfilaments to the postsynaptic plasma membrane (Schmitz *et al.*, 1993). The distribution of actin in the postsynaptic elements resembles that in conventional synaptic arrays, e.g. dendritic spines (see below).

7.2.3 PLASTICITY OF THE PHOTORECEPTOR–HORIZONTAL CELL SYNAPTIC COMPLEX

The photoreceptor–HC synapses have been extensively studied for signs of morphological changes associated with light and dark adaptation and hence different functional states of the retina. In rod spherules of the rat, Brandon and Lam (1983) showed originally that the invaginating processes of the HC axon terminals have smooth contours during the light phase. In the dark, however, finger-like protrusions of photoreceptor cytoplasm were found to extend into the HC processes on either side of the synaptic ribbon underneath the HC membrane surrounding the protrusions, and there was an electron-dense band of fuzzy material. These morphogenetic changes have been interpreted as a result of the dynamics of increased membrane traffic during synaptic activity. In cone pedicles of the teleost retina, light adaptation induces spinules to arise from HC dendrites positioned laterally to synaptic ribbons; spinules are degraded at the beginning of the dark phase (Wagner, 1980). Since synaptic activity of photoreceptors is decreased during the light and thus membrane turnover may be assumed to be reduced, spinules have been ascribed a more specific role. The time course of their formation and degradation, as well as their pharmacology, correlates closely with physiological feedback activity recorded from HCs. Therefore, they have been proposed as sites of the feedback synapses (Wagner and Djamgoz, 1993).

In other vertebrates, and especially amphibians, the morphological pattern of HC processes branching from a parent axon has been compared to the spine apparatus found in many other neurons of the CNS (Winslow *et al.*, 1989; Witkovsky *et al.*, 1989). Mathematical modeling of the electrical properties of the spine head and stem would suggest that morphological changes similar to those found in teleosts (as well as neurochemical modulation of the conductances of the photoreceptor–HC synapses) could also contribute to the differences in signal transmission and feedback activity associated with light and dark adaptation.

Another conspicuous component of the photoreceptor–HC synapse subject to morphological changes related to the state of adaptation is the synaptic ribbon. In rat, turtle and fish, ribbons show changes in shape and/or number, being longest and most numerous during the day, and shortest and least numerous during the night (Wagner, 1973; Spadaro *et al.*, 1978; Abe and Yamamoto, 1984). These changes, typically, are not only light-dependent but governed by a circadian rhythm. In rat and turtle they are most prominent in rods, whereas in fish they are restricted to cones. The implication of synaptic ribbons in the process of vesicle transport and exocytosis has been demonstrated in pharmacological experiments indicating that the high levels of intracellular Ca^{2+} required for tonic release of transmitter during the dark is responsible for their shortening or even disappearance in teleost cone pedicles (Schmitz *et al.*, 1989); furthermore, the phosphoinositide metabolism is also involved in the regulation of the structural changes of these synaptic ribbons (Schmitz and Drenckhahn, 1993a).

7.3 PHOTORECEPTOR–HORIZONTAL CELL SYNAPTIC TRANSMISSION

7.3.1 BASIC MODEL

The main mechanism of synaptic transmission from photoreceptors to HCs was originally proposed by Trifonov (1968) in the fish retina. Accordingly, an excitatory neurotransmitter is released from photoreceptors in the dark (Chapter 6) and exerts a depolarizing action on the HC membrane potential. When photoreceptors respond to light stimuli by membrane hyperpolarization, the amount of neurotransmitter released is reduced, so the HC membrane hyperpolarizes. The light-evoked HC response is known as the 'S-potential' in recognition of Svaetichin who was first to record this response (Svaetichin, 1953). Over the years, a substantial body of experimental evidence has been gathered, as follows, to support the basic proposal of Trifonov (1968).

1. Depolarization of photoreceptor synaptic terminals by extrinsic current results in depolarization of the HC membrane potential (Byzov and Trifonov, 1968).
2. Suppression of photoreceptor synaptic input to HCs by application of divalent cations like Co^{2+} or Mg^{2+} results in hyperpolarization of the HC membrane potential (Dowling and Ripps, 1973; Kaneko and Shimazaki, 1975).
3. The input resistance of the HC membrane increases during light-evoked hyperpolarization (Byzov and Trifonov, 1981). This effect is also seen under voltage clamp condition (Miyachi and Murakami, 1989; Low *et al.*, 1991).
4. The photoreceptor neurotransmitter candidate glutamate depolarizes HCs both in intact retina (e.g. Murakami *et al.*, 1972;

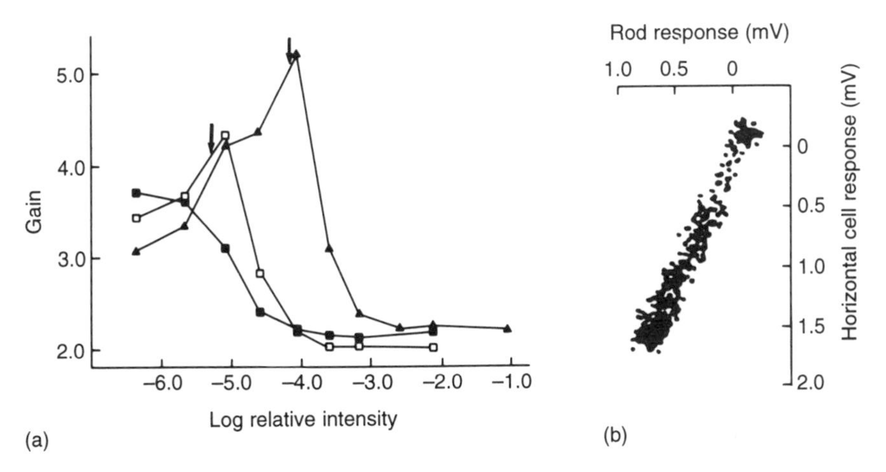

Figure 7.6 Voltage gain in the synaptic transmission from cones (a) and rods (b) to horizontal cells in turtle and toad retinas respectively. (a) The relationship between gain (amplitude of horizontal cell response divided by cone response amplitude) and the intensity of the light stimulus (relative values). Filled squares, dark-adapted state; open squares and triangles, light-adapted states induced by background illumination of −5.27 and −4.17, respectively (values relative to test light intensity indicated by vertical arrows. (Modified from Normann and Perlman (1979).) (b) Amplitude of light-evoked rod responses plotted against corresponding horizontal cell responses. (Modified from Belgum and Copenhagen (1988).)

Rowe and Ruddock, 1982a) and after dissociation (Lasater and Dowling, 1982; Ishida *et al.*, 1984). In the former case, the effect of glutamate is also seen after blocking synaptic input to HCs.
5. Glutamate has been shown to be released from photoreceptors (Copenhagen and Jahr, 1989). Glutamatergic antagonists block both the effects of light and exogenous glutamate on HCs (Rowe and Ruddock, 1982b).

The input–output relationships at photoreceptor–HC (mostly, rod–HC) synapses have been studied in a few species (e.g. Normann and Perlman, 1979; Attwell *et al.*, 1987; Belgum and Copenhagen, 1988; Wu, 1988). Experimental and theoretical analyses of the synaptic transfer have been found to be quite complex, especially after the discovery of light/dark adaptation-dependent neuromodulation at these synapses (section 7.4). On the whole, the gain in the transmission of voltage signals has been reported to be in the range 2–20. Importantly, synaptic gain appears to be non-linear, being highest for photoreceptor activity near the 'dark' resting level, i.e. for dim stimuli (Figure 7.6).

7.3.2 GLUTAMATE RECEPTORS ON HORIZONTAL CELLS

In the vertebrate central nervous system (CNS), glutamate receptors fall into two broad categories, as follows:

1. **Ionotropic receptors**. These receptors are characterized by the ligand-binding site and the ionophore being integral parts of the same protein. In turn, two subtypes exist: those that are selectively activated by *N*-methyl-D-aspartate (NMDA receptors) and those that are not (non-NMDA receptors). The latter may prefer as agonist kainate (KA) or quisqualate (QUIS)

/α-amino-3-hydroxy-5-methyl-4-isoxazole-propionate (AMPA).

2. **Metabotropic receptors**. These receptors do not possess an ionophore. Instead, ligand binding is signaled to the ion channel(s), which are separate proteins, by second messenger molecules. Activation of metabotropic glutamate receptors have been shown either to stimulate or to inhibit turn-over of second messenger(s).

Horizontal cells of a variety of vertebrates have been shown to possess glutamate receptors of both major types.

The main receptor mechanism mediating the photoreceptor synaptic input to HCs is ionotropic and of the non-NMDA type. Thus, KA and QUIS have been shown to depolarize HCs in several vertebrate classes: fish (Lasater and Dowling, 1982; Rowe and Ruddock, 1982a; Ishida *et al.*, 1984; Ishida and Neyton, 1985); amphibia (Slaughter and Miller, 1983, 1985; Yang and Wu, 1989, 1991; Krizaj *et al.*, 1994); reptiles (Perlman *et al.*, 1987) and mammals (Massey and Miller, 1987; Hankins and Ikeda, 1991). An extensive study of excitatory amino acid receptors on HC subtypes in the white perch retina was performed by Zhou *et al.* (1993). Four different (H1–H4) subtypes of HC were found to respond to glutamate, KA, QUIS and AMPA with constant relative amplitudes of response, suggesting that there might be a single population of non-NMDA receptors on these cells. On the other hand, in the salamander retina, the non-NMDA receptor antagonist 6-cyano-7-nitroquinoxaline-2, 3-dione (CNQX) blocked the action of KA and AMPA, but not those of QUIS and glutamate, indicating the existence of KA/AMPA and CNQX-resistant QUIS receptors on HCs (Yang and Wu, 1991). In voltage clamp studies, the membrane current elicited in HCs by KA appeared sustained, whereas that evoked by AMPA or QUIS was desensitizing (O'Dell and Christensen, 1989; Zhou *et al.* 1993; Krizaj *et al.*, 1994), consistent with what is found generally in the CNS (e.g. Mayer *et al.*, 1991).

Two studies have also suggested the presence of NMDA receptors on HCs. Hals *et al.* (1986) and O'Dell and Christensen (1989) showed that HCs isolated from catfish retina respond to NMDA (Chapter 14, section 14.6). In the salamander retina, Yang and Wu (1989) suggested that the NMDA receptors on HCs may be extrasynaptic since CNQX was found to block the light-evoked responses as well as the KA-induced depolarization of HCs, while having no effect on the action of glutamate.

There is also evidence to suggest that HCs of fish and amphibian retinas possess functional metabotropic receptors. Yang and Wu (1989) noted that prolonged application of *trans*-1-aminocyclopentane-1, 3-dicarboxylic acid (ACPD), an agonist of metabotropic glutamate receptors linked to phosphatidylinositol metabolism, hyperpolarized the HCs of salamander retina and enhanced their light-evoked responses. Yang and Wu (1989) suggested, therefore, that these receptors may be involved in adjusting the sensitivity of HC responses. Another type of metabotropic glutamate receptor is selectively activated by 2-amino-4-phosphonobutyrate (APB) and linked negatively to cGMP via a G-protein. The functioning of this receptor is understood best from work on ON-center bipolar cells (Chapter 12). Thus, activation of APB receptors by glutamate leads to stimulation of a phosphodiesterase and a consequent reduction in the intracellular concentration of cGMP. In turn, cGMP-gated cation channels close and the cell hyperpolarizes. In the case of ON-center bipolar cells, reduction in the amount of glutamate released from photoreceptors during light means that ON-center bipolar cells depolarize in response to light. Takahashi and Copenhagen (1992) found that solitary HCs of fish (catfish and goldfish) retinas respond to APB (section 7.4.4

includes further discussion of basic aspects). Yasui *et al.* (1990) showed in the carp retina that the APB receptors H1 type HCs are associated with short-wavelength-sensitive cone input to these cells. Thus, application of APB hyperpolarized the 'dark' membrane potential and significantly enhanced the cells' short wavelength sensitivity. It was suggested that the underlying APB-sensitive receptor mechanism could have a subtle role in chromatic signaling in the outer retina.

7.3.3 IONIC BASIS OF GLUTAMATE ACTION UPON HCS

Kaneko and Shimazaki (1975) showed originally in the isolated carp retina that Na^+ is the major ion gated by glutamate through ionotropic receptors on HCs. This view was confirmed in the axolotl retina by Waloga and Pak (1976). On the other hand, Miller and Dacheux (1976) suggested that Cl^- is the most important ion mediating the light-evoked responses of HCs in the mudpuppy retina. A subsequent study on goldfish by Tachibana (1985) involving patch clamp recording from solitary HCs showed that responses to glutamate involved K^+ and Ca^{2+}, in addition to Na^+. The reversal potential of glutamate- and light-evoked responses were found to be in the range −14 to +5 mV, consistent with the involvement of non-selective cation channels (Trifonov *et al.*, 1974; Tachibana, 1985; Murakami and Takahashi, 1987; Miyachi and Murakami, 1989; Low *et al.*, 1991).

When HCs normally respond to light with membrane hyperpolarization, voltage-dependent conductances may also come into play. In particular, the non-synaptic K^+ conductance is enhanced by hyperpolarization in a non-linear manner and can significantly affect the waveform and amplitude of the light-evoked responses (e.g. Byzov *et al.*, 1977; Perlman *et al.*, 1993). HCs are also capable of generating Ca^{2+} spikes during glutamate-induced depolarization but these are seen only after blocking the 'background' K^+ channels (Murakami and Takahashi, 1987).

Kaneko and Tachibana (1985) initially suggested that the depolarizing action of glutamate on HCs could involve reduction in a K^+ current normally generated by anomalous rectifier K^+ channels. Perlman *et al.* (1988) subsequently showed, however, that this effect was more likely to be due to a glutamate-induced build-up of K^+ around (non-perfused) HCs (but see section 7.4.4). Djamgoz (1988) found that the microenvironment is important in glutamate action in the retina. In the isolated retina of the cyprinid fish, the roach, application of glutamate caused a millimolar rise in the extracellular K^+ activity; also the field potential changed in the negative direction by some 8 mV for a 10-fold increase in the K^+ activity. Light-evoked activity of the outer retina is also accompanied by a number of extracellular ionic changes including a reduction in K^+ concentration (Oakley, 1977). Furthermore, the subcellular space of the retina may change during illumination (Huang and Karwoski, 1992). The regulation of such extracellular ionic fluxes depends on metabolic pumping (Djamgoz, 1988) and spatial buffering (Karwoski *et al.*, 1989). Ionic fluxes within the retinal microenvironment can reach abnormal levels causing excessive release of neurotransmitters, and the whole effect can propagate by spreading depression. Such events can contribute to visual dysfunction during pathological conditions such as ischemia (Chapter 16).

7.3.4 ROD AND CONE-DRIVEN ACTIVITY IN HORIZONTAL CELLS

Rod- and cone-driven signals are integrated in HCs to varying extents in retinas of different vertebrates. In addition to perikarya and axon terminals of HCs contacting rod and/or cones directly in varying proportions, further integration occurs by means of interphotoreceptor signaling (Chapter 6).

Retinas of teleost fish are an extreme case where there is little, if any, direct rod–cone interaction (Scholes, 1975; Stell, 1980). This segregation is maintained at the HC level. Electrophysiological aspects of photoreceptor inputs to HCs have mostly been studied in the cyprinid fish retina. Thus, the quantum spectral sensitivity of intermediate HCs matches the absorption spectrum of porphyropsin and shows no Purkinje shift (Kaneko and Yamada, 1972). External (H1, H2 and H3) HCs generate a variety of spectral types of cone-driven S-potentials (Figure 7.7). H1 type HCs generally hyperpolarize in response to light stimuli of all wavelengths. Although

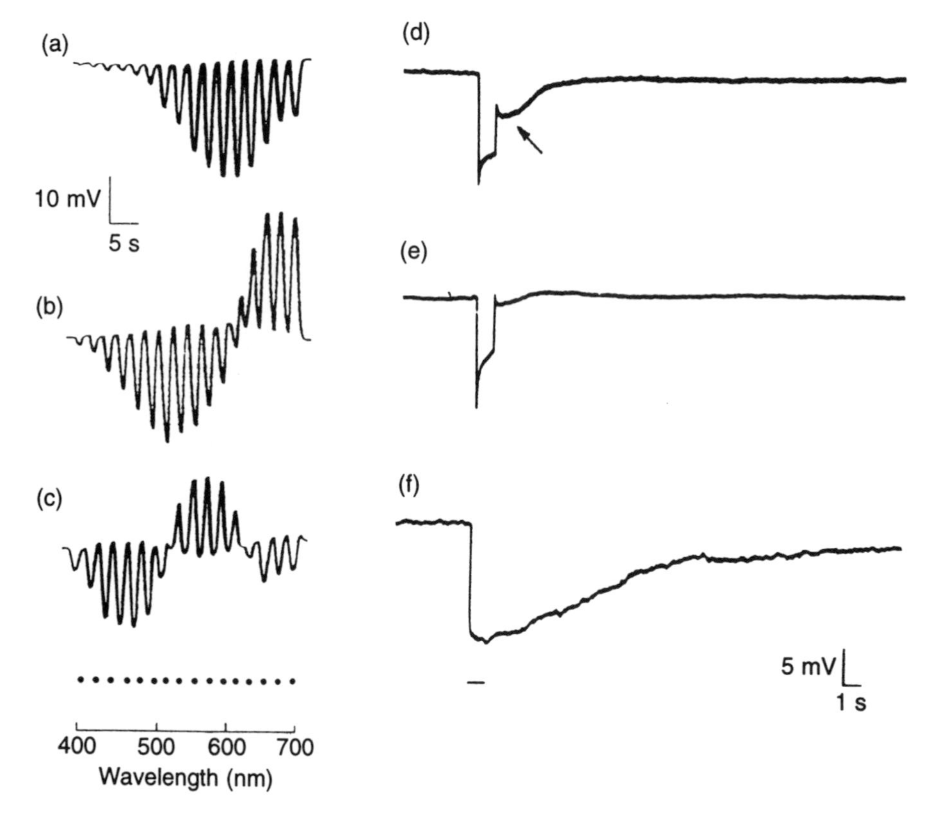

Figure 7.7 Light-evoked responses of horizontal cells in the retina of the cyprinid fish, the roach (a–c) and the rabbit (d–f) showing cone- and/or rod-driven activity. (a) Red-sensitive luminosity (H1); (b) biphasic, chromaticity (H2); (c) triphasic, chromaticity type responses, all driven by cones with no rod input. Sixteen different spectral stimuli (20 nm intervals) of near-equal quantum content were presented in sequence (dots at bottom of figure) to cover the spectral raise from 400 to 700 nm (inclusive). (Modified from Djamgoz and Yamada (1990).) (d–f) Horizontal cell responses to high intensity stimulation with white light (duration indicated by short horizontal bar under the bottom trace). (d) Axonless horizontal cell, showing mixed cone- and rod-driven activity (transition indicated by small arrow head). (e) Perikaryon of axon-bearing horizontal cell showing primarily cone-driven response. (f) Axon terminal response showing mainly rod-driven response. (Modified from Dacheux and Raviola (1982).)

their spectral sensitivity follows the absorption system of the R-cone pigment in the middle- and long-wavelength regions of the spectrum some discrepancies have been noted for short-wavelengths (e.g. Djamgoz, 1984). H2 type HCs which do not contact R-cones generate biphasic, chromaticity (blue–green-hyperpolarizing; red-depolarizing) type responses (Downing and Djamgoz, 1989). H3 type HCs on the other hand, which selectively contact B-cones generate triphasic, chromaticity response, i.e. blue-hyperpolarizing/green-depolarizing/red-hyperpolarizing (Greenstreet and Djamgoz, 1994). Finally, some HCs may receive additional input from ultraviolet-sensitive cones (Hashimoto *et al.*, 1988). Interestingly, this input may exist only transiently in small fish, where the ocular medium would have some UV transmittance (Bowmaker and Kunz, 1987).

The electrophysiological spectral response characteristics of H1–H3 HCs described above agree with the hierarchical system of cone feed-forward and feedback connectivity originally suggested from Golgi studies by Stell and Lightfoot (1975) and Stell *et al.* (1975). However, the basic model has been extended to include a number of features which enable fine tuning of its functional outputs. First, recent work has suggested that the H1 HCs of cyprinid fish also receive a minor depolarizing input from B- and/or G-cones. This input appears to be mediated by metabotropic, conductance-decreasing, APB-sensitive receptors (Yasui *et al.*, 1990). Thus, when the predominant R-cone input is suppressed by selective adaptation, short wavelength-sensitive depolarizing response can be revealed (Yamada *et al.*, 1991). Second, the feedback connections responsible for chromaticity type S-potentials are 'plastic' being strong when the retina is light-adapted and weak in the dark-adapted state (Djamgoz *et al.*, 1988; Wagner and Djamgoz, 1993). Such plasticity of HC-cone feedback probably contributes significantly to the functional expediency of the HC system. It should be noted that a different concept of cone-HC connectivity and signaling has been proposed from a modeling study by Kamermans *et al.* (1989). Although this model is interesting, it has no neuroanatomical correlate.

Amphibian and reptilian HCs have also been shown to generate luminosity and chromaticity type S-potentials. The internal HCs of the frog retina were found to generate hyperpolarizing S-potentials that were due to a mixture of rod and cone inputs to the cells' somatic dendrites and axon terminals (Ogden *et al.*, 1984, 1985). External HCs, which appeared axonless, generated biphasic chromaticity type S-potentials with a hyperpolarizing component that was driven by the 433 nm rods and a depolarization due to an input from the 580 nm cones (Stone *et al.*, 1990). Generation of biphasic chromaticity type S-potentials by mixed rod/cone input was suggested earlier in the mudpuppy retina by Fain (1975). An initial study on the turtle retina found a system of hierarchical connectivity with cones, as in the cyprinid fish retina (Leeper, 1978a, b). However, a subsequent study involving intracellular recording and staining suggested that HCs were not so specific in their connectivity. In particular, a type of cone incorporating a pale-green oil droplet, which was thought to be R-sensitive, contacted biphasic and triphasic chromaticity HCs (Ohtsuka and Kouyama, 1986). On the other hand, if this particular cone type were B-sensitive, the scheme of Leeper (1978a, b) would be upheld. The spectrally non-specific or limited cone connectivity of HCs in mammalian (cat, rhesus and macaque monkey and human retinas) appear to be reflected in S-potentials that are purely hyperpolarizing and non-chromatic (Nelson *et al.*, 1975; Nelson, 1977, 1985; Gouras and Zrenner, 1981; Bloomfield and Miller, 1982; Dacheux and Raviola, 1982) (Figure 7.7).

7.4 MODULATION OF FUNCTION IN HORIZONTAL CELLS

A few years ago, Kaczmarek and Levitan (1987) defined neuromodulation as 'the ability of neurons to alter their electrical properties in response to intracellular biochemical changes resulting from synaptic or hormonal stimulation', a definition which continues to encompass the latest results concerning neuromodulation of retinal HC function. Neuromodulation is mediated by a complex array of intracellular biochemical machinery, including a host of membrane receptors, soluble second messengers such as cAMP, Ca^{2+} and Ca^{2+}-regulating proteins, a variety of protein kinases and the presence of multiple allosteric sites on voltage- and synaptically-gated channels. Not only are many, if not all, of these present in retinal HCs, but there is good evidence that the three main categories of channel in the HC (glutamate-gated, voltage-gated and gap junction) are all subject to modulation. The best understood, and perhaps the main modulatory substance is dopamine, which as far as is known works only through second messengers. However, glutamate, the fast, ionotropic, excitatory transmitter can also function as a modulator, both through metabotropic receptors, and also by influencing the degree of Ca^{2+} entry into the cell (Linn and Christensen, 1992).

A related theme in HC physiology is 'plasticity', or the ability to reorganize circuitry in response to external or internal changes. This may not be a question of forming connections *de novo*, or breaking off existing ones, but rather strengthening or weakening connections as changing environmental conditions may demand. Retinal function is influenced strongly by two general environmental modifiers: an internal one, the circadian clock, and an external one, light. Cahill and Besharse (1993) have provided strong evidence for the location within photoreceptors of a circadian oscillator governing melatonin production. Melatonin is a potent inhibitor of dopamine release (Dubocovich, 1983). Light and dark adaptation result in important modifications of retinal function, including retinomotor movements, spinule growth and retraction and a rearrangement of retinal circuitry corresponding to rod- and cone-mediated vision. Each of these is associated with structural/biochemical changes, which are beginning to be elucidated. For example, Janssen–Bienhold *et al.* (1993) have shown that light adaptation increased the phosphorylation in five protein bands derived from HCs, ranging from 15 to 68 kDa. The phosphorylation of three of these bands was increased by calmodulin. One of these, the 28–30 kDa protein, is in the same size range as the gap junctional protein. Levitan (1994) has summarized the evidence that modulation of ion channel activity through protein phosphorylation is a ubiquitous mechanism in the vertebrate CNS.

7.4.1 MODULATION OF AMINO ACID-MEDIATED SYNAPTIC TRANSMISSION

(a) Dopamine–glutamate interactions

The seminal study by Knapp and Dowling (1987) provided a link between the major excitatory neurotransmitter, glutamate, and the neuromodulator, dopamine. They showed, in isolated fish HCs, that dopamine by itself elicited no membrane current, but it enhanced the current evoked by glutamate, or one of its analogs, KA. The effect was mediated by a D1 dopamine receptor, which is linked to adenylate cyclase, resulting in a rise in intracellular cAMP. Liman *et al.* (1989) found that cAMP acted through a protein kinase, suggesting that the glutamate channel, or some closely related protein, is modulated through phosphorylation. Single channel recordings (Knapp *et al.*, 1990) indicated that dopamine increases the frequency of channel opening and extends slightly the open time, without changing channel conductance.

The desensitization property of glutamate receptors on HCs (section 7.3.2) is of great potential interest to synaptic function in the outer retina, due to the tonic nature of transmitter release by photoreceptors (Trifonov, 1968). At excitatory synapses in hippocampus and cerebellum, activation by rapid perfusion of QUIS results in a fast inward current that decays within a few milliseconds to a reduced, maintained plateau current, the relative magnitude of which varies with the particular type of glutamate receptor, and with mRNA editing (Sommer *et al.*, 1990). Under conditions of tonic release, it is probable, therefore, that glutamate receptors on HCs will be in a desensitized state.

Desensitization is subject to modulation by a variety of substances, including concanavalin A (O'Dell and Christensen, 1989; Mayer *et al.*, 1991) and aniracetam (Ito *et al.*, 1990), both of which affect the kinetics of AMPA-induced currents. Kruse and Schmidt (1993) showed, in isolated fish HCs, that glutamate evokes a desensitizing current. Dopamine enhanced the maximum glutamate-induced current without shifting the EC_{50} value, but the dopamine effect was abolished by pretreatment with concanavalin A. Aniracetam greatly enhanced the L-glutamate current, and also prevented its further modulation by dopamine.

Ishida and Neyton (1985) found that currents elicited by KA in isolated goldfish HCs were reduced in the presence of QUIS. The inhibitory interaction of these two glutamate agonists has been observed in many systems (Patneau and Mayer, 1991). Since glutamate, not QUIS or KA, is the natural transmitter, the significance of QUIS/KA interactions is in exploring the subunit composition and desensitization properties of AMPA-type glutamate receptors. Krizaj *et al.* (1994) examined the effects of glutamate and its non-NMDA agonists on HCs as a function of the adaptational state of the retina. Glutamate, AMPA or QUIS elicited desensitizing responses under light-adapted conditions, but larger and more sustained responses when the retina was dark-adapted. Dopamine or its D1 agonist potentiated KA-evoked responses, but inhibited those elicited by QUIS. These data suggest that light and dark adaptation, or in cellular terms, a shift in the balance of rod- vs. cone-initiated signals, are associated with changes in the response properties of the glutamate receptors on HCs.

The non-independence of rod and cone inputs to retinal neurons is well established. In human vision, rods inhibit cone responses, an inhibition relieved by a weak green background field (Alexander and Fishman, 1984). In amphibian eyes, in which second-order cells receive mixed rod and cone input (Hare *et al.*, 1986; Witkovsky *et al.*, 1988a), responses to weak cone-effective stimuli are enhanced when rods are partially suppressed (Frumkes and Eysteinsson, 1987). In reciprocal fashion, cone activity inhibits rod to HC transmission (Witkovsky *et al.*, 1989; Krizaj *et al.*, 1994). Perhaps more than one mechanism is responsible for this reciprocal inhibition, but a very important one is located on the HC in association with the glutamate-gated channels. Under mesopic conditions, application of the AMPA blocker, CNQX, results in a suppression of cone input, and a simultaneous growth of rod input (Krizaj *et al.*, 1994). On the other hand, injection of cAMP into a dark-adapted HC results in the resuscitation of a latent cone input (Witkovsky *et al.*, 1993a). Although much remains to be learned, these results indicate strongly that modulation of glutamate receptors underlies some aspects of light and dark adaptation.

(b) GABA–dopamine interactions

HCs in many vertebrate retinas utilize GABA as a neurotransmitter (Chapter 10). In addition, they may have both a Na^+-dependent transporter for GABA (Malchow and Ripps, 1990; Kamermans and Werblin, 1992) and a GABA-gated ion channel. Dong and Werblin (1994) noted that dopamine reduced the GABA-gated channel current, but was without effect on the transporter. The dopamine action was mimicked by a D1 dopamine agonist, or by forskolin, indicating mediation by cAMP. It is of interest that Veruki and Yeh (1992, 1994) reported that vasointestinal polypeptide (VIP) potentiates $GABA_A$ -mediated currents in bipolar and ganglion cells of the rat retina through a cAMP-dependent mechanism involving protein kinase A. These studies thus also indicate the generality of second messenger-mediated modulation of ligand-gated membrane channels operating through similar intracellular machinery.

7.4.2 CONTROL OF DOPAMINE RELEASE

A light-driven circuit from rod photoreceptors (Besharse and Witkovsky, 1992), acting through APB-sensitive bipolar cells (Besharse, 1992) increases dopamine outflow. In the mudpuppy, Myhr *et al.* (1994) present evidence that cones also can stimulate dopamine release. In turn, dopamine, acting through a D2-like autoreceptor, inhibits its own release. Harsanyi and Mangel (1992) found that quinpirole, a D2 dopamine agonist, increased HC coupling in goldfish retina, but its action was blocked by pretreatment with a D1 dopamine agonist. They drew the reasonable inference that quinpirole acted by decreasing dopamine outflow, an inference supported by studies of $[^3H]$dopamine release using D2 dopamine ligands (Rashid *et al.*, 1993).

The dopaminergic neuron additionally receives synaptic input from amacrine cells and centrifugal fibers (Zucker and Dowling, 1987). In fish, the centrifugals arise from the nucleus olfactoretinalis and reach the retina via the terminal nerve (Stell *et al.*, 1984). Centrifugal fibers release the peptides gonadotropin-releasing hormone (GnRH) and FMRFamide. Umino and Dowling (1991) showed that GnRH had dopamine-like effects on HCs in whole retinas, but was without effect in 6-OHDA-treated retinas. Behrens *et al.* (1993) noted that GnRH increased spinule formation, an effect blocked by haloperidol. These results indicate that GnRH works through a dopamine pathway. FMRFamide, on the other hand, antagonized the actions of GnRH (Umino and Dowling, 1991). In turtle retinas, Kolbinger and Weiler (1993) documented that dopamine release is tonically inhibited by GABA and certain peptides, with the result that addition of exogenous GABA has little effect, whereas application of its antagonist, bicuculline, induces a large increase in dopamine release. The physiology of the HC bears this out: GABA causes little change in HC coupling, whereas bicuculline application results in a sharp reduction of coupling (Piccolino *et al.*, 1982). Similarly, application of the opiate antagonist, naloxone, increased dopamine release 8-fold, whereas enkephalin was without effect. However, Weiler and Ball (1989) showed that enkephalin antagonized the release of $[^3H]$dopamine in response to elevated extracellular K^+.

The question whether, in fish retinas, dopamine increases in light or darkness now seems resolved. Dopamine release *per se* increases during light. Dearry and Burnside (1989) demonstrated that medium taken from light-adapted retinas caused light-adapted pigment dispersion in isolated retinal pigment epithelium. This effect was blocked by a D2 dopamine antagonist, sulpiride. Direct measures showed that light-adaptation increased retinal dopamine overflow by 3–4-fold over dark-adapted retinas. Similar findings were obtained by Kirsch and Wagner (1989) and, in amphibian retinas, by Boatright *et al.* (1989). Light also stimulates retinal tyrosine hydroxylase activity in the green sunfish (Dearry, 1991).

The problem has been that certain physiological results led to the speculation that dopamine levels were low in light and maximal only after prolonged dark-adaptation (Mangel and Dowling, 1985, 1987). The main finding was a loss of responsiveness in HCs following either prolonged dark-adaptation or dopamine application. More recently, however, Umino *et al.* (1991) showed that flickering light induced a 5–10 mV depolarization in perch HCs associated with an increased response to small spot stimulation; both effects were blocked by haloperidol, indicating their dependence on dopamine. The depolarization is explained by an enhanced action of glutamate released by cones (Knapp and Dowling, 1987; Witkovsky *et al.*, 1988b) and the increased small spot response to uncoupling of HCs (Piccolino *et al.*, 1984).

The loss of responsiveness in prolonged darkness appears to derive from an uncoupling of R-cones from their postsynaptic targets, as first shown by Raynauld *et al.* (1979) in goldfish retina. Yang *et al.* (1994) studied H1 type HC responses in intact carp. As the eye dark-adapted, the spectral input to the cells shifted from red to green and the depolarizing responses of chromatic HCs disappeared. Destruction of dopaminergic cells with 6-OHDA only partially suppressed the loss of responsivity to red light stimuli during dark-adaptation, whereas that to dim green light actually increased, indicating a preferential loss of R-cone signals, as Raynauld *et al* (1979) found for ganglion cells. The role of dopamine, if any, in this loss remains to be determined, for other physiological evidence indicates strongly that dopamine in fact enhances R-cone to HC communication. For example, spinule formation and the increase in the depolarizing responses of chromatic cells are stimulated both by light-adaptation (Kirsch *et al.*, 1990) and by dopamine, acting through a D1 mechanism (Kirsch *et al.*, 1991; Wagner and Djamgoz, 1993). The consensus is that the depolarizing responses of chromatic HCs depend on a feedback signal which originates in R-cones (Chapter 9).

7.4.3 MODULATION OF HC COUPLING

This topic is covered in Chapter 8; only some essential points related to neuromodulation are given here. The degree of HC coupling is modulated by dopamine through a D1 receptor (Lasater and Dowling, 1985; DeVries and Schwartz, 1989) which acts in the classical way to increase cAMP production and thereby to activate a protein kinase (Lasater, 1987). At a single channel level, dopamine reduces by fourfold the open probability of gap junction channels (McMahon *et al.*, 1989) without altering single channel conductance.

HC gap junctions can also be modulated by a pathway involving cGMP (Miyachi and Murakami, 1991; DeVries and Schwartz, 1992). Miyachi *et al.* (1990) found that arginine, which activates soluble guanylate cyclases, produced a very sharp reduction in gap junctional conductance when injected intracellularly. Parenthetically this method is a useful tool for examining synaptic conductances of HCs, which normally cannot be estimated accurately due to current flow through the low resistance gap junctions. Another activator of soluble guanylate cyclase is nitric oxide, which is produced during metabolism of L-arginine. Weiler and Kewitz (1993) reported that NADPH-diaphorase (a marker for nitric oxide synthase) was localized histochemically to a variety of retinal neurons in carp retina, including HCs. McMahon (1994) noted that cGMP reduced gap junction channel conductance by reducing the frequency of channel openings, without altering channel open time. Thus cGMP and cAMP each mediate a reduction in macroscopic gap junctional permeability, but through different pathways and mechanisms. In that regard it is worth noting that Baldridge and Ball (1991) found that background light still decreased HC coupling even in the presence of a D1 dopamine antagonist or when dopaminergic neurons were lesioned, thus implicating another mediator besides dopamine.

There is evidence also that the number and/or the distribution of connexons is subject to modulation. Schmitz and Wolburg (1991) noted that reduced pH caused gap junctional particles in goldfish HCs to aggregate in crystalline arrays, a state associated with uncoupling. Kurz-Isler and Wolburg (1986) earlier noted a clustering of connexons associated with dark-adaptation, and Baldridge *et al.* (1987) determined that connexon density was decreased by exposure to dopamine.

Recent data document the presence of inositol triphosphate (ITP), a soluble second messenger, in the retina. Peng *et al.* (1991) showed that the ITP receptor protein is localized to the synaptic layers of rat retina. HCs isolated from salamander were negative, although photoreceptor synaptic terminals stained strongly. Another product of phosphatidylinositol phosphate hydrolysis is diacylglycerol, a non-soluble messenger, which together with Ca^{2+}, activates protein kinase C (PKC). Akopian *et al.* (1992) found that phorbol esters, which activate PKC, caused changes in HC receptive fields consistent with uncoupling. The phorbol ester effects, however, were largely reduced by dopamine antagonists, suggesting that the PKC may not be located primarily in the HC. On the other hand, Rodrigues and Dowling (1990) reported that the retraction of HC terminal neurites induced by dopamine or a D1 dopamine agonist was mediated by PKC. The evidence was that phorbol esters or a diacylglycerol analog did cause neurite retraction, and, staurosporine, a PKC inhibitor, blocked the action of dopamine on neurites. Cyclic AMP analogs were ineffective. This finding raises the interesting possibility that D1 dopamine receptor agonists are not invariably coupled to adenylate cyclase. Further work is required to characterize this pathway.

7.4.4 MODULATION OF TIME AND VOLTAGE-DEPENDENT ION CHANNELS IN HCs

Whole-cell patch clamp studies have identified a number of intrinsic channels in HCs, including three K^+ currents (an A-current, a delayed rectifier, and an inward (anomalous) rectifier); two calcium currents (T-type and L-type); and a TTX-sensitive Na^+ current (reviewed in Lasater, 1991). At least some of these currents are subject to modulation. Kaneko and Tachibana (1985) noted that glutamate reduced the inward rectifier current of goldfish HCs, an effect seen also in turtle HCs (Golard *et al.*, 1992). Although Perlman *et al.* (1988) considered the effect of an artefact due to the pressure puff of glutamate altering the ionic environment of the test cell, Priddy *et al.* (1992) obtained the same result on neurons of the solitary tract with a superfusion system. It will be interesting to see whether this action of glutamate is mediated by a metabotropic receptor.

The 4-AP sensitive, transient A-current is modulated by intracellular GTP, as well as by Ca^{2+} and glutamate (Akopian and Witkovsky, 1994). GTP shifts the activation range of the A-current by up to +10 mV, which brings it into the normal operating range of the cell. The action of GTP is mediated by a G-protein.

Pfeiffer-Linn and Lasater (1993) characterized two Ca^{2+} currents, a T-type and an L-type, in cone-driven HCs of a fish retina.

Dopamine, at low micromolar doses inhibited the T-current, but enhanced the L-current. The dopamine effect was mediated by a D1 receptor, coupled through cAMP to a protein kinase. It is noteworthy that the Ca^{2+} current of the rod-driven HC was unaffected by dopamine.

The function of intrinsic channels in HCs is only partly understood. HCs in many species are thought to release transmitter in a Ca^{2+}-independent way. However, Ca^{2+} evidently is implicated in multiple biochemical pathways, e.g., in PKC activation. It also contributes to the oscillatory behavior of the HC membrane, and to a depolarizing spike that appears at light 'off' (Akopian *et al.*, 1991; Perlman *et al.*, 1993).

The metabotropic glutamate analog, APB, was found to reduce the light responses of goldfish HCs and to hyperpolarize them, but this action occurred only in HEPES-buffered, not in bicarbonate-buffered, medium (Takahashi and Copenhagen, 1992). The mechanism of APB action is unclear; in isolated HCs it antagonized glutamate-induced depolarizations even in a cocktail that blocked K^+ and Ca^{2+} channels, but it does depend on pH, being active in acidic media. Hare and Owen (1992) postulated that in salamander retina, APB worked presynaptically on cones to decrease transmitter release. In relation to pH, Dixon *et al.* (1993) found that glutamate suppressed an L-type (high voltage-activated) Ca^{2+} current in isolated catfish HCs by lowering intracellular pH. More information is needed on the control of intracellular pH in order to evaluate the possible significance of these findings in an *in vivo* situation.

7.4.5 NON-UNIVARIANCE OF PHOTORECEPTOR ACTIVITY: ROLE OF NEUROMODULATORS

The principle of univariance (Naka and Rushton, 1966) states that the output of a photoreceptor depends only on its quantum catch. A clear violation of univariance is given by the feedback circuit which provides depolarizing input to cones (Chapter 9). Rods are also known to violate univariance. Yang and Wu (1989) found that rod–cone coupling was increased by background lights and that the increment threshold curve of the rod was shallower for 700 nm test flashes than for 500 nm flashes, indicating cone input. The kinetics of the rod photoresponse also reflect cone input (Hare and Owen, 1992), particularly in the initial hyperpolarizing transient peak. Witkovsky *et al.* (1993a and work in progress) noted that the D2 dopamine agonist, quinpirole, elicited similar light-adapted changes in the rod. The significance of rod to HC transmission still has to be determined, for D2 agonists also greatly increase cone input to the amphibian HC beyond what is transmitted through rods, as shown by simultaneous recordings from rods and HCs (Krizaj and Witkovsky, 1993).

Within the invagination of the cone, processes of horizontal, bipolar and interplexiform cells come into very close proximity. The Muller glial cell is excluded, and this is significant in view of its active uptake of glutamate (Brew and Attwell, 1987). Photoreceptors contribute glutamate; HCs or interplexiform cells (Nakamura *et al.*, 1980) contribute GABA, and glycine is emitted by a glycinergic interplexiform cell (Smiley and Yazulla, 1990). The local concentrations of amino acids within the cone pedicle are determined by a balance between release and re-uptake by transporters. Glutamate is taken up by photoreceptors (Tachibana and Kaneko, 1988) and GABA by the HC (Malchow and Ripps, 1990; Kamermans and Werblin, 1992). Presumably glycine is transported by the interplexiform cells, but this has yet to be demonstrated. The factors governing transporter activity are just beginning to be elucidated. They depend on Na^+ and Cl^- gradients (Eliasof and Werblin, 1993; Dong *et al.*, 1994) and possibly are subject to

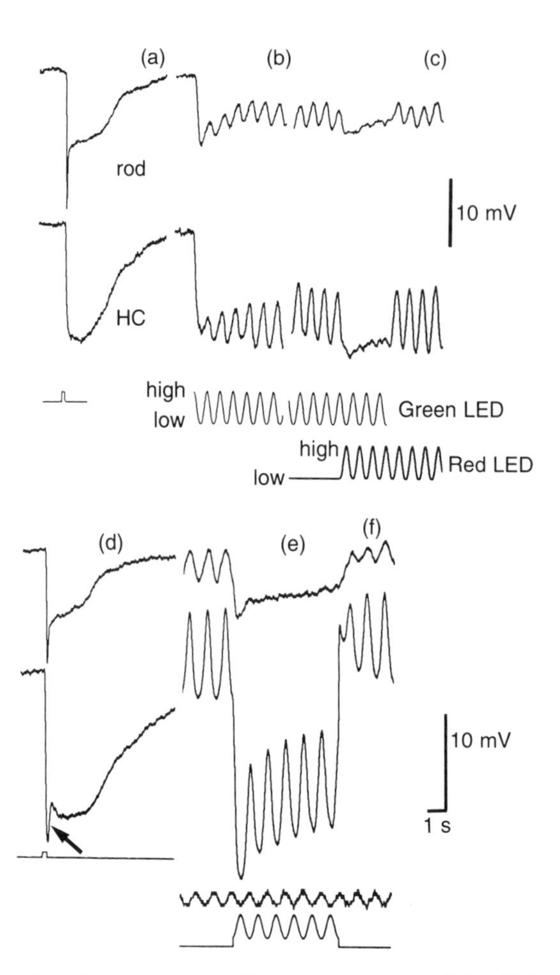

Figure 7.8 The 'null' method for detecting cone input to HCs. (a–c) Recordings from *Xenopus* retina bathed in normal Ringer solution. (a) Simultaneous recording from a rod and a horizontal cell (HC) in dark-adapted retina. Responses to a 200 ms flash of 527 nm light. Stimulus marker at bottom. (b) Responses of rod and HC to a 1 Hz sinusoidal flicker from a green light-emitting diode (LED). Stimulus marker at bottom. (c) When a red LED is illuminated in counterphase, and its intensity adjusted to stimulate rods equally as the green LED, a 'null' is seen, i.e., a loss of the time-modulated signal. HC also shows null, indicating input from rods only. (d–f) Recordings from retinas bathed in Ringer solution containing 5 μM dopamine. (d) Cell responses and stimulus markers as for upper panel (a). After a few minutes exposure to 0.5 μM dopamine, the HC flash response acquires an initial transient (arrow) due to cone input. (e and f) In response to 1 Hz counterphase flicker, the rod produces a null, while the HC evinces a prominent flicker response (stimulus conditions in (e) and (f) correspond to those in (c) and (d), respectively). (Adapted from Krizaj and Witkovsky (1993).)

modulation. Kamermans and Werblin (1992) have outlined a complex relation between GABA transport and GABA release, and there is some indication that GABA release may increase during dark-adaptation (Yang and Wu, 1993).

Thus, within the cone base neuronal activity is dictated only in part by synaptic geometry. For example, the amphibian HC responds to glycine (Stone and Witkovsky, 1984), even though the glycinergic interplexiform cell does not make a defined synaptic contact onto it (Smiley and Yazulla, 1990). Nevertheless, the glycinergic terminal is positioned optimally to short circuit cone to HC synaptic transfer. Tonic glycine release also damps the oscillatory behavior of the HC by tending to clamp the membrane near E_{Cl}, which in HCs is positive to the dark potential (Djamgoz and Laming, 1987).

Other modulatory substances such as dopamine diffuse to distal retinal targets from the border of the inner nuclear and inner plexiform layer, at least 25 μm distant. Witkovsky *et al.* (1993b) provided voltametric data indicating that dopamine levels around photoreceptors vary in the range 0.1–0.7 mM, being highest in the light near dawn and lowest in darkness, and Krizaj and Witkovsky (1993) found that 0.5 mM exogenous dopamine was sufficient to induce a cone input to the HC (Figure 7.8).

7.5 CONCLUSIONS AND FUTURE PERSPECTIVES

On the whole, our understanding of the photoreceptor connectivity of HCs in a variety of vertebrates, from fish to human, is good, although different patterns have been observed in different species. Our understanding is also good for the basic mechanism of synaptic transmission from photoreceptors to HCs involving glutamate as the neurotransmitter acting on non-NMDA type ionotropic receptors on HCs. However, we know much less about ultrastructural and molecular aspects of neuronal interactions within photoreceptor synaptic terminals and the role of metabotropic receptors in the transmission process.

Importantly, synaptic transmission from photoreceptors to HCs and elsewhere in the retina and brain, can no longer be considered to be 'hard-wired', occurring only at morphologically well-defined sites and dictated solely by the properties of ligand-gated, ionotropic channels. Both the internal and external milieux of the retinal neuron are in constant flux and communication. The modes of action of one of the neuromodulators, dopamine, have been studied in detail. However, there is increasing evidence to suggest that physiological changes that retinal neurons undergo during different visual conditions, including light/dark adaptation, involve additional neuromodulators. Thus, a host of neuromodulatory substances, working through a number of second messenger pathways, appear to contrive to alter subtly the excitability of channels, and in ways yet to be spelled out fully, they must alter visual information processing.

The quality and quantity of visual information transmitted from photoreceptors to HCs are likely to be affected in diseases (e.g. Parkinsonism) where a neurochemical loss may occur. Also, following photoreceptor loss occurring during ageing and in degenerative diseases (e.g. retinitis pigmentosa) there will be a general weakening of the visual signals received by HCs. Since HCs provide a major part of the synaptic drive of the bipolar cells, a malfunction in HC activity is likely to affect visual signals reaching the brain. An in-depth understanding of photoreceptor–HC synaptic transmission and identification of the neurochemicals involved in modulating this pathway have generated the promise that improvement of vision in such clinical conditions may ultimately be possible.

REFERENCES

Abe, H. and Yamamoto, T.Y. (1984) Diurnal changes in synaptic ribbons of rod cells of the turtle. *Journal of Ultrastructural Research*, **86**, 246–51.

Ahnelt, P. and Kolb, H. (1994) Horizontal cells and cone photoreceptors in human retina: a Golgi EM study of spectral connectivity. *Journal of Comparative Neurology*, **343**, 406–427.

Akopian, A. and Witkovsky, P. (1994) Modulation of transient outward potassium current by GTP, calcium and glutamate in horizontal cells of the *Xenopus* retina. *Journal of Neurophysiology*, **71**, 1661–71.

Akopian A., McReynolds, J. and Weiler, R. (1991) Short term potentiation of off-responses in the turtle horizontal cells. *Brain Research*, **546**, 132–8.

Akopian, A., McReynolds, J. and Weiler, R. (1992) Activation of protein kinase C modulates light responses in horizontal cells of the turtle retina. *European Journal of Neuroscience*, **4**, 745–9.

Alexander, K.R. and Fishman, G.A. (1984) Rod–cone interaction in flicker perimetry. *British Journal of Ophthalmology*, **68**, 303–9.

Attwell, D. (1986) Ion channels and signal processing in the outer retina. *Quarterly Journal of Physiology*, **71**, 497–536.

Attwell, D. (1990) The photoreceptor output synapse. *Progress in Retinal Research*, **9**, 337–62.

Attwell, D., Borges, S., Wu, S.M. and Wilson, M. (1987) Signal clipping by the rod output synapse. *Nature*, **328**, 522–4.

Baldridge, W.H. and Ball, A.K. (1991) Background illumination reduces horizontal cell receptive-field size in both normal and 6-hydroxy-dopamine-lesioned goldfish retinas. *Visual Neuroscience*, **7**, 441–50.

Baldridge, W.H., Ball, A.K. and Miller, R.G. (1987) Dopaminergic regulation of horizontal cell gap junction particle density in goldfish retina. *Journal of Comparative Neurology*, **265**, 428–36.

Behrens, U.D., Douglas, R.H. and Wagner, H-J. (1993) Gonadotropin-releasing hormone, a neuropeptide of efferent projections to the teleost retina induces light-adaptive spinule formation on horizontal cell dendrites in dark-adapted preparations kept *in vitro*. *Neuroscience Letters*, **164**, 59–62.

Belgum, J.H. and Copenhagen D.F. (1988) Synaptic transfer of rod signals to horizontal and bipolar cells in the retina of the toad. *Journal of Physiology*, **396**, 225–45.

Besharse, J.C. (1992) The 'ON'-bipolar agonist, L-2-amino-4-phosphonobutyrate, blocks light-evoked cone contraction in *Xenopus* eye cups. *Neurochemical Research*, **17**, 75–80.

Besharse, J.C. and Witkovsky, P. (1992) Light-evoked contraction of red absorbing cones in the *Xenopus* retina is maximally sensitive to green light. *Visual Neuroscience*, **8**, 243–9.

Bloomfield, S.A. and Miller, R.F. (1982) A physiological and morphological study of the horizontal cell types of the rabbit retina. *Journal of Comparative Neurology*, **208**, 288–303.

Boatright, J.H., Hoel, M.J. and Iuvone, P.M. (1989) Stimulation of endogenous dopamine release and metabolism in amphibian retina by light- and K^+-evoked depolarization. *Brain Research*, **482**, 164–8.

Bowmaker, J.K. and Kunz, Y.W. (1987) Ultraviolet receptors, tetrachromatic colour vision and retinal mosaics in the brown trout (*Salmo trutta*): age-dependent changes. *Vision Research*, **27**, 2101–8.

Boycott, B.B. and Kolb, H. (1973) The horizontal cells of the rhesus monkey retina. *Journal of Comparative Neurology*, **148**, 115–39.

Boycott, B.B., Peichl, L. and Wässle, H. (1978) Morphological types of horizontal cells in the retina of the domestic cat. *Proceedings of the Royal Society of London B*, **203**, 247–67.

Boycott, B.B., Hopkins, J.M. and Sperling, H.G. (1987) Cone connections of the horizontal cells of the rhesus monkey's retina. *Proceedings of the Royal Society of London B*, **186**, 317–31.

Brandon, C. and Lam, D. M-K. (1983) The ultrastructure of rat rod synaptic terminals: effects of dark adaptation. *Journal of Comparative Neurology*, **217**, 167–175.

Brew, H. and Attwell, D. (1987) Electrogenic glutamate uptake is a major current carrier in the membrane of axolotl retinal glial cells. *Nature*, **327**, 707–9.

Burkhardt, D.A. (1993) Synaptic feedback, depolarization, and color opponency in cone photoreceptors. *Visual Neuroscience*, **10**, 981–89.

Byzov, A.L. and Shura–Bura, T.M. (1986) Electrical feedback mechanisms in the processing of signals in the outer plexiform layer of the retina. *Vision Research*, **86**, 33–4.

Byzov, A.L. and Trifonov, Y.U. (1968) The response to electric stimulation of horizontal cells in the carp retina. *Vision Research*, **8**, 817–22.

Byzov, A.L. and Trifonov, Yu. A. (1981) Ionic mechanisms underlying the nonlinearity of horizontal cell membrane. *Vision Research*, **21**, 1573–81.

Byzov, A.L., Trifonov, Y.A., Chailahian, L.M. and Golubtzov, K.W. (1977) Amplification of graded potentials in horizontal cells of the retina. *Vision Research*, **17**, 265–273.

Cahill, G.M. and Besharse, J.C. (1993) Circadian clock functions localized in *Xenopus* retinal photoreceptors. *Neuron*, **10**, 573–7.

Copenhagen, D.R. and Jahr, C. (1989) Release of endogenous excitatory amino acids from turtle photoreceptors. *Nature*, **341**, 536–9.

Dacheux, R.F. and Raviola, E. (1982) Horizontal cells in the retina of the rabbit. *Journal of Neuroscience*, **2**, 1486–93.

Dearry, A. (1991) Light onset stimulates tyrosine hydroxylase activity in isolated teleost retinas. *Vision Research*, **31**, 395–9.

Dearry, A. and Burnside, B. (1989) Light-induced dopamine release from teleost retinas as a light-adaptive signal to the retinal pigment epithelium. *Journal of Neurochemistry*, **53**, 870–8.

De Vries, S.H. and Baylor, D.A. (1993) Synaptic circuitry of the retina and olfactory bulb. *Neuron*, **10** (suppl), 139–49.

DeVries, S.H. and Schwartz, E.A. (1989) Modulation of an electrical synapse between solitary pairs of catfish horizontal cells by dopamine and second messengers. *Journal of Physiology*, **414**, 351–75.

DeVries, S.H. and Schwartz, E.A. (1992) Hemi-gap-junction channels in solitary horizontal cells of the catfish retina. *Journal of Physiology*, **445**, 201–30.

Dixon, D.B., Takahashi, K-I. and Copenhagen, D.R. (1993) L-glutamate suppresses HVA calcium current in catfish horizontal cells by raising intracellular proton concentration. *Neuron*, **11**, 267–77.

Djamgoz, M.B.A. (1984) Electrophysiological characterization of the spectral sensitivities of horizontal cells in cyprinid fish retina. *Vision Research*, **24**, 1677–87.

Djamgoz, M.B.A. (1988) Effect of glutamate on extracellular potassium ion activity in the isolated retina of cyprinid fish. *Neurochemistry International*, **13**, 439–48.

Djamgoz, M.B.A. and Downing, J.E.G. (1988) A horizontal cell selectively contacts blue-sensitive cones in cyprinid fish retina: intracellular staining with horseradish peroxidase. *Proceedings of the Royal Society of London B*, **235**, 281–7.

Djamgoz, M.B.A. and Laming, P.J. (1987) Microelectrode and functional aspects of chloride activity in cyprinid fish retina: extracellular activity and intracellular activities of L- and C-type horizontal cells. *Vision Research*, **27**, 1481–9.

Djamgoz, M.B.A. and Yamada, M. (1990) Electrophysiological characteristics of retinal neurones: synaptic interactions and functional outputs, in *The Visual System of Fish* (eds R.H. Douglas and M.B.A. Djamgoz), Chapman & Hall, London, pp.159–210.

Djamgoz, M.B.A., Downing, J.E.G. and Wagner, H-J. (1987) Retinal neurones of cyprinid fish: intracellular marking with horseradish peroxidase and correlative morphological analysis. *Experimental Biology*, **46**, 203–16.

Djamgoz, M.B.A., Downing, J.E.G., Kirsch, M. *et al.* (1988) Light-dependent plasticity of horizontal cell functioning in cyprinid fish retina: effects of background illumination of moderate intensity. *Journal of Neurocytology*, **17**, 701–10.

Dong, C-J. and McReynolds, J.S. (1994) Cones contribute to light-evoked, dopamine-mediated uncoupling of horizontal cells in the mudpuppy retina. *Journal of Neurophysiology* (in press).

Dong, C-J. and Werblin, F.S. (1994) Dopamine modulation of GABA receptor function in an isolated retinal neuron. *Journal of Neurophysiology*, **71**, 1258–60.

Dong, C-J., Picaud, S.A. and Werblin, F.S. (1994) GABA transporters and GABA-like receptors on catfish cone- but not rod-driven horizontal cells. *Journal of Neuroscience*, **14**, 2648–58.

Dowling, J.E. (1987) *The Retina: An Approachable Part of The Brain*. Harvard University Press, Cambridge, MA.

Dowling, J.E. and Ripps, H. (1973) Neurotransmission in the distal retina: the effect of magnesium on horizontal cell activity. *Nature*, **242**, 101–3.

Downing, J.E.G. and Djamgoz, M.B.A (1989) Quantitative analysis of cone photoreceptor–horizontal cell connectivity patterns in the retina of a cyprinid fish: electron microscopy of functionally identified and HRP-labelled horizontal cells. *Journal of Comparative Neurology*, **289**, 537–53.

Dubocovich, M.L. (1983) Melatonin is a potent modulator of dopamine release in the retina. *Nature*, **306**, 782–4.

Eliasof, S. and Werblin, F. (1993) Characterization of the glutamate transporter in retinal cones of the tiger salamander. *Journal of Neuroscience*, **13**, 402–11.

Fain, G.L. (1975) Interactions of rod and cone signals in the mudpuppy retina. *Journal of Physiology*, **252**, 735–69.

Falk, G. (1989) Signal transmission from rods to bipolar and horizontal cells: a synthesis. *Progress in Retinal Research*, **8**, 255–79.

Famiglietti, E.V. (1990) A new type of wide-field horizontal cell, presumably linked to the cones, in rabbit retina. *Brain Research*, **535**, 174–9.

Fisher, S.K. and Boycott, B.B. (1974) Synaptic connections made by horizontal cells within the outer plexiform layer of the retina of the cat and the rabbit. *Proceedings of the Royal Society of London B*, **186**, 317–31.

Frumkes, T.E. and Eysteinsson, T. (1987) Suppressive rod–cone interaction in distal vertebrate retina: intracellular records from *Xenopus* and *Necturus*. *Journal of Neurophysiology*, **57**, 1–23.

Gallego, A. (1985) Advances in horizontal cell terminology since Cajal, in *Neurocircuitry of The Retina, A Cajal Memorial* (eds A. Gallego and P. Gouras), Elsevier, Amsterdam, pp. 122–40.

Gallego, A., Baron, M. and Gayoso, M. (1975) Horizontal cells of the avian retina. *Vision Research*, **15**, 1029–30.

Golard, A., Witkovsky, P. and Tranchina, D. (1992) Membrane currents of horizontal cells isolated from turtle retina. *Journal of Neurophysiology*, **68**, 351–61.

Gouras, P. and Zrenner, E. (1981) Color coding in primate retina. *Vision Research*, **21**, 1591–8.

Greenstreet, E.H. and Djamgoz, M.B.A. (1994) Triphasic chromaticity type horizontal cells selectively contact short wavelength-sensitive cone photoreceptors in the retina of a cyprinid fish, *Rutilus rutilus*. *Proceedings of the Royal Society of London B*, **256**, 227–30.

Hals, G., Christensen, B.T., O'Dell, T. *et al.* (1986) Voltage clamp analysis of currents produced by glutamate and some glutamate analogues on horizontal cells isolated from the catfish retina. *Journal of Neurophysiology*, **56**, 19–31.

Hankins, M.W. and Ikeda, H. (1991) Non-NMDA type excitatory amino acid receptors mediate rod input to horizontal cells in the isolated rat retina. *Vision Research*, **31**, 609–17.

Hare, W.A. and Owen, W.G. (1992) Effects of 2-amino-4-phosphonobutyric acid on cells in the distal layers of the tiger salamander's retina. *Journal of Physiology*, **445**, 741–57.

Hare, W.A., Lowe, J.S. and Owen, W.G. (1986) Morphology of physiologically identified bipolar cells in the retina of the tiger salamander, *Ambystoma tigrinum*. *Journal of Comparative Neurology*, **252**, 130–8.

Harsanyi, K. and Mangel, S.C. (1992) Activation of a D2 receptor increases electrical coupling between retinal horizontal cells by inhibiting dopamine release. *Proceedings of the National Academy of Sciences USA*, **89**, 9220–4.

Hashimoto, Y., Haros, F.I., Ueki, K. and Fukurotani, I-I. (1988) Ultraviolet-sensitive cones in the color-coding systems of cyprinid retinas. *Neuroscience Research*, *Suppl.* **8**, S81–S95.

Hidaka, S., Christensen, B.N. and Naka, K. (1986) The synaptic ultrastructure in the outer plexiform layer of the catfish retina: a three-dimensional study with HVEM an conventional EM of Golgi-impregnated bipolar and horizontal cells. *Journal of Comparative Neurology*, **247**, 181–99.

Huang, B. and Karwoski, C.J. (1992) Light-evoked expansion of subretinal space volume in the retina of the frog. *Journal of Neuroscience*, **12**, 4243–52.

Ishida, A.T. and Neyton, J. (1985) Quisqualate and L-glutamate inhibit retinal horizontal-cell responses to kainate. *Proceedings of the National Academy of Sciences USA*, **82**, 1837–41.

Ishida, A.T., Kaneko, A. and Tachibana, M. (1984) Responses of solitary retinal horizontal cells from *Carassius auratus* to L-glutamate and related amino acids. *Journal of Physiology*, **358**, 169–82.

Ito, I., Tanabe, S., Kohda, A. and Sugiyama, H. (1990) Allosteric potentiation of quisqualate receptors by a inotropic drug aniracetam. *Journal of Physiology*, **424**, 533–43.

Janssen-Bienhold, U., Nagel, H. and Weiler, R. (1993) In vitro phosphorylation in isolated horizontal cells of the fish retina: effects of the state of light adaptation. *European Journal of Neuroscience*, **5**, 548–93.

Kaczmarek, L.K. and Levitan, I.B. (1987) *Neuromodulation. The Biochemical Control of Neuronal Excitability*. Oxford University Press, New York.

Kamermans, M. and Werblin, F. (1992) GABA-mediated positive autofeedback loop controls horizontal cell kinetics in tiger salamander retina. *Journal of Neuroscience*, **12**, 2451–63.

Kamermans, M., van Dijk, B.W. and Spekreijse, H. (1989) Lateral feedback from monophasic horizontal cells to cones in carp retina. II A quantitative model. *Journal of General Physiology*, **93**, 695–714.

Kaneko, A. (1987) The functional role of retinal horizontal cells. *Japanese Journal of Physiology*, **37**, 578–80.

Kaneko, A. and Shimazaki, H. (1975) Effects of external ions on the synaptic transmission from photoreceptors to horizontal cells in the carp retina. *Journal of Physiology*, **252**, 509–22.

Kaneko, A. and Tachibana, M. (1985) Effect of L-glutamate on the anomalous rectifier potassium current in horizontal cells of *Carassius auratus* retina. *Journal of Physiology*, **358**, 169–92.

Kaneko, A. and Yamada, M. (1972) S-potentials in the dark adapted retina of the carp. *Journal of Physiology*, **227**, 261–73.

Karwoski, C.J., Lu, H-K. and Newman, E.A. (1989) Spatial buffering of light-evoked potassium increases by retinal Müller (glial) cells. *Science*, **244**, 578–80.

Kim, H.G. and Miller, R.F. (1991) Rods and cones activate different excitatory amino acid receptors on the mudpuppy retinal horizontal cell. *Brian Research*, **538**, 141–6.

Kirsch, M. and Wagner, H-J. (1989) Release pattern of endogenous dopamine in teleost retinae during light adaptation and pharmacological stimulation. *Vision Research*, **29**, 147–54.

Kirsch, M., Djamgoz, M.B.A. and Wagner, H-J. (1990) Correlation of spinule dynamics and plasticity of the horizontal cell spectral response in cyprinid fish retina: quantitative analysis. *Cell and Tissue Research*, **260**, 123–30.

Kirsch, M., Wagner, H-J. and Djamgoz, M.B.A. (1991) Dopamine and plasticity of horizontal cell function in the teleost retina: regulation of a spectral mechanism through D1-receptors. *Vision Research*, **31**, 401–12.

Kivela T., Tarkkamen, A. and Virtanen, I. (1989) Synaptophysin in the human retina and retinoblastoma, *Investigative Ophthalmology and Visual Science*, **30**, 212–19.

Knapp, A.G. and Dowling, J.E. (1987) Dopamine enhances excitatory amino acid-gated conductances in retinal horizontal cells. *Nature*, **325**, 437–9.

Knapp, A.G., Schmidt, K.F. and Dowling, J.E. (1990) Dopamine modulates the kinetics of ion channels gated by excitatory amino acids in retinal horizontal cells. *Proceedings of the National Academy of Sciences USA*, **87**, 767–71.

Kolb, H. (1974) The connections between horizontal cells and photoreceptors in the retina of the cat: electron microscopy of Golgi preparations. *Journal of Comparative Neurology*, **155**, 1–14.

Kolb, H., Mariani, A. and Gallego, A. (1980) A second type of horizontal cell in the monkey retina. *Journal of Comparative Neurology*, **189**, 31–44.

Kolbinger, W. and Weiler, R. (1993) Modulation of endogenous dopamine release in the turtle retina: effects of light, calcium, and neurotransmitters. *Visual Neuroscience*, **10**, 1035–42.

Krause, W. (1868) Die Retina I. Die membrana fenestrata der retina. *Int. Mschr. Anat. Hist.*, **1**, 225–54,

Kretz, R., Ishida, A.T. and Stell, W.K. (1982) Ratfish retina – intracellular recordings and HRP injections in an isolated, superfused all-rod retina. *Vision Research*, **22**, 857–61.

Krizaj, D. and Witkovsky, P. (1993) Effects of submicromolar concentrations of dopamine on photoreceptor to horizontal cell communication. *Brain Research*, **627**, 122–8.

Krizaj, D., Akopian, A. and Witkovsky, P. (1994) The effects of L-glutamate, AMPA, quisqualate and kainate on retinal horizontal cells depend on adaptational state: implications for rod–cone interactions. *Journal of Neuroscience*, **14**, 5661–71.

Kruse, M. and Schmidt, K-F. (1993) Studies on the dopamine-dependent modulation of amino acid-gated currents in cone horizontal cells of the perch (*Perca fluviatilis*). *Vision Research*, **33**, 2031–42.

Kurz–Isler, G. and Wolburg, H. (1986) Gap junctions between horizontal cells in the cyprinid fish alter rapidly their structure during light and dark adaptation. *Neuroscience Letters* **67**, 7–12.

Lasanky, A. (1979) Lateral contacts and interactions of horizontal cell dendrites in the retina of the larval tiger salamander. *Journal of Physiology*, **301**, 59–68.

Lasater, E.M. (1987) Retinal horizontal cell gap junctional conductance is modulated by dopamine through a cyclic AMP-dependent protein kinase. *Proceedings of the National Academy of Science USA*, **84**, 7319–23.

Lasater, E.M. (1991) Membrane properties of distal retinal neurones. *Progress in Retinal Research*, **11** 215–46.

Lasater, E.M. and Dowling, J.E. (1982) Carp horizontal cells in culture respond selectively to L-glutamate and its agonists. *Proceedings of the National Academy of Science USA*, **79**, 936–40.

Lasater, E.M. and Dowling, J.E. (1985) Dopamine decreases conductance of the electrical junctions between cultured retinal horizontal cells. *Proceedings of the National Academy of Science USA*, **82**, 3025–9.

Leeper, H.F. (1978a) Horizontal cells of the turtle retina 1. Light microscopy of golgi preparations. *Journal of Comparative Neurology*, **182**, 777–93.

Leeper, H.F. (1978b) Horizontal cells of the turtle retina. 2. Analysis of interconnections between photoreceptor cells and horizontal cells by light microscopy. *Journal of Comparative Neurology*, **182**, 795–809.

Levitan, I.B. (1994) Modulation of ion channels by protein phosphorylation and dephosphorylation. *Annual Review of Physiology*, **56**, 193–212.

Liman, E.R., Knapp, A.G. and Dowling, J.E. (1989) Enhancement of kainate-gated currents in retinal horizontal cells by cyclic AMP-dependent protein kinase. *Brain Research*, **481** 399–402.

Linberg, K.A. and Fisher, S.K. (1988) Ultrastructural evidence that horizontal cells axon terminals are presynaptic in the human retina. *Journal of Comparative Neurology*, **268**, 281–97.

Linn, C.P. and Christensen, B.N. (1992) Excitatory amino acid regulation of intracellular Ca^{2+} in isolated catfish cone horizontal cells measured under voltage- and concentration-clamp conditions. *Journal of Neuroscience*, **12**, 2156–64.

Low, J.C., Yamada, M. and Djamgoz, M.B.A. (1991) Voltage clamp study of electrophysiologically-identified horizontal cells in carp retina. *Vision Research*, **31**, 437–49.

Malchow, R.P. and Ripps. H. (1990) Effects of gamma-aminobutyric acid on skate retinal horizontal cells: evidence for an electrogenic uptake mechanism. *Proceedings of the National Academy of Sciences USA*, **87**, 8945–9.

Malchow, R.P., Qian, H.H., Ripps, H. and Dowling, J.E. (1990) Structural and functional properties of two types of horizontal cells in the skate retina. *Journal of General Physiology*, **95**, 177–89.

Mangel, S.C. and Dowling, J.E. (1985) Responsiveness and receptive field size of carp horizontal cells are reduced by prolonged darkness and dopamine. *Science*, **229**, 1107–9.

Mangel, S.C. and Dowling, J.E. (1987) The interplexiform–horizontal cell system of the fish retina: effects of dopamine, light stimulation and time in the dark. *Proceedings of the Royal Society of London B*, **231**, 91–121.

Mariani, A.P. and Leure-du Pree, A.E. (1978) Photoreceptors and oil droplet colors in the red area of the pigeon retina. *Journal of Comparative Neurology*, **182**, 821–37.

Massey, S.C. and Miller, R.F. (1987) Excitatory amino acid receptors of rod and cone-driven horizontal cells in the rabbit retina. *Journal of Neurophysiology*, **57**, 645–59.

Mayer, M.L., Vyklicky, L., Benveniste, M. *et al.* (1991) Desensitization at NMDA and AMPA-kainate receptors, in *Excitatory Amino Acids and Synaptic Function* (eds H. Wheal and A. Thomson), Academic Press, New York, pp. 123–40.

McMahon, D.G. (1994) Modulation of zebrafish horizontal cell electrical synapses by nitroprusside and cGMP. *Investigative Ophthalmology and Visual Science*, **35**, 1821.

McMahon, D.G., Knapp, A.G. and Dowling, J.E. (1989) Horizontal cell gap junctions: single-channel conductance and modulation by dopamine. *Proceedings of the National Academy of Sciences USA*, **86**, 7639–43.

Miller, R.F. and Dacheux, R.F (1976) Synaptic organization of ionic basis of ON and OFF channels in mudpuppy retina. III. A model of ganglion cell reception field organization based on chloride free experiments. *Journal of General Physiology*, **67**, 679–90.

Miyachi, E-I. and Murakami, M. (1989) Decoupling of horizontal cells in carp and turtle retinae by intracellular injection of cAMP. *Journal of Physiology*, **419**, 213–24.

Miyachi, E-I. and Murakami, M. (1991) Synaptic inputs to turtle horizontal cells analyzed after blocking of gap junctions by intracellular injection of cyclic nucleotides. *Vision Research*, **31**, 631–5.

Miyachi, E-I., Murakami, M. and Nakaki, T. (1990) Arginine blocks gap junctions between retinal horizontal cells. *NeuroReport*, **1**, 107–10.

Müller, B. and Peichl, L. (1993) Horizontal cells in the cone-dominated tree shrew retina: morphology, photoreceptor contacts, and topographical distribution. *Journal of Neuroscience*, **13**, 3628–46.

Murakami, M., Ohtsu, K. and Ohtsuka, T. (1972) Effects of chemicals on receptors and horizontal cells in the retina. *Journal of Physiology*, **227**, 899–913.

Murakami, M. and Takahashi, K. (1987) Calcium action potential and its use for measurement of reversal potentials of horizontal cell responses in carp retina. *Journal of Physiology*, **386**, 165–80.

Myhr, K.L., Dong, C-J. and McReynolds, J.S. (1994) Cones contribute to light-evoked, dopamine-mediated uncoupling of horizontal cells in the mudpuppy retina. *Journal of Neurophysiology*, **72**, 56–62.

Naka, K.I. and Rushton, W.A.H. (1966) An attempt to analyse colour reception by electrophysiology. *Journal of Physiology*, **185**, 556–85.

Nakamura, Y., McGuire, B.A. and Sterling, P. (1980) Interplexiform cell in cat retina: identification by uptake of gamma-[^{3}H]aminobutyric acid and serial reconstruction. *Proceedings of the National Academy of Sciences USA*, **77**, 658–61.

Negishi, K., Salas, R., Parthe, V. and Drujan, B.D. (1988) Identification of horizontal cells generating different spectral responses in the retina of a teleost fish (*Eugerres plumieri*). *Journal of Neuroscience Research*, **20**, 246–56.

Nelson, R. (1977) Cat cones have rod input: a comparison of the response properties of cones and horizontal cell bodies in the retina of the cat. *Journal of Comparative Neurology*, **172**, 109–36.

Nelson, R. (1985) Spectral properties of cat horizontal cells. *Neuroscience Research Suppl.*, **2**, S167–S183.

Nelson, R., Lützow, A.S.C., Kolb, H. and Gouras, P. (1975) Horizontal cells in the cat retina with independent dendritic systems. *Science*, **189**, 137–9.

Normann, R.A. and Perlman, I. (1979) Signal transmission from red cones to horizontal cells in the turtle retina. *Journal of Physiology*, **286**, 509–24.

Oakley, B. II (1977) Potassium and the photoreceptor-dependent pigment epithelial hyperpolarization. *Journal of General Physiology*, **70**, 405–25.

O'Dell, T.J. and Christensen, B.G. (1989) Horizontal cells isolated from catfish retina contain two types of excitatory amino acid receptors. *Journal Neurophysiology*, **61**, 1097–109.

Ogden, T.E., Pierantoni, R. and Citron, M.C. (1982) Studies of horizontal cell processes in frogs and primates; use of computer reconstruction; in *The Structure of the Eye* (ed. J.G. Hollyfield), Elsevier, Amsterdam, pp.141–9.

Ogden, T.E., Mascetti, G.G. and Pierantoni, R. (1984) The internal horizontal cell of the frog. Analysis of receptor input. *Investigative Ophthalmology and Visual Science*, **25**, 1382–94.

Ogden, T.E., Mascetti, G.G. and Pierantoni, R. (1985) The outer horizontal cell of the frog retina. Morphology, receptor input and function. *Investigative Ophthalmology and Visual Science*, **26**, 643–56.

Ohtsuka, T. and Kouyama, N. (1986) Electron microscopic study of synaptic contacts between photoreceptors and HRP filled horizontal cells in the turtle retina. *Journal of Comparative Neurology*, **250**, 141–56.

Patneau, D.K. and Mayer, M.L. (1991) Kinetic analysis of interactions between kainate and AMPA: evidence for activation of a single receptor in mouse hippocampal neurons. *Neuron*, **6**, 785–98.

Peichl, L. and Gonzalez-Soriano, J. (1993) Unexpected presence of neurofilaments in axon-bearing horizontal cells of the mammalian retina. *Journal of Neuroscience*, **13**, 4091–100.

Peng, Y–W., Sharp, A.H., Snyder, S.H. and Yau, K–W. (1991) Localization of the inositol 1,4,5-trisphosphate receptor in synaptic terminals in the vertebrate retina. *Neuron*, **6**, 525–31.

Perlman, I., Normann, R.A. and Anderson, P.J. (1987) The effects of prolonged superfusions with acidic neurotransmitter antagonists. *Neuroscience Letters*, **30**, 257–62.

Perlman, I., Knapp, A.G. and Dowling, J.E. (1988) Local superfusion modifies the inward rectifying potassium conductance of isolated retinal horizontal cells. *Journal of Neurophysiology*, **60**, 1322–32.

Perlman, I., Sullivan, J.M. and Normann, R.A. (1993) Voltage and time-dependent potassium conductances enhance the frequency response of horizontal cells in the turtle retina. *Brain Research*, **619**, 89–97.

Pfeiffer-Linn, C. and Lasater, E.M. (1993) Dopamine modulates in a differential fashion T- and L-type calcium currents in bass retinal horizontal cells. *Journal of General Physiology*, **102**, 277–94.

Piccolino, M., Neyton, J., Witkovsky, P. and Gerschenfeld, H.M. (1982) γ-aminobutyric acid antagonists decrease junctional communication between L-horizontal cells of the retina. *Pro-*

ceedings of the National Academy of Sciences USA, **79**, 3671–5.

Piccolino, M., Neyton, J. and Gerschenfeld, H.M. (1984) Decrease of gap junction permeability induced by dopamine and cyclic adenosine 3′:5′-monophosphate in horizontal cells of turtle retina. *Journal of Neuroscience*, **4**, 2477–88.

Priddy, M., Drewe, J.A. and Kunze, D.L. (1992) L-Glutamate inhibition of an inward potassium current in neonatal neurons from the nucleus of the solitary tract. *Neuroscience Letters*, **136**, 131–5.

Rashid, K., Baldridge, W.H. and Ball, A.K. (1993) Evidence for D2 receptor regulation of dopamine release in the goldfish retina. *Journal of Neurochemistry*, **61**, 2025–33.

Raviola, E. and Dacheux, R.F. (1990) Axonless horizontal cells of the rabbit retina: synaptic connections and origin of the rod aftereffect. *Journal of Neurocytology*, **19**, 73–6.

Raynauld, J-P., Laviolette, J.R. and Wagner, H-J. (1979) Goldfish retina: a correlate between cone activity and morphology of the horizontal cell in cone pedicules. *Science*, **204**, 1436–8.

Rodieck, R.W. (1988) The primate retina, in *Comparative Primate Biology, Volume 4, Neurosciences*, Alan R. Liss, New York, pp.203–78.

Rodrigues, P.d S. and Dowling, J.E. (1990) Dopamine induces neurite retraction in retinal horizontal cells via diacylglycerol and protein kinase C. *Proceedings of the National Academy of Sciences USA*, **87**, 9693–7.

Röhrenbeck, J., Wässle, H. and Heizmann, C.W. (1987) Immunocytochemical labelling of horizontal cells in mammalian retina using antibodies against calcium-binding protein. *Neuroscience Letters*, **77**, 255–60.

Rowe, J.S. and Ruddock, K.H. (1982a) Hyperpolarization of retinal horizontal cells by excitatory amino-acid neurotransmitter antagonists. *Neuroscience Letters*, **30**, 251–6.

Rowe, J.S. and Ruddock, K.H. (1982b) Depolarization of retinal horizontal cells by excitatory amino-acid neurotransmitter agonists. *Neuroscience Letters*, **30**, 257–62.

Sakai, H.M. and Naka, K. (1986) Synaptic organization of the cone horizontal cells in the catfish retina. *Journal of Comparative Neurology*, **245**, 107–15.

Schaeffer, S.F., Raviola, E. and Heuser, J.E. (1982) Membrane specializations in the outer plexiform layer of the turtle retina. *Journal of Comparative Neurology*, **204**, 253–67.

Schmitz F. and Drenckhahn, D. (1993a) Distribution of actin in cone photoreceptor synapses. *Histochemistry*, **100**, 35–40.

Schmitz F. and Drenckhahn, D. (1993b) Li^+-induced structural changes of synaptic ribbons are related to the phosphoinositide metabolism in photoreceptor synapses. *Brain Research*, **604**, 142–8.

Schmitz, F., Kirsch, M. and Wagner, H-J. (1989) Calcium modulated synaptic ribbon dynamics: a pharmacological and electron spectroscopic study. *European Journal of Cell Biology*, **49**, 207–12.

Schmitz, F., Holbach, M. and Dreckhahn, D. (1993). Colocalization of retinal dystrophin and actin in postsynaptic dendrites of rod and cone photoreceptor synapses. *Histochemistry*, **100**, 473–9.

Schmitz, Y. and Wolburg, H. (1991) Gap junction morphology of retinal horizontal cells is sensitive to pH alterations in vitro. *Cell and Tissue Research*, **263**, 303–10.

Scholes, J.H. (1975) Colour receptors and their synaptic connections, in the retina of a cyprinid fish. *Philosophical Transactions of the Royal Society of London B*, **270**, 61–118.

Slaughter, M.M. and Miller, R.F. (1983) An excitatory amino-acid antagonist blocks cone input to sign conserving second order retinal neurons. *Science*, **219**, 1230–2.

Slaughter, M.M. and Miller, R.F. (1985) Characterization of an extended glutamate receptor of the on bipolar neuron in the vertebrate retina. *Journal of Neuroscience*, **5**, 224–33.

Smiley, J.F. and Yazulla, S. (1990) Glycinergic contacts in the outer plexiform layer of the *Xenopus laevis* retina characterized by antibodies to glycine, GABA, and glycine receptors. *Journal of Comparative Neurology*, **299**, 375–88.

Sommer, B., Keinanen, K, Verdoorn, T.A. *et al.* (1990) Flip and flop: a cell-specific functional switch in glutamate-operated channels of the CNS. *Science*, **249**, 1580–5.

Spadaro, A., de Simone, I. and Puzzolo, D. (1978) Ultrastructural data and chronobiological patterns of the synaptic ribbons in the outer

plexiform layer in the retina of albino rats. *Acta, Anatomica* (Basel) **102**, 365–73.
Stell, W.K. (1972) The morphological organization of the vertebrate retina, In *Handbook of Sensory Physiology* vol. VII/2. Physiology of Photoreceptor Organs (ed. M.G.F. Fourtes), Springer-Verlag, Berlin, pp.111–214.
Stell, W.K. (1975) Horizontal cell axon terminals in goldfish retina. *Journal of Comparative Neurology*, **159**, 513–20.
Stell, W.K. (1980) Photoreceptor-specific synaptic pathways in goldfish retina: a world of colour, a wealth of connections, in *Colour Vision Deficiencies V* (ed. G. Verriest), Adam Hilger, Bristol, pp.1–14.
Stell, W.K., and Lightfoot, D.O. (1975) Color-specific interconnections of cones and horizontal cells in the retina of the goldfish. *Journal of Comparative Neurology*, **159**, 473–502.
Stell, W.K. and Witkovsky, P. (1973) Retinal structure in the smooth dogfish, *Mustelus canis*: light microscopy of photoreceptor and horizontal cells. *Journal of Comparative Neurology*, **148**, 33–45.
Stell, W.K., Lightfoot, D.O., Wheeler, T.G. and Leeper, H.F. (1975), Goldfish retina; functional polarization of cone horizontal cell dendrites and synapses. *Science*, **190**, 989–90.
Stell, W.K., Kretz, R. and Lightfoot, D.O. (1982) Horizontal cell connectivity in goldfish, in *The S-Potential* (eds B. Drujan and M. Laufer), Alan R. Liss, New York, pp.51–75.
Stell, W.K., Walker, S.E., Chohan, K.S. and Ball, A.K. (1984) The goldfish *nervus terminalis*: a luteinizing hormone-releasing hormone and molluscan cardioexcitatory peptide immunoreactive olfactoretinal pathway. *Proceedings of the National Academy of Sciences USA*, **81**, 940–4.
Stone, S. and Witkovsky, P. (1984) The actions of γ-aminobutyric acid, glycine and their antagonists upon horizontal cells of the *Xenopus* retina. *Journal of Physiology*; **353**, 249–264.
Stone, S., Witkovsky, P. and Schütte, M. (1990) A chromatic horizontal cell in the *Xenopus* retina: intracellular staining and synaptic pharmacology. *Journal of Neurophysiology*, **64**, 1683–95.
Svaetichin, G. (1953) The cone action potential. *Acta Physiologica Scandinavica*, **299**, Suppl. **106**, 565.
Tachibana, M. (1985) Permeability changes induced by L-glutamate in solitary retinal horizontal cells isolated from *Carassius auratus*. *Journal of Physiology*, **358**, 153–67.
Tachibana, M., and Kaneko, A. (1988) L-Glutamate induced depolarization in solitary photoreceptors: a process that may contribute to the interaction between photoreceptors in situ. *Proceedings of the National Academy of Sciences USA*, **85**, 5315–19.
Takahashi, K-I. and Copenhagen, D.R. (1992) APB suppresses synaptic input to retinal horizontal cells in fish: a direct action on horizontal cells modulated by intracellular pH. *Journal of Neurophysiology*, **67**, 1633–42.
Tarres, M.A., Baron, M. and Gallego, A. (1986) The horizontal cells in the retina of the owl, *Tyto alba*, and owlet, *Carinae noctua*. *Experimental Eye Research*, **41**, 315–21.
Trifonov, Y.A. (1968) Study of synaptic transmission between photoreceptors and horizontal cells by means of electrical stimulation in the retina. *Biofizika*, **13**, 809–17.
Trifonov, Y.A., Byzov, A.L. and Chailahan, L.M. (1974) Electrical properties of subsynaptic and nonsynaptic membranes of horizontal cells in the fish retina. *Vision Research*, **14**, 229–41.
Tsukamoto, Y., Yamada, M. and Kaneko, A. (1987) Morphological and physiological studies of rod-driven horizontal cells with special reference to the question of whether they have axons and axon terminals. *Journal of Comparative Neurology*, **255**, 305–16.
Umino, O. and Dowling, J.E. (1991) Dopamine release from interplexiform cells in the retina: effects of GnRH, FMRFamide, bicuculline, and enkephalin on horizontal cell activity. *Journal of Neuroscience*, **11**, 3034–46.
Umino, O., Lee, Y. and Dowling, J.E. (1991) Effects of light stimuli on the release of dopamine from interplexiform cells in the white perch retina. *Visual Neuroscience*, **7**, 451–8.
Van Haesendonck, E. and Missotten, L. (1979) Synaptic contacts of the horizontal cells in the retina of the marine teleost, *Callionymus lyra*. *Journal of Comparative Neurology*, **184**, 167–92.
Veruki, M.L. and Yeh, H.H. (1992) Vasoactive intestinal polypeptide modulates $GABA_A$ receptor function in bipolar cells and ganglion cells of the rat retina. *Journal of Neurophysiology*, **67**, 791–7.

Veruki, M.L. and Yeh, H.H. (1994) Vasoactive intestinal polypeptide potentiates $GABA_A$ receptor function through cyclic AMP. *Investigative Ophthalmology*, **35**, 1365.

Wagner, H-J. (1973) Darkness-induced reduction of the number of synaptic ribbons in fish retina. *Nature*, **246**, 54–5.

Wagner, H-J. (1980) Light dependent plasticity of the morphology of horizontal cell terminals in cone pedicles of fish retinas. *Journal of Neurocytology*, **9**, 575–90.

Wagner, H-J. (1990) Retinal structure of fishes, in *The Visual System of Fish* (eds R.H. Douglas and M.B.A. Djamgoz), Chapman and Hall, London, pp.109–58.

Wagner, H-J. and Djamgoz, M.B.A. (1993) Spinules: a case for retinal synaptic plasticity. *Trends in Neuroscience*, **16**, 201–6.

Waloga, G. and Pak, W.L. (1976) Horizontal cell potentials: dependence on external sodium ion concentration. *Science*, **191**, 964–6.

Wässle, H. and Boycott, B.B. (1991) Functional architecture of the mammalian retina. *Physiological Reviews*, **71**, 447–80.

Weiler, R. and Ball, A.K. (1989) Enkephalinergic modulation of the dopamine system in the turtle retina. *Visual Neuroscience*, **3**, 455–61.

Weiler, R. and Kewitz, B. (1993) The marker for nitric oxide synthase, NADPH-diaphorase, colocalizes with GABA in horizontal cells and cells of the inner retina in the carp retina. *Neuroscience Letters*, **158**, 151–4.

Weiler, R. and Wagner, H-J. (1984) Light-dependent change of cone–horizontal cell interactions in carp retina. *Brain Research*, **298**, 1–9.

West, R.W. (1978) Bipolar and horizontal cells of the gray squirrel retina: Golgi morphology and receptor connections. *Vision Research*, **18**, 129–36.

Winslow, R.L., Miller, R.F. and Ogden, T.E. (1989) Functional role of spines in the retinal horizontal cell network. *Proceedings of the National Academy of Sciences USA*, **86**, 387–91.

Witkovsky, P. and Powell, C.C. (1981) Synapse formation and modification between distal retinal neurons in larval and juvenile *Xenopus*. *Proceedings of the Royal Society of London B*, **211**, 373–89.

Witkovsky, P., Stone S. and MacDonald, E.D. (1988a) Morphology and synaptic connections of HRP-filled axon-bearing horizontal cells in the *Xenopus* retina. *Journal of Comparative Neurology*, **275**, 29–38.

Witkovsky, P. Stone, S. and Besharse, J. (1988b) Dopamine modifies the balance of rod and cone inputs to horizontal cells of the *Xenopus* retina. *Brain Research*, **449**, 332–6.

Witkovsky, P., Stone, S. and Tranchina, D. (1989) Photoreceptor to horizontal cell synaptic transfer in the *Xenopus* retina: modulation by dopamine ligands and a circuit model for interactions of rod and cone inputs. *Journal of Neurophysiology*, **62**, 864–81.

Witkovsky, P., Akopian, A., Krizaj, D. and Zhang, J. (1993a) A cellular basis for light and dark adaptation in the *Xenopus* retina. *Great Basin Visual Science Symposium*, **1**, 49–57.

Witkovsky, P., Nicholson, C., Rice, M.E. *et al.* (1993b) Extracellular dopamine concentration in the retina of the clawed frog, *Xenopus laevis*. *Proceedings of the National Academy of Sciences USA*; **90**, 5667–71.

Wu, S.M. (1988) Synaptic transmission from rods to horizontal cells in dark-adapted tiger salamander retina. *Vision Research*, **28**, 1–8.

Wu, S.M. (1994) Synaptic transmission in the outer retina. *Annual Reviews of Physiology*; **56**, 141–68.

Yamada, M., Djamgoz, M.B.A., Low, J.C. *et al.* (1991) Conductance-decreasing cone input to H1 horizontal cells in carp retina. *Neuroscience Research, Supplement* **15**, S51–S65.

Yang, X–L. and Wu, S.M. (1989) Modulation of rod-cone coupling by light. *Science*, **224**, 352–4.

Yang, X.-L. and Wu, S.M. (1991) Co-existence and function of glutamate receptor subtypes in the horizontal cells of the tiger salamander retina. *Visual Neuroscience*, **7**, 377–82.

Yang, X.-L. and Wu, S.M. (1993) Effects of GABA on horizontal cells in the tiger salamander retina. *Vision Research*, **33**, 1339–44.

Yang, X.-L., Fan, T.-X. and Shen, W. (1994) Effects of prolonged darkness on light responsiveness and spectral sensitivity of cone horizontal cells in carp retina, *in vivo*. *Journal of Neuroscience*, **14**, 326–34.

Yasui, S., Yamada, M. and Djamgoz, M.B.A. (1990) Dopamine and 2-amino-4-phosphonobutyrate differentially affect spectral responses of H1 horizontal cells in carp retina. *Experimental Brain Research*, **83**, 79–84.

Zhou, Z.J., Fain, G.L. and Dowling, J.E. (1993) The excitatory and inhibitory amino acid receptors on horizontal cells isolated from the white perch retina. *Journal of Neurophysiology*, **70**, 8–19.

Zucker, C.L. and Dowling, J.E. (1987) Centrifugal fibers synapse on dopaminergic interplexiform cells in the teleost retina. *Nature*, **330**, 166–8.

8

Horizontal cell coupling and its regulation

MARK W. HANKINS

8.1 INTRODUCTION

Studies of the physiology of retinal horizontal cells in numerous vertebrate species have revealed a number of characteristic functions indicative of their role in visual processing in the outer plexiform layer. One such important characteristic is the extended nature of horizontal cell receptive field response properties. Horizontal cells in general respond to light stimuli that are presented well beyond their individual dendritic field. The origin of such expansive summation is now accepted to reside in the extensive gap-junction coupling between adjacent horizontal cells of a given functional class (Svaetichin *et al.*, 1961; Yamada and Ishikawa, 1965; Naka and Rushton, 1967; Kaneko, 1971). We can therefore regard each morphological, or corresponding physiological, class of horizontal cell as a functional syncytium which may, under certain conditions, extend across the whole retina. These syncytia form the functional 'S-space' originally proposed and modeled by Naka and Rushton (1967).

It is now well established that horizontal cells provide the principal antagonistic surround input to bipolar cells through feed-forward and feedback synapses in the outer plexiform layer (Kaneko, 1973; Toyoda and Tonasaki, 1978). The strength of horizontal cell coupling therefore has important implications to visual processing in the retina, particularly since it has become apparent that the extent of the coupling is not a hard-wired feature of the outer plexiform layer. Such coupling appears to be under the regulation of complex endogenous control mechanisms, the activity of which may adjust a number of features of the retina output in accordance with prevailing light or adaptational conditions. Horizontal cell coupling has thus been examined in relation to the pharmacology of its controlling mechanisms, in order to determine which endogenous pathways act to regulate this functional property. Other studies have probed the second messenger systems of horizontal cells, which translate the pharmacological inputs into changes in gap-junction conductance. More recently, the emphasis has been on the study of light and dark adaptational regulation of horizontal cell coupling, the goal of such studies being to provide an insight into the functional modification of visual processing in the outer retina.

Neurobiology and Clinical Aspects of the Outer Retina
Edited by M.B.A. Djamgoz, S.N. Archer and S. Vallerga
Published in 1995 by Chapman & Hall, London
ISBN 0 412 60080 3

8.2 ESTIMATION OF HORIZONTAL CELL COUPLING

Studies of the coupling status of horizontal cells have encompassed a range of methodologies, including electrophysiological and anatomical approaches. These methods and how they are applied are examined in the following sections, prior to evaluating the information that they have provided.

8.2.1 ELECTROPHYSIOLOGICAL APPROACHES

Intracellular recording from single horizontal cells provides the primary functional assessment of horizontal cell receptive fields. Such measurements usually involve the recording of light-evoked potentials elicited by light stimuli with a variety of spatial configurations. Perhaps the simplest configuration is that of a paired spot and annulus, in which the cell is stimulated alternately with a small spot confined to an area covered by the dendritic field (~ 100–150 μm diameter) and an annulus covering retinal areas outside the dendritic field (~ 150–200 μm inside diameter). Such an arrangement gives an indirect assessment of horizontal cell coupling by measuring the direct dendritic response, relative to that which is mediated primarily via adjacent coupled cells (Figure 8.1). This approach has been used extensively in pharmacological investigations of horizontal cell coupling, for example in the initial studies of dopamine on horizontal cell coupling (Negishi and Drujan, 1978). Horizontal cell receptive fields can also be assessed by measuring light-evoked responses with spots of increasing diameters centered on the impaled cell (Figure 8.2); such an approach was adopted by Cohen and Dowling (1983). Other studies have used a small light spot stimulus which is deflected left and right of the receptive field center (see for example Teranishi *et al.*, 1984). A refinement of the displaced stimulus method utilizes a small slit of light, moved stepwise across the receptive field. Using this approach, Lamb (1976) demonstrated in the turtle retina that the analysis of the horizontal cell spatial response could be simplified into a one-dimensional problem where the decline in the peak response (at receptive field center) is described by a single exponential function:

$$V(x) = V_o \exp(-x/\lambda) \qquad (8.1)$$

where x is the distance between the slit and the recording electrode; $V(x)$ is the response amplitude to the slit at position x; V_o is the maximum response amplitude at receptive field center; and λ is the length constant.

Within the normal physiological range, the length constant for a given cell is defined as:

$$\lambda = D\,(g_s/g_m)^{0.5} \qquad (8.2)$$

where D is the mean spacing between cells; g_s is the coupling conductance between a pair of adjacent cells; and g_m is the single horizontal cell plasma membrane conductance.

Thus, the spatial length constant for a given cell varies both as a function of the coupling conductance, together with any changes in the effective 'shunt' conductance of the horizontal cell plasma membrane. Therefore, changes in the apparent length constant of cells may occur quite independently of gap-junction conductance. This is an important factor in interpreting of experimental data.

In studies of other species, the single exponential function model of Lamb (1976) has been found to hold for rod-driven horizontal cells in the carp, which anatomically have no axon terminals (Shigematsu and Yamada, 1988; Yamada *et al.*, 1992). A complication has arisen from studies of the cone-driven horizontal cells in the teleost fish retina, in that they possess both a soma (cell body) and an axon terminal. This effectively introduces a secondary functional syncytium so that the spatial model for the decline of the

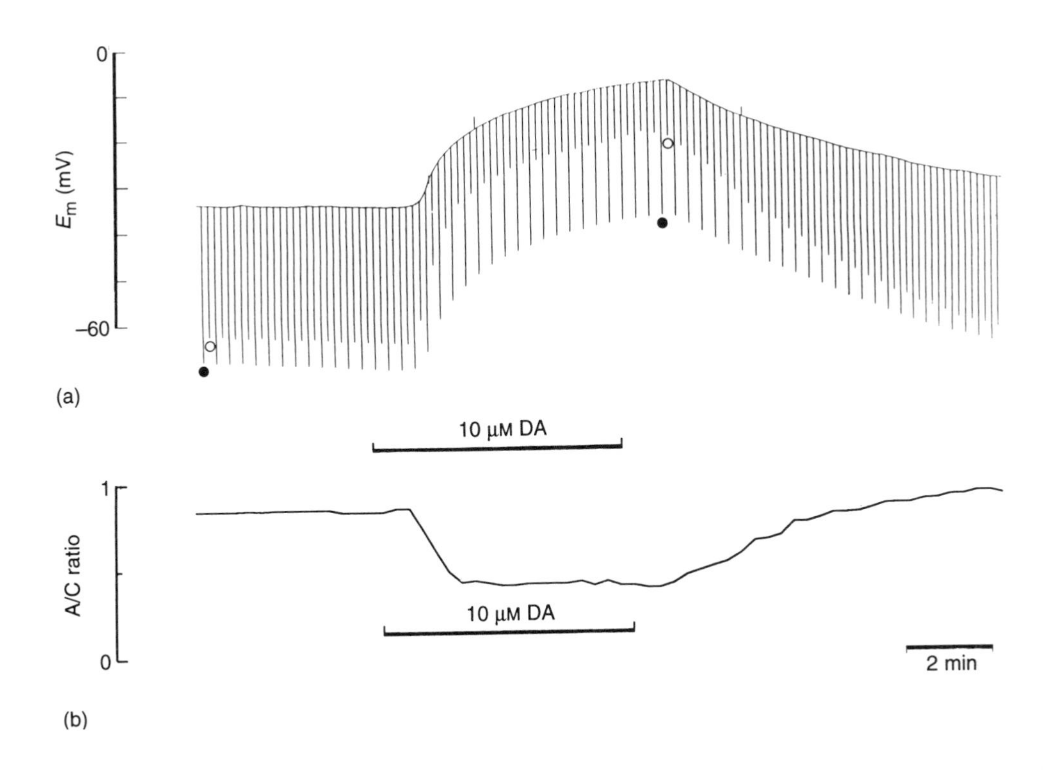

Figure 8.1 The differential effect of dopamine on a rat rod-driven horizontal cell response to peripheral or central light stimuli reveals a dopaminergic reduction in receptive field. (a) An electrophysiological recording from a single rod-driven cell. Membrane potential (E_m, mV) is displayed with respect to time and each downward deflection on the resting potential is a hyperpolarizing light response (S-potential) to a 520 nm light stimulus. Dopamine is applied during the times indicated by the bar. The cell is stimulated alternately by a central spot (100 μm diameter; denoted by the filled circle) and an annulus (internal diameter 150 μm, external diameter 4 mm; denoted by the open circle), both centered on the recording electrode. Note that 10 μM dopamine (DA) depolarizes the cell and selectively reduces the amplitude of the annulus response, whereas the spot-evoked response remains relatively unaffected. (b) A plot of the concurrent changes in the annulus/central (A/C) ratio, showing the rapid and selective loss of surround sensitivity evoked by dopamine and its subsequent recovery. (Modified from Hankins and Ikeda (1993).)

response at the cell soma in this case is described as the sum of two independent exponential functions (Yagi, 1986; Shigematsu and Yamada, 1988).

Whereas electrophysiological assessment of horizontal cell receptive fields and coupling are largely confined to recordings from single horizontal cells within the syncytia of the intact retina, an important technique for recording from identified coupled pairs of horizontal cells has been developed using partially dissociated cells in culture (Lasater and Dowling, 1985). Thus, by recording from one cell (the follower) and injecting current into the other (the driver), it is possible to assess directly changes in the coupling con-

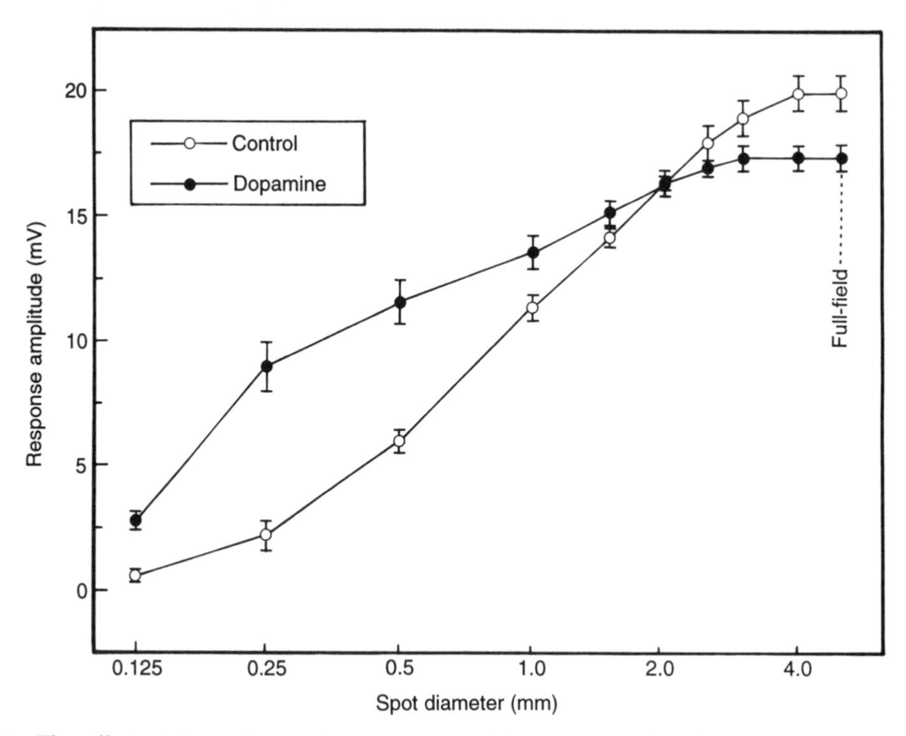

Figure 8.2 The effect of dopamine on the receptive field properties of rat horizontal cells assessed by the spot size method. Averaged peak amplitudes of horizontal cell S-potentials plotted as a function of increasing stimulus spot diameter. Cells were recorded either from normal control retinas (n=8; open circle), or during the application of dopamine (n=8; solid circle). Note that dopamine (10 μM) significantly increased the amplitude of responses to small spots (<1.0 mm), but also decreased the responses to large and full-field stimuli (>3.0 mm). Each data point represents the mean ± SEM.

ductance (g_s) while eliminating any changes in the general membrane conductance (g_m).

8.2.2 DYE-COUPLING STUDIES

This approach originates from the observation that fluorescent dyes such as Lucifer Yellow, when injected into single horizontal cells, normally diffuse to adjacent horizontal cells via gap junctions (Kaneko and Stuart, 1980; Piccolino *et al.*, 1982; Teranishi *et al.*, 1983). The technique has demonstrated functional cytoplasmic continuity between adjacent cells via gap junctions and by using controlled protocols of fixed injection time and current it provides a semiquantitative index of cell coupling by comparing the diffusion of dye from cell to cell (Teranishi *et al.*, 1983). In this way, comparisons can be made from one preparation to another under different conditions and under pharmacological manipulation. Given the experimental variability of the technique however, this application is limited to clearly defined extremes of coupling or uncoupling. Another limitation also arises from the potential heterogeneity of gap-junction subtypes. Observations in the inner retina suggest that some retinal gap-junction channels are impermeant to dyes such as Lucifer Yellow, but readily allow the passage of markers such as neurobiotin (Hampson *et al.*, 1992). Therefore, the impermeance of Lucifer Yellow alone cannot negate the presence of functional gap-junction channels. In summary, dye-injection studies measure dye-specific coupling, rather than functional electrotonic coupling.

8.2.3 VISUALIZATION OF GAP JUNCTIONS

A number of techniques have been employed to assess gap-junctional conductance through the visualization of single gap-junction channels. The structure of gap junctions follows the general motif seen in other channel proteins, in which a number of protein subunits are assembled around a central pore (Bennett *et al.*, 1991). Generally, in the case of gap junctions, it appears that the two hexameric half channels of connexin proteins are joined across adjacent cell membranes to form a continuous pore. However, the connexin proteins have several homologous forms with typical masses in the range of 30–43 kDa. Given the large size of the assembled channels, it has been possible to create freeze-fracture replicas from the horizontal cell membranes of the goldfish retina and, thereby, to measure the density of gap-junction particles (Kurtz–Isler and Wolberg, 1986). The technique has been proposed as a semiquantitative index of cell coupling and used to compare horizontal cells under various conditions or pharmacological states, on the assumption that the direct control of gap-junction channel density plays a role in the regulation of horizontal cell coupling (Baldridge *et al.*, 1987). A more recent study using conventional transmission electron microscopy suggests, however, that the presence of gap-junction particles cannot in itself provide an index of junctional conductance, since gap-junction channels undergo periods of renewal and removal, during which times they may be functionally inactive (Vaughan and Lasater, 1990). Furthermore, the conductance of gap-junction channels is under the regulation of second messenger systems (section 8.3.4).

8.3 PHARMACOLOGY OF HORIZONTAL CELL COUPLING: NEUROANATOMICAL BASIS

The functional machinery of horizontal cell coupling, namely the presence of gap-junction channels between adjacent cells, is now well established (Yamada and Ishikawa, 1965; Vaughan and Lasater, 1990). As to the neuroanatomical evidence for a neuromodiolatory synaptic pathway that may regulate the coupling, there are limited possibilities. In general, retinal horizontal cells receive synaptic inputs from photoreceptors mediated by L-glutamate (Chapter 7). In many species, however, some types of horizontal cells receive a secondary synaptic input from interplexiform cells (Dowling and Ehinger, 1978). Interplexiform cells form an intraretinal centrifugal pathway from the inner to the outer plexiform layer, where they form extensive synapses primarily upon cone-driven horizontal cells (Chapter 15). The interplexiform cells receive inputs in the inner plexiform layer from multiple amacrine cell types and, in the teleost fish, from centrifugal fibers (Zucker and Dowling, 1987). In most species, the principal interplexiform cells appear to be dopaminergic neurons (reviewed by Djamgoz and Wagner, 1992). However, although dopaminergic interplexiform cells make contact with certain classes of horizontal cells in many species groups, including teleosts, cat, rabbit, rat, New World monkeys and humans, they are notably absent from other groups such as the elasmobranchs, frogs, turtles and birds (Djamgoz and Wagner, 1992). Furthermore, recent immunohistochemical studies have revealed an apparent diversity of neurotransmitter types among subcategories of interplexiform cells in various species. In addition to the dopaminergic interplexiform cells (I1 type), glycinergic interplexiform cells (I2 type) have been observed in fish and cat (Marc and Lui, 1984), GABAergic interplexiform cells in cat (Nakamura *et al.*, 1980) and most recently 'I3-adrenergic' cells reported in teleosts (Baldridge and Ball, 1993). Such findings raise the possibility of multiple regulatory pathways affecting horizontal cells via these diverse interplexiform cell populations.

8.4 DOPAMINERGIC MODULATION OF HORIZONTAL CELL COUPLING

In many vertebrate species retinal dopamine-containing cells are not a single category of neuron, rather they consist of at least two principal classes, the dopamine–amacrine cells and the dopamine–interplexiform cells. Some species possess only one of these classes. For example, the majority of the teleost fish retinas appear to contain only the dopamine–interplexiform type, whereas in the turtle only dopamine–amacrine cells are found. In contrast, most mammalian retinas possess both cell types (Djamgoz and Wagner, 1992).

The dopamine cells represent a sparse population of retinal neurons, with typical cell densities which rarely exceed 25 cells/mm^2. In contrast, dopamine receptors occur on virtually all morphological classes of neuron and this is reflected by electrophysiological and biochemical studies in a broad range of species which reveal a multiplicity of dopamine effects on the activity of retinal neurons (recent reviews Witkovsky and Dearry, 1991; Djamgoz and Wagner, 1992).

Dopamine was the first endogenous substance to be investigated in relation to the regulation of the receptive field properties of horizontal cells. The first studies, in the teleost retina, examined the effects of exogenous dopamine on cone-driven horizontal cells (Negishi and Drujan, 1978). These experiments established that dopamine increased the horizontal cell response to small central spot stimuli, but decreased the horizontal cell response to concentric annular stimuli. This was the first demonstration that dopamine was capable of restricting the receptive field properties of horizontal cells. It has since been reported that dopamine has broadly similar effects on the majority of horizontal cell types studied in numerous vertebrate species. Thus dopamine not only restricts the receptive field properties of cone-driven horizontal cells in numerous teleost fish species (Negishi and Drujan, 1978) but also rod-driven horizontal cells in the carp (Yamada *et al.*, 1992), the axon terminals of horizontal cells in the turtle (Gerschenfeld *et al.*, 1982; Piccolino *et al.*, 1984), horizontal cells in the mudpuppy (Dong and McReynolds, 1991) and mammalian rod-driven horizontal cells in the rat (Hankins and Ikeda, 1991b). The general effects of dopamine on the receptive field properties of horizontal cells are illustrated by recordings and data obtained from rod-driven horizontal cells in the rat retina in Figures 8.1 and 8.2.

The evidence suggests that dopaminergic control of the horizontal cell receptive field is a common feature in the vertebrate retina, but it is also apparent that the effects of dopamine are not restricted to those horizontal cells which receive direct synaptic input from dopamine–interplexiform cells. For example, dopaminergic interplexiform cells do not make contact with rod-driven horizontal cells in the fish retina, and there are no dopamine–interplexiform cells in the turtle retina. This has led to the proposal that in some retinas, the modulatory effects of dopamine might be mediated via a paracrine release of dopamine into the neural retina (Piccolino *et al.*, 1987).

8.4.1 ARE THE PRINCIPAL EFFECTS OF DOPAMINE ON RECEPTIVE FIELDS MEDIATED VIA GAP-JUNCTION COUPLING?

The suggestion that the effects of dopamine on horizontal cell receptive fields occur through the modulation of gap-junction conductance gained some support from the observation that dopamine also reduces significantly the diffusion of Lucifer Yellow between adjacent horizontal cells (Gerschenfeld *et al.*, 1982; Teranishi *et al.*, 1983). Direct evidence for this is provided by the results of measurements on pairs of horizontal cells in culture (Lasater and Dowling, 1985), which established that whereas dopamine dramatically reduces the junctional conductance between paired cells, it has no direct effect on the input resistance of isolated single cells.

8.4.2 UNCERTAINTIES IN THE DOPAMINE-RECEPTIVE FIELD MODEL

It would appear that not all of the effects of dopamine on the receptive field properties of horizontal cells can be explained by changes in gap-junction coupling. More specifically, there are a number of studies in which dopamine is reported to reduce the response to large spots or full-field stimuli (Hedden and Dowling, 1978; Teranishi *et al.*, 1984; Hankins and Ruddock, 1984a; Hankins and Ikeda, 1991b; Figure 8.2). This cannot be reconciled with changes in coupling alone, since in response to full-field stimuli all horizontal cells would be driven equally, therefore uncoupling them should not reduce the central horizontal cell light response. It is only possible to model such changes in receptive field properties by incorporating a secondary effect of dopamine on the horizontal cell membrane, or shunt conductance (g_m). It has been suggested that this may occur through the dopamine-mediated potentiation of the photoreceptor glutamate input to horizontal cells (Knapp and Dowling, 1987). However, such a mechanism appears to be species dependent since, for example, dopamine has no apparent effect on horizontal cell full-field light evoked responses in the turtle retina (Piccolino *et al.*, 1984). Further, exogenous dopamine does not always evoke horizontal cell depolarization in the intact retina, which might suggest the effect of dopamine on the L-glutamate-mediated current is not a general phenomenon in all species. In the rat retina, dopamine consistently depolarizes horizontal cells in conjunction with its effects on the receptive field and furthermore, the depolarizing effect of dopamine does not persist when the photoreceptor input is abolished during synaptic block by cobalt (Hankins and Ikeda, 1991a,b). However, recent attempts in our laboratory to reveal an effect of dopamine on excitatory amino acid-mediated depolarization of horizontal cells *in situ*, failed to reveal any dopamine-mediated potentiation (Figure 8.3). The mechanisms through which dopamine depolarizes horizontal cells in the rat retina have yet to be clarified.

Another mechanism for a dopamine-mediated change in horizontal cell membrane conductance has been suggested by Djamgoz and Laming (1987), who reported that exogenous dopamine applied *in situ*, evoked an increase in the primary Cl^- conductance of roach horizontal cells. This might also indicate a possible source of variability in the dopamine-evoked depolarization of horizontal cells, since under some conditions the dark resting potential (E_m) approaches the Cl^- equilibrium potential. It has also been shown in the roach that the depolarizing effects of dopamine persist after synaptic blockade with cobalt (Hankins and Ruddock, 1984a, 1986), thus in some species, the effects of dopamine on the receptive field properties

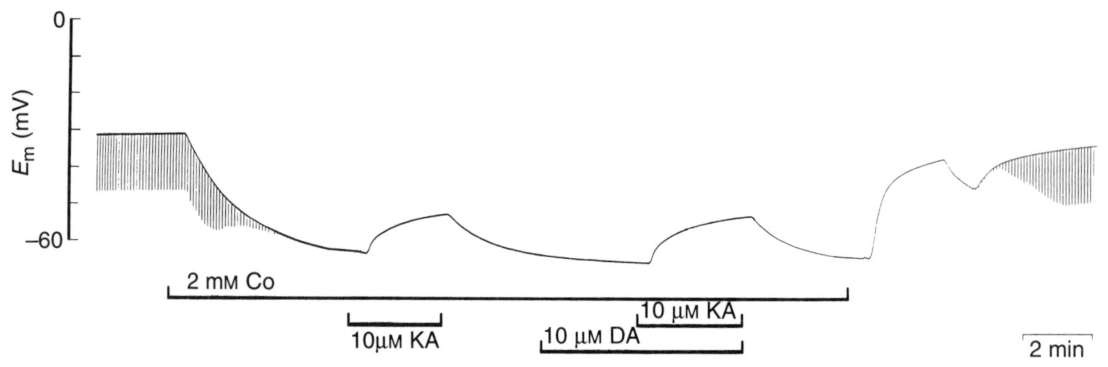

Figure 8.3 Dopamine does not facilitate the excitatory amino acid (EAA) induced depolarization of rat retinal horizontal cells. In this recording the cell is initially hyperpolarized by the addition of 2 mM cobalt (Co), the resulting hyperpolarization and abolition of the S-potential reflecting the blockade of photoreceptor transmitter release. The subsequent addition of 10 μM kainate (KA), an agonist at the EAA receptors on these cells (Hankins and Ikeda, 1991a), then evokes a clear postsynaptic depolarization. Note that the subsequent addition of dopamine (DA) has no effect on the resting potential in the presence of cobalt, neither does it affect the magnitude of a subsequent co-application of 10 μM kainate. These results suggest that dopamine, at concentrations which affect the receptive field properties of horizontal cells in this preparation, has no potentiating effect on the EAA-evoked depolarization. (S-potential responses evoked by repeated full-field stimuli, 520 nm).

of horizontal cells may include direct or indirect modulation of the horizontal cell membrane conductance.

8.4.3 PHARMACOLOGY OF THE DOPAMINE–HORIZONTAL CELL COUPLING PATHWAY

The effects of dopamine are mediated by membrane receptors which have been broadly divided into two categories, the D1and D2 types (Kebabian and Calne, 1979). Molecular biological studies have, however, demonstrated that the D2 type in particular, consists of a heterogeneous group of subclasses identified as D2s, D21, D3 and D4 receptors (Sibley and Monsma, 1992). There are, at present, no potent and selective pharmacological agonists and antagonists which reliably differentiate these receptors *in situ*.

Our knowledge of the dopamine receptors involved in retinal pathways is primarily derived from the use of pharmacological agonists and antagonists selective for the D1 and D2 receptor families. One functional distinction between D1 and D2 receptors is revealed by differential second messenger coupling. D1 receptor activation enhances intracellular cAMP formation, whereas D2 receptor activation either inhibits or is without effect on cAMP. In addition, the general affinity of D1 sites for dopamine is in the micromolar range, whereas that of the D2 sites is in the nanomolar range. The electrophysiological studies of dopamine on horizontal cells by many authors consistently reveal that the concentration of dopamine required *in situ* to evoke changes in receptive field and cell coupling is in the 10–100 μM range. Furthermore, the effects of dopamine on horizontal cell coupling also appear to involve an increase in intracellular cAMP. For example in the fish retina, the application of dibutyryl cAMP *in situ* (Teranishi *et al.*, 1983)

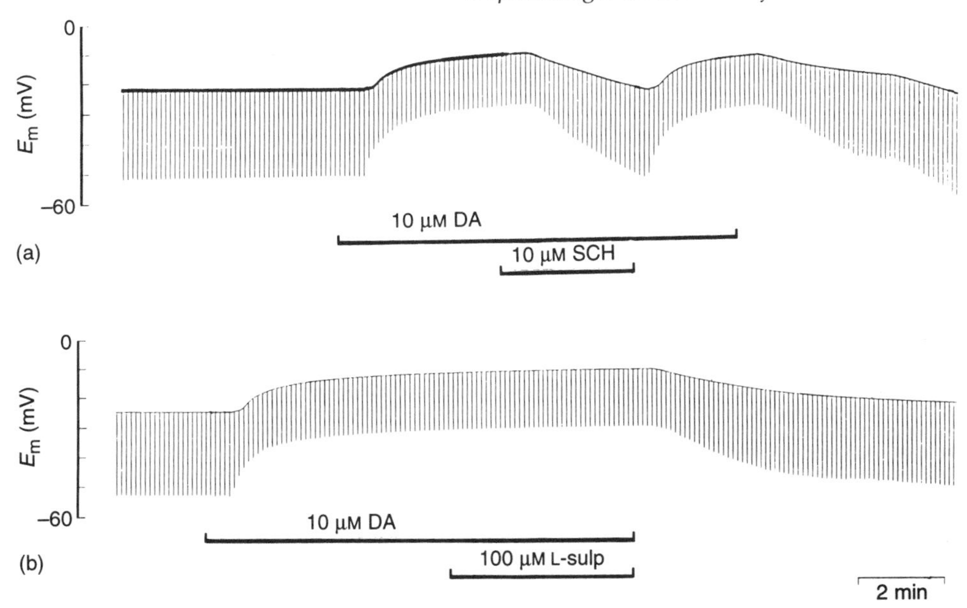

Figure 8.4 Pharmacological studies of the actions of dopamine at horizontal cells in the rat retina. Effects of a D1 antagonist, SCH 23390 (a) and the D2 antagonist, L-sulpiride (b) on the exogenous DA-evoked effects on horizontal cells. In both recordings the cells are stimulated by repeating full-field light flash of 520 nm wavelength. Note that in (a) dopamine (DA) evokes a pronounced depolarization of the membrane potential, an effect associated with a reduction in the full-field light response. Both effects of dopamine are reversed by the subsequent addition of 10 μM SCH 23390 (SCH); whereas in (b) 100 μM L-sulpiride (L-sulp) has no effect on the DA-evoked response. These data suggest a pronounced effect of dopamine on horizontal cells that is mediated by D1 type receptors. (Redrawn from Hankins and Ikeda, (1993).)

or intracellularly applied cAMP (Lasater and Dowling, 1985) also uncouples horizontal cells (section 8.4.4). This has led to a generalized hypothesis that dopamine released from dopamine–interplexiform cells acts primarily upon D1 receptors in fish retinal horizontal cells to increase intracellular cAMP. In the rat retina, both the dopamine-evoked depolarization of horizontal cells and the changes in receptive field properties are reversed by application of the selective D1 antagonist SCH 23390, but are unaffected by the D2 antagonist L-sulpiride (Hankins and Ikeda, 1991b, 1993; see Figure 8.4). These results are consistent with the effects of exogenous dopamine on horizontal cells being mediated by D1 receptors. In addition, horizontal cells are hyperpolarized and their light responses to broad-field stimuli enhanced by D1, but not D2 antagonists (Hankins and Ikeda, 1993). These experiments confirm the presence of an active endogenous dopamine pathway mediated by D1 receptor activation.

To date, relatively few studies have examined the direct actions of agonists selective for D1 or D2 receptors and currently no such data are available for the mammalian retina. Interestingly, it appears that in the turtle retina D2 receptor agonists may promote a small increase, rather than a decrease of horizontal cell coupling (Demontis *et al.*, 1988). In other species, it has been suggested that such changes in coupling evoked by D2 agonists may reflect the activation of presynaptic D2 autoreceptors known to inhibit endogenous dopamine release in the CNS (Harsanyi and Mangel, 1992). Whatever the mechanism, functional evidence indicates that D2 receptor activation, presynaptic or postsynaptic, promotes an increase in horizontal cell coupling.

8.4.4 SECOND MESSENGER INVOLVEMENT IN THE REGULATION OF HORIZONTAL CELL COUPLING

This section deals with the specific intracellular mechanisms which act to link pharmacological receptor activation with changes in horizontal cell gap-junctional conductance. The first studies to address this question involved the application of cAMP and its membrane permeant analogs to horizontal cells both *in situ* and in isolated cell preparations. The application of dibutyryl-cAMP was found to mimic the effects of dopamine on horizontal cells in the carp retina, restricting receptive fields and dye-coupling in preparations previously depleted of dopaminergic neurons (Teranishi *et al.*, 1983). Similarly, the incorporation of excess intracellular cAMP in experiments on paired horizontal cells, also resulted in the uncoupling of horizontal cells (Lasater and Dowling, 1985). These and other studies have established that exogenous dopamine and intracellular cAMP have analogous effects. Other studies have examined the biochemical stages beyond cAMP activation, and it has been reported that the injection of the catalytic subunit of cAMP-dependent protein kinase also uncouples horizontal cells (Lasater, 1987). DeVries and Schwartz (1989), studying paired catfish horizontal cells, established that the uncoupling effect of dopamine was not only mimicked by cAMP and forskolin, but was also blocked by an inhibitor of cAMP-dependent protein kinase.

The activation of cAMP-dependent protein kinase may provide the important link to the phosphorylation of specific target proteins, perhaps even the connexin gap-junction channel itself. In this case, phosphorylation of the channel would lead to a decrease in gap-junction conductance. These steps beyond the cAMP-dependent protein kinase remain speculative and await additional investigation to reveal the entire second messenger pathway.

Evidence now exists that additional second messenger systems, other than cAMP, are also involved in the regulation of gap-junction conductance. In their studies of the catfish, DeVries and Schwartz (1989) reported that both cAMP and cGMP independently decreased the coupling of horizontal cells. Furthermore, lowering the intracellular pH within the physiological range by 0.2–0.3 units was found to profoundly reduce cell coupling. Similarly, in the turtle, intracellular injection of either cAMP or cGMP to horizontal cells blocked gap-junctional conductance (Miyachi, 1992). This particular study showed that L-arginine, a precursor of nitric oxide, also had an uncoupling effect. This raises the possibility of a pathway in which nitric oxide may activate a soluble guanylate cyclase, raising cGMP and uncoupling horizontal cells. Such a pathway may be acting independently of the dopamine–cAMP system to provide secondary control of horizontal cell gap-junction conductance. In the skate retina, where rod-driven horizontal cells are coupled through gap junctions, the junctional conductance between cells is not affected by dopamine or the second messengers cAMP and cGMP, but is dramatically reduced following intracellular acidification (Qian *et al.*, 1993). It remains to be seen whether there are any neuromodulatory regulatory pathways, analogous to dopamine, in this species.

The second messenger systems discussed here appear to be directed at the control of single gap-junction channel conductance, and they suggest that channel phosphorylation and dephosphorylation may represent the regulatory endpoint. Indeed, it has been shown by noise analysis in paired cultured cells that the response to dopamine is accompanied by a reduction in the open state probability of gap-junction channels (McMahon *et al.*, 1989). A complication to this interpretation has arisen from investigations of the density of gap-junctional particles (Baldridge *et al.*, 1987) which suggest that the application of dopamine results in a rapid decline in their number. Consequently, it was proposed that changes in the junctional conductance between horizontal cells may be regulated by changes in the number of channels, rather than the single channel conductance. However, given the evidence suggesting that changes in horizontal cell gap-junction particle density may be independent of functional channels (Vaughan and Lasater, 1990; section 8.2.3), it is now assumed that the measured changes in number of channel particles probably reflect an additional dopamine-dependent effect on the channel renewal processes (Baldridge and Ball, 1991).

8.5 EVIDENCE FOR OTHER PUTATIVE MODULATORS OF HORIZONTAL COUPLING

8.5.1 DIRECT PATHWAYS

Although the dopaminergic system represents the best established neuromodulatory pathway acting to regulate horizontal cell coupling in the vertebrate retina, recent studies indicate the presence of additional independent pathways with similar functional roles.

We have already noted the independent second messenger pathways involving nitric oxide and cGMP, neither of which appears to be regulated by dopamine. Further, it has been shown that changes in illumination continue to modulate the receptive field properties of horizontal cells following the destruction of retinal dopamine cells (section 8.6; Baldridge and Ball, 1991). It therefore seems probable that other, non-dopaminergic, modulatory signals are involved in the regulation of horizontal cell coupling and the apparent diversity of neurotransmitters within interplexiform cell types (section 8.3) may provide clues to the identity of such pathways. In the goldfish retina, a new class of interplexiform cell has been described, the 'I3' adrenergic-containing cell (Baldridge and Ball, 1993). It was also shown that exogenous epinephrine, like dopamine, reduced the receptive fields of horizontal cells. Although this effect was blocked by the D1 antagonist SCH 23390, the possibility remains that the adrenergic pathway may represent an independent regulatory system, albeit one which acts through the same receptors as the dopamine system. It would be of considerable interest to examine the independence of this system by applying adrenergic agonists to retinas, depleted of dopaminergic cells. Additional independent regulation of horizontal cell receptive field may also originate from the retinal GABAergic system. GABA-containing horizontal cells release GABA when they are depolarized by darkness (Marc *et al.*, 1978) and GABA also usually depolarizes synaptically isolated horizontal cells by a direct action on horizontal cell ion channels (Hankins and Ruddock, 1984b; Stockton and Slaughter, 1991). Although this pathway would act on membrane conductance (g_m), rather than the coupling conductance (g_s), it may well represent a light-dependent local mechanism capable of regulating the receptive field responses of horizontal cells in the absence of dopamine signals. Furthermore, the GABA receptors located on horizontal cells may also represent a target site for transmitter release from the GABAergic interplexiform cells reported in some species (Nakamura *et al.*, 1980).

8.5.2 INDIRECT PATHWAYS

We have noted the evidence regarding the dopaminergic regulation of horizontal cell receptive field properties. In the fish retina, at least, such a pathway appears to involve the synaptic release of dopamine onto horizontal cells from dopaminergic interplexiform cells. Although such dopamine–interplexiform cells are a sparse population of neurons which have not been examined directly by neuropharmacological approaches, a number of studies suggest that various endogenous pathways act presynaptically to regulate the release of dopamine. The rationale of such studies is that an applied substance affects the receptive field properties of horizontal cells, but does so via the dopamine system and is thus both sensitive to dopaminergic antagonists and ineffective in retinas depleted of dopaminergic cells. With this approach it is possible to analyze the convergence of transmitter pathways presynaptic to the dopamine–interplexiform system.

The involvement of a presynaptic GABA pathway was suggested by the finding that the $GABA_a$ receptor antagonist, bicuculline, acts to restrict the normal receptive field pattern of horizontal cells in the teleost retina (Negishi *et al.*, 1983) and the turtle retina (Piccolino *et al.*, 1982). Furthermore, this effect of GABA is blocked by selective dopamine antagonists. These experimental data are best explained by a tonic inhibitory GABA–amacrine cell input, capable of directly regulating dopamine–interplexiform cell or dopamine–amacrine cell activity (Witkovsky and Dearry, 1991).

In the teleost retina, an additional regulatory amacrine cell input to the dopamine cells mediated by enkephalin has been proposed (Umino and Dowling, 1991). The application of met-enkephalin evoked a similar effect on horizontal cells as that caused by dopamine, but it remains to be established whether this excitatory enkephalinergic input acts directly on the dopamine–interplexiform cells.

It has been suggested that dopamine–interplexiform cells in the fish retina are regulated by the synaptic influence of centrifugal fibers (Zucker and Dowling, 1987). Electrophysiological evidence for this has been provided by experiments involving the application of the putative centrifugal fiber transmitters, in particular gonadotrophin releasing hormone (GnRH). GnRH was found to mimic the effect of dopamine on horizontal cells and, furthermore, GnRH had no effect on retinas in which the dopamine cells had been lesioned by 6-hydroxydopamine (6-OHDA) (Umino and Dowling, 1991).

A potentially important retinal mechanism implicated in the regulation of dopamine activity involves an interaction with melatonin. Retinal melatonin synthesis is restricted primarily to photoreceptors (Wiechman and Hollyfield, 1988), from where it appears to be passively released (section 8.7). One of the established effects of melatonin is as a potent inhibitor of dopamine release in the mammalian retina (Dubocovich, 1983). We have shown that the application of nanomolar concentrations of melatonin abolishes the endogenous effects of dopamine on horizontal cells in the rat retina (Hankins and Ikeda, 1993). Furthermore, the effects of melatonin were found to be associated with an expansion of the normal horizontal cell receptive field.

Recent studies have started to unravel some of the complexities of pathways and mechanisms which appear to regulate the activity of the retinal dopamine system. Experimental difficulties with recording from single identified dopamine-containing cells *in situ* still pose a barrier to a comprehensive understanding of the integration of these apparently complex pathways. Figure 8.5 summarizes the current understanding of the direct and indirect mechanisms implicated in the regulation of horizontal cell coupling.

8.6 LIGHT AND DARK REGULATION OF HORIZONTAL CELL COUPLING

Our knowledge regarding the pathways and subcellular mechanisms regulating horizontal cell coupling is well advanced. In particular, studies in a wide range of species suggest that dopaminergic regulation of the horizontal cell receptive field and of cell coupling are general features of the vertebrate retina (section 8.4). Given that such properties of horizontal cells are modulated pharmacologically, this raises the question as to what are the physiological conditions which initiate such regulation. This is an important question, the answer to which should provide an insight into the physiological modulation of synaptic plasticity in the outer plexiform layer and also elucidate the role of horizontal cell coupling in retinal processing. Recent attention has therefore been focused on the general effects of light or darkness on horizontal cell receptive field properties. However, the results of these studies, from various species and preparations, have provided apparently contradictory evidence. The current problem is thus to formulate a coherent visual model, rather than produce a number of interpretations based solely on species difference.

Two general approaches have been applied to these studies. The most direct being to measure the effect of various light and dark levels and/or adaptation states on the receptive field characteristics of horizontal cells. Other studies have presumed that dopamine is the primary signal regulating horizontal cell receptive fields and have examined aspects of dopaminergic activity in relation to the light/dark status. This latter approach carries with it the inherent danger that other independent regulatory pathways will be overlooked.

8.6.1 THE EFFECT OF PROLONGED DARKNESS

Electrophysiological studies of the effect of prolonged darkness have been of two general forms, either they involve comparative studies of horizontal cell responses recorded from retinas taken from animals which have previously been dark adapted for either short or long periods. Alternatively, they have examined horizontal cell responses in long-term dark-adapted retinas in order to follow, *in vitro*, the changes in responses of horizontal cells to adapting light backgrounds. Experiments on the fish retina by Mangel and Dowling (1985) were the first to indicate that prolonged dark adaptation had profound effects on the responses of cone-driven horizontal cells. They showed that in comparison to the responses of horizontal cells obtained following short-term dark adaptation (20 min), the responses of cells from long-term adaptation (90 min) were suppressed in amplitude and that this effect was associated with a reduction in the receptive field. The effect of prolonged darkness thus appeared similar to that evoked by exogenous dopamine on horizontal cells taken from short-term dark-adapted retinas. Later studies in the white perch retina provided strong evidence in support of prolonged dark adaptive changes in horizontal cell response and coupling, consistent with an increase in the activity of dopaminergic pathways (Yang *et al.*, 1988a,b; Tornqvist *et al.*, 1988). These experiments, which involved both electrophysiological and dye-coupling studies, established that, following prolonged dark adaptation, horizontal cell responses were indeed strongly inhibited and their dye-coupling reduced. In addition, it was shown that exogenous dopamine produced dual and time-dependent effects. The application of dopamine for brief periods was associated with an increase in the cell's response to small spot stimuli, but a suppression of the full-field responses.

In contrast, long-term applications led to a pronounced suppression of horizontal cell responses to all types of stimuli. This latter condition is broadly similar to that which results from long-term dark adaptation in the white perch retina.

Further evidence in support of this effect of prolonged darkness has been obtained from rod-driven horizontal cells in the fish retina. Retinas taken from animals subjected to prolonged dark-adaption were exposed to a background light *in vitro*, and the extent of horizontal cell receptive fields were found to increase markedly (Villa *et al.*, 1991). Such results suggest that during prolonged darkness, interplexiform cells become active, releasing dopamine and uncoupling horizontal cells (Dowling, 1989). This conclusion forms the foundation for a functional hypothesis, that the uncoupling of horizontal cells during prolonged darkness acts to abolish the surround input to bipolar cells, promoting a greater sensitivity in the responses of second-order neurons.

Prolonged dark adaptation may, however, involve an additional effect, since under dark conditions, horizontal cells are maximally depolarized and thus GABA release from cone horizontal cells is also maximal (Marc *et al.*, 1978; Kamermans *et al.*, 1990; Chapter 10, Section 10.7). It is known that GABA has a pronounced and direct effect on horizontal cells (Hankins and Ruddock, 1984b; Stockton and Slaughter, 1991), acting to increase the membrane conductance (g_m), which may reduce the response to full-field stimuli. However, the relative contribution of this mechanism remains difficult to ascertain pharmacologically, as GABA has a presynaptic influence on DA–interplexiform cells at the inner retina (section 8.5.2). Such difficulties could be eliminated by measuring the effects of GABA on cell coupling in the presence of dopamine antagonists, or in retinas depleted of dopamine neurons. None-the-less, although the GABA hypothesis cannot explain the enhanced horizontal cell responses to small spots observed during prolonged darkness

Figure 8.5 Schematic representation of the principal mechanisms, together with direct and indirect factors, involved in the regulation of horizontal cell (HC) receptive field properties in the vertebrate retina. The figure summarizes the findings of many research laboratories, as discussed in the text. The conductance through single gap-junction channels (Gs) at horizontal cells in the outerplexiform layer (OPL) is modulated principally through cyclic-AMP (cAMP) and cAMP dependent protein kinase ($cAMP_{PK}$). However, there appears to be a variable expression of gap-junction conductance modulated independently via cyclic-GMP (cGMP) or by changes in intracellular pH. Changes in HC receptive field properties may also be regulated through changes in the plasma membrane conductance (Gm), possibly mediated through the dark release of GABA ($GABA_{DARK}$) acting at $GABA_a$ receptors (G_A). The level of cAMP in horizontal cells is regulated through postsynaptic dopamine, D1 type receptors. These receptors are activated primarily via the presynaptic release of dopamine (DA) from interplexiform cells IPC, although in some species release occurs from DA–amacrine cells. The DA–IPCs are influenced by a variety of inhibitory pathways, including: GABA release from amacrine cells acting at $GABA_a$ receptors (G_A); dopamine autoreceptor activation (D2); paracrine melatonin (MEL) release acting at melatonin receptors (M). Facilitatory pathways affecting dopamine release also appear to include feedback excitation from centrifugal fibers releasing gonadotrophin releasing hormone (GnRH) and direct or indirect excitation mediated by enkephalinergic (ENK) amacrine cells. Melatonin synthesis is shown within the photoreceptors, this proceeds maximally in darkness, but is inhibited by light through either a reduction in free calcium (Ca^{2+}) or dopamine D4 receptor coupled reductions in cAMP.

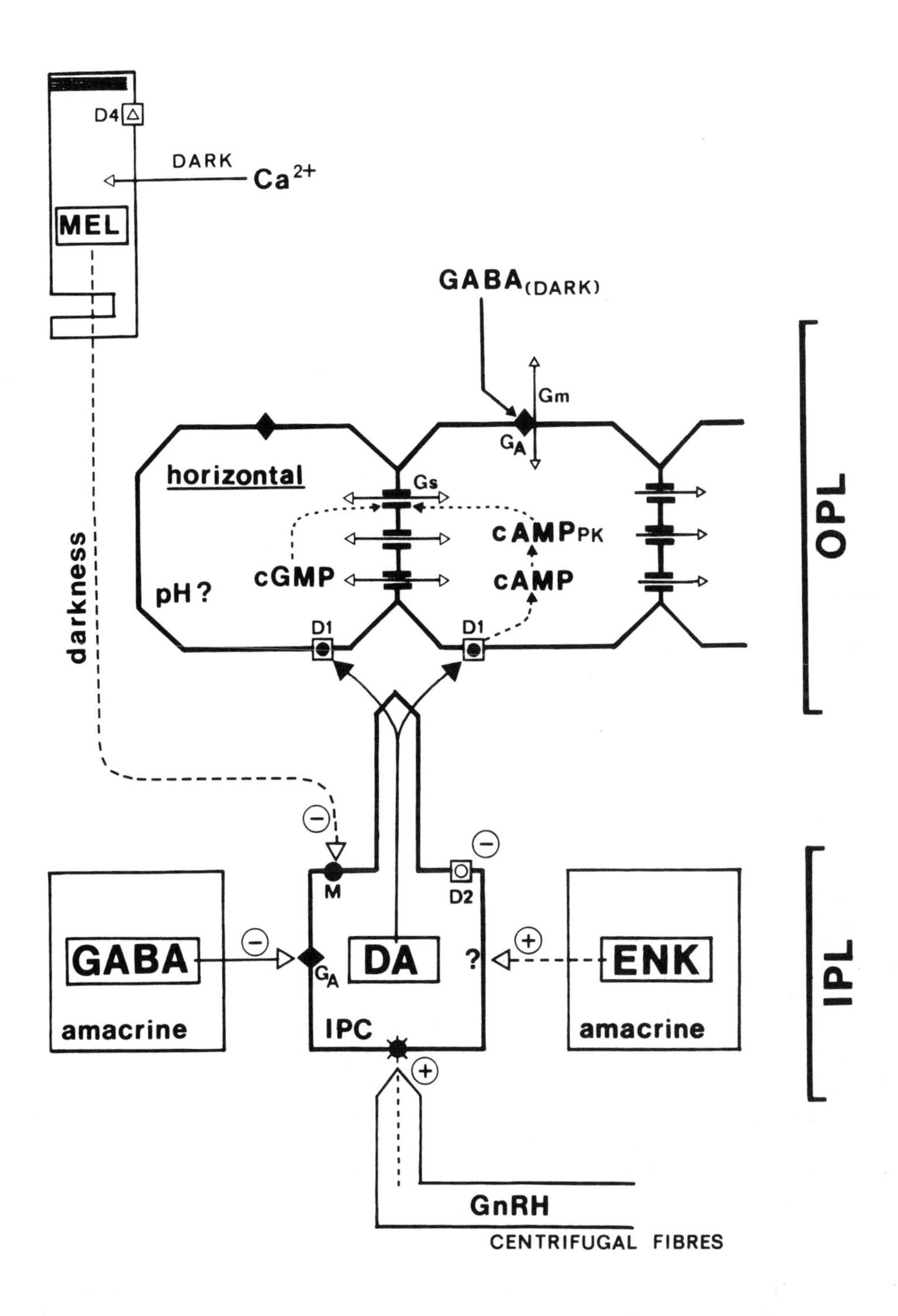
D4
DARK
Ca2+
MEL
GABA(DARK)
Gm
GA
horizontal
Gs
cAMPPK
cGMP
cAMP
pH ?
D1
D1
darkness
OPL
M
D2
GABA
GA
DA
?
ENK
amacrine
IPC
amacrine
IPL
GnRH
CENTRIFUGAL FIBRES

(Mangel and Dowling, 1985), it may provide an explanation of the decrease in response to full-field stimuli, which are not readily interpreted in terms of dopamine activity.

8.6.2 THE EFFECT OF LIGHT ON HORIZONTAL CELL COUPLING

In apparent contrast to the effects of darkness outlined in the previous section, there are other studies which suggest that exposure to adapting background light *in vitro* can also initiate a constriction of the horizontal cell receptive field, associated with a reduction in cell coupling. Thus, it has been proposed that in the carp retina light adaptation mimics the effects of dopamine on horizontal cell receptive fields (Shigematsu and Yamada, 1988). By recording from the soma of H1 type horizontal cells and measuring the receptive fields using displaced slit stimuli, it was shown that, whereas the overall light-sensitivity was reduced following light adaptation, the receptive field was narrowed and a discrete central component revealed. Other studies have examined the effects of the transient presentation of background illumination. Thus, Dong and McReynolds (1991) reported that a brief (2.5 min) exposure to a moderate adapting background light caused a reduction of the receptive field of horizontal cells. These experiments assessed the general extent of receptive fields by using alternating spot and annulus stimuli, and it was found that, whereas the response to the annulus was suppressed, the response to the spot was enhanced following the presentation of background light. Furthermore, these effects were blocked by the application of the D1 antagonist SCH 23390, suggesting that they are mediated by a light-induced increase in retinal dopaminergic activity. We have recently examined the effect of a transient light adaptation on the responses of rod-driven horizontal cells in the rat retina. These experiments demonstrated that the presentation of a background adapting field resulted in a rapid increase in the horizontal cell response to small spot stimuli together with a suppression of the annular response (Figure 8.6).

Experiments on the goldfish retina have also demonstrated that background illumination reduces the horizontal cell receptive field size and the extent of dye-coupling (Baldridge and Ball, 1991). In these studies, however, the effects of background light were not reversed by dopaminergic antagonists, and persisted in retinas lesioned by 6-OHDA pretreatment. The effects of light on the receptive field properties of horizontal cells have also been examined in the white perch retina (Umino *et al.*, 1991). These experiments compared the effect of flickering and steady background light. Both types of stimuli were found to reduce the receptive field sizes of horizontal cells, but only the effects of flickering light were found to be blocked by the dopamine antagonist, haloperidol. These authors proposed that the effects of steady background light may involve changes in the general membrane conductance (g_m). Collectively this evidence suggests that in some species the steady state light-induced uncoupling of horizontal cells may not be regulated exclusively by dopamine.

8.6.3 ARE THE STUDIES OF LIGHT AND DARK-MEDIATED EFFECTS CONTRADICTORY?

The experimental evidence reviewed in the previous section implies that horizontal cell receptive field changes evoked by either prolonged dark adaptation or, in some cases, by light backgrounds are mediated, at least in part, by the activity of retinal dopaminergic pathways. This raises the question as to whether the effects of both darkness and light can be regulated by the same retinal DA pathway. There are very few data regarding

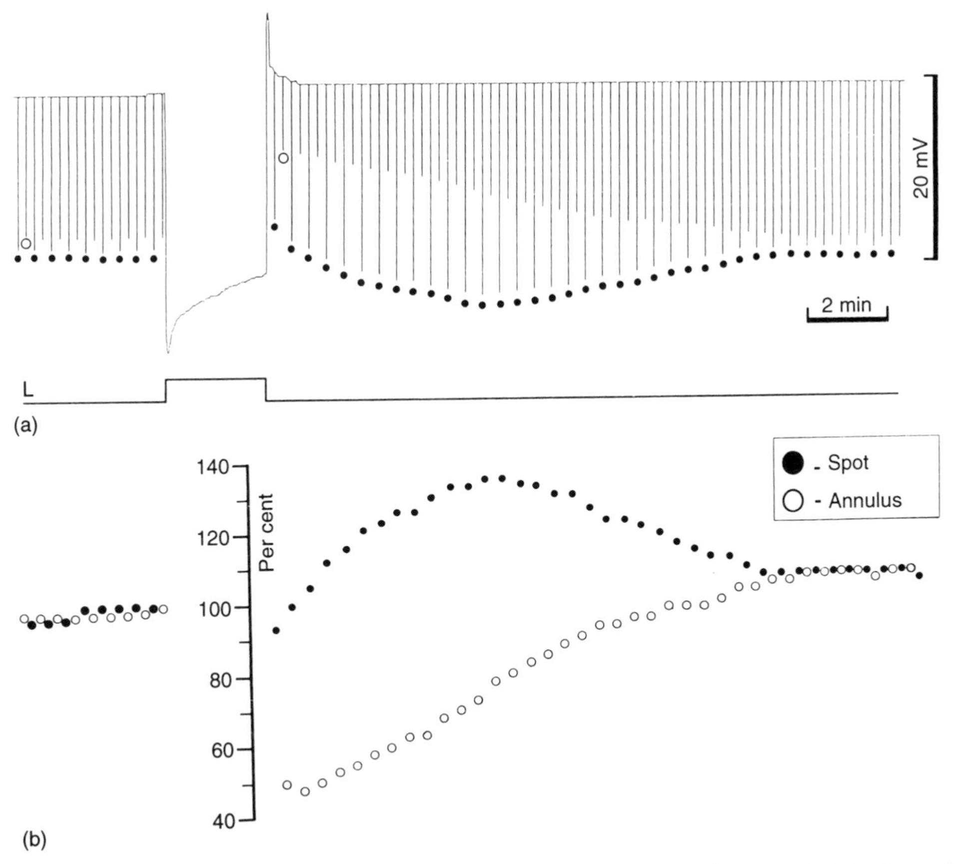

Figure 8.6 The effect of an adapting background light on the responses of a rat retinal horizontal cell to peripheral or central stimulation. (a) An electrophysiological recording of a cell receiving alternate small spot (solid circle) and annular (open circle) stimuli, identical to the parameters used in Figure 8.1. The presentation of the adapting field (irradiance 4000 quanta/μm/s; wavelength 520 nm), denoted by the trace 'L', results in a hyperpolarizing response followed by a slow relaxation. After 2.5 min adaptation, note the significant reduction in the response to annular stimuli and the progressive enhancement of the response to small spots. (b) A concurrent plot of the spot and annulus responses as a percentage of their pre-adaptation control levels. The data suggest that light adaptation results in a functional constriction of horizontal cell receptive field.

the effects of light levels on the electrophysiological responses of single DA-containing cells and therefore the measurement of endogenous dopamine release, using HPLC techniques, remains an important method in the examination of light-dependent activity of the retinal dopamine pathways. Numerous studies have examined the effects of light on the basal release of DA, and this topic has been reviewed extensively (Witkovsky and Dearry, 1991; Djamgoz and Wagner, 1992). The general conclusion from these studies in fish, amphibian and mammalian retinas, is that DA release is low in the dark-adapted state and increased by the presentation of flickering light and/or steady background light (Godley and Wurtman, 1988; Boatright *et al.*, 1989; Weiler *et al.*, 1989; Kirsch and

Wagner, 1989). Thus, the release studies generally support the notion that light-induced changes in horizontal cell receptive field and coupling status can be explained by a light-evoked increase in retinal DA release.

There remain difficulties however, in reconciling the electrophysiological evidence for a DA-mediated effect during prolonged darkness, since this condition is associated with low basal levels of DA release. It is possible that the global retinal release, or 'overflow', measured by HPLC does not reflect changes in the local synaptic release of DA from the synaptic terminals of DA–interplexiform cells. This may be particularly relevant in retinas where the population of DA-containing cells is not homogeneous (Section 8.4), but this does not apply in the case of the fish retina. Another, as yet unexplored, possibility is that the enhanced activity of DA pathways during prolonged dark adaptation may reflect changes in parameters other than DA release. For example, in the adult rat retina it has been shown that the density of D1 type dopamine receptors is dramatically increased, or up-regulated, during dark adaptation (De Montis *et al.*, 1988). A possible model, consistent with the current data, is one in which dopamine release is maintained at low basal levels during prolonged dark adaptation, but the subsequent up-regulation of D1 receptors in darkness increases the synaptic efficacy of the dopamine pathway and uncouples horizontal cells. Other parameters could also regulate the efficacy of DA pathways; for example, we currently have little experimental knowledge of the mechanisms and regulation of selective retinal DA-uptake.

Finally, it should be emphasized that although there is evidence that DA pathways may regulate both dark-adaptional and light-evoked uncoupling of horizontal cells, there remains in some species clear evidence that other, as yet unidentified, pathways may be acting independently of DA.

8.6.3 IMPLICATIONS FOR VISUAL PROCESSING IN THE OUTER PLEXIFORM LAYER

Although there may be some degree of species and cell specific variability in the expression of the light- and dark-regulated mechanisms involved, there is good evidence that both light and prolonged darkness can initiate the uncoupling of retinal horizontal cells. Furthermore, in the retina of at least one species (*Roccus americana*), both light- and dark-evoked mechanisms are operative (Umino *et al.*, 1991). What then are the functional implications of these systems for visual processing at the outer plexiform layer? On the basis of the studies of prolonged darkness in the fish retina, Dowling (1989) has proposed that the uncoupling of horizontal cells represents a mechanism by which overall retinal light sensitivity may be enhanced under scotopic conditions. Thus, uncoupling horizontal cells minimizes the inhibitory surround input to bipolar cells. Under these conditions, the efficacy of synaptic transfer from photoreceptors to bipolar cells would be enhanced. In this respect, the functional role of horizontal cells at the outer plexiform layer would be restricted to the gain control at the first retinal synaptic stage. The physiological significance of this mechanism remains to be clarified, since the dark-adaptational changes are initiated only after prolonged periods in darkness, by which time cone-driven systems are unlikely to be operative. Interestingly, in the white perch, where prolonged darkness has a profound effect both on horizontal cell sensitivity and coupling, dark-adaptational changes are restricted to cone-driven horizontal cells and not expressed by the rod-driven horizontal cells (Yang *et al.*, 1988a). It remains possible therefore, that the dark-related effects of DA are part of a functional pathway involved in the general suppression of the cone system under scotopic conditions.

The suggestion that light backgrounds can act to uncouple horizontal cells has a profound physiological significance in relation to spatial visual processing in the retina. A reduction in the size of horizontal cell receptive fields under photopic conditions would restrict the extent of bipolar cell antagonistic surround input and potentially enhance spatial resolution at the outer retina, since the center response of bipolar cells would be referred only to a limited surround field. The suggestion that DA pathways play a role in the regulation of spatial acuity gains support from the spatial deficits reported in both the dopamine-deficient Parkinsonian retina, and in the MPTP-treated primate (Chapter 20).

Although the number of pharmacological studies of horizontal cells in the mammalian retina is relatively few, light responses have been recorded in many species including: rat, cat, dog, rabbit and primate. From these studies, there is currently no evidence for any mammalian equivalent to the depolarizing color type horizontal cells seen in lower vertebrates. This may suggest that in the mammalian retina, color opponency functions are largely confined to the inner retina, as has also been proposed on the basis of anatomical evidence (Boycott, 1988). If this is the case, then the functional significance of horizontal cell outputs, fed forward to bipolar cells and back to photoreceptors, may be restricted to the regulation of light sensitivity or spatial acuity.

8.7 DIURNAL AND CIRCADIAN REGULATION

This chapter has focused considerable attention on the effects of the retinal dopamine system on the coupling of horizontal cells. There is, however, evidence that the paracrine release of dopamine in the retina may act as a general signal regulating a multiplicity of diverse retinal functions including; cell coupling, retinomotor movements, pigment migration, the balance of rod/cone inputs, spinule formation and gap-junction channel density (reviews: Witkovsky and Dearry, 1991; Djamgoz and Wagner, 1992; Wagner and Djamgoz, 1993). Viewed collectively, such regulatory mechanisms are consistent with the view that the retinal dopamine system plays a pivotal role, integrating a number of adaptational functions involved in the transition of the retina from the dark to the light state. There is also considerable evidence that links the retinal dopamine pathway to the activity of the retinal melatonin system (review: Besharse and Iuvone, 1992; section 8.5.2). Melatonin synthesis in the retinas of numerous species, including humans, is confined to photoreceptors and some bipolar cells (Wiechmann and Hollyfield, 1988). It appears that melatonin synthesis proceeds according to a circadian rhythm in the absence of light cues, with release being highest in the night phase (Cahill and Besharse, 1993). The synthesis and passive release of melatonin are, however, also rapidly inhibited by light, because the activity of *N*-acetyltransferase (NAT), one of the rate-limiting enzymes in melatonin synthesis, is dependent both on intracellular Ca^{2+} and cAMP (Iuvone and Besharse, 1986; Redburn and Mitchell, 1989). Thus, either a light-dependent decrease in photoreceptor free Ca^{2+} or an increase in the paracrine activation of outer segment dopamine (D4) receptors (which are negatively coupled to cAMP) will inhibit melatonin synthesis. Conversely, the dark-release of melatonin is a potent inhibitor of dopamine release in the mammalian retina (Dubocovich, 1983). Indeed, our studies have shown that nanomolar concentrations of melatonin abolish the electrophysiological effects of endogenous dopamine on horizontal cells in the rat retina (Hankins and Ikeda, 1993). Such evidence is consistent with the proposal that dopamine and melatonin may have contra-regulatory roles in the light- and dark-adaptational process (Besharse and

Iuvone, 1992), and the possibility remains that long-term changes in retinal adaptation are under the neuromodulatory 'push–pull' control of the dopamine and melatonin systems.

The circadian regulation of synaptic plasticity in the retina may represent a system which plays a role in the preparation of the retina for expected day (light) or expected night (dark). How, then, would such mechanisms function in the case of horizontal cell coupling? It has already been shown in this chapter that dopamine rapidly regulates horizontal cell coupling through the activation of second messenger systems, but it must be remembered that both light and dopamine have an additional long-term effect on gap-junction particle density (Baldridge *et al.*, 1987; Baldridge and Ball, 1991). Although it is now thought that the gap-junction particle density does not reflect absolute changes in functional cell coupling (Vaughan and Lasater, 1990; Baldridge and Ball, 1991), such density changes may, none-the-less, reflect an underlying change in synaptic plasticity involved in the preparation of the retina for the light phase. These mechanisms may be important in signaling long-term modifications to retinal sensitivity and, furthermore, since melatonin synthesis can proceed under circadian control, such functional plasticity may persist in the absence of visual cues.

8.8 CLINICAL SIGNIFICANCE OF HORIZONTAL CELL COUPLING

The significance of the regulation of horizontal cell coupling in relation to visual processing in the outerplexiform layer has already been examined (section 8.6.3). The dark- and light-induced uncoupling of horizontal cells suggest that there are two quite distinct functional advantages conferred by the uncoupling of horizontal cells. In darkness, the uncoupling of horizontal cells may enhance absolute retinal light sensitivity, whereas in light, the uncoupling of horizontal cells appears to accord greater spatial acuity to second order neurons. What are the clinical implications of such functional regulation? A potential key to understanding the role of horizontal cell coupling changes in human vision is the study of retinas in which the regulatory dopaminergic pathways are abnormal. The obvious examples of this type are Parkinson's disease patients, the retinas from whom have been shown to be dopamine deficient (Harnois and DiPaolo, 1990). In Chapter 20, Bodis–Wollner and Antal give a detailed account of the visual dysfunction associated with Parkinson's disease. The functional deficit in contrast sensitivity and visual acuity ascribed to Parkinson's patients can be modeled, in part, according to subnormal light-induced uncoupling of horizontal cells at the outer plexiform layer. Of course the retinal dopamine system controls a multiplicity of discrete regulatory mechanisms (section 8.4 and 8.7; Djamgoz and Wagner, 1992), and there will certainly be additional factors to include in such a model. Further evidence that retinal dopamine pathways have a role in human spatial vision originates from studies of the effects of dopaminergic drugs on normal subjects. Thus, Domenici *et al.* (1985), have shown that the DA-selective uptake inhibitor, nomifensine, significantly increases contrast sensitivity for spatial frequencies above 2 cycles/degree in the normal observer. A number of subsequent studies, using a range of dopamine agonists, report broadly similar results. Once again, it is tempting to associate the spatial sensitivity changes evoked by dopaminergic drugs, with their effects on the coupling of horizontal cells. Certainly, the better the understanding of the synaptic pathways involved in the regulation of retinal spatial acuity, the better the potential for the treatment of abnormalities in spatial contrast sensitivity.

Evidence from animal studies suggests the presence of a number of dopamine-related abnormalities in the senescent retina. For example, dopamine receptor binding studies in the rat retina show that D1 receptor expression is enhanced in senile rats, through a reduction in the rate of receptor degradation (Giorgi *et al.*, 1992), whereas general dopamine specific cell markers are reduced. What role then, might differences in retinal dopaminergic pathways play in age-related abnormalities in spatial vision? This area remains largely unexplored.

The evidence derived from animal studies for a contra-regulatory system involving dopamine and melatonin has considerable implications for the apparent diurnal and adaptational changes that occur in retinal sensitivity (section 8.7). It is therefore of considerable clinical importance to establish whether such mechanisms are functional in the human retina. Some indication that this may be so, is provided by the results of an electroretinographic study on humans examining the effects of psoralen analogs, which are known to enhance melatonin release (Souetre *et al.*, 1989). It was reported that clinical doses of 5-methoxypsoralen given to normal volunteers caused a pronounced enhancement of retinal b-wave amplitude under photopic conditions, and furthermore enhanced the b-wave amplitude during the early stages of scotopic dark adaptation. Unfortunately, no studies have been performed on the spatial contrast sensitivity of subjects receiving psoralen drugs.

8.9 CONCLUSIONS AND FUTURE PERSPECTIVES

This chapter has examined the modulatory pathways and the subcellular machinery that regulate horizontal cell coupling in the retinas of a wide range of vertebrate species. The evidence that the status of horizontal cell coupling is a dynamic phenomenon controlled, at least in part, by the light- or dark-adaptational state of the retina has also been reviewed.

One uncertainty with the interpretation of the current data regarding the effects of light and dark, is the relationship between the experimental conditions in a given study and the actual physiological adaptational status. It is perhaps naive to assume that light or dark adaptation of the retina *in vitro* will proceed in a manner equivalent to that which occurs *in vivo*. For one thing, most of the experimental studies are performed on the isolated neural retina and are thus devoid of interactions with the pigment epithelium. Furthermore, it is now clear that primary photoreceptor light adaptation is itself regulated by intracellular Ca^{2+} (Matthews, 1991), thus variations in experimental Ca^{2+} concentrations would have a profound influence on photoreceptor light adaptation. Certainly horizontal cell responses in the fish retina are highly sensitive to changes in the extracellular free Ca^{2+} concentration (Hankins *et al.*, 1985). In addition, if nanomolar concentrations of melatonin released from photoreceptors represent an important regulatory signal expressed during dark-adaptation, then the leaching of melatonin from perfused preparations may shift the adaptational status of the retina independently of the experimental light conditions or light-history. The use of terms such as scotopic, mesopic and photopic cannot, therefore, be applied with any degree of certainty to the current experimental studies. In recent years we have begun to unravel some of the complexities of neuromodulatory pathways in the retina. The effects of these pathways are often quite subtle, thus it is particularly important that experiments are carefully designed. Apparent contradictions in the results of experimental studies may revolve around small differences in experimental techniques.

There are a number of specific areas on which our current knowledge regarding the regulation of horizontal cell coupling is limited. The identity of the specific protein subunits that constitute the single gap-junction channels in horizontal cells remains unknown. These are most likely to be members of the connexin family. In other preparations, the identity of specific channel subunits and their associated regulatory binding sites are being correlated with the diversity of conferred second messenger sensitivities (Bennett *et al.*, 1991). It remains to be established whether the apparent diversity of second messenger systems in retinal horizontal cells may reflect multiple subunit families of connexin proteins. Furthermore, it is also possible that horizontal cell coupling may be independently regulated by non-connexin channel proteins such as the synaptophysins.

As knowledge of the molecular diversity of dopamine receptors increases (Sibley and Monsma, 1992), it can be expected that appropriate selective agonists and antagonists for the newly identified families of DA receptors will emerge. An understanding of the detailed contributions of such specific receptors to spatial visual processing will be important for the treatment of clinical abnormalities in spatial vision, when the new generation of dopaminergic drugs emerge into the therapeutic area.

One of the critical regulatory systems in the dopaminergic pathway is the removal of DA from the extracellular space by active uptake, and this may be particularly important in the regulation of the proposed paracrine effects of dopamine in the retina. In other regions of the CNS this is achieved through a specific DA-transporter which has been cloned (Giros and Caron, 1993). It appears that a selective DA transporter functions in the retina, since the drug nomifensine has been shown to increase the basal dark release of dopamine in the rabbit (Godley and Wurtman, 1988). There are, however, no direct electrophysiological studies which have examined the contribution of retinal reuptake on the efficacy of the dopamine pathway. It is important that the dynamics of the uptake system are elucidated since this may also provide an alternative target for clinical therapy.

Finally, the evidence has been presented for the existence of additional, non-dopaminergic, regulatory pathways which modulate the coupling of horizontal cells, and whilst these pathways remain unidentified, there will be an additional 'hidden hand' in the adaptational process which regulates synaptic plasticity in the neural retina. Greater understanding of the mechanisms which subserve retinal plasticity may, in the long term, provide a wider range of therapeutic strategies designed for treatment of retinal visual dysfunction.

ACKNOWLEDGEMENTS

The author's current research in the Gunnar Svaetichin Laboratory is supported by a Fellowship grant from the Wellcome Foundation. I also thank Professor K.H. Ruddock for useful discussions and Jane Yeats for her help in preparing the manuscript.

REFERENCES

Baldridge, W.H. and Ball, A.K. (1991) Background illumination reduces horizontal cell receptive-field size in both normal and 6-hydroxy-dopamine-lesioned goldfish retinas. *Visual Neuroscience*, **7**, 441–50.

Baldridge, W.H. and Ball, A.K. (1993) A new type of interplexiform cell in the goldfish retina is PNMT-immunoreactive. *NeuroReport*, **4**, 1015–18.

Baldridge, W.H., Ball, A.K. and Miller, R.G. (1987) Dopaminergic regulation of horizontal cell gap-junction density in goldfish retina. *Journal of Comparative Neurology*, **265**, 428–36.

Bennett, M.V.L., Barrio, L.C., Bagiello, T.A. *et al.* (1991) Gap junctions: new tools, new answers, new questions. *Neuron*, **6**, 305–20.

Besharse, J.C. and Iuvone, P.M. (1992) Is dopamine a light-adaptive or a dark-adaptive modulator in the retina? *Neurochemistry International*, **20**, 193–9.

Boatright, J.H., Hoel, M.J. and Iuvone, P.M. (1989) Stimulation of endogenous dopamine release and metabolism in amphibian retina by light- and K^+-evoked depolarization. *Brain Research*, **482**, 164–8.

Boycott, B.B. (1988) Horizontal cells of mammalian retinae. *Neuroscience Research Suppl.*, S97–S111.

Cahill, G.M. and Besharse, J.C. (1993) Circadian clock functions localised in xenopus photoreceptors. *Neuron*, **10**, 573–7.

Cohen, J.L. and Dowling, J.E. (1983) The role of retinal interplexiform cells: effects of 6–hydroxydopamine on the spatial properties of carp horizontal cells. *Brain Research*, **264**, 307–10.

DeMontis, G., Porceddu, M.L., Pepitoni, S. *et al.* (1988) D1 dopamine receptors in the retina of adult and senescent rats: physiological and pharmacological modulation, in *Dopaminergic Mechanisms in Vision*. Alan R. Liss, New York, pp.71–93.

DeMontis, G.C., Beani, L., Bianchi, C. and Piccolino, M. (1988) Influence of D2 dopamine drugs on horizontal cell (HC) electrical coupling and [^{3}H]-dopamine release in turtle retina. *Neuroscience Letters, Suppl.* **33**, S72.

DeVries, S.H. and Schwartz, E.A. (1989) Modulation of an electrical synapse between pairs of catfish horizontal cells by dopamine and second messengers. *Journal of Physiology, London*, **414**, 351–75.

Djamgoz, M.B.A. and Laming, P.J. (1987) Microelectrode measurements and functional aspects of chloride activity in cyprinid fish retina: extracellular activity and intracellular activities of L- and C-type horizontal cells. *Vision Research*, **27**, 1481–9.

Djamgoz, M.B.A. and Wagner, H.-J. (1992) Localization and function of dopamine in the adult retina. *Neurochemistry International*, **20**, 139–91.

Domenichi, L., Trimarchi, C., Piccolino, M. *et al.* (1985) Dopaminergic drugs improve human visual contrast sensitivity. *Human Neurobiology*, **4**, 195–7.

Dong, C.-J. and McReynolds, J.S. (1991) The relationship between light, dopamine release and horizontal cell coupling in the mudpuppy retina. *Journal of Physiology, London*, **440**, 291–309.

Dowling, J.E. (1989) Neuromodulation in the retina: the role of dopamine. *Seminars in the Neurosciences*, **1**, 35–43.

Dowling, J.E. and Ehinger, B. (1978) The interplexiform cell system. I. Synapses of the dopaminergic neurons of the goldfish retina. *Proceedings of the Royal Society London B*, **201**, 7–26.

Dubocovich, M.L. (1983) Melatonin is a potent modulator of dopamine release in the retina. *Nature*, **306**, 782–4.

Gerschenfeld, H.M., Neyton, J., Piccolino, M. and Witkovsky, P. (1982) L-horizontal cells of the turtle: network organization and coupling modulation. *Biomedical Research*, Suppl. **3**, 21–32.

Giorgi, O., Pibiri, M.G., Toso, R.D. and Ragatzu, G. (1992) Age related changes in the turnover rates of D1-dopamine receptors in the retina and in distinct areas of the rat brain. *Brain Research*, **569**, 323–9.

Giros, B. and Caron, M.G. (1993) Molecular characterization of the dopamine transporter. *Trends in Pharmacological Sciences*, **14**, 43–9.

Godley, B.F. and Wurtman, R.J. (1988) Release of endogenous dopamine from superfused rabbit retina in vitro: effects of light stimulation. *Brain Research*, **452**, 393–5.

Hampson, E.C.G.M., Vaney, D.I. and Weiler, R. (1992) Dopaminergic modulation of gap-junction permeability between amacrine cells in mammalian retina. *Journal of Neuroscience*, **12**, 4911–22.

Hankins, M.W. and Ikeda, H. (1991a) Non-NMDA type excitatory amino acid receptors mediate rod input to horizontal cells in the isolated rat retina. *Vision Research*, **31**, 609–17.

Hankins, M.W. and Ikeda, H. (1991b) The role of dopaminergic pathways at the outerplexiform layer of the mammalian retina. *Clinical Vision Sciences*, **6**, 87–93.

Hankins, M.W. and Ikeda, H. (1993) Dopamine pathways in the outer mammalian retina – and abnormalities expressed in hereditary retinal dystrophy. *Zdravniski vestnik*, **62**, Suppl.1, 33–8.

Hankins, M.W. and Ruddock, K.H. (1984a) Hyperpolarization of fish horizontal cells by kainate and quisqualate. *Nature*, **308**, 360–2.

Hankins, M.W. and Ruddock, K.H. (1984b) Electrophysiological effects of GABA on fish retinal horizontal cells are blocked by bicuculline but not by picrotoxin. *Neuroscience Letters*, **44**, 1–6.

Hankins, M.W. and Ruddock, K.H. (1986) Neuropharmacological actions of kynurenic and quinolinic acids on horizontal cells of the isolated fish retina. *Brain Research*, **380**, 297–302.

Hankins, M.W., Rowe, J.S. and Ruddock, K.H. (1985) Properties of amino acid binding sites on horizontal cells determined by electrophysiological studies on the isolated roach retina, in *Neurocircuitry of the Retina A Cajal Memorial* (ed. A. Gallego and P. Gouras), Elsevier, New York, pp.99–108.

Harnois, C. and DiPaolo, T. (1990) Decreased dopamine in the retinas of patients with Parkinson's disease. *Investigative Ophthalmology and Visual Science*, **31**, 2473–5.

Harsanyi, K. and Mangel, S.C. (1992) Activation of a D2 receptor increases electrical coupling between retinal horizontal cells by inhibiting dopamine release. *Proceedings of the National Academy of Sciences, USA*, **89**, 9220–4.

Hedden, W.L. and Dowling, J.E. (1978) The interplexiform cell system. II Effects of dopamine on goldfish retinal neurones. *Proceedings of the Royal Society of London B*; **201**, 27–51.

Iuvone, P.M. and Besharse, J.C. (1986) Dopamine receptor-mediated inhibition of serotonin-*N*-acetyltransferase activity in the retina. *Brain Research*, **369**, 168–76.

Kamermans, M., Van Dijk, E.W. and Spekreijse, H. (1990) An evaluation of dopamine and light induced changes in carp horizontal cells. *Investigative Ophthalmology and Visual Science*, **31** (4), 177.

Kaneko, A. (1971) Electrical connections between horizontal cells in the dogfish retina. *Journal of Physiology, London*, **213**, 95–105.

Kaneko, A. (1973) Receptive field organization of bipolar and amacrine cells in the goldfish retina. *Journal of Physiology, London*, **235**, 133–53.

Kaneko, A. and Stuart, A.E. (1980) Coupling between horizontal cells in the carp retina examined by diffusion of lucifer yellow. *Biology Bulletin*, **159**, 486.

Kebabian, J.W. and Calne, D.B. (1979) Multiple receptors for dopamine. *Nature*, **277**, 93–6.

Kirsch, M. and Wagner, H.-J. (1989) Release pattern of endogenous dopamine in teleost retinae during light adaptation and pharmacological stimulation. *Vision Research*, **29**, 147–54.

Knapp, A.G. and Dowling, J.E. (1987) Dopamine enhances excitatory amino acid-gated conductances in cultured retinal horizontal cells. *Nature*, **325**, 437–9.

Kurtz-Isler, G. and Wolberg, H. (1986) Gap junctions between horizontal cells in the cyprinid fish alter rapidly their structure during light and dark adaptation. *Neuroscience Letters*, **67**, 7–12.

Lamb, T.D. (1976) Spatial properties of the horizontal cell in the turtle retina. *Journal of Physiology, London*, **263**, 239–55.

Lasater, E.M. (1987) Retinal horizontal cell gap junctional conductance is modulated by dopamine through a cyclic AMP-dependent protein kinase. *Proceedings of the National Academy of Sciences, USA*, **84**, 7319–23.

Lasater, E.M. and Dowling, J.E. (1985) Dopamine decreases conductance of the electrical junctions between cultured retinal horizontal cells. *Proceedings of the National Academy of Sciences, USA*, **82**, 3025–9.

Mangel, S.C. and Dowling, J.E. (1985) Responsiveness and receptive field size of carp horizontal cells are reduced and prolonged by darkness and dopamine. *Science*, **229**, 1107–9.

Marc, R.E. and Lui, W.-L.S. (1984) Horizontal cell synapses onto glycine accumulating interplexiform cells. *Nature*, **311**, 266–9.

Marc, R.E., Stell, W.K., Bok, D. and Lam, D.M.-K. (1978) GABAergic pathways in the goldfish retina. *Journal of Comparative Neurology*, **182**, 221–45.

Matthews, H.R. (1991) Incorporation of chelator into guinea-pig rods shows that calcium mediates photoreceptor light adaptation. *Journal of Physiology, London*, **436**, 93–105.

McMahon, D.G., Knapp, A.G. and Dowling, J.E. (1989) Horizontal cell gap junctions: single channel conductance and modulation by dopamine. *Proceedings of the National Academy of Sciences, USA*, **86**, 7639–43.

Miyachi, E. (1992) Intracellular messengers and

their roles in retinal gap junctions. *Japanese Journal of Psychopharmacology*, **12**, 129–34.

Naka, K.-I. and Rushton, W.A.H. (1967) The generation and spread of S-potentials in fish (Cyprinidae). *Journal of Physiology, London*, **192**, 437–61.

Nakamura, Y., McGuire, B.A. and Sterling, P. (1980) Interplexiform cell in the cat retina: identification by uptake of [^{3}H]GABA and serial reconstruction. *Proceedings of the National Academy of Sciences, USA*, **77**, 658–61.

Negishi, K. and Drujan, B.D. (1978) Effects of catecholamine on horizontal cell membrane potential in the fish retina. *Sensory Processes*, **2**, 388–95.

Negishi, K., Teranishi, T. and Kato, S. (1983) A GABA antagonist, bicuculline, exerts its uncoupling action on external horizontal cells through dopamine cells in carp retina. *Neuroscience Letters*, **37**, 261–6.

Piccolino, M., Neyton, J., Witkovsky, P. and Gerschenfeld, H.M. (1982) Y-aminobutyric acid antagonists decrease junctional communication between horizontal cells of the retina. *Proceedings of the National Academy of Sciences, USA*, **79**, 3671–5.

Piccolino, M., Neyton, J. and Gerschenfeld, H.M. (1984) Decrease of gap junction permeability induced by dopamine and cyclic adenosine 3′:5′-monophosphate in horizontal cells of turtle retina. *Journal of Neuroscience*, **4**, 2477–88.

Piccolino, M., Witkovsky, P. and Timarchi, C. (1987) Dopaminergic mechanisms underlying the reduction of the electrical coupling between horizontal cells of the turtle retina induced by D-amphetamine, bicuculline and veratridine. *Journal of Neuroscience*, **7**, 2273–84.

Qian, H., Malchow, R.P. and Ripps, H. (1993) Gap-junctional properties of electrically coupled skate horizontal cells in culture. *Visual Neuroscience*, **10**, 287–95.

Redburn, D.A. and Mitchell, C.K. (1989) Darkness stimulates rapid synthesis and release of melatonin in rat retina. *Visual Neuroscience*, **3**, 391–403.

Shigematsu, Y. and Yamada, M. (1988) Effects of dopamine on spatial properties of horizontal cell responses in the carp retina. *Neuroscience Research*, Suppl.**8**, S69–80.

Sibley, D.R. and Monsma, F.J. (1992) Molecular biology of dopamine receptors. *Trends in Pharmacological Sciences*, **13**, 61–9.

Souetre, E., De Galeani, B., Gastaud, P. *et al.* (1989) 5-methoxypsoralen increases the sensitivity of the retina to light in humans. *European Journal of Clinical Pharmacology*, **36**, 59–61.

Stockton, R.A. and Slaughter, M.M. (1991) Depolarizing actions of GABA and glycine on amphibian retinal ganglion cells. *Journal of Neurophysiology*, **65**, 680–92.

Svaetichin, G., Laufer, M., Mitarai, G. *et al.* (1961) Glia control of neuronal networks and receptors, in *The Visual System: Neurophysiology and Psychophysics* (eds R. Jung and H. Kornhuber), Springer, Berlin, pp. 445–56.

Teranishi, T., Negishi, K. and Kato, S. (1983) Dopamine modulates S-potential amplitude and dye-coupling between external horizontal cells in carp retina. *Nature*, **301**, 243–6.

Teranishi, T., Negishi, K. and Kato, S. (1984) Regulatory effect of dopamine on spatial properties of horizontal cells in carp retina. *Journal of Neuroscience*, **4**, 1271–80.

Tornqvist, K., Yang, X.-L. and Dowling, J.E. (1988) Modulation of cone horizontal cell activity in the teleost retina. III. Effects of prolonged darkness and dopamine on electrical coupling between horizontal cells. *Journal of Neuroscience*, **8**, 2259–68.

Toyoda, J.-I. and Tonasaki, K. (1978) Effect of polarization of horizontal cells on the ON-center bipolar cell of the carp retina. *Nature*, **276**, 399–400.

Umino, O. and Dowling, J.E. (1991) Dopamine release from interplexiform cells in the retina: effects of GnRH, FMRFamide, bicuculline, and enkephalin on horizontal cell activity. *Journal of Neuroscience*, **11**, 3034–46.

Umino, O., Lee, Y. and Dowling, J.E. (1991) Effects of light stimuli on the release of dopamine from interplexiform cells in the white perch retina. *Visual Neuroscience*, **7**, 451–8.

Vaughan, D.K. and Lasater, E.M. (1990) Evidence for renewal of electrical synapses in teleost horizontal cells. *Investigative Ophthalmology and Visual Science*, **31** (4), 285.

Villa, P., Bedmar, M.D. and Baron, M. (1991) Studies on rod horizontal cell S-potential in dependence of the dark/light adapted state:

a comparative study in *Cyprinus carpio* and *Scyliorhinus canicula* retinas. *Vision Research*, **31**, 425–35.

Wagner, H.-J. and Djamgoz, M.B.A. (1993) Spinules: a case for retinal synaptic plasticity. *Trends in Neurosciences*, **6**, 201–6.

Weiler, R., Kolbinger, W. and Kohler, K. (1989) Reduced light responsivness of the cone pathway during prolonged darkness does not result from an increase in dopaminergic activity in the fish retina. *Neuroscience Letters*, **99**, 214–18.

Wiechmann, A.F. and Hollyfield, J.G. (1988) HIOMT-like immunoreactivity in the vertebrate retina: a species comparison. *Experimental Eye Research*, **49**, 1079–95.

Witkovsky, P. and Dearry, A. (1991) Functional roles of dopamine in the vertebrate retina. *Progress in Retinal Research*, **11**, 247–92.

Yagi, T. (1986) Interaction between soma and the axon terminal of horizontal cells in *Cyprinus carpio*. *Journal of Physiology, London*, **375**, 121–35.

Yamada, E. and Ishikawa, T. (1965) Some observations on the fine structure of the vertebrate retina. *Cold Spring Harbor Symposia on Quantitative Biology*, **30**, 383–92.

Yamada, M., Shigematsu, Y., Umetani, Y. and Saito, T. (1992) Dopamine decreases receptive field size of rod-driven horizontal cells in carp retina. *Vision Research*, **32**, 1801–7.

Yang, X.-L., Tornqvist, K. and Dowling, J.E. (1988a) Modulation of cone horizontal cell activity in the teleost retina. I. Effects of prolonged darkness and background illumination on light responsiveness. *Journal of Neuroscience*, **8**, 2259–68.

Yang, X.-L., Tornqvist, K. and Dowling, J.E. (1988b) Modulation of cone horizontal cell activity in the teleost retina. II. Role of interplexiform cells and dopamine in regulating light responses. *Journal of Neuroscience*, **8**, 2269–78.

Zucker, C.L. and Dowling, J.E. (1987) Centrifugal fibres synapse on dopaminergic interplexiform cells in the teleost retina. *Nature*, **330**, 166–8.

9

Cross-talk between cones and horizontal cells through the feedback circuit

MARCO PICCOLINO

9.1 INTRODUCTION: FROM 'THE PARADOX OF HORIZONTAL CELLS' TO THE DISCOVERY OF FEEDBACK

More than a century ago Santiago Ramón Y Cajal published his landmark study on the structure of the vertebrate retina, *La Rétine des Vertébrés* (Ramón Y Cajal, 1893), a work which still remains a fundamental reference for any study of retinal circuits. Cajal had a 'photographic' conception of retinal organization i.e. one-to-one communication from photoreceptors to bipolar and ganglion cells, with minimum convergence and divergence and with virtual absence of lateral communication in the main line of centripetal transmission. Productive as it was, however, this approach proved to be inadequate for the understanding of the role of neural elements like horizontal and amacrine cells which connect distant visual neurons over a tangential retinal plan (Piccolino, 1988).

The 'operative' and 'integrative' views of the functional organization of the retina took a long time to emerge, but slowly retinal circuits were shown to be designed to carry out complex operations on the visual image, aimed at extraction and transmission of aspects of rich adaptive value, such as form, contrast, color and movement. These operations require complex interactions among the visual pathways in order to generate a multiplicity of neural 'transforms' which may appear only remotely related to the optical image focused on the retina.

The notion of lateral inhibition, leading to the concept of center-surround antagonism of the receptive fields of visual neurons was introduced some 40 years ago in the *Limulus* eye and in mammalian retinas (Hartline, 1949; Barlow, 1953; Kuffler, 1953; Ratliff, 1965). Inhibition was absent from Cajal's conceptual framework interpreting the functioning of nerve circuits in the retina, as well as in other regions of the central nervous system. This is one of the reasons why Cajal could not account for the lateral flow of signals along the 'tangential' neurons of the retina, a limit of his theories that he explicitly recognized in the last period of his life referring to '. . . the paradox of retinal horizontal cells . . .' and '. . . the enigma of amacrine cells . . .' (Ramón Y Cajal, 1933).

A further period was necessary to under-

Neurobiology and Clinical Aspects of the Outer Retina
Edited by M.B.A. Djamgoz, S.N. Archer and S. Vallerga
Published in 1995 by Chapman & Hall, London
ISBN 0 412 60080 3

stand the functional role of retinal horizontal cells (HCs), even after the introduction of the notion of lateral inhibition in retinal physiology and although HCs were the first retinal neurons to be studied by intracellular recording (Svaetichin, 1953). The difficulty arose because, in the retina of many species, the only obvious morphological contacts of HCs appeared to be with photoreceptors, and, on morphological grounds, these contacts seemed exclusively to be sites of transmission from photoreceptors to HCs. 'The paradox of retinal horizontal cells' was only solved in 1971 when Baylor, Fuortes and O'Bryan revealed the existence of a negative feedback loop from HCs to cones in the turtle retina. Hyperpolarization of HCs, achieved by injecting inward current through the intracellular recording electrode, induced a depolarizing deflection in cones. Through this negative feedback, illumination of the receptive field periphery could induce in cones a response of opposite sign to that elicited by central illumination. Cones respond to small central light spots with tonic graded hyperpolarizations the amplitude of which depends on the intensity of light stimulus. These hyperpolarizations are transmitted to HCs through a sign-conserving synapse (Trifonov, 1968; Dowling and Ripps, 1973; Cervetto and Piccolino, 1974; Kaneko and Shimazaki, 1975) (Chapter 7). However, HCs have a large summation area within their receptive field, due to their strong electrical coupling (Chapter 8), and generate large-amplitude hyperpolarizations in response to light stimuli of large area. Such stimuli flashed on the retina elicit, through the feedback synapse, a depolarization of cones, i.e. an effect opposite in sign, 'antagonistic', to the direct response elicited by central illumination.

Even though center-surround antagonistic organization of the receptive field has been found to be common in more proximal retinal neurons (Kuffler, 1953), it came as a rather unexpected finding in photoreceptors, the first neural elements of the visual pathways. The experiments of Baylor *et al* (1971) showed that the retina functions as an 'operational network' from the very beginning, carrying out even at its initial stage rather complex processing of visual information.

HC feedback has been shown to be involved in retinal mechanisms analyzing temporal, spatial and chromatic features of the visual signal as well as 'neural' control of light sensitivity (Fuortes *et al.*, 1973; O'Bryan, 1973; Werblin, 1974; Pasino and Marchiafava, 1976; see Wu, 1992 and Burkhardt, 1993 for reviews). In particular, HC feedback is thought to be the primary mechanism responsible, at least in part, for the color-opponent responses elicited in certain HCs by stimuli of different wavelengths (Gouras, 1972; Fuortes and Simon, 1974; Stell *et al.*, 1975; Lipetz, 1978) and for the surround responses of the bipolar cells (Kaneko, 1973; Schwartz, 1974).

9.2 CONE MEMBRANE MECHANISMS INVOLVED IN HC FEEDBACK

9.2.1 CONDUCTANCE CHANGES

O'Bryan (1973) first investigated the conductance changes associated with feedback responses in cones in the turtle retina. He found that the amplitudes of the voltage deflections induced by short current pulses injected through the cone membrane decreased during feedback depolarization, thus confirming the earlier suggestion that feedback involves an increase in cone membrane conductance. However, other results suggested that feedback responses were associated with more complex changes in cone membrane properties. When long-duration, depolarizing currents of large amplitude were injected, the late phase changed polarity (i.e. became hyperpolarizing), whereas the initial phase did not reverse. Depolarizing currents of moderate magnitude produced spike-like deflections in the initial phase, sometimes up to 20 mV in amplitude. Feedback responses disappeared with strong hyperpolarizing currents. Thus, it was clear that feedback depolarizations involved a multiplicity of membrane mechanisms that could be associated with regenerative phenomena.

9.2.2 Ca^{2+} AND Ca^{2+}-DEPENDENT RESPONSES

Piccolino and Gerschenfeld (1980) demonstrated that an important component of HC feedback was due to an increase of Ca^{2+} conductance in the cone membrane. This conclusion was reached by studying spike-like feedback responses (Figure 9.1) similar to those recorded by O'Bryan (1973). These were shown to be Ca^{2+}-dependent action potentials, being facilitated by perfusing the retina with high-Ca^{2+} Ringer or by adding Sr^{2+} or Ba^{2+} to extracellular medium; they were not affected by tetrodotoxin, the blocker of the voltage-dependent Na^+ channels, responsible for the 'classical' (i.e. Hodgkin–Huxley type) action-potentials. The feedback spikes were associated with an increase of membrane conductance and exhibited a

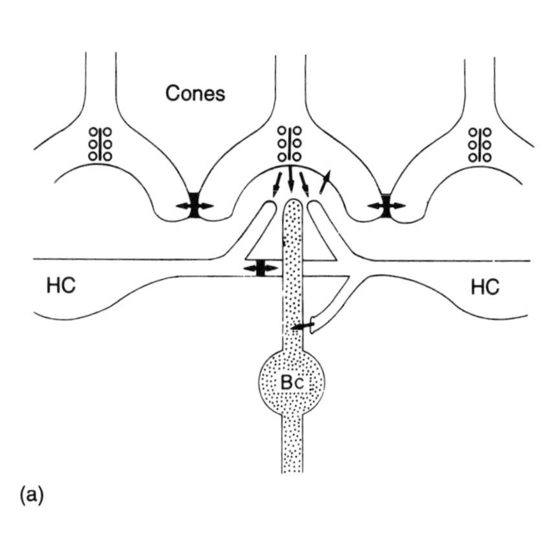

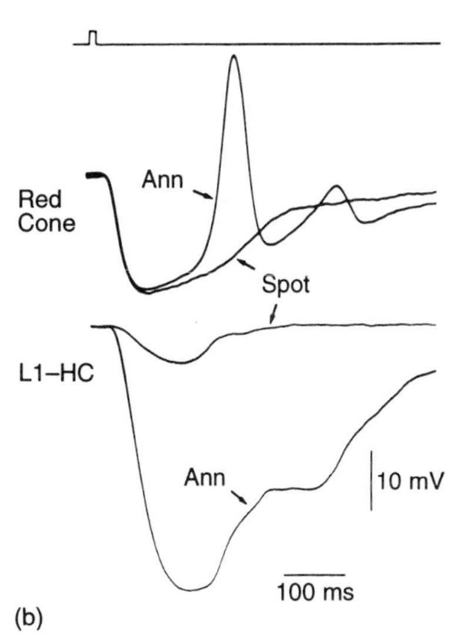

Figure 9.1 Horizontal cell (HC) feedback circuit and feedback responses in cones. (a) Schematic representation of the synaptic pathways involving cones, horizontal cells (HC) and bipolar cells (Bc) in the outer plexiform layer of the retina. Single arrows indicate sites of presumptive chemical transmission, whereas the double-arrow contact represents an electrical synapse. The direct synapse from horizontal to bipolar cells has been described only in some species. (b) Intracellular recording of the feedback response of a cone in the turtle retina. A red-sensitive cone and a horizontal cell axon terminal (L1-HC) of the turtle retina were simultaneously penetrated with two independent electrodes, and the retina was stimulated with either a 250 μm-diameter spot of white light or with a large white-light annulus (3600 and 430 μm outer and inner diameter). Notice the large spike-like depolarization followed by oscillatory wavelets in the late phase of the cone response to the annulus, after the initial hyperpolarization (due to the light scattering to the receptive field center). Light was attenuated by 1.2 log units with reference to the maximum available intensity (about 1.5×10^{-5} μW/μm^2). In the cone the initial hyperpolarizing phase of the annulus response is largely due to light scattering to the receptive field center. (M. Piccolino, J. Neyton and H. Gerschenfeld, unpublished results.)

threshold behavior. Moreover, and importantly, the effect of Sr^{2+} or Ba^{2+}-containing media were seen in virtually all cones of the turtle retina, even in cones which had not shown any sign of feedback influence prior to the application of the divalent cations. This suggested that feedback could affect all cones in the turtle retina. In the presence of Sr^{2+} or Ba^{2+}, prolonged stimulation of the receptive field periphery resulted in the discharge of repetitive trains of spikes. This indicated that the increase of cone Ca^{2+}-conductance produced by prolonged stimuli was sustained (Piccolino and Gerschenfeld, 1980).

An important functional property of the HC feedback emerging from these studies (Gerschenfeld and Piccolino, 1980; Piccolino and Gerschenfeld, 1980; Piccolino *et al.*, 1981) concerns the effect of the cone membrane potential on the feedback response. It was consistently found that feedback depolarizations could be delayed, reduced or suppressed by hyperpolarizing the cone membrane either by intracellular current injection, as also reported by O'Bryan (1973), or by central illumination. This hyperpolarization block of cone responses could account for some unexpected trigger features of feedback responses. For instance, bright light annuli were usually more effective in activating feedback depolarizations, compared with large spots of the same intensity and outer diameter. If feedback responses depended only on the amplitude of the light-induced hyperpolarization in HCs, they should be evoked more efficiently by large spots than by large annuli, and by more intense light, since such stimuli are more effective in driving HCs. It appeared as if a light stimulus was more powerful in eliciting cone feedback responses if it induced large HC responses and, at the same time, it did not induce a large hyperpolarization of cone membrane (directly by light stimulation or cone–cone coupling; Chapter 6). The suppressive influence of cone hyperpolarization on feedback responses has also been shown for salamander cones (Lasansky and Vallerga, 1975; Skrzypek and Werblin, 1983; Wu, 1991).

The inflow of Ca^{2+} in the cone induced by activation of HC feedback may, in turn, activate other membrane conductances, in particular Ca^{2+}-dependent K^{+} and Cl^{-} conductances. It has been shown (Burkhardt *et al.*, 1988; Thoreson and Burkhardt, 1990, 1991) that the activation of a Ca^{2+}-dependent Cl^{-} conductance is the main mechanism underlying the long-duration depolarizing feedback response frequently observed in turtle cones and referred to as 'prolonged depolarizations'. When the feedback response spikes, these prolonged depolarizations became regenerative and depended on the characteristics of the light stimulus in a highly non-linear way. Prolonged depolarizations are reduced or blocked by treatments leading to a decreased concentration of Ca^{2+} in cones (e.g. by extracellular application of low-Ca^{2+} saline, blockade of Ca^{2+} inflow, intracellular application of Ca^{2+} buffers) whereas they are facilitated by high-Ca^{2+} media. The prolonged depolarizations are also facilitated by experimental procedures leading to an increase of Cl^{-} efflux through the cone membrane, but not appreciably modified by changes in the extracellular concentration of Na^{+} or K^{+}. In order to explain how activation of a Ca^{2+}-dependent Cl^{-} conductance may lead to a depolarizing response, Burkhardt and coworkers assumed that the Cl^{-} equilibrium potential is positive with respect to cone membrane potential in darkness. However, this assumption is at variance with experiments carried out on dissociated turtle cones (section 9.2.1).

Although spikes and prolonged depolarizations are normally induced in cones by activation of the HC feedback mechanism with large spots or annuli of light, they can also be elicited in the absence of light stimulation under favorable conditions. Moreover, after prolonged treatment with

Sr^{2+} or Ba^{2+}, the spikes and, less commonly, the prolonged depolarizations, can appear in darkness in the absence of any current stimulus (Gerschenfeld and Piccolino, 1980; Thoreson and Burkhardt, 1991). Comparable responses can also be elicited by current injection into dissociated cones (Maricq and Korenbrot, 1988; Barnes and Deschênes, 1992). These data clearly indicate that the membrane mechanisms underlying both types of feedback response are intrinsic to cones; feedback only facilitates their activation through a synaptic process. In fact, studies on isolated cones have shown the presence of a voltage-dependent Ca^{2+} and Ca^{2+}-dependent Cl^- current thought to generate spikes and prolonged depolarizations (Maricq and Korenbrot, 1988; Barnes and Hille, 1989; Barnes and Deschênes, 1992; Lasater and Witkovsky, 1991; see Lasater, 1991 for review). Only a small component (often difficult to detect) of the overall feedback response reflects the true synaptic potential resulting from activation of the HC feedback itself (Burkhardt *et al.*, 1988; Burkhardt, 1993).

Forward transmission from cones to the second-order neurons occurs through chemical synapses (Chapters 7 and 12) The contacts between HCs and cones supposed to underlie the feedback transmission are located in close proximity to the sites of release of the cone neurotransmitter. In most chemical synapses, the transmitter release depends critically on Ca^{2+} inflow in the synaptic terminal. Since feedback controls the Ca^{2+} current in cone endings, it appears that it is strategically located to control the release of the cone transmitter. In our opinion, most of the cone membrane potential changes caused by the feedback are an 'epiphenomenon' of this main control mechanism. Voltage responses in cones appear clearly only when the activation of Ca^{2+} current triggers the regenerative potential in one membrane. On these grounds, we can conceive that HC feedback may affect second-order neurons heavily, even though it may not induce substantial changes in the cone membrane potential.

9.3 TRANSMISSION AT THE FEEDBACK SYNAPSE

9.3.1 THE 'CLASSICAL' CHEMICAL SYNAPTIC HYPOTHESIS OF FEEDBACK TRANSMISSION: IS GABA INVOLVED?

Baylor *et al.* (1971) originally suggested that the feedback transmission involved a chemical synapse since it was sign-inverting and led to an increase of cone membrane conductance following peripheral illumination. The transmitter at chemical synapses is normally released at a high rate when the presynaptic membrane is depolarized. The membrane of the HCs (presynaptic elements) is depolarized in darkness and is hyperpolarized by light (resulting in cone feedback depolarizations). Thus, feedback depolarizations in cones should result from a reduction of the HC transmitter release. It would follow, therefore, that the feedback transmitter should hyperpolarize cones and, moreover, it should decrease cone membrane conductance when released at a high rate in darkness. However, HCs may be unconventional since the release of their synaptic transmitter appears to be mostly through a non-vesicular, Ca^{2+}-independent, carrier-mediated mechanism (Schwartz, 1982, 1987; Yazulla and Kleinschmidt, 1983; Attwell *et al.*, 1993; and Chapter 10). Still, the electrical characteristics of the carrier are such that the release is favored when HCs are depolarized and, therefore, the aforementioned conclusions on the polarity of feedback transmitter action remain valid.

A large body of experimental evidence (reviewed by Yazulla, 1986; Brecha, 1992; and Chapter 10) indicates that γ-aminobutyric acid (GABA) is the transmitter of HCs, particularly in lower vertebrates. Moreover, immunohistochemical studies indicate that $GABA_A$ receptors are localized on cone pedicles (Yazulla *et al.*, 1989; Vardi *et al.*, 1992; Yang *et al.*, 1992). Thus it seems safe to assume that GABA is the transmitter at the feedback synapse. Consistently with this view, Tachibana and Kaneko (1984) have shown that iontophoretic application of GABA close to the synaptic pedicle of isolated cones of turtle activates a Cl^- current. This action of GABA was mimicked by muscimol and blocked by bicuculline, thus suggesting the involvement of $GABA_A$ receptors. The polarity of the observed effect would correspond to that of the feedback transmitter, if Cl^- equilibrium potential (E_{Cl}) was negative with respect to the cone resting potential in physiological conditions. This view was supported by subsequent work (Kaneko and Tachibana, 1986), but it is in contrast with studies on the prolonged depolarizations of cones *in situ* by Burkhardt and coworkers (section 9.2.2).

Direct physiological support for the involvement of GABA in HC feedback transmission comes also from studies on slice preparation of the salamander retina. In a preliminary study, Eliasof and Werblin (1989) showed that salamander, like turtle, cones respond to GABA with an increase in Cl^- conductance and that the effect is likely to be mediated by $GABA_A$ receptors. Moreover, they found that GABA also activated another membrane mechanism with a reversal potential near E_K, involving $GABA_B$ receptors. It was suggested that the input from GABA-ergic HCs could activate the two mechanisms in similar proportions under physiological conditions.

'Pure' feedback responses in the absence of the 'contaminating' effect of direct light stimuli were recorded from salamander cones *in situ* after mechanically removing their outer segment (Wu, 1991). Feedback responses in such 'truncated' cones were drastically reduced by application of bicuculline. The reversal potential of the responses was consistent with GABA affecting Cl^- channels, assuming E_{Cl} to be more negative than the dark resting potential. Wu (1991) also concluded therefore, that the feedback transmission in salamander cones is mediated by GABA and involves mainly $GABA_A$ receptors.

Thoreson and Burkhardt (1990) applied GABA and $GABA_A$ or $GABA_B$ agonists and antagonists to cones *in situ* and could not detect any consistent effect on the cone membrane potential and feedback response. The drugs were ineffective also at high concentrations and for long incubations, and also when applied to a slice preparation. Moreover, GABA was ineffective also in the presence of the uptake blocker nipecotic acid. The reasons of the discrepancy between these studies and the work of Tachibana and Kaneko (1988) and Wu (1991) remain elusive. If GABA is not the feedback transmitter, then the effects of GABA on isolated cones reported by Tachibana and Kaneko (1988) might represent some regulatory action of GABA on cones, not related directly to feedback transmission. On the other hand, if GABA is the feedback transmitter, one can still suppose that GABAergic drugs were ineffective in the retina *in situ* due to some diffusion barrier, notwithstanding the experimental precautions of Burkhardt and collaborators.

9.3.2 HETERODOX HYPOTHESIS OF FEEDBACK TRANSMISSION: THE 'ELECTRICAL' HYPOTHESIS

Byzov and coworkers (Byzov *et al.*, 1977; Byzov and Shura-Bura, 1986) proposed an 'electrical' theory for the HC feedback based on the fact that the contact region between

HCs and cones is almost isolated from extracellular current flow. This could explain the efficacy of feedback on second-order neurons in spite of the small amplitude of feedback potentials in most cones. This theory assumes that the extracellular current flow generated by the HC membrane in response to chemical synaptic input from photoreceptors could influence directly the nearby cone membranes, and lead to sign-inverting transmission via an extracellular field effect similar to that operating in fish Mauthner cells (Furukawa and Furshpan, 1963; Korn and Faber, 1975). An electrical field effect has also been supposed to underlie lateral inhibitory interaction in some insect eyes (Shaw, 1984).

The contacts between HCs and cones are located in the synaptic terminals of cones, near the transmitter release sites. Thus, the local current flow induced by 'electrical' feedback in the cone membrane could modify significantly the release of the cone synaptic transmitter, even in the absence of any substantial changes in the cone membrane potential. The hypothesis requires that the synaptic current of the HCs does not have free access to the 'open' extracellular space and is 'forced' to go through the cone membrane. There is indeed anatomical support for a restricted current flow in the extracellular space nearby HC–cone contacts since HC processes penetrate deeply into the cone pedicles at invaginating synapses. Moreover, in teleost retinas, changes in the efficacy of feedback transmission have been related to morphological modifications at these synapses. The contact area increases with the development of HC 'spinules', finger-like protrusions of the HC membrane, which indent the cone membrane making the synaptic cleft less accessible to the open extra cellular space (Wagner, 1980; Weiler and Wagner, 1984; Djamgoz *et al.*, 1988; see Wagner and Djamgoz, 1993 for review).

9.3.3 THE 'UPTAKE MODULATION' HYPOTHESIS

Another theory of HC feedback transmission is based on data (obtained from the turtle retina) suggesting a glutamate uptake mechanism dependent on the HC membrane potential. The uptake would occur at a high rate in darkness, when the HC membrane is depolarized, and it would slow down when HCs hyperpolarize in response to diffuse illumination (Schütte and Schlemermeyer, 1993). Such voltage dependence is unusual, since glutamate in other nerve and glial cells is taken up at a higher rate when the cell membrane is hyperpolarized (Attwell *et al.*, 1993). Glutamate release from cone endings is decreased directly by illumination of photoreceptors (Cervetto and MacNichol, 1972; Murakami *et al.*, 1972; Ayoub *et al.*, 1989; Copenhagen and Jahr, 1989). Thus, the level of glutamate in the synaptic cleft would be modulated in opposite ways by central or peripheral illumination: center illumination decreasing the release and peripheral illumination decreasing the uptake from the synaptic cleft. The actual changes of glutamate concentration would be more prominent in synaptic spaces relatively separated from the free extracellular medium. This could account for the feedback responses of cones and the opposite polarities of bipolar cell responses to center or surround illumination. Glutamate applied near the synaptic pedicle depolarizes cones both in salamander and in turtle retina (Sarantis *et al.*, 1988; Tachibana and Kaneko 1988; Eliasof and Werblin, 1993). Therefore, an increase of glutamate level in cone endings caused by peripheral illumination (via inhibition of HC uptake) could produce depolarizing responses in cones. Up to now, however, the involvement of such a mechanism in cone feedback is only speculative. There is no information on the kinetics of the proposed uptake process, the evidence for glutamate uptake in HCs is indirect, and it is not clear if this process can cause a significant change of glutamate concentration in the synaptic cleft. On the other hand, a mechanism based on the modulation of the extracellular level of glutamate, as well as the electrical field effect, could easily explain the apparent lack of effects of GABAergic drugs on feedback responses (as observed by Thoreson and Burkhardt, 1990). Nevertheless, on balance the chemical hypothesis still appears as the reference model for transmission at the HC feedback synapse.

9.4 HC FEEDBACK AND TEMPORAL PROPERTIES OF CONES

The activation of the HC feedback circuit may have an important effect on the time course of the cone response. The discovery of feedback came from the observation that the recovery phase of cone responses to a brief flash of light are faster if the surround of the receptive field is illuminated. For prolonged, step-wise illumination, the activation of the feedback circuit may lead to a more pronounced sag. In some cones with 'large' feedback responses, step illumination may evoke only a hyperpolarizing transient at the light onset (Gerschenfeld and Piccolino, 1980; Burkhardt *et al.*, 1988).

The temporal influence of the HC feedback may change with light intensity. At high intensity, the feedback fails to modify the early peak phase due to a suppressive influence of the large, direct light-induced hyperpolarization. For dim light, feedback may also affect the initial phase of the cone response decreasing the peak amplitude (Baylor *et al.*, 1971; Piccolino *et al.*, 1981; M. Piccolino, unpublished observations).

The temporal transfer functions of salamander and catfish cones are modified, by enlarging the illuminated retinal area (Pasino and Marchiafava, 1976; Lasater, 1982); this is consistent with the hypothesis of feedback affecting the properties of the cones. In both species, large stimuli improved the responses to high frequency stimulation. However, whereas in catfish retina responses to low temporal frequencies increased, in the salamander retina they decreased. These differences may be species specific or due to the different experimental procedures used, i.e. sinusoidal light stimuli in salamander and white-noise stimuli in catfish. The problem is also complicated by the difficulty of eliciting stable feedback responses (Baylor *et al.*, 1971; Piccolino and Gershenfeld, 1980); this may explain the reported lack of significant difference in cone temporal transfer functions for small or large stimuli (Kawasaki *et al.*, 1984).

9.5 CONES, FEEDBACK AND SECOND-ORDER NEURONS

9.5.1 FEED FORWARD AND FEEDBACK INTERACTIONS BETWEEN CONES AND HORIZONTAL CELLS

The variety of neural operations involving feedback mechanisms is particularly evident in the turtle retina, where the connections of the cells in the outer plexiform layer are known in detail (Chapter 7). In this retina, there is one type of axon-bearing horizontal cells (H1) and at least two classes of axonless horizontals (H2 and H3; Leeper 1978a). The axon-bearing cell gives hyperpolarizing responses for any wavelength of the stimulus (luminosity type, L-HC; Figure 9.2a), whereas the response polarity of axonless HCs changes with the stimulus wavelength (chromaticity type; C-HC). The response of the H2-type is hyperpolarizing for green lights and depolarizing for red lights (red–green cells, RG-HC; Figure 9.2b). The response of H3-type HCs is hyperpolarizing for blue lights and depolarizing for green–orange stimuli (green–blue cells, GB-HC; Figure 9.2c).

Red-sensitive (R-) cones are the main input to L-HCs (Fuortes *et al.*, 1973). The summa-

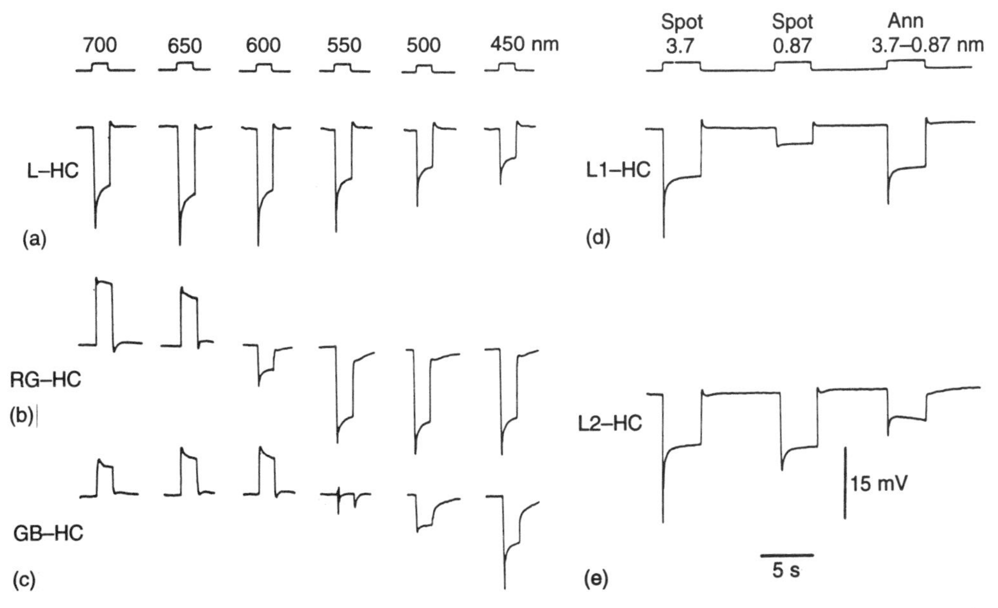

Figure 9.2 (a–c) Intracellular recordings of the light responses induced by monochromatic lights of different wavelengths in (a) a luminosity horizontal cell (L-HC), (b) a red–green chromaticity horizontal cell (RG-HC), and (c) a green–blue chromaticity horizontal cell (GB-HC). Numbers near the stimulus traces in (a) give the stimulus wavelength in nanometers. The stimulus was a spot of 3.7 mm diameter delivering about 1.7×10^5 quanta/$\mu m^2/s^1$. (d,e) Simultaneous recordings of the responses of (d) a H1-axon terminal (L1-HC) and (e) a H1-cell body (L2-HC), elicited by white light stimuli covering different areas of their receptive field as indicated. The numbers near the stimulus traces give the outer diameter of the spot or the outer and inner diameter of the annulus. Notice that the response elicited by the annulus in the H1-axon terminal is larger than the response elicited by the small spot, whereas the opposite occurs in the H1-cell body. The light was attenuated by 2.7 log units.

tion area of the receptive field is different for axon terminals (H1-ATs) and for cell bodies (H1-CBs) of L-HCs (Simon, 1973; Leeper, 1978b; Piccolino *et al.*, 1982). Responses of rather large amplitude are recorded in H1-CBs (L2-HC in Figure 9.2e) with relatively small light spots, which elicit very small responses in H1-ATs (L1-HC in Figure 9.2d). Moreover, in H1-CBs (and not in H1-ATs), large light annuli may elicit small depolarizing responses, suggesting a weak surround antagonism (Byzov, 1975; Piccolino *et al.*, 1981). The difference in response properties between H1-CBs and H1-ATs is due to three main reasons.

1. The difference in electrical coupling, strong in the H1-ATs network and weak in the H1-CBs network (Byzov, 1975; Piccolino *et al.*, 1982; Witkovsky *et al.*, 1983);
2. The absence of significant communication between H1-ATs and H1-CBs through the thin axon fiber (Piccolino *et al.*, 1981; Golard *et al.*, 1992);
3. The arrangement of their feedforward and feedback connections with photoreceptors.

Axon terminals contact R-cones and rods, whereas H1-CBs contact R- and green-sensitive (G-) cones (Leeper, 1978b). Anatomical and physiological evidence suggests that

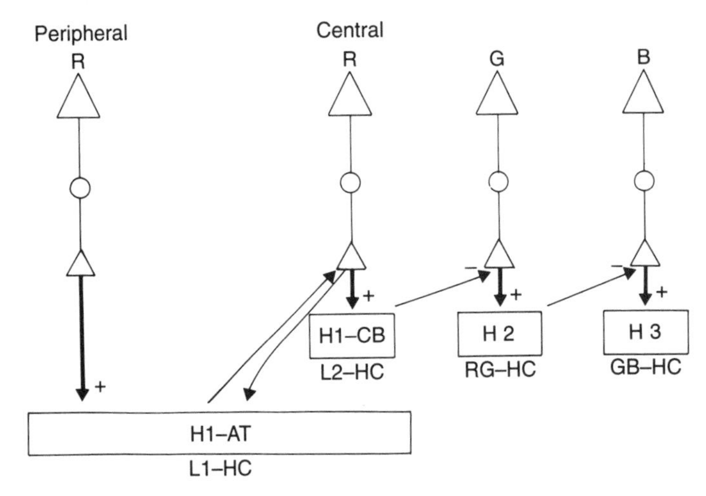

Figure 9.3 Simplified diagram of the feedforward and feedback connections between cones and horizontal cells in the turtle retina. (+) and (−) indicate sign-conserving and sign-inverting synapses, respectively. R, G and B indicate red, green and blue-sensitive cones, respectively. H1 stands for the axon-bearing horizontal cell, and H2 and H3 indicate the two types of axonless horizontal cells according to the classification of Leeper (1978a). Under each horizontal cell element is indicated the response-type it generates. In the forward pathway the thickness of the line represents the relative weight of cone input. 'CENTRAL' or 'PERIPHERAL' refer, respectively, to the pool of cones inside or outside a circular region of the receptive field of about 1 mm diameter.

H1-ATs receive dominant input from R-cones and rods and provide the main feedback input to R-cones; H1-CBs receive the dominant input from R-cones and provide the main feedback input to G-cones (Leeper, 1978b; Leeper and Copenhagen, 1982; Neyton *et al.*, 1981; Piccolino *et al.*, 1981).

As far as the axonless HCs are concerned, H2 HCs receive the dominant input from G-cones and feedback onto blue-sensitive (B-) cones; H3 HCs receive the main input from B-cones but its feedback target, if any, is unknown (Fuortes and Simon, 1974; Leeper, 1978b, Figure 9.3). The functional connections illustrated in Figure 9.3 define some response properties of photoreceptors and HCs in the turtle retina (Fuortes *et al.*, 1973: Fuortes and Simon, 1974; Leeper, 1978b; Piccolino *et al.*, 1981; Neyton *et al.*, 1981). Surround antagonistic responses of R-cones depends mainly on feedback from H1-ATs. Surround antagonism of R-cones causes the surround antagonism of H1-CBs. The spatial and chromatic antagonism of G-cone responses (Figure 9.4a) are explained by feedback from H1-CBs. Long wavelength stimulis, poorly absorbed by G-cones, can cause depolarizing responses (Figure 9.4b) through the following pathway: R-cones → H1-CBs → G-cones (Figure 9.3). In G-cones, long wavelength light spots of moderate size (about 1 mm diameter) are more effective in eliciting feedback depolarizations than large light annuli (Figure 9.4c, d), since such stimuli are optimal for covering the summation area of H1-CBs, but not of H1-ATs (Figure 9.4e–h). Large annuli might reduce the depolarizing influence of feedback due to red light spots stimulating G-cones via the following pathway: peripheral R-cones → H1-ATs → central R-cones → H1-CBs. Such an 'anti-antagonistic' effect of far-surround

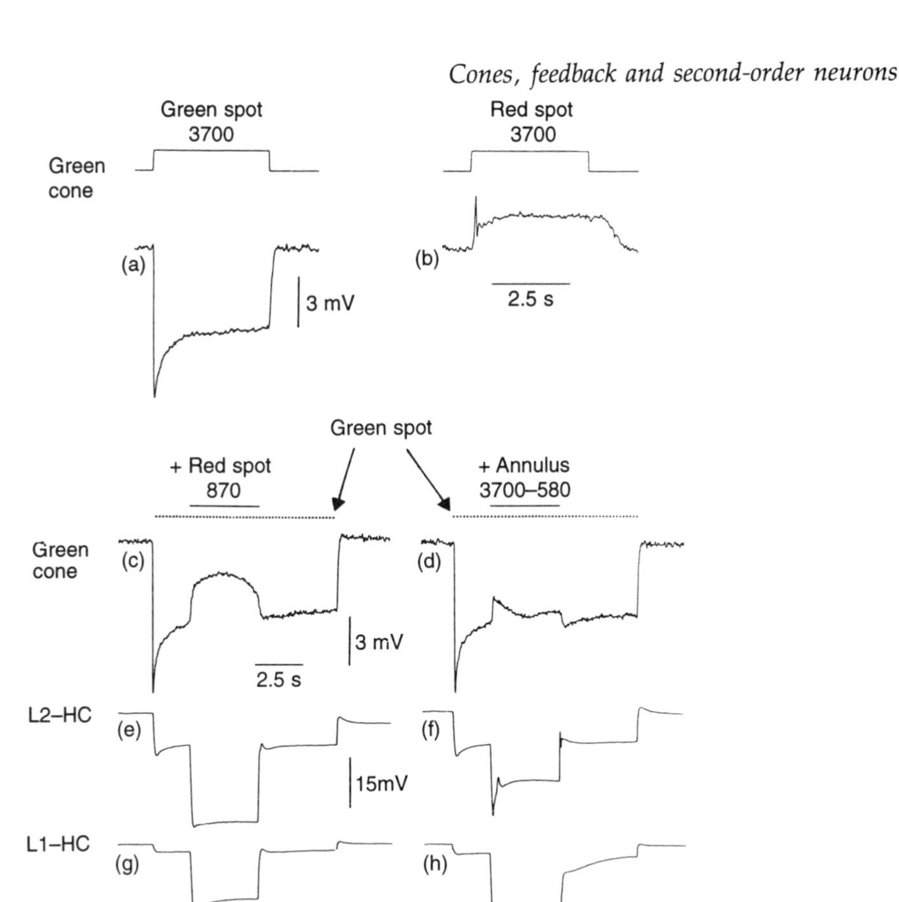

Figure 9.4 Intracellular recordings from cones (a–d) and horizontal cells (e–h) in the turtle retina. (a,b) Light responses elicited in a green-sensitive (G-) cone by a 3700 μm light spot of monochromatic light of 550 nm (green spot) and 700 nm (red spot) delivering approximately the same photon flux of about 1.2×10^5 quanta/μm² s, respectively. (c–h) Intracellular recordings of the light evoked responses from a G-cone (c,d), L2-HC cell body and (e,f) L1-HC axon terminal (g,h). Monochromatic red light stimuli (700 nm; 4.2×10^6 quanta/μm²/s) applied in the presence of a background illumination with green light (550 nm; 1.2×10^5 quanta/μm²/s) were used. The duration of the red light stimuli (spot or annulus with diameters indicated in μm near the stimulus traces) was monitored by the full line trace. The dotted line trace indicates the duration of the green spot (250 μm spot for all the recordings). Notice that amplitude of the depolarization induced in G- cones by the red light is larger with the spot than with the annulus, and is correlated with the L2-HC response amplitude, but not with the amplitude of the L1-HC response (From M. Piccolino, J. Neyton and M. Gerschenfeld, unpublished observations.)

stimulation has indeed been observed in G-cones (Neyton *et al.*, 1981).

In conclusion, the receptive field of G-cones is arranged in three mutually antagonistic concentric areas: (1) a central one, sensitive to green light, responsible for the central hyperpolarizing responses; (2) a near-surround field sensitive to red lights, responsible for the feedback depolarizations; and (3) a red-sensitive far-surround, whose stimulation reduces the antagonistic influence of near-surround illumination (Figure

9.5). The complexity of the chromatic and spatial interactions involved in G-cone responses in turtle retina illustrates the great operational capability of the retinal network at the first synaptic stage in the visual pathway.

Feedback is also involved in the chromatic opponency of HCs in fish retina, where luminosity and chromatic-type HC responses were first recorded (Svaetichin, 1953, 1956; Stell *et al.*, 1975; Lipetz, 1978). The organization of feedforward and feedback connections between cones and HCs in fish was outlined by Stell *et al.* (1975) (Chapter 7). The 'Stell model' is still valid even if some of the assumptions have been questioned (Kamermans *et al.*, 1989, 1991). The essential hypothesis of this model is the specific connectivity between the different chromatic types of cones and HCs. As in turtle retina, the dominating hyperpolarizing response of any HC subtype would depend primarily on direct input from a specific class of cone, while the opponent spectral responses would be mediated through the feedback circuit. Although the model was originally based primarily on anatomical observations (Stell and Lightfoot, 1975; Stell *et al.*, 1975) it has been substantiated by electrophysiological experiments (Downing and Djamgoz, 1989; Greenstreet and Djamgoz, 1994). Direct physiological evidence for feedback is scarce in fish, due to the difficulty of obtaining stable electrical recording from cones (but see Murakami *et al.*, 1982a, b) Interestingly, in fish retina, the efficacy of the feedback pathways involved in chromatic opponency of HCs increases during light and decreases in dark

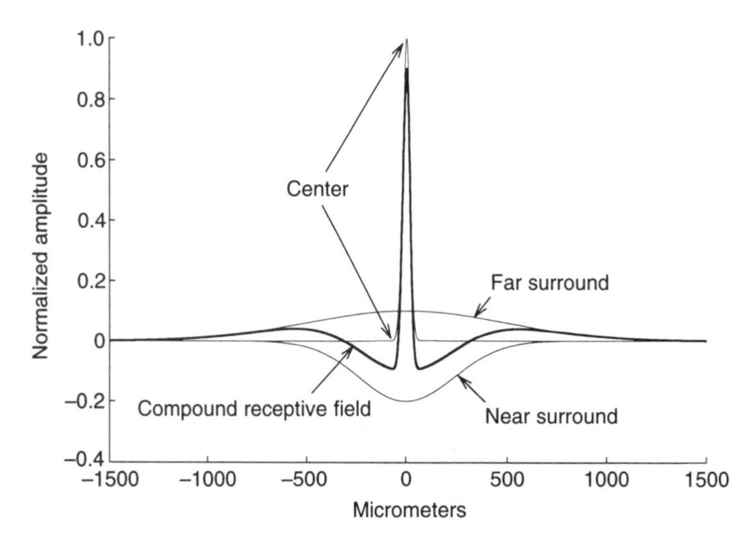

Figure 9.5 Model of the overall receptive field profile of a green-sensitive cone of the turtle retina (thick line) with its different components (thin lines). For computational convenience, the different components have been approximated by Gaussian functions whose widths (standard deviations) correspond to the space constants measured with thin, linear light stimuli (Smith and Sterling, 1990). Space constants values: 26 μm for the center mechanism (Detwiler and Hodgkin, 1979), 334 μm for the near surround and 700 μm for the far surround. These last two values are the average of the space constants of H1-CBs and H1-ATs, calculated from the data of Leeper and Copenhagen (1982). The amplitude of the near and far surround, relative to the center profile, are exaggerated for the purpose of illustration.

adaptation. As already mentioned, the adaptation-dependent effect of feedback action probably correlates with the appearance of spinules indenting the cone pedicles membrane. These functional and morphological changes are controlled by dopamine, which, in turn, depends on the level of retinal illumination and also follows a circadian rhythm (see Wagner and Djamgoz, 1993, for review).

The feedback effect on the chromatic opponency in C-HCs has been questioned in both turtle and fish on the basis of anatomical and physiological studies (Burkhardt and Hassin, 1978, 1983; Kolb and Jones, 1984; Ohtsuka and Kouyama, 1986; Burkhardt, 1993). It has been suggested that chromatic opponency in C-HCs could result from sign-inverting input from specific spectral classes of cones to different HC subtypes, or it could be the consequence of direct interactions between HCs. Nevertheless the feedback model is supported by rather sound experimental evidence and accounts for many properties of C-HC light responses. In particular, it explains why the reversal potential of the hyperpolarizing and depolarizing opponent responses of C-HCs is the same (Miyachi and Murakami, 1989, 1991; Yamada *et al.*, 1991); this would indeed be expected to be the case if both response components would result from modulation of the release of the same cone transmitter. The feedback also explains why in bright light the 'direct' hyperpolarizing response of C-HCs is dominating (Naka and Rushton, 1966a, b; Yazulla, 1976; Burkhardt and Hassin, 1978; Kato, 1979; Gottesman and Burkhardt, 1987; M. Piccolino and A. Pignatelli, unpublished observations). This would be the consequence of the blockade of depolarizing feedback responses elicited by the cone hyperpolarization induced by direct light. The modifications of C-HC responses induced by Sr^{2+} have also easily been accounted for by Piccolino *et al.* (1980) on the basis of the feedback model. Finally, the basic assumption of the feedback model of C-HC responses, i.e. that chromatic opponency is present already in cones, is well documented at least in turtle retina (Fuortes *et al.*, 1973; Fuortes and Simon, 1974; Piccolino *et al.*, 1980; Neyton *et al.*, 1981). Although other interactions could contribute to shaping the spectral behavior of C-HCs, they cannot be the main mechanism of chromatic opponency at this level (see Piccolino, 1995 for further discussion).

It is not known if HC feedback occurs in the mammalian retina. Selective connectivity between different chromatic classes of cones and HCs apparently does not occur in mammals. Only luminosity-type HC responses have been recorded in different mammalian species, including primates (e.g. Nelson, 1977; Dacheux and Raviola, 1982, 1990; Leeper and Charlton, 1985). A morphological study in monkey retina indicates that a HC subtype (H1) contacts all cones in its receptive field, thus excluding chromatic selectivity (Wässle *et al.*, 1989). On the other hand, a role of HCs in chromatic signaling cannot be excluded. Some subtypes of HCs in primates may establish relatively selective connections with different chromatic classes of cones and they could contribute to the chromatic opponency of bipolar cells (Kolb, 1991; Ahnelt and Kolb, 1994). In the monkey retina the functional opponency of red and green stimuli is reduced by local administration of GABA antagonist (Mills and Sperling, 1990; Sperling and Mills, 1991). Although in the mammalian retina there are many possible targets for GABAergic drugs (see Yazulla, 1986, and Brecha 1992 for review), the possibility cannot be excluded that the effect involved blockage of a feedback synapse. Furthermore, GABA receptors have been shown to be present in the synaptic pedicles of mammalian cones whereas the GABA-synthesizing enzyme is present in mammalian HCs (Vardi *et al.*, 1992).

9.5.2 HC FEEDBACK AND BIPOLAR CELLS

Bipolar cell receptive fields are characterized by center-surround antagonistic organization; the 'surround' field is thought to be mediated by HCs (Werblin and Dowling, 1969; Kaneko, 1970; Schwartz, 1974; Richter and Simon, 1975; Skrzypek and Werblin, 1983; Hare and Owen, 1990). This hypothesis is supported by experiments carried out on turtle and fish retinas (Marchiafava, 1978; Toyoda and Tonosaki, 1978) showing that hyperpolarization of HC membrane by extrinsic current injection evokes in bipolar cells a response similar to that induced by peripheral illumination. In principle, HCs could influence bipolar cells either via a direct synaptic action or via the feedback circuit. There is morphological evidence of a direct HC–bipolar cell synapse in some amphibia (Dowling and Werblin, 1969; Lasansky, 1973; Witkovsky and Powell, 1981) and in catfish (Sakai and Naka, 1986). On the other hand, since feedback causes spatial antagonism in cones, it must contribute to the surround antagonism of bipolar cells, at least for cone-driven responses (Kaneko, 1973; Schwartz, 1974). Some feedback effect has been observed in fish bipolar cells where the reversal potential of the surround response evoked by current injected into HCs was the same as the reversal potential of the bipolar cell response to central illumination (Saito *et al.*, 1981). This is consistent with involvement of the feedback since in such cases both central and peripheral responses would be due to the action of the same transmitter. Moreover, feedback responses in cones, as well as surround responses of bipolar cells in retinas of salamander, turtle and fish are reduced or suppressed by intense central illumination (Skrzypek and Werblin, 1983; Schwartz, 1974; M. Piccolino and A. Pignatelli, unpublished observations).

The relative contributions of the direct synaptic input and of HC feedback mechanism in the surround responses of bipolar cell have been evaluated in the salamander retina, where direct HC–bipolar cell contacts have been demonstrated (Yang and Wu, 1991; Hare and Owen, 1992). Cone input to ON-center bipolar cells can be blocked selectively by L-α-amino phosphonobutyric acid (APB), without affecting cone transmission to HCs and to OFF-center bipolar cells. In the presence of APB, the ON-center bipolar cells did not respond to central illumination, but showed a small hyperpolarization to surround stimulation. This residual peripheral response was assumed to be driven directly by HCs. These experiments suggested that 30–50% of peripheral responses of ON-center bipolar cells was mediated by this direct action. On the other hand Mangel (1991) found in the rabbit retina that surround response in ON-center bipolar cells depends entirely on the feedback circuit. This conclusion was based on the finding that APB suppressed all light responses of ON-center ganglion cells and abolished the peripheral responses elicited in these cells by current-induced hyperpolarization of HCs. A theoretical study in cat retina suggested that the surround response of a subset of bipolar cells may wholly depend on the feedback circuit (Smith and Sterling, 1990).

9.5.3 FEEDBACK RESPONSES IN HORIZONTAL-CELL AXON-TERMINALS

(a) Transient responses

In principle, an easy way to identify the transient and sustained components due to the feedback action in the response of any retinal cell would be to compare the response to a given stimulus under normal, physiological conditions and after blocking the feedback pathway. However, up to now no pharmacological agent capable of blocking selectively the feedback synapse has been found (Thoreson and Burkhardt, 1990). If one considers the operation of the feedback loop (Figure 9.3), it would appear that even if feedback cannot cause a sustained depolarizing influence on H1-ATs, it could, in principle, produce oscillatory responses through some transmission delay in the feedback circuit and an adequate gain in the recurrent synapse. Oscillatory deflections can sometimes be observed in the recovery phase of the response to short light flashes in M1-ATs (Fuortes *et al.* 1973; O'Bryan 1973; Piccolino and Neyton, 1982), and, more rarely, in the transient phase of the response to light steps. These oscillations are more prominent with large stimuli, which are optimal for evoking feedback depolarizations in R-cones. The H1-AT response oscillations are correlated with the oscillatory deflections induced in cones by feedback. However, the oscillations in H1-ATs can be short-lived, whereas prolonged light stimuli may evoke in cones maintained oscillations or other sustained feedback influences (Gerschenfeld and Piccolino, 1980).

(b) Sustained oscillations

Sustained oscillatory potentials can be elicited in turtle H1-ATs by prolonged illumination of a large retinal area. The trigger features of these responses suggest that they depend on the feedback mechanism, probably maintained between R-cones and H1-ATs (Piccolino and Belluzzi, 1990; M. Piccolino, unpublished). Figure 9.6a illustrates the response elicited in a H1-AT by prolonged illumination of the retina with a large annulus of light. As the response 'sags' gradually from the initial peak to the less hyperpolarized plateau, the baseline becomes progressively more noisy and eventually (~ 5 min), the noise amplitude attains a large steady-state value. In general, the development of noise during peripheral illumination has a slow time course, sometimes up to 20 min, but this can be shortened by prior exposure of the H1-AT to peripheral stimulation (Figure 9.6a,b). Illumination of the retina with a small light spot (adjusted in size and intensity to elicit the same level of maintained hyperpolarization as the annulus) was ineffective in inducing the appearance of the noise (Figure 9.6b).

The membrane 'noise' in H1-AT elicited by annul or stimulus consisted of a high-frequency oscillatory component modulated irregularly by slower waves (Figure 9.6c). Fourier analysis of this response revealed the presence of a major component at about 45 Hz, with side bands extending to more than 4 Hz from the peak frequency (Figure 9.7a, continuous line). In some cases, oscillations were relatively constant in amplitude and frequency with a spectrum characterized by a narrow band without secondary peaks (Figure 9.7b). A peak of small amplitude but clearly distinguishable from the mains interference could sometimes be detected even in 'quiet' H1-ATs, i.e., in cells which did not develop overt oscillations during peripheral illumination (Figure 9.7c). Sustained oscillations elicited by prolonged peripheral illumination of H1-ATs had a higher frequency than transient oscillations elicited by a flash of light. However, if light was flashed over a large steady background, the frequency of the transient oscillations increases and could approach the level of the sustained oscillations (Figure 9.8).

In almost all H1-ATs in which peripheral illumination induced detectable oscillations (~ 30% of the units studied), the main component of the Fourier spectrum was between 40 and 50 Hz. The proximity of the main frequency of peripheral oscillations to the 50 cycle interference may be one reason why these oscillations have been overlooked. Oscillations somewhat similar in time course and trigger features have been reported for HCs of fish retina but only after treatment with catecholaminergic or cholinergic agents

(Negishi and Drujan, 1979). Unlike the turtle, however, the high-frequency oscillations of fish could be maintained in darkness.

Feedback responses in cones involve the activation of a regenerative Ca^{2+} conductance; their kinetics are profoundly modified by divalent cations such as Ba^{2+} or Sr^{2+} which can permeate through Ca^{2+} channels (Piccolino and Gerschenfeld, 1980; Gerschenfeld and Piccolino, 1980). Divalent cations could also increase the amplitude and change the time course of H1-AT oscillations. In the presence of Ba^{2+}, oscillations reached an amplitude of several millivolts, and their frequency decreased to values as low as 1 Hz. These slow oscillations had a similar time course to the slow spikes observed in cones in the presence of Ba^{2+}. This was attributed to Ba^{2+}-induced modification of the conductance change(s) in the cone membrane, probably block of K^+ conductance (Piccolino and Gerschenfeld, 1980). A further reason for the decreased rate of oscillations could be the slowing down of the feedforward synaptic transmission between cones and H1-ATs (Piccolino and Gerschenfeld, 1980).

(c) A feedback hypothesis for the sustained oscillations in H1-ATs

The sustained peripheral oscillations of H1-ATs have several similarities to the feedback responses of cones, suggesting a common origin. Both are evoked exclusively by illumination of a large retinal area and suppressed by stimulation of the receptive field centre (Gerschenfeld and Piccolino, 1980). In the case of cone feedback responses, it was possible to show that the suppression by small central stimuli was due to the hyperpolarization of the cone membrane. The assumption that the H1-AT oscillations depend on the feedback circuit is consistent with the greater efficacy of large, peripheral annuli in eliciting the responses compared with small central spots generating the same level of hyperpolarization. A further similarity is the effect on amplitude and time course following application of Ba^{2+} or Sr^{2+}. Transient activation of the feedback circuit by flash illumination of large retinal areas tends to produce short-lived oscillations in both cones and HCs. Therefore, the sustained oscillations in H1-ATs induced by long-duration peripheral stimuli could be due to an electrical signal generated by feedback leading to strong activation and underdamping of the feedback mechanism. The need for prolonged illumination to induce sustained oscillations in H1-ATs could be due to stochastic feedback fluctuations; thus, a long time may be necessary to synchronize signals circulating between a multitude of cones and electrically coupled H1-ATs. Modifications to the membrane characteristics of either H1-ATs and cones and the synaptic transmission properties might also occur during constant peripheral illumination.

Figure 9.6 Intracellular recordings obtained in a H1-type horizontal cell axon terminal (H1-AT) during stimulation with an annulus (5 mm and 1.8 mm outer and inner diameter, respectively) (a), or a spot (1.2 mm diameter) of fixed intensity light (b) respectively. The light stimulus, monitored by the traces above the recordings, was attenuated by 0.9 log units in (a) and by 0.6 log units in (b) with respect to the maximum available intensity. (c) Recording of membrane potential in darkness; (d) after 22 min from the onset of fixed intensity illumination with a light annulus (5 mm and 8 mm outer and inner diameter, respectively); and (e) after changing the stimulus from the annulus to a small light spot (0.8 mm diameter), 28 min after the light stimulation had begun. In (d) and (e) the light was attenuated by 0.9 log units, and the mean level of maintained hyperpolarization of the membrane was -17 mV in both cases.

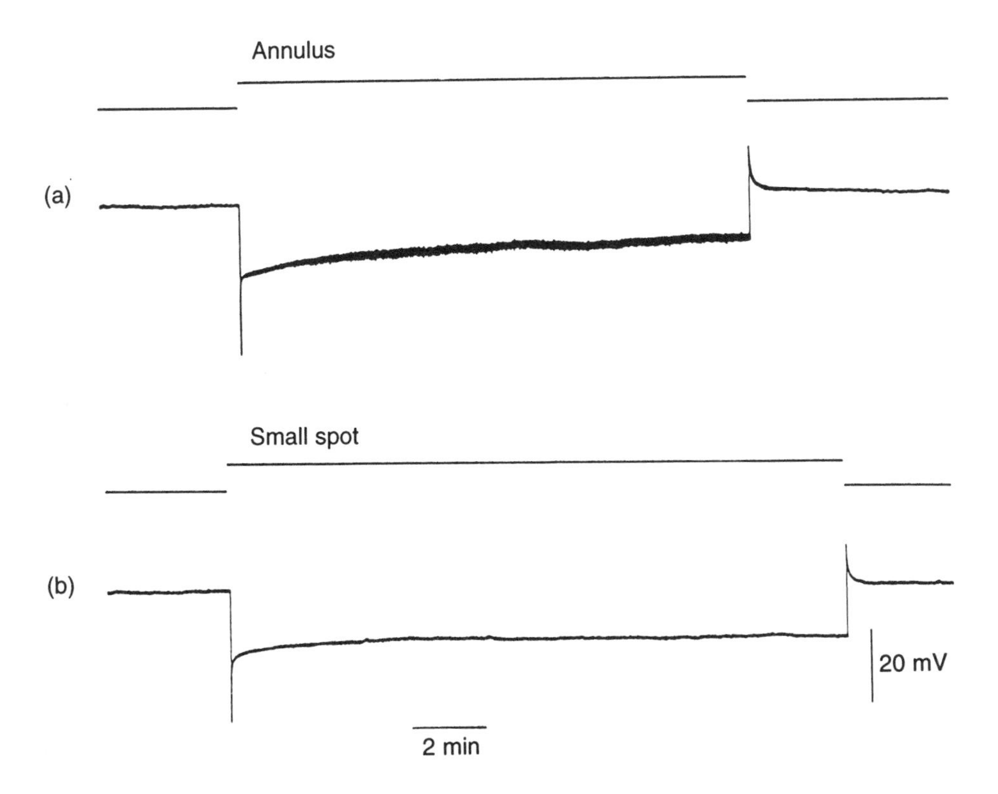
Annulus
(a)
Small spot
(b)
20 mV
2 min

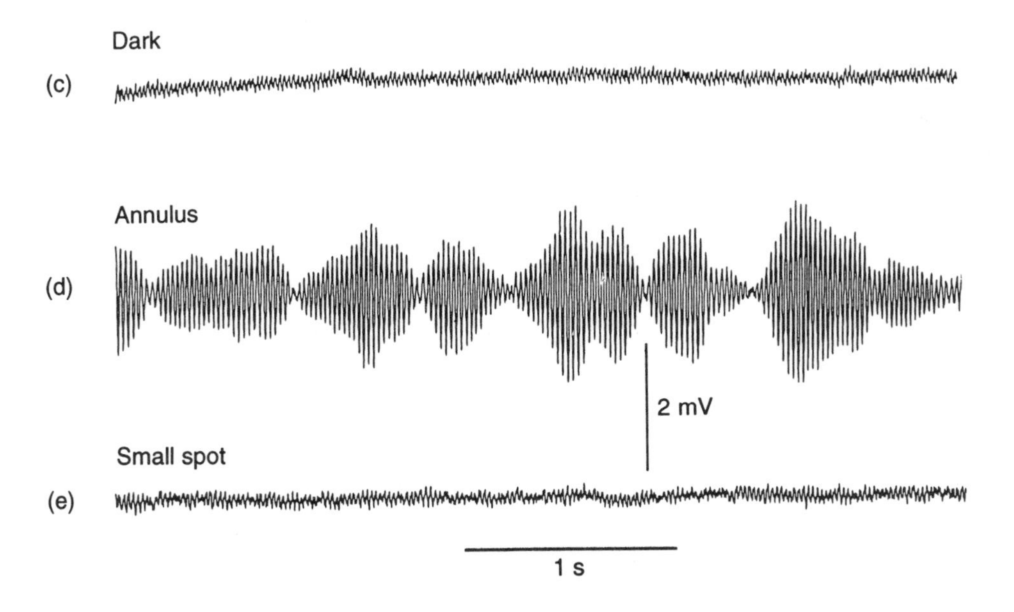
Dark
(c)
Annulus
(d)
2 mV
Small spot
(e)
1 s

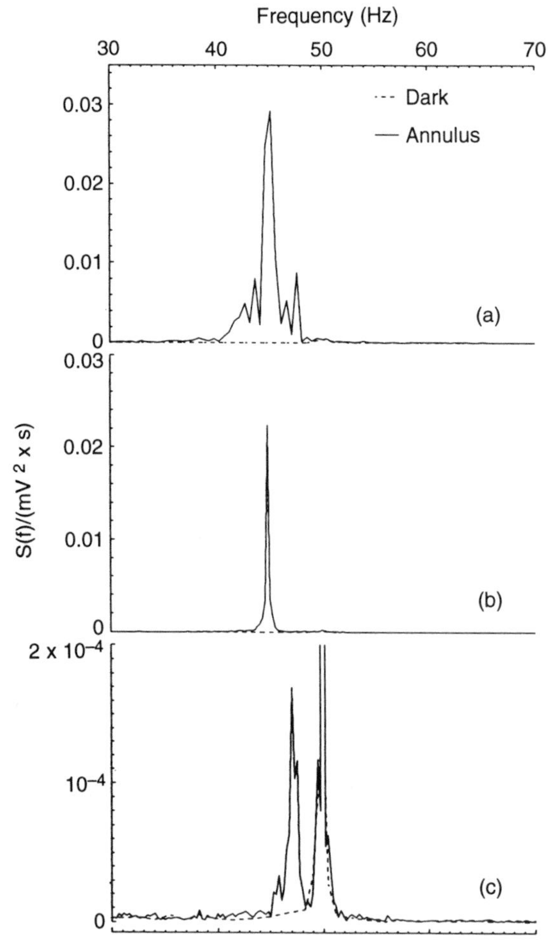

Figure 9.7 Power spectra of membrane potential recordings of two different H1-type horizontal cell axon terminals (H1-ATs) obtained in darkness or in the presence of a background illumination with a light annulus (5 mm and 8 mm outer and inner diameter, respectively). Light was attenuated by 0.9 log units. (a) and (b) refer to H1-ATs which developed large amplitude oscillations in the presence of the annulus, whereas (c) illustrates the power spectrum of a 'quiet' H1-AT. Note the difference in ordinate scale between (c) and (a) or (b).

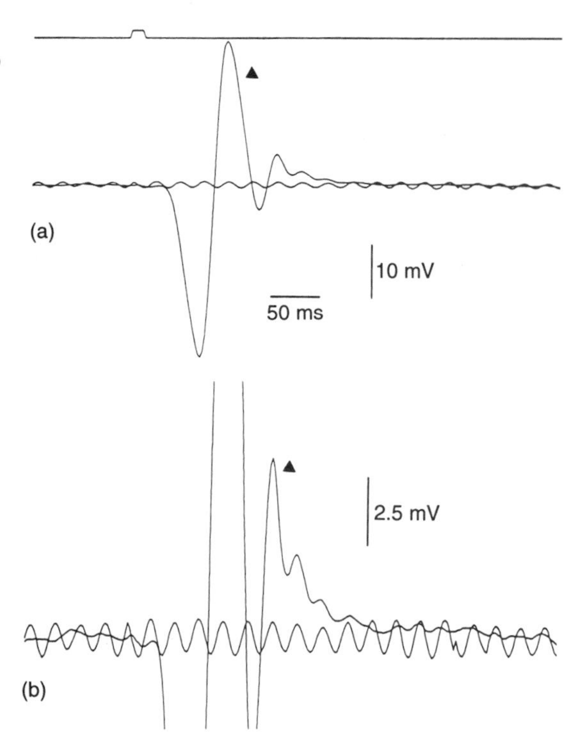

Figure 9.8 Transient response of a H1-AT to flash illumination in the presence of a peripheral background recorded 10 min after the onset of the background (tracings, marked by arrows), compared with the sustained oscillations developed by the same H1-AT 15 min later, in the continued presence of the background alone (unmarked tracings). The stimulus was an annulus of the same dimensions for both the flash and the background (5 mm and 1.8 mm diameter annulus), and the light was attenuated by 0.9 and 0.3 log units for the steady annulus and the flash, respectively. The flash recording is the average of 12 traces. (a) and (b) illustrate the same recordings, but the voltage axis was expanded in (b) for better comparison of the tracings, showing similar frequency of the transient and sustained oscillations in the two stimulus conditions.

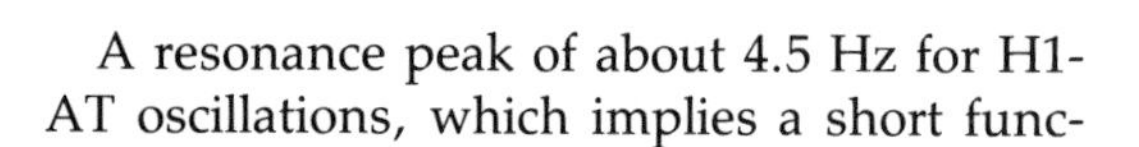
A resonance peak of about 4.5 Hz for H1-AT oscillations, which implies a short functional delay (~ 22 ms) in the feedback loop, seems inconsistent with the physiological evidence that H1-AT responses lag behind

cones' by several tens of milliseconds. However, transmission between cones and H1-ATs is fast for stimuli eliciting response amplitudes comparable to H1-AT oscillations observed in the present experiments (Schnapf and Copenhagen, 1982). The delay in a single stage of the synaptic mechanisms involved in the color-opponent responses of HCs has been estimated to be 25 ms or less (Spekreijse and Norton, 1970; Yamada *et al.*, 1985). This would appear to be compatible with the 22 ms functional delay implied in the generation of sustained oscillations in H1-ATs.

(d) H1-AT sustained oscillations: a model for high frequency oscillations in the retina and other regions of the nervous system?

Oscillations of around 50 Hz are present in other regions of the central nervous system, including the visual cortex, and have been postulated also to result from recurrent excitation and inhibition among populations of interacting neurons (Freeman, 1987; Gray and Singer, 1989). Feedback inhibition has also been implicated in the origin of oscillatory waves in the second-order neurons of the fly visual system (Kirschfeld, 1992). Fast oscillatory wavelets are present in the flash electroretinogram of the vertebrate eye (Ogden, 1974); in cold-blooded species these wavelets have frequencies comparable to the turtle H1-AT oscillations (Alcayaga *et al.*, 1989). Oscillations in H1-AT responses arise at the initial stage of visual processing through interaction of a limited number of cell types with well-known anatomical and functional connectivity. The study of oscillatory responses in H1-ATs can, therefore, serve as a useful model for understanding the mechanisms of generation of oscillatory potentials in the nervous system generally.

9.6 FEEDBACK VERSUS FEEDFORWARD: WHY FEEDBACK?

9.6.1 FEEDBACK, A SPECIAL CASE OF LATERAL INHIBITION

HC feedback represents a special case of lateral inhibitory interaction, which introduces spatial, temporal and chromatic opponency in the cone response. It is generally assumed that lateral interactions in the visual system enable efficient detection of changes in the visual image normally rich in information on borders, edges, moving patterns, intensities, chromatic contrasts etc. (Ratliff, 1965; Marr, 1982; De Valois and De Valois, 1990). HC feedback makes cones more responsive to local and transient illumination as compared to diffuse and steady light stimuli, and sharpens their chromatic signaling, contributing to the overall selectivity of the visual system in the spatial, temporal and chromatic domains. In particular, feedback may contribute to the low-frequency attenuation of sensitivity to both temporally and spatially modulated visual stimuli observed in most vertebrates including humans (De Valois and De Valois, 1990).

9.6.2 THE 'GLOBAL' CHARACTER OF FEEDBACK ACTION

A remarkable characteristic of HC feedback is that it acts at an early stage of retinal processing, at the cone level, and can thus affect all retinal neurons in the main 'centripetal' line of retinal transmission. This 'global' character of feedback may be important for optimization of the information-carrying capacity of retinal subcircuits. Feedback is essentially a mechanism for subtracting from the response of an individual cone the responses of neighboring photoreceptors, thus reducing the possibility of saturation and extending the operating range of the synapse. Saturation of synaptic signaling at the cone level could not be prevented by feedforward transmission. Feedback also improves the signal-to-noise ratio of the transmitting pathway (Barlow, 1981; Srinivasan *et al.*, 1982; Laughlin, 1994).

Feedback can induce a variety of effects in different neurons through a single synaptic mechanism. This can be seen in the bipolar cell surround responses which are hyperpolarizing for ON-center cells and depolarizing for OFF-center cells. By means of the feedback scheme, both surround responses can be achieved by the same mechanism. In contrast a feedforward model would require different synaptic receptors or transmitters, or even different HC types. On the other hand, feedforward could have the advantage of a greater specificity of actions, e.g. specific HC subtypes could exert their effect on specific subsets of second- or higher-order neurons (Sakai and Naka, 1985; Marshak and Dowling, 1987), and thus bring about visual signal modification limited to the neurons acted upon.

9.7 CONCLUSIONS AND FUTURE PERSPECTIVES

In view of the differences in the functional implications of the feedback versus the feedforward mode of HC action, they may coexist and operate in a complementary way in retinal processing of visual information. Future work is necessary in order to characterize this and other important aspects of the physiology of neural circuits at the very beginning of the visual pathways.

HC feedback studies have been limited almost exclusively to lower vertebrates due to the difficulty in obtaining intracellular recordings from cones in mammals. However, spatial antagonism similar to that observed in turtle cones has been observed in the tree squirrel in a rare study of cone responses in a mammalian species (Leeper and Charlton, 1985). This finding and some fragments of indirect evidence gathered in monkey (Smith and Sterling, 1990) and rabbit (Mangel, 1991), indicate that feedback may be operative in higher vertebrates, although it may not play a similar role in chromatic perception as in lower vertebrates. Although it may be unsafe to make inferences about the physiology and pathology of the human retina from experimental results obtained from lower vertebrates the possibility that some abnormalities of human vision may depend, at least in part, on altered functioning of the HC feedback circuit could be relevant for clinical studies. This applies especially to the selective spatiotemporal deficiency associated with abnormal functioning of the GABAergic and the dopaminergic systems, which are both implied in the feedback circuit. Since GABA is probably the transmitter at the feedback synapse, feedback action may be altered by drugs interfering with the GABAergic function (see below). Moreover, GABAergic agents may also modify feedback physiology indirectly, since GABA normally exerts an inhibitory control on the release of dopamine in the retina, and, in turn, dopamine can influence feedback function by multiple mechanisms (Dowling, 1991; Witkovsky and Dearry, 1991; Wagner and Djamgoz 1993 for reviews). Dopamine may modulate directly the efficiency of feedback transmission by acting at the HC–cone synapses, as suggested by studies in fish (see Wagner and Djamgoz, 1993 for review). Dopamine receptors are present at the photoreceptor level also in higher vertebrates including humans (Brann and Young, 1986; Denis *et al.*, 1990). Dopamine reduces the summation area of the HC receptive field (Piccolino *et al.*, 1984; Teranishi *et al.*, 1984; and Chapter 8); this effect is also present in mammals (Hankins and Ikeda, 1991; Hampson *et al.*, 1994). In addition, dopamine may influence the responsiveness of HCs by regulating their sensitivity to the photoreceptor transmitter (Knapp and Dowling, 1987) and controlling the balance of cone and rod input to HCs, as demonstrated in the *Xenopus* retina (Witkovsky and Shi, 1990; Krizaj and Witkovsky, 1993) (Chapter 7).

Several reports have demonstrated selective deficits in spatial and temporal visual performance after administration of benzodiazepines, a class of drugs which potentiates GABAergic transmission by acting on the $GABA_A$ receptor complex (Smith and Misiak, 1976; Parrot, 1982; MacNab *et al.*, 1985; Harris and Philipson, 1990; Blin *et al.*, 1993). However, benzodiazepines have widespread effects on the nervous system, and thus these defects may be the consequence of actions on higher visual centers (Chun and Artola, 1989; Hendry *et al.*, 1990; Müller and Singer, 1989). The possible involvement of retinal mechanisms is suggested by the changes of electroretinographic responses induced by benzodiazepines (Schültze and Appel, 1977; Jaffe *et al.*, 1986). It is also interesting to recall that, although mammals differ somewhat from lower vertebrates for the color selectivity of cone–HCs connections, the $GABA_A$ antagonist bicuculline reduces red/green opponency in monkey retina (Mills and Sperling, 1990; section 9.5.1).

Studies of Parkinson's disease in humans, neurotoxic states of both humans and animals, and the effects of dopamine agonists and antagonists on visual functions strongly suggest that the dopaminergic system is involved in selective abnormalities of the vertebrate visual system (Chapter 20). It would suffice to note here that it was the demonstration of the dopaminergic effects on the spatial interactions of HCs in lower vertebrates that prompted the research on spatial vision changes associated with dysfunctions of the dopaminergic system in humans. A selective impairment of the spatial visual performance is, in fact, a prominent aspect of the visual anomalies observed in Parkinson's disease and in related alterations of dopaminergic transmissions. Thus, experimental studies on animals, even when they explore basic physiologic mechanisms, can potentially continue to provide important information of clinical relevance including the possibility of developing treatment for pathological conditions in humans.

ACKNOWLEDGEMENTS

I thank G. Rispoli for critically reading the manuscript and P. Taccini for preparing the illustrations. This work was supported by Human Frontiers Science Program and by the Italian National Research Council and by the Ministry for the Scientific and Technological Research.

REFERENCES

Ahnelt, P. and Kolb, H. (1994) Horizontal cells and cone photoreceptors in human retina: a Golgi EM study of spectral connectivity. *Journal of Comparative Neurology*, **343**, 406–27.

Alcayaga, J., Bustamante, S. and Gutierrez, C.O. (1989) Fast activity and oscillatory potentials of carp retina in the frequency domain. *Vision Research*, **29**, 949–55.

Attwell, D., Barbour, B., and Szatkowski, M. (1993) Nonvescicular release of neurotransmitters. *Neuron*, **11**, 401–7.

Ayoub, G.S., Korenbrot, J.I., and Copenhagen, D.R. (1989) Release of endogenous glutamate from isolated cone photoreceptors of the lizard. *Neuroscience Research*, **7**, Suppl. 10, S47–S56.

Barlow, H.B. (1953) Summation and inhibition in the frog's retina. *Journal of Physiology*, **119**, 69–88.

Barlow, H.B. (1981) The Ferrier Lecture, 1980. Critical limiting factors in the design of the eye and visual cortex. *Proceedings of the Royal Society of London B*, **212**, 1–34.

Barnes, S. and Deschènes, M.C. (1992) Contribution of Ca and Ca-activated Cl channels to regenerative depolarization and membrane bistability of cone photoreceptors. *Journal of Neurophysiology*, **68**, 745–55.

Barnes, S. and Hille, B. (1989) Ionic channels of the inner segment of tiger salamander *Journal of General Physiology*, **94**, 719–43.

Baylor, D.A., Fuortes, M.G.F. and O'Bryan, P.M. (1971) Receptive field of cones in the retina of the turtles. *Journal of Physiology*, **214**, 265–94.

Blin, O., Mestre, D., Paut, O. *et al.* (1993) GABA-ergic control of visual perception in healthy volunteers – Effects of midazolam, a benzodiazepine, on spatio-temporal contrast sensitivity. *British Journal of Clinical Pharmacology*, **36**, 117–24.

Brann, M.R. and Young, W.S. (1986) Dopamine receptors are located in rods in bovine retina. *Neuroscience Letters*, **69**, 221–6.

Brecha, N.C. (1992) Expression of GABA(A) receptors in the vertebrate retina. *Progress in Brain Research* **90**, 3–28.

Bressler, S.L. (1990) The gamma wave: a cortical information carrier? *Trends in Neurosciences*, **13**, 61–162.

Buchsbaum, G. and Gottschalk, A. (1983) Trichromacy, opponent colours coding and optimum colour information transmission in the retina. *Proceedings of the Royal Society of London B*, **220**, 89–113.

Burkhardt, D.A. (1993) Synaptic feedback, depolarization, and color opponency in cone photoreceptors. *Visual Neuroscience*, **10**, 981–9.

Burkhardt, D.A. and Hassin, G. (1978) Influences of cones upon chromatic and luminosity-type horizontal cells in pike–perch retina. *Journal of Physiology*, **281**, 125–37.

Burkhardt, D.A. and Hassin, G. (1983) Quantitative relations between color-opponent response of horizontal cells and action spectra of cones. *Journal of Neurophysiology*, **49**, 961–75.

Burkhardt, D.A., Gottesman, J., and Thoreson, W.B. (1988) Prolonged depolarization in turtle cones evoked by current injection and stimulation of the receptive field surround. *Journal of Physiology*, **407**, 329–48.

Byzov, A.L. (1975) Interaction between the horizontal cells of the turtle retina. *Neurofiziologia*, **7**, 279–86 (in Russian).

Byzov, A.L. and Shura-Bura, T.M. (1986) Electrical feedback mechanism in the processing of signals in the outer plexiform layer of the retina. *Vision Research*, **26**, 33–44.

Byzov, A.L., Golubtzov, K.V. and Trifonov, Yu. (1977) The model of mechanism of feedback between horizontal cells and photoreceptors in the vertebrate retina, in *Vertebrate Photoreception* (eds H.B. Barlow and P. Fatt), Academic Press, New York, pp. 265–74.

Cervetto, L. and MacNichol, E.F.J. (1972) Inactivation of horizontal cells in turtle retina by glutamate and aspartate. *Science*, **178**, 767–8.

Cervetto, L. and Piccolino, M. (1974) Synaptic transmission between photoreceptors and horizontal cells in the turtle retina. *Science*, **183**, 417–19.

Chun, M.H. and Artola, A. (1989) GABA-like activity in *in vitro* slices of the rat visual cortex – immunocytochemistry and electrophysiology. *Brain Research*, **494**, 401–6.

Copenhagen, D.R. and Jahr, C.E. (1989) Release of endogenous excitatory amino acids from turtle photoreceptors. *Nature*, **341**, 536–9.

Dacheux, R.F. and Raviola, E. (1982) Horizontal cells in the retina of the rabbit. *Journal of Neuroscience*, **2**, 1486–93.

Dacheux, R.F. and Raviola, E. (1990) Physiology of HI horizontal cells in the primate retina. *Proceedings of the Royal Society of London B*, **239**, 213–30.

De Valois, R.L. and De Valois, K.K. (1990) *Spatial Vision*, Oxford University Press, Oxford, pp. 1–381.

Denis, P., Elena, P.-P., Nordmann, J.-P. *et al.* (1990) Autoradiographic localization of D_1 and D_2 dopamine binding sites in the human retina. *Neuroscience Letters*, **116**, 81–6.

Detwiler, P.B. and Hodgkin, A.L. (1979) Electrical coupling between cones in turtle retina. *Journal of Physiology*, **291**, 75–100.

Djamgoz, M.B.A., Downing, J.E.G., Kirsch, M. *et al.* (1988) Plasticity of cone horizontal cell functioning in cyprinid fish retina: effects of background illumination of moderate intensity. *Journal of Neurocytology*, **17**, 701–10.

Dowling, J.E. (1991) Retinal neuromodulation: the role of dopamine. *Visual Neuroscience*, **7**, 87–97.

Dowling, J.E. and Ripps J.H. (1973) Effects of magnesium on horizontal cell activity in the skate retina. *Nature*, **242**, 101–3.

Dowling, J.E. and Werblin, F. (1969) Organization of retina of the mudpuppy, *Necturus maculosus* I Synaptic structure. *Journal of Neurophysiology*, **32**, 315–38.

Downing, J.E.G. and Djamgoz, M.B.A. (1989) Quantitative analysis of cone photoreceptor–horizontal cell connectivity patterns in the retina of a cyprinid fish: electron microscopy of functionally identified and HRP-labelled horizontal cells. *Journal of Comparative Neurology*, **289**, 537–53.

Eliasof, S. and Werblin, F. (1989) GABA(A) and GABA(B) mediated synaptic transmission to cones in the tiger salamander retina. *Investigative Ophthalmology and Visual Science, suppl.*, **30**, 163.

Eliasof, S. and Werblin, F. (1993) Characterization of the glutamate transporter in retinal cones of the tiger salamander. *Journal of Neuroscience*, **13**, 402–11.

Freeman, W.J. (1987) Simulation of chaotic EEG patterns with a dynamic model of the olfactory system. *Biological Cybernetics*, **56**, 139–50.

Fuortes, M.G.F. and Simon, E.J. (1974) Interactions leading to horizontal cell responses in the turtle retina. *Journal of Physiology*, **240**, 177–80.

Fuortes, M.G.F., Schwartz, E.A. and Simon, E.J. (1973) Colour dependence of cone responses in the turtle retina. *Journal of Physiology*, **234**, 199–216.

Furukawa, T. and Furshpan, E.J. (1963) Two inhibitory mechanisms in Mauthner neurons of goldfish. *Journal of Neurophysiology*, **26**, 140–76.

Gerschenfeld, H.M. and Piccolino, M. (1980) Sustained feedback effects of L-horizontal cells on turtle cones. *Proceedings of the Royal Society of London B*, **206**, 465–80.

Golard A., Witkovsky, P. and Tranchina, D. (1992) Membrane currents of horizontal cells isolated from turtle retina. *Journal of Neurophysiology*, **68**, 351–61.

Gottesman, J. and Burkhardt, D.A. (1987) Response properties of C-type horizontal cells in the retina of the bowfin. *Vision Research*, **27**, 179–89.

Gouras, P. (1972), S-potentials, in *Handbook of Sensory Physiology*, Vol. VII/2 (ed. M.G.F. Fuortes), Springer, Berlin, pp.609–34.

Gray, C.M. and Singer, W. (1989) Stimulus-specific neuronal oscillations in orientation columns of cat visual cortex. *Proceedings of the National Academy of Sciences of the USA*, **86**, 1698–702.

Greenstreet, E.H. and Djamgoz, M.B.A. (1994) Triphasic chromaticity-type horizontal cells selectively contact short wavelength-sensitive cone photoreceptors in the retina of a cyprinid fish, *Rutilus rutilus*. *Proceedings of the Royal Society of London B*, **256**, 227–30.

Hampson, E.C.G.M., Weiler, R. and Vaney, D.I. (1994) Ph-Gated dopaminergic modulation of horizontal cell gap junctions in mammalian retina. *Proceedings of the Royal Society of London B*, **255**, 67–72.

Hankins, M.W. and Ikeda, H. (1991) The role of dopaminergic pathways at the outer plexiform layer of the mammalian retina. *Clinical Vision Sciences*, **6**, 87–93.

Hare, W.A. and Owen, W.G. (1990) Spatial organization of the bipolar cell's receptive field in the retina of the tiger salamander. *Journal of Physiology*, **421**, 223–45.

Hare, W.A. and Owen, W.G. (1992) Effects of 2-amino-4-phosphonobutyric acid on cells in the distal layers of the tiger salamander's retina. *Journal of Physiology*, **445**, 741–57.

Harris, J. and Philipson, O. (1990) The role of GABA-ergic mechanisms in human contrast sensitivity. *Perception*, **19**, 395 (A 73).

Hartline, H.K. (1949) Inhibition of activity of visual receptors by illuminating nearby retinal areas in the *Limulus* eye. *Federation Proceedings*, **8**, 69.

Hendry, S.H.C., Fuchs, J. Deblas, A.L. and Jones, E.G. (1990) Distribution and plasticity of immunocytochemically localized $GABA_A$ receptors in adult monkey visual cortex. *Journal of Neuroscience*, **10**, 2438–50.

Jaffe, M.J., Hommer, D., Caruso, R.C. and Karson, C.N. (1986) Effects of diazepam on the human ganzfeld electroretinogram. *Investigative Ophthalmology and Visual Science, suppl.*, **27**, 232.

Kamermans, M., van Dijk, B.W., Spekreijse, H. and Zweypfenning, R.C.V.J. (1989) Lateral feedback from monophasic horizontal cells to cones in carp retina. I. Experiments. *Journal of General Physiology*, **93**, 681–94.

Kamermans, M., van Dijk, B.W. and Spekreijse, H. (1991) Color opponency in cone-driven horizontal cells in carp retina. Aspecific pathways between cones and horizontal cells. *Journal of General Physiology*, **97**, 819–43.

Kaneko, A. (1970) Physiological and morphological identification of horizontal, bipolar and amacrine cells in goldfish retina. *Journal of Physiology*, **207**, 623–33.

Kaneko, A. (1973) Receptive field organization of

bipolar and amacrine cells in the goldfish retina. *Journal of Physiology*, **235**, 133–53.
Kaneko, A. and Shimazaki, H. (1975) Effects of external ions on the synaptic transmission from photoreceptors to horizontal cells in the carp retina. *Journal of Physiology*, **252**, 509–22.
Kaneko, A. and Tachibana, M. (1986) Effects of γ-aminobutyric acid on isolated cone photoreceptors of the turtle retina. *Journal of Physiology*, **373**, 443–61.
Kato, S. (1979) C-type horizontal cell responses to annular stimuli. *Experimental Eye Research*, **28**, 627–39.
Kawasaki, M., Aoki, K. and Naka, K.-I. (1984) Effects of background and spatial pattern on incremental sensitivity of catfish horizontal cells. *Vision Research*, **24**, 1197–204.
Kirschfeld, K. (1992) Oscillations in the insect brain: do they correspond to the cortical gamma-waves of vertebrates? *Proceedings of the National Academy of Sciences of the USA*, **89**, 4764–8.
Knapp, A.G. and Dowling, J.E. (1987) Dopamine enhances excitatory amino acid-gated conductances in retinal horizontal cells. *Nature*, **325**, 437–9.
Kolb, H. (1991) Anatomical pathways for color vision in the human retina. *Visual Neuroscience*, **7**, 61–74.
Kolb, H. and Jones, J. (1984) Synaptic organization of the outer plexiform layer of the turtle retina: an electron microscope study of serial sections. *Journal of Neurocytology*, **13**, 567–91.
Korn, H. and Faber, D.S. (1975) An electrically mediated inhibition in goldfish medulla. *Journal of Neurophysiology*, **38**, 452–71.
Krizaj, D. and Witkovsky, P. (1993) Effects of submicromolar concentrations of dopamine on photoreceptor to horizontal cell communication. *Brain Research*, **627**, 122.
Kuffler, S.W. (1953) Discharge patterns and functional organization of the mammalian retina. *Journal of Neurophysiology*, **16**, 37–68.
Lasansky, A. (1973) Organization of the outer synaptic layer in the retina of the larval tiger salamander. *Philosophical Transactions of the Royal Society of London B*, **265**, 471–89.
Lasansky, A. and Vallerga, S. (1975) Horizontal cell responses in the retina of the larval tiger salamander. *Journal of Physiology*, **251**, 145–65.
Lasater, E.M. (1982) A white-noise analysis of responses and receptive fields of catfish cones. *Journal of Neurophysiology*, **47**, 1057–68.
Lasater, E.M. (1991) Membrane properties of distal retinal neurons. *Progress in Retinal Research*, **11**, 215–46.
Lasater, E.M. and Witkovsky, P. (1991) The calcium current of turtle cone photoreceptor axon terminals. *Neuroscience Research, suppl.*, **15**, S165–S173.
Laughlin, S.B. (1994). Matching coding, circuits, cells, and molecules to signals: general principles of retinal design in the fly's eye. *Progress in Retinal and Eye Research*, **13**, 165–96.
Leeper, H.F. (1978a) Horizontal cells of the turtle retina: I. Light microscopy of Golgi preparations. *Journal of Comparative Neurology*, **182**, 777–94.
Leeper, H.F. (1978b) Horizontal cells of the turtle retina: II. Analysis of the interconnections between photoreceptors and horizontal cells by light microscopy. *Journal of Comparative Neurology*, **182**, 795–810.
Leeper, H.F. and Charlton, J.S. (1985) Response properties of horizontal cells and photoreceptor cells in the retina of the tree squirrel, *Sciurus carolinensis*. *Journal of Neurophysiology*, **54**, 1157–66.
Leeper, H.F. and Copenhagen, D.R. (1982) Horizontal cells in turtle retina: structure, synaptic connections, and visual processing, in *The S-Potential* (eds M. Laufer and B. Drujan), Alan Liss, New York, pp.77–104.
Lipetz, L.E. (1978) A model of function of the outer plexiform layer of the cyprinid retina, in *Frontiers in Visual Science* (eds S.J. Cool and E.L. Smith III), Springer, Berlin, pp.471–82.
MacNab, M.W., Foltz, E.L. and Sweitzer, J. (1985) Evaluation of signal detection theory on the effects of psychotropic drugs on critical flicker-fusion frequency in normal subjects. *Psychopharmacology*, **85**, 431–5.
Mangel, S.C. (1991) Analysis of the horizontal cell contribution to the receptive field surround of ganglion cells in the rabbit retina. *Journal of Physiology*, **442**, 211–34.
Marchiafava, P.L. (1978) Horizontal cells influence membrane potential of bipolar cells in the retina of the turtle. *Nature*, **275**, 141–2.
Maricq, A.V. and Korenbrot, J.I. (1988) Calcium

and calcium-dependent chloride currents generate action potentials in solitary cone photoreceptors. *Neuron*, **1**, 503–15.

Marr, D. (1982) *Vision*, Freeman, San Francisco.

Marshak, D.W. and Dowling, J.E. (1987) Synapses of cone horizontal cell axons in goldfish retina. *Journal of Comparative Neurology*, **256**, 430–43.

Mills, S.L. and Sperling, H.G. (1990) Red/green opponency in the rhesus macaque ERG spectral sensitivity is reduced by bicuculline. *Visual Neuroscience*, **5**, 217–21.

Miyachi, E.-I. and Murakami, M. (1989) Decoupling of horizontal cells in carp and turtle retinae by intracellular injection of cyclic AMP. *Journal of Physiology*, **419**, 213–24.

Miyachi, E.-I. and Murakami, M. (1991) Synaptic inputs to turtle horizontal cells analyzed after blocking of gap junctions by intracellular injection of cyclic nucleotides. *Vision Research*, **31**, 631–5.

Muiller, C.M. and Singer, W. (1989) Acetylcholine-induced inhibition in the cat visual cortex is mediated by a GABAergic mechanism. *Brain Research*, **487**, 335–42.

Murakami, M., Ohtsu, K. and Ohtsuka, T. (1972) Effects of chemicals on receptors and horizontal cells in the retina. *Journal of Physiology*, **227**, 899–913.

Murakami, M., Shimoda, Y., Nakatani, K. *et al.* (1982a) GABA-mediated negative feedback from horizontal cells to cones in carp retina. *Japanese Journal of Physiology*, **32**, 911–26.

Murakami, M., Shimoda, Y., Nakatani, K. *et al.* (1982b) GABA-mediated negative feedback and colour opponency in carp retina. *Japanese Journal of Physiology*, **32**, 927–35.

Naka, K.-I. and Rushton, W.A.H. (1966a) An attempt to analyse colour reception by electrophysiology. *Journal of Physiology*, **185**, 556–86.

Naka, K.-I. and Rushton, W.A.H. (1966b) S-potentials from colour units in the retina of fish (*Cyprinidae*). *Journal of Physiology*, **185**, 536–55.

Negishi, K. and Drujan, B.D. (1979) Similarities in effects of acetylcholine and dopamine on horizontal cells in the fish retina. *Journal of Neuroscience Research*, **4**, 335–49.

Nelson, R. (1977) Cat cones have rod input: a comparison of response properties of cones and horizontal cell bodies in the retina of the cat. *Journal of Comparative Neurology*, **172**, 109–36.

Neyton, J., Piccolino, M. and Gerschenfeld, H.M. (1981) Involvement of small-field horizontal cells in feed-back effects on green cones of turtle retina. *Proceedings of the National Academy of Sciences of the USA*, **78**, 4616–19.

O'Bryan, P. M. (1973) Properties of the depolarizing synaptic potential evoked by peripheral illumination in cones of the turtle retina. *Journal of Physiology*, **235**, 207–23.

Ogden, T.E. (1974) The oscillatory waves of primate electroretinogram. *Vision Research*, **13**, 1059–74.

Ohtsuka, T. and Kouyama, N. (1986) Electron microscopic study of synaptic contacts between photoreceptors and HRP-filled horizontal cells in the turtle retina. *Journal of Comparative Neurology*, **250**, 141–56.

Parrot, A.C. (1982) The effects of clozabam upon critical flicker fusion threshold: a review. *Drug Development Research*, **suppl. 1**, 57–66.

Pasino, E. and Marchiafava, P.L. (1976) Transfer properties of rod and cone cells in the retina of the tiger salamander. *Vision Research*, **16**, 381–6.

Piccolino, M. (1988) Cajal and the retina: a 100-year retrospective. *Trends in Neurosciences*, **11**, 521–5.

Piccolino, M. (1995) The feedback synapse from horizontal cells to cone photoreceptors in the vertebrate retina. *Progress in Retinal and Eye Research*, **14**, 141–196.

Piccolino, M. and Belluzzi, O. (1990) Oscillatory potentials induced by steady peripheral illumination in horizontal cell axon terminals HCATs of the isolated Pseudemys turtle retina. *Journal of Physiology*, **427**, 52P.

Piccolino, M. and Gerschenfeld, H.M. (1980) Characteristics and ionic processes involved in feedback spikes of turtle cones. *Proceedings of the Royal Society of London B*, **201**, 309–15.

Piccolino, M. and Neyton, J. (1982) The feedback effect from luminosity horizontal cells to cones in the turtle retina, in *The S-potential*; (eds M. Laufer and B. Drujan), Alan Liss, New York, pp.161–79.

Piccolino, M., Neyton, J. and Gerschenfeld, H.M. (1980) Synaptic mechanisms involved in responses of chromaticity horizontal cells of turtle retina. *Nature*, **284**, 58–60.

Piccolino, M., Neyton, J. and Gerschenfeld, H.M. (1981) Center surround antagonistic organization in the small field L-horizontal cells of turtle retina. *Journal of Neurophysiology*, **45**, 361–73.

Piccolino, M., Neyton, J., Witkovsky, P. and Gerschenfeld, H.M. (1982) GABA antagonists decrease junctional communication between L-horizontal cells of the retina. *Proceedings of the National Academy of Sciences of the USA*, **79**, 3671–5.

Piccolino, M., Neyton, J. and Gerschenfeld, H.M. (1984) Decrease of gap junction permeability induced by dopamine and cyclic adenosine 3′:5′-monophosphate in horizontal cells of turtle retina. *Journal of Neuroscience*, **4**, 2477–88.

Ramón Y Cajal, S. (1893) La Rétine des Vertebrés. *La Cellule*, **9**, 119–257.

Ramón Y Cajal, S. (1933) Los problemas histophysiologicos de la retina. *Proceedings of the XIV Concilium Ophthalmologicum Hispaniae*, pp.1–19.

Ratliff, F. (1965) *Mach Bands: Quantitative Studies on Neural Network in the Retina*, Holden-Day, San Francisco.

Richter, A. and Simon, E.J. (1975) Properties of centre-hyperpolarizing, red-sensitive bipolar cells in the turtle retina. *Journal of Physiology*, **248**, 317–34.

Saito, T., Kondo, H. and Toyoda J.-I. (1981) Ionic mechanisms of two types of on-center bipolar cells in the carp retina. II. The responses to annular illumination. *Journal of General Physiology*, **78**, 569–89.

Sakai, H.M. and Naka, K.-I. (1985) Novel pathway connecting outer and inner vertebrate retina. *Nature*, **315**, 570–1.

Sakai, H.M. and Naka, K.-I. (1986) Synaptic organization of the cone horizontal cells in the catfish retina. *Journal of Comparative Neurology*, **245**, 107–15.

Sarantis, M., Everett, K. and Attwell, D. (1988) A presynaptic action of glutamate and the cone output synapse. *Nature*, **332**, 451–3.

Schnapf, J.L. and Copenhagen, D.R. (1982) Difference in kinetics of rod and cone synaptic transmission. *Nature*, **296**, 862–4.

Schültze, J. and Appel, E. (1977) Drug effects on b-wave amplitude and on readaptation after glare. *Documenta Ophthalmologica. Proceedings Series*, **15**, 39–44.

Schütte, M. and Schlemermeyer, E. (1993) Depolarization elicits, while hyperpolarization blocks uptake of endogenous glutamate by retinal horizontal cells of the turtle. *Cell Tissue Research*, **274**, 553–8.

Schwartz, E.A. (1974) Responses of bipolar cells in the retina of the turtle. *Journal of Physiology*, **236**, 211–24.

Schwartz, E.A. (1982) Calcium-independent release of GABA from isolated horizontal cells of the toad retina. *Journal of Physiology*, **323**, 211–27.

Schwartz, E.A. (1987) Depolarization without calcium can release γ-aminobutyric acid from a retinal neuron. *Science*, **238**, 350–5.

Shaw, S.R. (1984) Early visual processing in insects. *Journal of Experimental Biology*, **112**, 225–51.

Simon, E.J. (1973) Two types of luminosity horizontal cells in the retina of the turtle. *Journal of Physiology*, **230**, 199–211.

Skrzypek, J. and Werblin, F. (1983) Lateral interactions in absence of feedback to cones. *Journal of Neurophysiology*, **49**, 1007–16.

Smith, J.M. and Misiak, H. (1976) Critical flicker frequency (CFF) and psychotrophic drugs in normal subjects – a review. *Psychopharmacology*, **47**, 175–82.

Smith, R.G. and Sterling, P. (1990) Cone receptive field in cat retina computed from microcircuitry. *Visual Neurosciences*, **5**, 453–61.

Spekreijse, H. and Norton, A.L. (1970) The dynamic characteristics of color-coded S-potentials. *Journal of General Physiology*, **56**, 1–15.

Sperling, H.G. and Mills, S.L. (1991) Red–green interactions in the spectral sensitivity of primates as derived from ERG and behavioral data. *Visual Neuroscience*, **7**, 75–86.

Srinivasan, M., Laughlin, S.B. and Dubs, A. (1982) Predictive coding: a fresh view of inhibition in the retina. *Proceedings of the Royal Society of London B*, **216**, 427–59.

Stell, W.K. and Lightfoot, D.O. (1975) Color-specific interconnection of cones and horizontal cells in the retina of the goldfish. *Journal of Comparative Neurology*, **159**, 473–502.

Stell, W.K., Lightfoot, D.O., Wheeler, T.G. and Leeper, H.F. (1975) Goldfish retina: functional polarization of cone horizontal cell dendrites and synapses. *Science*, **190**, 989–90.

Svaetichin, G. (1953) The cone action potential. *Acta Physiologica Scandinavica*, **29** (Suppl. 106), 565–99.

Svaetichin, G. (1956) Spectral responses from single cones. *Acta Physiologica Scandinavica*, **39** (Suppl. 134), 17–46.

Tachibana, M. and Kaneko, A. (1984) γ-Aminobutyric acid acts at axon terminals of turtle photoreceptors: difference in sensitivity among cell types. *Proceedings of the National Academy of Sciences of the USA*, **81**, 7961–4.

Tachibana, M. and Kaneko, A. (1988) L-glutamate-induced depolarization in solitary photoreceptors: a process that may contribute to the interaction between photoreceptors *in situ*. *Proceedings of the National Academy of Sciences of the USA*, **85**, 5315–19.

Teranishi, T., Negishi, K. and Kato, S. (1984) Regulatory effect of dopamine on spatial properties of horizontal cells in carp retina. *Journal of Neuroscience*, **4**, 1271–80.

Thoreson, W.B. and Burkhardt, D.A. (1990) Effects of synaptic blocking agents on the depolarizing responses of turtle cones evoked by surround illumination. *Visual Neuroscience*, **5**, 571–83.

Thoreson, W.B. and Burkhardt, D.A. (1991) Ionic influences on the prolonged depolarization of turtle cone in situ. *Journal of Neurophysiology*, **65**, 96–110.

Toyoda, J.-I. and Tonosaki, K. (1978) Effect of polarization of horizontal cells on the on-centre bipolar cell of carp retina. *Nature*, **276**, 399–400.

Trifonov, Yu. (1968) Study of synaptic transmission between photoreceptors and horizontal cells by means of electrical stimulation of the retina. *Biophizica*, **13**, 809–17 (in Russian).

Vardi, N., Masarachia, P. and Sterling, P. (1992) Immunoreactivity to $GABA_A$ receptor in the outer plexiform layer of the cat retina. *Journal of Comparative Neurology*, **320**, 394–7.

Wagner, H.-J. (1980) Light dependent plasticity of the morphology of horizontal cell terminals in cone pedicles of fish retinas. *Journal of Neurocytology*, **9**, 575–90.

Wagner, H.J. and Djamgoz, M.B.A. (1993) Spinules – a case for retinal synaptic plasticity. *Trends in Neurosciences*, **16**, 201–6.

Wässle, H, Boycott, B.B. and Röhrenbeck, J. (1989) Horizontal cells in the monkey retina: cone connections and dendritic network. *European Journal of Neuroscience*, **1**, 421–35.

Weiler, R. and Wagner, H.J. (1984) Light-dependent change of cone–horizontal cell interactions in carp retina. *Brain Research*, **298**, 1–9.

Werblin, F.S. (1974) Control of retinal sensitivity. II: Lateral interactions at the outer plexiform layer. *Journal of General Physiology*, **40**, 23–62.

Werblin, F. and Dowling, J.E. (1969) Organization of the retina of the mudpuppy, *Necturus maculosus*: II Intracellular recording. *Journal of Neurophysiology*, **32**, 339–55.

Witkovsky, P. and Dearry, A. (1991) Functional roles of dopamine in the vertebrate retina. *Progress in Retinal Research*, **11**, 247–92.

Witkovsky, P. and Powell, C.C. (1981) Synapse formation and modification between distal retinal neurons in larval and juvenile *Xenopus*. *Proceedings of the Royal Society of London B*, **211**, 373–89.

Witkovsky, P. and Shi, X.-P. (1990) Slow light and dark adaptation of horizontal cells in the *Xenopus* retina: a role for endogenous dopamine. *Visual Neuroscience*, **5**, 405–13.

Witkovsky, P., Owen, D.G. and Woodworth, M. (1983) Gap junctions among perikarya, dendrites, and axon terminals of the luminosity-type horizontal cell of the turtle retina. *Journal of Comparative Neurology*, **216**, 359–68.

Wu, S.M. (1991) Input–output relations of the feedback synapse between horizontal cells and cones in the tiger salamander retina. *Journal of Neurophysiology*, **65**, 1197–206.

Wu, S.M. (1992) Feedback connections and operations of the outer plexiform layer of the retina. *Current Opinion in Neurobiology*, **2**, 462–78.

Yamada, M., Shigematsu Y. and Fuwa, M. (1985) Latency of horizontal cell response in the carp retina. *Vision Research*, **25**, 767–74.

Yamada, M., Djamgoz, M.B.A., Low, J.C. *et al.* (1991) Conductance-decreasing cone input to H1 horizontal cells in carp retina. *Neuroscience Research Supplement* **15**, S51–S65.

Yang, C.-Y., Lin, Z.-S. and Yazulla, S. (1992) Localization of $GABA_A$ receptor subtypes in the tiger salamander retina. *Visual Neuroscience*, **8**, 57–64.

Yang, X.-L. and Wu, S.M. (1991) Feedforward lateral inhibition in retinal bipolar cells: input–output relation of the horizontal cell-depolarizing bipolar cell synapse. *Proceedings of*

the National Academy of Sciences of the USA **88**, 3310–13.
Yazulla, S. (1976) Cone input to horizontal cells in the turtle retina. *Vision Research*, **16**, 727–35.
Yazulla, S. (1986) GABAergic mechanisms in the retina. *Progress in Retinal Research*, **5**, 1–52.
Yazulla, S. and Kleinschmidt, J. (1983) Carrier-mediated release of GABA from retinal horizontal cells. *Brain Research*, **263**, 63–75.
Yazulla, S., Studholme, K.M., Vitorica, J. and de Blas, A.L. (1989) Immunocytochemical localization of $GABA_A$ receptors in goldfish and chicken retinas. *Journal of Comparative Neurology*, **280**, 15–26.

10

Neurotransmitter release from horizontal cells

STEPHEN YAZULLA

10.1 INTRODUCTION

Perhaps the most important function of a neuron is the integration of input stimuli as expressed in the modulation of its neurotransmitter release. The measurement of neurotransmitter release from specific neurons has been the most difficult aspect of identifying a given chemical as 'neurotransmitter' of that neuron. It is here that retinal horizontal cells have several advantages that make them attractive for use in studying neurotransmitter release. Horizontal cells are neurons in the central nervous system (CNS), diencephalon to be exact, that are located within the neural retina. They are easily identified interneurons that are part of a well-characterized neural circuit; they respond with graded potentials and can be studied *in situ* or in isolation (Dowling, 1987) (Chapter 7). Studies on horizontal cells from the retinas of non-mammalian vertebrates have greatly increased our understanding of the mechanisms controlling the release of amino acid neurotransmitters, not only in the retina but throughout the vertebrate CNS. This chapter reviews transmitter release from horizontal cells and describes an effort that confronted the accepted view of Ca^{2+}-dependent vesicular release and emerged to establish Na^{+}-dependent transport carriers as an alternative neurotransmitter release mechanism.

In general, horizontal cells receive synaptic input from photoreceptors and transmit lateral inhibitory signals within the outer plexiform layer (OPL) to other photoreceptors and/or bipolar cells. Proposed functions of this lateral inhibition include temporal tuning of the photoreceptor response (Marmarelis and Naka, 1973), formation of the bipolar cell receptive field surround (Werblin and Dowling, 1969) and generation of color-coding particularly in non-mammalian vertebrates (Stell *et al.*, 1975). As with other retinal neurons, horizontal cells consist of multiple subtypes that differ in their connectivity and presumptively in their function. Horizontal cells typically comprise only two to four subtypes for any given species, and may be either axon-bearing or axonless (Gallego, 1986; for review). The synaptic organization of horizontal cells is not the same across species which complicates efforts to discern a unifying theme regarding their function and mechanism of transmitter release (Chapter 7).

Synaptic inputs to horizontal cells are found on their dendrites from photoreceptors

Neurobiology and Clinical Aspects of the Outer Retina
Edited by M.B.A. Djamgoz, S.N. Archer and S. Vallerga
Published in 1995 by Chapman & Hall, London
ISBN 0 412 60080 3

and, depending on the species, on the soma from interplexiform cells (Gallego, 1986; and Chapter 7). However, description of the output of horizontal cells has been more problematic. There is general agreement from electrophysiological evidence that horizontal cells affect the responses of cone photoreceptors and bipolar cells by feedback to cones, or by direct input to bipolar cells (Baylor *et al.*, 1971; Gerschenfeld *et al.*, 1980; Wu, 1992; Burkhardt, 1993; and Chapter 9). Unlike other areas of the CNS, the synaptic ultrastructure within the OPL is atypical. Although there are numerous reports of conventional vesicular synapses made by horizontal cell dendrites onto bipolar cell dendrites and photoreceptor terminals within the neuropil of the OPL ((e.g., cat and rabbit, Dowling *et al.* (1966); mudpuppy (Dowling and Werblin, 1969); turtle (Kolb and Jones, 1984); catfish (Sakai and Naka, 1983, 1986) and human (Lindberg and Fisher, 1988)), except perhaps for catfish and mudpuppy, conventional synapses in horizontal cells are not very common and probably could not account for synaptic feedback. Previous reports of conventional synapses in the OPL of *Rana* and *Xenopus* frogs (Dowling, 1968; Witkovsky and Powell, 1981) involved processes of glycine interplexiform cells rather than horizontal cells as the presynaptic element (Kleinschmidt and Yazulla, 1984; Smiley and Yazulla, 1990). Other morphological specializations in horizontal cell dendrites within the photoreceptor invagination lack the components expected for Ca^{2+}-dependent vesicular transmitter release (Stell, 1967; Witkovsky and Dowling, 1969; Lasansky, 1971, 1980; Chen and Witkovsky, 1978; Wagner, 1980; Wagner and Djamgoz, 1993). Although all agree that horizontal cells release neurotransmitter, the precise locus for the bulk of that release is as yet unresolved.

Despite the presence of multiple subtypes of horizontal cell, only one neurotransmitter used by these cells has been identified, γ-amimoburyric acid (GABA) (Yazulla, 1986). Much histochemical data demonstrating high affinity ^{3}H-labeled GABA (^{3}H-GABA) uptake, immunoreactivities for GABA and glutamic acid decarboxylase (GAD; the biosynthetic enzyme for GABA), indicate that at least one class of GABAergic horizontal cell is ubiquitous among non-mammalian vertebrates (see Yazulla, 1986, for review) and is also found among some mammals (e.g., cat; Pourcho and Owczarzak, 1989; Wässle and Chun, 1989) but not others (e.g., rat) or only in early development (e.g., mouse; Schnitzer and Russof, 1984). GABA/GAD-immunoreactive horizontal cells in adult rabbit (Mosinger and Yazulla, 1987) and primate (Nishimura *et al.*, 1985; Agardh *et al.*, 1987) appear to be regionally localized (Grünert and Wässle, 1990; M.T.R. Perez, personal communication; Rowe-Rendleman and D. Redburn, personal communication). Electrophysiological studies showing direct effects of GABA on cone photoreceptors in fish, salamander and turtle (e.g., Lam *et al.*, 1978; Murakami *et al.*, 1982a; Lasater and Lam, 1984; Tachibana and Kaneko, 1984; Wu, 1986), indicate that cones are postsynaptic to GABAergic horizontal cells. Although glutamate-immunoreactivity and nucleoside uptake (adenosine and inosine) have been localized to horizontal cells (Ehinger and Perez, 1984; Marc *et al.*, 1990; Blazynski and Perez, 1992), there is insufficient evidence presently that these substances serve as horizontal cell transmitters. As a result, discussion in this chapter will be restricted to GABA. Furthermore, with the exception of neonatal rabbit retina (Moran *et al.*, 1986; Messersmith and Redburn, 1993) all other studies regarding GABA release from horizontal cells have been performed on retinas of non-mammals.

Perhaps the most difficult criterion to satisfy in establishing the identity of a transmitter is to demonstrate the release of that transmitter from the neuron in question in response to the appropriate stimulus, which in the case of the visual system is light. For horizontal cells, in particular, the appropriate stimulus is decrements in light intensity, a condition that increases release of the depolarizing photoreceptor transmitter. The most widely used depolarizing stimuli are elevated [K^+] and acidic amino acids (L-glutamate, L-aspartate) and their agonists, as these latter are the most likely candidates for the photoreceptor transmitter (Massey, 1990) (Chapter 6).

10.2 EXPERIMENTAL PROCEDURES

10.2.1 ^{3}H-GABA VERSUS ENDOGENOUS GABA

Most studies have measured the release of newly accumulated ^{3}H-GABA, presumably following high-affinity uptake of ^{3}H-GABA into neurons. Fewer studies have measured the release of endogenous GABA and related these data to those obtained with ^{3}H-GABA (Ayoub and Lam, 1984; Campochiaro *et al.*, 1984, 1985; Yazulla *et al.*, 1985; Cunningham *et al.*, 1988). Overall, the pharmacology of evoked ^{3}H-GABA release is very similar to evoked endogenous GABA release. However, two differences have been reported. First, when expressed as a percentage of the amount of the compound in the tissue, up to 10 times more ^{3}H-GABA is released than endogenous GABA under comparable stimulating conditions (Ayoub and Lam, 1984; Yazulla *et al.*, 1985; Cunningham *et al.*, 1988). It is clear that there is a large pool of endogenous GABA that is not available for evoked release. In addition to cytoplasmic pools of GABA, postembedding immunocytochemistry has shown intense GABA-immunoreactivity in the nuclei of GABA-stained neurons (e.g., Yazulla and Yang, 1988; Wässle and Chun, 1989; Grünert and Wässle, 1990; Sherry and Yazulla, 1993), perhaps accounting for a portion of the non-releasable pool of GABA. Second, Campochiaro *et al.* (1984) compared ^{3}H-GABA and GABA release from chick retina following excitotoxic lesioning and found a corresponding decrease in GAD activity and GABA release, whereas ^{3}H-GABA release was relatively unaffected. They concluded that ^{3}H-GABA may be an invalid probe to study transmitter GABA release. These data raise concern for the similarly perturbed preparation used by Schwartz (1982) in *Bufo* toad. However, in the latter study autoradiography was used to localize ^{3}H-GABA uptake to horizontal cells, thereby demonstrating viability in the outer retina following excitotoxic lesioning. In addition, pharmacological data derived from isolated horizontal cells or intact fish and *Xenopus* retinas have yielded similar conclusions regardless of whether ^{3}H-GABA or GABA release were measured (Ayoub and Lam, 1984; Yazulla *et al.*, 1985; Cunningham *et al.*, 1988). Since the two probes have yielded very similar but not identical results in the various preparations studied, caution should be used when generalizing conclusions drawn from these data.

10.2.2 TISSUE PREPARATIONS

Three types of preparation have been used to study GABA release from the retina: synaptosomes, isolated cells and whole retinas. A problem with all such release studies has been to localize GABA release to horizontal cells. For synaptosomal preparations, differential centrifugation is used to separate small synaptosomes of the inner plexiform layer (P_2 fraction) from photoreceptor terminals of the outer plexiform layer (P_1 fraction) (Neal and Atterwill, 1974; Redburn, 1977). A problem with this procedure is that horizontal cell synapses may not fall discretely into the P_1/P_2 categories. Although horizontal cells contain small numbers of conventional vesicular-type synapses, the presynaptic structure may be located in a dendrite, soma or an expansive axon terminal (Marc and Liu, 1984; Sakai and Naka, 1985; Marshak and Dowling, 1987). Furthermore, many horizontal cells (e.g., in teleost fish) do not appear to have specialized junctions at presumed release sites in the OPL. This uncertainty of the location of presynaptic release sites in horizontal cells would compromise interpretation of data obtained with differential centrifugation.

Enzymatic dissociation of the retina produces isolated neurones of all varieties. Enriched fractions of various cell types, including horizontal cells, were produced using velocity sedimentation gradients (Lam, 1972), thus enabling the study of GABA release from these identified horizontal cells (Lam and Ayoub, 1983). Tissue culture of embryonic chick retina has been used to study ^{3}H-GABA release; however, identification of the embryonic cell types in the culture is less certain (de Mello *et al.*, 1988; do Nascimento and de Mello, 1985).

The use of whole retinas to study ^{3}H-GABA release is complicated by the fact that, in non-mammalian vertebrates, GABA is taken up by and contained in several subtypes of amacrine cell as well as a sub-type of horizontal cell (Yazulla, 1986, 1991). To circumvent the problem of multiple potential sources of GABA release, autoradiography was used to compare the distribution of ^{3}H-GABA in the tissue before and after experimentation. Reduction of grain density over any population of neurons indicated release of ^{3}H-GABA from that cell type. Localization of ^{3}H-GABA release to horizontal cells was accomplished by either arranging the incubation conditions to favor uptake into horizontal cells (i.e., Schwartz, 1982; Cunningham and Neal, 1985), or by the relative specificity of evoked release from horizontal cells compared to amacrine cells (i.e., Ayoub and Lam, 1984; Cunningham and Neal, 1985; Moran *et al.*, 1986; Yazulla and Kleinschmidt, 1982, 1983).

10.3 RELEASE OF GABA FROM HORIZONTAL CELLS

10.3.1 HISTORICAL PERSPECTIVE

Early studies on GABA release from horizontal cells were directed more toward understanding the properties of the photoreceptor neurotransmitter rather than the properties of GABA release. At that time, ^{3}H-GABA uptake autoradiography showed very dense label over a subtype of horizontal cell in the retinas of non-mammalian vertebrates (e.g., Lam and Steinman, 1971; Marshall and Voaden, 1974), whereas GAD-immunoreactivity (IR) was localized to horizontal cells in fish and frog (Lam *et al.*, 1979; Brandon *et al.*, 1980). In electrophysiological experiments, negative feedback from horizontal cells to cones was demonstrated in turtle and fish (Baylor *et al.*, 1971; Marmarelis and Naka, 1973; Burkhardt, 1977) and was shown to be due to GABA in fish (Lam *et al.*, 1978; Djamgoz and Ruddock, 1979; Murakami *et al.*, 1982a,b; Wu and Dowling, 1980). Although there was compelling evidence for GABAergic horizontal cells, the identity of the photoreceptor neurotransmitter that provided the input to these horizontal cells could be either L-glutamate or L-aspartate. The ability of glutamate or aspartate to isolate the photoreceptor response from second-order neurons (Sillman et al., 1969; Cervetto and MacNicol, 1972) was thought to be due to their mimicking the photoreceptor neurotransmitter. Exogenously applied L-glutamate and L-aspartate not only depolarized horizontal cells and OFF-center bipolar cells, but, surprisingly, they hyperpolarized ON-center bipolar cells (Murakami *et al.*, 1972; 1975), exactly the opposing physiological effects expected for second-order neurons in response to the photoreceptor transmitter.

Considering that a major effect of an input to a neuron is to modulate the transmitter release of the target neuron, our strategy in

1980 (see also Miller and Schwartz, 1983) was to use ^{3}H-GABA release from goldfish horizontal cells as an assay to study the properties of the cone photoreceptor transmitters (i.e., glutamate versus aspartate). The mechanism underlying GABA release from horizontal cells was not an issue during that time. Calcium-dependent exocytosis was an entrenched concept, despite reports that conventional synaptic specializations were not present in goldfish horizontal cell dendrites (Stell, 1967; Witkovsky and Dowling, 1969) and indications that the high-affinity transport carrier for GABA in synaptosomes of the CNS could operate bidirectionally, resulting in GABA efflux (Martin, 1976). Experiments performed in low Ca^{2+}, high Mg^{2+} were the expected controls to eliminate indirect excitatory effects of the acidic amino acids and high-K^+. It was, therefore, surprising to find that ^{3}H-GABA release from horizontal cells in *Bufo* toad and goldfish was Ca^{2+} independent when evoked by acidic amino acids and only partially Ca^{2+}-dependent when evoked by K^+ (Schwartz 1982; Yazulla and Kleinschmidt, 1982, 1983; Lam and Ayoub, 1983).

The relative amount of Ca^{2+}-dependent release of GABA from horizontal cells depends on two broad factors: the depolarizing agents used, and the species. The next three sections contain an overview of data obtained in birds, amphibians and fish, followed by general topics regarding GABA release that transcend species.

10.3.2 BIRDS

^{3}H-GABA release was studied in retinal synaptosomes and isolated retinal preparations from adult chickens (López-Colomé *et al.*, 1978; Tapia and Arias, 1982; Morán and Pasantes-Morales, 1983; Campochiaro *et al.*, 1984; Morán *et al.*, 1986) and in cultured retinal neurons of chick embryos (do Nascimento and de Mello, 1985; de Mello *et al.*, 1988, 1993). Of these, only Morán *et al.* (1986) localized ^{3}H-GABA release to horizontal cells; but the Ca^{2+} dependence of this release was not determined. L-glutamate, L-aspartate and kainate, as determined by autoradiography, depleted ^{3}H-GABA labeling over chick horizontal cells compared to neurons in the inner retina (Morán *et al.*, 1986). Considering that such evoked release of ^{3}H-GABA from intact retina and the P_1 fraction was independent of extracellular Ca^{2+}- and Na^+-dependent (Tapia and Arias, 1982; Morán and Pasantes-Morales, 1983; Campochiaro *et al.*, 1984), it can be inferred that ^{3}H-GABA release from chick horizontal cells is Ca^{2+} independent when evoked by acidic amino acids. Campochiaro *et al.* (1985) reported similar properties of endogenous GABA release from intact chick retina. With regard to high-K^+, there was an overall reduction of ^{3}H-GABA labeling over horizontal cells as well as amacrine cells (Morán *et al.*, 1986). Thus, we cannot interpret the varied data on the Ca^{2+} dependence of K^+-evoked ^{3}H-GABA from the chick retina since the source of that release is not known (López-Colomé *et al.*, 1978; Tapia and Arias, 1982; Morán and Pasantes-Morales, 1983; Campochiaro *et al.*, 1984).

10.3.3 AMPHIBIANS

Localization of evoked ^{3}H-GABA release to horizontal cells has been reported in *Xenopus* and *Rana* frogs and *Bufo* toads (Hollyfield *et al.*, 1979; Schwartz, 1982; Cunningham and Neal, 1985; Cunningham *et al.*, 1988). There are considerable species differences among the frogs and toad in the Ca^{2+} dependence of the evoked ^{3}H-GABA release from horizontal cells. In *Xenopus*, ^{3}H-GABA release is Ca^{2+} dependent when evoked by elevated K^+ (Hollyfield *et al.*, 1979; Cunningham *et al.*, 1988) but Ca^{2+} independent when evoked by L-glutamate and kainate (Cunningham *et al.*, 1988). In *Rana*, ^{3}H-GABA release is Ca^{2+} dependent regardless of how it is evoked

(Cunningham *et al.*, 1985), whereas in *Bufo*, ^{3}H-GABA release is Ca^{2+} independent regardless of how it is evoked (Schwartz, 1982). Despite these differences there is some common ground regarding voltage and sodium dependence of the ^{3}H-GABA release. For example, K^+-evoked ^{3}H-GABA release is not affected by Na^+ substitution with choline$^+$, Tris$^+$ or Li^+ (Schwartz, 1982; Cunningham *et al.*, 1985) suggestive of voltage-dependent release. However, 'glutamate-evoked' ^{3}H-GABA release is reduced by choline$^+$ or Tris$^+$ substitution, indicating Na^+ dependence (Schwartz, 1982; Cunningham *et al.*, 1985); but not by Li^+ substitution, indicating an interaction with voltage dependence as well (Schwartz, 1982). This issue of Na^+ versus voltage dependence of ^{3}H-GABA and endogenous GABA release was investigated in *Xenopus* retina by Cunningham *et al.* (1988), who found that incubation in various pharmacological agents that caused the same membrane potential in horizontal cells (determined by intracellular recording) evoked very different levels of GABA release from the horizontal cells. Those conditions that increased intracellular Na^+ as well as depolarized horizontal cells (i.e., L-glutamate), evoked more GABA release than conditions of elevated K^+ that only passively depolarized the cell (Cunningham *et al.*, 1988).

10.3.4 FISH

The properties of ^{3}H-GABA release from H1 horizontal cells of teleost fish have received the most extensive investigation and, as such have been the best characterized. Most studies have been carried out in goldfish retina, including intact retina (Marc *et al.*, 1978; Yazulla and Kleinschmidt, 1982, 1983) Yazulla, 1983, 1985a; Ayoub and Lam, 1984; O'Brien and Dowling, 1985; Cha *et al.*, 1986;) and isolated horizontal cells (Lam and Ayoub, 1983; Ayoub and Lam, 1985). Comparisons of endogenous GABA release with newly accumulated ^{3}H-GABA release have been performed in intact retina (Yazulla *et al.*, 1985) and isolated horizontal cells (Ayoub and Lam, 1985). Studies on the catfish retina (Lasater and Lam, 1984; Schwartz, 1987) and the retina of a marine teleost *Eugerres plumieri* (Jaffe *et al.*, 1984) also have been reported. These studies have produced a rather coherent picture of transmitter GABA release from teleost horizontal cells, which is very similar to that suggested for toad horizontal cells (Schwartz, 1982).

Teleost H1 horizontal cells (Figure 10.1) differ from axon-bearing horizontal cells of other vertebrate classes in that the axon terminal does not contact photoreceptors, but rather descends into the inner nuclear layer where relatively few conventional synaptic contacts are made with cell bodies and/or processes of amacrine, bipolar, or glycine I2 interplexiform cells (Sakai and Naka, 1983; Marc and Lui, 1984; Marshak and Dowling, 1987). Considering that the presumed transmitter release sites at H1 horizontal cell dendrites do not contain synaptic vesicles (Stell, 1967; Witkovsky and Dowling, 1969), the anatomical evidence indicates that two mechanisms operate to mediate transmitter release: vesicular and non-vesicular. The pharmacological studies support this notion.

There is massive cytochemical evidence supporting the GABAergic nature of H1 horizontal cells. Both the soma and axon terminal accumulate ^{3}H-GABA with high affinity (Agardh and Ehinger, 1982; Marc *et al.*, 1978; Yazulla and Kleinschmidt, 1982, 1983), and are immunoreactive for GAD (Lam *et al.*, 1979; Zucker *et al.*, 1984; Brandon, 1985), GABA-T (Yazulla, 1986; Prince *et al.*, 1987) and GABA antisera (Mosinger *et al.*, 1986). Autoradiographic experiments showed that ^{3}H-GABA was released selectively from the H1 horizontal cell bodies by 1–5 mM L-aspartate or L-glutamate (Yazulla and Kleinschmidt, 1982, 1983; Figure 10.2); the ^{3}H-GABA content of the H1 axon terminals

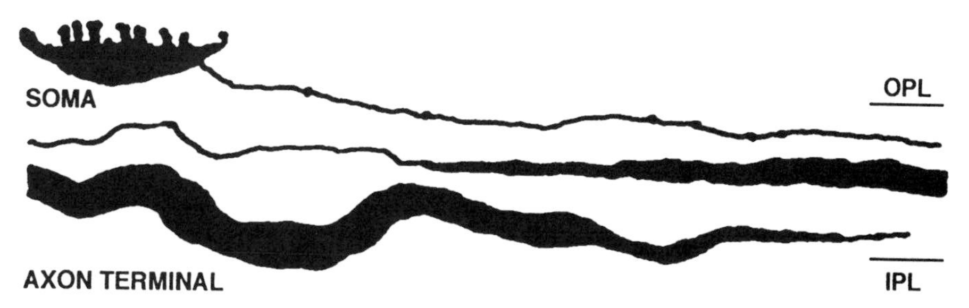

Figure 10.1 Scale drawing of a goldfish H1 GABAergic horizontal cell, adapted from Stell (1975), illustrating the relative sizes, positions within the inner nuclear layer and distance separating the soma and axon terminal. OPL, outer plexiform layer; IPL, inner plexiform layer. Calibration bar = 25 μm.

appeared unchanged. Although the electrical responses of H1 cell axon terminals are similar to those recorded from the cell body (Weiler and Zettler, 1979), the axon terminals do not release ^{3}H-GABA in quantities that are detected by autoradiography in response to cone transmitter candidates. It was also found that diffusion of ^{3}H-GABA from the axon terminal back to the cell body was very slow (Yazulla and Kleinschmidt, 1983), indicating that GABA taken up by the axon terminal is not available to the cell body for subsequent synaptic release.

^{3}H-GABA is released from H1 horizontal cells in a dose-dependent manner by K^+, L-glutamate, L-aspartate (Figure 10.3), quisqualic acid, kainic acid, L-cysteate, *N*-methyl aspartate and veratridine. With the exception of K^+ and veratridine in *Eugerres* retina (Jaffe *et al.*, 1984), it is generally agreed that ^{3}H-GABA release evoked by elevated K^+ is only partially Ca^{2+}-dependent whereas ^{3}H-GABA release evoked by the acidic amino acids and analogs is independent of extracellular Ca^{2+} (Lam and Ayoub, 1983; Yazulla, 1983; Yazulla and Kleinschmidt, 1983; Ayoub and Lam, 1984; Lasater and Lam, 1984; O'Brien and Dowling, 1985). The same results were reported for evoked endogenous GABA release from goldfish retina (Yazulla *et al*, 1985) and isolated horizontal cells with respect to K^+-evoked GABA release (Ayoub and Lam, 1985). Curiously, ^{3}H-GABA and GABA release evoked by L-glutamate and L-aspartate were actually larger in a Ca^{2+}-free medium (Figure 10.3; Yazulla and Kleinschmidt, 1983; O'Brien and Dowling, 1985; Yazulla *et al*, 1985). Similar results have been reported in chick retina (Morán and Pasantes-Morales, 1983) as well as rat striatal slices (Bernath and Zigmond, 1988; Bernath *et al.*, 1989). O'Brien and Dowling (1985) suggested that the enhanced GABA release was due to the removal of a Ca^{2+}- dependent inhibitory input, which in the case of H1 horizontal cells would be derived from dopaminergic interplexiform cells (DA-IPC; see below for more discussion of dopamine input).

Although extracellular Ca^{2+} was not required for evoked GABA efflux from horizontal cells, it was possible that intracellular Ca^{2+} stores in mitochondria or endoplasmic reticulum could be released in response to Na^+ influx following 'glutamate' stimulation (Carafoli *et al.*, 1974) and play a role in the release of neurotransmitters (Alnaes and Rahamimoff, 1975; Sandoval, 1980). The first attempt to address this issue in retina was based on observations that mitochondrial transport of Ca^{2+} was inhibited by lanthanides and the mucopolysaccharide dye ruthenium red (Reed and Bygrave, 1974; Moore, 1971). If mitochondrial stores of Ca^{2+} were involved in the Na^+-dependent GABA release, then inclusion of ruthenium red should potentiate that efflux, as shown for veratridine-evoked ^{3}H-GABA release from cortical slices (Cunningham and Neal, 1981). Yazulla (1985b) found no enhancement of ^{3}H–GABA release from goldfish retina evoked by either L-glutamate or elevated K^+ in the presence of ruthenium red, suggesting that mitochondrial Ca^{2+} stores were not a factor in ^{3}H-GABA release from horizontal cells.

The definitive study that eliminated intracellular Ca^{2+} as a factor in GABA release from teleost fish horizontal cells was provided by Schwartz (1987). Pairs of isolated retinal neurons were used, each pair comprising: (1) an mb bipolar cell from goldfish used as a GABA detector because its synaptic terminal receives input from numerous GABAergic synapses and is responsive to GABA application (Marc *et al.*, 1978; Schwartz, 1987); and (2) a cone horizontal cell from catfish retina used as a source of GABA (Lam *et al.*, 1978; Lasater and Lam, 1984). Cone horizontal cells were loaded with the Ca^{2+}-sensitive dye fura-2 and impaled with a micropipette. Ion substitution experiments showed that, in a medium containing 10 μM Ca^{2+} and 10 mM Mg^{2+}, the horizontal cell could be depolarized from −60 to +20 mV without increasing the intracellular Ca^{2+} concentration. Under these conditions, a bipolar cell was whole-cell patched and maneuvered into close proximity to the horizontal cell. Depolarizing steps of the horizontal cell evoked increasing outward currents in the bipolar cell that were due to a substance released by the horizontal cell rather than passage of current between the cells. In a series of experiments, Schwartz (1987, p.353) concluded that the released compound was GABA because 'the released compound opened a channel whose permeant ion, amplitude and mean open time were the same as the channel opened by GABA, and the released compound had an effect that was desensitized by exogenous GABA and was decreased by a pharmacological antagonist of GABA action' (picrotoxin). Schwartz (1987) also provided evidence that the GABA release occurred via a transporter. He showed that 10 μM nipecotic acid inhibited currents recorded from the bipolar cell in response to depolarization of the horizontal cell, whereas 10 μM nipecotic acid had no effect on currents induced by direct application of GABA to the bipolar cell terminal. Because nipecotic acid has a much higher affinity for blocking GABA transporters than postsynaptic receptors, Schwartz (1987) concluded that GABA was released from horizontal cells by a transporter in a voltage-dependent manner without changes in intracellular Ca^{2+}. A small Ca^{2+}-dependent release of GABA from conventional synapses probably also is involved. It is likely that the same

Figure 10.2 Light microscopic autoradiographs of ^{3}H-GABA uptake in goldfish retina. Isolated retinas were incubated in 0.72 μM ^{3}H-GABA in normal Ringer in room light for 15 min, followed by 15 min incubation without ^{3}H-GABA in: (a) normal Ringer as a control, (b) 5 mM L-glutamate, (c) 5 mM L-glutamate plus 100 μM dopamine plus 2 mM IBMX. Retinas were fixed in mixed aldehydes, embedded in EPON, sectioned at 1.5 μm, dipped in Kodak NTB-2 emulsion and exposed for 7 days. Compared to control (a), L-glutamate selectively reduced grain density over horizontal cell somas (b, HS), but had little effect on grain density over horizontal cell axon terminals (AT) and inner plexiform layer (IPL). Dopamine plus IBMX (c) blocked the effect of L-glutamate on horizontal cell somas. Two heavily labeled cell bodies of pyriform Ab amacrine cells are indicated with arrowheads in (a) and (c). Calibration bar = 25 μm. (Figures adapted from Yazulla and Kleinschmidt (1982).)

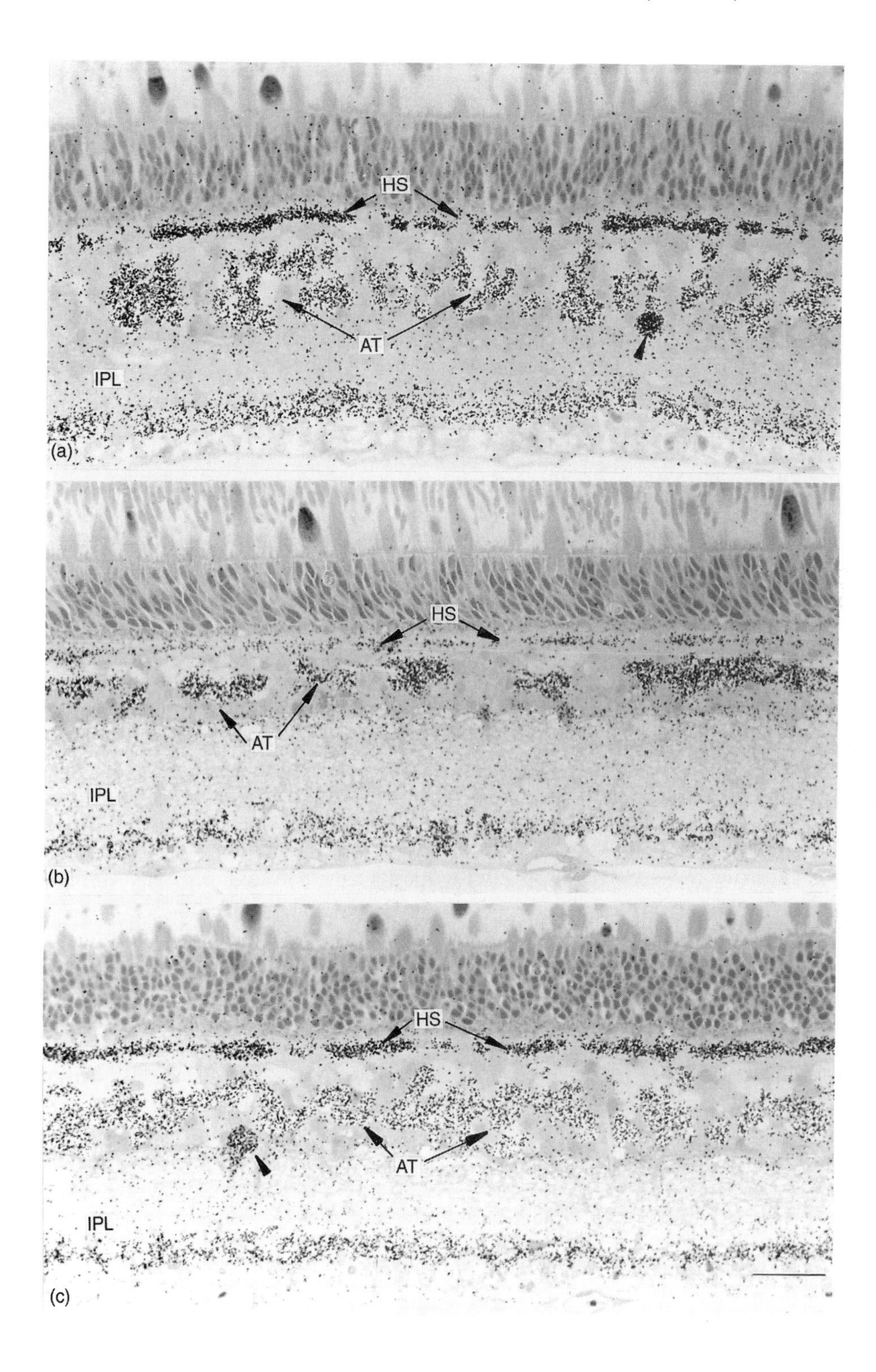
HS
AT
IPL
(a)
HS
AT
IPL
(b)
HS
AT
IPL
(c)

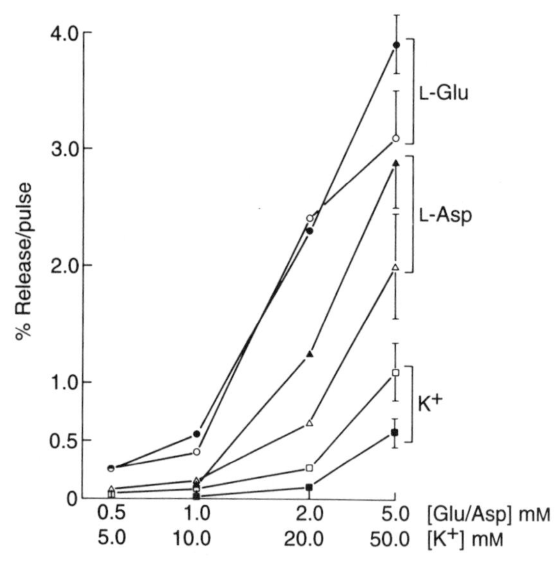

Figure 10.3 Dose-response functions for L-glutamate, (L-Glu), L-aspartate, (L-Asp) and K^+ on ^{3}H-GABA release from the superfused isolated goldfish retina in a normal HEPES-buffered Ringer solution (open symbols) and in a zero Ca^{2+}/20 mM Mg^{2+} solution (closed symbols). Isolated retinas were perfused in a closed chamber at 1 ml/min. Stimuli were presented in increasing concentrations for 2 min at 10 min intervals. Responses are the amount of ^{3}H-GABA released/2 min pulse as a percentage of the total ^{3}H-GABA in the retina. Note that, in the zero Ca^{2+}/high Mg^{2+} Ringer, evoked ^{3}H-GABA release was less for K^+ but higher for L-glutamate and L-aspartate compared to control. (Figure redrawn from Yazulla (1983).)

conclusions may be applied to horizontal cells of birds, *Bufo* toads and *Xenopus* frogs, but not *Rana* frogs in which Ca^{2+} independence has not been demonstrated (Cunningham and Neal, 1985).

10.4 Na^+ AND VOLTAGE DEPENDENCE OF GABA RELEASE

Further evidence for the involvement of a transporter is that both the high affinity uptake and 'glutamate'-evoked release of GABA from horizontal cells are Na^+ dependent. Substitution of Na^+ with impermeant cations choline$^+$ or Tris$^+$ greatly reduces L-glutamate or veratridine-evoked ^{3}H-GABA release in fish, toad and frog, suggestive of voltage and/or Na^+ dependence (Schwartz, 1982; Yazulla and Kleinschmidt, 1983; Ayoub and Lam, 1984; Jaffe *et al.*, 1984; Cunningham and Neal, 1985; Cunningham *et al.*, 1988). Li^+, a permeant cation will substitute for Na^+ in toad retina when ^{3}H-GABA release is evoked by elevated K^+ indicating voltage dependence (Schwartz, 1982). However, Li^+ will not substitute for Na^+ for L-glutamate evoked ^{3}H-GABA release in fish and toad retinas (Schwartz, 1982; Yazulla and Kleinschmidt, 1983; Jaffe *et al.*, 1984), indicating a greater dependence on Na^+ compared to voltage.

GABA is co-transported with 2 Na^+ and 1 Cl^-, thus making it electrogenic and subject to membrane potential and the Na^+ gradient (Kanner *et al.*, 1983; Malchow and Ripps, 1990). The relative importance of Na^+ in GABA release from horizontal cells was suggested by observations that acidic amino acids that increased intracellular Na^+ evoked much greater ^{3}H-GABA and GABA release than elevated K^+ that only passively depolarized the cell (Yazulla, 1983; Yazulla *et al.*, 1985). This was further illustrated in experiments in fish and *Xenopus* retinas which measured ^{3}H-GABA and/or GABA release under conditions that produced the same horizontal cell membrane potential. For an equivalent membrane potential, L-glutamate and kainate evoked up to an order of magnitude more ^{3}H-GABA/GABA release than high-K^+ (Yazulla, 1983; Cunningham *et al.*, 1988). Furthermore, glycine, which also depolarizes *Xenopus* horizontal cells failed to evoke ^{3}H-GABA or GABA release (Cunningham *et al.*, 1988), presumably because glycine gates a chloride rather than Na^+ conductance. Campochiaro *et al.* (1985) suggested that the relative ineffectiveness of K^+ stimulation was due to K^+-evoked release of an inhibitory agent of ^{3}H-GABA release, most likely dopamine, which inhibits ^{3}H-GABA release, from goldfish horizontal cells (Yazulla and Kleinschmidt, 1982; O'Brien and Dowling, 1985; Yazulla, 1985a). Their suggestion could account, in part, for differences observed in intact retina. The end result of these studies was that the properties of evoked GABA release were very similar to those of GABA uptake indicating that the same mechanism was responsible for both events.

10.5 GLUTAMATE RECEPTORS MEDIATING GABA RELEASE: NON-NMDA VERSUS NMDA

Depolarizing input to horizontal cells comes from photoreceptors and as such, the photoreceptor transmitter and its agonists should stimulate GABA release from horizontal cells. L-Glutamate is the most likely transmitter for long-wavelength absorbing cones (Massey, 1990, for review). Experiments dealing with the relative efficacies of L-glutamate and its agonists and antagonists on ^{3}H-GABA release have produced several agreements and some disagreements regarding the properties of receptors evoking GABA release from horizontal cells, whether isolated or in intact tissue, in fish (Yazulla, 1983; Ayoub and Lam, 1984; Yazulla *et al.*, 1985; Cha *et al.*, 1986) and and *Rana* frog (Cunningham and Neal, 1985). In general, (1) kainate and quisqualate are about 100-fold more effective than either L-glutamate or L-aspartate; (2) although comparable in initial potency, L-glutamate evokes about 1.5 to 4 times the GABA release as L-aspartate; (3) kainate evokes about twice the GABA release as quisqualate. Differences relate to the effects of D-aspartate and N-methyl D-aspartate (NMDA).

D-Aspartate inhibits the uptake of both L-glutamate and L-aspartate in goldfish retina (Marc and Lam, 1981). It is agreed that D-aspartate potentiates the depolarizing action of L-glutamate on horizontal cells (Ishida and Fain, 1981; Rowe and Ruddock, 1982; Mangel *et al.*, 1989) as well as the ability of L-glutamate to evoke ^{3}H-GABA and endogenous GABA release from the intact retina (Cha *et al.*, 1986; Yazulla, 1983; Yazulla *et al.*, 1985). The disagreement relates to the manner in which D-aspartate interacts with L-aspartate. First reports were that 1–3 mM D-aspartate inhibited release of ^{3}H-GABA (Ayoub and Lam, 1984; Yazulla, 1983, 1985a) and endogenous GABA (Yazulla *et al.*, 1985) when evoked by L-aspartate. However, there were later reports that D-aspartate potentiated the depolarizing effects of both L-aspartate and L-glutamate in carp horizontal cells (Mangel *et al.*, 1989) and the release of ^{3}H-GABA from goldfish horizontal cells (Cha *et al.*, 1986). Clearly, these latter studies are consistent with the uptake inhibitory effects of D-aspartate. Yet apart from invoking 'procedural differences' among studies, there is no satisfactory explanation for the reported inhibitory action of D-aspartate on effects of L-aspartate. The conceptual controversy involves the hypothesis of Yazulla (1985a) who suggested that L-aspartate could not be the photoreceptor transmitter in fish because dark-evoked release of ^{3}H-GABA occurred in the presence of D-aspartate, which potentiated L-glutamate but inhibited L-aspartate. However this hypothesis has been contested (Ariel *et al.*, 1986; Chan *et al.*, 1986) and the issue has yet to be resolved.

N-Methyl-DL-aspartate (NMDLA) evoked ^{3}H-GABA release from goldfish retina at low concentrations (< 50 μM); this release was inhibited by Mg^{2+} and D-α-aminoadipic acid (Yazulla, 1983). However, in electrophysiological studies, NMDLA hyperpolarized carp horizontal cells in intact retina and *in vitro* (Ariel *et al.*, 1984), and NMDA evoked little or no response in mudpuppy horizontal cells (Slaughter and Miller, 1983) or isolated sting ray and perch horizontal cells (O'Dell and

Christensen, 1989; Zhou *et al.*, 1993). Rather, NMDA appears to act as an antagonist of the photoreceptor transmitter on horizontal cells in carp and *Rana* (Ariel *et al.*, 1986; Cunningham and Neal, 1985). In the absence of evidence for direct depolarizing action of NMDA on goldfish horizontal cells, it is possible that the NMDLA-evoked ^{3}H-GABA release in the Yazulla (1983) study derived from amacrine cells rather than horizontal cells. The issue is not closed because NMDA depolarizes horizontal cells in catfish (O'Dell and Christensen, 1986) and turtle (Anderton and Millar, 1989; Perlman *et al.*, 1987). It seems curious that NMDA would evoke such diverse responses in horizontal cells given the overall conservation in regard to GABA content and response to glutamate.

Although L-aspartate evokes ^{3}H-GABA release from isolated horizontal cells (Ayoub and Lam, 1984) and horizontal cells in intact retina of goldfish (Yazulla, 1983; Yazulla and Kleinschmidt, 1983; Cha *et al.*, 1986), *Rana* and *Xenopus* (Cunningham and Neal, 1985; Cunningham *et al.*, 1988), and potently depolarizes horizontal cells in intact retina (Wu and Dowling, 1978; Ariel *et al.*, 1984), L-aspartate has no electrophysiological effect on isolated horizontal cells (e.g., Ishida *et al.*, 1984; Lasater and Dowling, 1982; Zhou *et al.*, 1993). It is unlikely that the potency of L-aspartate in intact tissue is due to heteroexchange with L-glutamate, which is actually the stimulant (Ariel *et al.*, 1984) because D-aspartate, a substrate for the same carrier (Marc and Lam, 1981), has very little effect on membrane potential (Mangel *et al.*, 1989) or ^{3}H-GABA release from horizontal cells (Yazulla, 1983, 1985a; Ayoub and Lam, 1984; Cha *et al.*, 1986). The most likely explanation is that enzymatic dissociation of the retina altered the membrane receptors in some manner to select between L-glutamate and L-aspartate. Curiously, isolated horizontal cells release ^{3}H-GABA (Ayoub and Lam, 1984) but do not depolarize in response to L-aspartate, perhaps a further illustration of the dissociation between membrane potential and transmitter release. Another possibility is suggested by work on cultured embryonic avian retina cells (de Mello *et al.*, 1988). These authors suggested that a glutamate/GABA exchange mechanism could account for Na^+-dependent, Ca^{2+}-independent ^{3}H-GABA release from the cultured cells. However, the conclusion drawn from studies on matured neuron is that GABA release from horizontal cells is evoked most effectively by activation of non-NMDA-type glutamate receptors, probably the AMPA (α-amino-3-hydroxy-5-methyl-4-isoxazole propionate) type.

10.6 DARKNESS – THE NATURAL STIMULUS OF GABA RELEASE

A critical criterion for any transmitter identification is that its release be evoked by the natural stimulus, which in the case of horizontal cells would be depolarization caused by an increase in photoreceptor transmitter release in response to a decrease in illumination. Such dark-induced release of ^{3}H-GABA has been measured from the goldfish retina by biochemical means (Figure 10.4; Ayoub and Lam, 1984; Yazulla, 1985a). In both studies, the amount of ^{3}H-GABA release was very small, about 1–2% of the retinal ^{3}H-GABA content after 6–15 min in the dark following dim red light exposure. Compare this with the 20% ^{3}H-GABA release evoked by 5 mM L-glutamate over the same time frame (Yazulla, 1983). Also, in both studies the ^{3}H-GABA release was localized to horizontal cells by showing that retinas, preloaded in ^{3}H-GABA in a way to favor uptake into amacrine cells, did not display dark-evoked ^{3}H-GABA release. In order to detect dark-induced ^{3}H-GABA release, Ayoub and Lam (1984) included 0.1 mM nipecotic acid to inhibit the reuptake of released ^{3}H-GABA, whereas Yazulla (1985a) used 3.2 mM D-aspartate to inhibit the uptake of photo-

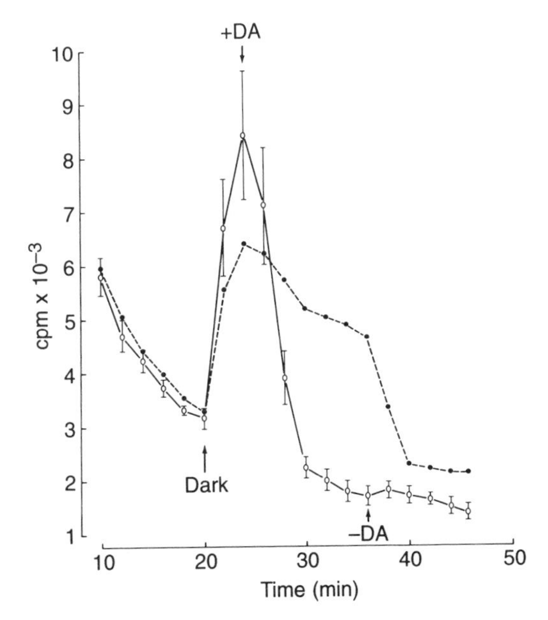

Figure 10.4 ^{3}H-GABA efflux from the isolated goldfish retina in response to darkness, light and dopamine (DA). Isolated retinas were incubated in dim red safety light for 15 min in 0.72 μM ^{3}H-GABA, perfused for 25 min in Ringer containing 3.2 mM D-aspartate, after which the light was turned off (Dark), resulting in increased ^{3}H-GABA efflux. The increased ^{3}H-GABA efflux was inhibited by turning the light back on (closed symbols, at min 36) and by application of freshly made 100 μM dopamine (open symbols, +DA). (Figure adapted from Yazulla (1985a).)

receptor transmitter, thereby potentiating stimulation by darkness. Whereas, at least for teleost fish, GABA could be released from horizontal cells by conditions that evoke and mimic photoreceptor neurotransmitter release, it can be concluded that GABA is a neurotransmitter of a class of horizontal cells.

10.7 INHIBITION OF GABA RELEASE BY DOPAMINE

The other major synaptic input to teleost H1 horizontal cells is a dopaminergic interplexiform cell, acting via D1 receptors to increase cAMP (Dowling and Ehinger, 1978; Van Buskirk and Dowling, 1981). Dopamine inhibits the release of ^{3}H-GABA from horizontal cells whether evoked by L-glutamate (Figure 10.2; O'Brien and Dowling, 1985; Yazulla and Kleinschmidt, 1982) or darkness (Figure 10.4; Yazulla, 1985a). Dopamine, however, did not block the ^{3}H-GABA release from horizontal cells evoked by ouabain (Yazulla and Kleinschmidt, 1982), although nipecotic acid did block effects of ouabain (Yazulla and Kleinschmidt, 1983). These findings led to the suggestion that dopamine, via cAMP, could stimulate the Na^+/K^+-ATPase which would reduce the level of Na^+ required for co-transport of GABA, thereby inhibiting GABA release (Yazulla, 1985b). The role of the Na^+/K^+-ATPase in modulating transmitter release is not a new idea (e.g., Vizi, 1978; O'Fallon *et al.*, 1980) and regulation of the Na^+/K^+-ATPase by neurotransmitters has received considerable attention (Hernández, 1992, for review). Inhibition of the Na^+/K^+-ATPase by dopamine has been reported in a variety of neural and non-neural preparations (i.e., Aperia *et al.*, 1987; Bertorello *et al.*, 1990; Horiuchi *et al.*, 1993), including chick retina (Laitinen, 1993). Electrophysiological and histochemical evidence also indicate the presence of the Na^+/K^+-ATPase in carp and goldfish horizontal cells (Yasui, 1987; Yazulla and Studholme, 1987). However, the hypothesized stimulation of the Na^+/K^+-ATPase by dopamine in horizontal cells (Yazulla, 1985b) has yet to be tested. Regardless of the mechanism of GABA inhibition by dopamine, the point is that GABAergic horizontal cells appear to be subjected to antagonistic influences from photoreceptors and dopamine interplexiform cells.

A hypothetical mechanism by which this interaction occurs was described by Yazulla and Kleinschmidt (1982) and is illustrated in Figure 10.5. A main feature of the hypothesis is that glutamate and dopamine act *via* different means to increase or decrease the local concentration of intracellular Na^+ that leads to increases or decreases in GABA efflux by the transport carrier.

10.8 IS CARRIER MEDIATED GABA RELEASE ADEQUATE?

Although there is massive evidence to demonstrate transport carrier mediated release of GABA from retinal horizontal cells, the questions arise as to whether the release is fast enough, in an adequate concentration and over a physiological range of horizontal cell membrane potential. The answer to these questions, addressed experimentally in isolated horizontal cells (Lam and Ayoub, 1983; Ayoub and Lam, 1985; Schwartz, 1987) and in a recent review (Attwell *et al.*, 1993) is 'yes'. First, GABA release is linear over a range of −60 mV to O mV, well within the operating range of horizontal cells. Second, the Ca^{2+}-independent release of GABA from a horizontal cell evokes a current in a juxtaposed bipolar cell within 100 ms (Schwartz, 1987). Considering that these responses took place in a Petri dish rather that the confined space of a synaptic pedicle, the response *in situ* should be much faster and within limits of the graded light-evoked responses of outer retinal neurons. Third, GABA uptake into horizontal cells is electrogenic (Malchow and Ripps, 1990) and probably has a stoichiometry of $2Na^+$ and 1 Cl^- per GABA molecule transported (Kanner *et al.*, 1983). Attwell *et al.* (1993) calculated that within the membrane potential range of −70mV to −20mV, such a transporter could maintain extracellular GABA concentrations of 4–28 μM, above the sensitivity of isolated turtle cones to GABA (Kaneko and Tachibana, 1986). It seems likely, therefore, that the GABA transporter could function in the physiological range to account for neurotransmitter release of GABA from horizontal cells.

10.9 HORIZONTAL CELL TRANSMITTERS OTHER THAN GABA?

Only GABA has been clearly identified as a neurotransmitter in horizontal cells. Although it was thought initially that there was only one type of GABAergic horizontal cell per retina, there is evidence that GABA may be used by multiple horizontal cell types. For example, Van Haesendonck and Missotten (1992) described GABA-immuno reactivity (IR) in all three types of cone horizontal cell in the dragonet, a marine teleost. Also, in cat and monkey retinas, both A- and B-type horizontal cells are GABA-IR (Wässle and Chun, 1989; Grünert and Wässle, 1990), suggesting that in some mammals, both types of horizontal cell are GABAergic. However, GABA is not likely to be the transmitter of all horizontal cells, i.e., rod horizontal cells in duplex retinas of non-mammals or horizontal cells in rodents.

There are relatively few alternatives to GABA as a transmitter for the other types of horizontal cells. Antisera against the biosynthetic enzymes, tyrosine hydroxylase and phenylethanolamine *N*-methyltransferase have been localized to axonless A-type horizontal cells in ferret retina, indicating the presence of epinephrine (Keyser *et al.*, 1987). However, this finding has not been replicated in any other mammal, to my knowledge. They could be more to this finding because it would seem unlikely that Keyser *et al.* (1987) would have been so lucky (or unlucky) to have studied the only mammal with potential catecholaminergic horizontal cells.

The strongest candidate for an additional horizontal cell transmitter is a purine, probably adenosine (Blazynski and Perez, 1992, for review). [^{3}H]-Adenosine derived uptake has been described in horizontal cells of birds, frog and fish (Ehinger and Perez, 1984; Perez and Bruun, 1987). [^{3}H]-Adenosine uptake and GABA-IR colocalize in some chick horizontal cells (Perez and Bruun, 1987).

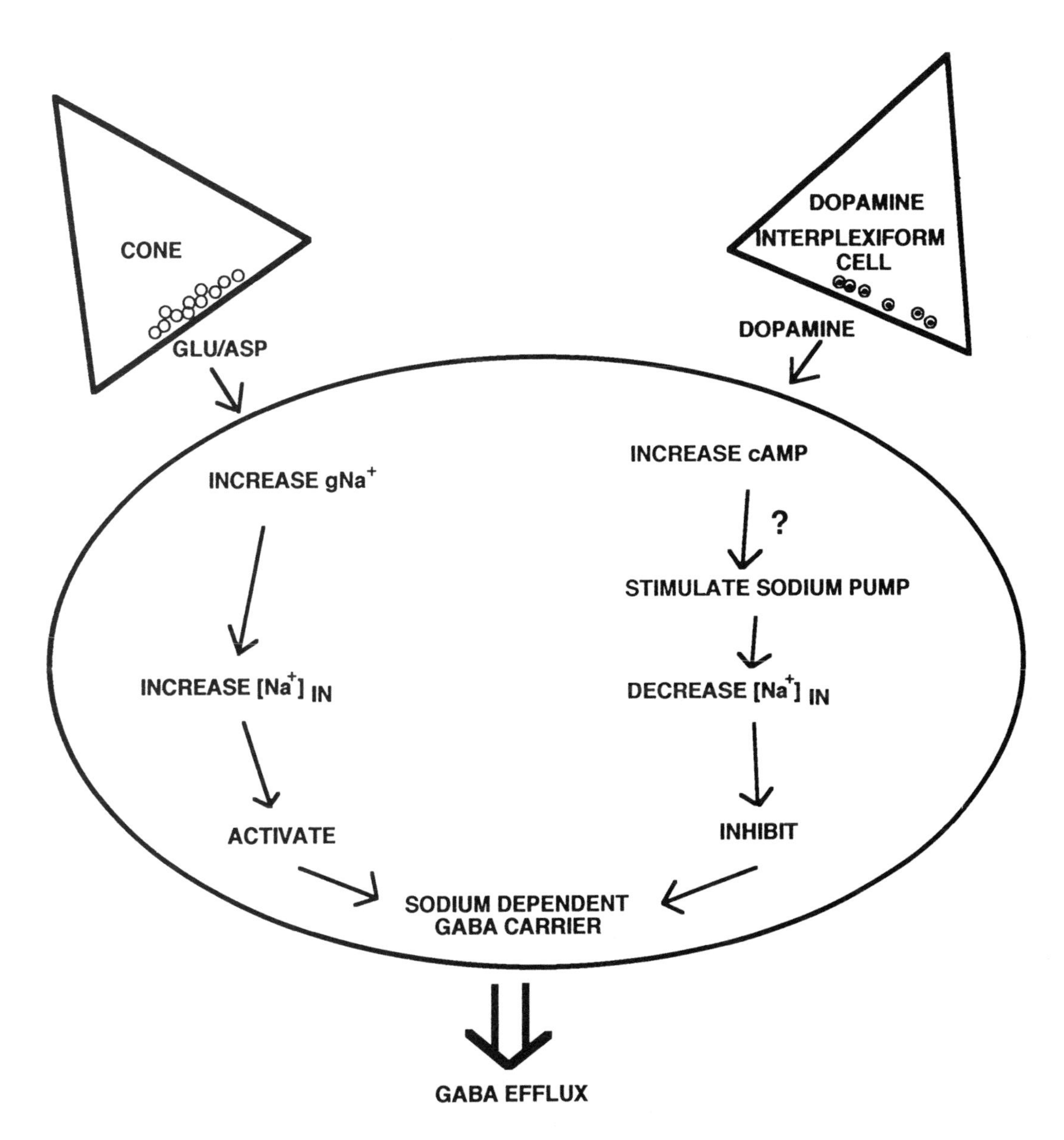

Figure 10.5 Schematic illustration of the hypothesized mechanism for the antagonistic modulation carrier-mediated release of GABA from teleost horizontal cells. Excitatory input is provided by cones that release an acidic amino acid, probably glutamate, that activates ionotropic receptors, increasing intracellular Na^+ and stimulating GABA efflux. Inhibitory input comes from dopamine I1 interplexiform cells that, via metabatropic receptors, increases cAMP that is hypothesized (?) to stimulate that Na^+/K^+ATPase, which would reduce intracellular Na^+ $[Na^+]_{IN}$ and inhibit GABA release by the transport carrier. This hypothesis was first presented by Yazulla and Kleinschmidt (1982). (Figure is adapted from Yazulla (1985b).)

However, in goldfish, these markers do not colocalize; [^{3}H]-adenosine uptake occurs in cells just proximal to the GABAergic H1 cells, and are likely to be rod horizontal cells (Studholme and Yazulla, 1994). Adenosine-like-IR has been localized to horizontal cells in a wide variety of mammals, including humans (Braas *et al.*, 1987; Blazynski *et al.*, 1989b). However, the distribution of adenosine A_1 receptors is not particularly enriched in the outer plexiform layer of mammals (Braas *et al.*, 1987; Blazynski, 1990). Evoked release of ^{3}H-labeled purines can be achieved with high-K^+, acidic amino acid agonists and GABA, glycine and dopamine antagonists. There was a large Ca^{2+}-dependent component for the excitatory agents and mixed results with the antagonists (Perez *et al.*, 1986; Perez and Ehinger, 1989). The mixed results regarding Ca^{2+} dependence is not surprising considering that several neuron types contain purines and the source of evoked release was not determined. There is electrophysiological evidence for purine transmission in the outer retina. The amplitude of the cat ERG b-wave is enhanced in the presence of micromolar concentrations of adenosine (Blazynski *et al.*, 1989a) and the A_1 agonist, CHA (Früh *et al.*, 1989). Also, Barnes and Hille (1989) reported that adenosine inhibited a Ca^{2+} activated Cl^- conductance in about 25% of salamander cone photoreceptors. These data are consistent with a role for purine transmission from horizontal cells. Perhaps purinergic function may be one of the more fruitful areas of future research on neuromodulation in the outer retina.

10.10 CONCLUSIONS AND FUTURE PERSPECTIVES

The general consensus is that the synaptically evoked release of GABA from horizontal cells occurs by two mechanisms. The first involves a Na^+-dependent transport carrier and cytoplasmic stores of GABA; it is Ca^{2+} independent and driven by both membrane potential and the Na^+ gradient. It is uncertain as to whether the carrier-mediated release occurs at focal sites within dendrites or is more globally distributed over the surface of the horizontal cell. The second mechanism involves Ca^{2+}-dependent vesicular release of GABA from conventional synapses located in the dendrites, soma and/or axon terminal. The release of GABA is increased by the photoreceptor transmitter (probably L-glutamate) and, at least in teleost fish, is inhibited by dopamine, the transmitter of I1 interplexiform cells (Figure 10.5).

The whole issue of vesicular versus non-vesicular (carrier-mediated) release of GABA is not unique to retinal horizontal cells, but has been addressed in photoreceptors (Schwartz, 1986) and rabbit starburst amacrine cells (O'Malley *et al.*, 1992). This issue is not even unique to retina. From 1982 to 1987, the properties of Ca^{2+}-independent GABA release from horizontal cells were described, corroborated and solidified. Although still immersed in controversy, the same properties of GABA release as described for non-mammalian horizontal cells have been proposed in other areas of the vertebrate CNS (Adam-Vizi, 1992; Bernath, 1992; Levi and Raiteri, 1993, for recent reviews). GABAergic horizontal cells, along with the motor neurons at the neuromuscular junction, are among the best characterized neurons with respect to transmitter release. Given the relatively few number of conventional synapses, the horizontal cell serves as an excellent preparation to study the accumulation and turnover of cytoplasmic stores of transmitter. This particular property provides an important advantage over other neurons that have significant stores of synaptic vesicles. These studies on GABA release from horizontal cells of non-mammalian vertebrates have yielded important clues on a variety of topics, including the identity of the photoreceptor transmitter (L-glutamate), the most important ion species gated by that transmitter (Na^+), an alternative to vesicular transmitter release (transport carrier), the importance of cytoplasmic stores of transmitter, dissociation of transmitter release from intracellular Ca^{2+} and membrane potential. It is very likely that mammalian horizontal cells will function similarly to those of non-mammals, although some differences are to be expected; but, probably no big surprises. Also, as evidenced by the recent flurry of primary and review articles, the prospect of carrier-mediated release of GABA, as well as other amino acid transmitters, throughout the CNS, is being taken far more seriously than it was five to ten years ago. Despite the early work on GABA transport in synaptosomes and the hypothesis that the GABA transporter could work in reverse (Martin, 1976), there is certain satisfaction in that much of the pioneering work on this idea was done successfully on retinal horizontal cells. It is clear that the dual mechanisms of vesicular- and carrier-mediated transmitter release operate under different rules (voltage, ionic dependence, release sites, etc.) that impart far more flexibility in postsynaptic response and consequent neuromodulation than could be achieved by either mechanism alone. Future work on the molecular diversity of transporters and their cellular regulation (e.g., Clark and Amara, 1993, for review) should reveal even greater flexibility and complexity of transmitter release than described here.

REFERENCES

Adam-Vizi, V. (1992) External Ca^{2+}-independent release of neurotransmitters. *Journal of Neurochemistry*, **58**, 395–405.

Agardh, E., Ehinger, B. and Wu, J.-Y. (1987) GABA and GAD-like immunoreactivity in the primate retina. *Histochemistry*, **86**, 485–90.

Agardh, E. and Ehinger, B. (1982) ^{3}H-Muscimol, ^{3}H-nipecotic acid and ^{3}H-isoguavacine as autoradiographic markers for GABA neurotransmission. *Journal of Neurotransmission*, **54**, 1–18.

Alnaes, E. and Rahamimoff, R. (1975) On the role of mitochondria in transmitter release from motor nerve endings. *Journal of Physiology (London)*, **248**, 285–306.

Anderton, P.J. and Millar, T.J. (1989) MK801-induced antagonism of NMDA-preferring excitatory amino acid receptors in horizontal cells of the turtle retina. *Neuroscience Letters*, **101**, 331–36.

Aperia, A., Bertorello, A. and Seri, I. (1987) Dopamine causes inhibition of Na^+-K^+ -ATPase activity in rat proximal convoluted tubule segments. *American Journal of Physiology*, **252**, F39–F45.

Ariel, M., Lasater, E.M., Mangel, S.C. and Dowling, J.E. (1984) On the sensitivity of H1 horizontal cells of the carp retina to glutamate, aspartate and their agonists. *Brain Research*, **295**, 179–83.

Ariel, M., Mangel, S.C. and Dowling, J.E. (1986) *N*-methyl D-aspartate acts as an antagonist of the photoreceptor transmitter in the carp retina. *Brain Research*, **372**, 143–8.

Attwell, D., Barbour, B. and Szatkowski, M. (1993) Nonvesicular release of neurotransmitter. *Neuron*, **11**, 401–7.

Ayoub, G.S. and Lam, D.M. (1985) The content and release of endogenous GABA in isolated horizontal cells of the goldfish retina. *Vision Research*, **25**, 1187–93.

Ayoub, G.S. and Lam, D.M.K. (1984). The release of γ-aminobutyric acid from horizontal cells of the goldfish (*Carassius auratus*) retina. *Journal of Physiology (London)*, **355**, 191–214.

Barnes, S. and Hille, B. (1989) Ionic channels of the inner segment of tiger salamander cone photoreceptors. *Journal of General Physiology*, **94**, 719–43.

Baylor, D.A., Fuortes, M.G.F. and O'Bryan, P.M. (1971) Receptive field of cones in the retina of the turtle. *Journal of Physiology (London)*, **214**, 265–94.

Bernath, S. (1992) Calcium-independent release of amino acid neurotransmitters: fact or artifact. *Progress in Neurobiology*, **38**, 57–91.

Bernath, S. and Zigmond, M.J. (1988) Characterization of [^{3}H]GABA release from striatal slices: evidence for a calcium-independent process via the GABA uptake system. *Neuroscience*, **27**, 563–70.

Bernath, S., Keller Jr., R.W. and Zigmond, M.J. (1989) Release of endogenous GABA can occur through Ca^{2+}-dependent and Ca^{2+}-independent processes. *Neurochemistry International*, **14**, 439–45.

Bertorello, A.M., Hopfield, J.F., Aperia, A. and Greengard, P. (1990) Inhibition by dopamine of (Na^+ +K^+)ATPase activity in neostriatal neurons through D_1 and D_2 dopamine receptor synergism. *Nature*, **347**, 386–8.

Blazynski, C. (1990) Discrete distributions of adenosine receptors in mammalian retina. *Journal of Neurochemistry*, **54**, 648–55.

Blazynski, C. and Perez, M.-T.R. (1992) Neuroregulatory functions of adenosine in the retina. *Progress in Retinal Research*, **11**, 293–332.

Blazynski, C., Cohen, A.I., Früh, B. and Niemeyer, G. (1989a) Adenosine: autoradiographic localization and electrophysiologic effects in the cat retina. *Investigative Ophthalmology and Visual Science*, **30**, 2533–6.

Blazynski, C., Mosinger, J.L. and Cohen, A.I. (1989b) Comparison of adenosine uptake and endogenous adenosine-containing cells in mammalian retina. *Visual Neuroscience*, **2**, 109–16.

Braas, K.M., Zarbin, M.A. and Snyder, S.H. (1987) Endogenous adenosine and adenosine receptors localized to ganglion cells of the retina. *Proceedings of the National Academy of Sciences of the United States of America*, **84**, 3906–10.

Brandon, C. (1985) Retinal GABA neurons: localization in vertebrate species using an antiserum to rabbit brain glutamate decarboxylase. *Brain Research*, **344**, 286–95.

Brandon, C., Lam, D.M.K., Su, Y.Y.T. and Wu, J.Y. (1980) Immunocytochemical localization of GABA neurons in the rabbit and frog retina. *Brain Research Bulletin*, **5** (Suppl.2), 21–9.

Burkhardt, D.A. (1977) Responses and receptive-field organization of cones in perch retinas. *Journal of Neurophysiology*, **40**, 53–62.

Burkhardt, D.A. (1993) Synaptic feedback, depolarization, and color opponency in cone photoreceptors. *Visual Neuroscience*, **10**, 981–9.

Campochiaro, P., Ferkany, J.W. and Coyle, J.T. (1984) The dissociation of evoked release of [^{3}H]-GABA and endogenous GABA from chick retina *in vitro*. *Experimental Eye Research*, **39**, 299–305.

Campochiaro, P., Ferkany, J.W. and Coyle, J.T. (1985) Excitatory amino acid analogs evoke release of endogenous amino acids and acetyl choline from chick retina *in vitro*. *Vision Research*, **25**, 1375–86.

Carafoli, E., Tiozzo, R., Crovetti, F. and Kratzing, C. (1974) The release of calcium from heart mitochondria by sodium. *Journal of Molecular and Cellular Cardiology*, **6**, 361–71.

Cervetto, L. and MacNicol, E.F., Jr (1972) Inactivation of horizontal cells in turtle retina by glutamate and aspartate. *Science*, **178**, 767–8.

Cha, J., O'Brien, D.R. and Dowling, J.E. (1986) Effects of D-aspartate on excitatory amino acid-induced release of ^{3}H-GABA from goldfish retina. *Brain Research*, **376**, 140–8.

Chen, F. and Witkovsky, P. (1978) The formation of photoreceptor synapses in the retina of larval *Xenopus*. *Journal of Neurocytology*, **7**, 721–40.

Clark, J.A. and Amara, S.G. (1993) Amino acid neurotransmitter transporters: structure, function, and molecular diversity. *BioEssays*, **15**, 323–32.

Cunningham, J. and Neal, M.J. (1981) On the mechanism by which veratridine causes a calcium-independent release of gamma-aminobutyric acid from brain slices. *British Journal of Pharmacology*, **73**, 655–67.

Cunningham, J.R. and Neal, M.J. (1985) Effect of excitatory amino acids on gamma-aminobutyric acid release from frog horizontal cells. *Journal of Physiology (London)*, **362**, 51–67.

Cunningham, J.R., Neal, M.J., Stone, S. and Witkovsky, P. (1988) GABA release from *Xenopus* retina does not correlate with horizontal cell membrane potential. *Neuroscience*, **24**, 39–48.

de Mello, M.C.F., Klein, W.L. and de Mello, F.G. (1988) L-glutamate evoked release of GABA from cultured avian retina cells does not require glutamate receptor activation. *Brain Research*, **443**, 166–72.

de Mello, M.C.F., Guerra-Peixe, R. and de Mello, F.G. (1993) Excitatory amino acid receptors mediate the glutamate-induced release of GABA synthesized from putrescine in cultured cells of embryonic avian retina. *Neurochemistry International*, **22**, 249–53.

Djamgoz, M.B.A. and Ruddock, K.H. (1979) Effects of picrotoxin and strychnine on fish retinal S-potentials: evidence for inhibitory control of depolarizing responses. *Neuroscience Letters*, 12, 329–34.

do Nascimento, J.L.M. and de Mello, F.G. (1985) Induced release of gamma-aminobutyric acid by a carrier-mediated, high-affinity uptake of L-glutamate in cultured chick retina cells. *Journal of Neurochemistry*, **45**, 1820–7.

Dowling, J.E. (1968) Synaptic organization of the frog retina: an electron microscopic analysis comparing the retinas of frogs and primates. *Proceedings of the Royal Society of London*, **170**, 205–28.

Dowling, J.E. (1987) *The Retina: An Approachable Part of the Brain*. Harvard University Press, Cambridge, MA.

Dowling, J.E. and Ehinger, B. (1978) The interplexiform cell system-I. Synapses of the dopaminergic neurons of the goldfish retina. *Proceedings of the Royal Society of London*, **201**, 7–26.

Dowling, J.E. and Werblin, F.S. (1969) Organization of retina of the mudpuppy, *Necturus maculosus*.I. Synaptic structure. *Journal of Neurophysiology*, **32**, 315–38.

Dowling, J.E., Brown, J.E. and Major, D. (1966) Synapses of horizontal cells in rabbit and cat retinas. *Science*, **153**, 1639–41.

Ehinger, B. and Perez, M.T.R. (1984) Autoradiography of nucleoside uptake into the retina. *Neurochemistry International*, **6**, 369–81.

Früh, B., Niemeyer, G. and Onoe, S. (1989) Adenosine enhances the ERG b-wave and depresses the light peak in perfused cat eyes. *Investigative Ophthalmology and Visual Science*, (Suppl.) **30**, 124.

Gallego, A. (1986) Comparative studies on horizontal cells and a note on microglial cells. *Progress in Retinal Research*, **5**, 165–206.

Gerschenfeld, H.M., Piccolino, M. and Neyton, J. (1980) Feedback modulation of cone synapses by L-horizontal cells of the turtle retina. *Journal of Experimental Biology*, **89**, 177–92.

Grünert, U. and Wässle, H. (1990) GABA-like immunoreactivity in the macaque monkey retina: a light and electron microscopic study. *Journal of Comparative Neurology*, **297**, 509–24.

Hernandez, -R.J. (1992) Na^+/K^+ -ATPase regulation by neurotransmitters. *Neurochemistry International*, **20**, 1–10.

Hollyfield, J.G., Rayborn, M.F., Sarthy, P.V. and Lam, D.M.K. (1979) The emergence, localization and maturation of neurotransmitter systems during development of the retina in *Xenopus laevis* I. γ-aminobutyric acid. *Journal of Comparative Neurology*, **188**, 587–98.

Horiuchi, A., Takeyasu, K., Mouradian, M.M. *et al.* (1993) D_{1A} dopamine receptor stimulation inhibits Na^+/K^+-ATPase activity through protein kinase A. *Molecular Pharmacology*, **43**, 281–5.

Ishida, A.T. and Fain, G.L. (1981) D-aspartate potentiates the effects of L-glutamate on horizontal cells in goldfish retina. *Proceedings of the National Academy of Sciences of the United States of America*, **78**, 5890–94.

Ishida, A.T., Kaneko, A. and Tachibana, M. (1984) Responses of solitary retinal horizontal cells from *Carassius auratus* to L-glutamate and related amino acids. *Journal of Physiology (London)*, **348**, 255–70.

Jaffe, E.H., Hernandez, N. and Holder, L.G. (1984) Study on the mechanism of release of [^{3}H]GABA from a teleost retina *in vitro*. *Journal of Neurochemistry*, **43**, 1226–35.

Kaneko, A. and Tachibana, M. (1986) Effects of gamma-aminobutyric acid on isolated cone photoreceptors of the turtle retina. *Journal of Physiology (London)*, **373**, 463–79.

Kanner, B.I., Bendahan, A. and Radian, R. (1983) Efflux and exchange of gamma-aminobutyric acid and nipecotic acid catalyzed by synaptic plasma membrane vesicles isolated from immature rat brain. *Biochimica et Biophysica Acta*, **731**, 54–62.

Keyser, K.T., Karten, H.J., Katz, B. and Bohn, M.C. (1987) Catecholaminergic horizontal and amacrine cells in the ferret retina. *Journal of Neuroscience*, **7**, 3996–4004.

Kleinschmidt, J. and Yazulla, S. (1984) Uptake of ^{3}H-glycine in the outer plexiform layer of the retina of the toad, *Bufo marinus*. *Journal of Comparative Neurology* **230**, 352–60.

Kolb, H. and Jones, J. (1984) Synaptic organization of the outer plexiform layer of the turtle retina: an electron microscope study of serial sections. *Journal of Neurocytology*, **13**, 567–91.

Laitinen, J.T. (1993) Dopamine stimulates K^+

efflux in the chick retina via D_1 receptors independently of adenyl cyclase activation. *Journal of Neurochemistry*, **61**, 1461–9.

Lam, D.M.K. (1972) The biosynthesis and content of gamma-aminobutyric acid in the goldfish retina. *Journal of Cell Biology* **54**, 225–31.

Lam, D.M. and Ayoub, G.S. (1983) Biochemical and biophysical studies of isolated horizontal cells from the teleost retina. *Vision Research*, **23**, 433–44.

Lam, D.M.K. and Steinman, L. (1971) The uptake of [γ-^{3}H] aminobutyric acid in the goldfish retina. *Proceedings of the National Academy of Sciences of the United States of America* **68**, 2777–81.

Lam, D.M.K., Lasater, E.M. and Naka, K.-I. (1978) γ-Aminobutyric acid: a neurotransmitter candidate for cone horizontal cells of the catfish retina. *Proceedings of the National Academy of Sciences of the United States of America*, **75**, 6310–13.

Lam, D.M.K., Su, Y.Y.T., Swain, L. *et al.* (1979) Immunocytochemical localization of L-glutamic acid decarboxylase in the goldfish retina. *Nature*, **278**, 565–7.

Lasansky, A. (1971) Synaptic organization of cone cells in the turtle retina. *Philosophical Transactions of the Royal Society of London*, **262**, 365–81.

Lasansky, A. (1980) Lateral contacts and interactions of horizontal cell dendrites in the retina of the larval tiger salamander. *Journal of Physiology (London)*, **301**, 59–68.

Lasater, E.M. and Dowling, J.E. (1982) Carp horizontal cells in culture respond selectively to L-glutamate and its agonists. *Proceedings of the National Academy of Sciences of the United States of America*, **79**, 936–40.

Lasater, E.M. and Lam, D.M. (1984) The identification and some functions of GABAergic neurons in the distal catfish retina. *Vision Research*, **24**, 497–506.

Levi, G. and Raiteri, M. (1993) Carrier-mediated release of neurotransmitters. *Trends in Neuroscience*, **16**, 415–19.

Lindberg, K.A. and Fisher, S.K. (1988) Ultrastructural evidence that horizontal cell axon terminals are presynaptic in the human retina. *Journal of Comparative Neurology*, **268**, 281–97.

Lopez-Colomé, A.M., Salceda, R. and Pasantes-Morales, H. (1978) Potassium-stimulated release of GABA, glycine, and taurine from the chick retina. *Neurochemical Research*, **3**, 431–41.

Malchow, R.P. and Ripps, H. (1990) Effects of gamma-aminobutyric acid on skate retinal horizontal cells: evidence for an electrogenic uptake mechanism. *Proceedings of the National Academy of Sciences of the United States of America*, **87**, 8945–9.

Mangel, S.C., Ariel, M. and Dowling, J.E. (1989) D-Aspartate potentiates the effects of both L-aspartate and L-glutamate on carp horizontal cells. *Neuroscience*, **32**, 19–26.

Marc, R.E. and Lam, D.M.K. (1981) Uptake of aspartic and glutamic acid by photoreceptors in goldfish retina. *Proceedings of the National Academy of Sciences of the United States of America*, **78**, 7185–9.

Marc, R.E. and Liu, W.L.S. (1984) Horizontal cell synapses onto glycine-accumulating interplexiform cells. *Nature*, **312**, 266–8.

Marc, R.E., Stell, W.K. Bok, D. and Lam, D.M.K. (1978) GABAergic pathways in the goldfish retina. *Journal of Comparative Neurology*, **182**, 221–46.

Marc, R.E., Liu, W.-L.S, Kalloniatis, M *et al.* (1990) Patterns of glutamate immunoreactivity in the goldfish retina. *Journal of Neuroscience*, **10**, 4006–34.

Marmarelis, P.Z. and Naka, K.-I. (1973) Nonlinear analysis and synthesis of receptive-field responses in the catfish retina. III. Two-input white-noise analysis. *Journal of Neurophysiology*, **36**, 634–48.

Marshak, D.W. and Dowling, J.E. (1987) Synapses of cone horizontal cell axons in goldfish retina. *Journal of Comparative Neurology*, **256**, 430–43.

Marshall, J. and Voaden, M.J. (1974) An autoradiographic study of the cells accumulating ^{3}H-gamma-aminobutyric acid in the isolated retinas of pigeons and chickens. *Investigative Ophthalmology and Visual Science*, **13**, 602–7.

Martin, D.L. (1976) Carrier-mediated transport and removal of GABA from synaptic regions, in *GABA in Nervous System Function*, eds. E. Roberts, T.N. Chase, and D.B. Tower, Raven Press, New York, pp. 347–86.

Massey, S.C. (1990) Cell types using glutamate as a neurotransmitter in the vertebrate retina. *Progress in Retinal Research*, **9**, 399–425.

Messersmith, E.K. and Redburn, D.A. (1993) The

role of GABA during development of the outer retina in the rabbit. *Neurochemical Research*, **18**, 463–70.

Miller, A.M. and Schwartz, E.A. (1983) Evidence for the identification of synaptic transmitters released by photoreceptors of the toad retina. *Journal of Physiology (London)*, **334**, 325–49.

Moore, C.L. (1971) Specific inhibition of mitochondrial Ca^{++} transport by ruthenium red. *Biochemical and Biophysical Research Communications*, **42**, 298–305.

Moran, J. and Pasantes-Morales, H. (1983) Effects of excitatory amino acids, and of their agonists and antagonists on the release of neurotransmitters from the chick retina. *Journal of Neuroscience Research*, **10**, 261–71.

Moran, J., Pasantes-Morales, H. and Redburn, D.A. (1986) Glutamate receptor agonists release ^{3}H. GABA preferentially from horizontal cells. *Brain Research*, **398**, 276–87.

Mosinger, J. and Yazulla, S. (1987) Double-label analysis of GAD- and GABA-like immunoreactivity in the rabbit retina. *Vision Research*, **27**, 23–30.

Mosinger, J.L., Studholme, K.M. and Yazulla, S. (1986) Immunocytochemical localization of GABA in the retina: a species comparison. *Experimental Eye Research*, **42** 631–44.

Murakami, M., Ohtsu, K. and Ohtsuka, T. (1972) Effects of chemicals on receptors and horizontal cells in the retina. *Journal of Physiology (London)*, **227**, 899–913.

Murakami, M., Ohtsuka, T. and Shimazaki, H. (1975) Effects of aspartate and glutamate on the bipolar cells in the carp retina. *Vision Research*, **15**, 456–8.

Murakami, M., Shimoda, Y., Nakatani, K. *et al.* (1982a) GABA-mediated negative feedback and color opponency in carp retina. *Japanese Journal of Physiology*, **32**, 927–35.

Murakami, M., Shimoda, Y., Nakatani, K. *et al* (1982b) GABA-mediated negative feedback from horizontal cells to cones in carp retina. *Japanese Journal of Physiology*, **32**, 911–26.

Neal, M.J. and Atterwill, C.K. (1974) Isolation of photoreceptor and conventional nerve terminals by subcellular fractionation of rabbit retina. *Nature*, **251**, 331–3.

Nishimura, Y., Schwartz, M.L. and Rakic, P. (1985) Localization of γ-aminobutyric acid and glutamic acid decarboxylase in rhesus monkey retina. *Brain Research*, **359**, 351–55.

O'Brien, D.R. and Dowling, J.E. (1985) Dopaminergic regulation of GABA release from the intact goldfish retina. *Brain Research*, **360**, 41–50.

O'Dell, T.J. and Christensen, B.N. (1986) *N*-methyl-D-aspartate coexist with kainate and quisqualate receptors on single isolated catfish horizontal cells. *Brain Research*, **381**, 359–62.

O'Dell, T.J. and Christensen, B.N. (1989) A voltage-clamp study of isolated stingray horizontal cell non-NMDA excitatory amino acid receptors. *Journal of Neurophysiology*, **61**, 162–72.

O'Fallon, J.V., Brosemer, R.W. and Harding, J.W. (1980) The Na^+, K^+ATPase: a plausible trigger for voltage independent release of cytoplasmic neurotransmitters. *Journal of Neurochemistry*, **36**, 369–78.

O'Malley, D.M., Sandell, J.H. and Masland, R.H. (1992) Co-release of acetylcholine and GABA by the starburst amacrine cells. *Journal of Neuroscience*, **12**, 1394–408.

Perez, M.T.R. and Bruun, A. (1987) Colocalization of [^{3}H]-adenosine accumulation and GABA immunoreactivity in the chicken and rabbit retinas. *Histochemistry*, **87**, 413–17.

Perez, M.T.R. and Ehinger, B. (1989) Multiple neurotransmitter systems influence the release of adenosine derivatives from the rabbit retina. *Neurochemistry International*, **15**, 411–20.

Perez, M.T.R., Ehinger, B., Lindstrom, K. and Fredholm, B.B. (1986) Release of endogenous and radioactive purines from the rabbit retina. *Brain Research*, **398**, 106–12.

Perlman, I., Normann, R.A. and Anderton, P.J. (1987) The effects of prolonged superfusions with acidic amino acids and their agonists on field potentials and horizontal cell photoresponses in the turtle retina. *Journal of Neurophysiology*, **57**, 1022–32.

Pourcho, R.G. and Owczarzak, M.T. (1989) Distribution of GABA immunoreactivity in the cat retina: a light- and electron-microscopic study. *Visual Neuroscience*, **2**, 425–35.

Prince, D.J., Djamgoz, M.B.A. and Karten, H.J. (1987) GABA transaminase in cyprinid fish retina: localization and effects of inhibitors on temporal characteristics of S-potentials. *Neurochemistry International*, **11**, 23–30.

Redburn, D.A. (1977) Uptake and release of [^{14}C]-

GABA from rabbit retina synaptosomes. *Experimental Eye Research*, **25**, 265–75.

Reed, K. C. and Bygrave, F.L. (1974) The inhibition of mitochondrial calcium transport by lanthanides and ruthenium red. *Biochemical Journal*, **140**, 143–55.

Rowe, J.S. and Ruddock, K.H. (1982) Depolarization of retinal horizontal cells by excitatory amino acid neurotransmitter agonists. *Neuroscience Letters*, **30**, 257–62.

Sakai, H. and Naka, K. (1983) Synaptic organization involving receptor, horizontal and on- and off-center bipolar cells in the catfish retina. *Vision Research*, **23**, 339–51.

Sakai, H. and Naka, K.-I. (1985) Novel pathway connecting the outer and inner vertebrate retina. *Nature*, **315**, 570–1.

Sakai, H.M. and Naka, K. (1986) Synaptic organization of the cone horizontal cells in the catfish retina. *Journal of Comparative Neurology*, **245**, 107–15.

Sandoval, M.E. (1980) Sodium-dependent efflux of [^{3}H]GABA from synaptosomes probably related to mitochondrial calcium mobilization. *Journal of Neurochemistry*, **35**, 915–21.

Schnitzer, J. and Rusoff, A.C. (1984) Horizontal cells of the mouse retina contain glutamic acid decarboxylase-like immunoreactivity during early developmental stages. *Journal of Neuroscience*, **4**, 2948–55.

Schwartz, E.A. (1982) Calcium-independent release of GABA from isolated horizontal cells of the toad retina. *Journal of Physiology (London)*, **323**, 211–27.

Schwartz, E.A. (1986) Synaptic transmission in amphibian retinae during conditions unfavourable for calcium entry into presynaptic terminals. *Journal of Physiology (London)*, **376**, 411–28.

Schwartz, E.A. (1987) Depolarization without calcium can release γ-aminobutyric acid from a retinal neuron. *Science*, **238**, 350–5.

Sherry, D.S. and Yazulla, S. (1993) A simple Golgi/immunocytochemical double-labelling technique to identify neurotransmitter content of Golgi impregnated neurons. *Journal of Neuroscience Methods*, **46**, 41–8.

Sillman, A.J., Ito, H. and Tomita, T. (1969) Studies on the mass receptor potential of the isolated frog retina. I. General properties of the response. *Vision Research*, **9**, 1435–42.

Slaughter, M.M. and Miller, R.F. (1983) The role of excitatory amino acid transmitters in the mudpuppy retina: an analysis with kainic acid and *N*-methyl aspartate. *Journal of Neuroscience*, **3**, 1701–11.

Smiley, J.F. and Yazulla, S. (1990) Glycinergic contacts in the outer plexiform layer of the *Xenopus laevis* retina characterized by antibodies to glycine, GABA, and glycine receptors. *Journal of Comparative Neurology*, **299**, 375–88.

Stell, W.K. (1967) The structure and relationships of horizontal cells and photoreceptor–bipolar synaptic complexes in goldfish retina. *American Journal of Anatomy*, **120**, 401–24.

Stell, W.K. (1975) Horizontal cell axons and axon terminals in goldfish retina. *Journal of Comparative Neurology*, **159**, 503–20.

Stell, W.K., Lightfoot, D.O., Wheeler, T.G., and Leeper, H.F. (1975) Goldfish retina: functional polarization of cone horizontal cell dendrites and synapses. *Science*, **190**, 989–90.

Studholme, K.M. and Yazulla, S. (1994) ^{3}H-adenosine derived uptake labels rod horizontal cells in the goldfish retina. *Investigative Ophthalmology and Visual Science*, **35** (Suppl.) 2153.

Tachibana, M. and Kaneko, A. (1984) gamma-Aminobutyric acid acts at axon terminals of turtle photoreceptors: difference in sensitivity among cell types. *Proceedings of the National Academy of Science of the United States of America*, **81**, 7961–4.

Tapia, R. and Arias, C. (1982) Selective stimulation of neurotransmitter release from chick retina by kainic and glutamic acids. *Journal of Neurochemistry*, **39**, 1169–78.

Van Buskirk, R. and Dowling, J.E. (1981) Isolated horizontal cells from carp retina demonstrate dopamine-dependent accumulation of cyclic AMP. *Proceedings of the National Academy of Sciences of the United States of America*, **78**, 7825–9.

Van Haesendonck, E. and Missotten, L. (1992) Three types of GABA-immunoreactive cone horizontal cells in teleost retina. *Visual Neuroscience*, **8**, 443–8.

Vizi, E.S. (1978) Na^+K^+-activated adenosinetriphosphatase as a trigger in transmitter release. *Neuroscience*, **3**, 367–84.

Wagner, H.-J. (1980) Light-dependent plasticity of the morphology of horizontal cell terminals in cone pedicles of fish retinas. *Journal of Neurocytology*, **9**, 573–90.

Wagner, H.-J. and Djamgoz, M.B.A. (1993) Spinules: a case for retinal synaptic plasticity. *Trends in Neuroscience*, **16**, 201–6.

Wässle, H. and Chun, M.H. (1989) GABA-like immunoreactivity in the cat retina: light microscopy. *Journal of Comparative Neurology*, **279**, 43–54.

Weiler, R. and Zettler, F. (1979) The axon bearing horizontal cells in the teleost retina are functional as well as structural units. *Vision Research*, **19**, 1261–8.

Werblin, F.S. and Dowling, J.E. (1969) Organization of the retina of the mudpuppy, II. Intracellular recording. *Journal of Neurophysiology*, **32**, 339–355.

Witkovsky, P. and Dowling, J.E. (1969) Synaptic relationships in the plexiform layers of carp retina. *Zeitschrift für Zellforschung und Mikroskopisne Anatomie*, **100**, 60–82.

Witkovsky, P. and Powell, C.C. (1981) Synapse formation and modification between distal retinal neurons in larval and juvenile *Xenopus*. *Proceedings of the Royal Society of London*, **211**, 373–89.

Wu, S.M. (1986) Effects of gamma-aminobutyric acid on cones and bipolar cells of the tiger salamander retina. *Brain Research*, **365**, 70–7.

Wu, S.M. (1992) Feedback connections and operation of the outer plexiform layer of the retina. *Current Opinions in Neurobiology*, **2**, 462–8.

Wu, S.M. and Dowling, J.E. (1978) L-Aspartate: evidence for a role in cone photoreceptor synaptic transmission in the carp retina. *Proceedings of the National Academy of Science of the United States of America*, **75**, 5205–9.

Wu, S.M. and Dowling, J.E. (1980) Effects of GABA and glycine on the distal cells of the cyprinid retina. *Brain Research*, **199**, 401–14.

Yasui, S. (1987) Ca and Na homeostasis in horizontal cells of the cyprinid fish retina: evidence for Na–Ca exchanger and Na–K pump. *Neuroscience Research* Suppl. 6, S133–S146.

Yazulla, S. (1983) Stimulation of GABA release from retinal horizontal cells by potassium and acidic amino acid agonists. *Brain Research*, **275**, 61–74.

Yazulla, S. (1985a) Evoked efflux of ^{3}H-GABA from goldfish retina in the dark. *Brain Research*, **325**, 171–80.

Yazulla, S. (1985b) Factors controlling the release of GABA from goldfish retinal horizontal cells. *Neuroscience Research* Suppl. 2, S147–S165.

Yazulla, S. (1986) GABAergic mechanisms in the retina. *Progress in Retinal Research*, **5**, 1–52.

Yazulla, S. (1991) The mismatch problem for GABAergic amacrine cells in goldfish retina: Resolution and other issues. *Neurochemical Research*, **16**, 327–39.

Yazulla, S. and Kleinschmidt, J. (1982) Dopamine blocks carrier-mediated release of GABA from retinal horizontal cells. *Brain Research*, **233**, 211–15.

Yazulla, S. and Kleinschmidt, J. (1983). Carrier-mediated release of GABA from retinal horizontal cells. *Brain Research*, **263**, 63–75.

Yazulla, S. and Studholme, K.M. (1987) Ultracytochemical distribution of ouabain-sensitive, K^+-dependent, *p*-nitrophenylphosphatase in the synaptic layers of goldfish retina. *Journal of Comparative Neurology*, **261**, 74–84.

Yazulla, S. and Yang, C.-Y. (1988) Colocalization of GABA- and glycine-immunoreactivities in a subset of retinal neurons in tiger salamander. *Neuroscience Letters*, **95**, 37–41.

Yazulla, S., Neal, M.J. and Cunningham, J. (1985). Stimulated release of endogenous GABA and glycine from the goldfish retina. *Brain Research*, **345**, 384–8.

Zhou, Z.J., Fain, G.L. and Dowling, J.E. (1993) The excitatory and inhibitory amino acid receptors on horizontal cells isolated from the white perch retina. *Journal of Neurophysiology*, **70**, 8–19.

Zucker, C., Yazulla, S. and Wu, J. (1984) Non-correspondence of ^{3}H-GABA uptake and GAD localization in goldfish amacrine cells. *Brain Research*, **298**, 154–8.

11

The organization of photoreceptor to bipolar synapses in the outer plexiform layer

HELGA KOLB and RALPH NELSON

11.1 INTRODUCTION

In the vertebrate retina, light, acting upon visual pigment molecules in the photoreceptor outer segment, is transduced into electrical signals, which then modulate the flow of neurotransmitter from photoreceptors to second-order neurons, the bipolar and horizontal cells in the outer plexiform layer (OPL). The photoreceptor neurotransmitter, glutamate, binds to receptor complexes located on bipolar-cell and horizontal-cell postsynaptic membranes, directly or indirectly gating ion channels which mediate either ON-center or OFF-center responses. Thus, information concerning brightness, darkness and color is transmitted proximally. Information is also transformed in this process, yielding second-order neurons with unique electrophysiological signatures. Horizontal cells integrate signals laterally over wide spatial regions and mediate antagonistic surround responses of bipolar cells. Signals from different photoreceptor types are combined, often synergistically to form broad-band, luminosity type responses, but sometimes antagonistically, to form color-opponent responses. Still other types of second-order neurons in the OPL remain faithful to single photoreceptor types, processing only rod or cone signals. Some of these unique bipolar-cell types can be identified by immunohistochemical markers, such as the presence of protein kinase C isozymes in rod and cone bipolar cells (Negishi *et al.*, 1989; Kolb *et al.*, 1993), or cholecystokinin in blue-sensitive cone bipolar cells (Kouyama and Marshak, 1992).

Bipolar cells send axons to the retinal inner plexiform layer (IPL) where they synapse on amacrine and ganglion cells. The pattern of connectivity is highly selective in this layer. Axons of ON- and OFF-center bipolar cells are confined to distinct sublaminae, and even within the sublaminae, synaptic contacts are further restricted to specific cell types. For example, mammalian rod bipolar cells appear to interact only with special amacrine cell types, and ganglion cells may receive input from only one of several different cone bipolar types branching within their dendritic arborizations. Thus, at the IPL level, information continues to be processed in synaptically specific pathways. However, it is the connec-

Neurobiology and Clinical Aspects of the Outer Retina
Edited by M.B.A. Djamgoz, S.N. Archer and S. Vallerga
Published in 1995 by Chapman & Hall, London
ISBN 0 412 60080 3

tions of photoreceptors to bipolar cells at the OPL that are the source of the two major pathways responsible throughout the whole visual system for the interpretation of the visual image as lighter or darker than background (Gouras, 1971; Famiglietti and Kolb, 1976; Schiller, 1992). It is these important connections at the OPL that is the subject of this chapter.

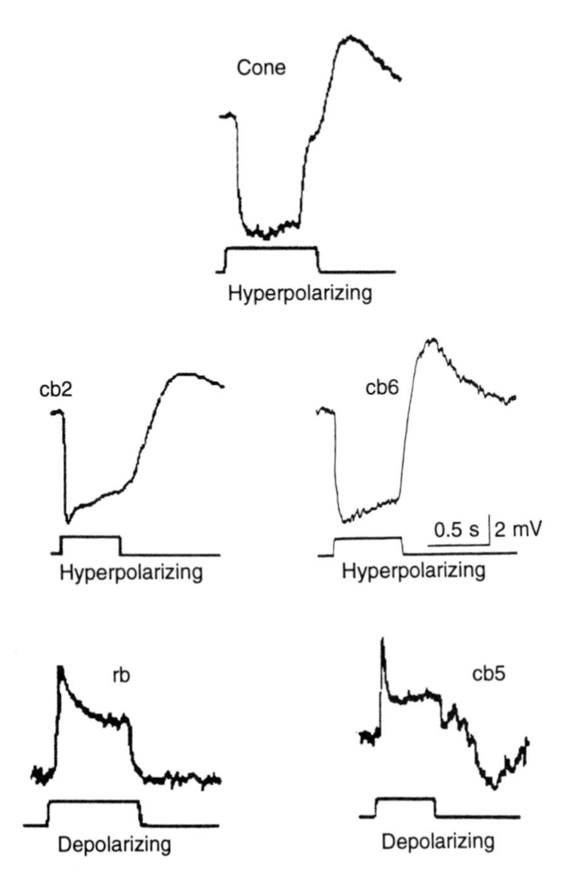

Figure 11.1 Sample intracellular recordings from a cone and four different bipolar cell types in the cat and rabbit retina. In cat, the cone and cone bipolar, cb2 and cb6 all respond to light stimuli (rectangular, photocell traces beneath each record) with hyperpolarizing, OFF-center slow potentials. cb5 and the rod bipolar (rb) respond with depolarizing, ON-center slow potentials to the light flash. (Rod bipolar response from rabbit, Dacheux and Raviola, 1986; others from cat, Nelson and Kolb, 1983).

11.2 ELECTROPHYSIOLOGY OF ON- AND OFF-CENTER BIPOLAR CELLS

In both mammalian and non-mammalian vertebrates, bipolar cells fall into two physiological classes: ON-center and OFF-center (Werblin and Dowling, 1969; Kaneko, 1970; Richter and Simon, 1975; Naka, 1976; Lasansky, 1978; Marchiafava and Weiler, 1980; Nelson and Kolb, 1983; Dacheux and Raviola, 1986). Example responses are shown in Figure 11.1. When a luminous spot is projected onto a bipolar-cell receptive field center, OFF-center cells hyperpolarize whereas ON-center cells depolarize, both in a sustained manner. Illumination of the receptive field periphery tends generally to repolarize the center response, antagonizing it. This is the 'center-surround' characteristic of bipolar-cell receptive fields. A further feature of bipolar-cell responses in some species is color coding (Kaneko, 1973; Yazulla, 1976; Hashimoto and Inokuchi, 1981; Kaneko and Tachibana, 1981, 1983). Thus, double opponent bipolar-cell types exists in carp retina (Kaneko and Tachibana, 1981, 1983).

Both ON-center and OFF-center responses are driven by ionic gradients with reversal potentials positive to the dark potential of the cell. This is accomplished by opposite conductance changes. Light causes an increase in conductance in ON-center bipolar cells, and a decrease in conductance in OFF-center cells (Toyoda, 1972; Nelson, 1973; Lasansky, 1992). One exception to this rule is known: cone-dominated responses in ON-center carp bipolar cells are accompanied by decreases in conductance to ions with a reversal potential more negative than the dark potential (Saito *et al.*, 1979).

Since vertebrate photoreceptors are in a depolarized state in darkness and are hyperpolarized by light, as can be seen in Figure 11.1 (Bortoff, 1964), release of the neurotransmitter, glutamate, occurs continuously in the dark and is suppressed by light (reviewed in Dowling, 1987; Massey, 1990). Different glutamate receptor types are associated with ON- and OFF-center bipolar cells (Miller and Slaughter, 1986). OFF-bipolar receptors appear related to the α-amino-3-hydroxy -5-methyl-isoxazole propionate (AMPA)–kainate type. Light responses of this cell type are suppressed by the antagonists *cis*-2, 5-piperidine dicarboxylate (PDA) (Slaughter and Miller, 1983), 6-cyano-7-nitroquinoxaline-2, 3-dione (CNQX) (Hensley *et al.*, 1993) or kynurenic acid (Kim and Miller, 1993). Thus the photoreceptor to OFF-bipolar synapse appears to be a common, excitatory, glutamatergic type. ON-bipolar receptors bind selectively the glutamate agonist APB (or AP4, 2-amino-4-phosphonobutyrate), and are insensitive to AMPA kainate ligands. Application of APB selectively hyperpolarizes the membrane potential and suppresses the light-responses of ON-center bipolar cells (Slaughter and Miller, 1981; Nawy and Copenhagen, 1987; Hensley *et al.*, 1993; Kim and Miller, 1993). This synaptic action is uncommon, involving glutamatergic blockage of cation channels, and sign inversion of the presynaptic signal. Exploration of ON-center bipolar cells by whole-cell recording with patch electrodes revealed that both APB-, and light-induced responses were labile. These responses 'washed out' with time, suggesting mediation by diffusible intracellular components (Nawy and Jahr, 1990; Shiells and Falk, 1990; Yamashita and Wässle, 1991; Lasansky, 1992; Kim and Miller, 1993). Receptor-activated G-proteins, and perhaps a cyclic-GMP cascade similar to that occurring in photoreceptors have been suggested (Nawy and Jahr, 1990; Shiells and Falk, 1990). Further details of the electrophysiology of bipolar cells are given in Chapter 12.

11.3 STRUCTURE OF ROD AND CONE SYNAPSES WITH BIPOLAR CELLS

The nature of bipolar cell contacts with the photoreceptor synaptic region has always aroused interest, for this synapse represents the first stage in the retina where integration of influences from different photoreceptors, and the lateral spatial information from horizontal cells can be combined and brought to bear on the vertical, bipolar-cell pathways. The division of signals originating from the photoreceptor into two bipolar cell 'channels' subserving brightness and darkness (ON- and OFF-center types, respectively) was appreciated at about the same time that electron microscope (EM) studies showed that bipolar cells could be classified according to the type of contacts made with photoreceptors. Examples of bipolar-cell contacts with rods and cones are shown in Figures 11.2 and 11.3. There was a concurrence of findings in a variety of species: human (Missotten, 1965), fish (Stell, 1967), monkey (Kolb, 1970), and turtle (Lasansky, 1971). Basically, two types of synaptic contact were seen. One type of bipolar penetrated rod spherule or cone pedicle to terminate as the central member of a triad of processes at the synaptic ribbon (Figure 11.2a, b, RB, IMB). These became known as 'invaginating bipolar cells'. The other variety had dendritic terminals ending on the surface of the photoreceptor base at a distance from the synaptic ribbons (Figure 11.3a,b, FB). These became known as 'flat bipolar cells'.

Fundamentally, synaptic contacts between bipolar and photoreceptor cells are classified according to two criteria: (1) the position of the contact with respect to the photoreceptor synaptic ribbon, and (2) the ultrastructure of the synaptic cleft. Position is the easiest to determine. In addition to the 'flat' and 'invaginating' contacts originally described, current evidence suggests yet a third category. This is the 'semi-invaginating' contact. This

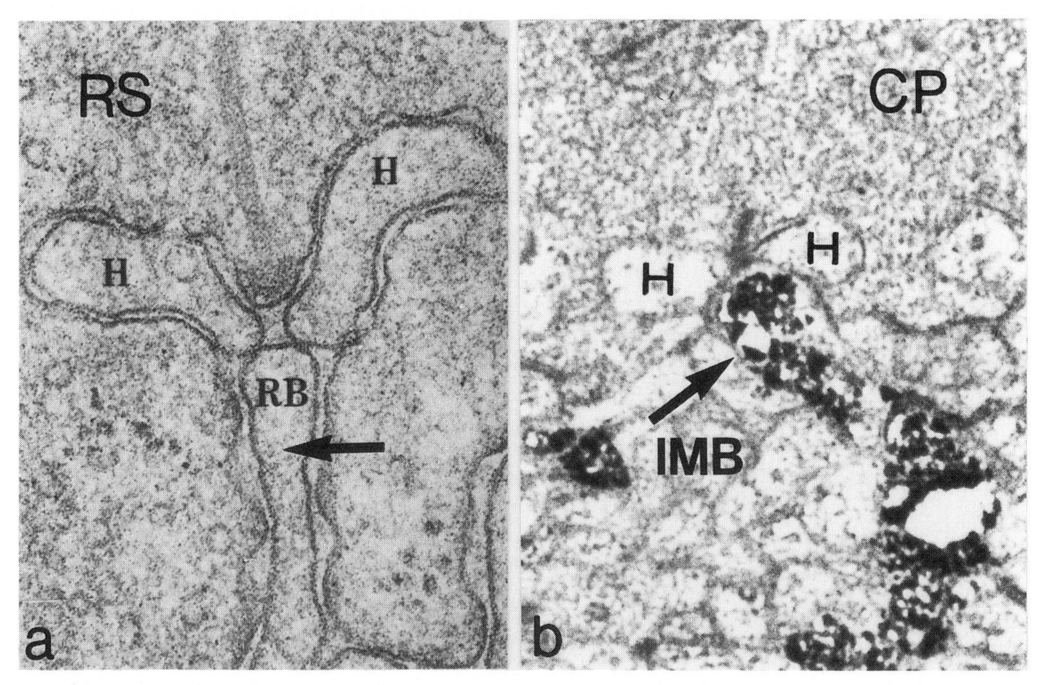

Figure 11.2(a) Electron micrograph of an invaginating ribbon-related contact made by a rod bipolar cell dendrite (rb, arrow) in a rod spherule (RS) of the cat retina. On either side of the RB are lateral elements, (H), processes arising from horizontal cell axon terminals. Magnification ×95 000. (b) Invaginating ribbon-related contact made by an invaginating midget bipolar cell dendrite (IMB, arrow) in a cone pedicle (CP) of the monkey retina. The Golgi stained IMB dendrite ends as a central element below the synaptic ridge of the ribbon and is flanked by two lateral elements contributed from horizontal cell (H) (triad arrangement). Magnification ×30 000.

was first reported in monkey retina as the exclusive synapse made between flat midget bipolar cells and cones (Kolb, 1970). Dendritic tips of flat midget bipolar cells were found always associated with photoreceptor synaptic ribbons, but not as central members of the triad. Rather, they ended around the mouth of the invagination at a synaptic density resembling that made by 'flat' contacts. This arrangement has also been found for cat cone bipolar type cb5 (Nelson and Kolb, 1983), and for the B6 cone bipolar of turtle (Kolb *et al.*, 1986).

In the turtle retina, there is an elaborate system of interphotoreceptor connections via telodendria. Photoreceptor telodendria form central elements of ribbon triads. The position normally occupied by invaginating bipolar-cell dendrites, might often be occupied by the telodendron of a neighboring photoreceptor instead (Kolb and Jones, 1985; Lasansky, 1971; Mariani and Lasansky, 1984). Thus, in the turtle retina, narrow-cleft 'semi-invaginating' bipolar-cell contacts, often described as exhibiting 'punctate' densities on the presynaptic membrane (Kolb and Jones, 1984), might represent the displacement of a true central invaginating bipolar dendrite by a photoreceptor process. In cat and monkey retinas, however, this explanation of the 'semi-invaginating' contact appears unlikely. If the 'semi-invaginating' bipolar-cell den-

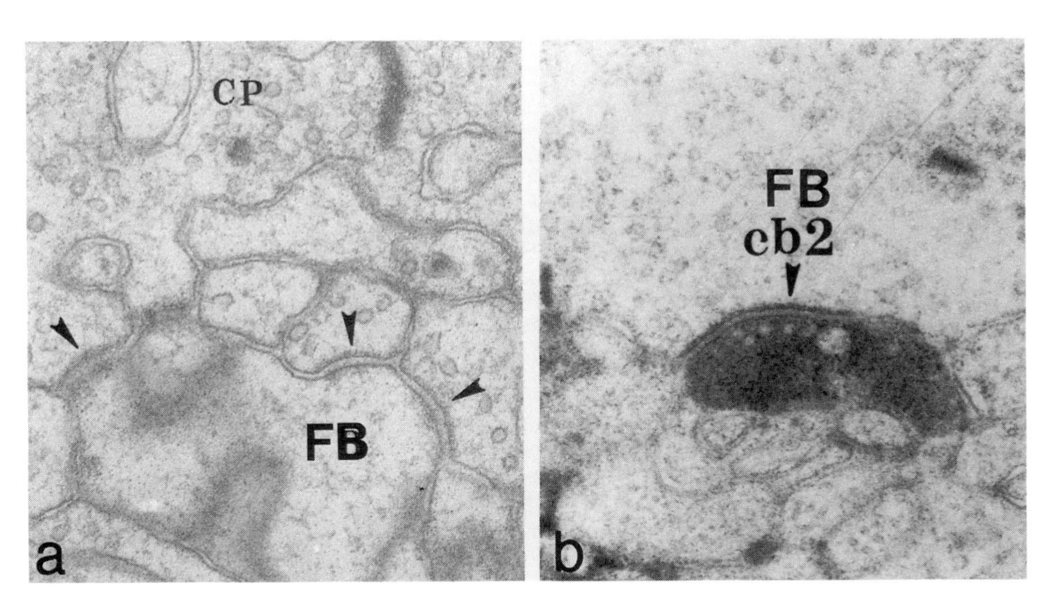

Figure 11.3 (a) Wide-cleft basal junction made by a flat bipolar cell dendrite (FB) upon the base of a cone pedicle (CP) in the turtle retina. A wide-cleft separates pre- and postsynaptic membranes, and striations of dense material cross the cleft (arrow). Magnification ×52 000. (b) Wide-cleft basal junction made by an HRP-injected and physiologically recorded flat cone bipolar cell (cb2) in the cat retina. Magnification ×50 000.

drite is displaced, it is by other bipolar cell dendrites.

Lasansky (1971) studying turtle retina suggested refinements of the simpler definitions of 'invaginating' and 'flat' (Dowling and Boycott, 1966; Kolb, 1970; Raviola and Gilula, 1975), and coined the phrase 'basal junction' to describe in elegant EM micrographs, the wide-cleft, striated flat bipolar contacts in the turtle. In addition, he described the 'ribbon-related' junctions as being actually narrow-cleft junctions, making synaptic contact along a length of apposition between cone-bipolar dendrite and cone pedicle membrane within the invagination. Narrow-cleft junctions were thus added to the ultrastructural descriptions of the types of contact bipolars can make with photoreceptors (Lasansky, 1972; Stell *et al.*, 1977; Kolb and Jones, 1984). Stell *et al.* (1977) studying goldfish retina used the terms 'wide-cleft' and 'narrow-cleft' to refer to the ultrastructural 'basal' and 'ribbon-related' contacts described by Lasansky (1971). This change in nomenclature provided the advantage that synaptic cleft structure need not be tied to the position of the contact with respect to the ribbon, as implied by the terms 'basal' and 'ribbon-related'. Nonetheless typical 'flat' contacts usually make 'wide-cleft' junctions. In wide-cleft contacts, presynaptic and postsynaptic membranes appear rather straight and smooth; there is a regular wide-space cleft between them with orthogonal striations in the cleft material. In 'narrow-cleft' contacts presynaptic and postsynaptic membranes approach more closely, and the spacing is somewhat irregular. These contacts are fre-

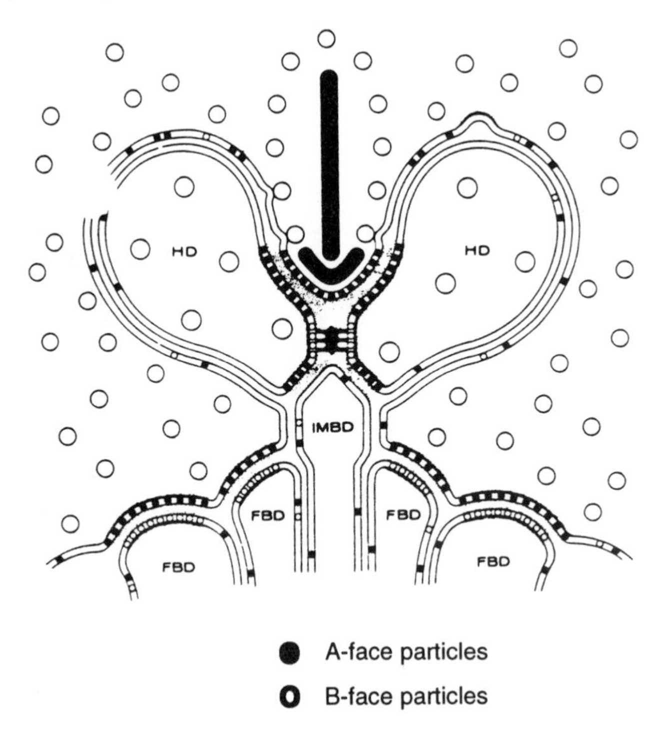

Figure 11.4 Summary drawing of A- and B-face particle distribution in the pre- and postsynaptic membranes of the cone to bipolar and cone to horizontal cell synapses in the monkey retina. Particles which remain preferentially associated with the cytoplasmic portion of the membrane (A-face) are represented by solid black spots: particles which remain with the external portion of the membrane (B-face) are represented as small open circles. HD, horizontal cell dendrite; IMBD, invaginating midget bipolar dendrite; FBD, dendrites of flat midget and diffuse cone bipolar cells. Large circles represent synaptic vesicles. ('hole' like appearance of HD and cone membranes to the left is an anomoly of the original figure). (Taken from Raviola and Gilula (1975) with permission.)

quently made *en passage* by the dendritic tips of invaginating bipolar cells. Stell *et al.* (1977), furthermore, proposed that it was the distinction in the structure of the cleft which was of fundamental physiological importance. Examining EM-reconstructed, Golgi-stained bipolar cells in goldfish, morphologically similar to ON- and OFF-center types (Kaneko, 1970), Stell *et al.* (1977) concluded that wide-cleft junctions mediated OFF-center responses, whereas narrow-cleft junctions mediated ON-center responses (some examples are illustrated in Figure 11.6c).

Raviola and Gilula (1975) performed a freeze fracture study on rabbit and monkey OPL in order to look for structural differences in particle distribution in pre- and postsynaptic membranes of photoreceptor synapses. Figure 11.4 summarizes the results of this type of analysis. The distribution of A-face and B-face particles at the flat bipolar basal junctions was described as being most like the situation in excitatory synapses in other parts of the vertebrate CNS, i.e. in cerebellum and olfactory bulb (Landis and Reese, 1974; Landis *et al.*, 1974). In contrast, the lack of B-face particles in the invaginating dendrite resembled inhibitory synaptic structure in the CNS. The interpretation of these data is that flat contacting bipolar cells make excitatory synapses with the photoreceptor (Figure 11.4) and thereby give the same response to light as the photoreceptor itself, i.e. a hyperpolarizing or OFF-center response, whereas invaginating ribbon-related bipolar types have inhibitory input (Figure 11.4) and show a sign-inverted response, i.e. they are depolarizing to light stimuli or ON-center (see also section 11.6).

11.4 STRATIFICATION OF ON- AND OFF-CENTER BIPOLAR CELL AXONS

An important clue in the attempted correlation between synaptic ultrastructure and physiological properties of bipolar cells is the observation that, in certain species like the monkey, stratification of bipolar cell axon terminal endings in the IPL, in addition to photoreceptor contacts, serve to distinguish different bipolar types. In the rhesus monkey retina, the axon terminals of flat midget bipolar cells end close to amacrine cell bodies (sublamina **a**) in distal neuropil of the IPL. The axon terminals of invaginating midget types end close to ganglion cell bodies (sublamina **b**), in the proximal neuropil of the IPL (Kolb, 1970). These bipolar endings correspond exactly to the level at which two different midget ganglion cells send their dendritic trees (Kolb, 1970; Kolb and DeKorver, 1991). Gouras (1971) speculated that flat-midget bipolar → midget-ganglion cell connections were responsible for OFF-center receptive fields, and that invaginating-midget bipolar → midget-ganglion cell connections underlay ON-center receptive fields, in these ganglion cells.

A schematic illustration of the co-stratification of flat and invaginating bipolar cell types with dendrites of ganglion cells in cat retina is shown in Figure 11.5. Nelson and coworkers (1978) proved conclusively, with intracellular recordings and dye injections of ganglion cells, that dendritic tree level in the IPL correlated with OFF- and ON-center physiology. Thus ganglion cells branching in sublamina **a** proved to be OFF-center and

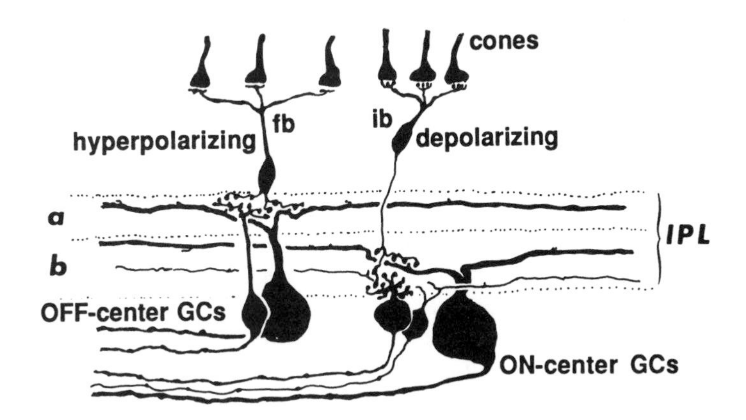

Figure 11.5 Summary drawing of the connections of cone bipolar cells and ganglion cells responsible for ON- and OFF-center transmission of the visual image in the cat retina. Modified from Nelson *et al.* (1978). Each cone connects to a center-hyperpolarizing flat bipolar (fb), and a center-depolarizing invaginating bipolar (ib) cell respectively. The center-hyperpolarizing bipolar contacts an OFF-center type of ganglion cell (GC) that branches in sublamina **a** of the IPL. The center-depolarizing bipolar contacts an ON-center ganglion cell that branches in sublamina **b** of the IPL.

ganglion cells branching in sublamina **b** proved to be ON-center (Nelson *et al.*, 1978). The conclusion followed, that OFF-center ganglion cells might be driven by center-hyperpolarizing bipolar cells, making flat contacts with photoreceptors, and that ON-center ganglion cells might be driven by center-depolarizing bipolar cells, making contacts at invaginating synapses (Famiglietti and Kolb, 1976), as shown in Figure 11.5.

In submammalian species where bipolar cells have been recorded and marked with dyes for morphological classification, the generalization that OFF-center bipolar cells send axonal endings to sublamina **a**, and ON-center bipolar cells send axonal endings to sublamina **b** appears to be true (Famiglietti *et al.*, 1977; Lasansky, 1978; Saito *et al.*, 1983, 1985; Sakai and Naka, 1983; Werblin, 1991). However, turtle, dace and cat retina show very obvious exceptions to this rule. Multistratified bipolar cells are common in turtle retina (Kolb, 1982) and there are even bistratified types in cat, monkey and human retinas (Kolb *et al.*, 1981; Mariani, 1983; Kolb *et al.*, 1992). Clearly, such bipolar cells are candidates to provide inputs of the 'wrong sign' either to the ON or to the OFF-layer. In turtle, it appears that hyperpolarizing bipolar cells can have portions of their bi- or tristratified axon terminals in sublamina **b** as long as one portion is in sublamina **a** (Weiler, 1981; Ammermüller and Kolb, 1995). In cat, the cb6 cone bipolar type is center-hyperpolarizing and its axon terminal is clearly monostratified in the middle of sublamina **b** with *no* branch in sublamina **a** (Nelson and Kolb, 1983). In dace retina, color-coded OFF-center bipolar cells (as defined by hyperpolarizing responses to white light) were found with axons in sublamina **b** (Hashimoto and Inokuchi, 1981).

The conclusion, concerning ganglion cells having a center response sign invariably related to the sublamina of the IPL in which their major dendrites branch, does appear to be correct for all vertebrate retinas, where unistratified or multistratified dendritic trees remain within *one* sublamina (Famiglietti *et al.*, 1977; Nelson *et al.*, 1978; DeMonasterio, 1979; Marchiafava and Weiler, 1980; Peichl and Wässle, 1981; Saito, 1983; Amthor *et al.*, 1989a; McReynolds and Lukasiewicz, 1989; Werblin, 1991; Ammermüller and Kolb, 1995). Ganglion cells with multistratified dendritic trees spanning both IPL sublaminae, however, are not included in the hypothesis and in fact can respond in any manner. ON–OFF-, ON-center and OFF-center responses have been found (Famiglietti *et al.*, 1977; Jensen and DeVoe, 1983; Djamgoz and Wagner, 1987; Amthor *et al.*, 1989b; Ammermüller and Kolb, 1995).

11.5 CIRCUITRY OF BIPOLAR CELL CONTACTS WITH GANGLION CELLS

The finding of anomalous branching in sublamina **b** of an OFF-center bipolar cell in cat (cb6) led to the push–pull theory of synaptic contacts to ganglion cells in cat retina (McGuire *et al.*, 1986). The idea is that, sublamina of dendritic branching notwithstanding, some ganglion cells, such as beta ganglion cells of cat, would receive both ON- and OFF-center bipolar inputs (Chapter 13). Thus, for an ON-center ganglion cell, an ON-center bipolar would provide ON-excitation through an excitatory synapse and on an OFF-center bipolar would provide OFF-inhibition through a sign-inverting synapse. This theory would of course require that an equivalent bipolar cell to hyperpolarizing cb6 (i.e. a depolarizing cone bipolar) be present in sublamina **a**. No such bipolar has been found physiologically yet, although morphological candidates exist (Kolb *et al.*, 1981). The 'push–pull' theory received no support from recent electrophysiological and pharmacological studies (Ikeda and Sheardown, 1983; Freed and Nelson, 1994). However, the complete range of bipolar inputs to beta ganglion cells in cat are not yet well understood. Both OFF- and ON-center beta cells appear to receive at least two different types of bipolar inputs, i.e. cb1 and cb2 synapse upon OFF-center beta cells (Kolb and Nelson, 1993) and cbb1 and cbb2 (according to Sterling and coauthors' nomenclature) synapse on ON-center beta cells (McGuire *et al.*, 1986; Cohen and Sterling, 1991). Even if push–pull is not physiologically significant, it seems likely that some oppositely signed input, for instance from some center-hyperpolarizing cb6 (Nelson and Kolb, 1983), equivalent to cbb2 (McGuire *et al.*, 1986) occurs upon ON-center beta cells even though excitatory input from ON-centre cb5 predominates (Nelson and Kolb, 1983; Freed and Nelson, 1994). Further information is required to resolve this question. Particularly intriguing will be determining the response sign of the cbl bipolar, which, in addition to the well-known OFF-center type cb2, makes synapses upon OFF-center beta cells (Nelson and Kolb, 1983; Kolb and Nelson, 1993).

Not all ganglion cells receive inputs from multiple bipolar-cell types. Freed and Sterling (1988) pointed out that ON-center alpha cells were contacted only by cbbl cone bipolar cells. Similarly Kolb and Nelson (1993) found that OFF-center alpha cells were contacted only by type cb2 cone bipolar cells. Recent work on midget ganglion cells in human retina has also revealed contacts only from single midget-type bipolar cells (Kolb and DeKorver, 1991).

11.6 LINKING SIGNALS TO SYNAPSES BY INTRACELLULAR STAINING OF BIPOLAR CELLS

The introduction of horseradish peroxidase (HRP) as a microelectrode stain suitable for electron microscopy (Graybiel and DeVor, 1974) provided a means to observe directly correlations between bipolar-to-photoreceptor contacts and ON- or OFF-center physiology. Electron microscopy of Golgi-stained bipolar cells also provided this opportunity in cases where responses of a morphologically recognized type were otherwise known (e.g. from intracellular fluorecent dye staining). The available results are summarized in Table 11.1 for 22 different bipolar cell types studied in nine species.

Lasansky (1978) was the first to cast doubt on seemingly well-founded expectations. Using the HRP-injection technique, he showed that center-depolarizing bipolar cells in salamander retina made contacts with rods and cones at both wide-cleft basal junctions and at invaginating ribbon junctions. Furthermore, it was the basal, wide-cleft junctions which accounted for about 80–90% of the contacts, suggesting that it was, in fact these contacts which were likely to mediate center-depolarizing responses. Center-hyperpolarizing bipolar cells also made both types of junction but in this case the invaginating ribbon-related variety was more numerous (about 80%) in both rods and cones (Figure 11.6a).

Figure 11.6 illustrates other exceptions to the expected rule in various species. As already noted, Stell *et al.* (1977) proposed that narrow-cleft junctions with ribbon related contacts were associated with center-depolarizing bipolar cells and that wide-cleft basal junctions were associated with center-hyperpolarizing types in goldfish (Figure 11.6c). Saito *et al.* (1985) confirmed this result in carp retina by electron-microscopic reconstruction of some of the HRP-injected cells (types ON B (II) and OFF B (II); Table 11.1). Additionally, however, two other bipolar cell types were found which did not conform. Type OFF B (I) formed ribbon contacts only with both rods and cones, thus becoming a clear example of an invaginating OFF-center type (Figure 11.6d). Furthermore, type ON B (I), while making ribbon contacts with rods, contacted cones at wide-cleft basal junctions (Saito *et al.*, 1983, 1985). Sakai and Naka

Table 11.1 Receptor contacts of ON- and OFF-center bipolar cells in vertebrate retinas

Species	Type	Resp	Rod contacts	Cone contacts	Reference
Salamander	ON type	ON	basal, narrow and wide-cleft	basal, narrow and wide-cleft	Lasansky (1978)
Catfish	ON type	ON	NONE	semi-invag.	Sakai and Naka (1983)
Goldfish	b1	ON	ribbon, narrow-cleft	narrow-cleft	Stell *et al.* (1977)
	b2	ON	ribbon, narrow-cleft	ribbon, narrow-cleft	
	b3	ON	ribbon	ribbon	
Carp	ON-B(I)	ON	ribbon	basal	Saito *et al.* (1983), (1985)
	ON-B (II)	ON	ribbon	semi-invag.	
Turtle	B6	ON	NONE	semi-invag., narrow-cleft	Kolb *et al.* (1986)
Cat	cb5	ON	NONE	semi-invag., narrow-cleft	Nelson and Kolb (1983)
	rod bipolar	ON?	ribbon	NONE	Kaneko *et al.* (1992)
Rabbit	rod bipolar	ON	ribbon	NONE	Dacheux and Raviola (1986)
Rat	rod bipolar	ON	ribbon	NONE	Yamashita and Wassle (1991)
Salamander	OFF type	OFF	ribbon, narrow and wide-cleft	ribbon, narrow and wide-cleft	Lasansky (1978)
Catfish	OFF type	OFF	NONE	ribbon	Sakai and Naka (1983)
Goldfish	a1	OFF	basal, wide-cleft	basal, wide-cleft	Stell *et al.* (1977)
	a2	OFF	basal, wide-cleft	basal, wide-cleft	
Carp	OFF-B (I)	OFF	ribbon	ribbon	Saito *et al.* (1983), (1985)
	OFF-B (II)	OFF	basal	basal	
Turtle	B4	OFF	NONE	basal, wide-cleft	Kolb *et al.* (1986)
	B10[a]	OFF	NONE	basal, wide and narrow-cleft	Zhang *et al.* (1992)
Cat	cb2	OFF	NONE	basal, wide-cleft	Nelson and Kolb (1983)
	cb6[a]	OFF	NONE	ribbon, wide-cleft	

[a]These disobey the sublamina axonal ending rule, i.e. B10 and cb6 are OFF-center units but send axon terminals only to sublamina **b** of the IPL.

(1983) studied catfish retinal bipolar cells by a similar technique. Although invaginating ON-center bipolar cells were found, so were invaginating OFF-center bipolar cells, which was not anticipated from the proposal of Stell *et al.* (1977).

In both fish species (goldfish and carp) there are purely invaginating, ON-center bipolar types. The cone contacts of this class of teleost ON-center bipolar appear somewhat similar to 'semi-invaginating' contacts of mammals and reptiles. They are ribbon associated, but there is always another bipolar process closer to the ribbon. This is the OFF-center bipolar (Sakai and Naka, 1983; Saito *et al.*, 1983, 1985). None of the available reports really defines the junctions with high enough resolution to enable clear definition of the membrane appositions. In other words, we cannot be sure whether the basal junctions that center-depolarizing bipolars make are narrow-cleft or whether center-hyperpolarizing bipolars make wide-cleft as well as ribbon-related synapses in these fish retinas.

The rod input to the depolarizing bipolar in fish retinas is mediated by an APB-sensitive glutamate receptor whereas the cone input to the same bipolar is mediated by a different glutamatergic mechanism (Nawy and Copenhagen, 1987). However, Saito *et al.* (1985) noted no correlation between glutamate receptor type and synaptic ultrastructure. The atypical glutamate receptor of the cone synapse is associated with both flat and invaginating contacts.

The original studies of Lasansky (1971) in turtle retina showed the important distinction between wide-cleft basal and narrow-cleft ribbon-related contacts. It is important to examine the physiology associated with these contacts in these turtle bipolar cells. Wide-cleft basal junctions would be expected to underlie OFF-center responses whereas narrow-cleft ribbon-related synapses would be expected to mediate ON-center responses. This turns out to be true for two bipolar-cell types of the turtle retina: B6 bipolar cells are known to be center-depolarizing (Marchiafava and Weiler, 1980) and EM study showed them to make narrow-cleft semi-invaginating contacts with cones (Figure 11.6b; Kolb *et al.*, 1986). In contrast, B4 bipolars are center-hyperpolarizing (Marchiafava and Weiler, 1980) and make primarily wide-cleft basal junctions (Figure 11.2a; Kolb *et al.*, 1986). A bipolar-cell type making both basal, and fully invaginating contacts with both rods and cones has also been described in turtle (Dacheux, 1982) but the responses of this cell type are not known.

Finally in the cat retina, cone bipolar cell cb6 appears to break both the sublaminal rule for axonal ending, and the rules for receptor contacts. Responses of this cell are OFF-center (Figure 11.1). When examined ultrastructurally for synaptic contacts with cones, cb6 was found to make *both* invaginating central contacts and wide-cleft basal junctions with cone pedicles (Figure 11.6e and f). The invaginating dendrite does not approach the ribbon completely. Instead a 'cap' of horizontal cell dendrites blocks the apex of the invaginating dendrite (Figure 11.6f). The types of contact that cb6 make with cones in the cat, are possibly analogous to hyperpolarizing units with invaginating ribbon-related synapses in salamander, catfish and carp (Lasansky, 1978; Saito *et al.*, 1983, 1985; Sakai and Naka, 1983). The cb6 axon terminal ends in sublamina **b**.

Table 11.1 summarizes present anatomical and physiological findings concerning photoreceptor-to-bipolar connections. In the case of bipolar cells with input from both rods and cones, each input was scored; however the following statistics take account only of the predominant synaptic types encountered, ignoring minority types. Considering the category of synaptic contact position, among basal (flat) contacts there are three examples mediating ON-center responses, and nine mediating OFF-center responses. Among invaginating contacts, 10 mediate ON-center responses, whereas six mediate OFF-center responses. All four examples of semi-invaginating contacts mediated ON-center responses. These data are sufficient to rule out much confidence in a predictive rule based solely on the position of photoreceptor-to-bipolar contact. Even the hopeful trend for semi-invaginating contacts is diminished when one considers that the flat midget bipolar cell of rhesus retina (Kolb, 1970), which is widely believed, though not yet proven, to be OFF-center, would fall into this category. Considering the classification according to ultrastructure of the synaptic cleft, this approach is less widely applicable since not all contacts have been, or perhaps can be, categorized in this way. Among the narrow-cleft synapses, seven were found associated with ON-center responses, whereas three appeared with OFF-center responses. Wide-cleft contacts occurred in ten OFF-center cases and two ON-center cases. Thus, although the tendencies predicted by the Stell *et al* (1977) hypothesis are evident, no fully secure predictive rule exists.

11.7 MAMMALIAN ROD BIPOLARS

In mammals, bipolar cells exist which selectively contact either rods or cones. Mammalian rod bipolars are a singular type and are expected to have a uniform physiology. They contact only rods, and do so with an invaginating type synapse (Kolb, 1970; Boycott and Kolb, 1973; Dacheux and Raviola, 1986). An example of such a synapse from cat is shown in Figure 11.2. Although invaginating, the axial apposition of bipolar–dendrite and cone–pedicle membrane is 'wavy' with irregular cleft width, failing to conform with either 'wide-cleft' or 'narrow-cleft' categories. Furthermore, in regard to the positional classification scheme, since only one bipolar cell dendrite is ever associated with triadic arrangements in rod spherules, the distinction between 'invaginating' and 'semi-invaginating' appears inappropriate.

Rod bipolars are easily identified in the light microscope. Dendrites are numerous and penetrate deeply into the OPL, passing the layer of cone pedicles to reach the multilayered band of rod spherules. At the opposite pole, the rod bipolar axon traverses the full thickness of the IPL to arborize in a band adjacent to ganglion cells. This pattern is found in all rod dominant mammals.

The first reports of rod bipolar responses came from cat retina (Nelson *et al.*, 1976). Intracellular stains of these cells suggested that they generate OFF-center responses. This result has been challenged on several grounds, however. First APB, a glutamate

Figure 11.6 Electron micrographs of bipolar synapses with photoreceptors which are 'exceptions to the rule' in a variety of different vertebrate retinas. (a) In the salamander retina, HRP-stained, center-hyperpolarizing bipolar cells make ribbon-related junctions with rods (b, arrow). R, rod; r, ribbon; H, horizontal cell dendrites (from Lasansky. 1978.) Magnification ×34 000. (b) In turtle retina, center-depolarizing bipolar cells (b) make narrow-cleft basal junctions (nbj, arrows) with puncta of stain on the cone pedicle pre-synaptic membranes. (From Kolb and Jones, 1984.). Magnification ×49 000. (c) In goldfish retina, center-hyperpolarizing bipolars (a) make wide-cleft basal junctions (large arrows) and center-depolarizing bipolars (b) make both narrow-cleft (small arrows) and ribbon-related (asterisk) junctions with the rod spherule (RSE) SR, synaptic ribbon, H, horizontal cell dendrites (From Stell *et al.*, 1977.) Magnification ×49 000. (d) In carp retina, center-hyperpolarizing responses are recorded in an HRP-stained bipolar cell (b, arrows) that makes ribbon-related contacts with the cone pedicle SR, synaptic ribbon; H, horizontal cell dendrites. (From Saito *et al.*, 1983.) Magnification ×50 000. (e) and (f) In cat retina, cb6 bipolar cells make both central element contacts (arrows) and wide-cleft basal junctions (wbj, arrow) with cone pedicles. The central element is shielded from the synaptic ribbon by a cap of horizontal-cell dendrites (H) in the triad. (From Nelson and Kolb, 1983.) Magnifications ×44 500 (e) and ×57 000 (f).

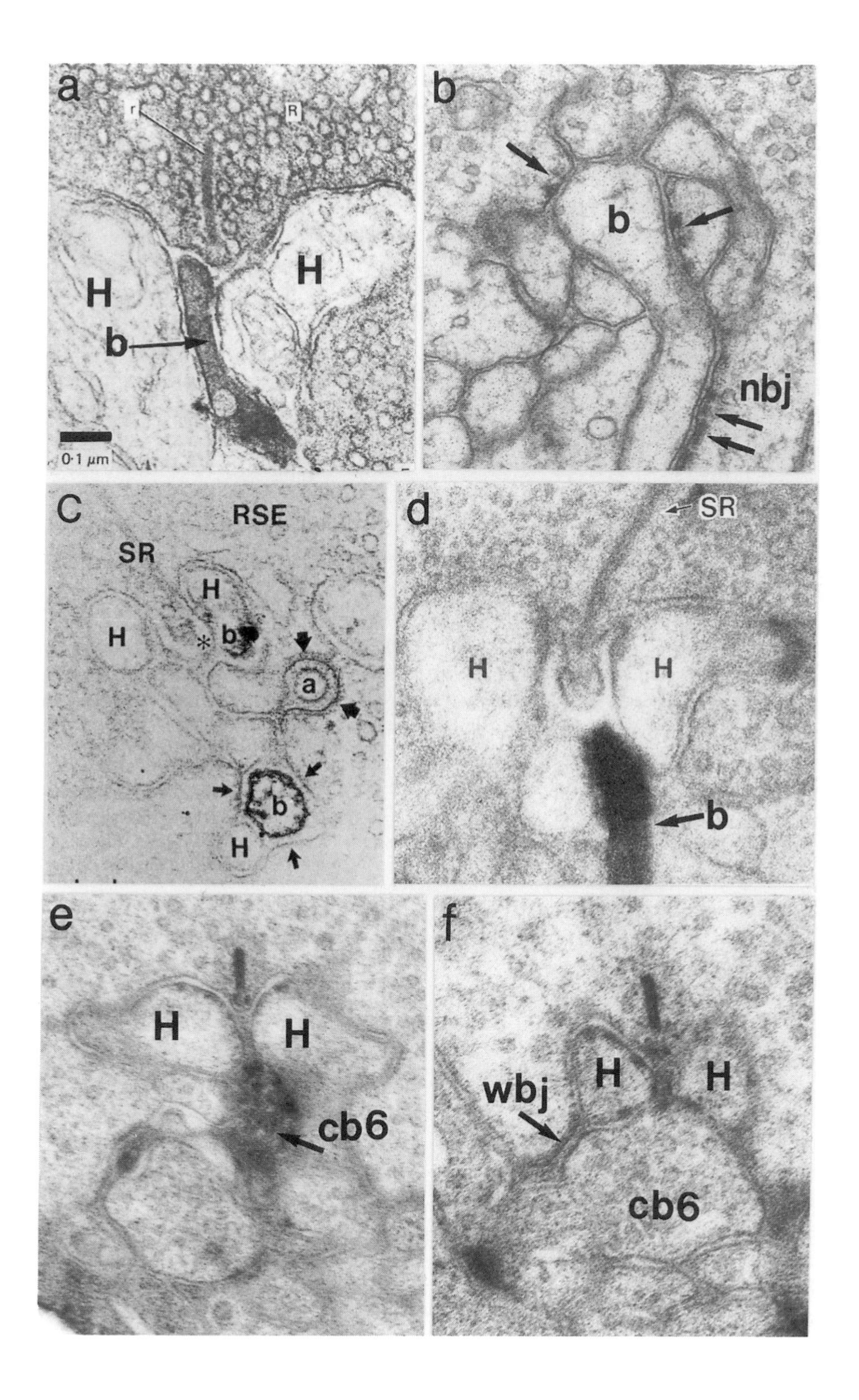
a
r
R
H
H
b
0·1 μm
b
b
nbj
C
RSE
SR
H
H
b
a
b
H
d
SR
H
H
b
e
H
H
cb6
f
wbj
H
H
cb6

agonist selective for ON-center bipolar cells (Slaughter and Miller, 1981), selectively blocks scotopic signals seen in cat ganglion cells (Müller *et al.*, 1988). Furthermore, dissociated cat bipolar cells acutely isolated in cell culture and studied with cGMP-doped cell-attached patch electrodes, respond as appropriate to ON-center bipolars when stimulated with either L-glutamate or APB (Kaneko *et al.*, 1992).

In rabbit retina, rod bipolars have been studied with intracellular microelectrodes and stained with HRP (Dacheux and Raviola, 1986). Figure 11.1 shows the depolarizing response of such a cell (rb). Bright stimuli evoke a characteristic rod component from such cells, the rod 'after effect': the cell remains depolarized long after the cessation of the stimulus (Dacheux and Raviola, 1986).

Dissociated rod bipolar cells, particularly of rodents, provide an elegant preparation in which to study the electrophysiology of rod bipolars in cell culture. A light microscopic view of such a cell and the associated response to APB are shown in Figure 11.7. Yamashita and Wässle (1991) using the nystatin patch technique, which helps to prevent the 'washout' of important cellular components, found that a net outward current was evoked by APB. There was also a reduction in noise, suggesting that the glutamate agonist APB blocked the 'noisiness' of tonically open channels. This is just the physiology expected of an ON-type bipolar cell.

11.8 PROTEIN KINASE C IMMUNOREACTIVITY OF BIPOLAR CELLS

Great interest was aroused when the second messenger, protein kinase C (PKC), was found in depolarizing rod-dominant bipolar cells of the fish retina (Negishi *et al.*, 1989). Since this initial report, PKC has almost been thought of as a specific marker for rod bipolar cells and/or ON-center bipolar-cell types (Wässle *et al.*, 1991). However, as attractive as this simplification is, it is not yet totally confirmed. The original PKC antisera used by early workers, was a mixture of α and β isozymes and provided confusing results that were very species-dependent. For example, in turtle retina, where there may not be a rod-specific bipolar cell, cone bipolar cells are stained and they may include hyperpolarizing types (by virtue of having axonal endings in IPL sublamina **a**; Cuenca *et al.*, 1990). By using monoclonal antibodies to the various isozymes of PKC it has been possible to stain a more diverse group of neurons than rod bipolar cells in the retina (Zhang *et al.*, 1992; Kolb *et al.*, 1993). Thus PKC-α stains a specific bipolar type in the turtle retina with a bistratified axon terminal with endings in sublamina **b** of the IPL. The same isozyme stains rod bipolar cells (Figure 11.8a) and the 'blue' cone bipolar cell in the human retina, whereas PKC-β stains bipolar cell types that end in sublamina **a** and make wide-cleft basal junctions (Figure 11.8b). Amacrine and ganglion cells, among them possible OFF-center types, also stain with PKC-β in both species (Zhang *et al.*, 1992; 1994; Kolb *et al.*, 1993).

The intriguing question of whether only depolarizing bipolar types are stained by PKC is as yet unanswered. In other words, even when one has specific contacts of the bipolars marked by PKC immunostaining as invaginating, ribbon-related or basal junctions, it does not mean that one can assume that the bipolar will be depolarizing or hyperpolarizing (see exceptions to the 'rule' in Figure 11.6 and Table 11.1). Similarly it does not appear possible to classify a bipolar cell as to response type simply on its axonal branching level in the IPL. Counter examples are found among turtle bipolar cells (Weiler, 1981; Ammermüller and Kolb, 1995) and cat cb6 (Figure 11.6 and f). An answer will only come from further combined intracellular recordings, HRP marking and ultrastructural examination of the contacts, looking particu-

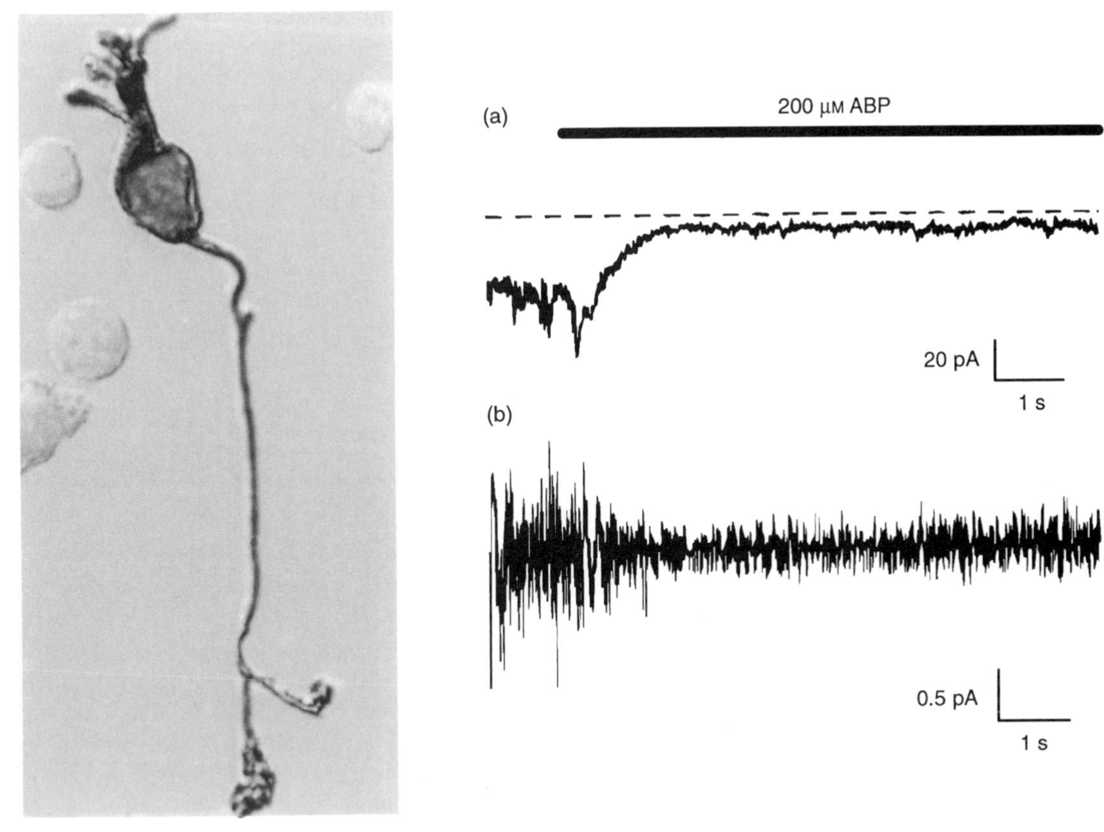

Figure 11.7 Glutamate agonist evokes 'ON-center membrane currents from dissociated rod bipolar cells of rat retina.' To the left is a light micrograph of a dissociated, rat, rod-bipolar cell, identified by dense binding of the antibody to PKC. To the right are currents evoked by APB, a glutamate agonist selective for ON bipolar cells. The currents were recorded with the nystatin patch technique which limits the 'wash-out' of delicate cytoplasmic chemistry involved in ON-center bipolar cell responses. Dissociated bipolar cells exhibit steady inward (depolarizing) currents in the absence of pharmacological stimulation. In (a) this current is blocked by APB. This is analogous to increased release of glutamate from photoreceptors at stimulus offset blocking depolarizing light responses in ON bipolars. In (b) APB is seen to decrease membrane noise, consistent with the closing of depolarizing channels. (From Yamishita and Wässle (1991) reproduced by permission of the authors.)

larly for the narrow- versus wide-cleft distinction. The bipolar type that stains with PKC-α (Zhang *et al.*, 1992) and type(s) that stain with the PKC-α/β combination (Cuenca *et al.*, 1990) in turtle have light microscopic morphologies of two color-coded bipolar cells recently recorded and marked in the same species (Ammermüller *et al.*, 1995). The latter are center-hyperpolarizing to bright white light (and one, at least, has axon terminals in sublamina **b**) but they have depolarizing components to either red or blue light in each case. Both types of junction with photoreceptors, namely narrow- and wide-cleft, have been seen by EM of PKC-α/β immunostaining (Figure 11.8c) (Zhang *et al.*, 1992).

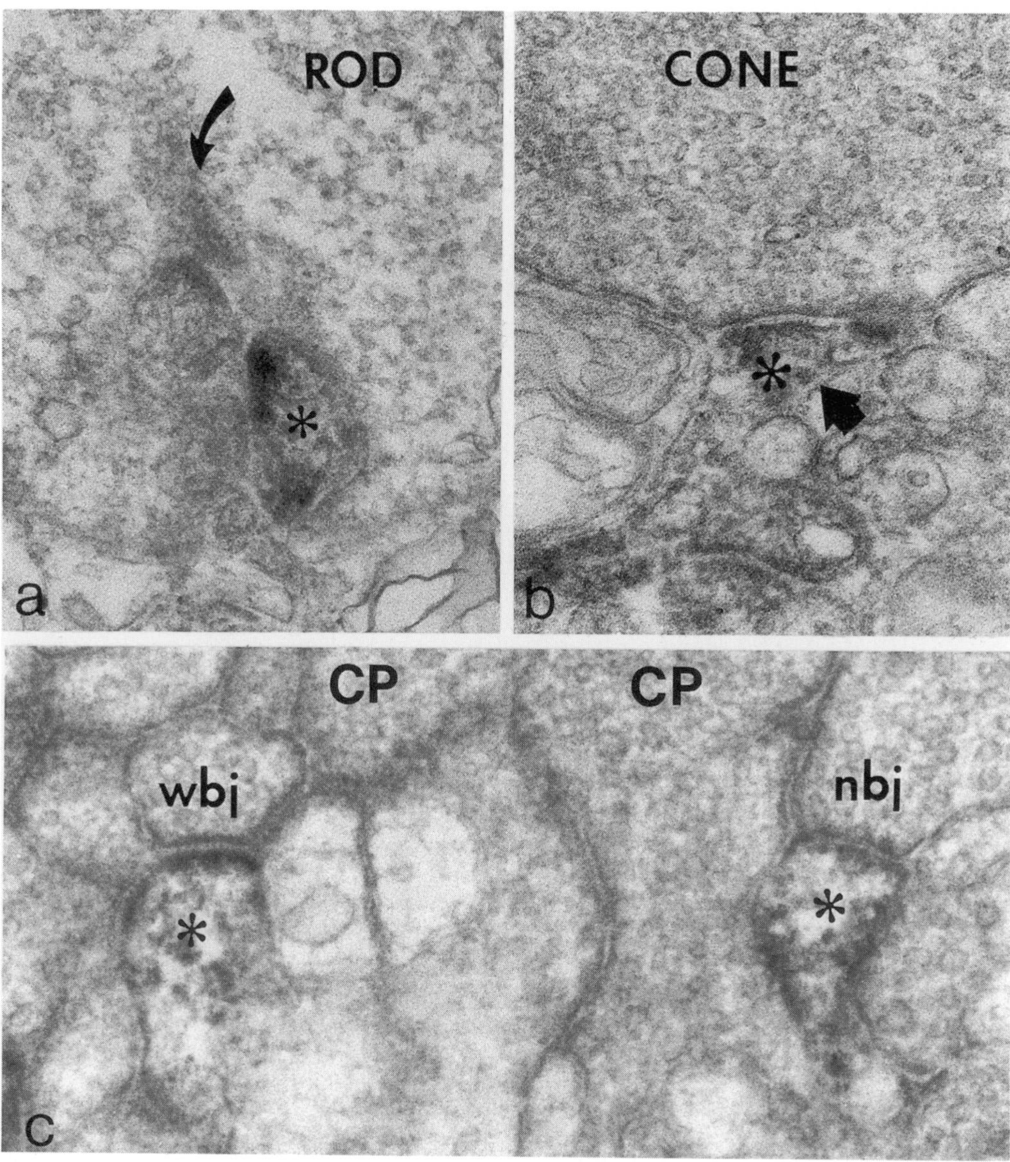

Figure 11.8 Electron micrographs of contacts made by PKC-IR bipolar cells in human and turtle retinas. (a) The antibody to PKC-α stains rod bipolar cells making typical invaginating ribbon-related contacts in rod spherules (ROD) in human retina. Curved arrow indicates synaptic ribbon. Magnification ×42 000. (b) PKC-β stains cone bipolar cells making wide-cleft basal junctions (asterisk, arrow) on cone pedicles (CONE) in human retina. Magnification ×50 000. (c) The mixed PKC-α/β antiserum stains bipolar cells that make both narrow-cleft, punctate (nbj, asterisk) and wide-cleft, striated junctions (wbj, asterisk) in two different cone pedicles in the turtle retina. The bipolars responsible for these synapses are bistratified in sublayers S2 and S4 of the IPL and may be color-coded. See text for further discussion. Magnification ×58 000.

Could it be that wide-cleft basal junction contacts to the specific spectral type of cone are associated with the hyperpolarizing component of the response, and that narrow-cleft semi-invaginating contacts in the other spectral type of cone are responsible for the depolarizing component of the response? In fact, different contacts associated with different spectral types of cone may be commoner among vertebrate retinas than we think. After all, color-coded bipolar cells have been described in goldfish, carp, dace, and turtle. Neither the contacts of these bipolar cells have been looked at from this perspective, nor have the extents of color-specific inputs in bipolar cells in general been analyzed adequately in physiological experiments. Only future investigation will be able to address this interesting possibility.

11.9 HORIZONTAL–CELL MEDIATION OF BIPOLAR-CELL SURROUND RESPONSES

Concerning color-coding in bipolar cells of vertebrate retinas, horizontal cells are thought to be responsible for providing color-antagonistic input to color-coded bipolar cells, thus producing double color-opponent bipolar cells. However, this type of bipolar cell has so far only been seen in goldfish retina (Kaneko and Tachibana, 1981; 1983; reviewed in Kolb and Lipetz, 1991) and the exact role of the chromatic horizontal cells has not yet been elucidated. In fact, many details about horizontal cells in the OPL are still not available. Particularly frustrating has been the lack of anatomical evidence for direct horizontal cell input to bipolar cells in fish and mammals. In general three pathways are considered possible for horizontal cell surround effects on bipolar cells: (1) feedback effects onto cones, which then indirectly influence bipolar cells; (2) direct interactions between horizontal cell and bipolar cell dendrites within the triadic complex; (3) direct synaptic contacts between horizontal cells and bipolar cells in the OPL.

The structure of the horizontal cell membrane at the invaginating, ribbon-related cone bipolar dendrite in the mammalian photoreceptor triad synapse seen in a freeze-fracture study (Raviola and Gilula, 1975) was somewhat suggestive of a direct horizontal cell inhibitory action on the bipolar (Figure 11.4). However, there do not appear to be any similar sites of interaction for the types of bipolar making wide-cleft basal junctions, of which there are a great number. Taking the freeze-fracture data on its face value, the implication is that invaginating bipolar cells should receive an antagonistic surround response from horizontal cells through a sign-inverting synapse, whereas flat bipolars should not have any surround responses at all. This interpretation is intriguing but may have the flaw that bipolar cells are known to have surround receptive fields at least in submammalian retinas (Werblin and Dowling, 1969; Kaneko, 1970; Naka, 1976; Lasansky, 1978; Marchiafava and Weiler, 1980; Saito *et al.*, 1983; McReynolds and Lukasiewicz, 1989). Although some of these cells may receive a mixture of basal and ribbon-related contacts (Lasansky, 1978; Dacheux, 1982) and so might receive surround signals from only the invaginating synaptic element, it seems likely that there are some types with exclusively basal contacts (Stell *et al.*, 1977; Kolb *et al.*, 1986) and these do exhibit surround responses.

So where does the bipolar cell surround come from in the vertebrate retina? Direct horizontal cell synapses upon bipolar cells were originally implicated (Dowling and Werblin, 1969) but are still not generally observed in vertebrate retinas. In turtle, salamander, frog, *Xenopus* and rabbit, synapses from horizontal cells to bipolar dendrites in the OPL have been reported (Dowling, 1970; Lasansky, 1972; Fisher and Boycott, 1974; Witkovsky and Powell, 1981;

Kolb and Jones, 1984), but not in carp, catfish, goldfish or most mammals. Interestingly, in the human retina, synaptic contacts from horizontal cell axon-terminal lateral elements within rod spherules, to rod bipolar cell dendrites have been seen (Linberg and Fisher, 1988).

Stimulation of horizontal cells by hyperpolarizing extrinsic currents evokes surround responses from fish retinal ganglion cells (Maximova, 1969; Naka, 1971), strongly implying that these cells are significant elements in ganglion-cell, and inferentially, bipolar-cell, surround responses. Physiological studies in carp retina (Saito *et al.*, 1981) provided convincing evidence that bipolar surrounds are generated in the feedback loop (Baylor *et al.*, 1971) from the horizontal cell to the cone itself. Furthermore, there is good anatomical evidence that horizontal cell feedback in fish retinas takes place through light induced control of spinules in horizontal cell lateral elements in cones (reviewed by Djamgoz and Kolb, 1993; Wagner and Djamgoz, 1993) or by direct synapses upon cone telodendria (Sakai and Naka, 1983). Thus, fish horizontal cells may form surrounds to bipolar cells via photoreceptor feedback circuits whereas anuran and reptilian horizontal cells may use direct chemical synapses upon bipolar dendrites in the OPL. With the exception of the human data on the rod axon terminal system mentioned above, however, there is little anatomical evidence for direct horizontal cell synapses upon bipolar dendrites or for feedback synapses to cones in mammalian retinas.

Congruent with the anatomical findings, physiological, 'surround' responses in bipolar cells have been rather difficult to evoke in the mammalian retina. Thus, recordings from cat bipolar cells have revealed weak surrounds for ON-center cone bipolar cb5, and no surround for any of the OFF-center types (Nelson and Kolb, 1983). Even the ON-center cone bipolar type exhibited only a slightly more transient wave form in distant regions of its receptive field (Nelson and Kolb, 1983) as did the depolarizing rod bipolar cell in the rabbit retina (Dacheux and Raviola, 1986). Hyperpolarizing cone bipolar types have, remarkably, simple, wide unimodal space constants of 85–100 μm (Nelson and Kolb, 1983). In explanation of this weakness of mammalian bipolar-cell surround responses, it has been suggested that surrounds to bipolar and other cell types are added in the IPL by amacrine cell circuits in the mammalian retina (Marc, 1989; Kolb and Nelson, 1993); particularly as surrounds in mammalian ganglion cells can be eliminated by pharmacological agents that affect neurotransmitters typical of amacrine cells (Ikeda, 1985; Jensen, 1991).

Nonetheless, the existence of a functional horizontal cell surround pathway has been clearly demonstrated in rabbit by Mangel (1991) by passing extrinsic current into horizontal cells to evoke antagonistic surround responses from ganglion cells. The pathway appears to involved horizontal cell feedbacks onto cones, since APB blocks the effect in ON-center ganglion cells. It should also be noted that horizontal cell surround effects on bipolar cells, under some conditions, reveal center-facilitating as well as center-antagonizing components. A good example of this is suppressive rod cone interaction (Goldberg *et al.*, 1983). In this paradigm, a flickering red spot is presented to the center of a bipolar-cell receptive field. This stimulus is cone selective both by virtue of color and flicker frequency. The amplitude of flicker evoked by this spot is greatly increased when surrounding areas are illuminated by dim blue or green rod-selective stimuli (Frumkes and Eysteinsson, 1987). Such effects are seen in a variety of cell types in both mammalian and submammalian species, and are observable behaviorally (Goldberg *et al.*, 1983; Frumkes and Eysteinsson, 1987; Pflug *et al*, 1990; Nelson *et al*; 1990). The mechanism appears to involve horizontal cell feedback through cones (Nelson *et al.*, 1990).

11.10 CONCLUSIONS AND FUTURE PERSPECTIVES

It is evident from this review of photoreceptor-to-bipolar cell synapses in the vertebrate retina that we are still, despite 30 years of anatomical and electrophysiological investigation, looking for better clues. We are hopeful, but yet not exactly sure, about correlations between morphology and physiology at this unique set of synapses, first in the visual system. Even the list of synaptic actions in distal retina may not yet be complete. We can, however, state some conclusions as follows:

1. The photoreceptor to bipolar synapse is where the visual message is split into center-hyperpolarizing (OFF-center) or center-depolarizing (ON-center) responses to light. Thus, the photoreceptor to bipolar synapse is either sign-conserving (excitatory, center-hyperpolarizing, OFF-center) or sign-inverting (inhibitory, center-depolarizing, ON-center).
2. There are no inviolate rules regarding the ultrastructure of synaptic contacts between photoreceptor and bipolar cells and the excitatory or inhibitory nature of the photoreceptor to bipolar synapse. However, OFF-center responses predominate among cells exhibiting wide-cleft basal junctions, and ON-center responses predominate among cells with narrow-cleft ribbon-related junctions.
3. The presence of bipolar-cell invaginating contacts appears to be among the weakest predictors of response sign.
4. Narrow-cleft, punctate junctions, typically associated with 'semi-invaginating' locations, are variations of the invaginating central element synapse and are common in submammalian rods and cones but are also present in mammalian cones. They are probably made by center-depolarizing bipolar cells in most species.
5. In anurans and reptiles, surrounds are added by direct chemical synapses between horizontal cell and bipolar cell dendrites in the OPL.
6. In carp and goldfish retinas, horizontal cell surrounds are added to bipolar cells by feedback loops through the photoreceptor, as evidenced by spinules.
7. In mammals only weak surround information is generated by horizontal cell pathways in the OPL. Surrounds may be formed instead by amacrine cell circuits in the IPL.
8. Monostratified ganglion cells are OFF-center if they branch purely in sublamina **a** and ON-center if they are restricted to sublamina **b** of the IPL.
9. OFF-center bipolar-cell axons usually end in sublamina **a**, and ON-center bipolar-cell axons usually end in sublamina **b**, though the stratification tendencies of ON- and OFF-center bipolar cell axons is weaker than that of ON- and OFF- centre ganglion cell dendrites. Coupled with the presence of multistratified bipolar cells, this assures that both ON and OFF- type bipolar cell axons coexist in the same IPL layers.
10. IPL sublamina-restricted ON- and OFF-centre ganglion cells may receive input from either single or multiple bipolar types. In the case of multiple types, not all need be of the same response sign. The same-signed input is clearly predominant, however, and determines the center response sign of the ganglion cell.

Although the relationship between ultrastructure of the synapses that bipolar cells engage in with photoreceptor synaptic endings is still unclear, it is hoped that future research will reveal such a relationship. The knowledge that we are achieving almost daily on the molecular biology of excitatory amino acid synaptic transmission in the CNS, indicates very precise molecular arrangements in receptor and channel architecture. These precisions must be related in some way to the anatomical structure of these pathways. We are fortunate that the retina is the most desirable CNS system for understanding structure underlying function, in a general sense as well as in this specific sense. With future ultrastructural investigations coupled with immunocytochemical markers for specific receptors as they become available, we will answer these important questions in the retina. Experiments in channel physiology and intracellular recordings will have to accompany the anatomy and molecular biology. Although future experiments are going to be difficult, particularly in the mammalian retina, we anticipate a more complete understanding of the synaptic events occurring at the first level of information processing for the whole visual system, namely, at the level of the photoreceptor to bipolar synapse, in the not too distant future.

REFERENCES

Ammermüller, J. and Kolb, H. (1995) The organization of the turtle inner retina I. The organization of on- and off-centre pathways. *J. Comp. Neurol.* (in press).

Ammermüller, J., Muller J. and Kolb, H. (1995) The organization of the turtle inner retina. II. Analysis of colour-coded and directionally selective cells. *J. Comp. Neurol.* (in press).

Amthor, F.R., Takahashi, E.S. and Oyster, C.W. (1989a) Morphologies of rabbit retinal ganglion cells with concentric receptive fields. *Journal of Comparative Neurology*, **280**, 72–96.

Amthor, F.R., Takahashi, E.S. and Oyster, C.W. (1989b) Morphologies of rabbit retinal ganglion cells with complex receptive fields. *Journal of Comparative Neurology*, **280**, 97–121.

Baylor, D.A., Fuortes, M.G.F. and O'Bryan, P.M. (1971) Receptive fields of the cones in the retina of the turtle. *Journal of Physiology. (London)*, **214**, 265–94.

Bortoff, A. (1964) Localization of slow potential responses in the Necturus retina. *Vision Research*, **4**, 627–35.

Boycott, B.B. and Kolb, H. (1973) The connection between the bipolar cells and photoreceptors in the retina of the domestic cat. *Journal of Comparative Neurology*, **148**, 91–114.

Cohen, E. and Sterling, P. (1991) Microcircuitry related to the receptive-field center of the ON-beta ganglion cell. *Journal of Neurophysiology*, **65**, 352–9.

Cuenca, N., Fernandez, E. and Kolb, H. (1990) Distribution of immunoreactivity to protein kinase C in the turtle retina. *Brain Research*, **532**, 278–87.

Dacheux, R.F. (1982) Connections of the small bipolar cells with the photoreceptors in the turtle. An electron microscope study of Golgi-impregnated, gold-toned retinas. *Journal of Comparative Neurology*, **205**, 55–62.

Dacheux, R.F. and Raviola, E. (1986) The rod pathway in the rabbit: a depolarizing bipolar and amacrine cell. *Journal of Neuroscience*, **6**, 331–45.

DeMonasterio, F.M. (1979) Asymmetry of on- and off-pathways of blue sensitive cones of the retina of macaques. *Brain Research*, **166**, 39–48.

Djamgoz, M.B.A. and Kolb, H. (1993) Ultrastructural and functional connectivity of intracellularly stained neurones in the vertebrate retina: correlative analyses. *Microscopy Research and Technique*, **24**, 43–66.

Djamgoz, M.B.A. and Wagner, H.-J. (1987) Intracellular staining of retinal neurones: applications to studies of functional organization. *Progress in Retinal Research*, **6**, 85–150.

Dowling, J.E. (1970) Organization of vertebrate retinas. *Investigative Ophthalmology*, **9**, 655–80.

Dowling, J.E. (1987) *The Retina: an Approachable Part of the Brain*. The Belknap Press, Harvard University Press, Cambridge, MA.

Dowling, J.E. and Boycott, B.B. (1966) Organization of the primate retina; electron microscopy. *Proceedings of the Royal Society of London B*, **166**, 80–111.

Dowling, J.E. and Werblin, F.S. (1969) Organization of the retina of the mudpuppy, *Necturus maculosus*. I. Synaptic structure. *Journal of Neurophysiology*, **32**, 315–38.

Famiglietti, E.V. Jr and Kolb, H. (1976) Structural basis for ON- and OFF-centre responses in retinal ganglion cells. *Science*, **194**, 193–5.

Famiglietti, E.V. Jr, Kaneko, A. and Tachibana, M. (1977) Neuronal architecture of on and off pathways to ganglion cells of the carp retina. *Science*, **198**, 1267–9.

Fisher, S.K. and Boycott, B.B. (1974) Synaptic connexions made by horizontal cells within the outer plexiform layer of the retina of the cat and rabbit. *Proceedings of the Royal Society of London B*, **186**, 317–31.

Freed, M. A. and Nelson, R. (1994) Conductances evoked by light in the ON-β ganglion cell of cat retina. *Visual Neuroscience*, **11**, 261–70.

Freed, M.A. and Sterling, P. (1988) The On-alpha ganglion cell of the cat retina and its presynaptic cell types. *Journal of Neuroscience*, **8**, 2303–20.

Frumkes, T.E. and Eysteinsson, T. (1987) Suppressive rod–cone interaction in distal vertebrate retina: Intracellular records from *Xenopus* and *Necturus*. *Journal of Neurophysiology*, **57**, 1361–82.

Goldberg, S.H., Frumkes, T.E. and Nygaard, R.W. (1983) Inhibitory influence of unstimulated rods in the human retina: evidence provided by examining cone flicker. *Science*, **221**, 180–2,

Gouras, P. (1971) The function of the midget system in primate colour vision. *Vision Research Suppl.*, **3**, 397–410.

Graybiel, A.M. and Devor, M.A. (1974) A microelectrode delivery technique for use with horseradish peroxidase. *Brain Research*, **68**, 167–73.

Hashimoto, Y. and Inokuchi, M. (1981) Characteristics of second order neurones in the dace retina: physiological and morphological studies. *Vision Research*, **21**; 1541–50.

Hensley, S.H., Yang, X.-L. and Wu, S.M. (1993) Identification of glutamate receptor subtypes mediating inputs to bipolar cells and ganglion cells in the tiger salamander retina. *Journal of Neurophysiology*, **69**, 2099–107.

Ikeda, H. (1985) Transmitter action at cat retinal ganglion cells. *Progress in Retinal Research*, **4**, 1–32.

Ikeda, H. and Sheardown, M.J. (1983) Functional transmitters at retinal ganglion cells in the cat. *Vision Research*, **23**, 1161–74.

Jensen, R.J. (1991) Involvement of glycinergic neurones in the diminished surround activity of ganglion cells in the dark-adapted rabbit retina. *Visual Neuroscience*, **6**, 43–54.

Jensen, R.J. and DeVoe, R.D. (1983) Comparisons of directionally selective with other ganglion cells of the turtle retina: intracellular recording and staining. *Journal of Comparative Neurology*, **217**, 271–87.

Kaneko, A. (1970) Physiological and morphological identification of horizontal, bipolar and amacrine cells in goldfish retina. *Journal of Physiology (London)*, **207**, 623–33.

Kaneko, A. (1973) Receptive field organization of bipolar and amacrine cells in the goldfish retina. *Journal of Physiology (London)*, **235**, 133–53.

Kaneko A., de la Villa, P. and Kurahashi, T. (1992) L-Glutamate-induced responses in isolated cat bipolar cells the subtype of which was identified by PKC-like immunoreactivity. *Investigative Ophthalmology and Visual Science*, **33**, 752.

Kaneko, A. and Tachibana, M. (1981) Retinal bipolar cells with double colour-opponent receptive fields. *Nature*, **293**, 220–2.

Kaneko, A. and Tachibana, M. (1983) Double colour opponent receptive fields of carp bipolar cells. *Vision Research*, **23**, 381–8.

Kim, H.G. and Miller, R.F. (1993) Properties of synaptic transmission from photoreceptors to bipolar cells in the mudpuppy retina. *Journal of Neurophysiology*, **69**, 352–60.

Kolb, H. (1970) Organization of the outer plexiform layer of the primate retina: electron microscopy of Golgi-impregnated cells. *Philosophical Transactions of the Royal Society of London B* **258**, 261–83.

Kolb, H. (1982) The morphology of the bipolar cells amacrine cells and ganglion cells in the retina of the turtle *Pseudemys scripta elegans*. *Philosophical Transactions of the Royal Society of London B*, **298**, 355–93.

Kolb, H. and DeKorver, L. (1991) Midget ganglion cells of the parafovea of the human retina: a study by electron microscopy and serial section reconstructions. *Journal of Comparative Neurology*, **303**, 617–36.

Kolb, H. and Jones, J. (1984) Synaptic organization of the outer plexiform layer of the turtle retina:

an electron microscope study of serial sections. *Journal of Neurocytology*, **13**, 567–91.
Kolb, H. and Jones, J. (1985) Electron microscopy of Golgi-impregnated photoreceptors reveals connections between red and green cones in the turtle retina. *Journal of Neurophysiology*, **54**, 304–17.
Kolb, H., Linberg, K.A. and Fisher, S.K. (1992) Neurones of the human retina: a Golgi study. *Journal of Comparative Neurology*, **318**, 147–87.
Kolb, H. and Lipetz, L.E. (1991) The anatomical basis for colour vision in the vertebrate retina, in *Vision and Visual Dysfunction*, volume 6: *The Perception of Colour* (ed. P. Gouras), Macmillan London, pp.128–45.
Kolb, H. and Nelson, R. (1993) OFF-alpha and off-beta ganglion cells in the cat retina. II. Neural circuitry as revealed by electron microscopy of HRP stains. *Journal of Comparative Neurology*, **329**, 85–110.
Kolb, H., Nelson, R. and Mariani, A. (1981) Amacrine cells, bipolar cells and ganglion cells of the cat retina: a Golgi study. *Vision Research*, **21**, 1081–114.
Kolb, H., Wang, H.-H. and Jones, J. (1986) Cone synapses with Golgi-stained bipolar cells that are morphologically similar to a center-hyperpolarizing and a center-depolarizing bipolar cell type in the turtle retina. *Journal of Comparative Neurology*, **205**, 510–20.
Kolb, H., Zhang, L. and DeKorver, L. (1993) Differential staining of neurones in the human retina with antibodies to protein kinase C isozymes. *Visual Neuroscience*, **10**, 341–51.
Kouyama, N. and Marshak, D.W. (1992) Bipolar cells specific for blue cones in the macaque retina. *Journal of Neurosciences*, **12**, 1233–52.
Landis, D.M.D. and Reese, T.S. (1974) Differences in membrane structure between excitatory and inhibitory synapses in the cerebellar cortex. *Journal of Comparative Neurology*, **155**, 93–126.
Landis, D.M.D., Reese, T.S. and Raviola, E. (1974) Differences in membrane structure between excitatory and inhibitory components of the reciprocal synapse in the olfactory bulb. *Journal of Comparative Neurology*, **155**, 67–92.
Lasansky, A. (1971) Synaptic organization of cone cells in the turtle retina. *Philosophical Transactions of the Royal Society of London B*, **262**, 365–81.
Lasansky, A. (1972) Cell junctions at the outer synaptic layer of the retina. *Investigative Ophthalmology*, **11**, 265–75.
Lasansky, A. (1978) Contacts between receptors and electrophysiologically identified neurones in the retina of the larval tiger salamander. *Journal of Physiology (London)*, **285**, 531–42.
Lasansky, A. (1992) Properties of depolarizing bipolar cell responses to central illumination in salamander retinal slices. *Brain Research*, **576**, 181–96.
Linberg, K.A. and Fisher, S.K. (1988) Ultrastructural evidence that horizontal cell axon terminals are presynaptic in the human retina. *Journal of Comparative Neurology*, **268**: 281–97.
Mangel, S.C. (1991) Analysis of the horizontal cell contribution to the receptive field surround of ganglion cells in the rabbit retina. *Journal of Physiology (London)*, **442**, 211–34.
Marc, R.E. (1989) The anatomy of multiple GABAergic and glycinergic pathways in the inner plexiform layer of the goldfish retina, in *Neurobiology of the Inner Retina* (eds. R. Weiler and N. Osborne), Springer-Verlag, Berlin., pp. 53–64.
Marchiafava, P.L. and Weiler, R. (1980) Intracellular analysis and structural correlates of the organization of inputs to ganglion cells in the retina of the turtle. *Proceedings of the Royal Society of London B.*, **208**, 103–13.
Mariani, A.P. (1983) Giant bistratified bipolar cells in monkey retina. *Anatomical Record*, **206**, 215–20.
Mariani, A.P. and Lasansky, A. (1984) Chemical synapses between turtle photoreceptors. *Brain Research*, **310**, 351–54.
Massey, S.C. (1990) Cell types using glutamate as a neurotransmitter in the vertebrate retina. *Progress in Retinal Research*, **9**, 399–425.
Maximova, E.M. (1969) Effect of intracellular polarization of horizontal cells on the activity of the ganglionic cells of the retina of fish. *Biofizika*, **14**, 537–44.
McGuire, B.A., Stevens, J.K. and Sterling, P. (1986) Microcircuitry of beta ganglion cells in cat retina. *Journal of Neuroscience*, **6**, 907–18.
McReynolds, J.S. and Lukasiewicz, P.D. (1989) Integration of synaptic input from on and off pathways in mudpuppy retinal ganglion cells, in *Neurobiology of the Inner Retina* (eds R. Weiler

and N.N. Osborne) NATO ASI Series, Springer-Verlag, Berlin, Heidelberg, pp.209–220.

Miller, R.F. and Slaughter, M.M. (1986) Excitatory amino acid receptors of the retina: diversity and subtype and conductive mechanisms. *Trends in Neurosciences*, **9**, 211–13.

Missotten, L. (1965) *The Ultrastructure of the Human Retina*. Arscia Uitgaven, Brussels, Ghent.

Müller, F., Wässle, H. and Voigt, T. (1988) Pharmacological modulation of the rod pathway in the cat retina. *Journal of Neurophysiology*, **59**, 1657–72.

Naka, K.-I. (1971) Receptive field mechanism in the vertebrate retina. *Science*, **171**, 691–3.

Naka, K.-I. (1976) Neuronal circuitry in the catfish retina. *Investigative Ophthalmology*, **15**, 926–35.

Nawy, S. and Copenhagen, D.R. (1987) Multiple classes of glutamate receptor on depolarizing bipolar cells in retina. *Nature*, **325**, 56–8.

Nawy, S. and Jahr, C.E. (1990) Suppression by glutamate of cGMP activated conductance in retinal bipolar cells. *Nature*, **346**, 269–71.

Negishi, K., Kato, S. and Teranishi, T. (1989) Immunocytochemical localization of protein kinase C in some vertebrate retinas, in *Neurobiology of the Inner Retina* (eds R. Weiler and N.N. Osborne), NATO ASI Series, Springer-Verlag, Berlin, Heidelberg, pp.425–30.

Nelson, R. (1973) A comparison of electrical properties of neurones in *Necturus* retina *Journal of Neurophysiology*, **36**, 519–35.

Nelson, R. and Kolb, H. (1983) Synaptic patterns and response properties of bipolar and ganglion cells in the cat retina. *Vision Research*, **23**, 1183–95.

Nelson, R., Kolb, H., Famiglietti, E.V. and Gouras, P. (1976) Neural responses in the rod and cone systems of the cat retina: intracellular records and procion stains. *Investigative Ophthalmology*, **18**, 946–53.

Nelson, R., Famiglietti, E.V. and Kolb, H. (1978) Intracellular staining reveals different levels of stratification for on-centre and off-centre ganglion cells in the cat retina. *Journal of Neurophysiology*, **41**, 427–83.

Nelson, R., Pflug, R. and Baer, S.M. (1990) Background-induced flicker-enhancement in cat retinal horizontal cells II: spatial properties. *Journal of Neurophysiology*, **64**, 326–40.

Peichl, L., and Wässle, H. (1981) Morphological identification of on- and off-centre brisk transient (Y) cells in the cat retina. *Proceedings of the Royal Society of London B*, **212**, 139–56.

Pflug, R., Nelson, R. and. Ahnelt, P.K. (1990) Background-induced flicker-enhancement in cat retinal horizontal cells I: temporal and spectral properties. *Journal of Neurophysiology*, **64**, 313–25.

Raviola, E. and Gilula, N.B. (1975) Intramembrane organization of specialized contacts in the outer plexiform layer of the retina: a freeze-fracture study in monkeys and rabbits. *Journal of Cell Biology*, **65**, 192–222.

Richter, A. and Simon, E.J. (1975) Properties of center-hyperpolarizing, red-sensitive bipolar cells in the turtle retina. *Journal of Physiology (London)*, **248**, 317–34.

Saito, H. (1983) Pharmacological and morphological differences between X- and Y-type ganglion cells in the cat's retina. *Vision Research*, **23**, 1299–308.

Saito, T., Kondo, H. and Toyoda, J.-I. (1979) Ionic mechanisms of two types of ON-centre bipolar cells in the carp retina: I. The responses to central illumination. *Journal of General Physiology*, **73**, 73–90.

Saito, T., Kondo, H. and Toyoda, J.-I. (1981) Ionic mechanisms of two types of ON-centre bipolar cells in the carp retina: II. The responses to annular illumination. *Journal of General Physiology*, **78**, 569–89.

Saito, T., Kujiraoka, T. and Yonaha, T. (1983) Connections between photoreceptors and horseradish peroxidase-injected bipolar cells in the carp retina. *Vision Research*, **23**, 353–62.

Saito, T., Kujiraoka, T., Yonaha, T. and Chino, Y. (1985) Reexamination of photoreceptor–bipolar connectivity patterns in carp retina: HRP-EM and Golgi-EM studies. *Journal of Comparative Neurology*, **236**, 141–60.

Sakai, H. and Naka, K.-I. (1983) Synaptic organization involving receptor, horizontal and on- and off-centre bipolar cells in the catfish retina. *Vision Research*, **23**, 339–51.

Schiller, P.H. (1992) The ON and OFF channels of the visual system. *Trends in Neurosciences*, **15**, 86–92.

Shiells, R.A. and Falk, G. (1990) Glutamate receptors of rod bipolar cells are linked to a cyclic GMP cascade via a G-protein. *Proceedings of the Royal Society of London B*, **242**, 91–4.

Slaughter, M.M. and Miller, R.F. (1981) 2-amino-4-phosphonobutyric acid: a new pharmacological tool for retina research. *Science*, **211**, 182–4.

Slaughter, M.M. and Miller, R.F. (1983) An excitatory amino acid antagonist blocks cone input to sign-conserving second-order retinal neurones. *Science*, **219**, 1230–2.

Stell, W.K. (1967) The structure and relationships of horizontal cells and photoreceptor–bipolar synaptic complexes in goldfish retina. *American Journal of Anatomy*, **121**, 401–24.

Stell, W.K., Ishida, A.T. and Lightfoot, D.O. (1977) Goldfish retina: structural basis for ON- and OFF-centre responses in retinal bipolar cells. *Science*, **198**, 1269–71.

Toyoda, J.-I. (1972) Membrane resistance changes underlying the bipolar cell response in the carp retina. *Vision Research*, **12**, 283–94.

Wagner, H.-J. and Djamjoz, M.B.A. (1993) Spinules: a case for retinal synaptic plasticity. *Trends in Neurosciences*, **16**, 201–6.

Wässle, H., Yamashita, M., Greferath, U. *et al.* (1991) The rod bipolar cell of the mammalian retina. *Visual Neuroscience*, **7**, 99–112.

Weiler, R. (1981) The distribution of center-depolarizing and center-hyperpolarizing bipolar cell ramifications within the inner plexiform layer of the turtle retina. *Journal of Comparative Physiology A*, **144**, 459–64.

Werblin, F. (1991) Synaptic connections, receptive fields, and patterns of activity in the tiger salamander retina. *Investigative Ophthalmology and Visual Science*, **32**, 459–83.

Werblin, F.S. and Dowling, J.E. (1969) Organization of the retina of the mudpuppy, *Necturus maculosus*. II. Intracellular recording. *Journal of Neurophysiology*, **32**, 339–55.

Witkovsky, P. and Powell, C.C. (1981) Synapse formation and modification between distal retinal neurones in larval and juvenile *Xenopus*. *Proceedings of the Royal Society of London B*, **211**, 373–89.

Yamashita, M. and Wässle, H. (1991) Responses of rod bipolar cells isolated from the rat retina to the glutamate agonist 2-amino-4-phophonobutyric acid (APB). *Journal of Neurosciences*, **11**, 2372–82.

Yazulla, S. (1976) Cone input to bipolar cells in the turtle retina. *Vision Research*, **16**, 737–44.

Zhang, L., DeKorver, L. and Kolb, H. (1992) Light and electron microscopy of immunostaining for protein kinase C and its isozymes in the turtle retina. *Journal of Neurocytology*, **21**, 833–45.

Zhang, L., DeKorver. L. and Kolb, H. (1994) Immunostaining of monkey retina with the PKC-β isozyme. *Investigative Ophthalmology and Visual Science Supplement*, **35**, 1582.

12

Photoreceptor–bipolar cell transmission

RICHARD SHIELLS

12.1 INTRODUCTION

Bipolar cells make direct synaptic contact with photoreceptors and form a 'through' pathway for information transmission in the retina before this is relayed to the higher visual centers of the brain. A high degree of processing of the visual signal occurs in the outer retina. For example, center-surround receptive field organization via lateral inhibitory inputs to bipolar cells from horizontal cells establishes the elements of contrast and edge detection at this first synaptic level in the visual pathway. The neurons of the outer retina are highly specialized in that they do not possess a regenerative Na^+ conductance and so do not generate action potentials. Instead, they respond to light with graded potential changes and photoreceptors, bipolar and horizontal cells all possess a linear range of response proportional to light intensity. The visual system is subdivided into ON-and OFF-pathways in the retina by two functionally distinct classes of bipolar cells, ON-center bipolar cells which depolarize and OFF-center bipolar cells which hyperpolarize in response to central illumination of their receptive fields (Werblin and Dowling, 1969; Kaneko, 1971). Bipolar cells may be further classified into rod, mixed rod–cone or cone ON-or OFF-bipolar cells depending on their type of photoreceptor input. How these cells generate responses of opposing polarities, given that they receive the same presynaptic signal from photoreceptors, has only recently been fully understood. The photoreceptors hyperpolarize in response to light, and this reduces the rate of release of the neurotransmitter, glutamate, from their synaptic terminals. This decrease in glutamate is sensed by different glutamate receptors on the ON- and OFF-bipolar cells. OFF-bipolar cells have ionotropic receptors which are directly linked to the opening of non-specific cation channels. Closure of these channels occurs when glutamate release falls with light, resulting in hyperpolarizing responses. On the other hand, rod ON-bipolar cells possess a novel class of metabotropic glutamate receptor (mGluR), which is indirectly linked via a cGMP cascade to the closure of cation channels (Shiells and Falk, 1990; Nawy and Jahr, 1990). The fall in glutamate release with light is transduced by rod ON-bipolar cells into a rise in intracellular cGMP, which opens cGMP-activated cation channels, generating depolarizing responses. This glutamate receptor-coupled cGMP cascade, which is

Neurobiology and Clinical Aspects of the Outer Retina
Edited by M.B.A. Djamgoz, S.N. Archer and S. Vallerga
Published in 1995 by Chapman & Hall, London
ISBN 0 412 60080 3

analogous to the cGMP cascade mediating phototransduction in photoreceptors (McNaughton, 1990; and Chapter 5), functions to amplify small rod signals in response to a few absorbed photons by about 100-fold (Ashmore and Falk, 1980a, b; Shiells and Falk, 1990). It is now known that in mammalian retina rod bipolar cells are predominately the ON-type (Dacheux and Raviola, 1986; Muller *et al.*, 1988; Daw *et al.*, 1990), and so this synaptic amplification is probably essential for the high sensitivity of the rod visual system.

The focus of this chapter is to illustrate how the functional elements involved in signal transmission to bipolar cells generate high synaptic voltage gain, with particular emphasis on the rod pathway. The temporal features of synaptic transmission from rods to bipolar cells are also considered.

12.2 GLUTAMATE RECEPTORS OF BIPOLAR CELLS

Generally in the central nervous system (CNS), the functional diversity of glutamate is determined by the possession of disparate glutamate receptors, which may be categorized into two distinct classes, ionotropic and metabotropic receptors (Nakanishi, 1992). In the retina, the formation of ON- and OFF-pathways at the bipolar cell level is determined by the differential expression of these glutamate receptor types. The ionotropic receptors contain integral cation-selective channels within their structure and are further subdivided into three main groups: *N*-methyl-D-aspartate (NMDA), α-amino-3-hydroxy-5-methyl-4-isoxazole propionate (AMPA) and kainate (KA) receptors (Seeburg, 1993 for review). The mGluRs are pharmacologically and functionally different from the ionotropic receptors. They are coupled via G-proteins to a variety of functions by mediating intracellular signal transduction (Sugiyama *et al.*, 1987; Schoepp *et al.*, 1990).

The glutamate receptors of ON-bipolar cells are distinct from those of OFF-bipolar cells as first demonstrated by Slaughter and Miller (1981) in mudpuppy retina and by Shiells *et al.* (1981) in the virtually all-rod dogfish retina. The glutamate analog 2-amino-4-phosphonobutyrate (APB) selectively activated the glutamate receptors of ON-bipolar cells, but had little or no action on OFF-bipolar or horizontal cells. On this basis, the ON-bipolar cell receptors were termed APB-receptors and were perhaps unique among other CNS glutamate receptors in that they were coupled to a non-specific cationic conductance decrease in rod-dominated retinas with a reversal potential near 0 mV (Shiells *et al.*, 1981). There is also evidence for a second type of ON-bipolar cell glutamate receptor which opens specific K^+ channels (Nawy and Copenhagen, 1987, 1990; Hirano and MacLeish, 1991) and since cone inputs to ON-bipolar cells are mediated by a K^+ conductance in teleost fish (Saito *et al.*, 1978; Saito and Kujiraoka, 1982), these receptors probably mediate cone transmission to ON-bipolar cells. One report suggested that these receptors were APB-insensitive (Nawy and Copenhagen, 1987) whereas evidence for their APB sensitivity was reported in salamander using higher doses of APB (Hirano and Macleish, 1991). This action would be consistent with evidence that APB blocks all on-pathways in the mammalian retina, from both rods and cones (Muller *et al.*, 1988; Schiller, 1992).

12.2.1 G-PROTEIN COUPLED GLUTAMATE RECEPTORS OF ROD ON-BIPOLAR CELLS

Evidence that APB receptors of ON-bipolar cells were linked via a G-protein to the control of a cGMP cascade was first presented by Nawy and Jahr (1990) using salamander retina and by Shiells and Falk (1990) in dogfish retina. Our study has identified this as the receptor type which mediates rod responses, since dogfish ON-bipolar cells have virtually no cone inputs. Using the whole-cell mode of the patch-clamp technique (Marty and Neher, 1983) applied to bipolar cells in retinal slices, it was possible to manipulate the intracellular biochemistry of these cells by the addition of compounds to the patch-pipette solution. Both groups found initially that soon after rupturing the membrane following giga-seal formation that the light or APB-induced responses were rapidly attenuated if GTP was not added to the patch-pipette solution. This suggested that internal dialysis of the ON-bipolar cells with the patch-pipette solutions effectively washed out some second messenger from the cytoplasm, and furthermore, that this second messenger was essential for gating the channels mediating their light or APB-responses. The omission of GTP from the patch-pipette solutions would explain why a conductance-decreasing action of APB was detected in only very few isolated mammalian ON-bipolar cells (Karschin and Wässle, 1990).

The identity of the second messenger was

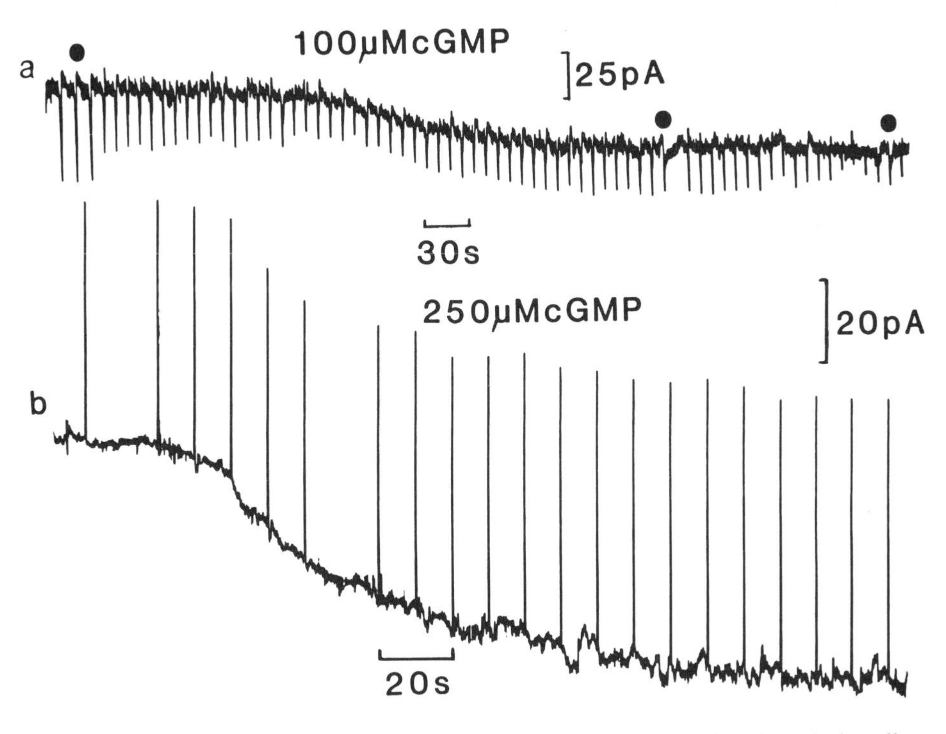

Figure 12.1(a) The action of 100 μM cGMP in the patch pipette on ON-bipolar whole-cell current (K soln). The record shows responses to dim light flashes, with brighter test flashes, giving maximal light responses, indicated by ●. Input resistance measurements showed a decrease from 27 to 20 MΩ with cGMP. Clamp potential (V_d) was −27 mV, seal resistance was 10 GΩ, and series resistance was 8 MΩ. (b) The more rapid effect of 250 μM cGMP (Cs soln) on whole cell current and input resistance of an ON-bipolar cell. Light responses were initially very small. The upward deflections are due to the application of 2 mV commands, showing a decrease in input resistance from 26 to 22 MΩ with cGMP. Clamp potential (V_d) was −24 mV, seal resistance was 4 GΩ, and series resistance was 7 MΩ. (Reproduced with permission from Shiells and Falk (1990).)

soon determined to be cGMP since in light-adapted salamander retinas treated with Co^{2+} to block synaptic transmission, cGMP added to the patch-pipette solution induced inward currents in ON-bipolar cells which were suppressed by APB (Nawy and Jahr, 1990). We reported a similar effect on dogfish ON-bipolar cells in dark-adapted retinal slices (Figure 12.1). The addition of 100 μM cGMP to the patch-pipette solution induced an inward current in ON-bipolar cells voltage-clamped to their dark potentials. This induced an increase in membrane conductance, and more rapid effects were seen with 250 μM cGMP. Complex effects were observed on the light responses. As the cells equilibrated with cGMP, a small increase in flash responses was observed, but the net effect was a decrease in flash response amplitude on equilibration with 100 μM cGMP. These experiments suggested that cGMP opened cation channels in the ON-bipolar cell membrane, and further, since cGMP suppressed the light responses, this second messenger was intimately associated with the generation of their light responses.

Evidence that a G-protein was involved in linking the APB-receptor to the control of intracellular cGMP was obtained by studying the effect of adding GTP-γ-S, a stable analog of GTP which activates G-proteins, to the patch-pipette solutions (Figure 12.2). This induced an outward current in ON-bipolar cells, due to the suppression of a tonic inward current present in the dark, which was accompanied by the suppression of light responses and a decrease in membrane conductance. These effects were similar to those

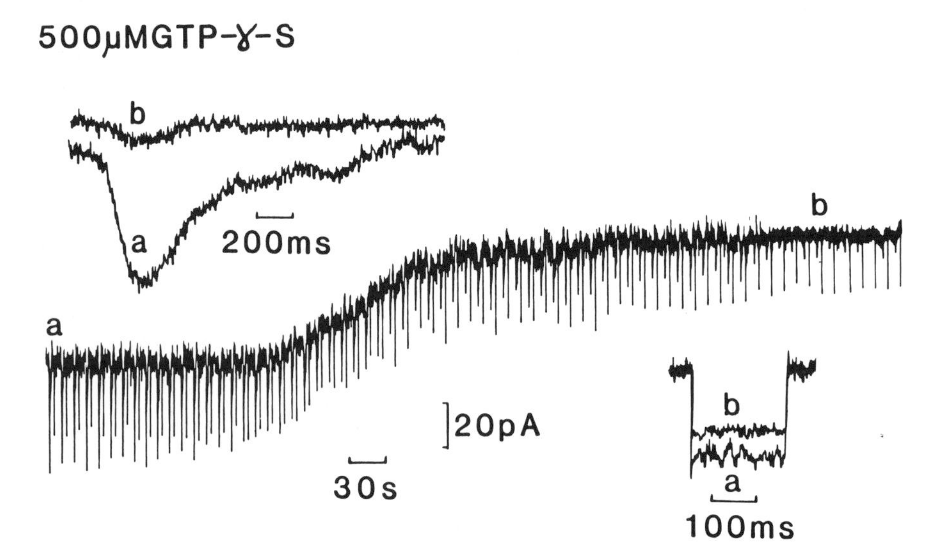

Figure 12.2 The action of 500 μM GTP-γ-S in the patch pipette on light responses, whole-cell current and input resistance of an on-bipolar cell (K soln). The continuous record shows the whole-cell current with responses to light flashes and 1 mV voltage steps applied every 10 s. Light responses (with the flash applied at the beginning of the traces) are shown on an expanded timescale (upper inset) before (a) and after (b) the action of GTP-γ-S. Responses to the 1 mV commands (lower inset) show an input resistance increase from 14 MΩ (a) to 22 MΩ (b) in the presence of GTP-γ-S (after correction for series resistance). The dark membrane potential (V_d) to which the cell was clamped, was −29 mV. Seal resistance was 8 GΩ and series resistance was 11 MΩ. (Reproduced with permission from Shiells and Falk (1990).)

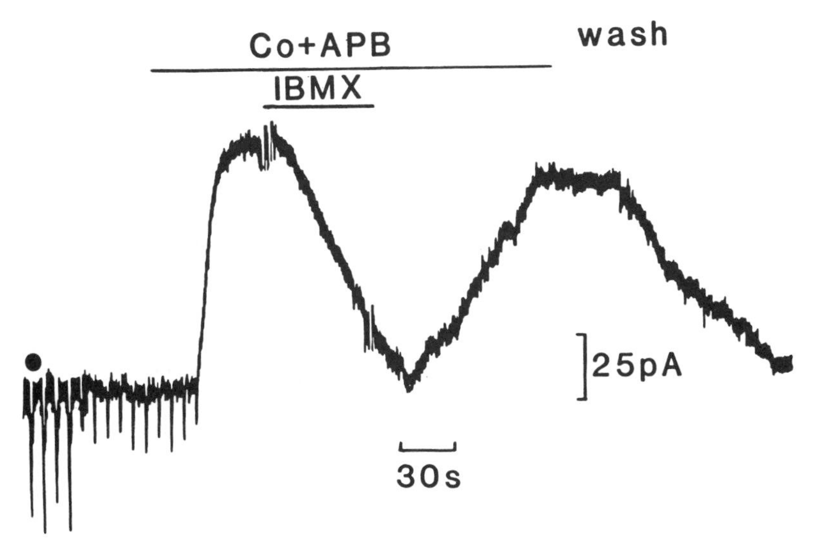

Figure 12.3 The action of the phosphodiesterase inhibitor IBMX in the superfusate on ON-bipolar whole-cell current (K soln). A similar flash protocol to Figure 12.2(a) was used. The perfusion fluid was switched to one containing 2mM $CoCl_2$ and 100 μM APB inducing an outward current and increase in input resistance from 16 to 35 MΩ. Ringer containing Co^{2+}, APB and 1 mM IBMX was then applied. This caused a decrease in input resistance from 35 to 17 MΩ accompanied by an inward current, similar to the effect of cGMP. On returning to Co^{2+} and APB alone, the inward current reversed and the input resistance returned to 32 MΩ. Clamp potential V_d was −24 mV, seal resistance was 8 GΩ, and the series resistance was 6 MΩ. (Reproduced with permission from Shiells and Falk (1990).)

induced by glutamate or APB, and so it seemed reasonable to propose that activation of the APB receptor in turn activated a G-protein to its GTP-bound state. Nawy and Jahr (1990) demonstrated this in salamander ON-bipolar cells, and additionally showed the converse effect with the G-protein inactivator, GDP-β-S.

To demonstrate the presence of a cGMP phosphodiesterase (PDE) in ON-bipolar cells, the action of the PDE inhibitor, isobutylmethylxanthine (IBMX) on the ON-bipolar cells was studied (Figure 12.3). Since the exposure of rods to IBMX was known to depolarize their membrane potential by inhibition of rod PDE allowing cGMP to rise, it was essential to study the external application of IBMX in the presence of the synaptic blocker Co^{2+}. APB and Co^{2+} were co-applied, inducing an outward current in the ON-bipolar cell due to the closure of cation channels by APB (due to a fall in cGMP). IBMX (1 mM) was then co-applied by superfusion of the retinal slice, and this induced an inward current accompanied by an increase in membrane conductance, consistent with the action of IBMX to inhibit PDE, resulting in a rise in cGMP-activated conductance. Other experiments ruled out the involvement of cAMP in this cascade. Membrane permeable analogs of cAMP were found to have no action on the ON-bipolar cells (Shiells and Falk, 1990), indicating that this was specifically a cGMP-PDE. The presence of a PDE was also demonstrated in salamander ON-bipolar cells (Nawy and Jahr, 1990, 1991).

These results led to the proposal of the reaction scheme in Figure 12.4. The transmit-

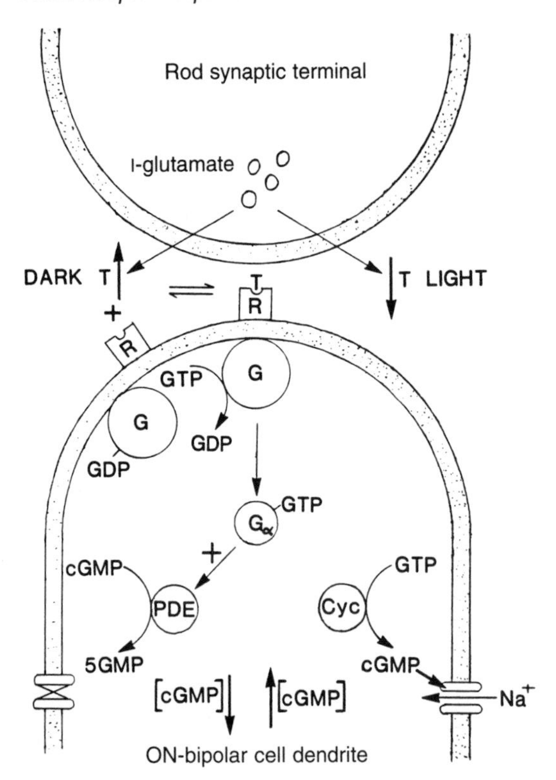

Figure 12.4 Proposed mechanism of the action of L-glutamate, the transmitter T, released from rod synaptic terminals. In the dark, when transmitter release is relatively high, glutamate binds to a postsynaptic receptor R, an interaction which catalyzes the exchange of GDP for GTP on a G-protein. The activated α-subunit of the G-protein bound to GTP (G_α-GTP) then activates a phosphodiesterase leading to a fall in cGMP and cation channel closure. With light, the rods hyperpolarize, decreasing glutamate release, resulting in G-protein inactivation, a rise in cGMP and to the opening of cation channels in the ON-bipolar cell dendritic membrane, the basis for depolarizing light responses. (Reproduced with slight modification from Shiells and Falk (1992a).)

ter T released from rods, glutamate, binds to the ON-bipolar cell membrane receptor R, and switches on an exchange of GTP for GDP bound to a G-protein. The GTP-bound form activates an effector enzyme cGMP-PDE, leading to a decrease in the concentration of the second messenger cGMP and the consequent closure of cation channels in the dark. The action of glutamate here is to induce a fall in cGMP, so that when the transmitter concentration in the synaptic cleft decreases with the light-induced hyperpolarization of rods, intracellular cGMP rises increasing cGMP-activated conductance, thus forming the basis of depolarizing light responses (Shiells and Falk, 1992a). This inverse coupling of external transmitter (glutamate) to a fall in internal transmitter (cGMP) is unusual, but a clear parallel may be drawn to other mGluRs which link to inhibitory cAMP cascades (Nakanishi, 1992). At other sites in the CNS, glutamate has been shown to increase intracellular cGMP by stimulating the release of nitric oxide (NO) which in turn activates soluble guanylate cyclase (Novelli *et al.*, 1987; Garthwaite, 1991).

The G-protein coupled glutamate receptor of ON-bipolar cells is now known to belong to a wider family of metabotropic glutamate receptors (mGluRs). Molecular cloning by cross-hybridization and polymerase chain reaction (PCR) techniques has revealed the presence of at least six subtypes of mGluRs, mGluR1 through mGluR6 (Tanabe *et al.*, 1992). One cloned species, mGluR6, was found to be highly homologous to the AP4-selective mGluR4 which has been shown to be linked to the suppression of synaptic transmission by the inhibition of glutamate release at a presynaptic site (Baskys and Malenka, 1991). The mGluR6 was found to be highly enriched in retina, which tentatively identified it as the mGluR of ON-bipolar cells (Nakanishi, 1992; Duvoisin, 1993). More recently, *in situ* hybridization studies have shown that mGluR6 is expressed specifically in the inner nuclear layer of the retina, where ON-bipolar cells are localized (Nakajima *et al.*, 1993). When expressed in Chinese hamster ovary cells, this receptor was found to link to an inhibitory cAMP cascade and had similar pharmacological properties to the rod ON-bipolar cell APB-receptor, with one exception. The expressed receptors did not respond to kainate, suggesting a slight difference from the native receptors which are activated by kainate (Shiells *et al.*, 1981; Slaughter and Miller, 1983).

12.2.2 IONOTROPIC GLUTAMATE RECEPTORS OF OFF-BIPOLAR CELLS

The ionotropic glutamate receptors of OFF-bipolar cells have been classified as AMPA/KA type in early studies on the basis that the excitatory amino acid antagonist *cis*-2,3-piperidine dicarboxylate (PDA) was found to block OFF-bipolar and horizontal cell light responses, and antagonize applied KA, in mudpuppy retina (Slaughter and Miller, 1983). PDA had little or no action on the ON-bipolar cells and thus has a complementary action to APB since it blocked the OFF- but not the ON-pathway. More recent reports on mammalian OFF-bipolar cells have identified the presence of AMPA/KA receptors, and NMDA receptors also seem to contribute to the glutamate response in these cells (Muller *et al.*, 1988; Sasaki and Kaneko, 1993). Furthermore, the KA receptor blocker, 6-cyano-7-nitroquinoxaline-2,3-dione (CNQX) blocked the majority of the glutamate-induced current in amphibian OFF-bipolar cells (Hensley *et al.*, 1993), but other work has isolated CNQX-sensitive and CNQX-insensitive receptors (Taylor and Copenhagen, 1993). Quisqualate-induced currents in OFF-bipolar cells decayed over a time course of tens of milliseconds, whereas the KA-induced currents showed no such desensitization (Gilbertson *et al.*, 1991), consistent with neuronal AMPA-receptor pharmacology (Seeburg, 1993). However, in other studies on isolated salamander OFF-bipolar cells, no desensitization to glutamate was observed (Attwell *et al.*, 1987). Concentration-jump experiments on isolated dogfish bipolar cells have demonstrated the complete absence of desensitization to glutamate in rod ON- and OFF-bipolar cells (Shiells and Falk, 1993b). Just as we have seen for the ON-bipolar cells, there may be two or more classes of glutamate receptors on OFF-bipolar cells mediating rod and cone inputs independently. The evidence would be consistent with a non-desensitizing glutamate receptor subserving rod inputs, whereas a desensitizing AMPA and/or NMDA receptor may subserve cone inputs to these cells. The inconsistencies in the literature may derive from differences in isolation procedures leading to the selection of bipolar cells expressing only one or both types of receptor. Recent experiments employing dual whole-cell recording from photoreceptors and OFF-bipolar cells have detected a differential sensitivity of rod and cone inputs to kynurenic acid (Kim and Miller, 1993), suggesting different glutamate receptor types. In amphibian off-bipolar cells, the reversal potentials for the action of quisqualate, kainate and glutamate were all close to 0 mV, as is the case for glutamate in rod OFF-bipolar cells (Shiells and Falk, 1993b). This suggests that these agonists gate a cation-selective channel for which Na^+ and K^+ have almost equal permeability (Attwell *et al.*, 1987; Gilbertson *et al.* 1991). On changing the external Ca^{2+} concentration, Gilbertson *et al.* (1991) detected a change in the reversal potential with high Ca^{2+}, indicating that these channels were much more permeable to Ca^{2+} ions than other non-NMDA channels. The only other known AMPA-receptor channels exhibiting this type of behavior are found on Bergmann glial cells, due to the absence of the subunit GluRB which restricts Ca^{2+} permeability (Seeburg, 1993). Glutamate receptors with similar properties to the non-desensitizing rod OFF-bipolar cell receptors have been reported in type-2 astrocytes (Usowicz *et al.*, 1989). It is probable that the glutamate receptors mediating rod vision followed a different evolutionary pathway to other known glutamate receptors, and have become specialized in their lack of desensitization to the continuous release of glutamate from rods in the dark. Small decreases in glutamate release at low light levels would simply not be detected by a desensitizing glutamate receptor, imposing a serious limitation on visual sensitivity. This would be consistent with the recent finding that the mGluR6 of rod ON-bipolar cells does not seem to be expressed elsewhere in the CNS (Nakajima *et al.*, 1993).

12.3 COMPONENTS OF THE cGMP CASCADE OF ON-BIPOLAR CELLS

To gain insight into how the glutamate receptor coupled cGMP cascade of ON-bipolar cells functions, we have primarily been interested in characterizing the components: the G-protein, cGMP-PDE, guanylate cyclase, and the cGMP-activated channels. Unlike photoreceptors, it is not possible to obtain a homogenous sample of ON-bipolar cells to perform bulk biochemical experiments on these components, and so we have relied on the patch-clamp technique to study their internal biochemistry at the single-cell level.

12.3.1 THE G-PROTEIN (G_b)

One method of characterizing G-proteins has been to determine their sensitivity to pertussis (PTX) and cholera (CTX) toxins. These toxins act by catalysing the transfer of adenosine diphosphate-ribose (ADPR) from intracellular nicotinamide adenine dinucleotide (NAD^+) to specific sites on the α-subunit of the G-protein. Transducin (G_T), the G-protein mediating phototransduction in rods, was unique among G-proteins in that it possessed ADP-ribosylation sites sensitive to both CTX and PTX (Van Dop *et al.*, 1984). The consequence of CTX action on G_T was activation, by prolonging the lifetime of the active GTP-bound complex, consistent with the activation of other CTX-sensitive G-proteins such as G_s (Cassel and Pfeiffer, 1978). Conversely, the action of PTX on G_T was inactivation, by prolonging the lifetime of the inactive GDP-bound complex, as is the case for other PTX-sensitive G-proteins such as G_i (Katada and Ui, 1982). The exposure of salamander rods to PTX resulted in depolarization of their membrane potential and suppression of their light responses, consistent with blocking G_T leading to a rise in cGMP via PDE inhibition (Falk and Shiells, 1988). By applying the activated toxin A-subunits via the patch-pipettes, we were able to show that G_b was also sensitive to both CTX and PTX (Shiells and Falk, 1992a). Figure 12.5 shows the effect of including the A-subunit of PTX, with excess NAD^+, in the patch-pipette solution on the input resistance, dark membrane current and light responses of an ON-bipolar cell recorded from the dark-adapted retinal slice of dogfish. It was important to obtain stable, long-term recordings from these cells because these toxins act relatively slowly. After a delay of a few minutes during which time the cell interior equilibrated with the patch-pipette solution, there was a slow decrease in the light responses in parallel with an increased inward current, and a rise in membrane conductance. This effect was consistent with the action of PTX in blocking G_b in the GDP-bound complex, resulting in PDE inactivation, thus increasing the cGMP-activated conductance. The converse effect was observed by including the A-subunit of CTX with excess NAD^+ in the patch-pipette solution (Figure 12.6). This induced an outward current due to the suppression of a tonic inward current present in the dark, a gradual decline in light responses, and a decrease in membrane conductance. An appropriate control for these experiments was to omit NAD^+ from the patch-pipette

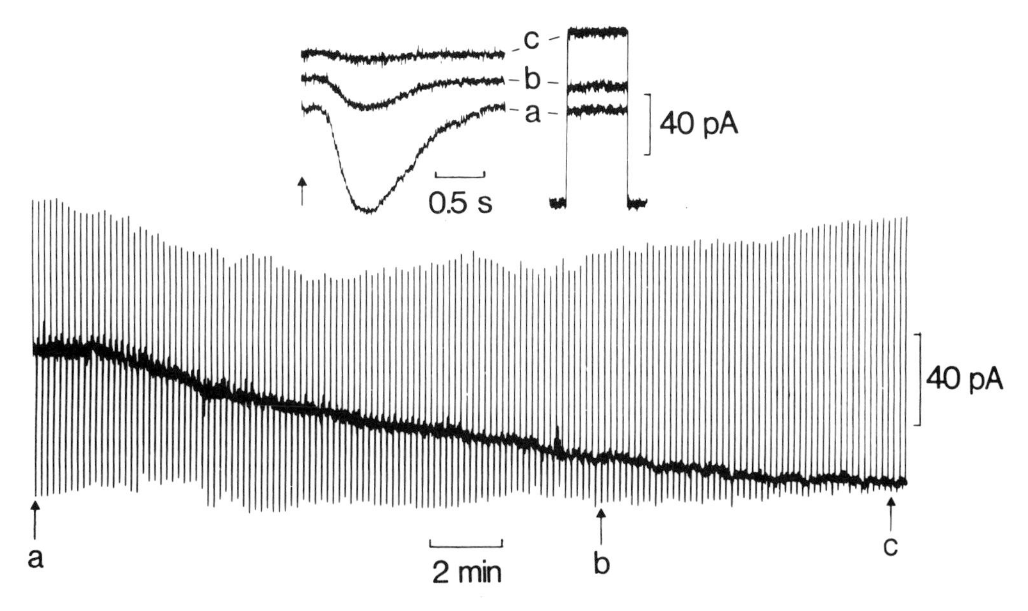

Figure 12.5 The effect of pertussis toxin A subunit (1 μg/ml) with 1 mM NAD^+ in the patch-pipette on an ON-bipolar cell. The dark membrane potential was measured initially in current clamp; the continuous record begins on voltage clamping the ON-bipolar cell to its dark potential of −26 mV. The downward deflections are inward currents evoked by flashes of light applied every 10 s. The upward deflections are current responses to 2 mV command pulses applied to monitor input resistance. These are shown inset on an expanded timescale (a) before, (b) during and (c) after pertussis toxin action. The input resistance decreased from 26 to 11 MΩ with pertussis toxin, and the inward current increased by 58 pA. The series resistance was 8 MΩ. (Reproduced with permission from Shiells and Falk (1992a).)

solutions containing the toxins, which would effectively wash out endogenous NAD^+. Without substrate NAD^+, these toxins had no effect on the ON-bipolar cells, ruling out possible non-specific actions of these toxins.

These results suggested that if a single G-protein coupled to the mGluR mediating the light responses of ON-bipolar cells, then this G_b might be structurally homologous to G_T in having ADP-ribosylation sites with similar amino acid sequences (Lerea *et al.*, 1986). It was unlikely that two G-proteins, consisting of inhibitory and stimulatory subunits (G_i and G_s) were involved, since in the adenylate cyclase system both CTX and PTX stimulate a rise in adenylate cyclase activity (Murayama and Ui, 1983). Blocking or activating other G-proteins present in ON-bipolar cells possibly linked to Ca^{2+} conductances or GABA-mediated Cl^- conductances would not be expected to modulate the light responses in this way. The presence of a cGMP cascade in ON-bipolar cells utilizing a similar G-protein to G_T would not be too surprising, given that both the photoreceptors and second-order cells appear to derive from the same precursor cells during development (Harris and Holt, 1990).

The identification of G_b using immunocytochemical techniques has thus far proved elusive. In a recent study, antibodies were raised to the components of the rod phototransduction cascade including G_T PDE, the cGMP-activated channel and arrestin (Vardi *et al.*, 1993). All of these reacted strongly with rods, but none reacted with bipolar cells.

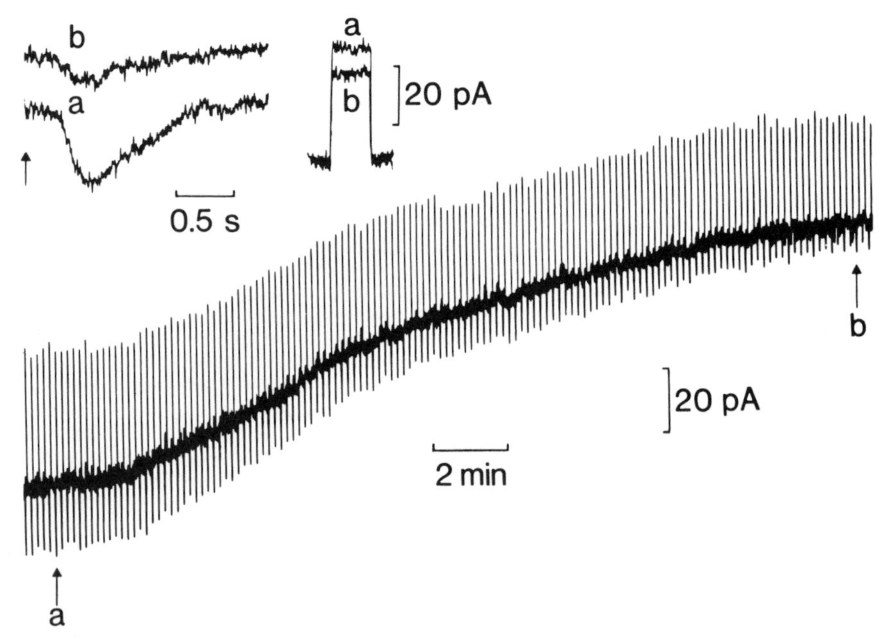

Figure 12.6 The effect of cholera toxin A subunit on an ON-bipolar cell when 10 μg/ml cholera toxin A subunit and 1 mM NAD^+ were added to the patch-pipette solution. The continuous record begins on whole-cell voltage clamping the ON-bipolar cell to its dark membrane potential of −27 mV. An outward current of 79 pA accompanied the suppression of light responses by cholera toxin and an increase in input resistance from 40 to 53 MΩ, shown inset on an expanded timescale (a) before and (b) after cholera toxin action. The series resistance was 10 MΩ. (Reproduced with permission from Shiells and Falk (1992a).)

However, antibodies raised to G_o did react strongly with rod bipolar cells, confirming previous studies (Terashima *et al.*, 1987). G_o is prominently expressed in the CNS and is usually associated with the regulation of Ca^{2+} conductance, so it seems unlikely that G_b could be a subspecies of G_o. Furthermore, antibodies raised against specific sequences found only on G_T expressed in rods did not cross-react with cone G_T (Lerea *et al.*, 1986), so it does not seem surprising that antibodies raised against G_T did not label G_b. More subtle molecular biological techniques will probably be necessary to identify G_b in the dendritic terminals of ON-bipolar cells, by screening cDNA libraries for the different G-proteins expressed in retina.

12.3.2 cGMP PHOSPHODIESTERASE (PDE)

The presence of a PDE in ON-bipolar cells was demonstrated by their responses to the PDE inhibitors, IBMX and dipyridamole (Shiells and Falk, 1990; Nawy and Jahr, 1990, 1991), which also inhibit the photoreceptor PDE (Cervetto and McNaughton, 1986). Membrane-permeable analogs of cAMP had no effect on the ON-bipolar cells, suggesting that this was a specific cGMP PDE. The PDE of rods has been shown to be activated by Ca^{2+}, an action mediated by a Ca^{2+}-binding protein, S-modulin (Kawamura and Murakami, 1991). Whether the PDE in ON-bipolar cells is also sensitive to regulation by Ca^{2+}

remains to be determined. If their light responses were accompanied by an influx of Ca^{2+}, this could function to accelerate the timecourse of PDE activation and restore cGMP to the dark level more rapidly. This does, however, seem unlikely since isolated ON-bipolar cells responded to step-changes in glutamate with sustained responses (Shiells and Falk, 1993b), as is the case for their responses to steps of light (Ashmore and Falk, 1980a, b).

12.3.3 GUANYLATE CYCLASE (GC)

Two principal forms of GC are known, a particulate transmembrane species and a soluble heme-containing form which is activated by nitric oxide (NO) (reviewed by Goy, 1991). Early biochemical experiments showed that both forms of GC could be isolated from whole retina (Troyer *et al.*, 1978). Two fractions were isolated, a soluble, Ca^{2+}-insensitive fraction and a particulate fraction which was inhibited by increases in Ca^{2+}. The particulate form corresponded to the GC expressed in photoreceptors (Fleischman and Denisevich, 1979), which is inhibited on elevation of intracellular Ca^{2+} (Koch and Stryer, 1988). The inhibitory action of Ca^{2+} on photoreceptor GC was mediated by the Ca^{2+}-sensitive protein recoverin, which activates the particulate form only at low Ca^{2+} levels (Dizhoor *et al.*, 1991), and light adaptation in rods is thought, in part, to be due to the Ca^{2+}-mediated inhibition of GC (McNaughton, 1990; and Chapter 5). Other particulate forms of GC are activated by peptides such as atrial natriuretic peptide or *E. coli* heat-stable enterotoxin (Schulz *et al.*, 1990; Yuen and Garbers, 1992), which activate the enzyme by binding to its extracellular domain.

Activation of soluble GC is thought to occur by the interaction of NO with the ferrous (reduced) form of the heme group associated with its regulatory subunit (Ignarro, 1990). Conversely, the inhibitors of soluble GC, ferricyanide or methylene blue are thought to act by oxidizing the heme group to its ferric form, resulting in an inactive state. By using these agents known to specifically activate or inhibit the different forms of GC, it was possible to determine the type of GC expressed in ON-bipolar cells (Shiells and Falk, 1992c).

The effect of NO on the ON-bipolar cells in dark-adapted retinal slices of dogfish was initially tested by adding a NO-donor, nitroprusside, to the patch-pipette solution (Figure 12.7a). As the cell equilibrated with the patch-pipette solution, an increase in inward current developed which was accompanied by a rise in dim and bright flash response amplitudes, and an increase in membrane conductance. Previously, we had observed small increases in light responses as ON-bipolar cells equilibrated with 100 μM cGMP, and furthermore, a similar effect was observed in salamander ON-bipolar cells (Nawy and Jahr, 1991). A lower dose of cGMP (20 μM) was included in the patch-pipette solution to test whether the increase in light responses was due to an elevation of cGMP due to GC activation (Figure 12.7b). The effects of NO on light responses, membrane conductance and inward current were reproduced by directly elevating the cGMP. Inclusion of the GC inhibitors, ferricyanide and methylene blue in the patch-pipette solution induced the converse effect, outward currents (due to the suppression of inward dark currents) which were accompanied by a decrease in light responses and membrane conductance (Figure 12.8). These effects were consistent with the presence of a soluble, NO-sensitive form of GC in ON-bipolar cells, a result confirmed in other recordings in which nitroprusside and methylene blue were applied by superfusion of the retinal slices, inducing reversible inward and outward changes in dark current, respectively. The inward currents induced by nitroprusside persisted in the presence of the

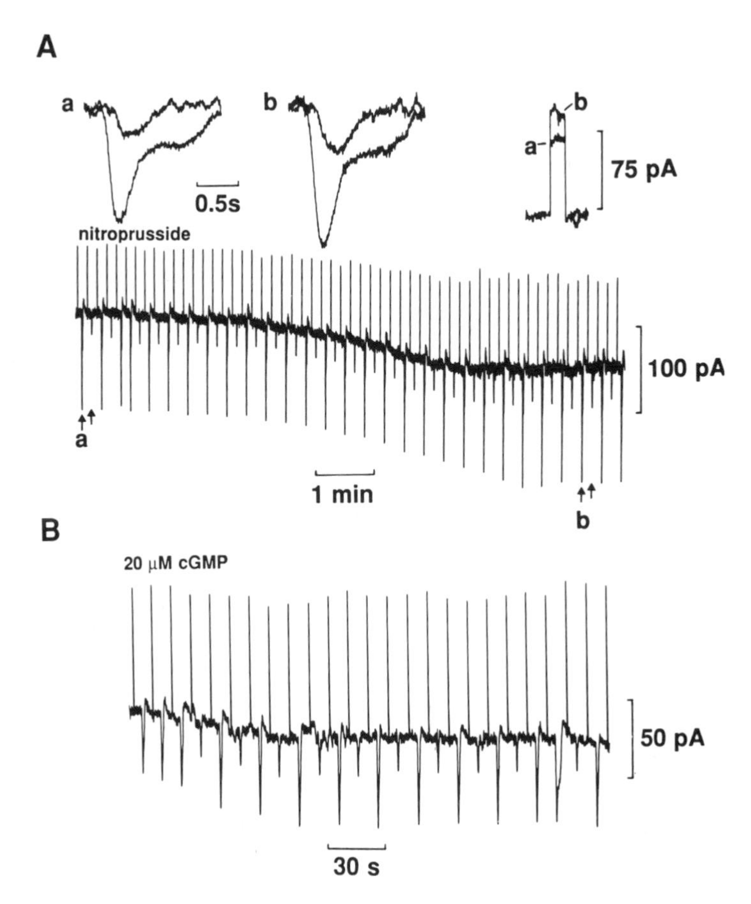

Figure 12.7 Whole-cell recordings from ON-bipolar cells voltage-clamped to their dark potentials with (A) 1mM nitroprusside (NP) or (B) 20 μM cGMP added to the patch-pipette solution. (A) Inward current light responses (downward deflections) to alternating dim and saturating flashes and outward current responses to voltage-command pulses (upward deflections) are shown on a slow time base (lower trace) and on expanded timescales (upper traces) before (a) and after equilibration with NP. NP induced an inward current of −73 pA, accompanied by an increase in peak saturated light response from 106 to 136 pA and a decrease in input resistance from 23 to 14 MΩ. Dark potential was −29 mV. (B) 20 μM cGMP induced an inward current of −22 pA, the peak saturated light response increased from 40 to 58 pA and the input resistance decreased from 26 to 20 MΩ. A supra-saturating flash was applied (black dot), and the dark potential was −24 mV. (Reproduced with permission from Shiells and Falk (1992c).)

synaptic blocker, Co^{2+}, confirming that the action of NO was localized to the ON-bipolar cell.

To determine whether a peptide-sensitive particulate form of GC was also expressed in ON-bipolar cells, *E. coli* heat-stable toxin (Sta) was applied to their synaptic regions from an external pipette by pressure injection, in the presence of Co^{2+} and APB. This had no effect on the ON-bipolar cells, nor did inclusion of Sta or recoverin in the patch-pipette solutions induce any significant responses. We therefore concluded that the principal form of GC expressed in ON-bipolar cells was the

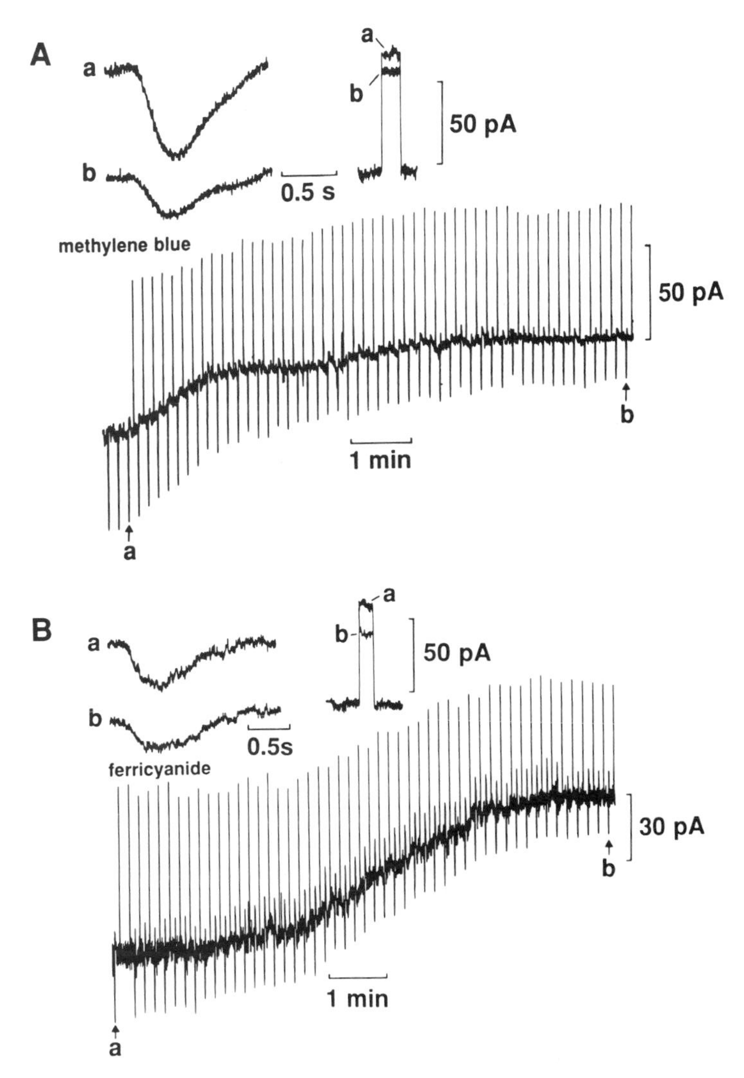

Figure 12.8 The actions of methylene blue (MeBlue) (A) and ferricyanide (B) included in the patch-pipette solution (1 mM) on the ON-bipolar cells voltage-clamped to their dark potentials. (A) MeBlue induced an outward current of 43 pA, a decrease in peak light response from 54 to 25 pA and an increase in input resistance from 21 to 27 MΩ. The dark potential was −24 mV. (B) Ferricyanide induced an outward current of 70 pA, a decrease in peak light response from 30 to 20 pA and an increase in input resistance from 20 to 33 MΩ. The dark potential was −32 mV. (Reproduced with permission from Shiells and Falk (1992c).)

NO-sensitive soluble form, excluding the presence of other particulate forms. This was consistent with recent molecular biological evidence, using *in situ* hybridization techniques, that the mRNA coding for the soluble form of GC was expressed in the inner nuclear layer (Barnstaple and Ahmad, 1992). The presence of other particulate forms of GC

(GC-A and GC-B) in the inner nuclear layer was excluded in the same studies, and the lack of ON-bipolar cell responses to *E. coli* Sta excluded the GC-C form (Schulz *et al.*, 1990). More recently, increases in cGMP stimulated by NO have been detected in rabbit ON-bipolar cells and horizontal cells using antibodies raised against a formaldehyde conjugate of cGMP (Massey *et al.*, 1993), confirming the predominance of the soluble, heme-containing form of GC in ON-bipolar cells.

The process of phototransduction in rods depends on the co-localization of the components rhodopsin, transducin, cGMP-PDE and GC to the disk or outer segment membranes (Pugh and Lamb, 1993). The presence of soluble GC in ON-bipolar cells would suggest that this enzyme is not compartmentalized within their dendritic regions, and so could not interact efficiently with G_b. This is probably coupled to PDE inhibition, consistent with the homology to transducin revealed by its CTX- and PTX-sensitivity. No other known G-protein coupled cascade operates by inhibition of soluble GC. The soluble GC of ON-bipolar cells and horizontal cells which is also activated by NO (Miyachi *et al.*, 1990; DeVries and Schwartz, 1992) may correspond to the Ca^{2+}-insensitive fraction isolated from whole retina in earlier studies (Troyer *et al.*, 1978). A Ca^{2+}-insensitive GC in ON-bipolar cells would be consistent with the absence of desensitization, or adaptation, of their responses to steps of light (Ashmore and Falk, 1980a) or to applied steps of glutamate (Shiells and Falk, 1993b), since one of the principal mechanisms mediating light adaptation in rods is via the inhibition of particulate GC by Ca^{2+}. It remains to be determined whether any light-stimulated release of NO from amacrine or horizontal cells modulates ON-bipolar cell responses by stimulation of soluble GC, since both of these cell types appear to possess NO-synthase (Sandell, 1985; Mills and Massey, 1993).

12.3.5 cGMP-ACTIVATED CHANNELS

To characterize the properties of ON-bipolar cell cGMP-activated channels, we used two well-known properties of the cGMP-activated channels of rods (Shiells and Falk, 1992a). These were, first, that rod cGMP-activated channels are blocked in a voltage-dependent manner by L-*cis*-diltiazem (Stern *et al.*, 1986; McLatchie and Matthews, 1992), and second, that the conductance of the photoreceptor channel is reduced by more than 100-fold by divalent cations (Hodgkin *et al.*, 1985; Lamb and Matthews, 1988). Whole-cell recordings were obtained from ON-bipolar cells in dogfish retinal slices with L-*cis*-diltiazem included in the patch-pipette solution (Figure 12.9). The continuous record (Figure 12.9a) shows the inward current responses to light flashes with the ON-bipolar cell voltage-clamped to its dark potential. As the cell equilibrated with the L-*cis*-diltiazem in the patch-pipette solution, there was a gradual development of an outward current due to the suppression of inward dark current. This was accompanied by a decrease in membrane conductance and a decrease in light responses. Light-evoked currents recorded before equilibration with L-*cis*-diltiazem at different voltage-clamp potentials are shown in the left-hand panel and after equilibration in the right-hand panel. At negative voltage-clamp potentials, the light-evoked peak current in the control increased linearly with membrane potential (Figure 12.9b) and the extrapolated reversal potential was near 0 mV, consistent with the opening of cation channels with an equal permeability to Na^+ and K^+ (Ashmore and Falk, 1980a, Attwell *et al.*, 1987). On equilibration with L-*cis*-diltiazem, the current-voltage relation became increasingly non-linear with hyperpolarization, indicating that L-*cis*-diltiazem blocks the ON-bipolar cell cGMP-activated channels in a voltage-dependent way. The D-*cis*-isomer was ineffective suggesting a stereospecificity sim-

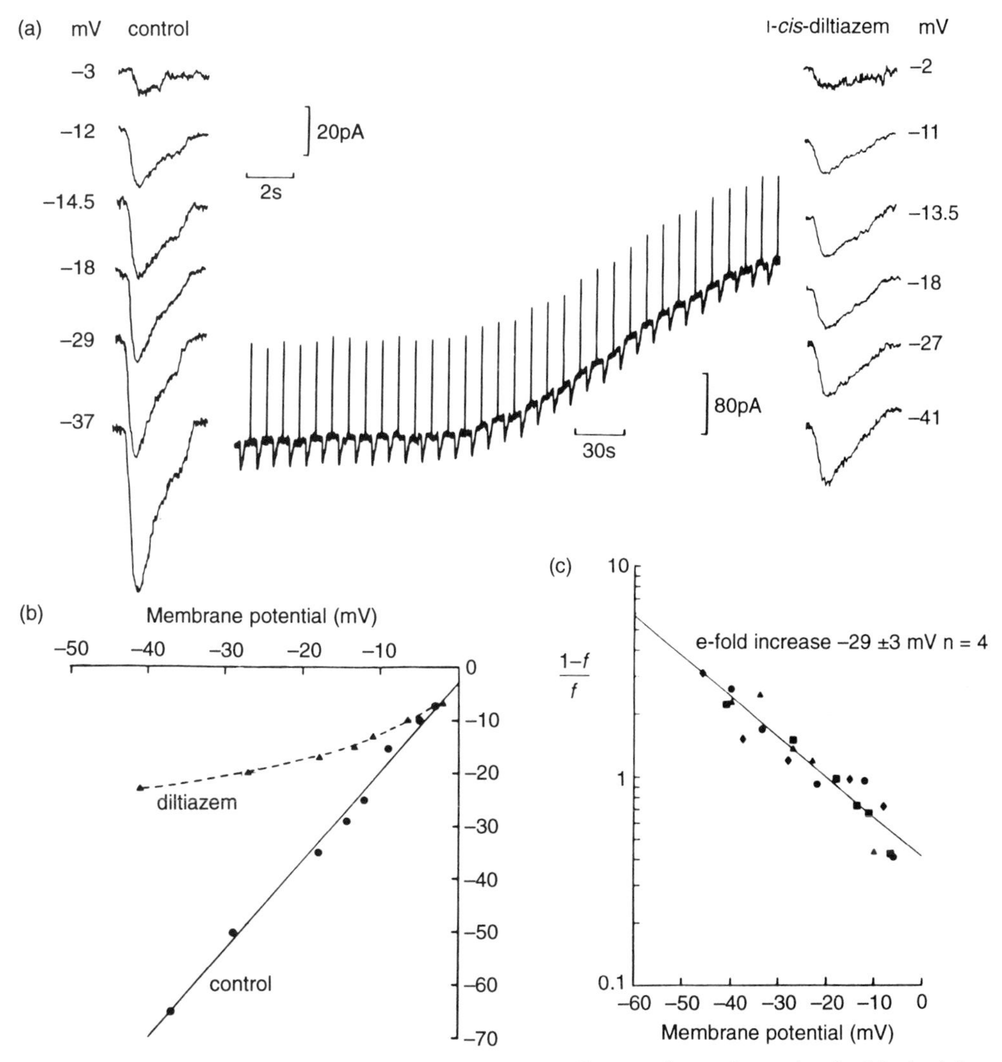

Figure 12.9 The light-evoked current of ON-bipolar cells is voltage-dependently blocked by L-*cis*-diltiazem, applied from the inside. (a) Whole-cell recording from an ON-bipolar cell in the retinal slice with 1 mM L-*cis*-diltiazem added to the patch-pipette solution. The control series of light current responses on the left was recorded at the voltage-clamp potentials shown, before the cell contents equilibrated with the patch-pipette solution. The cell was then clamped to its dark potential of −18 mV (center panel). The downward deflections are inward current responses to light and the upward delections are current responses to 5 mV command pulses, used to monitor input resistance. Diffusion of L-*cis*-diltiazem into the cell resulted in an outward current, which was accompanied by an increase in input resistance from 28 to 36 MΩ and a reduction of the light currents. The series resistance was 8 MΩ. Another series of light-evoked currents, at the clamp potentials shown on the right, was then recorded on equilibration with L-*cis*-diltiazem. (b) The current-voltage relation plotted from the light currents before (control) and on equilibration with L-*cis*-diltiazem. (c) The ratio of channels blocked to unblocked (1-*f*)/*f* plotted on a log scale against membrane potential, showing data pooled from four ON-bipolar cells. Since the extent of block varied among cells, but only the slope was of interest, a normalization procedure was used to represent the pooled data. The points for each cell were translated along the *y*-axis by an amount equal to the deviation from the mean value at each membrane potential. The squares are data from the cell in (a). (Reproduced with permission from Shiells and Falk (1992b).)

ilar to that of the photoreceptor channel (Stern *et al.*, 1986). The voltage-dependence of the channel block was analyzed by using a semilogarithmic plot against membrane potential of the ratio of light-sensitive channels blocked to those not blocked $(1\text{-}f)/f$, where f is the ratio of the light response with L-*cis*-diltiazem over the control at each membrane potential. The results from four cells (Figure 12.9c) gave an e-fold increase in the fraction of channels blocked over a mean value (± SEM) of 29 mV ± 3 mV hyperpolarization.

To determine whether external applications of L-*cis*-diltiazem could also block the ON-bipolar cell cGMP-activated channels, cells were isolated from the retinal slices by slowly raising the patch-electrodes in the cell-attached conformation. It was essential to use isolated cells since L-*cis*-diltiazem blocks the rod cGMP-activated conductance and so it would have been impossible to discriminate between pre- and postsynaptic effects in an intact preparation. Loosely connected bipolar cells on the surface of the retinal slices were selected, and were exposed to pressure applications of the selective glutamate agonist APB. This induced transient outward currents in isolated ON-bipolar cells voltage-clamped to their dark potentials, which reversed polarity to inward currents near 0 mV at positive potentials. Application of 100 μM L-*cis*-diltiazem by superfusion of isolated cells induced outward currents as occurred on internal dialysis, and a voltage-dependent decrease in the amplitude of the APB-induced currents. The amplitude of the currents suppressed by APB were used to determine the voltage-dependence of the block, and this was found to be the same whether L-*cis*-diltiazem was applied internally or externally (Shiells and Falk, 1992b).

The voltage-dependence of the block of ON-bipolar cell cGMP-activated channels with L-*cis*-diltiazem inside the cell was clearly anomalous, since diltiazem has a positive charge. If the blocking site within the channel was accessible from the inside of the cell, the charged blocking particle would be driven to the blocking site by the membrane electric field (Woodhull, 1973) and so the block should increase with depolarization as observed for the rod cGMP-activated channel (McLatchie and Matthews, 1992). The anomalous behavior could be resolved by proposing that L-*cis*-diltiazem could only gain access to the channel from the outside, to which access could be gained from the inside via the lipid phase of the membrane since diltiazem also exists in an uncharged, lipid-soluble form. From the voltage-dependence of block, and assuming that access to the channel was solely from the outside, we calculated that the blocking site lay at 0.86 of the electrical distance, that is close to the intracellular side of the channel. In comparison, the voltage-dependence of L-*cis*-diltiazem block of cGMP-activated channels in rods was weaker than that in ON-bipolar cells, increasing e-fold over 54 mV depolarization. It was also of the opposite polarity suggesting that L-*cis*-diltiazem blocks the rod channels by entering from the inside and blocks at a site half-way along the channel (McLatchie and Matthews, 1992). This led to the conclusion that the cGMP-activated channels of ON-bipolar cells were not the same channels as those expressed in rod photoreceptors.

A further difference between rod and ON-bipolar cell cGMP-activated channels was revealed in experiments designed to test the effect of removing divalent cations (Figure 12.10). If divalent cations blocked the cGMP-activated channels of ON-bipolar cells, as is the case in rods (Lamb and Matthews, 1988), then one would expect their removal from the external medium to induce a large inward current in cells held at negative voltage-clamp potentials. No such current was observed in the isolated cell on removal of divalent cations from the superfusate, nor was there any change in membrane conduct-

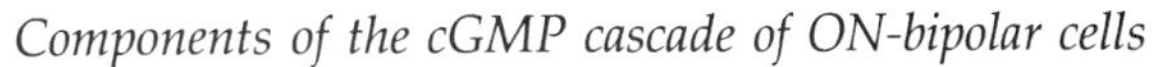

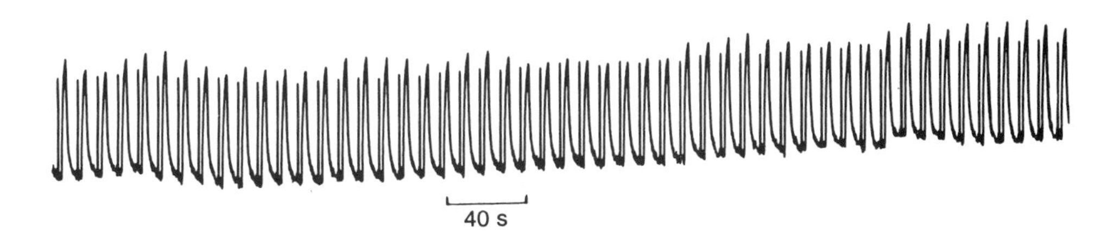

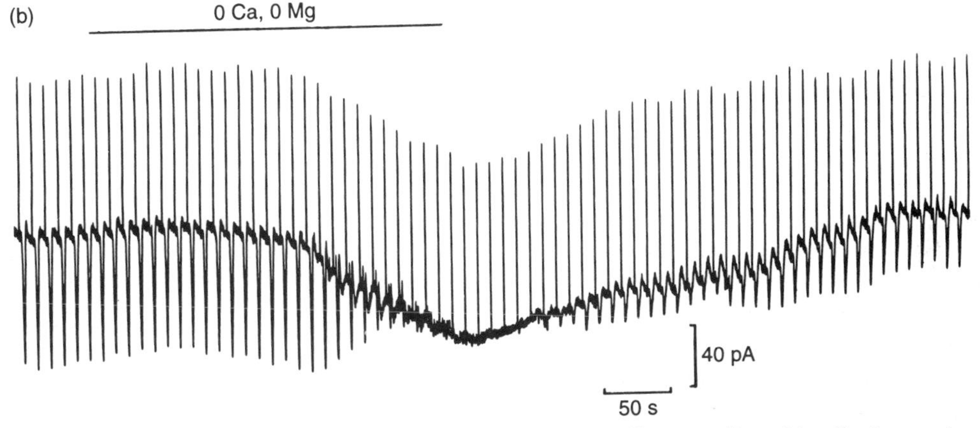

Figure 12.10 The cGMP-activated conductance of ON-bipolar cells is unaffected by divalent cations. (a) shows a whole-cell recording from an isolated ON-bipolar cell voltage clamped to its initial dark potential of −28 mV. Outward current responses were evoked in response to 400 ms pressure pulse applications of APB, preceded by command pulses. Input resistance was 21 MΩ. Normal Ringer solution was replaced by one containing no divalent cations with 0.25 mM EGTA and EDTA. The series resistance was 12 MΩ. (b) shows the effect of removal of divalent cations on light responses recorded from an ON-bipolar cell in the retinal slice. On removal of divalent cations, light responses (downward deflections) were suppressed due to failure of synaptic transmission, resulting in an inward current and a decrease in input resistance from 41 to 33 M Ω. The series resistance was 9 M Ω (Reproduced with permission from Shiells and Falk (1992b).)

ance or change in the amplitude of transient outward currents evoked by pressure applications of APB (Figure 12.10). The same experiment was repeated using an ON-bipolar cell recorded from the dark-adapted retinal slice (Figure 12.10b). On removal of external Ca^{2+}, the synaptic release of glutamate from rods was blocked resulting in a failure of the light responses, and a shift in the dark current towards the peak light current level. This inward current evoked by the removal of divalent cations never exceeded the peak light current level, consistent with the current shift being due to the suppression of transmitter release, rather than being due to a rise in single channel conductance. In the presence of divalent cations, rod cGMP-activated channels rectify strongly (Stern *et al.*, 1986; Baylor and Nunn, 1986) whereas the current-voltage relations of ON-bipolar cell light or APB-induced responses were linear, consistent with the absence of block by divalent cations from the outside. This difference in cGMP-activated channels in rods and ON-bipolar cells has recently been confirmed in immunocytochemical studies, in which antibodies raised to the rod channel protein did not label ON-bipolar cells (Vardi *et al.*, 1993).

cGMP-activated channels of rods are composed of several subunits which combine to form their characteristic functional properties. The human rod cGMP-activated channel has been cloned (Dhallan *et al.*, 1992) but there were prominent differences between the resulting expressed channel and the native channel. Co-expression of the cloned channel with another cloned subunit (hRCNCa) restored the native properties of the channel, including brief flickery openings to subconductance states and a high sensitivity to block by L-*cis*-diltiazem (Chen *et al.*, 1993; and Chapter 5). Another alternatively spliced variant (hRCNCb) was also found in human retina. This subunit, from immunocytochemistry, did not seem to be expressed in photoreceptors, but was localized in the outer plexiform layer. It will be interesting to see whether this is one of the functional subunits which combine to form the ON-bipolar cell channel.

The single-channel conductance of ON-bipolar cell cGMP-activated channels has been determined from the analysis of noise suppressed by glutamate in isolated salamander ON-bipolar cells (Attwell *et al.*, 1987). The decrease in membrane current noise induced by glutamate was described by the sum of two Lorentzian spectra and the analysis suggested a single-channel conductance of 11 pS. More recently, internal dialysis of isolated cat ON-bipolar cells with cGMP induced inward currents which were suppressed by glutamate. The inward currents induced by cGMP were accompanied by a rise in current noise (Kaneko *et al.*, 1993). Analysis of the difference spectra revealed a similar single-channel conductance to that determined in salamander ON-bipolar cells of 12 pS. Since the single-channel conductances determined from noise analysis are usually underestimates, the single-channel conductance of ON-bipolar cell cGMP-activated channels probably approaches the small conductance state of rod outer segment channels in the absence of divalent cations, which is of the order of 15–18 pS at negative potentials (Haynes *et al.*, 1986; Yau and Baylor, 1989; Torre *et al.*, 1992). The only other known cGMP-activated channel which is not blocked by divalent cations is found in invertebrate photoreceptors (Bacigalupo *et al.*, 1991). These have both large (40–43 pS) and small (15–18 pS) conductance states, reversal potentials near 0 mV, and similar linear current-voltage relations to both the rod (in the absence of divalent cations) and ON-bipolar cell cGMP-activated channels (reviewed by Yau and Chen, in press). The single-channel conductances determined from noise analysis await confirmation by direct recordings from isolated patches of ON-bipolar cell membrane. These studies will also resolve the question of whether the action of cGMP on the channels is direct, or whether cGMP-dependent protein kinases are required to phosphorylate the channel. A direct action seems likely, because agents which activate or inhibit protein kinase activity had no effect on light- or APB-induced currents whether applied internally or externally (Yamashita and Wässle, 1991; Nawy and Jahr, 1991; Shiells and Falk, 1992a).

The single photon event in dogfish ON-bipolar cells has been resolved from the analysis of intracellular responses to low light levels as 200 pS (Ashmore and Falk, 1980b). This would correspond to the opening of only 10–20 channels assuming the single-channel conductance of 12 pS. However, since cGMP-activated channels are known to have subconductance states, which may reflect the progressive binding of cGMP to the channels, a smooth 200 pS response probably derives from an increase in the open probability of a much larger number of cGMP-activated channels.

12.4 AMPLIFICATION OF PHOTORECEPTOR SIGNALS

The amplitude of the presynaptic voltage change in rods in response to the absorption of a single photon is very small, of the order of only 0.1 mV or less in the electrically coupled rod network (Fain, 1975; Detwiler *et al.*, 1980). Ganglion cells essentially function as threshold devices to generate action potentials requiring input signals of the order of 10 mV depolarization. A high degree of amplification is thus required to boost the rod signals sufficiently for mammalian ganglion cells to signal the absorption of a few photons within their receptive fields (Barlow *et al.*, 1971), and for the dark-adapted human observer to detect such low levels of light (Hecht *et al.*, 1942).

High gain in the rod pathway was first quantified by recording from a retina known to contain a relatively pure population of rod bipolar cells, the virtually all-rod retina of the dogfish. Measurements of ON-bipolar cell flash sensitivities suggested mean gains of the order of 100-fold accompanying synaptic transmission from rods (Ashmore and Falk, 1976, 1979, 1980a,b). The mean value agreed well with previous observations made on the electroretinogram (ERG). The b-wave of the ERG results directly from K^+ fluxes deriving from the population of ON-bipolar cell light responses (Falk and Shiells, 1986 and unpublished results; Stockton and Slaughter, 1989) whereas the a-wave is a reflection of photoreceptor activity. The scotopic b-wave is recruited at light intensities of the order of 100-times lower before the appearance of the a-wave, agreeing well with the mean amplification of rod signals by 100-fold. The apparent difference in sensitivity between a- and b-waves cannot be explained by the a-wave simply being a smaller signal, since normalizing for this difference in amplitude reveals that the intensity–response relation for the a-wave is shifted to 100-fold higher light intensities (Cone, 1963; Ashmore and Falk, 1976, 1980a; Fulton and Rushton, 1978). The scotopic b-wave in human ERG recordings therefore reliably reflects rod ON-bipolar cell activity, which, in order to generate such high gain, must also possess the mGluR-coupled cGMP cascade. This is fully substantiated by the recent discovery that mammalian rod ON-bipolar cells possess the mGluR-coupled cGMP cascade (Kaneko *et al.*, 1993). Amplification of rod signals by 100-fold also explains why rises in cGMP concentration were measured from whole retinas at low light intensities (Ames *et al.*, 1986). This would stimulate rises in cGMP in rod ON-bipolar cells at light intensities a factor of 100-times lower than the expected decrease in cGMP in rods.

Given that the overall mean gain in synaptic transmission from rods to ON-bipolar cells is of the order of 100-fold and that we have a unique postsynaptic mechanism, how is this synaptic gain generated? It has been proposed that some specialized mechanism must be operating here to explain not only, the high gain, but also the unusual filter characteristics of synaptic transmission at this synapse (Ashmore and Falk 1980a). Other workers in the field perhaps viewed such a high gain with skepticism, simply because in other mixed rod–cone retinas such as salamander the measured synaptic gains were much lower (Wu, 1985; Capovilla *et al.*, 1987). These retinas probably do contain some pure rod ON-bipolar cells, but the majority probably receive mixed rod–cone inputs. The effect of cone convergence onto ON-bipolar cells, and how this drastically reduces synaptic voltage gain due to electrical shunting of the rod input signals has been dealt with quantitatively (Falk, 1989). To address the problem of how such a high voltage gain is achieved in rod transmission to ON-bipolar cells, biophysical models applied to the transmitter release process from rods, the electrical properties of the rod ON-bipolar cell, and additional components incorporating the mGluR-coupled cGMP cascade must be considered.

12.4.1 THE SYNAPTIC VOLTAGE TRANSFER FUNCTION

The release of transmitter is known to rise very steeply with presynaptic membrane depolarization, ΔV_{pre} so that the transmitter concentration T in the synaptic cleft will be determined by the expression:

$$T = A \exp (b\Delta V_{pre}) \qquad (12.1)$$

where A and b are constants and the reciprocal of b determines the presynaptic voltage change which produces an exponential change in transmitter release.

The amplification associated with the synaptic transfer of photoreceptor light responses to ON- and OFF-bipolar cells is defined as the differential of the postsynaptic voltage (V_{post}) in the bipolar cell with respect to the presynaptic photoreceptor voltage (V_{pre}). The factors which determine synaptic gain or the input–output relation of the synapse have been considered in two recent reviews (Falk, 1989; Attwell, 1990). Four physical parameters were involved in determining the maximum possible gain. First the presynaptic component b, as defined in equation 12.1. Second, n, the Hill coefficient associated with the degree of cooperativity in the action of transmitter on postsynaptic glutamate receptors. Third, the driving potential for the generation of light responses given by the difference between the reversal potential of the transmitter-gated channels and that of non-synaptic channels ($V_r - V_o$). Fourth, the ratio of non-synaptic to synaptic conductance (G_o/G_r). The maximum synaptic voltage gain may be derived (Falk and Fatt, 1972, 1974a,b) as:

$$\frac{dV_{post}}{dV_{pre}} = \frac{bn(V_r - V_0)}{4[1 + (G_0/G_r)]} \qquad (12.2)$$

This expression could be further simplified by assuming that the synaptic conductance was much greater than the non-synaptic conductance, effectively removing the term (G_o/G_r) so that:

$$\text{gain}_{max} = bn(V_r - V_o)/4 \qquad (12.3)$$

For both ON- and OFF-bipolar cells, the reversal potential for transmitter action is close to 0 mV (Attwell *et al*, 1987; Shiells and Falk, 1993b) whereas the reversal potential of the non-synaptic channels may be assumed to be close to the K^+ reversal potential, around −70 mV. The value of b will be the same for both ON- and OFF-bipolar cells, since this parameter depends on the transmitter release mechanism. The maximum gain therefore, assuming a value for b of 0.25 mV^{-1} and cooperativity n of 2 (Shiells *et al.*, 1986; Shiells and Falk, 1993b), would be only about 5-fold if the glutamate receptors were directly linked to channels, as would apply to the ionotropic receptors of OFF-bipolar and horizontal cells. Much higher synaptic voltage gains could be obtained if it was assumed that no saturation of receptors by the transmitter occurred in the case of the channel closing action as applicable to the ON-bipolar cells (Falk, 1989). The limitation to synaptic gain posed by receptor saturation was rather paradoxical, because we knew that the gain was highest when the ON-bipolar cell was sitting at its dark potential, when most of the postsynaptic channels were closed, and yet to close most of the channels the transmitter concentration had to be at near-saturating levels, and thus operating in the least sensitive part of the dose–response relation. The assumption of no receptor saturation was in itself untenable because receptor–transmitter interaction must follow Michaelis–Menten kinetics. The advantage of linking mGluRs inversely to the control of many cGMP-activated channels is that it may be regarded as a mechanism capable of transcending the problem of receptor saturation.

The relation between postsynaptic conductance and transmitter concentration has been derived and verified experimentally by measuring the dose–response relation of isolated rod ON-bipolar cells to concentration-jumps of glutamate (Shiells and Falk, 1993a,b,c and unpublished results). This showed no saturation at higher glutamate concentrations, but instead was linear with a sharp cut-off at 200 μM glutamate. By amplifying small changes in glutamate binding to receptors, the cGMP cascade effectively linearizes the dose–response relation in what would otherwise be the saturated region of the curve (if the receptors were ionotropic). Assuming that, for simplicity, there is no cooperativity in the action of glutamate on receptors or of cGMP on channels, and that the reversal potential for transmitter action, $V_r = 0$, then the dependence of synaptic gain on postsynaptic potential V will be governed by the expression (Shiells and Falk, unpublished):

$$\frac{dV_{post}}{dV_{pre}} = \frac{-bV_0}{(a-1)}\frac{(1-aV)}{V_0} \times \left\{\frac{p}{r}\left(1-\frac{aV}{V_0}\right) + (a-1)\frac{V}{V_0}\right\} \quad (12.4)$$

where the ratio p/r is determined by:

$$\frac{p}{r} = \frac{cG_{max}}{K_{cG}}\left[\frac{(cG_{max} + K_{cG})}{BQ} - 1\right] \quad (12.5)$$

where $a = 1 + (G_r/G_o)$, K_{cG} is the dissociation constant for the action of cGMP on channels, cG_{max} is the peak cGMP concentration in bright light, B is a biochemical gain constant defined by the change in cGMP induced per activated mGluR, and Q is the total concentration of mGluRs. If we let $cG_{max} = 10^{-4}$M, $K_{cG} = 10^{-5}$M and put $BQ = 10^{-4}$, 1.1×10^{-4} and 2×10^{-4}M then this corresponds to p/r values of 1, 0 and −4.5 respectively. Solutions of equation 12.4 assuming a relatively low value of 5 for G_r/G_o (Falk, 1989) are shown in Figure 12.11.

This shows that at the dark potential of −40mV, if $p/r = 1$, there is a maximum gain of about 5-fold which decreases on hyperpolarizing or depolarizing the membrane potential. This is analogous to the model proposed by Falk (1989) which is limited by receptor saturation having a maximum gain of between 5 and 10. With $p/r = 0$, gains of the order of 25-fold are obtained around the dark potential of −40 mV and this situation is analogous to the model assuming no receptor saturation (Falk, 1989). With $p/r = -4.5$ this increases to the 100-fold range, and would account for the most sensitive ON-bipolar cells if these cells were more hyperpolarized in the dark (Ashmore and Falk, 1980a,b).

12.5 TEMPORAL PROPERTIES OF ROD–BIPOLAR CELL TRANSMISSION

Previous studies have revealed delays in the time course of rod ON-bipolar cells when compared to those of OFF-bipolar cells or horizontal cells (Ashmore and Falk, 1979, 1980a), and the filter characteristics of synaptic transmission to ON-bipolar cells were complex involving the selective amplification of rod signals at about 4 Hz, with a steep roll-off at higher frequencies. The question arises as to whether we can now account for these unusual properties in terms of the difference in glutamate receptors expressed on the ON- and OFF-bipolar cells. The principal difference between glutamate responses mediated by metabotropic and ionotropic receptors

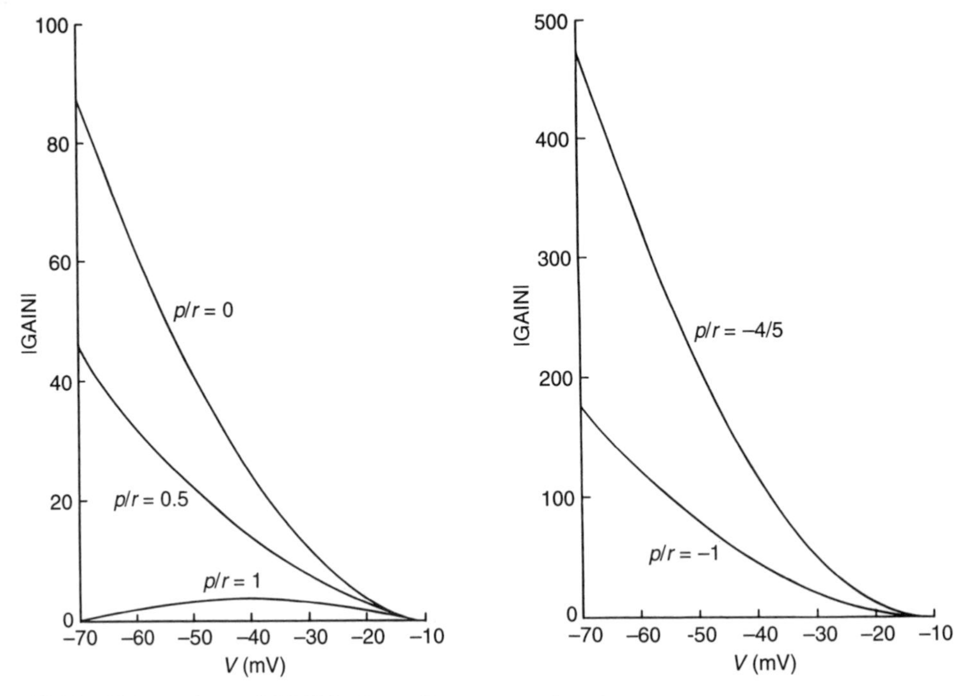

Figure 12.11 Gain at the rod ON-bipolar cell synapse. The absolute value of gain has been calculated from equation 12.4 using values for p/r ranging from 0 to 1 (left) and -1 and -4.5 (right) using values for the other parameters as given in the text. ON-bipolar cells had a mean dark potential of -45 mV, with a range from -35 to -55 mV (Ashmore and Falk, 1980a).

would be a delay associated with the activation kinetics of the cGMP cascade. Photoreceptors possess a similar system, and here the time course of cGMP reduction induced by flashes or steps of light has been well characterized. Rod responses have a waveform which is consistent with the concept that the cGMP concentration is reduced by a sequential chain of four or five first-order reactions, and then is restored by a further first-order reaction (McNaughton, 1990). To compare directly the time course of responses of isolated bipolar cells to concentration-jumps of glutamate, responses from ON- and OFF-bipolar cells which were isolated sequentially from the same retinal slice were lined up and plotted on an expanded time-scale (Figure 12.12). This revealed a delay of about 10 ms before the ON-bipolar cell responses rose above the baseline noise, whereas there was no delay in the OFF-bipolar cell responses. This delay is consistent with an inherent time-lag in the activation of the glutamate receptor-coupled cGMP cascade and is similar to that of rod photoreceptors exposed to bright light flashes (Cobbs and Pugh, 1987). The unusual frequency response characteristics of synaptic transmission to ON-bipolar cells may be accounted for by effectively linking two cGMP cascade processes together in series via the synapse. This probably functions to filter out high frequency noise components generated by the quantal nature of transmitter release (Shiells and Falk, unpublished results).

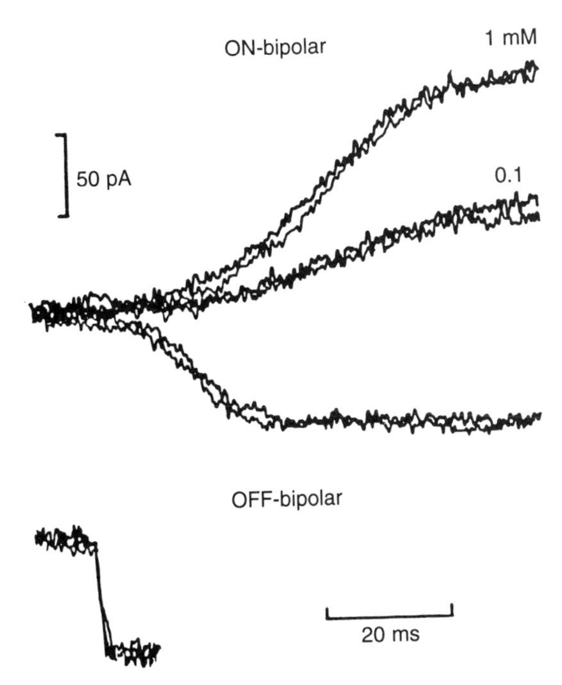

Figure 12.12 Responses from isolated ON- and OFF-bipolar cells shown synchronized on an expanded timescale to concentration-jumps of 1mM glutamate (the timing of the jump determined by stepping an open patch-pipette between two Ringer solutions of different ionic strengths, shown below). Two superimposed sweeps are shown in each case, with responses of the ON-bipolar cell to 0.1 mM glutamate. Lining up the responses in this way revealed a delay of about 10 ms before the ON-bipolar cell response rises above baseline, whereas, there is no delay in the OFF-bipolar cell response.

12.6 CONCLUSIONS AND FUTURE PERSPECTIVES

We have only recently determined how high amplification is generated at the rod–bipolar cell first synapse in the rod visual pathway, and it seems that the processes generating the inherent amplification in phototransduction are conserved in ON-bipolar cells making synaptic contact with rods (Shiells and Falk, 1990). The rod ON-bipolar cells possess metabotropic glutamate receptors (mGluRs) which are coupled via a G-protein to the control of a cGMP cascade, just as in rods the photon receptor molecule rhodopsin is coupled via transducin to a parallel cascade. Such biochemical cascade systems are capable of producing gains in terms of particle multiplication of the order of a million-fold (Stryer, 1986). In principle, high synaptic gain could be achieved by such a system if single glutamate receptors are coupled to the control of a large number of synaptic (cGMP-gated) channels (Shiells and Falk, 1993a). Amplification of photoreceptor signals is perhaps not so important in the OFF-pathway or indeed in the cone system, since these pathways signal light reduction from some ambient level and operate at 1000-fold higher light intensities in color vision, respectively. Not surprisingly, then, the expression of mGluRs and the cGMP cascade may be restricted to ON-bipolar cells making synaptic contact with rods, whereas, OFF-bipolar cells (and probably cone ON-bipolar cells) possess different forms of ionotropic receptors which by their very nature restrict receptor coupling to channels in a 1:1 ratio.

Research over the next few years should provide new insights into bipolar cell function. Direct recording from single cGMP-activated channels using inside-out patches of ON-bipolar cell membrane will resolve their single-channel conductances and determine their sensitivity to cGMP in the same detail as has been determined for photoreceptor cGMP-activated channels. The role of the novel neuromodulator, NO, in the retina is the subject of intensive research since there is now evidence that photoreceptors, horizontal cells, amacrine cells and even some ON-bipolar cells contain NO-synthase. One possibility may be that the release of NO from amacrine cells may form part of a positive feedback loop via stimulation of soluble guanylate cyclase in ON-bipolar cells. This would induce a rise in cGMP, and thus increase ON-bipolar cell light responses, a mechanism analogous to long-term potentiation in the hippocampus. Finally, the recent advances in molecular biological techniques (e.g. Goulding *et al.*, 1992) should resolve the identity of glutamate receptors mediating synaptic transmission from cones to ON-bipolar cells. It may be that different types of glutamate receptors mediate rod, blue-, red-, and green-cone inputs to the second order retinal neurons, providing the functional basis for the segregation of the visual system into dark-adapted (scotopic) and color (photopic) visual pathways respectively.

ACKNOWLEDGEMENTS

I would like to thank the Wellcome Trust and the Medical Research Council for financial support.

REFERENCES

Ames III, A., Walseth, T.F., Heyman, R.A. *et al.* (1986) Light-induced changes in cGMP metabolic flux correspond with electrical responses of photoreceptors. *Journal of Biological Chemistry*, **261**, 13034–42.

Ashmore, J.F. and Falk, G. (1976) Absolute sensitivity of rod bipolar cells in a dark-adapted retina. *Nature*, **263**, 248–9.

Ashmore, J.F. and Falk, G. (1979) Transmission of visual signals to bipolar cells near absolute threshold. *Vision Research*, **19**, 419–23.

Ashmore, J.F. and Falk. G. (1980a) Responses of rod bipolar cells in the dark-adapted retina of the dogfish, *Scyliorhinus canicula*. *Journal of Physiology*, **300**, 115–50.

Ashmore, J.F. and Falk, G. (1980b) The single-photon signal in rod bipolar cells of the dogfish retina. *Journal of Physiology*, **300**, 151–66.

Attwell, D. (1990) The photoreceptor output synapse. *Progress in Retinal Research*, **9**, 337–62.

Attwell, D., Mobbs, P., Tessier-Lavigne, M. and Wilson, M. (1987) Neurotransmitter-induced currents in retinal bipolar cells of the axolotl, *Ambystoma mexicanum*. *Journal of Physiology*, **387**, 125–61.

Bacigalupo, J., Johnson, E.C., Vergara, C. and Lisman, J.E. (1991) Light-dependent channels from excised patches of *Limulus* ventral photoreceptors are opened by cGMP. *Proceedings of the National Academy of Sciences USA*, **88**, 7938.

Barlow, H.B., Levick, W.R. and Yoon, M. (1971) Responses to single quanta of light in retinal ganglion cells of the cat. *Vision Research*, Suppl. **3**, 87–102.

Barnstaple, C.J. and Ahmad, J. (1992) Molecular cloning and differential laminar expression of soluble and particulate guanylate cyclases in rat retina. *Investigative Ophthalmology and Visual Science*, **33**, 1104.

Baskys, A. and Malenka, R.C. (1991) Agonists at metabotropic glutamate receptors presynaptically inhibit EPSCs in neonatal rat hippocampus. *Journal of Physiology*, **444**, 687–701.

Baylor, D.A. and Nunn, B.J. (1986) Electrical properties of the light-sensitive conductance of rods of the salamander *Ambystoma mexicanum*. *Journal of Physiology*, **371**, 115–45.

Capovilla, M., Hare, W.A. and Owen, W.G. (1987) Voltage gain of signal transfer from retinal rods to bipolar cells in the tiger salamander. *Journal of Physiology*, **391**, 125–140.

Cassel, D. and Pfeiffer, T. (1978) Mechanism of cholera toxin action: covalent modification of the guanyl nucleotide-binding protein of the adenylate cyclase system. *Proceedings of the National Academy of Sciences USA*, **75**, 2669–73.

Cervetto, L. and McNaughton, P.A. (1986) The effects of phosphodiesterase inhibitors and lanthanum ions on the light-sensitive current of toad retinal rods. *Journal of Physiology*, **370**, 91–109.

Chen, T.-Y., Peng, Y.-W., Dhallan, R.S. *et al.* (1993) A new subunit of the cyclic nucleotide-gated cation channel in retinal rods. *Nature*, **362**, 764–7.

Cobbs, W.H. and Pugh, E.N. Jr (1987) Kinetics and components of the flash photocurrent of isolated retinal rods of the larval salamander, *Ambystoma tigrinum*. *Journal of Physiology*, **394**, 529–72.

Cone, R.A. (1963) Quantum relations of the rat electroretinogram. *Journal of General Physiology*, **46**, 1267–86.

Dacheux, R.F. and Raviola, E. (1986) The rod pathway in the rabbit retina: a depolarizing bipolar and amacrine cell. *Journal of Neuroscience*, **6**, 331–45.

Daw, N.W., Jensen, R.J. and Brunken, W.J. (1990) Rod pathways in mammalian retinae. *Trends in Neuroscience*, **13**, 110–15.

Detwiler, P.B., Hodgkin, A.L. and McNaughton, P.A. (1980) Temporal and spatial characteristics of the voltage response in rods in the retina of the snapping turtle. *Journal of Physiology*, **300**, 213–50.

DeVries, S.H. and Schwartz E.A. (1992) Hemi-gap-junction channels in solitary horizontal cells of the catfish retina. *Journal of Physiology*, **445**, 201–30.

Dhallan, R.S., Macke, J.P., Eddy, R.L. *et al.* (1992) Human rod photoreceptor cGMP-gated channel: amino-acid sequence, gene structure, and functional expression. *Journal of Neuroscience*, **12**, 3248–60.

Dizhoor, A.M., Ray, S., Kumar, S. *et al.* (1991) Recoverin: a calcium sensitive activator of retinal rod guanylate cyclase. *Science*, **251**, 915–18.

Duvoisin, R. (1993) Isolation of a novel member of the metabotropic glutamate receptor gene family and its pattern of expression in mouse retina. *Investigative Ophthalmology and Visual Science*, **34**, 1381.

Fain, G.L. (1975) Quantum sensitivity of rods in the toad retina. *Science*, **187**, 838–41.

Falk, G. (1989) Signal transmission from rods to bipolar and horizontal cells: a synthesis. *Progress in Retinal Research*, **8**, 255–79.

Falk, G. and Fatt, P. (1972) Physical changes induced by light in the rod outer segments of vertebrates, in *Handbook of Sensory Physiology*, Vol. VII/I, *Photochemistry of Vision* (ed. H.J.A. Dartnall), Berlin, Springer, pp.235–44.

Falk, G. and Fatt, P. (1974a) Limitations to single-photon sensitivity in vision, in *Lecture Notes in Biomathematics*, Vol. 4: *Physics and Mathematics of the Nervous System* (eds M. Conrad, W. Guttinger and M. Dal Cin), Springer, Berlin, pp.171–204.

Falk, G. and Fatt, P. (1974b) The dynamic voltage-transfer function for rod–bipolar cell transmission. *Vision Research*, **14**, 739–41.

Falk, G. and Shiells, R.A. (1986) Do horizontal cell responses contribute to the electroretinogram (ERG) in dogfish? *Journal of Physiology*, **381**, 113P.

Falk, G. and Shiells, R.A. (1988) Block of light responses of salamander rods by pertussis toxin and reversal by nicotinamide. *Federation of the European Biochemical Society Letters*, **229**, 131–4.

Fleischman, D. and Denisevich, M. (1979) Guanylate cyclase of isolated bovine retinal rod axonemes. *Biochemistry*, **18**, 5060–6.

Fulton, A.B. and Rushton, W.A.H. (1978) The human rod ERG: correlation with psychophysical responses in light and dark adaptation. *Vision Research*, **18**, 785–92.

Garthwaite, J. (1991) Glutamate, nitric oxide and cell–cell signalling in the nervous system. *Trends in Neuroscience*, **14**, 60–7.

Gilbertson, T.A., Scobey, R. and Wilsom, M. (1991) Permeation of calcium ions through non-NMDA glutamate channels in retinal bipolar cells. *Science*, **251**, 1613–15.

Goulding, E.H., Ngai, J., Kramer, R.H. *et al.* (1992) Molecular cloning and single-channel properties of the cyclic nucleotide-gated channel from catfish olfactory neurons. *Neuron*, **8**, 45–58.

Goy, M.F. (1991) cGMP: the wayward child of the cyclic nucleotide family. *Trends in Neuroscience*, **14**, 293–9.

Harris, W.A. and Holt, C.E. (1990) Early events in the embryogenesis of the vertebrate visual system: cellular determination and pathfinding. *Annual Review of Neuroscience*, **13**, 155–69.

Haynes, L.W., Kay, A.R. and Yau, K-W. (1986) Single cyclic GMP-activated channel activity in excised patches of rod outer segment membrane. *Nature*, **321**, 66–70.

Hecht, S., Shlaer, S. and Pirenne, M.H. (1942) Energy, quanta and vision. *Journal of General Physiology*, **25**, 819–40.

Hensley, S.H., Yang, X-L. and Wu, S.M. (1993) Identification of glutamate receptor subtypes mediating inputs to bipolar cells and ganglion cells in the tiger salamander retina. *Journal of Neurophysiology*, **69**, 2099–107.

Hirano, A.A. and MacLeish (1991) Glutamate and 2-amino-4-phosphonobutyrate evoke an increase in potassium conductance in retinal bipolar cells. *Proceedings of the National Academy of Sciences USA*, **88**, 805–9.

Hodgkin, A.L., McNaughton, P.A. and Nunn, B.J. (1985) The ionic selectivity and calcium dependence of the light-sensitive pathway in toad rods. *Journal of Physiology*, **358**, 447–68.

Ignarro, L.J. (1990) Haem-dependent activation of guanylate cyclase and cGMP formation by endogenous nitric oxide: a unique transduction mechanism for transcellular signalling. *Pharmacological Toxicology*, **67**, 1–7.

Kaneko, A. (1971) Physiological studies of single retinal cells and their morphological identification. *Vision Research*, Suppl. **3**, 17–26.

Kaneko, A., De La Villa, P. and Sasaki, T. (1993) Glutamate-induced responses in identified cat bipolar cells. *International Union of Physiological Sciences Congress (Glasgow)*, Abstract 3.4/0 p.4.

Karschin, A. and Wassle, H. (1990) Voltage- and transmitter-gated currents in isolated bipolar cells of rat retina. *Journal of Neurophysiology*, **63**, 860–76.

Katada, T. and Ui, M. (1982) ADP-ribosylation of the specific membrane protein of C6 cells by islet activating protein associated with modification of adenylate cyclase activity. *Journal of Biological Chemistry*, **257**, 7210–16.

Kawamura, S. and Murakami, M. (1991) Calcium-dependent regulation of cyclic GMP phosphodiesterase by a protein from frog retinal rods. *Nature*, **349**, 420–3.

Kim, H.G. and Miller, R.F. (1993) Properties of synaptic transmission from photoreceptors to bipolar cells in the mudpuppy retina. *Journal of Neurophysiology*, **69**, 352–60.

Koch, K.-W. and Stryer, L. (1988) Highly cooperative feedback control of retinal rod guanylate cyclase by calcium ions. *Nature*, **334**, 64–6.

Lamb, T.D. and Matthews, H.R. (1988) External and internal actions in the response of salamander retinal rods to altered external calcium concentration. *Journal of Physiology*, **403**, 473–94.

Lerea, C.L., Somers, D.E., Hurley, J.B. *et al.* (1986) Identification of specific transducin α-subunits in retinal rod and cone photoreceptors. *Science*, **234**, 77–80.

Marty, A. and Neher, E. (1983) Tight seal whole-cell recording, in *Single Channel Recording* (eds B. Sakmann and E. Neher), Plenum, New York, pp.107–21.

Massey, S.C., Mills, S.L. and de Vente, J. (1993) Nitric oxide: identification of target cells in the rabbit retina. ARVO Abstract 3354, p.1382.

McLatchie, I.M. and Matthews, H.R. (1992) Voltage-dependent block by L-*cis*-diltiazem of the cyclic GMP-activated conductance of salamander rods. *Proceedings of the Royal Society of London B*, **247**, 113–16.

McNaughton, P.A. (1990) Light responses of vertebrate photoreceptors. *Physiological Reviews*, **70**, 847–83.

Mills, S.L. and Massey, S.C. (1993) L-Arginine uncouples A-type horizontal cells in rabbit retina. *Investigative Ophthalmology and Visual Science*, **34**, 1382.

Miyachi, E., Murakami, M. and Nakaki, T. (1990) Arginine blocks gap junctions between retinal horizontal cells. *Neuroreport*, **1**, 107–10.

Muller, F., Wassle, H. and Voigt, T. (1988) Pharmacological modulation of the rod pathway in the cat retina. *Journal of Neurophysiology*, **59**, 1657–72.

Murayama, T. and Ui, M. (1983) Loss of the inhibitory function of the guanine nucleotide regulatory component of adenylate cyclase due to its ADP-ribosylation by islet activating protein, pertussis toxin, in adipocyte membranes. *Journal of Biological Chemistry*, **258**, 3319–26.

Nakajima, Y., Iwakabe, H., Akazawa, C. *et al.* (1993) Molecular characterization of a novel retinal metabotropic glutamate receptor mGluR6 with a high agonist selectivity for L-2-amino-4-phosphonobutyrate. *Journal of Biological Chemistry*, **268**, 11868–73.

Nakanishi, S. (1992) Molecular diversity of glutamate receptors and implications for brain function. *Science*, **258**, 597–603.

Nawy, S. and Copenhagen, D.R. (1987) Multiple classes of glutamate receptor on depolarizing bipolar cells in retina. *Nature*, **325**, 56–8.

Nawy, S. and Copenhagen, D.R. (1990) Intracellular cesium separates two glutamate conductances in retinal bipolar cells of goldfish. *Vision Research*, **30**, 967–72.

Nawy, S. and Jahr, C.E. (1990) Suppression by glutamate of cGMP-activated conductance in retinal bipolar cells. *Nature*, **346**, 269–71.

Nawy, S. and Jahr, C.E. (1991) cGMP-gated conductance in retinal bipolar cells is suppressed by the photoreceptor transmitter. *Neuron*, **7**, 677–83.

Novelli, A., Nicoletti, F., Wroblewski, J.T. *et al.* (1987) Excitatory amino acid receptors coupled with guanylate cyclase in primary cultures of cerebellar granule cells. *Journal of Neuroscience*, **7**, 40–7.

Pugh, E.N. Jr and Lamb (1993) Amplification and kinetics of the activation steps in phototransduction. *Biochimica et Biophysica Acta*, **1141**, 111–49.

Saito, T. and Kujiraoka, T. (1982) Physiological and morphological identification of two types of on-center bipolar cells in the carp retina. *Journal of Comparative Neurology*, **205**, 161–70.

Saito, T., Kondo, H. and Toyoda, J.I. (1978) Rod and cone signals in the on-centre bipolar cell: their different ionic mechanisms. *Vision Research*, **18**, 591–5.

Sandell, J.H. (1985) NADPH diaphorase cells in the mammalian inner retina. *Journal of Comparative Neurology*, **238**, 466–72.

Sasaki, T. and Kaneko, A. (1993) Glutamate-induced responses in off-type bipolar cells of the cat retina. *Investigative Ophthalmology and Visual Science*, **34**, 1381.

Schiller, P.H. (1992) The ON and OFF channels of the visual system. *Trends in Neuroscience*, **15**, 86–92.

Schoepp, D., Bockaert, J. and Sladeczek (1990) Pharmacological and functional characteristics of metabotropic excitatory amino acid receptors. *Trends in Pharmacological Science*, **11**, 508–15.

Schulz, S., Green, C.K., Yuen, P.S.T. and Garbers, D.L. (1990) Guanylyl cyclase is a heat-stable enterotoxin receptor. *Cell*, **63**, 941–8.

Seeburg, P.H. (1993) The molecular biology of mammalian glutamate receptor channels. *Trends in Neuroscience*, **16**, 359–64.

Shiells, R.A. and Falk, G. (1990) Glutamate receptors of rod bipolar cells are linked to a cyclic GMP cascade via a G-protein. *Proceedings of the Royal Society of London B*, **242**, 91–4.

Shiells, R.A. and Falk, G. (1992a) The glutamate receptor linked cGMP cascade of retinal on-bipolar cells is pertussis and cholera toxin-sensitive. *Proceedings of the Royal Society of London B*, **247**, 17–20.

Shiells, R.A. and Falk, G. (1992b) Properties of the cGMP-activated channel of retinal on-bipolar cells. *Proceedings of the Royal Society of London B*, **247**, 21–5.

Shiells, R.A. and Falk, G. (1992c) Retinal on-bipolar cells contain a nitric oxide-sensitive guanylate cyclase. *Neuroreport*, **3**, 845–8.

Shiells, R.A. and Falk, G. (1993a) The glutamate receptor-coupled cyclic GMP cascade of rod bipolar cells. *International Union of Physiological Science Congress (Glasgow)*, abstract 3.3/0 p.4.

Shiells, R.A. and Falk, G. (1993b) Responses of isolated bipolar cells to concentration jumps of L-glutamate. *International Union of Physiological Science Congress (Glasgow)*, abstract 278.3/P p. 161.

Shiells, R.A. and Falk, G. (1993c) Responses of bipolar cells isolated from retinal slices of dogfish to concentration jumps of glutamate. *Journal of Physiology*, **473**, 9P

Sheills, R.A., Falk, G. and Naghshineh, S. (1981) Action of glutamate and asparate analogous on rod horizontal and bipolar cells. *Nature*, **294** 592–594.

Shiells, R.A., Falk, G. and Naghshineh, S. (1986) Ionophoteric study of the action of excitatory amino acids on rod horizontal cells of the dogfish retina. *Proceedings of the Royal Society of London B*, **227**, 121–135.

Slaughter, M. M. and Miller, R.F. (1981) 2-amino-4-phosphonobutyric acid: a new pharmacological tool for retina research. *Science*, **211**, 182–15.

Slaughter, M.M. and Miller, R.F. (1983) An excitatory amino acid antagonist blocks cone input to sign-conserving second-order retinal neurons. *Science*, **219**, 1230–2.

Stern, J.H., Kaupp, U.B. and Macleish, P.R. (1986) Control of the light-regulated current in rod photoreceptors by cyclic GMP, calcium, and *L-cis* diltiazem. *Proceedings of the National Academy of Science USA* **83**, 1163–7.

Stockton, M. and Slaughter, M.M. (1989) B-wave of the electroretinogram. A reflection of ON-bipolar cell activity. *Journal of General Physiology*, **93**, 101–22.

Stryer, L. (1986) Cyclic GMP cascade of vision. *Annual Review of Neuroscience*, **9** 87–119.

Sugimaya, H., Ito, I. and Hirono, C. (1987) A new type of glutamate receptor linked to inositol phospholipid metabolism. *Nature*, **325**, 531–3.

Tanabe, Y., Masu, M., Ishii, T. *et al.* (1992) A

family of metabotropic glutamate receptors. *Neuron*, **8**, 169–79.

Taylor, W.R. and Copenhagen, D.R. (1993) Analysis of light-evoked responses in off-bipolar cells of tiger salamander retina. *Investigative Ophthalmology and Visual Science*, **34**, 1381.

Terashima, T., Katada, T., Okada, E. *et al* (1987) Light microscopy of GTP-binding protein (G_0) immunoreactivity within the retina of different vertebrates. *Brain Research*, **436**, 384–9.

Torre, V., Straforini, M., Sesti, F. and Lamb, T.D. (1992) Different channel-gating properties of two classes of cGMP-activated channel in vertebrate photoreceptors. *Proceedings of the Royal Society of London B*, **250**, 209–213.

Troyer, E.W., Hall, I.A. and Ferrendelli, J.A. (1978) Guanylate cyclase in the CNS: enzymatic characteristics of soluble and particulate enzymes from mouse cerebellum and retina. *Journal of Neurochemistry*, **31**, 825–33.

Usowicz, M.M., Gallo, V. and Cull Candy, S.G. (1989) Multiple conductance channels in type-2 astrocytes activated by excitatory amino acids. *Nature*, **339**, 380–3.

Van Dop, C., Yamanaka, G., Steinberg, F. *et al.* (1984) ADP-ribosylation of transducin by pertussis toxin blocks the light-stimulated hydrolysis of GTP and cGMP in retinal photoreceptors. *Journal of Biological Chemistry*, **259**, 23–36.

Vardi, N., Matesic, D.F., Manning, D.R. *et al.* (1993) Identification of a G-protein in depolarizing rod bipolar cells. *Visual Neuroscience*, **10**, 473–8.

Werblin, F.S. and Dowling, J.E. (1969) Organisation of the retina of the mudpuppy, *Necturus maculosus*: II. Intracellular recording. *Journal of Neurophysiology*, **32**, 339–55.

Woodhull, A.M. (1973) Ionic blockade of sodium channels in nerve. *Journal of General Physiology*, **61**, 687–708.

Wu, S.M. (1985) Synaptic transmission from rods to bipolar cells in the tiger salamander retina. *Proceedings of the National Academy of Science USA*, **82**, 3944–7.

Yamashita, M. and Wässle, H. (1991) Responses of rod bipolar cells isolated from the rat retina to the glutamate agonist 2-amino-4-phosphonobutyric acid (APB). *Journal of Neuroscience*, **11**, 2372–82.

Yau, K.-W. and Baylor, D.A. (1989) Cyclic GMP-activated conductance of retinal photoreceptor cells. *Annual Review of Neuroscience*, **12**, 289–327.

Yau, K.-W. and Chen, T.-Y. (in press) Cyclic nucleotide-gated channels, in *CRC Handbook of Receptors and Channels* Vol. II *Ligand-gated and voltage-gated ion channels*

Yuen, P.S.T. and Garbers, D.L. (1992) Guanylyl cyclase-linked receptors. *Annual Review of Neuroscience*, **15**, 193–225.

13

Functional architecture of mammalian outer retina and bipolar cells

PETER STERLING, ROBERT G. SMITH,
RUKMINI RAO and NOGA VARDI

13.1 INTRODUCTION

'Function' in the outer retina has mainly been studied by recording *in situ* from single neurons. In lower vertebrates this approach to bipolar cells has been extremely fruitful (e.g. Chapter 12), but in mammals bipolar cell recordings can be counted on the fingers of (at most) two hands (Nelson and Kolb, 1983; Dacheux and Raviola, 1986). And, considering that the recordings include both rod bipolar and multiple types of cone bipolar cell (Chapter 11), the electrophysiological data regarding mammalian bipolar neurons are thinly spread. On the other hand, in lower vertebrates information essential to understanding the contribution of the outer retina to image processing (such as optics, sampling frequencies, and synaptic circuitry) hardly exists. So, in lower vertebrates how single neuron responses in the outer retina contribute to vision remains unclear.

Yet, single cell recording is not the only possible approach to understanding retinal function. An alternative strategy is to determine complete circuit structure ('wiring diagram') plus the chemical architecture and to incorporate this information, together with the optics and ganglion cell electrophysiology, into various computational models. Then, one might calculate backwards to the properties of the bipolar cells and photoreceptors. Such an effort leads to specific predictions regarding the photoreceptor and bipolar cell function, and with this approach a little electrophysiology goes a surprisingly long way. At least, that is our argument in this review. We emphasize cat retina, which is known in most detail, but also note recent data from rabbit and primate that indicate conservation of certain basic circuits and functions.

In the exact center of the cat area centralis, cone density reaches 30 000–40 000/mm^2 (Wässle and Riemann, 1978, Williams *et al.*, 1993), but the circuitry has been studied slightly off center (1° eccentricity) where the cone density is about 24 000/mm^2 and rod density is about 350 000/mm^2 (Sterling *et al.*, 1988). Here, due to the natural blur of the cat's optics (Wässle, 1971; Robson and Enroth–Cugell, 1978), the minimum number

Neurobiology and Clinical Aspects of the Outer Retina
Edited by M.B.A. Djamgoz, S.N. Archer and S. Vallerga
Published in 1995 by Chapman & Hall, London
ISBN 0 412 60080 3

of receptors stimulated from a point source is about 10 cones and 140 rods (Figure 13.1a,2b). This is many more receptors than that converging on a single bipolar cell, so even the finest spatial stimulus falling on either type of receptor will affect many bipolar cells. We review first the circuits for daylight that lead from cones because various portions of this circuit are parasitized by the circuits for twilight and starlight that lead from rods (Figure 13.2).

13.2 INTRINSIC CIRCUITS OF THE OUTER PLEXIFORM LAYER

The cone pedicles couple electrically to each other via small gap junctions (Raviola and Gilula, 1973; Kolb, 1977; Smith *et al.*, 1986; and Chapter 6). This is an apparently universal feature of the mammalian retina, described in cat, rabbit and monkey. In human retina, too, there are desmosomes between cones (Cohen, 1965; Missotten, 1965) that probably bracket gap junctions (Tsukamoto *et al.*, 1992). Surprisingly, gap junctions are found between cone pedicles even in the fovea where they couple essentially all adjacent pedicles without apparent selectivity for spectral type (Raviola and Gilula, 1973; Tsukamoto *et al.*, 1992). Pedicles in the human fovea send processes toward each other that probably also form gap junctions (Ahnelt and Pflug, 1986).

The cone pedicle also connects to two types of horizontal cell (Boycott and Kolb, 1973; Kolb, 1977; Boycott *et al.*, 1978). The types differ mainly in spatial scale: one is narrow-field (H_I in primate; B in cat and rabbit), and the other is wide-field (H_{II} in primate; A in cat and rabbit). The wide-field type is strongly coupled, and the narrow-field type weakly coupled (Müller and Peichl, 1993; Vaney, 1993). Compartmental models suggest that the difference of spatial scale between the

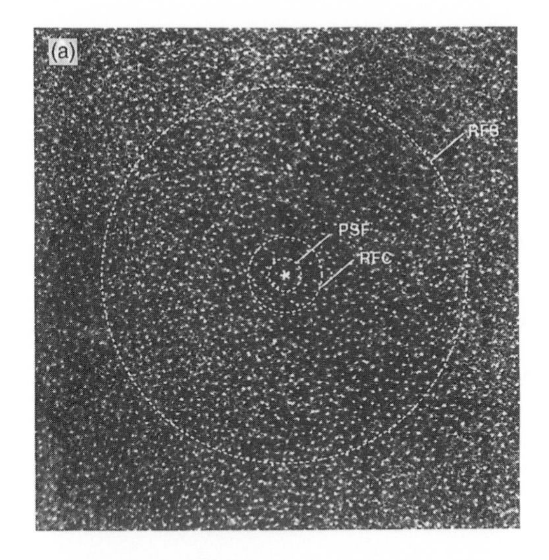

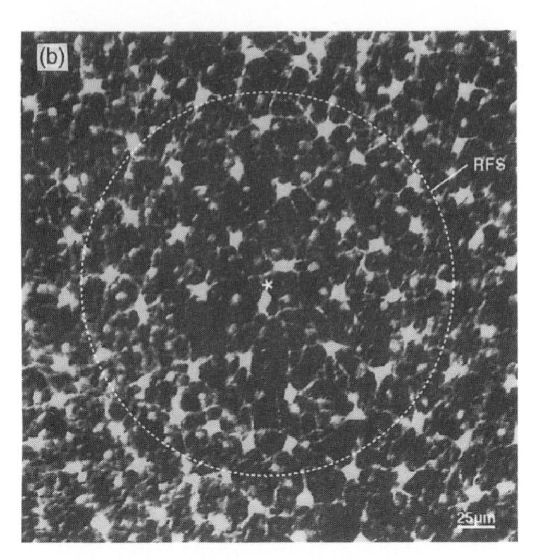

Figure 13.1 (a) Cone array in cat area centralis (tangential, 1 μm section stained with toluidine blue). Cone density is ~24 000/mm². Point spread function of cat's optics (PSF) includes about 10 cones; receptive field center (RFC) of one cone (*) includes about 50 cones. Receptive field surround (RFS) of one cone includes about 1200 cones. See text and Figure 13.4 (b) Horizontal cell arrays in cat area centralis stained with antibody to calbindin. Dark cells are type A; pale cells are type B. Both arrays contribute to the RFS of the cone (*). See text and Figure 13.5. Tissue prepared by S. Kumar and photographed by P. Auerbach. (Reprinted from Smith, 1995). Scale bar, same for (a) and (b).

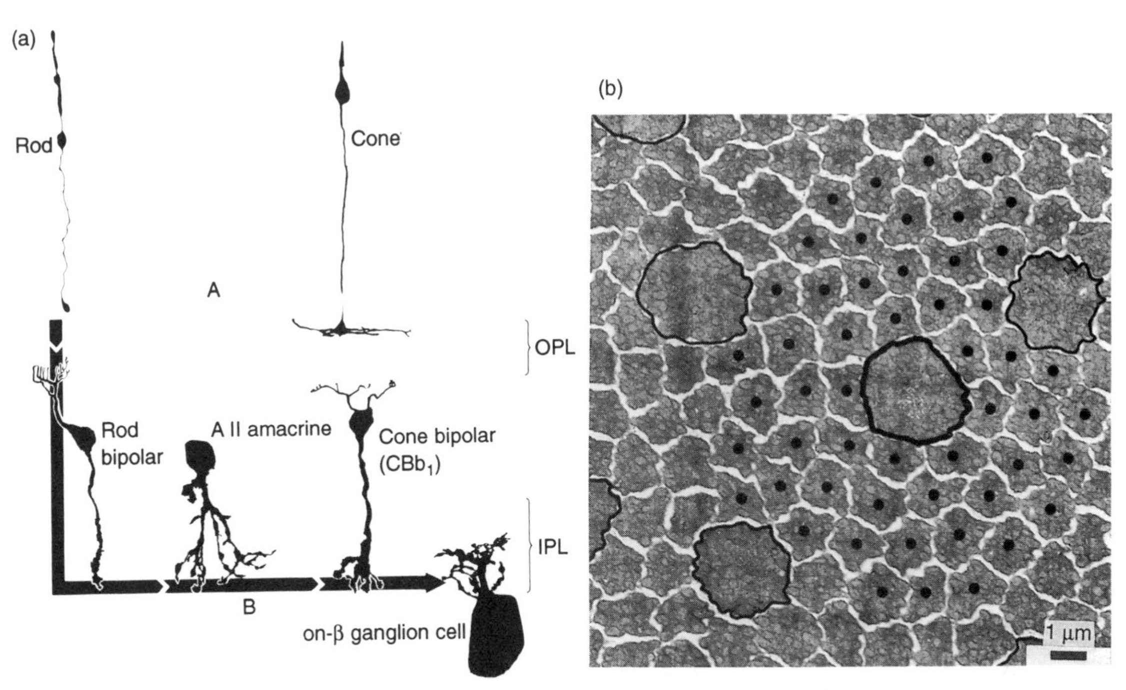

Figure 13.2 (a) Two rod pathways to the ON β (X) ganglion cell. A, via rod–cone gap junctions is thought to serve twilight (mesopic range); B, via rod bipolar cell is thought to serve starlight (scotopic range). (b) Electron micrograph of tangential section through photoreceptor inner segments. At this level, cones (outlined) are widely separated and the spaces are filled by rods. Forty-eight rods (dotted) surround the central cone. (a, b are reprinted from Smith *et al.* (1986).)

two types should be reflected in the extents of their receptive fields (Smith, 1994). Although in mammals this has, so far, not been clearly established by electrophysiology (Nelson, 1977; Bloomfield and Miller, 1982; Raviola and Dacheux 1983), in turtle it seems quite clear (Piccolino *et al.*, 1981). Both types are present in all mammals that forage in daylight and twilight, but in strongly nocturnal mammals the wide-field type is lacking (Peichl and Gonzáles-Soriano, 1993). This is probably related to the fact that the narrow-field horizontal cell has an axon that contacts only rods and thus provides a mechanism for lateral integration in the scotopic range (section 13.6).

In primate retina, there is evidence to suggest that both types of horizontal cell contact every cone within reach (Boycott *et al.*, 1987; Wässle *et al.*, 1989; reviewed by Wässle and Boycott, 1991). This implies that the signals generated by both types of horizontal cell would be spectrally broad band. This is supported by the direct demonstration in tree shrew that both blue-sensitive and red-sensitive cones contact both types of horizontal cell (Müller and Peichl, 1993). However, it has also been suggested that primates have three types of horizontal cell that collect somewhat selectively from different spectral types of cone (e.g. Kolb *et al.*, 1992). This issue is of some theoretical importance and should be technically possible to resolve.

13.2.1 MAMMALIAN HORIZONTAL CELLS ARE GABAERGIC

Mammalian horizontal cells do not readily show selective uptake of GABA or muscimol, as do their counterparts in lower vertebrates (Chapter 10), so, for some time, it was doubted whether they are GABAergic (reviewed by Freed, 1992). However, recently both types of horizontal cell were shown to contain endogenous GABA in cat (Pourcho and Owczarzak, 1989; Wässle and Chun, 1989) and primate (Agardh *et al.*, 1987; Grünert and Wässle, 1990). Furthermore, mRNA for glutamic acid decarboxylase (GAD), the synthetic enzyme of GABA, was demonstrated in cat (Sarthy and Fu, 1989), and immunostaining for GAD was shown in cat and monkey (Figure 1.3.3a; Agardh *et al.*, 1987; Vardi *et al.*, 1994). Interestingly, within a species, both horizontal cell types express the same isoform: GAD_{67} in cat and GAD_{65} in macaque monkey (Vardi *et al.*, 1994). Finally, immunostaining for the $GABA_A$ receptor was demonstrated in association with the structures postsynaptic to the horizontal cells (Vardi *et al.*, 1993; Vardi *et al.*, 1993; Vardi and Sterling, 1994). Almost certainly then, both types of mammalian horizontal cell are GABAergic, and thus inhibitory. This fits with the demonstration in rabbit that depolarizing a horizontal cell antagonizes the effect of light on the ganglion cell (Mangel, 1991).

(a) Where does the GABA act?

In lower vertebrates there is ample evidence that GABA secreted by the horizontal cell feeds back onto the cone terminal (Chapter 9) and also feeds forward onto the bipolar dendrites (Wu, 1994). It is natural to hypothesize similar actions for the mammalian circuit, and there is some evidence for both. Feedback onto the cone terminal was suggested by the demonstration that inhibition of the OFF ganglion cell, evoked by depolarizing the horizontal cell, survived blockage of

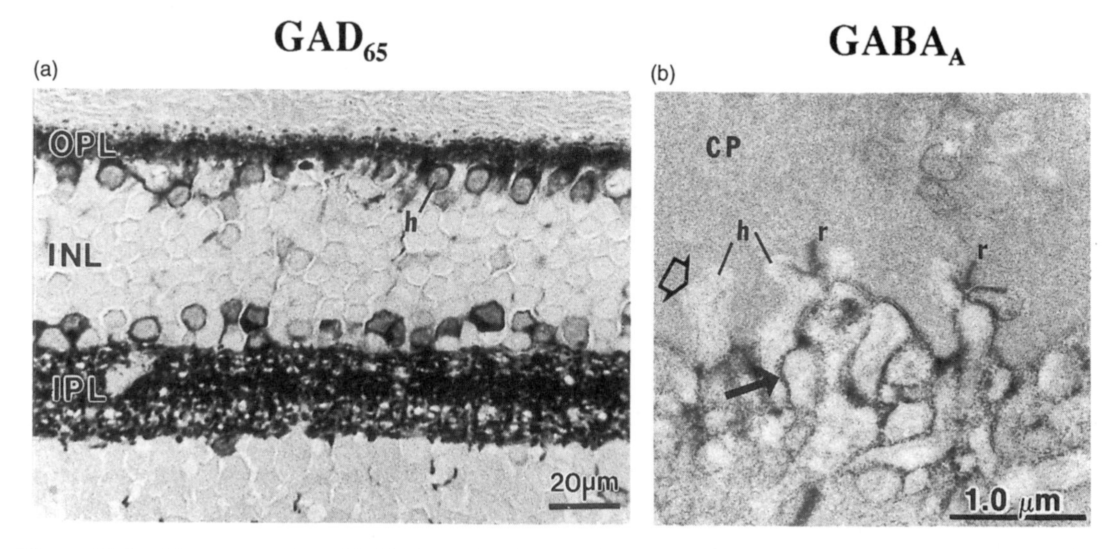

Figure 13.3 Macaque retina. (a) Horizontal cells (h) are stained for GAD_{65} (b) Bipolar dendrite (dark arrow) in apposition to horizontal cell terminal is stained by antibody to α_1 subunit of the $GABA_A$ receptor. Cone membrane (open arrow) in apposition to the horizontal cell terminal is unstained. OPL, outer plexiform layer; INL, inner nuclear layer; IPL, inner plexiform layer; cp, cone pedicle; r, ribbon. (a, reprinted from Vardi *et al.* (1994), b, reprinted from Vardi and Sterling (1994).)

the ON-bipolar cells with APB (Mangel, 1991). This suggested that at least some of the inhibition must act prior to the bipolar cells. Feedforward inhibition was also suggested by strong immunocytochemical staining for $GABA_A$ receptor $alpha_1$ and $beta_{2-3}$ subunits on the bipolar dendrites (Figure 13.3b; Vardi *et al.*, 1992; Vardi and Sterling, 1994). Staining for these subunits is absent on the cone pedicle itself, and so is the message for $alpha_{1-3}$ and $beta_{2-3}$ (Greferath *et al.*, 1993), but the cone might well express subunits of $GABA_A$ that are not detected by the particular antibodies employed. For example, although the rod terminal is also negative to these antibodies, the rat rod has been found by molecular methods to express the $beta_{1-2}$ subunit of $GABA_A$ (Grigorenko and Yeh, 1994).

13.2 MODEL OF THE CONE RECEPTIVE FIELD

How do the many factors, such as optical blur, cone coupling and inhibitory feedback from two types of coupled horizontal cells, contribute to the cone receptive field? A first step was to calculate the dimensions of the cone receptive field from a linear model based on the beta cell physiology and the known circuitry. Knowing the difference-of-Gaussians receptive field for the beta (X) ganglion cell at 1 ° eccentricity (Cleland *et al.*, 1979; Linsenmeier *et al.*, 1982), and knowing that 36 cones converge via bipolar circuits onto this cell (Figure 13.4a) (Cohen and Sterling, 1991, 1992), we computed the cone receptive field by deconvolution (Figure 13.4a; Smith and Sterling, 1990). The cone receptive field center proved to be much broader than the optical blur function and only slightly narrower than the beta ganglion cell center (compare Figures 13.1a and 13.5a). To assess the contributions of multiple factors we next constructed a multi-neuronal compartmental model that included an array of cones and their associated narrow-field and wide-field horizontal cells (Smith, 1992; Sterling *et al.*, 1992; Smith, 1995).

The compartmental model assumed a difference-of-Gaussians cone receptive field (obtained from the 'deconvolution model'), purely passive properties for the horizontal cells, and a feedback gain to the cone terminals adequate to hold their steady potentials near the midpoint of their operating range (~ -38mV). 'Tuning' the various parameters of the compartmental model to this cone receptive field showed that both optical blur and cone coupling are required to produce the Gaussian-like center. Either blur or coupling alone produce a too narrow, exponential center (Figure 13.5b). Also, the broad cone center is essential as input to the horizontal cells, for without it the horizontal cells fail to generate a cone surround. Finally, the narrow-field horizontal cells contribute the deep, proximal part of the cone surround, and the wide-field horizontal cells contribute the shallow, distal part of the surround (Figure 13.5c). Thus, from a purely mechanistic perspective, all these features together create a bandpass spatial filter at the cone terminal (Sterling *et al.*, 1992; Smith, 1995). Of course, temporal filtering also occurs but its mechanistic basis cannot be assessed from anatomical studies.

(a) Why the cone receptive field should be bandpass

Although 40 years have passed since Kuffler (1953) demonstrated the center-surround structure of the mammalian ganglion cell receptive field, there has been surprisingly little discussion of where in the retina this filter shape first arises. The repeated demonstration in lower vertebrates of a broad, center-surround receptive field in the cone (Baylor *et al.*, 1971) has made astonishingly little impression on investigators of mammal-

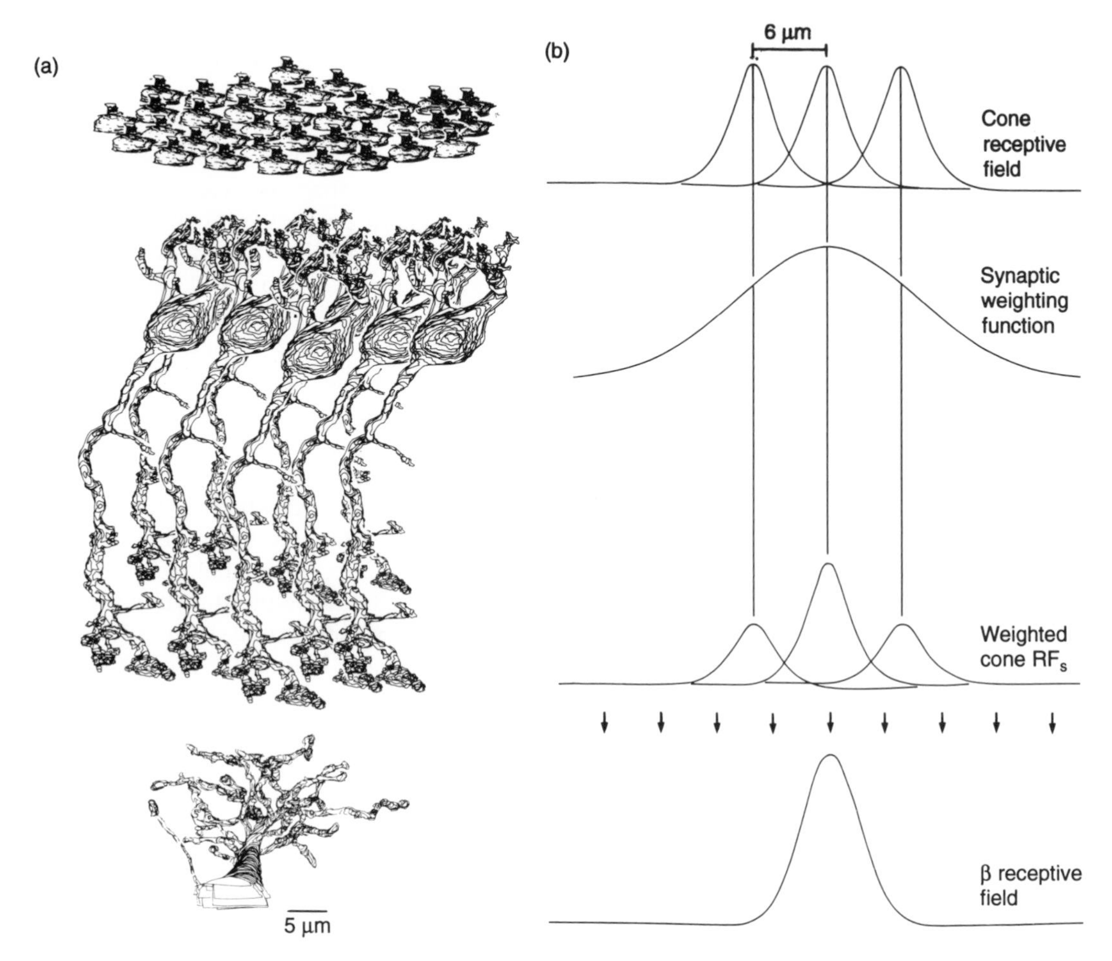

Figure 13.4 (a) 30 cones converge via seven b_1 bipolar cells to ON beta ganglion cell (Cohen and Sterling, 1992). (b) Method of deconvolution used to compute cone receptive field from beta receptive field. An initial cone receptive field template was copied 30 times; each copy was multiplied by its synaptic weight, spatially offset, and summed at the ganglion cell. (a,b reprinted from Smith and Sterling 1990).

ian retina, and doubt is still widely expressed that the cone or even the cone bipolar cells show this organization (e.g. Chapter 11). Yet, for efficient image-processing, this is exactly where a filter should be.

Receptive field center

The cone receptive field center should be lowpass in order to remove as much photon noise and transduction noise as possible before amplifying the signal at the first synapse. The basis for this improvement of signal/noise (S/N) ratio is that the signals in adjacent cones are partially correlated (due to correlations in the visual scene and to optical blur), whereas the noise in adjacent cones is uncorrelated. Consequently, spreading components of a given cone's voltage across the terminals of adjacent cones reduces the amplitude of the uncorrelated component

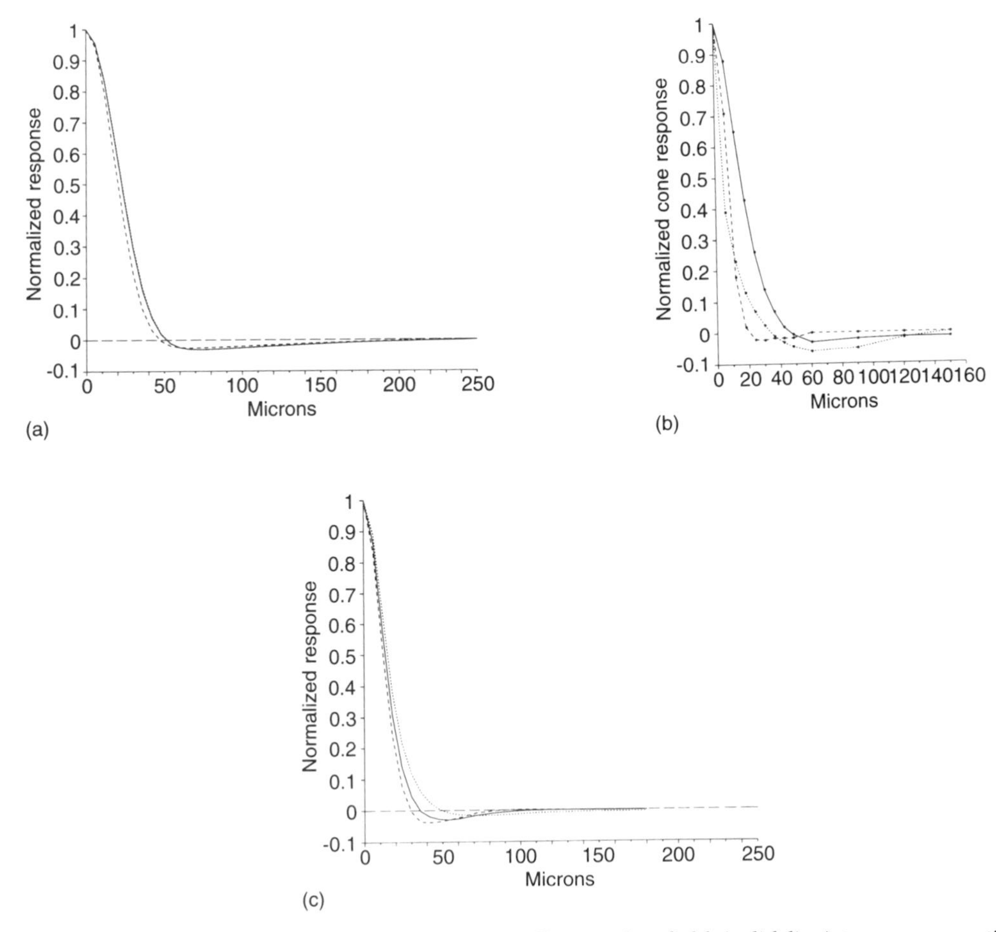

Figure 13.5 (a) Comparison of measured beta cell receptive field (solid line) to cone receptive field (dashed line). Beta cell receptive field computed by convolution of 30 cone receptive fields (dotted). Note that cone receptive field center is almost as broad as the ganglion cell center. (b) Sensitivity profiles of cone receptive field computed with the compartmental model. Solid curve: receptive field matches that from 'deconvolution model' shown in Figure 13.4. Dotted curve: receptive field computed without optical blur. Dashed curve: receptive field computed without cone coupling. Note that optical blur and cone coupling are both needed to give the center a Gaussian-like rather than exponential-like weighting. (c) Solid curve: cone receptive field matches 'deconvolution model'. Dotted curve: gain of wide-field horizontal cell (type A) increased. Note that the shallow, distal surround is enhanced at the expense of the deep, proximal surround. Dashed curve, gain of narrow-field horizontal cell (type B) increased. Note that the deep, proximal surround is enhanced at the expense of the shallow, distal surround. These computations suggest that narrow-field and wide-field horizontal cells are both needed to give the surround the appropriate, Gaussian-like weighting. (Reprinted from (a) Smith and Sterling (1990), (b) Sterling *et al.*, (1992), (c) Smith (1995).)

more than the correlated components. When Lamb and Simon (1976) made this point for turtle cones *in vitro*, they noted that '. . . this situation will not pertain for a small spot activating only one cone in this network . . .'. However, because of optical blur (Figure 13.1a), this situation does not arise (except for the fovea), therefore, with this exception, cone coupling must always improve the S/N ratio.

The finding of gap junctions between foveal cones was a surprise to psychophysicists, who are aware that we discriminate spatial gratings down to the resolution limit of the cone array (Hirsch and Curcio, 1989; Williams, 1992). In fact, they would claim that all spatial summation along the visual pathway before 'the first non-linearity' is accounted for by summation across the cone aperture (MacLeod *et al.*, 1992; Chen *et al.*, 1993). However, the experiments leading to this conclusion were conducted in bright light. At lower luminance, where photon noise sufficiently degrades the optical image, spatial summation across cones would become advantageous, and so, presumably, would coupling between cone pedicles.

The domed shape calculated for the cone receptive field center is probably no accident. Dome-like weighting of partially correlated signals optimally improves the S/N ratio (Tsukamoto *et al.*, 1990). The difference between domed weighting versus a flat or exponential weighting is modest, on the order of 10%. However, this may be enough to have provided the necessary selective pressure for the subtle matching of optics and cone coupling to have evolved (Sterling *et al.*, 1992; Smith, 1994). Another advantage of coupling cone terminals is that spreading the signal from one cone to many synaptic terminals increases the number of active zones used for transmission (Sterling *et al.*, 1988). Thus, the cat cone receptive field center at 1° that engages about 50 cone terminals for its transmission (Figure 13.1; Smith and Sterling, 1990), reduces the synaptic noise by $\sqrt{50}$ (i.e., 7-fold) over that of a single pedicle.

Receptive field surround

The cone receptive field should have an inhibitory surround (highpass) to remove the low spatial frequency components of the signal that are shared across adjacent cones. To postpone the removal of this redundant information to a later stage would be to force the cone synapse to encode it. This would reduce the gain that could be otherwise devoted to the signal components that differ between adjacent cones. Barlow made the point long ago that the ganglion cell spike train cannot encode enough levels to carry the redundant information (Barlow, 1961). We can now appreciate that the optimal stage to remove the redundancy is before the first forward synapse, and that much of the ganglion cell surround is probably accounted for by linear summation of the cone surrounds (Smith and Sterling, 1990; Smith, 1994).

The image-processing operation through which redundant information is removed by lowpass filtering has been termed 'predictive coding' (Srinivasan *et al.*, 1982). The principle is to predict the luminance at the center cone by means of a weighted average across the center and then subtract the prediction, leaving the difference to be transmitted forward. In this scheme the optimal weighting of the filter depends on the S/N ratio of the image, which is proportional to the square root of the luminance (Rose, 1973). Consequently in dim light, the cone surround should flatten and broaden, as it has been shown to do for single second order neurons in the fly's eye (Van Hateren, 1992; reviewed by Laughlin, 1994).

The outer plexiform model shows that appropriate adjustment of the surround weighting could emerge by increasing the effect of the wide-field horizontal cells at the expense of the narrow-field cells (Sterling *et al.*, 1992; Smith, 1995). This could occur through adjustments of their relative feedback gains. It could also occur through adjustments to their degrees of homotypic coupling, since greater coupling of the narrow-field cell effectively converts it into a wide-field type and vice versa for the wide-field type. Whether one or both mechanisms are used remains to be established. However, transmitter-gated changes in horizontal cell coupling have long been known for lower vertebrates and there is some evidence that this is true for mammals as well (Chapter 8).

13.3 CIRCUITS TO CONE BIPOLAR CELLS

Although there are only a few direct recordings from mammalian cone bipolar cells, no one doubts their fundamental division into depolarizing (ON) and hyperpolarizing (OFF) classes. This confidence rests on the juxtaposition of two major sets of observations. First, most mammalian bipolar axons arborize either in the outer or inner level of the inner plexiform layer (sublamina **a** or **b**), and these levels correspond to the respective arborizations of the OFF and ON ganglion cell dendrites (Nelson *et al.*, 1978). Second, numerous bipolar recordings in lower vertebrates show hyperpolarizing and depolarizing responses in accord with this pattern. So this is a case where insights from lower vertebrate physiology were quickly adopted.

The significance of this step, wherein at the very first junction neurons become specialized to encode signals brighter than or dimmer than the mean level, can hardly be overstated. This evolutionary 'decision' in circuit design represents a major commitment of neural resources, doubling the number of bipolar neurons, but also the number of ganglion cells, geniculate cells, and cortical cells, until the stage in V1 where the signals reconverge. It has been argued that the paired ON and OFF pathways are important for ganglion cells because spike trains are encoded best by increasing frequencies (Rodieck, 1973). However, the division originates in the outer retina, one stage before the ganglion cells. Apparently, halving the dynamic range to be encoded by a neuron offers a powerful advantage, the obvious one being that gain could be doubled. Although sensory neurons in other modalities use only single-ended outputs, their signals have already achieved major mechanical amplification. The hint that the tradeoff between gain and dynamic range is an issue for retinal neurons might have prepared us for the next big surprise regarding cone bipolar cells, namely their diversity of 'type'.

13.3.1 MANY TYPES OF CONE BIPOLAR CELL

The first report that mammalian cone bipolar cells are morphologically diverse was that by West (1976), who found many different morphologies based on Golgi impregnations of the ground squirrel retina. This report was followed-up by similar studies on cat and rabbit also suggesting about 8–10 morphological categories, divided about equally between those arborizing in sublaminae **a** and **b** (Kolb *et al.*, 1981; Famiglietti, 1981; Pourcho and Goebel, 1987a). The same approach has been applied to human retina with similar conclusions (Kolb *et al.*, 1992).

An important concern is whether the reported morphological differences represent continuous variation within a single population or discontinuous variations that represent natural types? If the variation is continuous, then the categories are not objective; 'splitters' will differ from 'lumpers' (Rodieck and Brening, 1983). This problem arises with any approach that divides a largely unseen population into many categories based on examples, and special precautions are needed to address it.

It helps somewhat to know whether neurons in different morphological categories form independent arrays across the retina. Boycott and Wässle (1991) suggested a scheme for six types of diffuse cone bipolar cell in monkey retina. The types were dis-

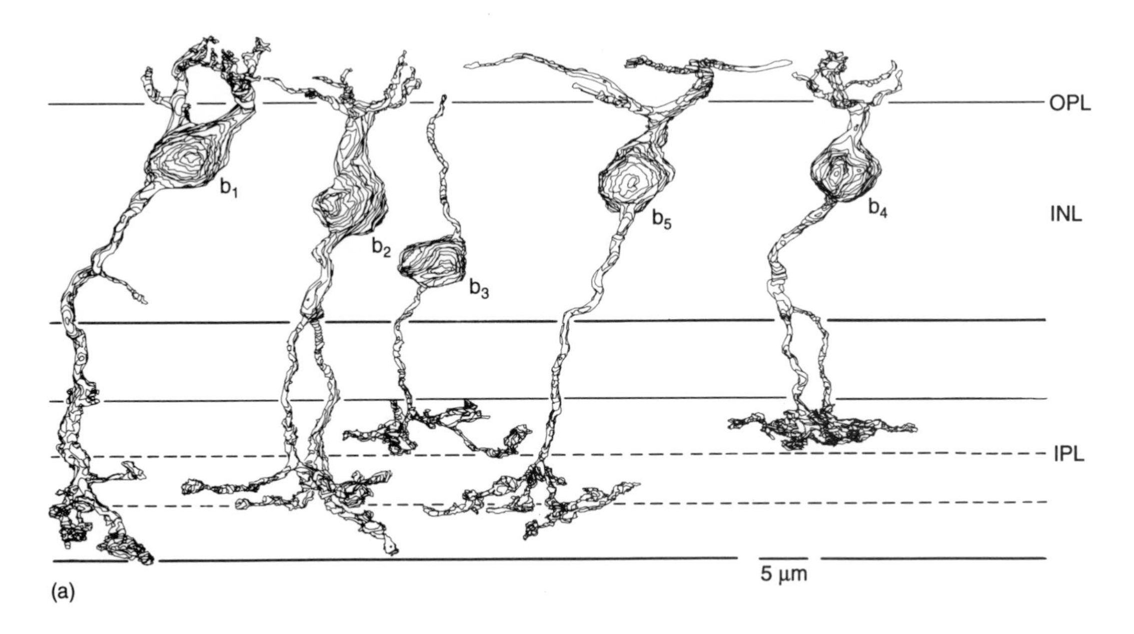
OPL
INL
IPL
b1
b2
b3
b4
b5
5 μm
(a)

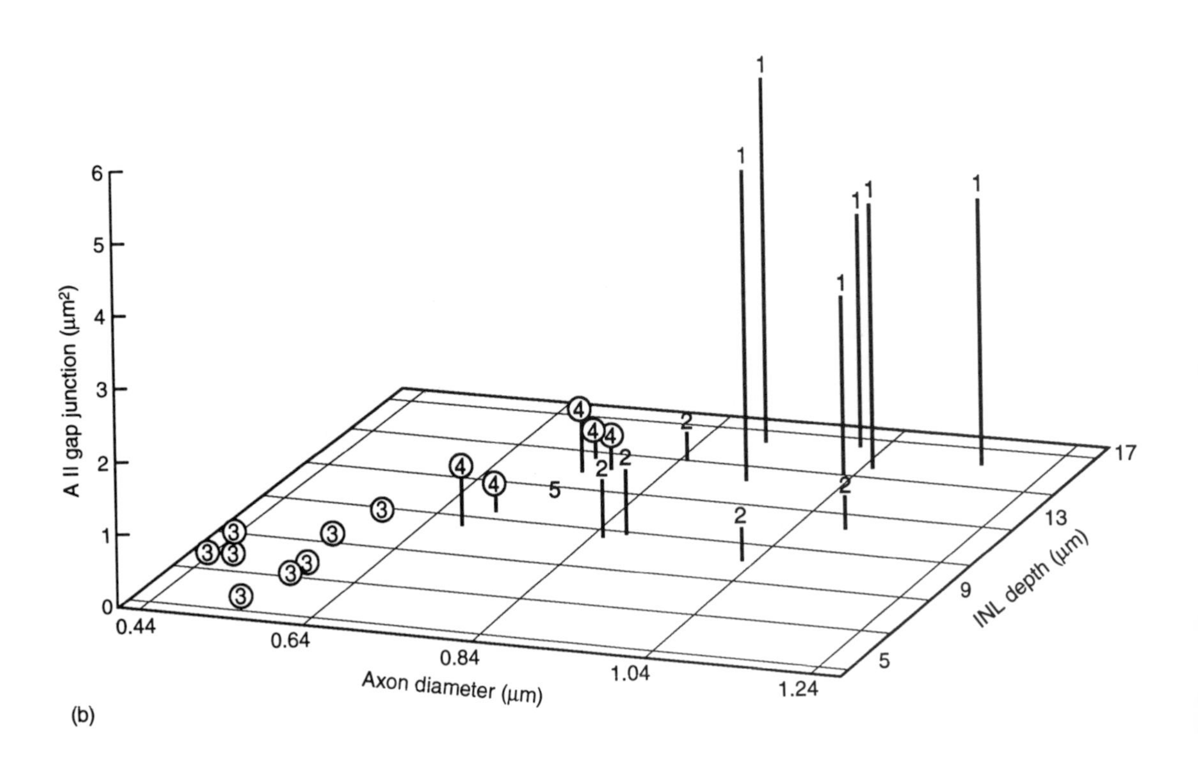
A II gap junction (μm²)
Axon diameter (μm)
INL depth (μm)
0.44
0.64
0.84
1.04
1.24
5
9
13
17
0
1
2
3
4
5
6
(b)

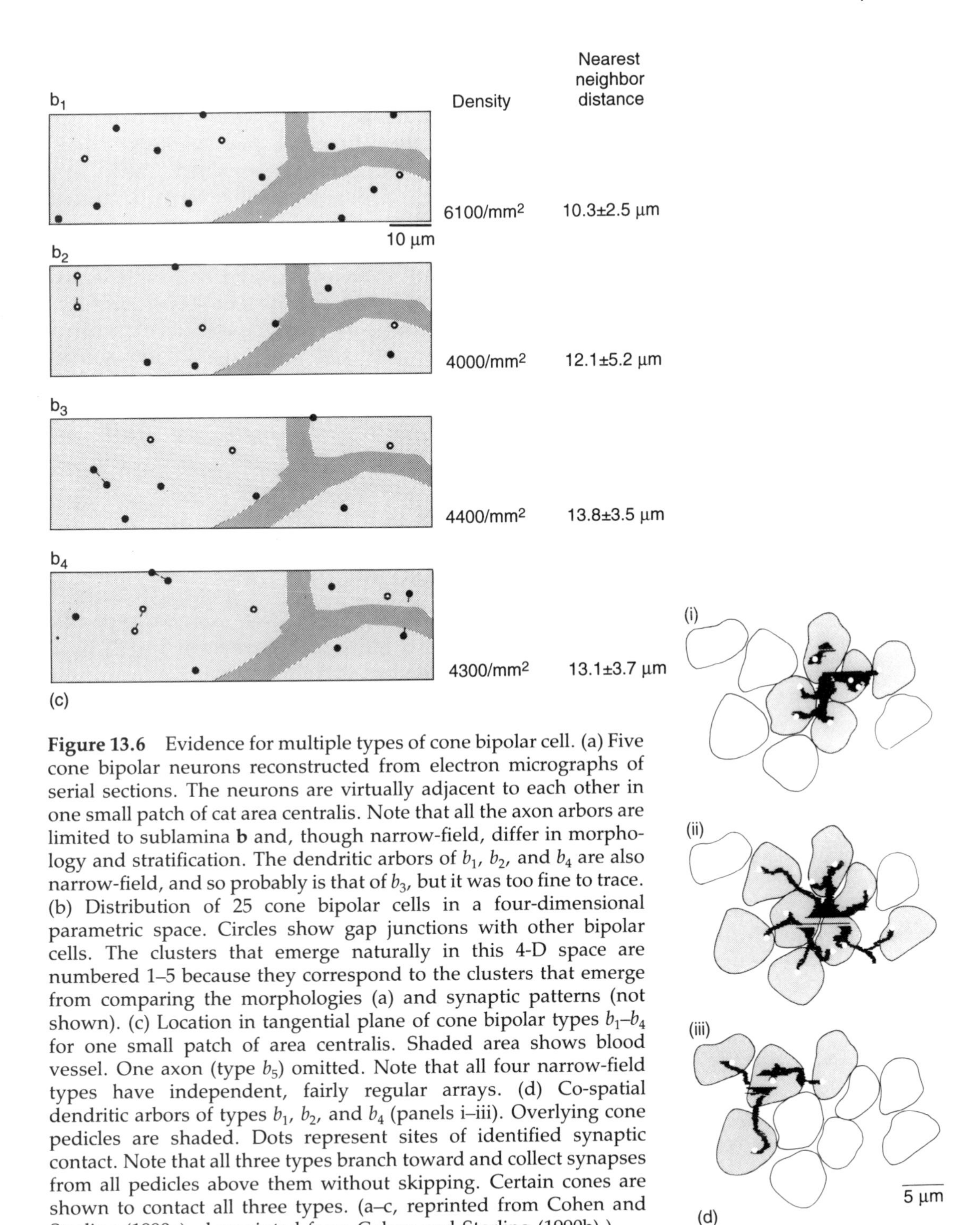

Figure 13.6 Evidence for multiple types of cone bipolar cell. (a) Five cone bipolar neurons reconstructed from electron micrographs of serial sections. The neurons are virtually adjacent to each other in one small patch of cat area centralis. Note that all the axon arbors are limited to sublamina **b** and, though narrow-field, differ in morphology and stratification. The dendritic arbors of b_1, b_2, and b_4 are also narrow-field, and so probably is that of b_3, but it was too fine to trace. (b) Distribution of 25 cone bipolar cells in a four-dimensional parametric space. Circles show gap junctions with other bipolar cells. The clusters that emerge naturally in this 4-D space are numbered 1–5 because they correspond to the clusters that emerge from comparing the morphologies (a) and synaptic patterns (not shown). (c) Location in tangential plane of cone bipolar types b_1–b_4 for one small patch of area centralis. Shaded area shows blood vessel. One axon (type b_5) omitted. Note that all four narrow-field types have independent, fairly regular arrays. (d) Co-spatial dendritic arbors of types b_1, b_2, and b_4 (panels i–iii). Overlying cone pedicles are shaded. Dots represent sites of identified synaptic contact. Note that all three types branch toward and collect synapses from all pedicles above them without skipping. Certain cones are shown to contact all three types. (a–c, reprinted from Cohen and Sterling (1990a), d reprinted from Cohen and Sterling (1990b).)

tinguished by differences in dendritic and axonal morphology observed by Golgi impregnation, as done previously. However, an additional effort was made to establish that each type forms an independent array. This was also done nicely with a different approach by Mills and Massey (1992) in rabbit. Using 4′,6-diamidino-2-phenylindole (DAPI) they observed several different staining patterns of bipolar cell nuclei. Upon injecting individual neurons with Lucifer yellow, they found that a given staining pattern was always associated with the same dendritic and axonal morphology, and in this way they distinguished three types of cone bipolar cell with axons in sublamina **a**. This effort satisfies the criterion, set out by Rodieck and Brening (1983), that the features used to classify be distributed discontinuously; and it satisfies the additional criterion set out by Wässle and Riemann (1978), that *a* type bipolar cells form an independent mosaic.

Our own approach to this problem in cat area centralis was to determine both morphology and synaptic connections of a complete population of bipolar neurons by reconstruction from electron micrographs of serial sections (Cohen and Sterling, 1990a). This permitted quantitative comparison of numerous parameters for every bipolar neuron in the population. Thus, we could determine whether the features formed discontinuous clusters (i.e., natural types), and also whether morphology and synaptic patterns were correlated. We could also determine the array structures for all types. This approach revealed for sublamina **b** five categories of cone bipolar cell (Figure 13.6a). Differing in morphology and synaptic connections, they cluster distinctly in multiparametric space (Figure 13.6b), and they form independent arrays (Figure 13.6c). Thus, the existence of five natural types innervating sublamina **b** is proven. Similar, though less complete studies of sublamina **a** (McGuire *et al.*, 1984) suggest at least three types, supporting the Golgi studies mentioned above.

Another critical point to emerge from these studies is that four of the five types of bipolar cell collect from all the overlying cones without exception (Figure 13.6d; Cohen and Sterling, 1990b). And, unlike the two types of horizontal cell, these four bipolar cell types have the same spatial scale, collecting from four to seven cones, and the same array structure, with densities of 4000–6500/mm^2. Thus, the four types of bipolar cell must carry the same spectral and spatial information to sublamina **b**. Their very existence implies, as for ON and OFF bipolar cells, that some important aspect of the cone's signals are being divided among parallel pathways, but which aspects?

13.3.2 PUSH–PULL HYPOTHESIS RE-EVALUATED

The first suggestion was that two types of bipolar cell, b_1 and b_2, connect to the ON beta (X) ganglion cell and form a 'push–pull' mechanism (McGuire *et al.*, 1986). Type b_1 (corresponding to Kolb *et al.*'s CB5) is reportedly depolarizing; type b_2 (corresponding to their CB6) is reportedly hyperpolarizing (Nelson and Kolb, 1983). Thus, it was natural to hypothesize that b_1 would excite the beta cell at light ON and b_2 would disinhibit. The idea seemed consistent with iontophoretic studies suggesting glycinergic inhibition to beta cells (Bolz *et al.*, 1985; Wässle *et al.*, 1986) and with the finding of endogenous glycine in the b_2 but not the b_1 cell (Pourcho and Goebel, 1987b). However, several observations have accumulated that cast considerable doubt on this hypothesis.

First, most cone bipolar terminals contain high levels of endogenous glutamate and are probably glutamatergic (Massey, 1990). Although certain bipolar cells in cat and monkey contain endogenous GABA and GAD (Wässle and Chun, 1989; Vardi *et al.*,

1994; Vardi and Auerbach, 1995), there are wide-field, sparsely distributed cells in cat that do not contact the beta ganglion cell (Cohen and Sterling, 1992). Second, the relatively intense immunocytochemical staining for glycine in the b_2 bipolar is hard to interpret. No difference in glycine accumulation was found between the various *b* types of bipolar cell (Cohen and Sterling, 1986), and all such types were shown to be coupled by gap junctions (directly or indirectly) to the glycine-accumulating AII amacrine cell (Cohen and Sterling, 1990a). Thus, glycine in cone bipolar cells might merely reflect this coupling (Cohen and Sterling, 1986). Third, the push–pull hypothesis predicts that a light stimulus to the receptive field center would decrease membrane resistance due to excitation and increase resistance due to disinhibition, leading to no net change. However, actual measurements show a clear decrease in resistance and thus are inconsistent with the hypothesis (Freed and Nelson, 1994). The only contradictory point is the report that type b_2 (=CB6) is hyperpolarizing. Since this report was apparently based on a single recording, one must now consider the possibility that the b_2 cell is actually depolarizing. If so, the push–pull hypothesis could finally be rejected.

13.3.3 WHY DOES A CONE NEED TEN PARALLEL PATHWAYS TO THE INNER PLEXIFORM LAYER?

Thus, considerable evidence now suggests that multiple, distinct pathways from the same cones transmit information to the distinct sublaminae of the inner retina devoted to increments and decrements of luminance. The arrangement represents a major investment of neural resources. For example, it requires the cat retina to have 50% more bipolar cells than cones (36 000/mm^2 versus 24 000/mm^2; Cohen and Sterling, 1986, 1990a). Further, since the densest spatial sampling of cones for sublamina **b** is accomplished by the $\mathbf{b}_1$ bipolar array (6500 cells/mm^2), the investment in three additional arrays ($\mathbf{b}_2$–$\mathbf{b}_4$) each at least 4000 cells/mm^2, means that the number of bipolar cells is 3-fold greater than what would be necessary merely for spatial sampling. Since our initial explanation, that two pathways to the same ganglion cell might provide a double-ended input, can be rejected, an hypothesis is needed to account for this investment.

Perhaps the simplest idea is that the different bipolar types divide the temporal bandwidth. Consistent with this suggestion, the beta ganglion cell (sustained-plus-transient response) collects from three types of bipolar cell (Cohen and Sterling, 1992), whereas the alpha ganglion cell (transient response only) collects from only one type of bipolar cell (Freed and Sterling, 1988). The hypothesis is also consistent with recordings from fish bipolar cells where different morphologies correlate with different temporal responses (Saito *et al.*, 1985). Only new recordings will tell us for sure whether different types divide the temporal bandwidth. However, the hypothesis might predict that glutamate receptors with different temporal properties would be expressed on different types, and conceiveably this will be studied first. Whatever the final answer(s) on the need for so many types of cone bipolar cell, it seems to be a remarkable reflection of how much information a single cone can transduce compared to how much a single bipolar cell can transmit; otherwise there would be fewer types.

13.3.4 MORPHOLOGICAL CORRELATES OF FUNCTION AT THE CONE SYNAPSE?

One apparent generalization of the 1970s was that the cone provides 'invaginating' contacts to bipolar cells innervating the ON ganglion cells and 'flat' contacts to bipolar cells innervating the OFF ganglion cells. The rule became fuzzier as 'semi-invaginating' contacts were observed (Nelson and Kolb, 1983; Cohen and Sterling, 1990a), and at least one frank exception was discovered, cone bipolar type a_2 with invaginating contacts (McGuire *et al.*, 1984). Whether the generalization is still useful or hopelessly riddled by exceptions, is treated thoroughly in Chapter 11.

13.3.5 POTENTIAL FOR ALIASING BY CONE BIPOLAR CELL ARRAYS

Although cone bipolar cells are more numerous than cones, the sampling array of each separate type is at most one-quarter of the cone sampling rate. This could cause severe aliasing by the cone bipolar cell arrays. However, the higher spatial frequencies (i.e., those which would aliased) are prefiltered by the combined optical and electrical blur at the cone pedicle. The ganglion cell arrays represent another stage of down-sampling from 4000–6000/mm^2 in a bipolar cell array (types b_{1-4} in cat), down to about 2000/mm^2 ON beta cells. Without further blurring by electrical coupling of the bipolar cells, this would place the beta ganglion cell arrays on the threshold of aliasing (Levitan and Buchsbaum, 1993).

Bipolar cells in lower vertebrates are extensively coupled (Hare and Owen, 1990; Borges and Wilson, 1987; Saito and Kujiraoka, 1988). Whether this serves to prevent aliasing is unclear because neither the optics nor the relative sampling frequencies of bipolar and ganglion cell arrays are known. In cat, cone bipolar types b_3 and b_4 form small gap junctions with each other, and types b_1–b_3 form gap junctions with the AII amacrine cell (Cohen and Sterling, 1990a). However, whether these electrical synapses actually conduct in daylight and thus help prevent aliasing at the ganglion cell remains to be determined.

13.4 BIPOLAR CIRCUIT FOR TWILIGHT

When the photon catch drops below about 100 Rh* (activated rhodopsin)/receptor/integration time, the cone signal declines in amplitude (Schnapf *et al.*, 1990), but the rod signal desaturates (Baylor *et al.*, 1984). About 40–50 rods surround each cone (Figure 13.2b), and small gap junctions connect the rod and cone synaptic terminals (Raviola and Gilula, 1973; Kolb, 1977; Smith *et al.*, 1986). Thus, 50 rods potentially couple to each cone. At a luminance where the cone outer segment collects 100 Rh*, its synaptic terminal sees a voltage generated by 5000 Rh* from the surrounding rods.

Thus, in the mesopic range (1–100 Rh*/receptor/integration time) the rod signals ought to be filtered by the intrinsic mechanisms described above for the daylight circuits (Figures 13.4 and 13.5) and they ought to be relayed forward by the cone bipolar circuits (Figures 13.2 and 13.6). Calculations from a compartmental model of the rod–cone coupled network suggests that most of the rod signal is conveyed to the cone terminal by this route (Smith *et al.*, 1986). The hypothesis also fits the observations that rod signals are indeed recorded in the cone itself (Nelson, 1977) and in the cell types postsynaptic to cones, including horizontal cells (Steinberg, 1969; Nelson, 1977; Raviola and Dacheux, 1990; Bloomfield and Miller, 1982; Lankheet *et al.*, 1991) and cone bipolar cells (Nelson and Kolb, 1983).

13.5 BIPOLAR CIRCUIT FOR STARLIGHT

It seems obvious from the geometry of the rod–cone array that the circuit carrying the rod signal to the cone must fail in scotopic luminances (Smith *et al.*, 1986). When the density of photoisomerizations falls much below 1 Rh*/rod/integration time, the rod's miniscule photocurrent (<1 pA; Baylor *et al.*, 1984) would dissipate into the local network. Thus, it would not effectively modulate transmitter release at cone terminals. Further, if the 50 rod terminals surrounding each cone were to remain coupled in the scotopic range (Figure 13.2), they would mainly transmit their continuous dark noise. In the cone terminal, the amplitude of this noise from 50 rods would exceed the amplitude of an Rh* event (Smith *et al.*, 1986), and to transmit this noise with amplification along the cone bipolar circuits, would be counterproductive.

Consequently, it has been suggested that at the mesopic→scotopic transition there is a switch-over from the cone bipolar to the rod bipolar circuit (Figure 13.2; Smith *et al.*, 1986). A key aspect of this switch should be the uncoupling of the rod and cone terminals. This would electrically isolate individual rod terminals, thus protecting each from the other's continuous dark noise. Equally important, it would dramatically improve transfer of the Rh* signal to the rod terminal (Smith *et al.*, 1986), from whence it can be transmitted to the rod bipolar cell. Evidence, as follows, has accumulated in support of this hypothesis.

First, the cone horizontal cell responds only in the mesopic range (Raviola and Dacheux, 1990) consistent with the idea that the rod→cone pathway uncouples in scotopic conditions. Second, the rod bipolar cell does respond in the scotopic range (Dacheux and Raviola, 1986). Also, the 'scotopic threshold response' (an extracellular potential that originates deep in the inner plexiform layer where rod bipolar cell terminals contact the AII amacrine cell) is detected at extremely low luminance (Sieving *et al.*, 1986). Third, APB, which blocks transmission to the rod bipolar dendrites, blocks the scotopic signal to ganglion cells (Müller *et al.*, 1988). If the signal were also carried by the rod→cone pathway, the scotopic signal should reach the OFF ganglion cells via the cone bipolar circuits. Behaviorally, too, a monkey with APB injected into the eye is blind in scotopic, but not in mesopic luminance (Dolan and Schiller, 1989). Fourth, psychophysical detection of flicker reveals a slow rod pathway at low intensities, plus a faster, less sensitive pathway at higher luminance (Stockman *et al.*, 1991).

13.6 FUNCTIONAL ARCHITECTURE OF THE MAMMALIAN ROD SYNAPSE

13.6.1 BINARY SIGNALING

The mammalian rod differs greatly from the amphibian rod most familiar to retinal physiologists. The differences seem critically reflected in the circuit architecture. The mammalian rod is about four fold faster than the amphibian rod (integration time of 250 ms versus 1 s).

To achieve this speed requires a short diffusion distance for cGMP (Lamb and Pugh, 1992), so the mammalian rod is thin (2 μm diameter vs. 10 μm in salamander). Due to the small cross-section and short integration time, the mammalian rod collects less than one photon per integration time over the full three log units of the scotopic range (Sterling *et al.*, 1987). Consequently, at scotopic intensities the rod detects a binary signal: 0 or 1 photon. Further, this signal apparently reaches the ganglion cell level, one Rh* triggering a burst of two or three spikes in several adjacent beta (X) cells (Barlow *et al.*, 1971; Mastronarde, 1983). The expression of single photon events in the ganglion cell implies that these events are not pooled at the level of the rods, so mammalian rods must not be electrically coupled. Consequently the signal presented to the rod synaptic terminal is binary, representing 0 or 1 Rh*.

13.6.2 HIGH RATE OF TONIC SIGNALING

The rod synapse has a single active zone that releases neurotransmitter quanta tonically in darkness and that responds to one Rh* with a brief pause in secretion (Figure 13.7a). Therefore, we asked: what mininum tonic rate would allow the bipolar cell to distinguish this pause from an extra-long interval between transmitter quanta due to the stochastic timing of release (Rao *et al.*, 1994)? We based the calculation on a model of the circuit that includes the rod convergence onto the bipolar cell and the bipolar cell's signal-to-noise ratio (Figure 13.7a). The result suggests that tonic release must be at least 40 quanta per second. This rate is consistent with the measurements at other 'ribbon' synapses (Ashmore and Copenhagen, 1983; von Gersdorff and Matthews, 1994).

The need to sustain a high tonic rate may explain the unique architecture of the rod active zone (Figure 13.7b). Sustained quantal release requires: (1) a substantial number of vesicle docking sites; (2) a mechanism for their prompt reloading and (3) a handy depot of fresh vesicles. The rod's vesicle-docking region is linear and extensive, with room for (~ 130 vesicles (Rao *et al.*, 1995) (versus ~20 at a conventional retinal synapse; Raviola and Raviola, 1982). The obvious candidate for a resupply depot is the plate-like synaptic ribbon to which ~640 vesicles are attached (Rao *et al.*, 1995). The ribbon's basal edge anchors parallel to the docking region at a distance of 30nm. Since this is about the diameter of one vesicle, the arrangement minimizes the distance for translocation of a vesicle to resupply an empty docking site.

13.6.3 SPACE EFFICIENCY

Although most neurons transfer a signal by employing a low quantal rate at multiple active zones, the rod employs a high rate at a single active zone. This design conserves space. For example, the cone terminal with 17 active zones (Sterling and Harkins, 1990) occupies ~8 times the volume of the rod terminal. Rods are the most numerous of retinal neurons (in cat, up to 500 000/mm^2; Steinberg *et al.*, 1973; Williams *et al.*, 1993) and their terminals occupy 66% of the retinal

Figure 13.7 Left: (a) Early stages of rod circuit in cat retina: *n* rods converge onto a bipolar cell, each contacting a separate dendrite. Right: diagrams of response at successive stages. Outer segment: single rhodopsin isomerization (Rh*) gives prolonged all-or-none hyperpolarizing response. Rod terminal: tonic quantal release (dots) and corresponding bipolar mpsps at bipolar dendrite. Bipolar soma: noise (σ) reflects temporal integration of mpsps from *n* rods with each rod contributing $\sigma/n^{1/2}$Rh* in one rod suppresses tonic release, causing depolarization of amplitude *S*. Bipolar terminal: depolarization by Rh* event triggers brief burst of quanta. Circuit illustrated here for cat is highly conserved across mammalian species (Rao *et al.*, 1994). (b) Rod spherule in cat contains a single active zone and four postsynaptic processes ('tetrad') (Rao *et al.*, 1995). The rod is the only vertebrate neuron whose total output is restricted to a single active zone. Consequently it is large and complex. Left: section perpendicular to the synaptic ribbon. Note that the synaptic vesicles tethered to the ribbon's basal edge contact the presynaptic membrane, thus forming an 'active zone'. Intramembrane particles, presumably including calcium channels, concentrate in the presysnaptic membrane convexity between the paired docking regions (Raviola and Gilula, 1973). Stippling on the horizontal cells (h) represents sites of postsynaptic densification and intramembrane particles (Raviola and Gilula, 1973), presumably kainate receptors. Stippling on the rod bipolar dendrites (b) represents presumed location of APB receptors. Right: section parallel to the face of the ribbon. Note that the active zone (darker vesicles at the presynaptic membrane) is arched and that the ribbon, conforming to the arch, is crescentic. The horizontal cell processes are extremely close to the release sites and the bipolar processes are far (see text).

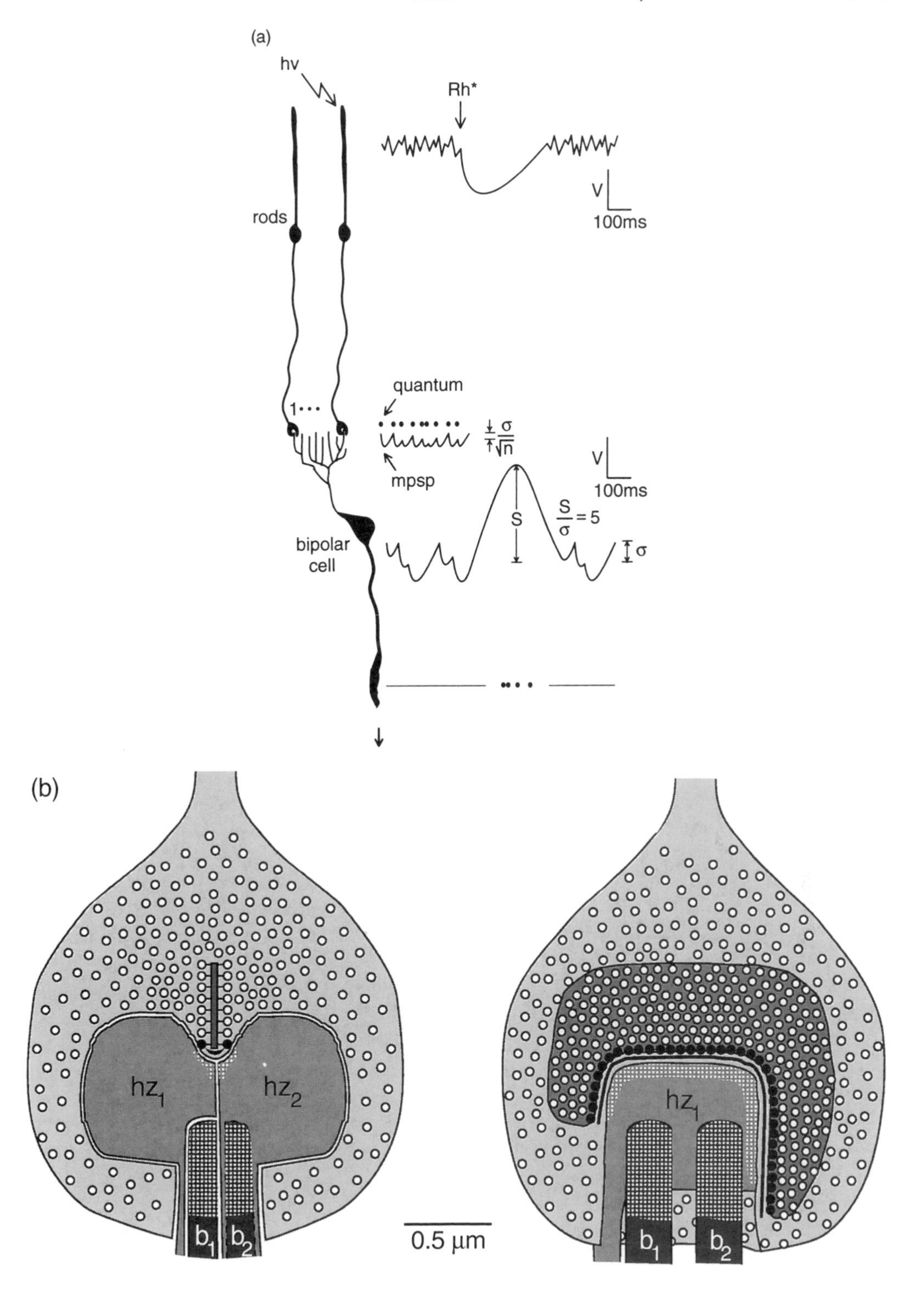
(a)
hv
Rh*
rods
V
100ms
quantum
1···
mpsp
bipolar
cell
S
$\frac{S}{\sigma} = 5$
$\frac{\sigma}{\sqrt{n}}$
σ
(b)
hz1
hz2
b1
b2
0.5 μm

volume devoted to photoreceptor terminals. Yet, rods are the most impoverished of signal: a rod in starlight absorbs a photon only about once in 10 minutes (Sterling *et al.*, 1987; MacLeod *et al.*, 1989). Therefore, the single active zone with a high quantal rate may reflect pressure to 'miniaturize' the apparatus needed for the reliable transfer of a binary signal.

Calculations from a two-dimensional diffusion model of this synapse suggest that one transmitter quantum released anywhere along the arch would affect all four postsynaptic processes. The rod horizontal cell terminals contain GABA (Wässle and Chun, 1989; Grünert and Wässle, 1990) and GAD (Vardi *et al.*, 1994). In humans, they form conventional, chemical synapses back onto the rod (Linberg and Fisher, 1988). In other mammals the locus of the presumed synapse is less obvious, but since the rod terminal in rat expresses mRNA for the $GABA_A$ receptor (Grigorenko and Yeh, 1994), it seems probable that all mammals have GABAergic feedback onto the rod terminal. The function of this feedback has not been demonstrated; however, a thresholding mechanism is needed at this stage to separate the Rh* event from the continuous dark noise. Otherwise, the continuous dark noise of the 20 or more rods that converge on the rod bipolar cell would swamp the Rh* signal (Baylor *et al.*, 1984; Freed *et al.*, 1987). We speculate that the GABA feedback onto the rod serves this thresholding function.

13.6.4 POSSIBLE THRESHOLDING AT BIPOLAR OUTPUT

The two bipolar members of the tetrad, being affected by the same pattern of transmitter quanta, must convey identical signals to the inner plexiform layer where each provides about 30 synapses to the next stage of the circuit (McGuire *et al.*, 1984; Sterling *et al.*, 1988). Consequently, when this pair of bipolar cells is depolarized by the same Rh* event, there will be a localized focus of about 60 active synapses on the processes of AII amacrine neurons (Sterling *et al.*, 1988) that form a coupled network in the inner plexiform layer (Hampson *et al.*, 1992). At this stage another thresholding mechanism is needed to separate the Rh* event from the noise of the 100 rod bipolar cells that converge (via the AII network) upon the beta (X) ganglion cell (Freed *et al.*, 1987; Sterling *et al.*, 1988). The AII cell, due to voltage-gated Na^+ channels (Boos *et al.*, 1993) apparently does have a threshold. We predict that this threshold would be exceeded by the coincident activation of 60 local rod bipolar synapses but not by the independent activation of a single rod bipolar axon at random locations within the field of 100 cells (Smith and Vardi, 1995). The circuit described here quantitatively for cat seems highly conserved across mammalian species: rat (Chun *et al.*, 1993), rabbit (Strettoi *et al.*, 1990; Young and Vaney, 1991; Vaney *et al.*, 1991), monkey (Grünert and Martin, 1991).

13.7 MAMMALIAN VERSUS AMPHIBIAN ROD CIRCUIT

The amphibian rod has a 25-fold greater cross-sectional area than the mammalian rod and a four fold longer integration time. Thus, it probably collects 100-fold more photons/integration time. Further, amphibian rods are strongly coupled (e.g. Attwell *et al.*, 1985; reviewed in Chapter 6) and thus pool their signals before forward transmission to the bipolar cell. Thus, the amphibian rod's synaptic terminal transmission is not a binary signal, but a graded one, representing 100s–1,000s of Rh*. In this respect the amphibian rod resembles a cone, and this seems to be reflected in its circuit design. Like a cone terminal, the amphibian rod terminal employs multiple active zones (Lasansky, 1973; Mariani, 1986); it also uses the same bipolar cells as cones (Lasansky, 1978), and it uses OFF- as well as ON-bipolar cells (Lasansky, 1978). Further, an individual mammalian rod does not adapt in the scotopic range, though it does so at mesopic luminances (Tamura *et al.*, 1989). Psychophysical experiments demonstrate that in the scotopic range, mammalian rod adaptation occurs at the level of the rod pool and that the size of the pool is in the order of a single rod bipolar cell dendritic field (MacLeod *et al.*, 1989). This makes perfect sense because, as MacLeod *et al.* (1989) point out, photons are too sparse for a single rod to estimate the local luminance; therefore a single rod lacks adequate data to set its own sensitivity.

13.8 CONCLUSIONS AND FUTURE PERSPECTIVES

Several themes emerge from this review. First, the circuitry for cones is vastly richer than for rods. At the cellular level, the cone employs ten types of bipolar neuron, whereas the rod employs only one type. And at the synaptic level, a cone provides about 20 presynaptic active zones and contacts about 100 postsynaptic processes, whereas a rod provides a single active zone and contacts only four processes. Second, this difference in circuit complexity apparently reflects a key functional difference: the photon shower upon the cones is dense and thus provides a rich source of spatial and temporal information; whereas the photon shower on the rods is sparse and thus provides irreducibly simple, binary signals. The cone's signal is so rich as to require compression (bandpass filtering) before the first forward synapse, plus multiple parallel paths, perhaps to divide responsibility for transmitting its full temporal bandwidth to the inner retina. The rod's signal requires, not compression, but rather thresholding to remove continuous dark noise. Third, the retina faces serious constraints on space. On the one hand, to transmit a really good optical image (rich in photons) apparently demands many parallel pathways and many synapses. On the other hand, to catch the last few photons requires an extremely dense distribution of rods and a correspondingly dense distribution of rod bipolar cells. In cat, for example, the sum of ten cone bipolar types is about equal to the number of rod bipolar cells (Cohen and Sterling, 1986; see also Grünert and Martin 1991 for monkey). The space constraints force elements that carry little information to be small (e.g., rod versus cone terminal). The same constraints force pathways to share space. Thus, the rod pathway for multiphoton signals shares the cone terminal and subsequent cone bipolar circuits (Figure 13.2). Similarly, the rod pathway for binary signals shares a final common node, the cone bipolar axon terminal (Figure 13.2). Thus, ganglion cells need not apportion additional dendritic space for an independent set of rod bipolar synapses.

The anatomical circuitry worked out in great detail over the last two decades has provided a rich source of models and hypotheses. And now, most fortunately, technical advances in whole cell recording *in situ* offer exciting prospects for testing them. First and most urgently, we need to record from cone bipolar cells in order to determine whether the different types really do divide the temporal bandwidth or serve some other purpose. Second, we need to identify the specific isoforms of transmitter-gated channels expressed by each identified cell type. One anticipates that the molecular and circuit properties will match, e.g., slow channels for the rod bipolar circuit and fast ones for the cone bipolar circuits.

Finally, we need to expand the quantitative, 'bottom-up' models of the identified circuits. Such models are simple tools to frame hypotheses regarding circuit function. For example, a model of current spread in the AII network followed by thresholding via the

AII's regenerative mechanism (Smith and Vardi, 1995) could predict how many **ON** beta ganglion cells are activated and how many **OFF** beta ganglion cells are suppressed by a single photon. Since one beta cell changes its firing by about three spikes in response to a photon (Barlow *et al.*, 1971; Mastronarde, 1983), one could estimate the total message to the brain (change in firing by both beta cell arrays) to one photon. We also need models to explore the transitions: photopic→mesopic and mesopic→scotopic, to grasp how the circuits serving each level interact.

ACKNOWLEDGEMENTS

We thank other members of the laboratory and collaborators, past and present: Barbara McGuire, Michael Freed, Ethan Cohen, Yoshihiko Tsukamoto and Gershon Buchsbaum who contributed much to the analysis summarized here. We also thank Patricia Masarachia, Sally Shrom and Peter Auerbach for superb technical assistance and Sharron Fina for secretarial help. The work was supported by EY00828, EY08124 and MH48168

REFERENCES

Agardh, E., Ehinger, B. and Wu, J.-Y. (1987) GABA and GAD-like immunoreactivity in the primate retina. *Histochemistry*, **86**, 485–90.

Ahnelt, P. and Pflug, R. (1986) Telodendrial contacts between foveolar cone pedicles in the human retina. *Experimentia*, **42**, 298–300.

Ashmore, J.F. and Copenhagen, D. (1983) An analysis of transmission from cones to hyperpolarizing bipolar cells in the retina of the turtle. *Journal of Physiology*, **340**, 569–97.

Attwell, D., Wilson, M. and Wu, S.M. (1985) The effect of light on the spread of signals through the rod network of the salamander retina. *Brain Research*, **343**, 79–88.

Barlow, H.B. (1961) Three points about lateral inhibition, in *Sensory Communication* (ed. W.A. Rosenblith), MIT Press/John Wiley, New York, pp.782–6.

Barlow, H.B., Levick, W.R. and Yoon, M. (1971) Responses to single quanta of light in retinal ganglion cells of the cat. *Vision Research*, **S3**, 87–101.

Baylor, D.A., Fuortes, M.G.F. and O'Bryan, P.M. (1971) Receptive fields of cones in the retina of the turtle. *Journal of Physiology*, **214**, 265–94.

Baylor, D.A., Nunn, B.J. and Schnapf, J.L. (1984) The photocurrent, noise and spectral sensitivity of rods of the monkey *Macaca fascicularis*. *Journal of Physiology*, **357**, 575–607.

Bloomfield, S.A. and Miller, R.F. (1982) A physiological and morphological study of the horizontal cell types of the rabbit retina. *Journal of Comparative Neurology*, **208**, 288–303.

Bolz, J., Thier, P., Voight, T. and Wässle, H. (1985) Action and localization of glycine and taurine in the cat retina. *Journal of Physiology*, **362**, 395–413.

Boos, R., Schneider, H. and Wässle, H. (1993) Voltage- and transmitter-gated currents of AII-amacrine cells in a slice preparation of the rat retina. *Journal of Neuroscience*, **13**, 2874–88.

Borges, S. and Wilson, M. (1987) Structure of the receptive fields of bipolar cells in the salamander retina. *Journal of Neurophysiology*, **58**, 1275–91.

Boycott, B.B. and Kolb, H. (1973) The horizontal cells of the Rhesus monkey retina. *Journal of Comparative Neurology*, **148**, 115–39.

Boycott, B.B. and Wässle, H. (1991) Morphological classification of bipolar cells of the primate retina. *European Journal of Neuroscience*, **3**, 1069–88.

Boycott, B.B., Peichl, L. and Wässle, H. (1978) Morphological types of horizontal cell in the retina of the domestic cat. *Proceedings of the Royal Society of London B*, **203**, 229–45.

Boycott, B.B., Hopkins, J.M. and Sperling, H.G. (1987) Cone connections of the horizontal cells of the rhesus monkey's retina. *Proceedings of the Royal Society of London B*, **229**, 345–79.

Chen, B., Makous, W. and Williams, D.R. (1993) Serial spatial filters in vision. *Vision Research*, **33**, 413–27.

Chun, M.H., Han, S.H., Chung, J.W. and Wässle, H. (1993) Electron-microscopic analysis of the rod pathway of the rat retina. *Journal of Comparative Neurology*, **332**, 421–32.

Cleland, B.G., Harding, T.H. and Tulunay-Keesey, U. (1979) Visual resolution and receptive-field size: examination of two kinds of cat retinal ganglion cell. *Science*, **205**, 1015–17.

Cohen, A.I. (1965) Some electron microscopic observations on inter-receptor contacts in the human and macaque retinae. *Journal of Anatomy*, **99**, 595–610.

Cohen, E. and Sterling, P. (1986) Accumulation of [^{3}H] glycine by cone bipolar neurons in the cat retina. *Journal of Comparative Neurology*, **250**, 1–7.

Cohen, E. and Sterling, P. (1990a). Demonstration of cell types among cone bipolar neurons of cat retina. *Philosophical Transactions of the Royal Society of London B*, **330**, 305–21.

Cohen, E. and Sterling, P. (1990b) Convergence and divergence of cones onto bipolar cells in the central area of cat retina. *Philosophical Transactions of the Royal Society of London B*, **330**, 323–8.

Cohen, E. and Sterling, P. (1991) Microcircuitry related to the receptive field center of the on-beta ganglion cell. *Journal of Neurophysiology*, **65**, 352–9.

Cohen, E. and Sterling, P. (1992) Parallel circuits from cones to the on-beta ganglion cell. *European Journal of Neuroscience*, **4**, 506–20.

Dacheux, R.F. and Raviola, E. (1986) The rod pathway in the rabbit retina: a depolarizing bipolar and amacrine cell. *Journal of Neuroscience*, **6**, 331–45.

Dolan, R.P. and Schiller, P.H. (1989) Evidence for only depolarizing rod bipolar cells in the primate retina. *Visual Neuroscience*, **2**, 421–4.

Famiglietti, E.V. Jr (1981) Functional architecture of cone bipolar cells in mammalian retina. *Vision Research*, **21**, 1559–63.

Freed, M.A. (1992) GABAergic circuits in the mammalian retina. *Progress in Brain Research*, **20**, 107–31.

Freed, M.A. and Nelson, R. (1994) Conductances evoked by light in the ON-β ganglion cell of the cat retina. *Visual Neuroscience*, **11**, 261–9.

Freed, M.A. and Sterling, P. (1988) The ON-alpha ganglion cell of the cat retina and its presynaptic cell types. *Journal of Neuroscience*, **8**, 2303–20.

Freed, M.A., Smith, R.G. and Sterling, P. (1987) Functional architecture of the rod bipolar neuron in cat retina. *Journal of Comparative Neurology*, **266**, 445–55.

Greferath, U., Müller, F., Wässle, H. *et al.* (1993) Localization of $GABA_A$ receptors in the rat retina. *Visual Neuroscience*, **10**, 551–61.

Grigorenko, E.V. and Yeh, H.H. (1994) Expression profiling of $GABA_A$ receptor β-subunits in the rat retina. *Visual Neuroscience*, **11**, 379–87.

Grünert, U. and Martin, P.R. (1991). Rod bipolar cells in the Macaque monkey retina: immunoreactivity and connectivity. *Journal of Neuroscience*, **11**, 2742–58.

Grünert, U. and Wässle, H. (1990) GABA-like immunoreactivity in the macaque monkey retina: a light and electron microscopic study. *Journal of Comparative Neurology*, **297**, 509–24.

Hampson, E.C., Vaney, D.I. and Weiler, R. (1992) Dopaminergic modulation of gap junction permeability between amacrine cells in mammalian retina. *Journal of Neuroscience*, **12**, 4911–22.

Hare, W.A. and Owen, W.G. (1990) Spatial organization of the biopolar cell's receptive field in the retina of the tiger salamander. *Journal of Physiology*, **421**, 223–45.

Hirsch, J. and Curcio, C. (1989) The spatial resolution capacity of human foveal retina. *Vision Research*, **29**, 1095–101.

Kolb, H. (1977) The organization of the outer plexiform layer in the retina of the cat: electron microscopic observations. *Journal of Neurocytology*, **6**, 131–53.

Kolb, H., Nelson, R. and Mariani, A. (1981) Amacrine cells, bipolar cells and ganglion cells of the cat retina: a Golgi study. *Vision Research*, **21**, 1081–114.

Kolb, H., Linberg, K.A. and Fisher, S.K. (1992) Neurons of the human retina: a Golgi study. *Journal of Comparative Neurology*, **318**, 147–87.

Kuffler, S.W. (1953) Discharge patterns and functional organization of mammalian retina. *Journal of Neurophysiology*, **16**, 37–68.

Lamb, T.D. and Pugh, E.N. Jr (1992) G-protein cascades: gain and kinetics. *Trends in Neuroscience*, **15**, 291–8.

Lamb, T.D. and Simon, E.J. (1976) The relation between intercellular coupling and electrical noise in turtle photoreceptors. *Journal of Physiology*, **263**, 257–86.

Lankheet, M.J.M., van Wezel, R.J.A. and van de Grind, W.A. (1991) Effects of background illumination on cat horizontal cell responses. *Vision Research*, **31**, 919–32.

Lasansky, A. (1973) Organization of the outer synaptic layer in the retina of the larval tiger salamander. *Philosophical Transactions of the Royal Society of London B*, **265**, 471–89.

Lasansky, A. (1978) Contacts between receptors

and electrophysiologically identified neurones in the retina of the larval tiger salamander. *Journal of Physiology*, **285**, 531–42.

Laughlin, S.B. (1994) Matching coding, circuits, cells, and molecules to signals: general principles of retinal design in the fly's eye. *Progress in Retinal and Eye Research*, **13**, 165–96.

Levitan, B. and Buchsbaum, G. (1993) Signal sampling and propagation through multiple cell layers in the retina: modeling and analysis with multirate filtering. *Journal of the Optical Society of America A*, **10**, 1463–79.

Linberg, K.A. and Fisher, S.K. (1988) Ultrastructural evidence that horizontal cell axon terminals are presynaptic in the human retina. *Journal of Comparative Neurology*, **268**, 281–97.

Linsenmeier, R.A., Frishman, L.J., Jakiela, H.G. and Enroth-Cugell, C. (1982) Receptive field properties of X and Y cells in the cat retina derived from contrast sensitivity measurements. *Vision Research*, **22**, 1173–83.

MacLeod, D.I.A., Chen, B. and Crognale, M. (1989) Spatial organization of sensitivity regulation in rod vision. *Vision Research*, **29**, 965–78.

MacLeod, D.I.A., Williams, D.R. and Makous, W. (1992) A visual nonlinearity fed by single cones. *Vision Research*, **32**, 347–63.

Mangel, S.C. (1991) Analysis of the horizontal cell contribution to the receptive field surround of ganglion cells in the rabbit retina. *Journal of Physiology*, **442**, 211–34.

Mariani, A.P. (1986) Photoreceptors of the larval tiger salamander retina. *Proceedings of the Royal Society of London B*, **227**, 483–92.

Massey, S.C. (1990) Cell types using glutamate as a neurotransmitter in the vertebrate retina, in *Progress in Retinal Research*, vol. 9 (eds N.N. Osborne and G. Chader), Pergamon Press, London, pp.399–425.

Mastronarde, D.N. (1983) Correlated firing of cat retinal ganglion cells. II. Responses of X- and Y-cells to single quantal events. *Journal of Neurophysiology*, **49**, 325–49.

McGuire, B.A., Stevens, J.K. and Sterling, P. (1984) Microcircuitry of bipolar cells in cat retina. *Journal of Neuroscience*, **4**, 2920–38.

McGuire, B.A., Stevens, J.K. and Sterling, P. (1986) Microcircuitry of beta ganglion cells in cat retina. *Journal of Neuroscience*, **6**, 907–18.

Mills, S.L. and Massey, S.C. (1992) Morphology of bipolar cells labeled by DAPI in the rabbit retina. *Journal of Comparative Neurology*, **321**, 133–49.

Missotten, L. (1965) *The Ultrastructure of the Human Retina*. Editions Arscia, Brussels.

Müller, B. and Peichl, L. (1993) Horizontal cells in the cone-dominated tree shrew retina: morphology, photoreceptor contacts, and topographical distribution. *Journal of Neuroscience*, **13**, 3628–46.

Müller, F., Wässle, H. and Voigt, T. (1988) Pharmacological modulation of the rod pathway in the cat retina. *Journal of Neurophysiology*, **59**, 1657–72.

Nelson, P., Famiglietti, E.V. and Kolb, H. (1978) Intracellular staining reveals different levels of stratification for On- and Off-center ganglion cells of the cat retina. *Journal of Neurophysiology*, **41**, 472–483.

Nelson, R. (1977) Cat cones have rod input: a comparison of the response properties of cones and horizontal cell bodies in the retina of the cat. *Journal of Comparative Neurology*, **172**, 109–36.

Nelson, R. and Kolb, H. (1983). Synaptic patterns and response properties of bipolar and ganglion cells in the cat retina. *Vision Research*, **23**, 1183–95.

Peichl, L. and González-Soriano, J. (1993) Unexpected presence of neurofilaments in axon-bearing horizontal cells of the mammalian retina. *Journal of Neuroscience*, **13**, 4091–100.

Piccolino, M., Neyton, J. and Gerschenfeld, H. (1981) Center-surround antagonistic organization in small-field luminosity horizontal cells of turtle retina. *Journal of Neuroscience*, **45**, 363–375.

Pourcho, R.G. and Goebel, D.J. (1987a) Visualization of endogenous glycine in cat retina: an immunocytochemical study with Fab fragments. *Journal of Neuroscience*, **7**, 1189–97.

Pourcho, R.G. and Goebel, D.J. (1987b) A combined Golgi and autoradiographic study of ^{3}H-glycine-accumulating cone bipolar cells in the cat retina. *Journal of Neuroscience*, **7** 1178–88.

Pourcho, R.G. and Owczarzak, M.T. (1989) Distribution of GABA immunoreactivity in the cat retina: a light- and electron-microscopic study. *Visual Neuroscience*, **2**, 425–435.

Rao, R., Buchsbaum, G. and Sterling, P. (1994) Rate of quantal transmitter release at the mammalian rod synapse. *Biophysical Journal*, **67**, 57–63.

Rao, R., Harkins, A., Buchsbaum, G. and Sterling,

P. (1995) Mammalian rod terminal architecture of a binary synapse *Neuron*, **14**, March.

Raviola, E. and Dacheux, R.F. (1983) Variations in structure and response properties of horizontal cells in the retina of the rabbit. *Vision Research*, **23**, 1221–7.

Raviola, E. and Dacheux, R.F. (1990) Axonless horizontal cells of the rabbit retina: synaptic connections and origin of the rod aftereffect. *Journal of Neurocytology*, **19**, 731–6.

Raviola, E. and Gilula, N.B. (1973) Gap junctions between photoreceptor cells in the vertebrate retina. *Proceedings of the National Academy of Sciences USA*, **70**, 1677–81.

Raviola, E. and Raviola, G. (1982) Structure of the synaptic membranes in the inner plexiform layer of the retina: a freeze-fracture study in monkeys and rabbits. *Journal of Comparative Neurology*, **209**, 233–248.

Robson, J.G. and Enroth-Cugell, C.E. (1978) Light distribution in the cat's retinal image. *Vision Research*, **18**, 159–73.

Rodieck, R.W. (1973) *The Vertebrate Retina: Principles of Structure and Function*. W.H. Freeman, San Francisco, CA.

Rodieck, R.W. and Brening, R.K. (1983) Retinal ganglion cells: properties, types, genera, pathways and trans-species comparisons. *Brain Behavior and Evolution*, **23**, 121–64.

Rose, A. (1973) *Vision: Human and Electronic*. Plenum Press, New York.

Saito, T. and Kujiraoka, T. (1988) Characteristics of bipolar–bipolar coupling in the carp retina. *Journal of General Physiology*, **91**, 275–287.

Saito, T., Kujiraoka, T., Yonaha, T. and Chino, Y. (1985) Reexamination of photoreceptor–bipolar connectivity patterns in carp retina: HRP-EM and Golgi-EM studies. *Journal of Comparative Neurology*, **236**, 141–60.

Sarthy, P.V. and Fu, M. (1989). Localization of L-glutamic acid decarboxylase mRNA in cat retinal horizontal cells by *in situ* hybridization. *Journal of Comparative Neurology*, **288**, 593–600.

Schnapf, J.L., Nunn, B.J., Meister, M. and Baylor, D.A. (1990) Visual transduction in cones of the monkey *Macaca fascicularis*. *Journal of Physiology*, **427**, 681–713.

Sieving, P.A., Frishman, L.J. and Steinberg, R.H. (1986) Scotopic threshold response of proximal retina in cat. *Journal of Neurophysiology*, **56**, 1049–61.

Smith, R.G. (1992) NeuronC: a computational language for investigating functional architecture of neural circuits. *Journal of Neuroscience Methods*, **43**, 83–108.

Smith, R.G. (1995) Simulation of an anatomically-defined local circuit: the cone-horizontal cell network in cat retina. *Visual Neuroscience* (in press).

Smith, R.G. and Sterling, P. (1990) Cone receptive field in cat retina computed from microcircuitry. *Visual Neuroscience*, **5**, 453–461.

Smith, R.G. and Vardi, N. (1993). Functional regenerative membrane and coupling in the AII amacrine cell of cat retina. *Neuroscience Abstracts*, **19**, 1415 (Abstr.).

Smith, R.G. and Vardi, N. (1995) Simulation of the AII amacrine cell of mammalian retina: functional consequences of electrical coupling and regenerative membrane properties. *Journal of Neuroscience* (in press).

Smith, R.G., Freed, M.A. and Sterling, P. (1986) Microcircuitry of the dark-adapted cat retina: functional architecture of the rod–cone network. *Journal of Neuroscience*, **6**, 3505–17.

Srinivasan, M.V., Laughlin, S.B. and Dubs, A. (1982) Predictive coding: a fresh view of inhibition in the retina. *Proceedings of the Royal Society of London B*, **216**, 427–59.

Steinberg, R.H. (1969) Rod and cone contributions to s-potenials from the cat retina. *Vision Research*, **9**, 1319–29.

Steinberg, R.H., Reid, M. and Lacy, P.L. (1973) The distribution of rods and cones in the retina of the cat (*Felis domesticus*). *Journal of Comparative Neurology*, **148**, 229–48.

Sterling, P. and Harkins, A.B. (1990) Ultrastructure of the cone pedicle in cat retina. *Investigative Ophthalmology and Visual Science*, **31**, 177.

Sterling, P., Cohen, E., Freed, M.A. and Smith, R.G. (1987) Microcircuitry of the on-beta ganglion cell in daylight, twilight and starlight. *Neuroscience Research*, (Suppl)6, 5269–85.

Sterling, P., Freed, M.A. and Smith, R.G. (1988) Functional architecture of the rod and cone circuits to the on-beta ganglion cell. *Journal of Neuroscience*, **8**, 623–42.

Sterling, P., Cohen, E., Smith, R.G. and Tsukamoto, Y. (1992) Retinal circuits for daylight: why ballplayers don't wear shades, in *Analysis and Modeling of Neural Systems* (ed. F.H. Eeckman), Kluwer Academic Publishers, Dordreeht, pp.143–62.

Stockman, A., Sharpe, L.T., Zrenner, E. and Nordby, K. (1991) Slow and fast pathways in the human rod visual system: electrophysiology and psychophysics. *Journal of the Optical Society of America*, **8**, 1657–65.

Strettoi, E., Dacheux, R.F. and Raviola, E. (1990) Synaptic connections of rod bipolar cells in the inner plexiform layer of the rabbit retina. *Journal of Comparative Neurology*, **295**, 449–66.

Tamura, T., Nakatani, K. and Yau, K.-W. (1989) Light adaptation in cat retinal rods. *Science*, **245**, 755–8.

Tsukamoto, Y., Smith, R.G. and Sterling, P. (1990) Collective coding of correlated cone signals in the retinal ganglion cell. *Proceedings of the National Academy of Sciences USA*, **87**, 1860–4.

Tsukamoto, Y., Masarachia, P., Schein, S.J. and Sterling, P. (1992) Gap junctions between the pedicles of macaque foveal cones. *Vision Research*, **32**, 1809–15.

van Hateren, J.H. (1992) Theoretical predictions of spatiotemporal receptive fields of fly LMCs and experimental validation. *Journal of Comparative Physiology*, **A171**, 157–70.

Vaney, D.I. (1993) The coupling pattern of axon-bearing horizontal cells in the mammalian retina. *Proceedings of the Royal Society of London B*, **252**, 93–101.

Vaney, D.I., Young, H.M. and Gynther, I.C. (1991) The rod circuit in the rabbit retina. *Visual Neuroscience*, **7**, 141–54.

Vardi, N. and Auerbach, P. (1994) Two forms of glutamic acid decarboxylase localize differently in cat retinal neurons. *Journal of Comparative Neurology*, **351**, 374–84.

Vardi, N. and Sterling, P. (1994) Subcellular localization of $GABA_A$ receptor on bipolar cells in macaque and human retina. *Vision Research*, **34**, 1235–46.

Vardi, N., Masarachia, P. and Sterling, P. (1992) Immunoreactivity to $GABA_A$ receptor in the outer plexiform layer of the cat retina. *Journal of Comparative Neurology*, **320**, 394–7.

Vardi, N., Matesic, D.F., Manning, D.R. *et al.* (1993) Identification of a G-protein in depolarizing bipolar cells. *Visual Neuroscience*, **10**, 473–8.

Vardi, N., Kaufman, D.L. and Sterling, P. (1994) Horizontal cells in cat and monkey retina express different isoforms of glutamic acid decarboxylase. *Visual Neuroscience*, **11**, 135–42.

von Gersdorff, H. and Matthews, G. (1994) Dynamics of synaptic vesicle fusion and membrane retrieval in synaptic terminals. *Nature*, **367**, 735–9.

Wässle, H. (1971) Optical quality of the cat eye. *Vision Research*, **11**, 995–1006.

Wässle, H., Boycott, B.B. (1991) Functional architecture of the mammalian retina. *Physiological Reviews*, **71**, 447–80.

Wässle, H., and Chun, M.H. (1989) GABA-like immunoreactivity in the cat retina: Light microscopy. *Journal of Comparative Neurology*, **279**, 43–54.

Wässle, H., and Riemann, H.J. (1978) The mosaic of nerve cells in the mammalian retina. *Proceedings of the Royal Society of London B*, **200**, 441–61.

Wässle, H., Schäfer–Trenkler, I. and Voigt, T. (1986) Analysis of a glycinergic inhibitory pathway in the cat retina. *Journal of Neuroscience*, **6**, 594–604.

Wässle, H. and Boycott, B.B. and Röhrenbeck, J. (1989) Horizontal cells in the monkey retina: cone connections and dendritic network. *European Journal of Neuroscience*, **1**, 421–35.

West, R.W. (1976) Light and electron microscopy of the ground squirrel retina: functional considerations. *Journal of Comparative Neurology*, **168**, 355–78.

Williams, D.R. (1992) Photoreceptor sampling and aliasing in human vision, in *Tutorials in Optics* (ed. D.T. Moore), Optical Society of America, Rochester, NY, pp.15–27.

Williams, R.W., Cavada, C. and Reinoso–Suárez, F. (1993) Rapid evolution of the visual system: a cellular assay of the retina and dorsal lateral geniculate nucleus of the Spanish wildcat and the domestic cat. *Journal of Neuroscience*, **13**, 208–228.

Wu, S.M. (1994) Synaptic transmission in the outer retina. *Annual Review of Physiology*, **56**,

Young, H.M. and Vaney, D.I. (1991) Rod-signal interneurons in the rabbit retina. 1. Rod bipolar cells. *Journal of Comparative Neurology*, **310**, 139–53.

Neurotransmitter release from bipolar cells

MASAO TACHIBANA

14.1 INTRODUCTION

Synaptic transmission is a key factor for the understanding of higher functions in the central nervous system (CNS). It has been demonstrated in the squid giant synapse that neurotransmitter is released upon activation of the Ca^{2+} current (Katz and Miledi, 1969; Llinas *et al.*, 1981; Augustine *et al.*, 1985; reviewed in Augustine *et al.*, 1987). Synaptic transmission in the vertebrate CNS seems to be more complicated than that at the squid giant synapse; CNS neurons are embedded in a complex network and receive multiple inputs (which may interact with each other via complicated intracellular processes) and the efficiency of the synaptic transmission is modified in various time ranges. Intensive studies have been carried out on the post-synaptic events associated with receptors, ion channels and second messenger cascades. However, the investigation of the presynaptic events, such as the Ca^{2+}-neurotransmitter release coupling and the modulation of neurotransmitter release, has been hindered largely due to the small size of presynaptic terminals.

One type of bipolar cell in the cyprinid retina (ON-type cells with inputs from rods and cones; Chapter 11) has an extraordinarily large axon terminal (approximately 10 μm in diameter, see Ishida *et al.*, 1980; Saito and Kujiraoka, 1982). Using this preparation we have been investigating how the excitation triggers the release of neurotransmitter. This chapter summarizes progress in this field (Tachibana and Okada, 1991; Okada *et al.*, 1992; Kobayashi *et al.*, 1993; Tachibana *et al.*, 1993).

14.2 PROPERTIES OF THE Ca^{2+} CURRENT IN GOLDFISH ON-TYPE BIPOLAR CELLS

Multiple subtypes of Ca^{2+} currents have been reported in various cells (reviewed in Bean, 1989; Carbone and Swandulla, 1989). These Ca^{2+} currents are different in many aspects, such as their voltage dependence, single-channel conductance, activation and inactivation kinetics, and sensitivity to pharmacological compounds (Hagiwara *et al.*, 1975; Carbone and Lux, 1984; Armstrong and Matteson, 1985; Nowycky *et al.*, 1985; Fox *et al.*, 1987; Kasai *et al.*, 1987). It has been suggested that each subtype of Ca^{2+} current may be related to a specific physiological function (Miller, 1987; Hirning *et al.*, 1988).

Neurobiology and Clinical Aspects of the Outer Retina
Edited by M.B.A. Djamgoz, S.N. Archer and S. Vallerga
Published in 1995 by Chapman & Hall, London
ISBN 0 412 60080 3

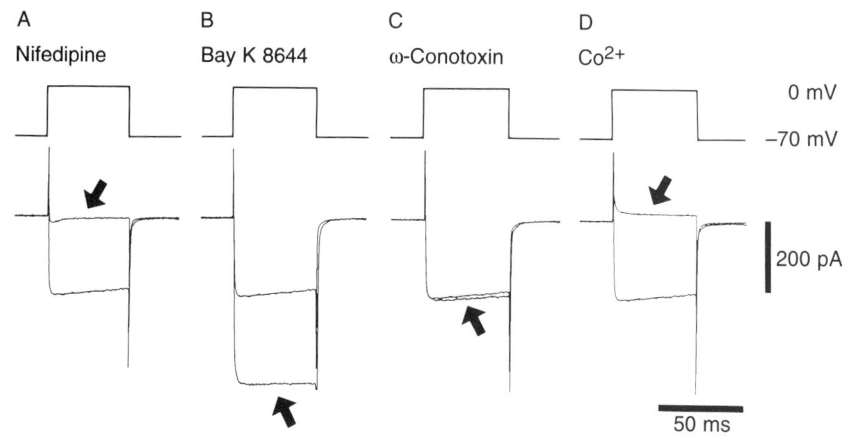

Figure 14.1 Effects of pharmacological agents on the Ca^{2+} current of goldfish ON-type bipolar cells. Superimposed current traces were obtained in the absence and presence (trace with arrow) of the pharmacological agent. A, 100 μM nifedipine; B, 0.2 μM Bay K 8644, C, 1 μM ω-conotoxin, D, 3.5 mM Co^{2+}. Each agent was applied from the Y-tube microflow system. The K^+ currents were blocked by Cs^+ and EGTA introduced via a patch pipette and by extracellular TEA^+. All records were obtained from the same bipolar cell.

The subtype(s) of Ca^{2+} currents present in goldfish ON-type bipolar cells was investigated first.

Bipolar cells were isolated enzymatically from the goldfish retina (Tachibana, 1981; Tachibana and Okada, 1991). The cells were voltage-clamped by a patch pipette in the whole-cell recording configuration (Hamill *et al.*, 1981). The voltage-activated and Ca^{2+}-activated K^+ currents (Kaneko and Tachibana, 1985) were almost completely suppressed by Cs^+ (120 mM) and EGTA (5 mM) introduced via the recording pipette. Various pharmacological agents were applied from a Y-tube microflow system (Suzuki *et al.*, 1990).

To examine whether goldfish ON-type bipolar cells were spatially equipotential, membrane potentials were recorded simultaneously from the axon terminal and the cell body of a single cell by using two patch pipettes in the whole-cell recording configuration (Tachibana *et al.*, 1993). There was no significant difference between the membrane potentials recorded from these two sites, which suggested that the space-clamp condition was well satisfied. In the following experiments the patch pipette was positioned either at the axon terminal or at the cell body.

When the membrane potential was depolarized from the holding potential of −70 mV to potentials more positive than −50 mV, an inward current was induced (Figure 14.1, traces without arrow). This current decayed very slowly during the depolarization. The current amplitude increased as the membrane potential was shifted up to approximately −10 mV, and decreased with larger depolarization. No inward current was detected when the membrane potential was depolarized from the holding potential of −100 mV to potentials below −50 mV.

The amplitude of the inward current was reduced in low extracellular Ca^{2+} concentration ($[Ca^{2+}]_o$), and was augmented in high $[Ca^{2+}]_o$. Depolarizations evoked the inward current when the extracellular Ca^{2+} was replaced with equimolar Sr^{2+} or Ba^{2+}. The order and relative value of the current amplitudes were: Ba^{2+} (1.9) $>$ Sr^{2+} (1.4) $>$ Ca^{2+} (1.0).

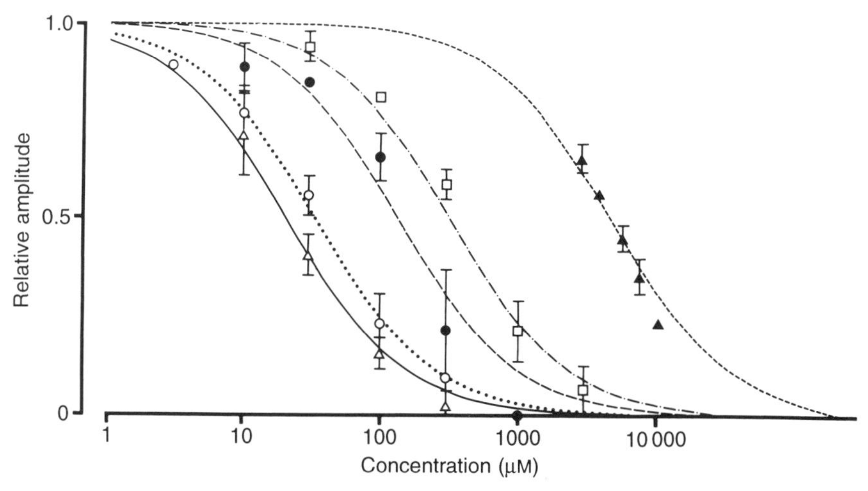

Figure 14.2 Blocking effects of various divalent cations on the Ca^{2+} current of goldfish ON-type bipolar cells. Dose–response curves were obtained in the presence of Cd^{2+} (open triangles), Ni^{2+} (open circles), Co^{2+} (filled circles), Mn^{2+} (open squares) and Mg^{2+} (filled triangles). Each point is the average ± SD (obtained from 3 to 5 cells) of the current amplitude recorded at different concentrations of the blocking ions and is normalized with respect to the control condition. The amplitude of the Ca^{2+} current was measured just before the cessation of 500-ms voltage pulses (from −70 mV to −20 mV). Curves were fitted by eye.

The inward current was almost completely blocked by various divalent cations, such as Co^{2+} (Figure 14.1D) and Cd^{2+} (Figure 14.9). The order of potency in blocking, estimated by the concentration that caused half-inhibition, was: Cd^{2+} (23 μM) > Ni^{2+} (50 μM) > Co^{2+} (120 μM) > Mn^{2+} (420 μM) > Mg^{2+} (5.3 mM) (Figure 14.2). These results demonstrated that the inward current after leakage subtraction was carried mainly by Ca^{2+} through Ca^{2+} channels. The underlying Ca^{2+} channel was of the high-voltage activated type.

In order to identify the Ca^{2+} current subtype, subtype-specific pharmacological agents were applied to goldfish ON-type bipolar cells. Dihydropyridine compounds were very effective in modifying the Ca^{2+} current. The Ca^{2+} current was strongly potentiated by Bay K 8644 (Figure 14.1B). On the other hand, nifedipine (Figure 14.1A) and nicardipine (Figure 14.3) suppressed the Ca^{2+} current almost completely. A lower concentration of these agents was required for the suppression of the Ca^{2+} current when the holding potential was less negative (Figure 14.3). Upon prolonged washout of these blocking agents, the Ca^{2+} current recovered partially. On the other hand, ω-conotoxin did not affect the Ca^{2+} current (Figure 14.1C).

This series of experiments demonstrated that the Ca^{2+} current of ON-type bipolar cells dissociated from the goldfish retina was of the high-voltage activated, dihydropyridine-sensitive type, the so-called 'L'-type (Nowycky *et al.*, 1985). There was no ω-conotoxin-sensitive ('N' type) component. In this preparation we did not observe the transient, low-voltage activated ('T'-type) Ca^{2+} current, which has been reported in bipolar cells dissociated from the mouse retina (Kaneko *et al.*, 1989). A similar result in goldfish ON-

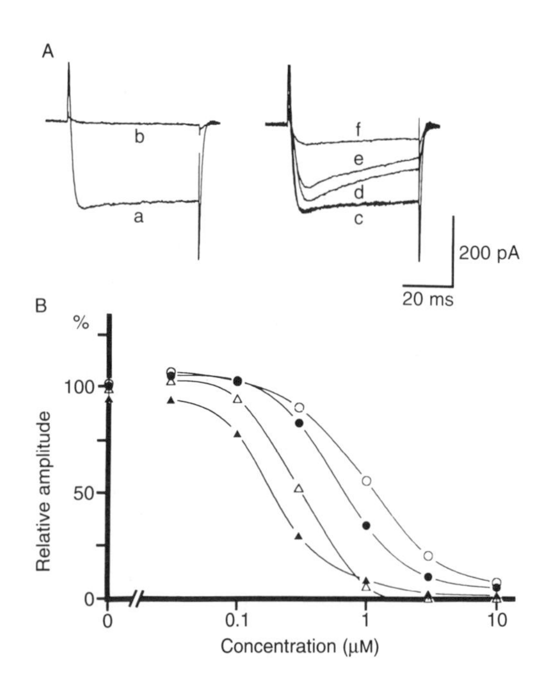

Figure 14.3 Holding-potential dependent blockage of the Ca^{2+} current by nicardipine. A, Superimposed current traces (left) were obtained in the absence (a) or in the presence of 10 μM nicardipine (b). The membrane potential of a goldfish ON-type bipolar cell was shifted from −50 mV to −20 mV for 50 ms. Superimposed current traces (right) were recorded in the absence (coinciding traces in c; holding potentials at −90 mV, −70 mV and −50 mV) and in the presence of 1 μM nicardipine at various holding potentials (d, −90 mV; e, −70 mV, f, −50 mV). The membrane potential was depolarized to −20 mV for 50 ms. B, Effect of holding potentials on the dose–response curve of the nicardipine blockage. Holding potentials were −50 mV (filled triangles), −60 mV (open triangles), −70 mV (filled circles) and −90 mV (open circles). Current amplitudes were measured immediately before the termination of pulses. Data were obtained from eight bipolar cells. Curves fitted by eye.

type bipolar cells was obtained by Heidelberger and Matthews (1992).

14.3 DISTRIBUTION OF Ca^{2+} CHANNELS OVER THE PLASMA MEMBRANE OF GOLDFISH ON-TYPE BIPOLAR CELLS

It is known that Ca^{2+} entering through Ca^{2+} channels can perform various physiological functions. If a consequence of the Ca^{2+} current of ON-type bipolar cells is to trigger the neurotransmitter release, at least a part of the Ca^{2+} current should flow into the presynaptic terminal. Ca^{2+} entering the cell body would not participate in the fast release of the neurotransmitter because it would take tens of seconds for Ca^{2+} to reach the presynaptic terminal through the axon fiber (ca. 30 μm in length) by diffusion. To gain insight into the physiological role of Ca^{2+} in goldfish ON-type bipolar cells, we measured the spatial and temporal changes of intracellular free Ca^{2+} concentration ($[Ca^{2+}]_i$) upon activation of the Ca^{2+} current (Tachibana *et al.*, 1993).

Bipolar cells were loaded with a fluorescent Ca^{2+} indicator, Fura-2 (Grynkiewicz *et al.*, 1985). The Fura-2 loaded cell was illuminated with UV lights (excitation wavelengths; 340 nm and 360 nm), and the emitted fluorescence (510 nm) was detected by a high sensitivity video camera (a SIT camera or an image-intensified CCD camera). The video images were fed into the image processor and were analyzed later (ARGUS system, Hamamatsu Photonics).

A puff of a high-K^+ solution induced a rapid and large increase of $[Ca^{2+}]_i$ in the axon terminal of the unclamped bipolar cell, whereas delayed, small increase of $[Ca^{2+}]_i$ was usually observed in the cell body. Intracellular free Ca^{2+} appeared to spread from the axon terminal toward the cell body through the axon fiber. Soon after the termination of the puff of the high-K^+ solution, $[Ca^{2+}]_i$ in the axon terminal started declining, whereas that in the cell body continued increasing for a while and then decreased more slowly.

When a Ca^{2+} ionophore (ionomycine or 4-bromo-calcium ionophore A 23187) was applied to the Fura-2 loaded bipolar cell, $[Ca2^{+}]_i$ in the cell body increased similarly to that in the axon terminal, suggesting that the Ca^{2+} buffering power would not be significantly different between two regions. In conclusion, the Ca^{2+} channels seemed to be highly localized to the axon terminals of ON-type bipolar cells.

In order to control the amount of Ca^{2+} entering through Ca^{2+} channels, the Ca^{2+} current and $[Ca^{2+}]_i$ were simultaneously monitored under voltage clamp. The pipette solution contained no Ca^{2+} buffer except 100 μM Fura-2 in order to minimize the disturbance of endogenous Ca^{2+} buffering power. A depolarizing pulse (< 500 ms in duration) activated the Ca^{2+} current and evoked a rapid and large increase of $[Ca^{2+}]_i$, which was usually restricted to the axon terminal (Figure 14.4). $[Ca^{2+}]_i$ reached a peak value near the pulse offset, and then decreased to the basal level in tens of seconds. A prolonged depolarization could evoke a slow and small increase of $[Ca^{2+}]_i$ in the cell body following a rapid and large increase in the axon terminal.

Application of 10 μM nicardipine, a light-resistant derivative of dihydropyridine, suppressed both the Ca^{2+} current and the increase of $[Ca^{2+}]_i$. Therefore, the Ca^{2+} entry occurred through dihydropyridine-sensitive Ca^{2+} channels.

The electrophysiological experiments also

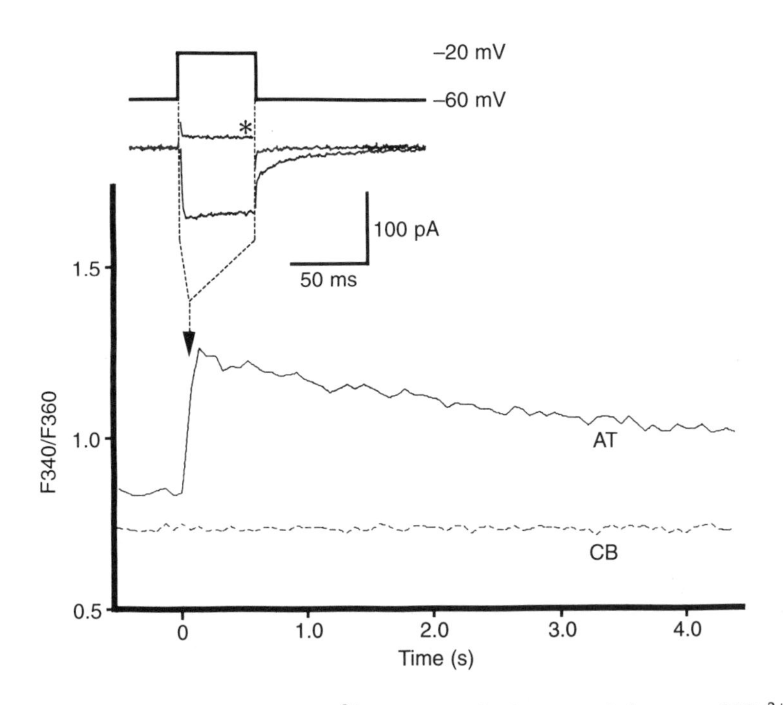

Figure 14.4 Simultaneous recordings of the Ca^{2+} current and of temporal changes of $[Ca^{2+}]_i$ in a voltage clamped bipolar cell. The membrane potential (top trace) was shifted from −60 mV to −20 mV for 50 ms. Superimposed current recordings (middle traces) were obtained in the absence and presence (*) of 10 μM nicardipine and are displayed on an expanded time scale. The main panel indicates the averaged fluorescence ratio (F340/F360) obtained in the control condition. Trace AT illustrates the data from the axon terminal region, and trace CB from the cell body region.

supported the idea that dihydropyridine-sensitive Ca^{2+} channels are localized to the axon terminal region. The Ca^{2+} current recorded from the bipolar cell, whose axon and axon terminal had been lost during the dissociation, was extremely small in maximal amplitude (< 2 pA). On the other hand, the maximal amplitude was more than 100 pA when the Ca^{2+} current was recorded from the axon terminal, which had been separated from the cell body during the dissociation procedure.

Heidelberger and Matthews (1992) estimated that approximately 40% of the total Ca^{2+} current would flow into the cell body of goldfish ON-type bipolar cells. The contradiction between our interpretation and theirs seems to be largely due to the differences in the experimental design; they measured the changes of $[Ca^{2+}]_i$, either in the axon terminal or in the cell body by a photomultiplier tube, and the cell was depolarized by applying a high-K^+ solution for tens of seconds without monitoring the Ca^{2+} current. It would be difficult to determine whether the increase of $[Ca^{2+}]_i$ in the cell body induced by a long depolarization was due to the influx of Ca^{2+} through Ca^{2+} channels located in the cell body, or due to the diffusion of intracellular free Ca^{2+} from the axon terminal.

14.4 REGULATION OF INTRACELLULAR FREE Ca^{2+} CONCENTRATION IN THE AXON TERMINALS OF GOLDFISH ON-TYPE BIPOLAR CELLS

The Ca^{2+} imaging of ON-type bipolar cells by a high-sensitivity video camera revealed the localization of dihydropyridine-sensitive Ca^{2+} channels at the axon terminal region. Since a satisfactory space-clamp condition had been achieved (section 14.2), the Ca^{2+} current recorded in the whole-cell clamp mode was actually the presynaptic Ca^{2+} current itself. The relation between the amount of Ca^{2+} influx and the increment of $[Ca^{2+}]_i$ has been analyzed quantitatively, and the mechanism of $[Ca^{2+}]_i$ regulation in the presynaptic terminal considered (Okada *et al.*, 1992; Tachibana *et al.*, 1993; Kobayashi *et al.*, 1993; Kobayashi and Tachibana, 1995).

Bipolar cells were voltage-clamped in the whole-cell recording configuration. The pipette solution contained 120 mM Cs^+, 100 μM Fura-2 and 50 mM HEPES. No additional Ca^{2+} buffers were included. The high concentration of HEPES was required to stabilize the intracellular pH during the extracellular application of a high pH solution. The averaged $[Ca^{2+}]_i$ in the axon terminal region was measured by a photomultiplier tube with the time resolution of 10 ms (Olympus OSP-3 system).

$[Ca^{2+}]_i$ in the axon terminal was approximately 60 nM at the holding potential of −70 mV. When the Ca^{2+} current was activated by a depolarizing pulse (from −70 mV to −10 mV), the Ca^{2+} transient was evoked. $[Ca^{2+}]_i$ increased rapidly without any discernible delay upon application of the pulse. For a short depolarizing pulse (ca. 50 ms), $[Ca^{2+}]_i$ increased at nearly a constant rate during the depolarization. $[Ca^{2+}]_i$ reached a peak value at the termination of the depolarization and then started declining monotonously. $[Ca^{2+}]_i$ recovered to the basal level in several seconds. The recovery phase of the Ca^{2+} transient could not be fitted by a single- or double-exponential function in most cases.

For a long depolarizing pulse, the rate of increase in $[Ca^{2+}]_i$ during the depolarization was almost constant initially and then slowed down with time. However, $[Ca^{2+}]_i$ continued increasing and never decreased as long as the Ca^{2+} current was activated by the depolarization. After the cessation of the pulse, $[Ca^{2+}]_i$ stayed at a plateau for a while and then declined to the basal level in a few tens of seconds.

The amount of the Ca^{2+} influx was calculated based on the assumption that Ca^{2+}

moving into the axon terminal during the pulse would distribute evenly in the spherical axon terminal. This value was plotted against the peak increment of the Ca^{2+} transient ($\Delta[Ca^{2+}]_i$), which was defined as the difference between the peak value of the Ca^{2+} transient and the basal level. The relation was almost linear when the amount of the Ca^{2+} influx was less than approximately 200 μM. Only a subtle fraction (< 1%) of the Ca^{2+} that entered the axon terminal could remain free in this range. As the amount of Ca^{2+} influx became larger, $\Delta[Ca^{2+}]_i$ did not increase proportionally any more and tended to reach a fixed level (approximately 1 μM). These results suggested that a powerful Ca^{2+} buffering system must control $[Ca^{2+}]_i$ in the axon terminal.

In various cells, a large gradient of Ca^{2+} across the plasma membrane is maintained by at least two kinds of Ca^{2+} transporters, the Na^+/Ca^{2+} exchanger and the Ca^{2+} pump (Dipolo and Beauge, 1983; Mcburney and Neering, 1987; Thayer and Miller, 1990; Tatsumi and Katayama, 1993). Extrusion of Ca^{2+} by the Na^+/Ca^{2+} exchanger is driven by the inward movement of Na^+ following its electrochemical gradient. The Ca^{2+} pump extrudes Ca^{2+} by hydrolyzing ATP. We investigated whether these mechanics contributed to the regulation of $[Ca^{2+}]_i$ in the axon terminals of goldfish ON-type bipolar cells.

Since extracellular Na^+ is essential to the operation of the Na^+/Ca^{2+} exchanger, we examined the effects of Na^+ removal on the Ca^{2+} transient. When extracellular Na^+ was replaced totally with Li^+ or TEA^+, the basal level of $[Ca^{2+}]_i$ remained constant. The Ca^{2+} transient evoked by a small amount of Ca^{2+} influx (i.e. by a short depolarizing pulse) did not change either in amplitude or in the time course. However, a notable change was observed when the Ca^{2+} transient was evoked by a large amount of the Ca^{2+} influx. The plateau phase, which appeared just after the offset of a long depolarizing pulse, was prolonged and it took much longer for $[Ca^{2+}]_i$ to recover to the basal level, although $\Delta[Ca^{2+}]_i$ was not affected at all by the removal of extracellular Na^+. These results suggested that the Na^+/Ca^{2+} exchanger contributes to the regulation of $[Ca^{2+}]_i$ in the axon terminals of goldfish ON-type bipolar cells once $[Ca^{2+}]_i$ reaches a relatively high level.

It has been reported that the plasma membrane Ca^{2+} pump transports H^+ in exchange for Ca^{2+} (Smallwood *et al.*, 1983), and its operation is inhibited by the reduction of extracellular H^+ concentration (i.e. by raising the extracellular pH) (Dipolo and Beauge, 1982). We examined whether $[Ca^{2+}]_i$ in the axon terminals of goldfish ON-type bipolar cells was affected by the elevation of extracellular pH.

When the extracellular pH was increased from 7.4 to 8.4, the basal level of $[Ca^{2+}]_i$ increased slightly but consistently. Upon application of a short depolarizing pulse, $[Ca^{2+}]_i$ jumped to a new level and, even after the termination of the voltage pulse, the elevated $[Ca^{2+}]_i$ did not decline obviously. For a long depolarizing pulse the rate of recovery from the increased $[Ca^{2+}]_i$ in the high pH solution was initially nearly as fast as that in the control solution (pH 7.4), but slowed down considerably as $[Ca^{2+}]_i$ was partially recovered. The treatment with the high-pH solution augmented the amplitude of the Ca^{2+} current but the relation between the amount of Ca^{2+} influx and $\Delta[Ca^{2+}]_i$ was not affected. These results indicated that the plasma membrane Ca^{2+} pump contributes to the extrusion of Ca^{2+} from the axon terminals of ON-type bipolar cells. The plasma membrane Ca^{2+} pump seemed to operate in the $[Ca^{2+}]_i$ range lower than that for the Na^+/Ca^{2+} exchanger.

We did not find clear evidence for the contribution of intracellular organelles to the regulation of $[Ca^{2+}]_i$ in the axon terminal. The peak amplitude of the Ca^{2+} transient may be determined by fast Ca^{2+} buffering substances with a high cooperativity, such as Ca^{2+} binding proteins.

14.5 CANDIDATES FOR THE NEUROTRANSMITTER OF BIPOLAR CELLS

In the axon terminals of ON-type bipolar cells a precisely arranged array of synaptic vesicles surrounds the synaptic ribbon (Kaneko *et al.*, 1980). We wished to know whether the increase of $[Ca^{2+}]_i$ in the axon terminal would actually trigger release of the neurotransmitter, and what kind of neurotransmitter is released. In this section, candidates for the neurotransmitter of bipolar cells are reviewed briefly.

Some substances have been proposed as the neurotransmitter of bipolar cells, such as glutamate, aspartate, homocysteic acid and 5-hydroxytryptamine (5-HT or serotonin). The former three substances are amino acids that interact with glutamate receptors, whereas the latter interacts with 5-HT receptors.

Several lines of evidence suggest that glutamate may be the neurotransmitter of bipolar cells. Glutamate immunoreactivity was high in bipolar cells among various retinal cells (Ehinger *et al.*, 1988; Ehinger, 1989; Marc *et al.*, 1990). The enzyme aminotransferase, which converts aspartate into glutamate, was included in bipolar cells (Mosinger and Altschuler, 1985). Retinal ganglion and amacrine cells, postsynaptic to bipolar cells, responded well to exogenously applied glutamate (Kato *et al.*, 1985; Barnes and Werblin, 1987; Aizenman *et al.*, 1988). The synaptic transmission from bipolar cells to ganglion cells was modified by the application of agonists and antagonists of glutamate (Slaughter and Miller, 1983; Bloomfield and Dowling, 1985; Lukasiewicz and McReynolds, 1985).

However, this evidence does not reject the possibility that the authentic neurotransmitter of bipolar cells may be aspartate. The presence of glutamate in bipolar cells was demonstrated by glutamate immunoreactivity but the role of aspartate has not yet been investigated in bipolar cells. It is very difficult to distinguish pharmacologically between glutamate and aspartate because both substances can interact with excitatory amino acid receptors.

Homocysteic acid has also been suggested to be a neurotransmitter of bipolar cells in the rabbit retina (Neal and Cunningham, 1989, 1992). Application of homocysteic acid resulted in the release of acetylcholine (ACh) from amacrine cells. *N*-Methyl-D-aspartic acid (NMDA) suppressed the ACh release induced by homocysteic acid or light stimulation, but did not affect the ACh release evoked by glutamate or aspartate. Homocysteic acid was present in the rabbit retina (Neal and Cunningham, 1989). However, it has not yet been demonstrated that homocysteic acid is contained in bipolar cells and that this substance is released from bipolar cells in a Ca^{2+}-dependent manner.

In the turtle retina, immunocytochemical techniques have revealed that a subpopulation of bipolar cells contains 5-HT (Weiler and Schutte, 1985a, b; Schutte and Weiler, 1987). Based on the morphological characteristics, these bipolar cells are assumed to be the OFF-type (Weiler and Schutte, 1985a; Hurd and Eldred, 1993). The intracellular concentration of 5-HT in bipolar cells was increased by preincubating the retina with a medium containing 5-HT, and decreased in a Ca^{2+}-dependent manner when bipolar cells were depolarized either by a high-K^+ solution or by kainate.

In the retinas of *Xenopus* and *Bufo*, bipolar cells were able to accumulate exogenous 5-HT, but did not synthesize 5-HT, perhaps due to the lack of 5-HT synthesizing enzyme (Zhu *et al.*, 1992). The accumulation of 5-HT in the bipolar cells of the *Xenopus* retina was suppressed by kainate, suggesting that these bipolar cells may be the OFF-type (Schutte and Witkovsky, 1990). In the goldfish retina, a subset of bipolar cells (OFF-type cells) accumulated 5-HT but did not display 5-HT-like immunoreactivity (Marc *et al.*, 1988).

It is probable that OFF-type bipolar cells in various species may use 5-HT as the neurotransmitter. However, it is necessary to show that 5-HT is synthesized and released from these bipolar cells, and that this substance has functional effects on postsynaptic neurons.

14.6 CATFISH CONE-TYPE HORIZONTAL CELLS AS A GLUTAMATE PROBE

As mentioned in the previous section, glutamate seems to be the leading candidate for ON-type bipolar cells. To investigate whether glutamate is released from goldfish ON-type bipolar cells in response to the activation of the dihydropyridine-sensitive Ca^{2+} current (section 14.2), we adopted an electrophysiological bioassay technique (Hume *et al.*, 1983; Young and Poo, 1983; Meriney *et al.*, 1989; Schwartz, 1987; Copenhagen and Jahr, 1989). A small amount of glutamate released from a single neuron could produce an electrophysiological response in another neuron containing a high density of glutamate receptors. Horizontal cells are second order neurons in the retina and are sensitive to glutamate, which is also the leading candidate for the neurotransmitter of photoreceptors (Chapter 6). Thus, we examined whether horizontal cells are suitable for the electrophysiological detection of glutamate.

Horizontal cells were isolated enzymatically from the retinas of goldfish and catfish (Tachibana, 1981). A horizontal cell was voltage-clamped by a patch pipette in the whole-cell recording configuration and various pharmacological substances were applied by a Y-tube microflow system (Tachibana and Okada, 1991).

Cone-type horizontal cells dissociated from the retinas of catfish and goldfish responded to glutamate, catfish horizontal cells being more sensitive to glutamate than those from goldfish. The dose that produced a half-maximal response was approximately 5 μM for catfish horizontal cells and approximately 30 μM for goldfish horizontal cells, the Hill coefficients being close to 1 and 2, respectively. Submicromolar concentrations of glutamate produced detectable responses in catfish horizontal cells but not in goldfish horizontal cells.

Properties of glutamate-induced responses were examined by applying various glutamate analogs. Horizontal cells were voltage-clamped at a positive potential (+30 mV or +40 mV) to avoid the blockage of NMDA receptor channels by extracellular Mg^{2+} at negative potentials (Nowak *et al.*, 1984). Catfish cone-type horizontal cells responded to all kinds of excitatory amino acids tested. The dose which produced a half-maximal response was 1.5 μM for quisqualate, 5 μM for glutamate, 30 μM for kainate, 35 μM for NMDA, 110 μM for aspartate and 135 μM for homocysteic acid. On the other hand, goldfish cone-type horizontal cells were sensitive to glutamate, quisqualate and kainate, but were insensitive to NMDA and aspartate. We concluded that catfish cone-type horizontal cells have both NMDA and non-NMDA receptors, whereas goldfish cone-type horizontal cells have only non-NMDA receptors.

It has been shown that the current through NMDA receptor channels is potentiated by glycine (Johnson and Ascher, 1987). Such potentiation could be profitable for detection of glutamate because larger responses would be more easily detected. When glycine was applied to catfish cone-type horizontal cells, the glutamate-induced current was obviously potentiated. Application of glycine alone did not produce any response. The amplitude of the glutamate-induced current was increased by glycine to approximately 150% at 0.1 μM and to approximately 180% at 10 μM (near maximal potentiation). The potentiation by glycine was also observed in the NMDA-induced current, but not in the quisqualate- or kainate-induced current. It is clear that glycine potentiated the current through NMDA receptor channels in catfish cone-type horizontal cells.

The voltage dependence of the glutamate-induced current in catfish cone-type horizontal cells was non-linear (Chapter 7). The reversal potential was close to 0 mV. The relation between the glutamate-induced current and the membrane potential was almost linear (or slightly outwardly rectifying) at potentials more positive than approximately −40 mV. However, when the membrane potential was shifted to more negative values than −40 mV, the glutamate-induced current decreased in amplitude contrary to the increased driving force. Thus the current versus voltage (I–V) relation was J-shaped. A similar J-shaped I–V relation was obtained when the current was induced by aspartate or NMDA, whereas a nearly linear I–V relation was obtained when the current was evoked by kainate or quisqualate. The non-linearity was due to the blockage of NMDA receptor channels by extracellular Mg^{2+} because the negative conductance region (< −40 mV) of the I–V curve disappeared after the removal of Mg^{2+} from the superfusate.

The glutamate-induced current in catfish cone-type horizontal cells was suppressed by various glutamate antagonists. The current induced by 5 μM glutamate was strongly suppressed by 50 μM D-2-amino-5-phosphonovaleric acid (APV), an antagonist specific to the NMDA receptor (Davies and Watkins, 1982), and the remaining current was abolished completely by 50 μM 6-cyano-7-nitroquinoxaline-2, 3-dione (CNQX), an antagonist specific to non-NMDA receptors (Honore *et al.*, 1988) (Figure 14.5A). The dose that was required to reduce the amplitude of the glutamate-induced current to half was 20 μM for APV, more than 100 μM for CNQX (Figure 14.5B) and 30 μM for kynurenic acid, a non-specific glutamate antagonist (Perkins and Stone, 1982). A high concentration of CNQX blocked the glutamate-induced current only partially, suggesting that the major glutamate receptors of catfish cone-type horizontal cells may be of the NMDA type.

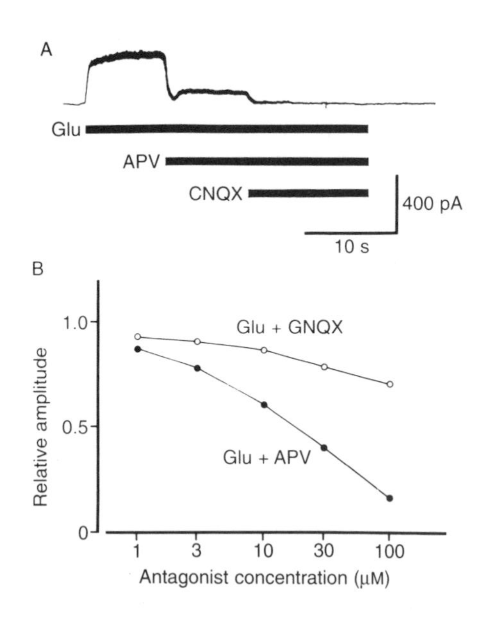

Figure 14.5 Effects of antagonists on the glutamate-induced responses in catfish cone-type horizontal cells. A, An outward current was evoked by the application of 5 μM glutamate in a horizontal cell maintained at +40 mV. The outward current was strongly suppressed by 50 μM APV. The remaining current was completely inhibited by the additional application of 50 μM CNQX. B, Dose-dependent blockage of the glutamate (5 μM)-induced current. Data obtained from three horizontal cells maintained at +30 mV.

The glutamate-induced current did not show obvious desensitization during continuous application of glutamate from the Y-tube microflow system (Figure 14.5A). Therefore, a temporal decrease in the response amplitude could be interpreted simply by the reduction of the local concentration of the released neurotransmitter.

We also examined the effects of other 'classical' neurotransmitters, such as ACh, dopamine, γ-aminobutyric acid (GABA), glycine, 5-HT, epinephrine, norepinephrine and ATP, on catfish horizontal cells (Tachibana and Okada, 1991). Among these substances only a high concentration of GABA produced a weak response. The relation between the GABA-induced current and the membrane potential was almost linear, and the reversal potential was close to 0 mV under our recording condition. The GABA-induced current was carried by Cl^- through $GABA_A$ receptors coupled with a Cl^- channel (Lasater *et al.*, 1984).

This series of experiments elucidated the conditions, whereby a small amount of glutamate could be detected electrophysiologically; catfish cone-type horizontal cells should be voltage-clamped at a positive membrane potential in the presence of 10 μM glycine. It must be kept in mind, however, that catfish cone-type horizontal cells respond not only to glutamate but also to aspartate, homocysteic acid and GABA, all of which would occur naturally in retinal tissues.

14.7 NEUROTRANSMITTER RELEASE FROM GOLDFISH ON-TYPE BIPOLAR CELLS

ON-type bipolar cells freshly dissociated from goldfish retina were dispersed into a culture dish, in which cone-type horizontal cells isolated from the catfish retina had been maintained in culture for one to six days (Tachibana and Okada, 1991). A bipolar cell was voltage-clamped at −60 mV in the whole-cell recording mode by a patch pipette positioned at the axon terminal, and was lifted up by manipulating the recording pipette. Then, a horizontal cell was voltage-clamped at +30 mV (or +40 mV) by another patch pipette. The bipolar cell was slowly lowered onto the horizontal cell until the axon terminal of the bipolar cell was slightly dimpled by the recording pipette to minimize the extracellular space between two cells. A solution containing 10 μM glycine was applied continuously to the cell pair to potentiate the NMDA receptor of the horizontal cell. In some cases, glutamate was applied ionophoretically from a microelectrode positioned near the horizontal cell after the bipolar cell was lifted up.

Upon application of a depolarizing pulse (from −60 mV to −20 mV for 500 ms) to the bipolar cell, the Ca^{2+} current was induced in the bipolar cell and an outward current was evoked in the apposed horizontal cell maintained at +40 mV (Figure 14.6A). The response recorded from the horizontal cell was approximately 20 pA in amplitude. Initiation of the horizontal cell response was usually observed a few milliseconds after the onset of the depolarization. Such delay suggests that the horizontal cell response may not be due to an electrical artifact caused by the voltage pulse. Hyperpolarizing pulses did not produce any responses in the horizontal cell. When the distance between the axon terminal of the bipolar cell and the horizontal cell was increased to more than 100 μm, the depolarization of the bipolar cell induced no response in the horizontal cell. Therefore, the response recorded from the horizontal cell was evoked by the neurotransmitter released from the axon terminal of the bipolar cell.

The reversal potential was examined to elucidate the ion-conductance mechanism of the neurotransmitter-induced response. Each time when the membrane potential of the apposed horizontal cell was shifted to a new potential level, the same depolarizing pulse was applied to the bipolar cell (Figure 14.7A). The neurotransmitter-induced current was outward at positive potentials. As the membrane potential became less positive, the outward current became smaller in ampli-

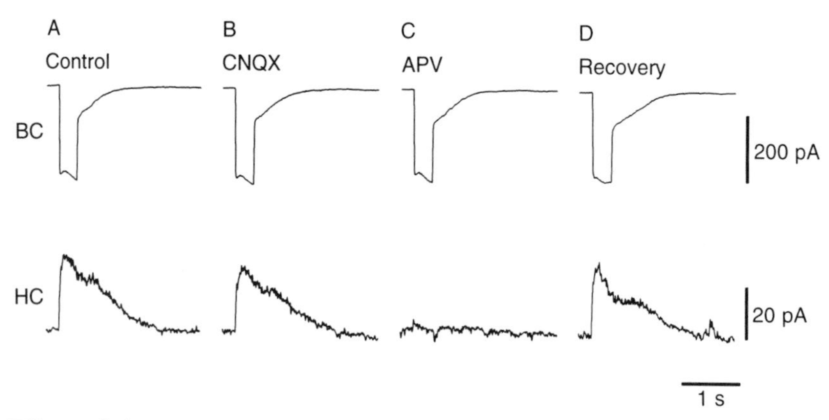

Figure 14.6 Effects of glutamate antagonists on the neurotransmitter-induced current recorded from a catfish cone-type horizontal cell. The axon terminal of a goldfish ON-type bipolar cell (BC) was closely apposed to a horizontal cell (HC). When BC was depolarized from −60 mV to −20 mV for 300 ms, the Ca^{2+} current was evoked in BC and the neurotransmitter-induced current was recorded from HC maintained at +40 mV (A). Simultaneous current recordings were then carried out in the presence of 50 μM CNQX (B) or 50 μM APV (C). After the washout of APV, the neurotransmitter-induced current recovered (D). All data were obtained from the same cell pair in the presence of 10 μM glycine.

tude. The neurotransmitter-induced current changed its polarity to inward at approximately 0 mV. At negative potentials, the current flowed inward. The relationship between the neurotransmitter-induced current and the membrane potential was nonlinear and J-shaped, suggesting that the neurotransmitter-induced current consisted mainly of the current through NMDA receptor channels. When 30% of the extracellular Na^+ was replaced with $choline^+$, the reversal potential shifted to −15 mV. This amount of shift was approximately half the value expected from the Nernst equation (−32 mV), suggesting that the neurotransmitter opens non-selective cation channels of the catfish horizontal cell.

After the bipolar cell was lifted up, glutamate was applied ionophoretically to the same horizontal cell (Figure 14.7B). Properties of the glutamate-induced response were almost identical to those of the neurotransmitter-induced response. The reversal potential of the glutamate-induced response was close to 0 mV in the control superfusate (Figure 14.7B), and shifted to a negative potential (ca. −15 mV) in the 30% Na^+ solution. The relation between the glutamate-induced cur-

Figure 14.7 Voltage-dependence of the neurotransmitter-induced current (A) and the glutamate-induced current (B). The membrane potential of the catfish cone-type horizontal cell was shifted to various values indicated on the left side of each current trace. Data were obtained in the presence of 10 μM glycine. A, The neurotransmitter was released from the apposed bipolar cell, which was depolarized from −70 mV to −20 mV for 500 ms. Evoked Ca^{2+} currents showed little run-down during the experiment (not illustrated). B, L-Glutamate was applied to the same horizontal cell by ionophoresis (intensity, 30 nA; duration, 100 ms) after the bipolar cell was lifted up. Transient deflections of each current trace were due to the artifact caused by ionophoretic current pulse.

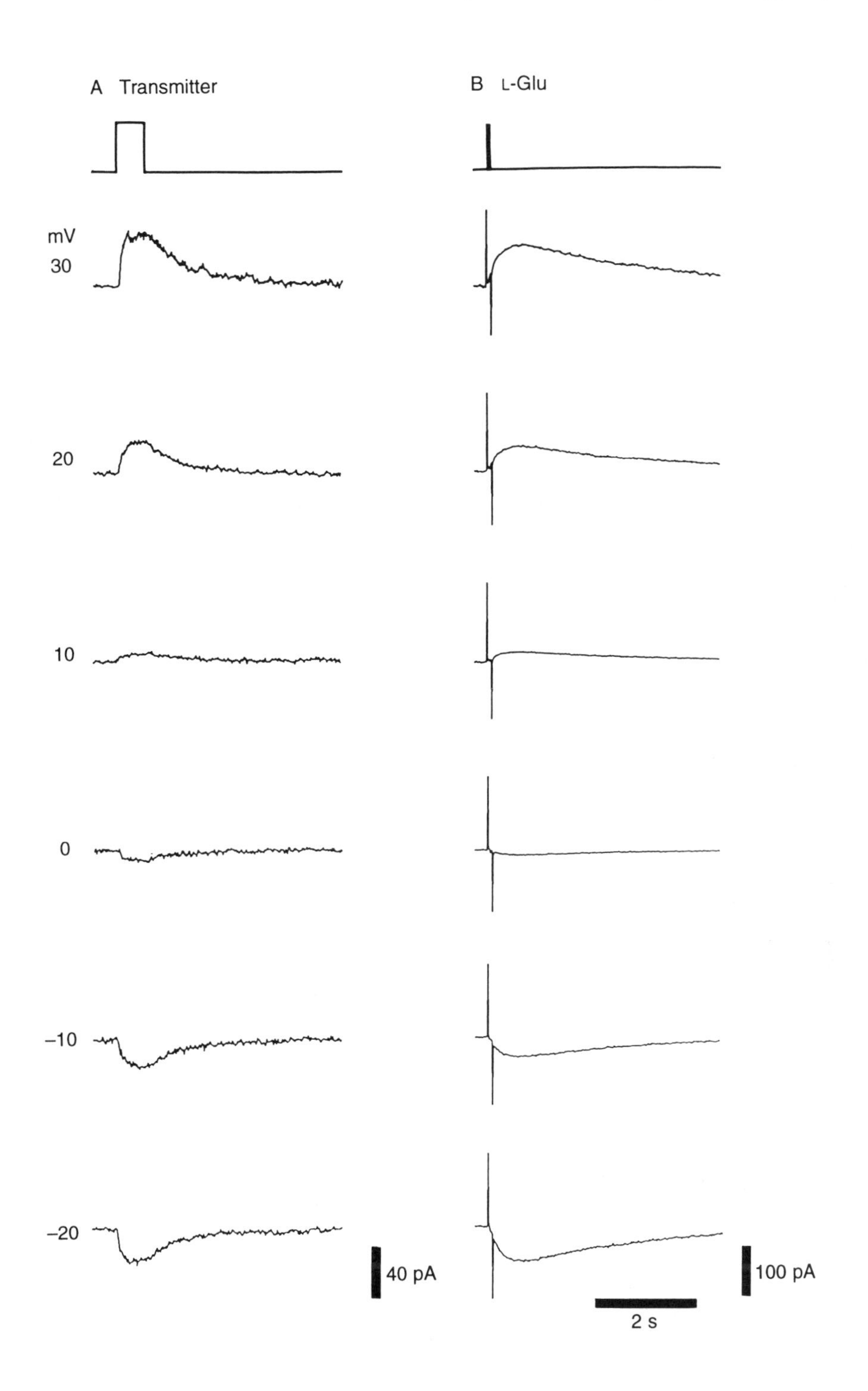
A Transmitter
B L-Glu
mV
30
20
10
0
−10
−20
40 pA
100 pA
2 s

rent and the membrane potential was nonlinear and J-shaped. These observations indicate that the neurotransmitter released from goldfish ON-type bipolar cells is glutamate or a closely related substance.

Activation of the NMDA receptor by the neurotransmitter was also suggested by the following observations. The amplitude of the neurotransmitter-induced current was increased in the presence of glycine to approximately 150% of the control value recorded in the absence of glycine. The neurotransmitter-induced current was suppressed strongly by 50 μM APV, an antagonist specific to NMDA receptors (Figure 14.6C). This concentration of APV suppressed the glutamate-induced current in catfish horizontal cells to approximately 20% (Figure 14.5).

In order to examine whether the non-NMDA receptor contributed to the production of the neurotransmitter-induced current in catfish cone-type horizontal cells, CNQX, an antagonist specific to non-NMDA receptors, was applied to the cell pair. The neurotransmitter-induced current was suppressed only slightly at a concentration of 50 μM (Figure 14.6B), which could reduce the glutamate-induced current to approximately 70% (Figure 14.5). Therefore, the neurotransmitter released from goldfish ON-type bipolar cells was mainly detected by the NMDA receptor in catfish cone-type horizontal cells. In conclusion, it is highly likely that the ON-bipolar cell neurotransmitter substance is glutamate, although we cannot reject the possibility that the authentic neurotransmitter may be aspartate or homocysteic acid.

Finally, we investigated the requirement of Ca^{2+} for the release of glutamate-like substance from goldfish ON-type bipolar cells. When the Ca^{2+} current of the bipolar cell was blocked by Co^{2+} or by Cd^{2+}, the neurotransmitter-induced current disappeared (Figure 14.9). Under this condition the direct application of glutamate evoked responses in the horizontal cell. The influx of Ca^{2+} into the bipolar cell through Ca^{2+} channels seems to be essential for the release of the glutamate-like substance.

Application of nifedipine suppressed both the Ca^{2+} current of the bipolar cell and the neurotransmitter-induced current of the apposed horizontal cell (Figure 14.8A). Iono-phoretically applied glutamate could evoke responses in the horizontal cell in the presence of nifedipine. On the other hand, ω-conotoxin did not affect either current (Figure 14.8B). It is obvious that the activation of the dihydropyridine-sensitive Ca^{2+} current mediates the release of glutamate-like substance from goldfish ON-type bipolar cells.

In order to gain insight into the Ca^{2+}-neurotransmitter release coupling in goldfish ON-type bipolar cells, the amount of the Ca^{2+} influx into the axon terminal of the bipolar cell was controlled by changing the intensity of voltage pulses (Figure 14.9). When a small depolarizing pulse induced a small Ca^{2+} current, the neurotransmitter-induced current was also small (Figure 14.9A). As the pulse intensity was increased, the neurotransmitter-induced current increased in amplitude concomitantly with the Ca^{2+} current (Figure 14.9B). For larger depolarization, the Ca^{2+} current decreased in amplitude due to the reduction of the driving force for Ca^{2+}, resulting in the decrease of the neurotransmitter-induced current (Figure 14.9C). During the application of a very large depolarization the Ca^{2+} current and the neurotransmitter-induced current were not obviously detected, but just after the termination of the depolarization, the Ca^{2+} tail current could evoke a neurotransmitter-induced current (an 'off' response) (Figure 14.9D). It is evident that the Ca^{2+} influx is essential for the release of the glutamate-like substance from goldfish ON-type bipolar cells.

The relation between the amount of the

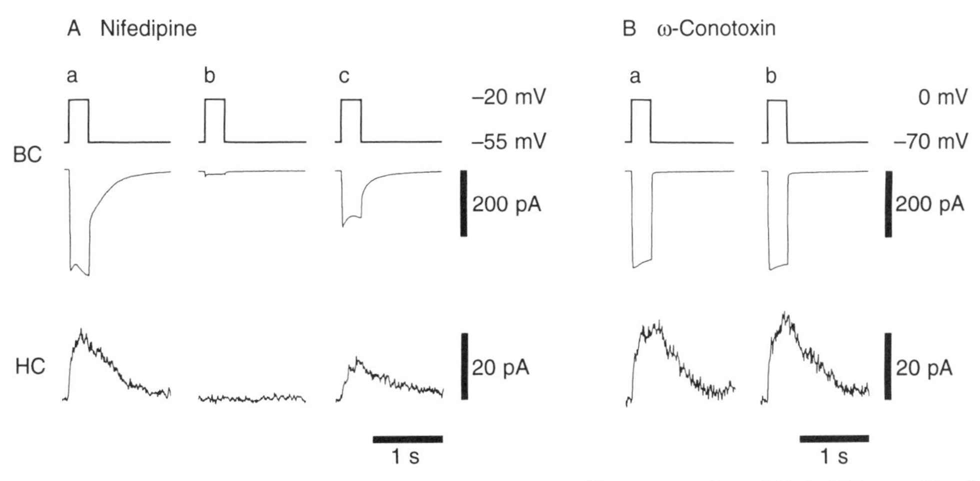

Figure 14.8 Effects of nifedipine and ω-conotoxin on the Ca^{2+} current of a goldfish ON-type bipolar cell and on the neurotransmitter-induced current of a catfish cone-type horizontal cell. A, A depolarizing pulse (300 ms pulse from −55 mV to −20 mV; top trace) applied to the bipolar cell (BC) evoked the Ca^{2+} current (middle trace) in BC and the neurotransmitter-induced current (bottom trace) in the apposed horizontal cell (HC) maintained at +30 mV (a). Application of 30 μM nifedipine suppressed both the Ca^{2+} current and the neurotransmitter-induced current (b). After washout both currents recovered partially (c). B, Recordings from another cell pair. The Ca^{2+} current and the neurotransmitter current were observed in the absence (a) and presence (b) of 1 μM ω-conotoxin. HC was voltage-clamped at +40 mV. (Reproduced with permission from Tachibana *et al.* (1993); published by Society for Neuroscience.)

Ca^{2+} influx and the peak amplitude of the neurotransmitter-induced current was examined. The larger the amount of the Ca^{2+} influx, the larger the peak amplitude of the neurotransmitter-induced current. The relation was a monotonous increasing function with a saturation. In this sense, this relation was similar to that between the amount of the Ca^{2+} influx and the peak increment of the Ca^{2+} transient (section 14.4). However, it would be too speculative to estimate the Ca^{2+}-neurotransmitter release coupling based on the present results. First, $\Delta[Ca^{2+}]_i$ measured by the fluorometric method would not reflect the changes of $[Ca^{2+}]_i$ just beneath the plasma membrane in the active zone. Second, the amount of the released neurotransmitter would not be estimated simply from the amplitude of the horizontal cell response without knowing the dilution factor in the extracellular space and the spatial profile of the released neurotransmitter over the apposed horizontal cell.

14.8 CONCLUSIONS AND FUTURE PERSPECTIVES

ON-type bipolar cells dissociated from the goldfish retina were used as a model system to investigate the mechanism of neurotransmitter release. The Ca^{2+} current was identified as the high-voltage activated, dihydropyridine-sensitive type. Activation of the Ca^{2+} current caused the increase of $[Ca^{2+}]_i$, which was mostly restricted to the axon terminal, and evoked the release of a glutamate-like substance. $[Ca^{2+}]_i$ in the axon terminal was regulated by a Na^+/Ca^{2+} exchanger, by the plasma membrane Ca^{2+} pump, and probably by cytoplasmic substances capable of fast Ca^{2+} buffering with a high cooperativity.

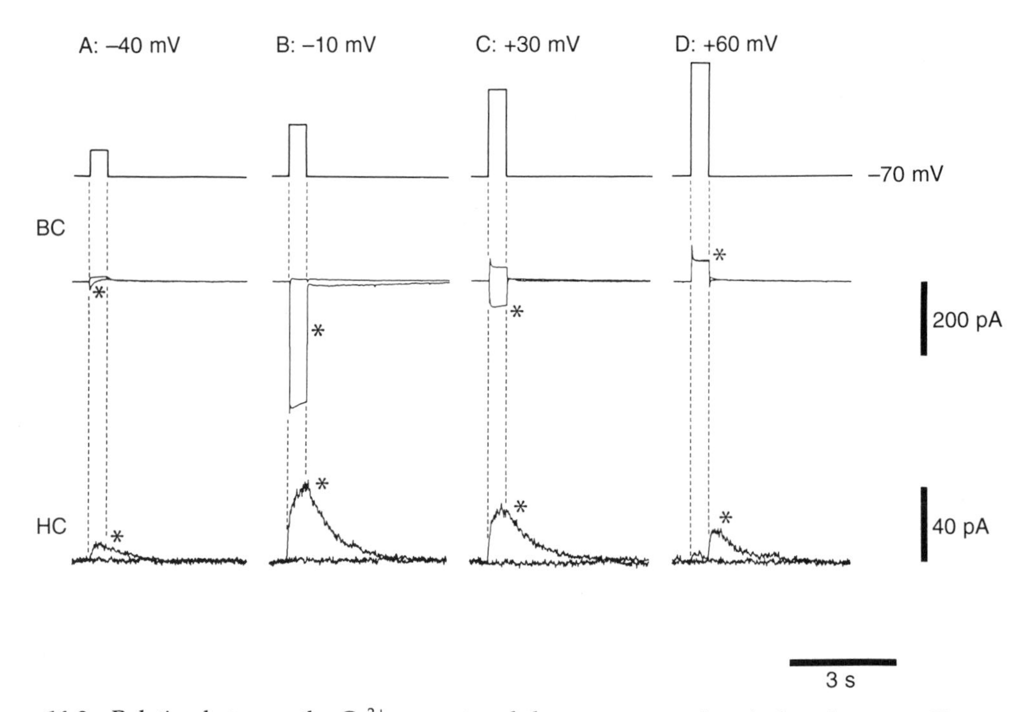

Figure 14.9 Relation between the Ca^{2+} current and the neurotransmitter-induced current. Top traces, membrane potential of a goldfish ON-type bipolar cell (BC). Middle traces, superimposed current traces were recorded from BC in the absence (*) and presence of 500 μM Cd^{2+}. Bottom traces, the currents recorded from the apposed horizontal cell (HC; holding potential at +30 mV) in the absence (*) and presence of Cd^{2+}. Data were obtained in the presence of 10 μM glycine. (Reproduced with permission from Tachibana and Okada (1991), published by Society for Neuroscience.)

The neurotransmitter of goldfish ON-type bipolar cells is very likely to be glutamate. However, we cannot reject the possibility that the authentic neurotransmitter would be aspartate or homocysteic acid because these amino acids could evoke responses in catfish cone-type horizontal cells at high concentrations. Furthermore, the substance which is not able to be detected by catfish cone-type horizontal cells might be co-released with the glutamate-like substance. The final conclusion should not be reached until the released substance is identified by its physical or chemical properties.

The electrophysiological bioassay technique was successfully applied to detect the release of the glutamate-like substance from ON-type bipolar cells. A similar technique has been applied to various cells (Hume *et al.*, 1983; Young and Poo, 1983; Schwartz, 1987; Copenhagen and Jahr, 1989; Meriney *et al.*, 1989). However, various technical limitations should be noted. The 'probe' cell may respond to a family of chemicals (e.g. excitatory amino acids) with low selectivity. It may be difficult to evaluate the factors of diffusion and dilution in the extracellular space. Furthermore, the dose–response relation of the 'probe' cell may not be linear (but usually a sigmoidal function). The time course of the neurotransmitter-induced current may not reflect faithfully the temporal change of the amount of neurotransmitter release if the receptors of the 'probe' cell desensitize with time.

Another way to monitor the transmitter release is to measure the membrane capacitance increase, which is caused by the fusion of synaptic vesicles (Lindau and Neher, 1988). The capacitance change seems to reflect the exocytosis and endocytosis of synaptic vesicles. Recently, using axon terminals of goldfish ON-type bipolar cells, von Gersdorff and Matthews (1994) succeeded in measuring the membrane capacitance change in response to the activation of the Ca^{2+} current. They suggested that the intracellular free Ca^{2+} concentration required for the exocytosis is larger than 50 μM. Since synaptic vesicles in neurons are very small, the detected capacitance change may be due to the fusion of thousands of vesicles. It is technically difficult to measure the capacitance changes during depolarizing pulses. On the other hand, combination of 'caged' Ca^{2+} compounds and the capacitance measurement could provide important information about the Ca^{2+} dependence of transmitter release.

The release of a glutamate-like substance from goldfish ON-type bipolar cells has been demonstrated and shown to depend critically on the activation of a dihydropyridine-sensitive Ca^{2+} current. This conclusion contrasts with the hypothesis that the activation of ω-conotoxin-sensitive Ca^{2+} current triggers the release of 'fast' neurotransmitters, whereas the activation of dihydropyridine-sensitive Ca^{2+} current induces the release of 'slow' neurotransmitters (Miller, 1987; Hirning *et al.*, 1988). It is likely that the neurotransmitter is released when $[Ca^{2+}]_i$ is increased by the Ca^{2+} influx through presynaptic Ca^{2+} channels, irrespective of their subtypes. The different presynaptic terminals may have different Ca^{2+} channels suitable for their functional requirements, such as the activation range, the activation and inactivation kinetics, the single-channel conductance, and the susceptibility to modulation by neurotransmitters and neuromodulators.

The fluorometric technique by fluorescent Ca^{2+} indicators is usually employed to measure the intracellular free Ca^{2+} concentration. However, this technique allows us to estimate the averaged changes of $[Ca^{2+}]_i$ in the deep parts of the cytoplasm rather than the local changes of $[Ca^{2+}]_i$ in the submembrane domain. To analyze the Ca^{2+}-neurotransmitter release coupling, it is essential to know the kinetics of Ca^{2+} just beneath the plasma membrane in the active zone where the exocytosis of neurotransmitter occurs. At present no technique is available to measure directly the kinetics of Ca^{2+} in the active zone. The measurement of Ca^{2+}-mediated currents may give insights into the kinetics of Ca^{2+} near the active zone (Roberts, 1993), if Ca^{2+}-mediated channels and Ca^{2+} channels are co-localized in this region.

ACKNOWLEDGEMENTS

I thank T. Okada, T. Arimura, K. Ono, K. Kobayashi, H. Odagiri and Dr M. Piccolino for their participation in some parts of the experiments. The work was supported by The Ministry of Education, Science and Culture (Japan) Grant-in-Aid for Scientific Research (634801111, 01640505, 01659503, 02223105, 02241203, 0345126, 0322526 and 03304026), by The Mitsubishi Foundation and by Human Frontier Science Program.

REFERENCES

Aizenman, E., Frosch, M.P. and Lipton, S.A. (1988) Responses mediated by amino acid receptors in solitary retinal ganglion cells from rat. *Journal of Physiology (London)*, **396**, 75–91.

Armstrong, C.M. and Matteson, D.R. (1985) Two distinct populations of calcium channels in a clonal line of pituitary cells. *Science*, **227**, 65–7.

Augustine, G.J., Charlton, M.P. and Smith, S.J. (1985) Calcium entry and transmitter release at voltage-clamped nerve terminals of squid. *Journal of Physiology (London)*, **369**, 163–81.

Augustine, G.J., Charlton, M.P. and Smith, S.J. (1987) Calcium action in synaptic transmitter release. *Annual Review of Neuroscience*, **10**, 633–93.

Barnes, S. and Werblin, F. (1987) Direct excitatory and lateral inhibitory synaptic inputs to amacrine cells in the tiger salamander retina. *Brain Research*, **406**, 233–7.

Bean, B.P. (1989) Classes of calcium channels in vertebrate cells. *Annual Review of Physiology*, **51**, 367–84.

Bloomfield, S.A. and Dowling, J.E. (1985) Roles of aspartate and glutamate in synaptic transmission in rabbit retina. II. Inner plexiform layer. *Journal of Neurophysiology*, **53**, 714–25.

Carbone, E. and Lux, H.D. (1984) A low voltage-activated, fully inactivating Ca channel in vertebrate sensory neurones. *Nature*, **310**, 501–2.

Carbone, E. and Swandulla, D. (1989) Neuronal channels: kinetics, blockade, and modulation. *Progress in Biophysics and Molecular Biology*, **54**, 31–58.

Copenhagen, D.R. and Jahr, C.E. (1989) Release of endogenous excitatory amino acids from turtle photoreceptors. *Nature*, **341**, 536–9.

Davies, J. and Watkins, J.C. (1982) Actions of D and L forms of 2-amino-5-phosphonovalerate and 2-amino-4-phosphonobutyrate in the cat spinal cord. *Brain Research*, **235**, 378–86.

Dipolo, R. and Beauge, L. (1982) The effect of pH on Ca^{2+} extrusion mechanisms in dialyzed squid axons. *Biochemica et Biophysica Acta*, **688**, 237–45.

Dipolo, R. and Beauge, L. (1983) The calcium pump and sodium–calcium exchange in squid axons. *Annual Review of Physiology*, **45**, 313–24.

Ehinger, B. (1989) Glutamate as a retinal neurotransmitter, in *Neurobiology of the Inner Retina, Series H, Cell Biology* Vol. 31 (eds R. Weiler and N.N. Osborne), Springer, Berlin, pp.1–14,

Ehinger, B., Ottersen, O.P., Storm-Mathisen, J. and Dowling, J.E. (1988) Bipolar cells in the turtle retina are strongly immunoreactive for glutamate. *Proceedings of the National Academy of Sciences of the USA*, **85**, 8321–25.

Fox, A.P., Nowycky, M.C. and Tsien, R.W. (1987) Kinetic and pharmacological properties distinguishing three types of calcium currents in chick sensory neurones. *Journal of Physiology (London)*, **394**, 149–72.

Grynkiewicz, G., Poenie, M. and Tsien, R.W. (1985) A new generation of Ca^{2+} indicators with greatly improved fluorescence properties. *Journal of Biological Chemistry*, **260**, 3440–50.

Hagiwara, S., Ozawa, S. and Sand, O. (1975) Voltage clamp analysis of two inward current mechanisms in the egg cell membrane of a starfish. *Journal of General Physiology*, **65**, 617–44.

Hamill, O.P., Marty, A., Neher, E. *et al.* (1981) Improved patch-clamp techniques for high-resolution current recording from cell and cell-free membrane patches. *Pflügers Archiv*, **391**, 85–100.

Heidelberger, R. and Matthews, G. (1992) Calcium influx and calcium current in single synaptic terminals of goldfish retinal bipolar neurons. *Journal of Physiology (London)*, **447**, 235–56.

Hirning, L.D., Fox, A.P., McCleskey, E.W. *et al.* (1988) Dominant role of N-type Ca^{2+} channels in evoked release of norepinephrine from sympathetic neurons. *Science*, **239**, 57–61.

Honore, T., Davies, S.N., Drejer, J. *et al.* (1988) Quinoxalinediones: potent competitive non-NMDA glutamate receptor antagonists. *Science*, **241**, 701–3.

Hume, R.I., Roll, L.W. and Fishbach, G.D. (1983) Acetylcholine release from growth cones detected with patches of acetylcholine receptor-rich membranes. *Nature*, **305**, 632–4.

Hurd, L.B. and Eldred, W.D. (1993) Synaptic microcircuitry of bipolar and amacrine cells with serotonin-like immunoreactivity in the retina of the turtle, *Pseudemys scripta elegans*. *Visual Neuroscience*, **10**, 455–71.

Ishida, A.T., Stell, W.K. and Lightfoot, D.A. (1980) Rod and cone inputs to bipolar cells in goldfish retina. *Journal of Comparative Neurology*, **191**, 315–35.

Johnson, J.W. and Ascher, P. (1987) Glycine potentiates the NMDA response in cultured mouse brain neurons. *Nature*, **325**, 529–31.

Kaneko, A. and Tachibana, M. (1985) A voltage-clamp analysis of membrane currents in solitary bipolar cells dissociated from *Carassius auratus*. *Journal of Physiology (London)*, **358**, 131–52.

Kaneko, A., Nishimura, Y., Tauchi, M. and Shimai, K. (1980) Distribution of afferent synapses along on-center bipolar cell axons in the carp retina. *Biomedical Research*, **1**, 345–48.

Kaneko, A., Pinto, L.H. and Tachibana, M. (1989) Transient calcium current of retinal bipolar cells of the mouse. *Journal of Physiology (London)*, **410**, 613–29.

Kasai, H., Aosaki, T. and Fukuda, J. (1987) Presynaptic Ca-antagonist ω-conotoxin irreversibly blocks N-type Ca channels in chick sensory neurons. *Neuroscience Research*, **4**, 228–35.

Kato, S., Teranishi, T. and Negishi, K. (1985) L-Glutamate depolarizes on–off transient type of amacrine cells in the carp retina: an ionophoretic study. *Brain Research*, **329**, 390–4.

Katz, B. and Miledi, R. (1969) Tetrodotoxin-resistant electric activity in presynaptic terminals. *Journal of Physiology (London)*, **203**, 459–87.

Kobayashi, K., Tachibana, M. and Okada, T. (1993) Ca^{2+} buffering in axon terminals of goldfish retinal bipolar cells. *Neuroscience Research*, Suppl. **18**, S28.

Kobayashi, K. and Tachibana, M. (1995) Ca^{2+} regulation in the presynaptic terminals of goldfish retinal bipolar cells. *Journal of Physiology* (London) (in press).

Lasater, E.M., Dowling, J.E. and Ripps, H. (1984) Pharmacological properties of isolated horizontal and bipolar cells from the skate retina. *Journal of Neuroscience*, **4**, 1966–75.

Lindau, M. and Neher, E. (1988) Patch-clamp techniques for time-resolved capacitance measurements in single cells. *Pflügers Archiv*, **411**, 137–46.

Llinas, R., Steinberg, I.Z. and Walton, K. (1981) Relationship between presynaptic calcium current and postsynaptic potential in squid giant synapse. *Biophysics Journal*, **33**, 323–52.

Lukasiewicz, P.D. and McReynolds, J.S. (1985) Synaptic transmission at *N*-methyl-D-aspartate receptors in the proximal retina of the mudpuppy. *Journal of Physiology (London)*, **367**, 99–115.

Marc, R.E., Liu, W.L., Scholz, K. and Muller, J.F. (1988) Serotonergic and serotonin-accumulating neurons in the goldfish retina. *Journal of Neuroscience*, **8**, 3427–50.

Marc, R.E., Liu, W.L., Kalloniatis, M. *et al.* (1990) Patterns of glutamate immunoreactivity in the goldfish retina. *Journal of Neuroscience*, **10**, 4006–34.

Mcburney, R.N. and Neering, I.R. (1987) Neuronal calcium homeostasis. *Trends in Neuroscience*, **10**, 164–9.

Meriney, S.D., Young, S.H. and Grinnell, A.D. (1989) Constraints on the interpretation of non-quantal acetylcholine release from frog neuromuscular junctions. *Proceedings of the National Academy of Sciences of the USA*, **86**, 2098–102.

Miller, R.J. (1987) Multiple calcium channels and neuronal function. *Science*, **235**, 46–52.

Mosinger, J.L. and Altschuler, R.A. (1985) Aspartate aminotransferase-like immunoreactivity in the guinea pig and monkey retinas. *Journal of Comparative Neurology*, **233**, 255–68.

Neal, M.J. and Cunningham, J.R. (1989) L-Homocysteic acid – a possible bipolar cell transmitter in the rabbit retina. *Neuroscience Letters*, **102**, 114–9.

Neal, M.J. and Cunningham, J.R. (1992) Effect of sulphur containing amino acids on [^{3}H]-acetylcholine release from amacrine cells of the rabbit retina. *British Journal of Pharmacology*, **105**, 563–8.

Nowak, L., Bregestovski, P., Ascher, P. *et al.* (1984) Magnesium gates glutamate-activated channels in mouse central neurones. *Nature*, **307**, 462–5.

Nowycky, M.C., Fox, A.P. and Tsien, R.W. (1985) Three types of neuronal calcium channel with different calcium agonist sensitivity. *Nature*, **316**, 440–3.

Okada, T., Tachibana, M., Arimura, T. and Kobayashi, K. (1992) The relationship between Ca current and intracellular Ca ion concentration in bipolar cells isolated from the goldfish retina. *Japanese Journal of Physiology*, **42** (Suppl), S120.

Perkins, M.N. and Stone, T.W. (1982) An iontophoretic investigation of the actions of convulsant kynurenines and their interaction with the endogenous excitant quinolinic acid. *Brain Research*, **247**, 184–7.

Roberts, W.M. (1993) Spatial calcium buffering in saccular hair cells. *Nature*, **363**, 74–6.

Saito, T. and Kujiraoka, T. (1982) Physiological

and morphological identification of two types of on-center bipolar cells in the carp retina. *Journal of Comparative Neurology*, **205**, 161–70.

Schutte, M. and Weiler, R. (1987) Morphometric analysis of serotoninergic bipolar cells in the retina and its implications for retinal image processing. *Journal of Comparative Neurology*, **260**, 619–26.

Schutte, M. and Witkovsky, P. (1990) Serotonin-like immunoreactivity in the retina of the clawed frog *Xenopus laevis*. *Journal of Neurocytology*, **19**, 504–18.

Schwartz, E.A. (1987) Depolarization without calcium can release γ-aminobutyric acid from a retinal neuron. *Science*, **238**, 350–5.

Slaughter, M.M. and Miller, R.F. (1983) Bipolar cells in the mudpuppy retina use an excitatory amino acid neurotransmitter. *Nature*, **303**, 537–8.

Smallwood, J.I., Waisman, D.M., Lafreniere, D. and Rasmussen, H. (1983) Evidence that the erythrocyte calcium pump catalyzes a Ca^{2+}: nH^{+} exchange. *Journal of Biological Chemistry*, **258**, 11092–7.

Suzuki, S., Tachibana, M. and Kaneko, A. (1990) Effects of glycine and GABA on isolated bipolar cells of the mouse retina. *Journal of Physiology (London)*, **421**, 645–62.

Tachibana, M. (1981) Membrane properties of solitary horizontal cells isolated from goldfish retina. *Journal of Physiology (London)*, **321**, 141–61.

Tachibana, M. and Okada, T. (1991) Release of endogenous excitatory amino acids from ON-type bipolar cells isolated from the goldfish retina. *Journal of Neuroscience*, **11**, 2199–208.

Tachibana, M., Okada, T., Arimura, T. *et al.* (1993) Dihydropyridine-sensitive calcium current mediates neurotransmitter release from bipolar cells of the goldfish retina. *Journal of Neuroscience*, **13**, 2898–909.

Tatsumi, H. and Katayama, Y. (1993) Regulation of the intracellular free calcium concentration in acutely dissociated neurones from rat nucleus basalis. *Journal of Physiology (London)*, **464**, 165–81.

Thayer, S.A. and Miller, R.J. (1990) Regulation of the intracellular free calcium concentration in single rat dorsal root ganglion neurones *in vitro*. *Journal of Physiology (London)*, **425**, 85–115.

von Gersdorff, H. and Matthews, G. (1994) Dynamics of synaptic vesicle fusion and membrane retrieval in synaptic terminal. *Nature*, **367**, 735–9.

Weiler, R. and Schutte, M. (1985a) Kainic acid-induced release of serotonin from OFF-bipolar cells in the turtle retina. *Brain Research*, **360**, 379–83.

Weiler, R. and Schutte, M. (1985b) Morphological and pharmacological analysis of putative serotonergic bipolar and amacrine cells in the retina of a turtle, *Pseudemys scripta elegans*. *Cell and Tissue Research*, **241**, 373–82.

Young, S.H. and Poo, M. (1983) Spontaneous release of transmitter from growth cones of embryonic neurones. *Nature*, **305**, 634–7.

Zhu, B., Gabriel, R. and Straznicky, C. (1992) Serotonin synthesis and accumulation by neurons of the anuran retina. *Visual Neuroscience*, **9**, 377–88.

15

Interplexiform cell connectivity in the outer retina

ROBERT E. MARC

15.1 INTRODUCTION

Interplexiform cells, often characterized as the 'sixth' form of retinal neuron, are multipolar cells with complex lateral arbors in both the outer and inner plexiform layers. The formal classifications of all other retinal neurons are likewise based solely on light microscopic form and this has also been adopted here. To my knowledge, the first published examples of interplexiform cells (likely to be glycinergic) were described by Ramón y Cajal (1892) in the retinas of perciform fishes: '. . . I have found cells whose morphological characteristics require us to regard them as a separate class of retinal elements.' Although Cajal leaned towards identifying them as a cone bipolar cell, he clearly understood their special and difficult morphologies. And as the densities of glycinergic interplexiform cells are so low (Marc and Lam, 1981; Marc, 1982), they have been detected but infrequently by Golgi methods (Wagner, 1976; Kalloniatis and Marc, 1990) and then usually as fragments. Cajal also described partial impregnations of probable glycinergic interplexiform cells in frogs, noting their similarity to perch interplexiform cells, and in avians as well (see the remarkable Figure 6 on page 87 of the Thorp and Glickstein translation (Cajal, 1972)). It is likewise evident that Cajal either failed to impregnate teleostean dopaminergic interplexiform cells or to recognize their unique status. And although Cajal also illustrates pieces of probable canine interplexiform cells, the work of Gallego (1971) contains both the formal introduction of the cell name and the clearest structural definition of the type. Without regard to neurotransmitter content or synaptic connectivity, these works show that form constitutes a necessary and sufficient definition of interplexiform cells: interplexiform cells are simply multipolar neurons with arbors in both plexiform layers. They are not bipolar in form and, although no interplexiform cell has been shown to be postsynaptic to photoreceptors, it ought to be acknowledged that there is no formal rule that precludes this.

The cell commonly considered the archetypal interplexiform cell is the dopaminergic neuron of cyprinid retinas (Ehinger *et al.*, 1969; Dowling and Ehinger, 1975). This is certainly an important neuron but it is likely to be a specialization of teleosts and not the

Neurobiology and Clinical Aspects of the Outer Retina
Edited by M.B.A. Djamgoz, S.N. Archer and S. Vallerga
Published in 1995 by Chapman & Hall, London
ISBN 0 412 60080 3

common vertebrate form (see later). With characteristic insight, Boycott *et al.* (1975) concluded that many mammalian interplexiform cells identified by Golgi impregnation were not aminergic. Indeed, the cells described by Gallego (1971) are probably GABAergic interplexiform cells. Interplexiform cells are proving more complex in form, neurochemistry and connectivity than previously believed and there is no basis for assuming the existence of a single 'homologous' form. We know the basic connectivities of two completely distinct forms of interplexiform cells in teleosts (there are at least as many types in most vertebrates) and evidence exists for the presence of at least two more forms. Though the objective of this brief review is to discuss connectivity in the outer plexiform layer, clarity demands some greater framework. So, some of the known neurochemical forms of interplexiform cells, some phylogenetic observations, and comparison of the connectivities of glycinergic and dopaminergic interplexiform cells are presented. There are many excellent reviews of retinal dopaminergic systems (e.g. Negishi *et al.*, 1990; Witkovsky and Dearry, 1991; Djamgoz and Wagner, 1992).

15.2 THE NEUROCHEMICAL FORMS OF INTERPLEXIFORM CELLS

15.2.1 DOPAMINERGIC INTERPLEXIFORM CELLS

The best-known interplexiform cells of the vertebrate retina are those from three disparate vertebrate classes (Osteichthyes, Amphibia and Mammalia) and the relationships among interplexiform types across taxa are still poorly explicated. Here both 'GABAergic' and 'dopaminergic' interplexiform cells are discussed, as they are one and the same cell in some vertebrates. The first strong evidence for a dopaminergic interplexiform cell system derived from observations on the vertebrate retina using the Falck–Hillarp or formaldehyde-induced fluorescence (FIF) technique for detecting cells containing high levels of heterocyclic amines. This literature has been thoroughly reviewed elsewhere (e.g. Boycott *et al.*, 1975; Ehinger and Dowling, 1987; Negishi *et al.*, 1990; Witkovsky and Dearry, 1991). The most dramatic observations were first made in advanced teleosts and these showed a dense plexus of amine-containing processes in the outer plexiform layer whereas the only obvious retinal sources of those processes resembled conventional amacrine cells (Ehinger and Falck, 1969). Subsequently, many analyses of the forms of catecholamine cells in cyprinid retinas have been published: extensive studies of FIF (Negishi *et al.*, 1990; Dowling and Ehinger, 1975); autoradiographic localizations of cells bearing dopamine transporters (Marc, 1982; Van Haesendonck *et al.*, 1993); and immunocytochemical localization of tyrosine hydroxylase immunoreactivity (Yazulla and Zucker, 1988; Van Haesendonck *et al.*, 1993) and dopamine immunoreactivity (Van Haesendonck *et al.*, 1993). Taken as a whole, these and other data indicate that dopamine is the prime catechole of the teleostean retina and dopaminergic interplexiform cells are the main dopaminergic neurons of the teleostean retina. The fundamental properties of teleostean dopaminergic interplexiform cells are (Figure 15.1):

1. large somata positioned in the amacrine cell layer proper;
2. multipolar morphologies with thick primary and secondary dendrites distributed throughout the inner plexiform layer, with preferential stratifications in sublayers 1,3,5;
3. multiple slender processes arising from the dendritic arbor and arborizing as a dense axonal field in the outer plexiform layer;

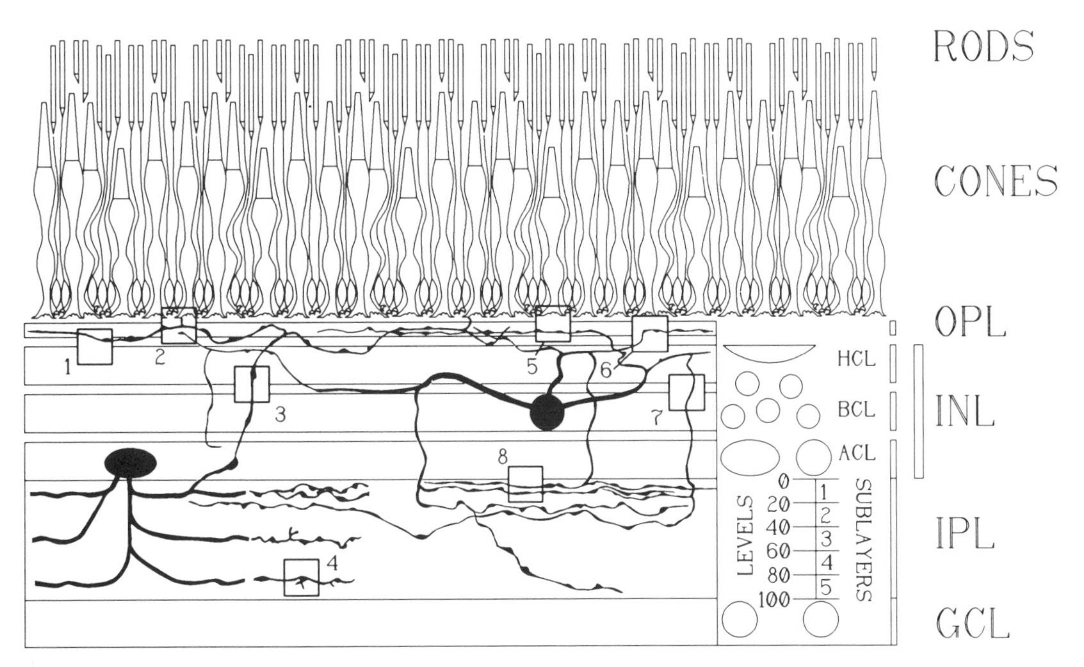

Figure 15.1 Schematic of the dopaminergic (left) and glycinergic (right) interplexiform cells and their arborizations in the outer and inner plexiform layers. Dopaminergic amacrine cells possess large ovoid somata and are located in the ACL (amacrine cell layer) with major dendritic arborizations in sublayers 1, 3 and 5 of the IPL (inner plexiform layer). Axonal extensions arise from primary or secondary dendrites and ascend to the OPL (outer plexiform layer) where they make contacts with horizontal cells and, occasionally, photoreceptors (boxes 1,2,3). The inputs to dopaminergic interplexiform cells originate in the inner plexiform layer and, in fishes, primarily involve GABAergic amacrine cells (box 4). The somata of glycinergic interplexiform cells are large, rounded and positioned in the middle to lower half of the INL (inner nuclear layer) in the nominal BCL (bipolar cell layer). They send dendrites through the INL to the distal border of the HCL (horizontal cell layer) where they receive direct synaptic inputs from H1 horizontal cells, are presynaptic to some horizontal cells; axons arise from these dendrites and descend into the inner plexiform layer (boxes 5,6,7). Contacts in the inner plexiform layer involve input to GABAergic amacrine cells (box 8).

4. a density of about 70–100 cells/mm^2 in a goldfish of 10–12 cm standard length (see Negishi *et al.*, 1990 for a comprehensive listing of references regarding cell density and size).

Recently, Baldridge and Ball (1993) reported the presence of yet another type of interplexiform cell in goldfish: one immunoreactive for PNMT (phenyl-*N*-methyltransferase). PNMT is generally considered be a marker for adrenergic (epinephrine-using) neurons. The extent to which this alters interpretations of the dopaminergic system remains to be clarified, but since epinephrine bears no free amino group, it is not fixable by glutaraldehyde as is dopamine. Thus immunocytochemistry with anti-dopamine immunoglobulins remains selective for the dopaminergic interplexiform cell (Van Haesendonck *et al.*, 1993).

Many other vertebrates (e.g. amphibians, turtles, primates, rodents) apparently possess populations of dopaminergic retinal neurons, where some neurons may be pure amacrine cells and others interplexiform cells, or where interplexiform processes arise from a subset of otherwise homogenous amacrine cells (Kolb *et al.*, 1987; Schütte and Witkovsky, 1991; Witkovsky and Dearry, 1991; Witkovsky and Schütte, 1991). Although these cells are formally classifiable as interplexiform cells, they clearly do not form the plexus characteristic of cyprinid interplexiform cells and do not share the same detailed format of arborization in the inner plexiform layer. This occasions a diversion into a brief consideration of amacrine cells and axons.

(a) The ascending processes of dopaminergic interplexiform cells are axons

Structural and physiological data indicate that many nominal amacrine cells possess bona fide axons. Of particular interest are the dopaminergic amacrine cells of the macaque retina. Dacey (1989, 1990) has shown that these and other primate amacrine cells possess a significant axon system beyond their basic dendritic arborizations. Famiglietti (1992a,b) described several types of polyaxonal amacrine cells in the rabbit retina and provisionally identified one Golgi type (PA4) as the rabbit dopaminergic amacrine cell. Likewise, dye injection studies of cyprinid amacrine cells have indicated that axon-like processes extend from the dendritic arbor of several types of otherwise conventional amacrine cells (e.g. Teranishi *et al.*, 1987). The 'association' amacrine cell of avian retinas (Cajal, 1892; Mariani 1982) is a classical Golgi type II interneuron with distinct dendritic and axonal fields. Some physiological counterparts of these structural considerations are (a) multiple spikes in mammalian amacrine cells (Bloomfield, 1990), (b) active spike systems in 'transient' amacrine cells (e.g. Werblin, 1977) and morphologically identified salamander interplexiform cells (Maguire *et al.*, 1990 and (c) blockade by tetrodotoxin of horizontal cell uncoupling in the turtle retina induced by veratridine and bicuculine (Piccolino *et al.*, 1987). Comparison of the general morphology of dopaminergic retinal cells with monaminergic systems in the brainstem leads to the obvious but unsubstantiated impression that these are similar, sparse systems with diffuse axonal targeting. The ascending processes of dopaminergic interplexiform cells resemble axons and likely function as axons. Moreover, the great similarities among dopaminergic amacrine cells and interplexiform cells suggest that they represent poles of a quantitative spectrum, as has been argued by Dacey (1990).

(b) Teleostean dopaminergic interplexiform cells are unique

Starting from the perspective of similarity across dopaminergic cell types, however, one can make both quantitative and qualitative distinctions between teleostean dopaminergic interplexiform cells systems that form dense, identifiable plexuses visible with even the least-sensitive probes (e.g. FIF) and instances of sparse plexi arising from occasional axons that ascend to the outer plexiform layer in mammals (e.g. Ngyuen–LeGros *et al.*, 1981; see also the discussion of the inner nuclear layer plexus by Boycott, *et al.*, 1975) and other tetrapods (Schütte and Witkovsky, 1991), and which require visualization methods of higher sensitivity. At one end of this presumed continuum are the homogeneous dopaminergic interplexiform cells of cyprinids. Each such cell sends several axons to the outer plexiform layer, thus forming a dense and uniform arbor around the horizontal cells (Kalloniatis and Marc, 1990; Van Haesendonck *et al.*, 1993). At the other pole are nominal dopaminergic interplexiform cells embedded in a large, otherwise indiscriminable population of dopaminergic amacrine cells which send a few axons to the outer plexiform layer and form a very sparse field. Many workers conclude that such cells are formal amacrine cells exhibiting a few eccentric processes (Voight and Wässle, 1987; Dacey, 1990; Vaney, 1990; Piechl, 1991). Such cells and certainly those of cat (Oyster *et al.*, 1985), which have a decidedly inhomogenous distribution, do not seem to provide a uniform targeting of an identifiable major cell population in the outer plexiform layer, if such targeting is even necessary. Implicit in this conclusion is the oft-unstated corollary that the processes are non-functional or adventitious. These axonal processes may be anomalous or mistargeted, or reflect a propensity of dopaminergic amacrine cells to develop extensive axonal arbors in regions of dopamine receptive elements, even though explicit synaptic contact may not be an absolute requirement to mediate the specific spatiotemporal features of dopaminergic transmission. The extent to which normal retinal cells form erroneous axonal fields is unknown. The axon terminals of goldfish horizontal cells can be observed in the outer nuclear layer, wandering among the rod nuclei, and these are undeniably misplaced, non-functional processes. In vascularized retinas, dopaminergic cells often send processes around retinal vessels even if they protrude well into the inner nuclear layer or inner plexiform layer (R. Marc, unpublished observations; Favard *et al.*, 1990; Kolb *et al.*, 1990), and though these processes may be truly functional, their presence nevertheless imparts the impression of a proclivity for wandering. The sense that specialized regions of mammalian retinas, such as in cat, seem to have an elevated level of interplexiform cell-like neurons and that they may indeed target specific sets of cells, must be balanced with the apparent variability among individual samples (Takahashi, 1988) and the extremely low frequency of such interplexiform cells (a few cells/mm^2). There is, as yet, no robust test of the hypothesis that the sparse outer plexiform layer processes of certain dopaminergic systems are adventitious. Based on the specificity and density of teleostean dopaminergic innervation in the outer plexiform layer and other issues detailed below, it may be difficult to substantiate the claim that teleostean dopaminergic interplexiform cells and mammalian dopaminergic amacrine cells are homologues in the sense of lineage and that the differences between them are only quantitative. Certainly, some aspects of phylogeny strongly argue against this (see below). Dacey (1990) has posited that the presence of intermediate forms of cells would constitute supporting evidence for the quantitative continuum hypothesis, but I would add that this would be true only if those forms were found in key species. Moreover, my data (R.E. Marc, unpublished observations) indicate that the density of dopaminergic innervation in the outer plexiform layer across species is bimodal – weak and strong – with no good evidence of graded forms in the appropriate taxa. In conclusion, it is not clear how to distinguish between adventitious and functional arbors or, without comprehensive ultrastructural data, how to define the density of outer plexiform layer arborization that represents functional innervation.

Two features of tetrapod dopaminergic interplexiform cells qualitatively distinguish them from cyprinid dopaminergic interplexiform cells. First, it appears that dopaminergic cells in fishes and amphibians contains no other known neurotransmitter substance (Wülle and Wagner, 1990; Watt and Florack, 1992), whereas the dopaminergic cells of the Amniota (Reptilia, Aves, Mammalia) are clearly GABA immunoreactive neurons (Wässle and Chun, 1988; Wülle and Wagner, 1990). Second, the dopaminergic interplexiform cells of goldfishes exhibit multiple ascending axon-like processes from the dendritic arbor, usually from primary or secondary dendrites (but occasionally high-order dendrites as well) and rarely if ever from the soma (Kalloniatis and Marc, 1990). In direct contrast, the interplexiform cells of turtle, rabbit, rat, mouse, cat and monkey often exhibit ascending processes arising from the soma as well as the arbor (e.g. Oyster *et al.* 1985, 1988; Ryan and Hendrickson, 1987; Witkovsky and Schütte, 1991). This seems true of both 'GABAergic' and 'dopaminergic' interplexiform cells of mammalian retinas (Nakamura *et al.*, 1980; Ryan and Hendrickson, 1987). Moreover, ultrastructural data (Kolb and West, 1977; Kolb *et al.*, 1990), Golgi impregnations of human retina (Kolb *et al.* 1992) and the cell size comparisons of dopaminergic and GABAergic interplexiform cells in primate retinas (Ryan and Hendrickson, 1987) indicate that there are almost certainly separate dopaminergic/GABAergic and GABAergic interplexiform cell populations in mammals, even if all 'dopaminergic' neurons also contain GABA. Do such interplexiform cells actually release dopamine, GABA or both in the outer plexiform layer? With the sole exception of the advanced teleosts, can we assert that any functions of interplexiform processes are dopaminergic? Probably not, at present. Although differences in innervation patterns in the outer plexiform layer have been reported for mammalian interplexiform cells viewed by autoradiography ([^{3}H]GABA: Nakamura *et al.*, 1980; dopamine: Frederick *et al.*, 1982), immunocytochemistry (Kolb *et al.*, 1990) or electron microscopic reconstruction (Kolb and West, 1977), there are as yet insufficient data from which to define unambiguously their selective connectivities in the outer plexiform layer.

15.2.2 GABAERGIC, PEPTIDERGIC AND OTHER INTERPLEXIFORM CELLS

Since the problem of GABAergic interplexiform cells has arisen, resolution of suspected populations of GABAergic interplexiform cells in non-mamalian vertebrates is seriously complicated by a background of horizontal cell somata and dendrites containing a strong GABA signal (Marc, 1992 for a review). For example, if all dopaminergic neurons in turtle are also GABA immunoreactive (Wagner and Wülle, 1990), then why are not these considered GABAergic rather than dopaminergic interplexiform cells? Thus, GABAergic interplexiform cells might well exist in many species and may be even more common than the dopaminergic form (section 15.3 on phylogenetic considerations). The serotonin immunoreactive S1 amacrine cells of the goldfish retina are also GABAergic (R. Marc, unpublished data; Ball and Tutton, 1990). However, serotonin immunoreactive neurons in teleosts can also show ascending processes (H.-J. Wagner, personal communication). This is also the case in goldfish (R. Marc, unpublished data). Since all serotonin immunoreactive cells in the goldfish are GABAergic, goldfish and other teleosts may thus possess a previously undisclosed population of GABAergic/serotoninergic interplexiform cells. Once again, the extent and density of the arbor is difficult to gauge, but it may be sparse, which would justify reviewing the same arguments just outlined regarding the discrimination of functional dopaminergic arbors. To this picture we must now add the novel PNMT interplexiform cell (Baldridge and Ball, 1993). It is plausible that the teleostean retina contains at least four types of interplexiform cells, though there are data on the connectivities of only the dopaminergic and glycinergic varieties. Peptides have not been considered in this review, and several immunoreactive peptide systems have been reported to possess processes in the outer plexiform layer (Tornqvist *et al.*, 1982). We know nothing of their possible roles in visual processing.

15.2.3 GLYCINERGIC INTERPLEXIFORM CELLS

(a) Teleostean fishes

Marc and Lam (1981) discovered the glycinergic interplexiform cells of the goldfish retina; these bore some similarity to the interplexiform cells of the perch described by Ramón y Cajal (1892). The somata of cyprinid glycinergic interplexiform cells are the largest found in the goldfish inner nuclear layer (>12 μm diameter) and they are typically located in the middle of the inner nuclear layer or just above the amacrine cell layer at a density of 25–50 cells/mm^2 (Figure 15.1). Each soma possesses between two and five primary dendrites that course laterally and obliquely in the inner nuclear layer before ascending to the outer plexiform layer where they form a fine, varicose dendritic arbor (Marc and Lam, 1981; Marc and Liu, 1984; Kalloniatis and Marc, 1990). Up to 20 slender axons descend from the outer plexiform arbor of a single glycinergic interplexiform cell into the inner plexiform layer where they branch in sublayers 1, 2 and 5 (Kalloniatis and Marc, 1990). On rare occasions, a somatic dendrite will enter the inner plexiform layer directly. Thus teleostean glycinergic interplexiform cells are fundamentally different from dopaminergic interplexiform cells not only in neurotransmitter content but also in basic neuronal form: dopaminergic interplexiform cells have their dendritic arbors in the inner plexiform layer and axonal fields in the outer plexiform layer; glycinergic interplexiform cells reverse the pattern by having their dendritic arbors in the outer plexiform layer and their apparent axonal fields in the inner plexiform layer. This presages their markedly different connectivities in the outer plexiform layer.

(b) Anuran amphibians

Glycinergic interplexiform cells have also been identified in the amphibian retina by Kleinschmidt and Yazulla (1984) and are apparently present in all anurans. A detailed study of these cells has been carried out by Smiley and Basinger (1988, 1990) and Smiley and Yazulla (1990). Roughly similar to goldfish interplexiform cells in form, the amphibian cells possess many dendrite-like profiles that course into the outer plexiform layer with various descending processes arising from the primary and secondary dendrites. They differ in detail as many processes arise from the soma and directly enter the inner plexiform layer (Smiley and Basinger, 1988). The polarity of these cells is indeterminate and may be mixed, with axonal and dendritic fields in both plexiform layers. Glycinergic interplexiform cells in *Rana* also contain a somatostatin-like peptide which has permitted the detailed study of their forms and distribution. It is worth noting that the glycinergic interplexiform cell of the goldfish retina does not contain somatostatin-like immunoreactivity.

15.3 PHYLOGENETICS OF INTERPLEXIFORM CELLS

15.3.1 MAMMALS AND TELEOSTEAN FISHES ARE 'SISTER' GROUPS

Few comparative studies of retinal neurochemistry have been published that examine key species in phylogeny. Teleostean fishes are often viewed as 'ancestral', 'primitive' or 'lower' and thus presented as a phylogenetic reference point. This is patently incorrect and deserves correction at the outset. Teleostean fishes are among the most modern and diverse of vertebrates, the earliest known of which appear in the late Triassic (Carroll, 1988) at the same time as the true mammals.

Comprising over 20 000 known living species (Nelson, 1984), they are the single largest vertebrate taxon. The last common ancestor of teleosts and mammals (representing the actinopterygian and sarcopterygian osteichthyan radiations respectively) must predate the early Devonian (≈ 400 MY BP). Thus, teleosts are in no way representative of an ancestral condition for mammals, and teleosts and mammals are formal sister groups among vertebrates that share several vertebrate synapomorphies and, in that sense, comparisons of mammalian and teleostean retinas allow some assessment of the status of their common ancestors. Some critical features of interplexiform cell 'evolution' are briefly reviewed here.

15.3.2 THE DOPAMINERGIC INTERPLEXIFORM CELLS OF TELEOSTS ARE APOMORPHIC

The dopaminergic interplexiform cell of teleosts is not ancestral to other vertebrate dopaminergic interplexiform cells and mammalian interplexiform cells with their low densities, inhomogeneous distributions and sparse arbors are not a vestigial form of the teleostean cell. Where does the teleostean dopaminergic interplexiform cell arise in phylogeny? Our data indicate that the rich arbor format of the teleostean dopaminergic interplexiform cell is synapomorphic for advanced, diurnal bony fishes (Marc, 1989). There are three major groups of modern euteleosts: the salmoniformes (e.g. pike, salmon, trout); the ostariophysians (cyprinids such as carp and goldfish); and the neoteleosts (including the perches and many advanced 'spiny' teleosts). Representatives of all of these groups can be shown to possess dopaminergic interplexiform cells (Ehinger *et al.*, 1969; Marc, 1989; Wagner and Wülle, 1990, 1992; Witkovsky and Schütte, 1991; Zaunreiter and Wagner, 1991; Wagner and Behrens, 1993), so the ancestral condition must arise earlier. However, modern isolates of diurnal fishes, such as (1) *Polypterus* and *Calamoichthyes* (Polypteriformes presumably related to primitive chondrosteans), (2) holostean grade gars such as the Florida spotted gar (Lepisosteiformes), as well as (3) *Protopterus* (an extant Dipnoan), lack the teleostean dopaminergic interplexiform cell based on [^{3}H]dopamine uptake autoradiography (Marc, 1989). In fact, the dopaminergic neurons of these fishes are predominantly highly stratified amacrine cells with a primary arborization in sublayer 1 and occasionally weak strata in sublayers 3 and 5, identical to those of urodele amphibians and very similar to those of reptiles, avians and most mammals. These data argue that the dopaminergic interplexiform cell of teleosts is apomorphic: a novel attribute developed late in the evolution of osteichthyans and not the homologue of mammalian dopaminergic interplexiform cells. It is not even a generic 'fish' feature, having arisen well after the sarcopterygian/actinopterygian divergence. This does not mean that dopaminergic interplexiform cells of fishes do not carry out some of the same functions as dopaminergic amacrine cells, but we really have no mechanistic, testable explanation for the morphological design of dopaminergic interplexiform cells. The dopaminergic systems of the other great branch of the early gnathostomes, the chondrichthyans, are largely amacrine cells (Brunken *et al.*, 1986). Other distantly related agnathan vertebrates, such as lampreys, do exhibit an interesting albeit sparse form of dopaminergic interplexiform cell (de Miguel and Wagner, 1990). The relationships among extant agnathans and the remaining vertebrates is currently poorly defined. It is however, likely that the immediate extant relatives of the last common ancestor of mammals and teleosts did not possess a teleostean interplexiform cell. Recently, Wagner *et al.* (1994) reported that tyrosine hydroxylase immunocytochemistry reveals interplexiform cells in both marine teleosts and chondrichthyans with cone rich retinas, but fails to demonstrate them in rod-dominated deep sea teleosts. They propose that interplexiform cells are part of cone vision. I would agree, but stress that the marine teleosts are the most recent of all the teleosts, and the absence of dopaminergic interplexiform cells from members of that group inhabiting scotopic regimes would also have to be considered apomorphic in view of the fact that cone-rich non-teleosts such as the gar, lungfishes and ropefishes lack dopaminergic interplexiform cells and closely resemble turtle or avian retinas. An examination of the cone-rich sting rays, the most advanced batoids, has shown that they too lack a distinct dopaminergic interplexiform cell system, although they could possess some sparse outer plexiform layer arbor like *Xenopus*, turtles and some mammals (R.E. Marc, unpublished observations). I conclude that the teleostean dopaminergic interplexiform cell represents a distinctive specialization of that vertebrate end-group, the like of which has not yet been described in any other vertebrate class or even osteichthyan order. In addition, the presence of habitat-specific expression (Wagner and Behrens, 1993) is representative of a strong niche selection acting in addition to the prime phylogenetic pattern of descent.

A strong argument for the phylogenetic independence of the sole known rich-arbor interplexiform cell in mammals, the platyrhine *Cebus*, is that its close relatives are notably impoverished in outer plexiform layer arbor. Moreover, even the Cebus monkey has nothing like the outer plexiform layer arbor of teleostean fishes. The point has not been articulated better than by Ehinger and Falck (1969) in their original work:

> Except in the Cebus monkey, the layer is never complete and is often represented by some scattered fibers only. The adrenergic fibers do not extend to the outer part of the outer plexiform layer, and do not seem to be in contact with any cone pedicles, but occasionally an adrenergic terminal was seen in the middle of the outer plexiform layer. In the Cebus monkey, adrenergic terminals were seen throughout the narrow outer plexiform layer. Some of the terminals came close to cone pedicles . . . No baskets of obvious synaptic character were seen around any of the horizontal cells, such as in the case in the teleost retina . . .

15.3.3 GABAERGIC NEURONS WITH INTERPLEXIFORM PROCESSES EXIST IN MOST VERTEBRATES

The type of interplexiform cell that is understood least is probably a very common form embedded in a matrix of GABAergic neuropils in the inner and outer plexiform layers. GABAergic interplexiform cells are present in chondrichthyans (Brunken *et al.*, 1986), probably in teleosts (section 15.2.1) and certainly in many mammals (e.g. Nakamura *et al.*, 1980; Ryan and Hendrickson, 1987). There is no evidence that they exist in extant agnathans, yet the phylogenetic distribution of GABAergic amacrine cells with ascending processes argues that they are possible sympleisiomorphic traits – shared, primitive features. However, that presumes that all GABAergic interplexiform cells are homologues and we have no evidence that such cells are equivalent across taxa. Since many of the dopaminergic interplexiform cells of amniotes may also be GABAergic, they clearly cannot be the same as any non-dopaminergic/GABAergic interplexiform cells. However, GABAergic interplexiform cells would be easier to comprehend than dopaminergic interplexiform cells, since spatial buffering of GABA by Müller's cells makes it impossible for GABA to act as a diffusional signal. GABAergic connectivity must involve close approximation between the source and target.

15.3.4 GLYCINERGIC INTERPLEXIFORM CELLS HAVE BEEN DOCUMENTED IN TELEOSTS, AMPHIBIANS, AND AVIANS, BUT ARE ABSENT IN MAMMALS

Had it not been for the very low level of intrinsic glycinergic label in the outer plexiform layer of the goldfish retina, it would have been difficult to resolve the processes of teleostean glycinergic interplexiform cells by autoradiography. However, they are distinctive and present in all diurnal osteichthyans so far examined (Marc, 1985). There is no strong evidence for chondrichthyan or agnathan glycinergic interplexiform cells, but both vertebrate classes do have numerous glycine-accumulating somata and glycine-immunoreactive somata in the inner nuclear layer (Marc, 1985; unpublished data). Many or all could be bipolar cells, but this would make glycinergic interplexiform cells very difficult to resolve. As mentioned above, both teleostean fishes and anuran amphibians exhibit glycinergic interplexiform cells of somewhat similar form. Yang and Yazulla (1988) provided immunochemical evidence for ascending processes of nominal glycinergic amacrine cells in the tiger salamander retina, indicating that such cells are widespread among the anamniota. Definitive evidence is not available, but glycinergic neurons of the inner nuclear layer are evident in both turtles (R. Marc, unpublished) and a glycinergic interplexiform cell 'has been described in the chicken retina (Kalloniatis and Fletcher, 1993), raising the likelihood that glycinergic interplexiform cells may be present in most non-mammalian vertebrates. There is no evidence for a glycinergic interplexiform cell in mammals (Marc, 1989).

15.4 CONNECTIVITY IN THE OUTER PLEXIFORM LAYER

The only interplexiform cells whose connectivities in the outer plexiform layer are sufficiently well known to justify physiological speculation are the dopaminergic and glycinergic interplexiform cells of the cyprinid retina and the glycinergic interplexiform cell of anurans. Both teleostean cells primarily target H1 horizontal cells and their functions apparently involve either the control of horizontal cell sensitivity, horizontal cell spatial integration or monitoring of an integrated flux signal from the horizontal cell layer (Chapter 7).

15.4.1 DOPAMINERGIC INTERPLEXIFORM CELLS

(a) Dopaminergic interplexiform cells primarily contact horizontal cells

Dowling and Ehinger (1975, 1978) first showed that the interplexiform cells marked by cytotoxic monoamine analogs made apparent synapses onto the somata of external or H1 horizontal cells in the goldfish retina (Figure 15.2). The nominally presynaptic contacts made by such aminergic neurons are often subtly different from standard Gray's type I or II synapses but are nevertheless specialized and worthy of recognition. Often, the postsynaptic specializations of the target profiles are indistinct, synaptic clefts are not of uniform width and presynaptic vesicle collections are small. Consequently dopaminergic 'synapses' deficient in conventional features but with nonetheless distinct specializations are often termed junctional appositions (Yazulla and Zucker, 1988). Subsequent work has shown that about 96% of the contacts formed by dopaminergic interplexiform cells in the goldfish target H1 horizontal cells (Van Haesendonck *et al.*, 1993). This is important because the remaining small number of contacts involve both bipolar cells and cone photoreceptors as targets. The densities of the major bipolar cell classes in the goldfish retina equal or exceed those of H1 horizontal cells (Marc, 1982), so

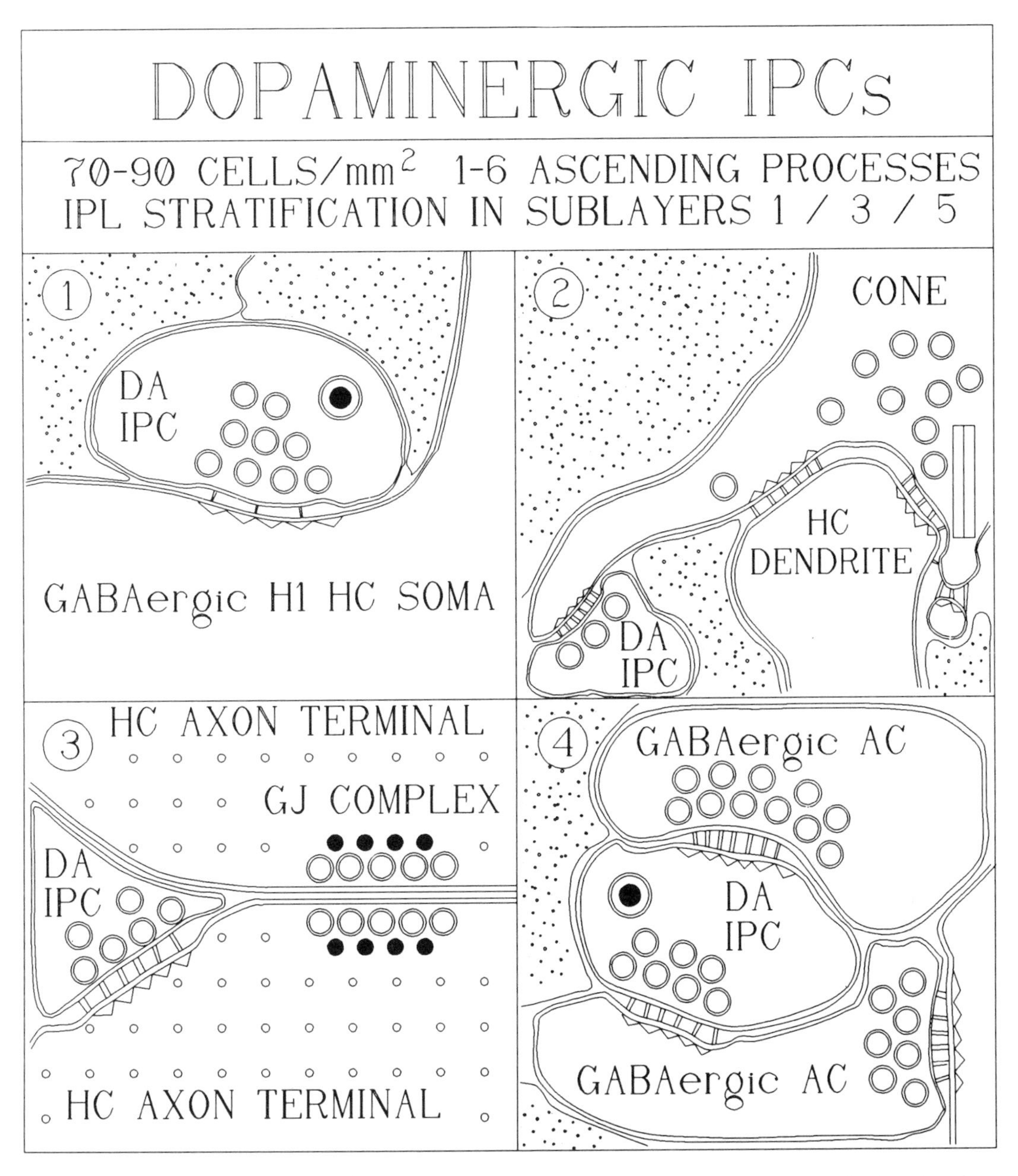

Figure 15.2 Summary data and schematic contacts of dopaminergic interplexiform cells. Boxes refer to locations in Figure 15.1. Box 1: DA IPC (dopaminergic interplexiform cell) varicosities directly contact H1 HC (horizontal cell) somata. Box 2: DA IPC varicosities form junctional appositions on cone pedicles. Box 3: DA IPC varicosities form junctional appositions on HC axon terminals near GJ (gap junction) complexes. Box 4: DA IPC varicosities receive conventional synaptic inputs from GABAergic AC (amacrine cell) terminals and, in turn, make junctional appositions onto GABAergic ACs.

the sparse dopaminergic contacts onto bipolar cells cannot represent uniform innervation of a major bipolar cell subset. Thus these contacts must either be adventitious or directed towards a specific and somewhat sparse bipolar cell variety. Similar concerns involve reported dopaminergic interplexiform cell contacts onto cone photoreceptors in cyprinids (see discussion in Wagner and Wülle, 1990; Van Haesendonck *et al.*, 1993). The evidence does not so far support a consistent innervation of a specific set of cone photoreceptors in these fishes. Alternatively, both silurid and cichlid fishes show a remarkably dense innervation of the photoreceptor layer with many or all cones (and some rods) possessing a close apposition (Wagner and Wülle, 1990, 1992). There may be a quantitative relationship between the degree of photoreceptor innervation by dopaminergic interplexiform cells and the speed or magnitude of retinomotor changes across species of teleosts.

As they course through the inner nuclear layer, processes of dopaminergic interplexiform cells occasionally contact axon terminals of horizontal cells (Marshak and Dowling, 1987), often near gap junctional complexes (Van Haesendonck *et al.*, 1993). It is tempting to posit that such apparent proximity is functional and involved in dopaminergic regulation of gap junctions. There is little evidence in fishes that the coupling between horizontal cell axon terminals can be modulated by dopamine in the same manner as somatic coupling (Teranishi *et al.*, 1984), unlike the situation in turtle retina where both the axon terminals and somata of horizontal cells show dopamine-modulated dye coupling and spatial integration (Piccolino *et al.*, 1982).

A problem yet to be resolved involves the fact that dopamine released from the inner plexiform layer can clearly diffuse throughout the retina, reaching pharmacologically active levels in the outer plexiform layer and even the outer nuclear layer (Witkovsky *et al.*, 1993). What, then, is the selection pressure underlying the development of complex outer plexiform layer arbors in teleosts? A number of possibilities exist, though we have no formal tests of them. First, there could be a component of dopaminergic action on fish horizontal cells that is time-dependent and must be kept in phase with other processes on a time scale faster than allowed by diffusion. This might be true if dopaminergic interplexiform cells are more than just scotopic/photopic adaptation switches. Second, the teleostean D1 dopamine receptor on horizontal cells may be so insensitive relative to the D2 receptor on cones that rapid and effective control of horizontal cell sensitivity demands direct, synaptic input (see discussion in Besharse and Iuvone, 1992). Third, dopaminergic interplexiform cells may release another agent from synaptic vesicles that will not diffuse through the retina to the outer plexiform layer. There is as yet, no evidence for such an agent, and it would seem that it cannot be GABA, glycine, glutamate, aspartate, acetylcholine, etc. In truth, there is no good explanation for the existence of the rich outer plexiform layer arbor of teleostean interplexiform cells.

On balance, though, one can reasonably argue that the primary purposes of interplexiform cell contacts involve direct, synaptic dopaminergic regulation of teleostean cone horizontal cell operations. There have been detailed reviews of this topic (e.g. Negishi *et al.*, 1990; Dowling, 1991; Witkovsky and Dearry, 1991; Besharse and Iuvone, 1992; Djangoz and Wagner, 1992). Micromolar levels of dopamine are able to (i) induce a large decrease in transjunctional conductance between horizontal cells (Lasater and Dowling, 1985) apparently by decreasing the open probability of single gap junctional channels (McMahon, *et al.*, 1989), (ii) enhance the apparent affinity of glutamate for kainate-type receptors on horizontal cells (Knapp and

Dowling, 1987), (iii) depress the maximal horizontal cell flash response (Mangel and Dowling, 1985, 1987), (iv) reduce T-type and potentiate L-type Ca^{2+} channels in bass horizontal cells (Pfeiffer–Linn and Lasater, 1993), and likely other effects, some of which are mediated through cAMP-dependent protein kinases (e.g. Lasater, 1987). There is evidence that dopamine causes receptive field changes in rod horizontal cells as well (Yamada *et al.*, 1992), though the pathway for this effect is unclear.

(b) Disinhibition and light-adaptation promote dopamine release

If and how these effects interact and the degree to which each is activated under normal conditions *in situ* remains unresolved. Though there are ongoing debates about whether dopamine release is primarily activated in response to light stimulation, dark onset or prolonged light deprivation (never biologically realized, in my view), it is likely that dopaminergic interplexiform cells have complex trigger features which will be more important for gating net dopaminergic potency than are simple light increments or decrements. There is anatomical (Yazulla and Zucker, 1988) and pharmacological evidence (O'Connor *et al.*, 1987, 1988) that the dominant proximal input to cyprinid dopaminergic interplexiform cells is a set of GABAergic amacrine cells. There is anatomical evidence for bipolar cell input to dopaminergic neurons in other vertebrates (Mariani and Hokes, 1988), but release and turnover studies indicate that the dominant input to fish and turtle dopaminergic neurons is GABAergic (e.g. O'Connor *et al.*, 1988; Critz and Marc, 1992; see reviews by Witkovsky and Dearry, 1991; Besharse and Iuvone, 1992). In fact, there is little evidence that dopaminergic neurons in turtles or fishes possess effective levels of glutamate receptors, as neither glutamate, kainate nor other analogs have any major impact on release (O'Connor *et al.*, 1988; Critz and Marc, 1992). The primary role of glutamate in these systems is to depolarize GABAergic amacrine cells, keeping the dopaminergic neurons under massive, tonic inhibition. In turtles, at least, these results are consistent with the primary signal path to dopaminergic amacrine cells being photoreceptors > hyperpolarizing bipolar cells > GABAergic amacrine cells > dopaminergic neurons. In this model, at least, light onset would promote dopamine release (see also Negishi *et al.*, 1990). Not all release studies are mutually consistent, but on balance, it seems that there is some form of light onset signal leading to an increase in dopamine release in the ectotherm outer plexiform layer (Boatwright *et al.*, 1989; Dearry and Burnside, 1989; Kirsch and Wagner, 1989) and that the magnitude of this signal could reach 100–1000 nM in the extracellular space of an animal such as *Xenopus laevis*, where the primary source of outer plexiform layer signal is probably diffusion from the inner plexiform layer (Witkovsky *et al.*, 1993). It is thus likely that supramicromolar levels of dopamine can also be achieved in the outer plexiform layer of teleostean fishes. However, the circuitry in fishes is likely more complex than this simple view of disinhibitions since the dopaminergic interplexiform cells have nearly as much arbor in the proximal layers of the inner plexiform layer as in the distal layers, and perhaps both depolarizing and hyperpolarizing bipolar cells feed into sets of GABAergic amacrine cells that ultimately converge on the dopaminergic interplexiform cell.

Studies in other tetrapod vertebrates indicate other stimulus selectivities underlying dopamine release. For example, Hamasaki *et al.* (1986) noted an increase in dopamine release in cat retina at light offset, consistent with a direct hyperpolarizing bipolar cell input to dopaminergic amacrine cells in sublayer 1 of the inner plexiform layer (see also Hokoc and Mariani, 1988a,b; Kolb *et al.*, 1990). This is not inconsistent with the dopaminergic signal representing a light adaptation event since such bipolar cells are all cone-driven. At mesopic–photopic background fluxes, stimulus polarity is irrelevant and it is modulation that concerns us. Either an ON-center or OFF-center path would suffice to initiate neuronal or network light adaptation provided the photopic modulation depth were adequate. Interestingly, the OFF-center path in mammals may be the *preferable* photopic monitor since those bipolar cells are not explicitly encumbered by the rod bipolar cell > AII amacrine cell > ON-center bipolar cell sequence for scotopic operation, and the gap junctions between ON-center cone bipolar cells and AII amacrine cells are a likely target for dopamine-activated decoupling (Voigt and Wässle, 1987; Hampson *et al.*, 1992). In amphibians, however, dopamine release can be evoked by virtually all classes of glutamate receptor agonists (Boatright *et al.*, 1992; Rubim *et al.*, 1993), consistent with the presence of significant bipolar cell input to tyrosine hydroxylase immunoreactive profiles throughout the anuran inner plexiform layer (Gabriel *et al.*, 1992). These varied circuitries are likely to reflect preferred paths for signaling scotopic/photopic transitions and represent either niche selections (fundamentally different modes and timings of photic transitions) or path selections forced by intrinsic network features, such as whether rod and cone pathways share bipolar cells (as in many ectotherms) or whether complex circuitries are gated to control rod signal flow (as in mammals). It should be borne in mind that these cells may have other fast functions, especially the mixed GABAergic/dopaminergic cells.

Whereas most physiological responses of retinas to dopamine application mimic nominal 'light adaptation' (see Besharse and Iuvone, 1992), both the robust impact of prolonged dark adaptation on horizontal cells and its similarity to the effects of exogenously applied dopamine are irrefutable findings (Mangel and Dowling, 1985, 1987; Yang *et al.*, 1988a,b). What might these effects imply? This has been debated many times by various principals; our impression would be that total 'prolonged' light deprivation leads to an indeterminate state very different from biologically meaningful dark adaptation. This is similar to Negishi *et al*'s (1990) explanation. True biological dark adaptation involves the graded diminution in photon density until a stable scotopic level is achieved; it is never light-free and involves preserving retinal responsiveness to the prevailing flux. Thus it is proposed that the prolonged dark experiments involve signal deprivation and that, at some point, the retina resolves its state in terms of a net dopaminergic interplexiform cell escape from tonic inhibition; that the normal transition to the scotopic state involves the explicit suppression of dopamine release from which the interplexiform cells do not escape until signals of an impending photopic regime occur.

(c) Dopaminergic interplexiform cells regulate spinule formation on horizontal cells

Evidence that endogenous patterns of dopamine release actually control defined physiological states has been difficult to obtain. Whether dopamine actually regulates horizontal cell receptive field size in light adaptation remains controversial, with published data supporting positive and negative conclusions. Even so, an intriguing set of results suggests that light-driven spinule formation by teleostean horizontal cells is primarily gated by dopamine release at light onset (Wagner *et al.*, 1992) whereas the dopamine-sensitive retinomotor activities of photoreceptors can be independent of dopaminergic interplexiform cell systems (Douglas *et al.*,

1992). Though dopamine has been shown to be a powerful initiator of cone contraction in fishes and effective in maintaining contraction, it is apparent that other pathways, including direct light input (Burnside *et al.*, 1993), can access the final common paths regulating retinomotor state. Dopamine must play some role in physiological regulation of retinomotor function, as dopamine D2 antagonists can diminish the magnitude of cone contraction (McCormack and Burnside, 1992), but dopaminergic regulation seems not to be the only state-signaling path for various retinal elements. One could legitimately argue that the only robust explanation for the rich outer plexiform layer arbor of teleostean dopaminergic interplexiform cells is that they directly control horizontal cell spinule formation (Djamgoz *et al.*, 1989). A more extensive bibliography of this research area can be found in Witkovsky and Dearry (1991) and Wagner and Djamgoz (1993).

15.4.2 GLYCINERGIC INTERPLEXIFORM CELLS

(a) H1 horizontal cells drive glycinergic interplexiform cells in teleosts

The operations of glycinergic interplexiform cells are totally different from those associated with the dopaminergic system and involve fast operations in retinal circuitry. The only known role for glycine release from glycinergic neurons involves opening strychnine-sensitive Cl^- channels. Though glycine can gate NMDA receptors, there is no evidence that glycinergic connectivity plays any role in that process. Glycinergic interplexiform cells send their dendrites to the neuropil of the outer plexiform layer just above the distal surfaces of the H1 horizontal cells (Marc and Lam, 1981; Marc and Liu, 1984) where they receive conventional synapses from the GABAergic horizontal cells (Figure 15.3). Each horizontal cell appears to provide 5–10 conventional synapses to each dendrite segment as it courses past (Marc and Liu, 1984) and, in sum, each glycinergic interplexiform cell could integrate signals from hundreds of GABAergic horizontal cell synapses (Kalloniatis and Marc, 1990). Within the limits of Golgi impregnation, glycinergic interplexiform cells seem not to provide uniform innervation of the outer plexiform layer and yet, since the space constants of even uncoupled horizontal cells are so large, the sparse array of glycinergic interplexiform cells can still uniformly sample the horizontal cell layer (Kalloniatis and Marc, 1990). Some of the inputs to glycinergic interplexiform cells come from horizontal cell axon terminals of unknown provenance (Marc and Liu, 1984; Marshak and Dowling, 1987) and they likely add little to the fundamental operations of such a cell. Presuming glycinergic interplexiforms cells maintain a low intracellular Cl^- level, their light responses are probably depolarizing.

(b) Teleostean glycinergic interplexiform cells copy the S-space to the inner plexiform layer

Glycinergic interplexiform cells in goldfish make sparse output synapses onto unknown horizontal cell dendrites, although it is certain that at least some of these targets are non-H1 horizontal cells (Marc and Lam, 1981). There is little specific evidence to resolve the potential role of glycine in the outer plexiform layer of the goldfish retina. Wu and Dowling (1990) demonstrated that some H2 horizontal cells were glycine sensitive and that responses driven by red-sensitive cones were diminished but not blocked by 300 μM glycine. In the roach, however, strychnine failed to elicit any major

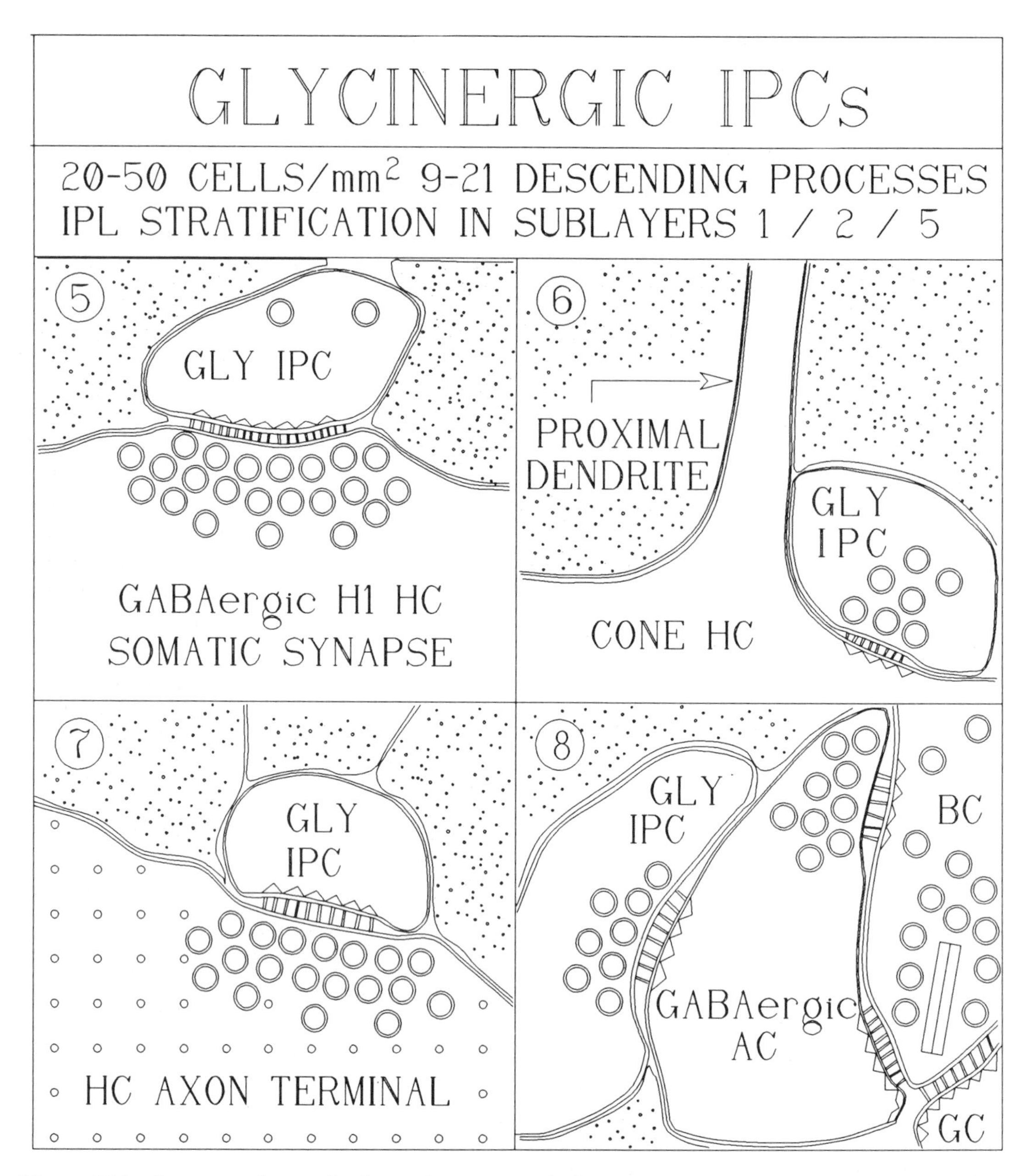

Figure 15.3 Summary data and schematic contacts of glycinergic interplexiform cells. Boxes refer to locations in Figure 15.1. Box 5: GLY IPC (glycinergic interplexiform cell) varicosities receive direct synaptic inputs from H1 HCs. Box 6: GLY IPC varicosities are sparsely presynaptic to unknown types of HCs. Box 7: GLY IPC axonal processes also receive conventional synaptic input from HC axon terminals. Box 8: GLY IPC axonal arborizations contact GABAergic ACs, which are in turn presynaptic to BC (bipolar cell) terminals.

change in color-opponent horizontal cell responses (Djamgoz and Ruddock, 1979), so the matter is still uncertain. Other workers report inconsistent hyperpolarizations of cone horizontal cells by glycine (Murakami *et al.*, 1972; Sugawara and Negishi, 1973; Negishi and Drujan, 1979). If the glycinergic interplexiform cells have a role to play in processing in the outer plexiform layer of cyprinids, it is an elusive one.

It has been proposed that glycinergic interplexiform cells in the goldfish function as devices to bypass the high spatial filter of the bipolar cells and present a copy of the integrated cone signal that exists in the aggregate of H1 horizontal cells, the 'S-space', to surround pathways of the inner plexiform layer. The evidence for this is purely morphological and there have been no pharmacological studies that could resolve the point since the path from horizontal cells to ganglion cells (the final element) would minimally involve a GABA>glycine chain and perhaps even a GABA>glycine>GABA chain prior to reaching its final target (Kalloniatis and Marc, 1990). Kalloniatis and Marc (1990) specifically argued that this pathway could provide a mechanism to reset ganglion cell thresholds in response to large, homogenous fields much greater in extent than bipolar cell surrounds. The evidence that these interplexiform cells target ganglion cells and GABAergic amacrine cells is based on the known patterns of glycinergic and GABAergic connectivity in the goldfish inner plexiform layer: that glycinergic cells, as a group, almost exclusively target ganglion cells and other non-glycinergic amacrine cells (Marc and Lam, 1981; Muller and Marc, 1990; R. Marc and W. Liu, unpublished). Yazulla and Studholme (1991) provided evidence that glycinergic interplexiform cells made axosomatic contacts with GABAergic amacrine cells of unknown type. A true picture of these connections awaits serial reconstructions and electron microscopic analysis of Golgi impregnated interplexiform cells.

(c) Anuran glycinergic interplexiform cells contact bipolar cells and may drive horizontal cells

The glycinergic interplexiform cell of anurans is a different cell. Although bearing features similar to cyprinid glycinergic interplexiform cells, these cells colocalize somatostatin, are an order of magnitude more frequent, and have different connectivities, forming conventional synapses primarily onto bipolar cell dendrites but not horizontal cells (Smiley and Basinger, 1988, 1990). Further, no evidence of direct horizontal cell input to the glycinergic interplexiform cells has been garnered, suggesting that these cells are predominantly presynaptic in the outer plexiform layer. Physiological evidence indicates that there is a direct, depolarizing, strychnine-sensitive glycinergic input to *Xenopus* horizontal cells (Stone and Witkovsky, 1984; Smiley and Basinger, 1990), and the presence of glycinergic interplexiform cells is an obvious correlate. The fact remains that neither autoradiography (Rayborn *et al.*, 1981), immunocytochemistry of somatostatin or glycine (Smiley and Basinger, 1988; Smiley and Yazulla, 1990), nor immunocytochemistry of gephyrin (Smiley and Yazulla, 1990; previously thought to be a glycine receptor subunit, but now known to be a cytoskeletal linker associated with various amino acid receptors) indicate the presence of classical glycinergic synapse onto horizontal cells. The mechanism of this effect remains a mystery and its significance in the physiology of horizontal cells is likewise difficult to assess. Unlike dopaminergic operations, glycinergic effects mediated by interplexiform cells are almost certain to be very fast and involve specific trigger features.

15.5 INTERPLEXIFORM CELLS EXHIBIT DIVERSE LIGHT-EVOKED RESPONSES

The light responses of interplexiform cells have been recorded infrequently and the conclusions regarding their stimulus preferences and response polarities are unclear. Most of the cells recorded with intracellular pipettes have probably been 'shunted' and any likelihood of observing good spikes compromised. The whole cell recordings (Maguire *et al.*, 1990) in tiger salamander interplexiform cells strongly argue that such cells use action potentials and, indeed, that is what one would expect of both dopaminergic and glycinergic interplexiform cells in fishes. One set of recordings from a probable dopaminergic interplexiform cell in goldfish (Djamgoz, *et al.*, 1991) exhibits phasic depolarizations in response to light flashes, consistent with the presence of regenerative mechanisms. Further, dopaminergic interplexiform cells are likely to be phasic in their frequency dependence since flickering light seems to mediate dopaminergic effects on horizontal cell receptive field size (Umino *et al.*, 1991). The first reported recordings of cyprinid interplexiform cells were made in dace (Hashimoto *et al.*, 1980) and came from cells that resemble neither the dopaminergic nor glycinergic cells of goldfish and that yielded either slow hyperpolarizing or depolarizing responses like bipolar cells. A large interplexiform cell also resembling the dopaminergic variety was reported to have slow hyperpolarizing responses (Shimoda *et al.*, 1992). Thus, none of the four publications on interplexiform cell physiology match well. However, on balance, the cell described by Djamgoz *et al.* (1991) most closely resembles the morphology of dopaminergic interplexiform cells with dendrites obliquely extending into the inner plexiform layer from sublayer 1. The neurons described by Shimoda *et al.* (1992) differ in some subtle ways. First, the primary dendrites seem strongly constrained to the distal half of sublayer 1 whereas autoradiography, FIF, tyrosine hydroxylase immunocytochemistry and dopamine immunocytochemistry all reveal large, oblique dendrites coursing towards sublayers 3 and 5 as well. Second, the axonal fields contact non-H1 and rod horizontal cells but not H1 horizontal cells, where the dopaminergic interplexiform cell sends 96% of its contacts to H1 horizontal cells. Contrary to the conclusions of Shimoda *et al.* (1992), therefore, I would suspect that this is not the dopaminergic interplexiform cell. On the other hand, it is remarkably similar to the S1 serotoninergic/GABAergic amacrine cell (Marc *et al.*, 1988; Ball and Tutton, 1990; Marc, 1992), which, as noted earlier (section 15.2.2), may possess interplexiform cell-like processes. Since S1 amacrine cells receive extensive and direct input from type Ma1 hyperpolarizing bipolar cells, this would explain their hyperpolarizing responses. Thus one might speculate from a broad synthesis of release data, connectivity and physiology, that in the goldfish retina there are three kinds of interplexiform cells with three distinct response waveforms, all of them exhibiting some spiking activity: (1) dopaminergic interplexiform cells with phasic depolarizing voltage responses (phasic or transient ON cells); (2) glycinergic interplexiform cells with tonic depolarizing responses (tonic or sustained ON cells); (3) serotoninergic/GABAergic interplexiform cells with tonic hyperpolarizing responses (tonic or sustained OFF cells).

15.6 CONCLUSIONS AND FUTURE PERSPECTIVES

The known interplexiform cells of vertebrates are a diverse cell group and are likely to have additional members join their ranks. The goldfish appears to have no less than four interplexiform cell varieties whereas the numbers in other vertebrates remain unknown, but there are at least two types in most mammals. All of these cells have varying degrees of axonal arborization, which suggests that they use action potentials. Some dopaminergic cells in mammalian retinas form such sparse outer plexiform fields that it is difficult to imagine that they could comprise a functional set, but we must remain cautious since, as yet, the hypothesis has not been tested. The teleostean dopaminergic interplexiform cell is quite distinctive in the density and specificity of its arbor in the outer plexiform layer; the cell is synapomorphic for diurnal teleosts but is absent in osteichthyans representing precursors to teleosts. The dopaminergic interplexiform cell is almost certainly responsible for signaling some but not all features of adaptation state and dopamine is one signal for photopic transitions. Glycinergic interplexiform cells are found in fishes, amphibians and avians, although their homologies remain to be established. In goldfish, the glycinergic interplexiform cell is a 'fast circuit' neuron involved in signaling the output of the horizontal cell layer or S-space to the inner plexiform layer. GABAergic neurons with interplexiform cell processes have been found in many vertebrate classes but nothing appears to be known of their functions.

A variety of experiments might be designed to clarify further the roles of various interplexiform cell types in the outer plexiform layer. One might surmise that quantitative differences may exist in the timing and magnitude of both dopamine activation and recovery from activation of effects in horizontal cells and or cones of those species possessing direct dopaminergic interplexiform cell input to the outer plexiform layer versus those signaling via diffusion. Perhaps a good pair of model systems would be teleosts and holeosts, as they are both cone-rich actinoptergian fishes with distinctive horizontal cell types differing in that only teleosts seem to possess a dopaminergic interplexiform cell system. There are also controversies yet to resolve about the environmental timings and degree of control exerted by dopaminergic systems on various outer retinal functions such as cone retinomotor operations. As noted above, factors in addition to dopamine seem to be involved in the regulation of cone length in teleosts, and one would like to know what those factors are and how they are regulated. Some of these queries may be resolved by more incisive mapping of dopamine receptor subtypes and genetic manipulation of receptor expression.

A more difficult problem is the prospect of deciphering the roles of GABAergic interplexiform cells in any species. In the goldfish, one might be able to uncover some of their functions in the outer plexiform layer (if they have any) (1) by destroying the serotoninergic S1 amacrine cells and observing whether any physiologically detectable GABA-mediated operations in the outer plexiform layer are contravened; or (2) by anatomically defining the targets of the S1 outer plexiform layer arbor. It is as likely that serotonin itself is the 'interplexiform' neurotransmitter in this case, and more serotoninergic pharmacology may open new doors. The tetrapod GABAergic systems remain completely enigmatic, but for a few anatomical clues (e.g. Nakamura *et al.*, 1980), and have simply been left out of the models of receptive field structure in mammals. It would be prudent to note that the functions of the inner plexiform layer involve a host of GABAergic amacrine cells whose synaptic innervation of any one target may be even less dense than the presumed outer plexiform layer targets of a GABAergic interplexiform cell. GABAergic interplexiform cells may have very potent and specific actions difficult to resolve from the activities of various concatenated GABAergic and glycinergic paths (Muller and Marc, 1990; Kalloniatis and Marc, 1990). Mapping of GABA receptor subtypes may further elucidate what these cells do in the outer plexiform layer.

Finally, glycinergic interplexiform cell systems are obviously not a popular topic as they are apparently absent from mammalian retinas. However, the possibility that they are widely spread among other vertebrate classes may implicate them as fundamental components of photopic processing. In our view, they are far from inconsequential components of ganglion cell receptive field construction and it might well be asked how mammals accomplish whatever it is that fishes achieve through glycinergic interplexiform cell functioning. We lack any strong pharmacological or molecular tools to subdivide types of glycinergic innervation and are still dependent on classical electrophysiological studies and further refinement of anatomic data to construct testable hypotheses of function. Hopefully, various forms of cell-specific markers or transgenic model systems will be developed, allowing the selective removal or modification of single neuronal types. Perhaps because they form diffuse connections between the plexiform layers, interplexiform cells remain rather underappreciated as components of visual processing. However, no formal understanding of vision is possible without the inclusion of every circuit component, and interplexiform cells give every evidence of playing pervasive roles in the control of signal flow through the retina.

ACKNOWLEDGEMENTS

I would like to thanks my colleagues, Michael Kalloniatis, Stuart Critz and W.-L. Sunny Liu for their invaluable contributions to the work cited from my laboratory. Errors contained herein are my responsibility alone. Portions of my work were supported by NIH grant EY02576 and a Jules and Doris Stein Research to Prevent Blindness Professorship.

REFERENCES

Baldridge, W.H. and Ball, A.K. (1993) A new type of interplexiform cell in the goldfish retina is PNMT immunoreactive. *Neuroreport*, **4**, 1015–18.

Ball, A.K. and Tutton, D.A. (1990) Contacts between S1 amacrine cells and 'I1' interplexiform cells in the goldfish retina. *Investigative Ophthalmology and Visual Science Supplement*, **27** 332.

Besharse, J.C. and Iuvone, P.M. (1992) Is dopamine a light-adaptive or dark-adaptive modulator in retina? *Neurochemistry International*, **20**, 193–9.

Bloomfield, S.A. (1990) Two types of orientation-sensitive responses of amacrine cells in the mammalian retina. *Nature*, **350**, 347–50.

Boatright, J.H., Hoel, M.J. and Iuvone, P.M. (1989) Stimulation of endogenous dopamine release and metabolism in amphibian retina by light- and K^+-evoked depolarization. *Brain Research*, **482**, 164–8.

Boatright, J.H., Gordon, J.R. and Iuvone, P.M. (1992) 2-Amino-4-phosphonobutyric acid (AP-4) blocks light-evoked dopamine release in frog retina. *Investigative Ophthalmology and Visual Science Supplement*, **33**, 1405.

Boycott, B.B., Dowling, J.E., Fisher, S.K. *et al.* (1975) Interplexiform cells of the mammalian retina and their comparison with catecholamine-containing retinal cells. *Proceedings of the Royal Society of London B*, **191**, 353–68.

Brunken, W.J., Witkovsky, P. and Karten, H.J. (1986) Retinal neurochemistry of three elasmobranch species: an immunohistochemical approach. *Journal of Comparative Neurology*, **243**, 1–12.

Burnside, B., Wang, E., Pagh–Roehl, K. and Rey, H. (1993) Retinomotor movements in isolated teleost cone inner–outer segment preparations (CIS–COS): effects of light, dark and dopamine. *Experimental Eye Research*, **57**, 709–22.

Cajal, S.R.Y. (1972) *The Structure of the Retina* (Compiled and translated by S.A. Thorpe and M. Glickson), C.C. Thomas, Springfield, IL.

Cajal, S.R.Y. (1892) La retine des vertebres. *La Cellule*, **9**, 121–246.

Carroll, R.L. (1988) *Vertebrate Paleontology and Evolution*. W.H. Freeman, New York, 698 pp.

Critz, S.D. and Marc, R.E. (1992) Glutamate antagonists that block hyperpolarizing bipolar cells increase the release of dopamine from turtle retina. *Visual Neuroscience*, **9**, 271–8.

Dacey, D.M. (1989) Axon-bearing amacrine cells of the macaque monkey retina. *Journal of Comparative Neurology*, **284**, 275–93.

Dacey, D.M. (1990) The dopaminergic amacrine cell. *Journal of Comparative Neurology*, **301**, 461–89.

Dearry, A. and Burnside, B. (1989) Regulation of cell motility in teleost retinal photoreceptors and pigment epithelium by dopaminergic D_2 receptors, in *Extracellular and Intracellular Messengers in the Vertebrate Retina* (eds D. Redburn and H. Pasantes-Morales), Alan R. Liss, New York, pp.229–56.

de Miguel, E. and Wagner, H-J. (1990) Tyrosine hydroxylase immunoreactive interplexiform cells in the lamprey retina. *Neuroscience Letters*, **113**, 151–5.

Djamgoz, M.B.A. and Ruddock, K.H. (1979) Effects of picrotoxin and strychnine on fish retinal S-potentials: evidence for inhibitory control of depolarizing responses. *Neuroscience Letters*, **12**, 329–34.

Djamgoz, M.B.A. and Wagner, H-J. (1992) Localization and function of dopamine in the adult vertebrate retina. *Neurochemistry International*, **20**, 139–91.

Djamgoz, M.B.A., Usai, C. and Vallerga, S. (1991) An interplexiform cell in the goldfish retina: light-evoked response pattern and intracellular staining with horseradish peroxidase. *Cell Tissue Research*, **264**, 111–16.

Douglas, R.H., Wagner, H-J., Zaunreiter, M. *et al.* (1992) The effect of dopamine depletion on light-evoked and circadian retinomotor movements in the teleost retina. *Visual Neuroscience*, **9**, 335–43.

Dowling, J.E. (1991) Retinal neuromodulation: the role of dopamine. *Visual Neuroscience*, **7**, 87–97.

Dowling, J.E. and Ehinger, B. (1975) Synaptic organization of the amine-containing interplexiform cells of the goldfish and *Cebus* monkey retinas. *Science*, **188**, 270–73.

Dowling, J.E. and Ehinger, B. (1978) The interplexiform cell system I. Synapses of the dopaminergic neurons of the goldfish retina. *Proceedings of the Royal Society of London B*, **201**, 7–26.

Ehinger, B. and Dowling, J.E. (1987) Retinal neurocircuitry and transmission, in *Handbook of Clinical Neuroanatomy*. V5. *Integrated Systems of the CNS*, Part I (eds A. Björklund, T. Hökfelt and L.W. Swanson), Elsevier, Amsterdam, pp.389–446.

Ehinger, B. and Falck, B. (1969) Adrenergic retinal neurons of some new world monkeys. *Zeitschrift fur Zellforschung*, **100**, 364–75.

Ehinger, B., Falck, B. and Laties, A.M. (1969) Adrenergic neurons in teleost retina. *Zeitschrift fur Zellforschung*, **97**, 285–97.

Famiglietti, E.V. Jr (1992a) Polyaxonal amacrine cells of rabbit retina: morphology of PA1 cells. *Journal of Comparative Neurology*, **316**, 391–405.

Famiglietti, E.V. Jr (1992b) Polyaxonal amacrine cells of rabbit retina: PA2, PA3, and PA4 cells: light and electron microscopic studies with a functional interpretation. *Journal of Comparative Neurology*, **316**, 422–46.

Favard, C., Simon, A., Vigny, A. and Nguyen-Legros, J. (1990) Ultrastructural evidence for a close relationship between dopamine cell processes and blood capillary walls in *Macaca* monkey and rat retina. *Brain Research*, **523**, 127–33.

Frederick, J.M., Rayborn, M.E., Laties, A.M. *et al.* (1982) Dopaminergic neurons in the human retina. *Journal of Comparative Neurology*, **210**, 65–79.

Gabriel, R., Zhu, B. and Straznicky, C. (1992) Synaptic contacts of tyrosine hydroxylase-immunoreactive elements in the inner plexiform layer of the retina of *Bufo marinus*. *Cell Tissue Research*, **276**, 525–34.

Gallego, A. (1971) Horizontal and amacrine cells in the mammal's retina. *Vision Research*, **3**, 33–50.

Hamasaki, D.I., Trattler, W.B. and Hajek, A.S. (1986) Light on suppresses and light off enhances the release of dopamine from the cat's retina. *Neuroscience Letters*, **68**, 112–66.

Hampson, E.C.G., Vaney, D.I. and Weiler, R. (1992) Dopaminergic modulation of gap junction permeability between amacrine cells in mammalian retina. *Journal of Neuroscience*, **12**, 4911–22.

Hashimoto, Y., Abe, M. and Inokuchi, M. (1980) Identification of the interplexiform cell in the dace retina by dye injection method. *Brain Research*, **197**, 331–40.

Hokoç, J.N. and Mariani, A.P. (1988a) Synapses from bipolar cells onto dopaminergic amacrine cells in cat and rabbit retinas. *Brain Research*, **461**, 17–26.

Hokoç, J.N. and Mariani, A.P. (1988b) Tyrosine hydroxylase immunoreactivity in the rhesus monkey reveals synapses from bipolar cells to dopaminergic amacrine cells. *Journal of Neuroscience*, **7**, 2785–93.

Kalloniatis, M.K. and Fletcher, E.L. (1993) Immunocytochemical localization of the amino acid neurotransmitters in the chicken retina. *Journal of Comparative Neurology*, **336**, 174–93.

Kalloniatis, M. and Marc, R.E. (1990) Interplexiform cells of the goldfish retina. *Journal of Comparative Neurology*, **297**, 340–58.

Kirsch, M. and Wagner, H-J. (1989) Release pattern of endogenous dopamine in teleost retinae during light adaptation and pharmacological stimulation. *Vision Research*, **29**, 147–54.

Kleinschmidt, J. and Yazulla, S. (1984) Uptake of [^{3}H]-glycine in the outer plexiform layer of the retina of the toad, *Bufo marinus*. *Journal of Comparative Neurology*, **230**, 352–60.

Knapp, A.G. and Dowling, J.E. (1987) Dopamine enhances excitatory amino acid-gated conductances in cultured retinal horizontal cells. *Nature*, **325**, 437–9.

Kolb, H. and West, R.W. (1977) Synaptic connections of the interplexiform cell in the retina of the cat. *Journal of Neurocytology*, **6**, 155–70.

Kolb, H., Linberg, K.A. and Fisher, S.K. (1992) Neurons of the human retina: A Golgi study. *Journal of Comparative Neurology*, **318**, 147–87.

Kolb, H., Cline, C., Wang, H.H. and Brecha, N. (1987) Distribution and morphology of dopaminergic amacrine cells in the retina of the turtle (*Pseudemys scripta elegans*). *Journal of Neurocytology*, **16**, 577–88.

Kolb, H., Cuenca, N., Wang, H.–H. and Dekorver, L. (1990) The synaptic organization of the dopaminergic amacrine cell in the cat retina. *Journal of Neurocytology*, **19**, 343–66.

Lasater, E.M. (1987) Retinal horizontal cell gap junctional conductance is modulated by dopamine through a cyclic-AMP dependent protein kinase. *Proceedings of the National Academy of Sciences, USA*, **84**, 7319–23.

Lasater, E.M. and Dowling, J.E. (1985) Dopamine decreases conductance of the electrical junctions between cultured retinal horizontal cells. *Proceedings of the National Academy of Sciences, USA*, **82**, 3025–9.

Maguire, G., Lukasiewicz, P. and Werblin, F. (1990) Synaptic and voltage-gated currents in interplexiform cells of the tiger salamander retina. *Journal of General Physiology*, **95**, 755–70.

Mangel, S.C. and Dowling, J.E. (1985) Responsiveness and receptive field size of carp horizontal cells are reduced by prolonged darkness and dopamine. *Science*, **229**, 1107–9.

Mangel, S.C. and Dowling, J.E. (1987) The interplexiform–horizontal cell system of the fish retina: effects of dopamine, light stimulation and time in the dark. *Proceedings of the Royal Society of London B*, **231**, 91–121.

Marc, R.E. (1982) Spatial organization of neurochemically classified interneurons in the goldfish retina. I. Local patterns. *Vision Research*, **22**, 589–608.

Marc, R.E. (1985) The role of glycine in retinal circuitry, in *Retinal Transmitters and Modulators: Models for the Brain*, Vol. 1 (ed. W. Morgan), CRC Press, Boca Raton, FL, pp.119–58.

Marc, R.E. (1989) Evolution of retinal circuits, in *Neural Mechanisms of Behavior*, Proceedings of the 2nd International Congress of Neuroethology (eds J. Erber, R. Menzel, H.–J. Pfluger and D. Todt), Thieme Medical Publishers, New York, pp.146–7.

Marc, R.E. (1992) The structure of GABAergic circuits in ectotherm retinas, in *GABA in the Retina and Central Visual System* (eds R. Mize, R.E. Marc, and A. Sillito), Elsevier, Amsterdam, pp. 61–92.

Marc, R.E. and Lam, D.M.K. (1981) Glycinergic pathways in the goldfish retina. *Journal of Neuroscience*, **1**, 152–65.

Marc, R.E. and Liu, W.–L. (1984) Horizontal cell synapses onto glycine-accumulating interplexiform cells. *Nature*, **311**, 266–9.

Marc, R.E., Liu, W.–L.S., Scholz, K. and Muller, J.F. (1988) Serotonergic pathways in the goldfish retina. *Journal of Neuroscience*, **8**, 3427–50.

Mariani, A.P. (1982) Association amacrine cells could mediate directional selectivity in pigeon retina. *Nature*, **298**, 654–55.

Mariani, A.P. and Hokoç, J.N. (1988) Two types of tyrosine hydroxylase immunoreactive amacrine cells in the rhesus monkey retina. *Journal of Comparative Neurology*, **276**, 81–91.

Marshak, D.W. and Dowling, J.E. (1987) Synapses of the cone horizontal cell axons of the goldfish retina. *Journal of Comparative Neurology*, **256**, 430–43.

McCormack, C.A. and Burnside, B. (1992) A role for endogenous dopamine in circadian regulation of retinal cone movement. *Experimental Eye Research*, **55**, 511–20.

McMahon, D.G., Knapp, A.G. and Dowling, J.E. (1989) Horizontal cell gap junctions: single-channel conductance and modulation by dopamine. *Proceedings of the National Academy of Sciences, USA*, **86**, 7639–43.

Muller, J.F. and Marc, R.E. (1990) GABA-ergic and glycinergic pathways in the inner plexiform layer of the goldfish retina. *Journal of Comparative Neurology*, **291**, 281–304.

Murakami, M., Ohtsu, K. and Ohtsuka, T. (1972) Effects of chemicals on receptors and horizontal cells in the retina. *Journal of Physiology*, **227**, 899–913.

Nakamura, Y., McGuire, B.A. and Sterling, P. (1980) Interplexiform cell in cat retina: identification by uptake of gamma-(^{3}H)-aminobutyric acid serial reconstruction. *Proceedings of the National Academy of Sciences, USA*, **77**, 658–61.

Negishi, K. and Drujan, B.D. (1979) Effects of catecholamines and related compounds on the horizontal cells of the fish retina. *Journal of Neuroscience Research*, **4**, 311–34.

Negishi, K., Teranishi, T. and Kato, S. (1990) The dopamine system of the teleost fish retina. *Progress in Retinal Research*, **9**, 1–48.

Nelson, J.S. (1984) *Fishes of the World*. Wiley, New York.

Ngyuen-LeGros, J., Berger, B., Vigny, A. and Alvarez, C. (1981) Tyrosine hydroxylase-like immunoreactive interplexiform cells in the rat retina. *Neuroscience Letters*, **27**, 255–9.

O'Connor, P., Zucker, C.L. and Dowling, J.E. (1987) Regulation of dopamine release from interplexiform cell processes in the outer plexiform layer of the carp retina. *Journal of Neurochemistry*, **49**, 916–20.

O'Connor, P., Dorisom, S.J., Watling, K.J. and Dowling, J.E. (1988) Factors affecting the release of [^{3}H] dopamine from perfused carp retina. *Journal of Neuroscience*, **6**, 1857–65.

Oyster, C.W., Takahashi, E.S., Cilluffo, M. and Brecha, N. (1985) Morphology and distribution of tyrosine hydroxylase-like immunoreactive neurons in the cat retina. *Proceedings of the National Academy of Sciences, USA*, **82**, 6335–9.

Oyster, C.W., Takahashi, E.S. and Brecha, N. (1988) Morphology of retinal dopaminergic neurons, in *Dopaminergic Mechanisms in Vision*, Alan R. Liss, New York, pp. 19–33.

Pfeiffer-Linn, C. and Lasater, E.M. (1993) Dopamine modulates in a differential fashion T- and L-type calcium currents in bass retinal horizontal cells. *Journal of General Physiology*, **102**, 277–94.

Piccolino, M., Neyton, J., Witkovsky, P. and Gerschenfeld, H.M. (1982) γ-Aminobutyric acid antagonists decrease junctional communication between L-horizontal cells of the turtle retina. *Proceedings of the National Academy of Sciences, USA*, **79**, 3671–5.

Piccolino, M., Witkovsky, P. and Trimarchi, C. (1987) Dopaminergic mechanisms underlying the reduction of electrical coupling between horizontal cells of the turtle retina induced by d-amphetamine, bicuculline and veratridine. *Journal of Neuroscience*, **7**, 2273–84.

Piechl, L. (1991) Catecholaminergic amacrine cells in the dog and wolf retina. *Visual Neuroscience*, **7**, 575–87.

Rayborn, M.E., Sarthy, P.V., Lam, D.M.-K. and Hollyfield, J.G. (1981) The emergence, localization and maturation of neurotransmitter systems during the development of the retina in *Xenopus laevis*: II. Glycine. *Journal of Comparative Neurology*, **195**, 585–93.

Rubim, N.M., Boatright, J.H., Gordon, J.R. and Iuvone, P.M. (1993) NMDA receptors regulate dopamine release in amphibian retina. *Investigative Ophthalmology and Visual Science Supplement*, **34**, 1770.

Ryan, M.K. and Hendrickson, A.E. (1987) Interplexiform cells in Macaque monkey retina. *Experimental Eye Research*, **45**, 57–66.

Schütte, M. and Witkovsky, P. (1991) Dopaminergic interplexiform cells and centrifugal fibers in the *Xenopus laevis* retina. *Journal of Neurocytology*, **20**, 195–207.

Shimoda, Y., Hidaka, S., Maehara, M. *et al.* (1992) Hyperpolarizing interplexiform cell of the dace retina identified physiologically and morphologically. *Visual Neuroscience*, **8**, 193–199.

Smiley, J.F. and Basinger, S.F. (1988) Somatostatin-like immunoreactivity and glycine high-affinity uptake colocalize to an interplexiform cell of the *Xenopus laevis* retina. *Journal of Comparative Neurology*, **274**, 608–18.

Smiley, J.F. and Basinger, S.F. (1990) Glycine stimulates calcium-independent release of [^{3}H]-GABA from isolated retinas of *Xenopus laevis*. *Visual Neuroscience*, **4**, 337–48.

Smiley, J.F. and Yazulla, S. (1990) Glycinergic contacts in the outer plexiform layer of the *Xenopus laevis* retina characterized by antibodies to glycine, GABA and glycine receptors. *Journal of Comparative Neurology*, **299**, 375–88.

Stone, S. and Witkovsky, P. (1984) The actions of γ-aminobutyric acid, glycine and their antagonists upon horizontal cells of the *Xenopus* retina. *Journal of Physiology*, **353**, 249–64.

Sugawara, K. and Negishi, K. (1973) Effects of some amino acids on horizontal cell membrane potential in the isolated carp retina. *Vision Research*, **13**, 977–81.

Takahashi, E.S. (1988) Dopaminergic neurons in the cat retina. *American Journal of Optometry and Physiological Optics*, **65**, 331–6.

Teranishi, T., Negishi, K. and Kato, S. (1984) Regulatory effects of dopamine on spatial properties of horizontal cells in carp retina. *Journal of Neuroscience*, **4**, 1271–80.

Teranishi, T., Negishi, K. and Kato, S. (1987) Functional and morphological correlates of amacrine cells in carp retina. *Neuroscience*, **20**, 935–50.

Tornquist, K., Uddman, R., Sundler, F. and Ehinger, B. (1982) Somatostatin and VIP neurons in the retina of different species, *Histochemistry*, **76**, 137–52.

Umino, O., Lee, Y. and Dowling, J.E. (1991) Effects of light stimuli on the release of dopamine from interplexiform cells in the white perch retina. *Visual Neuroscience*, **7**, 451–8.

Vaney, D.I. (1990) The mosaic of amacrine cells in the mammalian retina. *Progress in Retinal Research*, **9**, 49–100.

Van Haesendonck, E., Marc, R.E. and Missotten, L. (1993) New aspects of dopaminergic interplexiform cell organization in the goldfish retina. *Journal of Comparative Neurology*, **333**, 503–18.

Voight, T. and Wässle, H. (1987) Dopaminergic innervation of AII amacrine cells in mammalian retina. *Journal of Neuroscience*, **7**, 4115–28.

Wagner, H-J. (1976) Patterns of Golgi-impregnated neurons in a predator-type fish retina, in *Neural Principles in Vision* (eds F. Zettler and R. Wieler), Springer-Verlag, Berlin, pp. 7–25.

Wagner, H-J. and Behrens, U.D. (1993) Microanatomy of the dopaminergic systems in the rainbow trout retina. *Vision Research*, **33**, 1345–58.

Wagner, H-J. and Djamgoz, M.B.A. (1993) Spinules: a case for retinal synaptic plasticity. *Trends in Neuroscience*, **16**, 201–6.

Wagner, H-J. and Wülle, I. (1990) Dopaminergic interplexiform cells contact photoreceptor terminals in catfish retina. *Cell Tissue Research*, **261**, 359–65.

Wagner, H-J. and Wülle, I. (1992) Contacts of dopaminergic interplexiform cells in the outer retina of the blue acara. *Visual Neuroscience*, **9**, 325–33.

Wagner, H-J., Behrens, U.D., Zaunreiter, M. and Douglas, R.H. (1992) The circadian component of spinule dynamics in teleost horizontal cells is dependent on the dopaminergic system. *Visual Neuroscience*, **9**, 345–51.

Wagner, H-J., Frölich, E., Negishi, K. and Miki, N. (1994) Differentiation of dopaminergic cells in fish retinae: only cone containing retinae have interplexiform cells with extensive telodendria in the outer plexiform layer. *Investigative Ophthalmology and Visual Science Supplement*, **35**, 512.

Wässle, H. and Chun, M.H. (1988) Dopaminergic and indoleamine-accumulating amacrine cells express GABA-like immunoreactivity in the cat retina. *Journal of Neuroscience*, **8**, 3383–94.

Watt, C.B. and Florack, V.J. (1992) A double-label analysis demonstrating the non-coexistence of tyrosine hydroxylase-like and GABA-like immunoreactivities in amacrine cells of the larval tiger salamander retina. *Neuroscience Letters*, **148**, 47–50.

Werblin, F.S. (1977) Regenerative amacrine cell depolarization and formation of on–off ganglion cell responses. *Journal of Physiology*, **264**, 767–85.

Witkovsky, P. and Dearry, A. (1991) Functional roles of dopamine in the vertebrate retina. *Progress in Retinal Research*, **11**, 247–92.

Witkovsky, P. and Schütte, M. (1991) The organization of dopaminergic neurons in vertebrate retinas. *Visual Neuroscience*, **7**, 113–24.

Witkovsky, P., Nicholson, C., Rice, M.E. *et al.* (1993) Extracellular dopamine concentration in the retina of the clawed frog *Xenopus laevis*. *Proceedings of the National Academy of Sciences, USA*, **90**, 5667–71.

Wu, S.M. and Dowling, J.E. (1990) Effects of GABA and glycine on the distal cells of the cyprinid retina. *Brain Research*, **199**, 401–14.

Wülle, I. and Wagner, H-J. (1990) GABA and tyrosine hydroxylase immunocytochemistry reveal different patterns of colocalization in retinal neurons of various vertebrates. *Journal of Comparative Neurology*, **296**, 173–8.

Yamada, M., Shigematsu, Y., Umetani, Y. and Saito, T. (1992) Dopamine decreases receptive field size of rod-driven horizontal cells in carp retina. *Vision Research*, **32**, 1801–7.

Yang, C.-Y. and Yazulla, S. (1988) Light microscopic localization of putative glycinergic neurons in the larval tiger salamander by immunocytochemical and autoradiographical methods. *Journal of Comparative Neurology*, **272**, 343–57.

Yang, X.-L., Tornqvist, K. and Dowling, J.E. (1988a) Modulation of cone horizontal cell activity in the teleost fish retina. I. Effects of prolonged darkness and background illumination on light responsiveness. *Journal of Neuroscience*, **8**, 2259–68.

Yang, X.-L., Tornqvist, K. and Dowling, J.E. (1988b) Modulation of cone horizontal cell activity in the teleost fish retina. II. Role of interplexiform cells and dopamine in regulating light responsiveness. *Journal of Neuroscience*, **8**, 2259–68.

Yazulla, S. and Studholme, K. (1991) Glycinergic interplexiform cells make synaptic contact with amacrine cell bodies in goldfish retina. *Journal of Comparative Neurology*, **310**, 1–10.

Yazulla, S. and Zucker, C.L. (1988) Synaptic organization of dopaminergic interplexiform cells in the goldfish retina. *Visual Neuroscience*, **1**, 13–29.

Zaunreiter, M. and Wagner, H-J. (1991) Dopaminergic interplexiform cells contact photoreceptor terminals in teleosts. *Investigative Ophthalmology and Visual Science Supplement*, **32**, 1260.

16

The involvement of Müller cells in the outer retina

ANDREAS REICHENBACH and
STEPHEN R. ROBINSON

16.1 INTRODUCTION

Neurons are always found in intimate apposition to glia, regardless of the region of the nervous system examined or the species of vertebrate concerned (Reichenbach and Robinson, 1995). This tight correspondence strongly suggests that these two cell types are interdependent; this suggestion has indeed been amply confirmed by a large and diverse body of evidence. Different types of glia are specialized for different purposes: astrocytes provide metabolic support to neurons and maintain homeostasis of the extracellular environment; microglia mediate response to tissue damage and phagocytose neuronal debris; while oligodendrocytes on the other hand wrap axons and neuronal somata in an insulating sheath to limit the lateral, and to accelerate the longitudinal, conduction of electrical impulses.

H. Müller (1851) was the first to describe the radial trunks of macroglia spanning the thickness of the retina. These cells now bear his name and have been found in the retinas of all vertebrates investigated so far, including Agnatha. Müller cells are the only type of glia found in the retinas of non-mammalian vertebrates. Additional types of glia are found in those mammals that have a retinal vasculature. Astrocytes in such species are always restricted to vascularized regions of the nerve fiber layer, as are oligodendrocytes, although the latter cells are only found in a few species (e.g. rabbits). Vascularized retinas also contain microglia and these cells form into arrays in each of the plexiform layers. Nevertheless, even in vascularized retinas, Müller cells are the numerically dominant form of glia and they are the only glia with processes that extend into the outer nuclear layer.

The fact that Müller cells are the exclusive form of glia in the outer retina has meant that they are required to perform tasks normally undertaken by several different types of glia. This impost is made all the greater by the outer retina having one of the highest metabolic rates of any tissue in the central nervous system (CNS). In fact, Müller cells are responsible for a wide range of tasks and without these hard-working cells it is unlikely that the outer retina would be able to function at all.

Neurobiology and Clinical Aspects of the Outer Retina
Edited by M.B.A. Djamgoz, S.N. Archer and S. Vallerga
Published in 1995 by Chapman & Hall, London
ISBN 0 412 60080 3

16.2 DEVELOPMENT OF MÜLLER CELLS

Müller cells begin their service to the retina at an early stage of its development. The genesis of cells in the developing mammalian retina can be divided into two phases. The earliest phase produces several types of 'primary' neurons (Figure 16.1; black) cf. Robinson 1991), whereas in the second phase, columnar clones are generated from a common progenitor (Figure 16.1; white), such clones consist of one Müller cell and a defined number of rod photoreceptors, bipolar cells and amacrine cells (see references in Reichenbach *et al.*, 1994b). All these neurons are 'born' at the apical margin of the retinal neuroepithelium, in the zone adjacent to the pigment epithelium, so their nuclei have to migrate 50–200 μm to their final resting position. By analogy with events in the developing cerebral cortex, one might expect that the newborn retinal neurons would use the radial trunks of their Müller cell 'sisters' to guide their migration. Regional expression of certain marker molecules on the cell membranes of the Müller cells could assist the neurons in locating their correct laminar position. This mechanism is obviously not used by newborn ganglion cells, horizontal cells, cones and first amacrine cells, since cell birth-date studies have consistently shown that Müller cells are not born until the second phase of cytogenesis (for review see Robinson, 1991). Furthermore, such neurons are born at a time when the retinal tissue is rather thin; thus, they can translocate their somata within the cytoplasmic 'tubes' spanning the width of the retina. In contrast, Müller cell guidance might be important for those neurons appearing later.

Müller cells help to direct the growth of neurites that extend from differentiating neurons. This has been demonstrated most convincingly in preparations of dissociated embryonic rat retinas in which the neurites of differentiating rod photoreceptors show a strong preference for Müller cells as a substrate (Kljavin and Reh, 1991). Conversely, contact with maturing neurons appears to trigger the maturation of Müller cells such as expression of glutamine sythetase. This interaction may be mediated by the 5A11 antigen, which is homologous to the HT7 antigen (a member of the immunoglobulin super gene family), and is predominantly associated with Müller cells (Fadool and Linser, 1993). Exposure of retinal cell cultures to antibodies against the 5A11 antigen results in (i) reduced expression of glutamine synthetase as a marker enzyme of glial cell maturation, and (ii) altered development of the stereotypic arrangement of neurons and glia characterized by a reduction in the number and complexity of neuronal processes growing on glia as well as by reduced neuronal cell aggreagation on Müller cells (Fadool and Linser, 1993). Thus, Müller cells express recognition molecules which may play important roles in retinal development.

16.3 STRUCTURAL FEATURES OF MÜLLER CELLS

In mature retinas, the somata of Müller cells are situated in the inner nuclear layer where they may form a distinct sublayer. Müller cells have a bipolar morphology and they always extend processes towards the inner and outer limiting membranes of the retina. Although this review is concerned primarily with the outer retina, we will give a brief overview of the inner (vitread) part of Müller cells because many of the functions served by these cells in the outer retina are only made possible by the collective features of the whole Müller cell.

The dominant feature of the Müller cell is its radial trunk which extends from the outer plexiform layer (OPL) to the nerve fiber, layer. In the cone-dominated retinas of most reptilians and birds, where the inner retinal

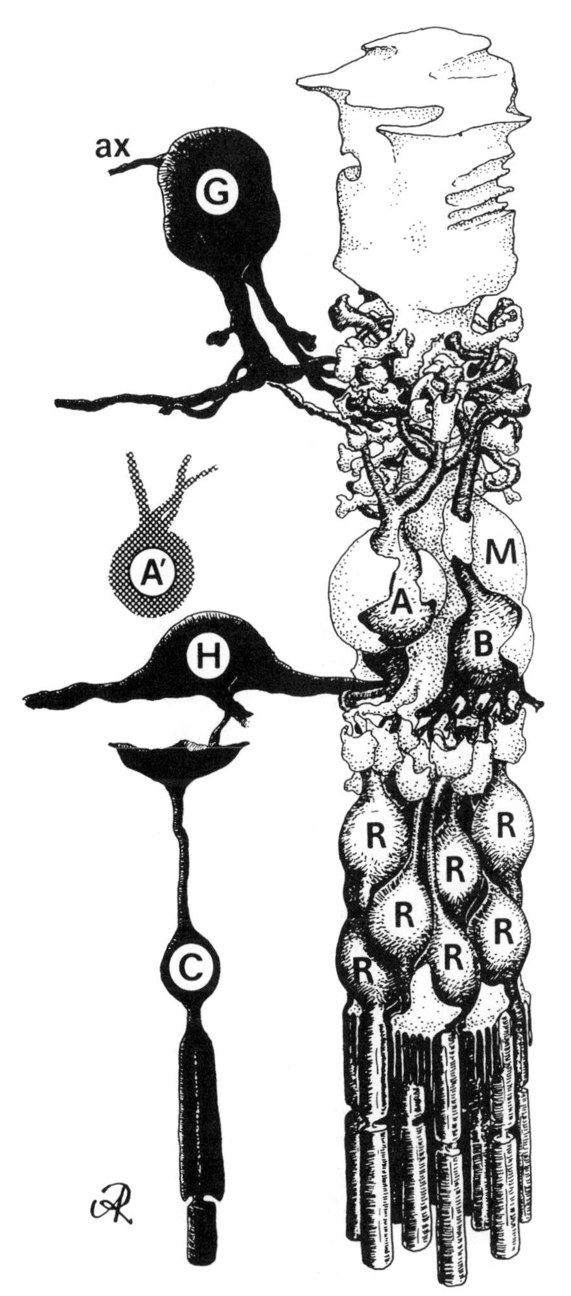

Figure 16.1 Semischematic drawing of the columnar arrangement of neuronal cells along Müller cells. ax, ganglion cell axon, A, amacrine cells; B, bipolar cell; C, cone photoreceptor cell; G, ganglion cell; H, horizontal cell, M, Müller cell; R, rod photoreceptor cell. All somata and processes of neuronal cells are extensively ensheathed by Müller cell processes. In black, the early-born types of neuronal cells are given; these cells are more rare than Müller cells, such that not every Müller cell contacts one of these cells. (Reproduced from Reichenbach *et al.* (1993) with permission.)

layers are very thick and complex, the trunk often splits into numerous fibrils as it enters the inner plexiform layer. In vascularized retinas, the trunks contain mitochondria (Uga and Smelser, 1973). In both vascular and avascular retinas, the trunks have a core of intermediate filaments that are immunoreactive for vimentin. This immunoreactivity is normally restricted to filaments in the inner half of the Müller cell, but following trauma to the retina (e.g. retinal detachment) vimentin immunoreactivity is massively upregulated and can be seen throughout the extent of the Müller cells, including those portions in the outer retina (Guèrin *et al.*, 1990). These filaments can also exhibit immunoreactivity for glial fibrillary acidic protein (GFAP). The extent of this immunoreactivity normally decreases as the cells mature, but it too is rapidly upregulated when Müller cells respond to retinal trauma (e.g. Bignami and Dahl, 1979; Guèrin *et al.*, 1990).

The trunk terminates in an expanded conical endfoot. These endfeet contain abundant endoplasmic reticulum, and lie adjacent to the basal lamina of the inner limiting membrane. In most species (with a few exceptions e.g. frogs), characteristic orthogonal arrays of particles (OAPs) are found in the region of the endfoot that abuts the basal lamina (Wolburg and Berg, 1988). In some species (e.g. mouse), OAPs are found in all parts of the cell membrane. Despite this variability, OAPs in all species are concentrated in regions of the cell membrane that lie adjacent to the vitreous humour or capillary walls, and it has been suggested that OAPs are sites of increased K^+ conductance (Wolburg and Berg, 1988). In species with vascularized retinas, Müller cells form 'en passant endfeet' onto retinal blood vessels (Wolburg and Berg, 1988); the trunks of a few Müller cells even terminate on large vessels instead of extending to the inner limiting membrane.

At their outermost end, Müller cells extend apical microvilli between the inner segments of photoreceptor cells in the subretinal space (Figure 16.1). The length and number of microvilli vary between species, possibly in inverse relation to the degree of retinal vascularization (Uga and Smelser, 1973; see however Dreher *et al.*, 1992). The scleradmost portion of Müller cell cytoplasm is densely filled with mitochondria, particularly in species with avascular retinas, where the choriocapillaris is the principal source of oxygen and nutrition. Apicolaterally, Müller cells are connected to neighboring Müller and photoreceptor cells by specialized junctions, thereby forming the outer limiting 'membrane'. In most vertebrates these junctions are zonulae adhaerentes or (in fish) tight junctions. However, amphibian Müller cells are connected by intermediate junctions or (in frogs and toads) gap junctions (Uga and Smelser, 1973; Fain *et al.*, 1976) that permit extensive electrical coupling (Attwell *et al.*, 1986). Neither gap nor tight junctions have been demonstrated in the apicolateral membranes of mammalian Müller cells *in situ*, and further doubt has been cast on their existence by the observation that Müller cells in the intact rabbit retina are never dye-coupled to each other (Robinson *et al.*, 1993).

Müller cells send side branches into both plexiform layers, where they ensheath dendrites and synapses, particularly the cone pedicles in the OPL (Figure 16.1). In the nuclear layers, Müller cells enclose neuronal somata within bubble-like lamellae. More detailed descriptions of the structure and ultrastructure of Müller cell processes are given elsewhere (Reichenbach, 1989; Reichenbach *et al.*, 1989; Dreher *et al.*, 1992).

The morphology of Müller cells and the way that they fit into the cytoarchitecture of the retina has several important implications for neuronal function. First, by contributing to both the inner and outer limiting membranes, Müller cells seal the retina from the external environment. Except for the outer segments of photoreceptors, all other points of entry into the retina are lined by the processes of Müller cells. Thus, passage of substances into and out of the retina is limited. Second, by spanning the thickness of the retina, Müller cells are able to provide a conduit for the passage of molecules, including ions, between the subretinal space and the vitreous humour. Third, Müller cells contribute to the blood: retina barrier by direct formation of apicolateral tight junctions, as in fish retinas, or through the induction of tight junctions between adjacent endothelial cells of retinal capillaries, as in mammals (Tout *et al.*, 1993). In either case, Müller cells are in a position to monitor the flow of substances into and out of the vasculature. Fourth, by completely ensheathing individual neuronal somata and synapses, Müller cells are able to isolate them from electrical or chemical interference. This isolation means that Müller cells are also able to control the extracellular microenvironment of individual neurons, particularly as the extracellular clefts in the retina are exceedingly narrow (Karwoski *et al.*, 1985). All of these features lead to the conclusion that Müller cells can potentially exert an extraordinary degree of control over the composition of the extracellular environment. Since neurons are exquisitely sensitive to even slight perturbations in the chemical composition of the fluid bathing them, the need for a detailed understanding of Müller cell function is self-evident. The following sections describe the various ways that Müller cells can influence neuronal function in the outer retina.

16.4 NUTRITIVE RELATIONSHIPS AND REGULATION OF pH

The retina is fueled exclusively by glucose and has one of the highest metabolic rates of any nervous tissue. Müller cells play a central role in the uptake of glucose, the synthesis and storage of glycogen and the provision of carbohydrate metabolites to neurons. Glucose is transported across the blood: retina barrier by facilitated diffusion that involves the Glut-1 glucose transporter (Mantych *et al.*, 1993). In the human retina, Glut-1 is almost entirely restricted to Müller cell and concentrated in their rootlets that entwine the energy-hungry photoreceptor somata in the outer nuclear layer (Mantych *et al.*, 1993). The human retina contains several networks of blood vessels that are potential sources of glucose; in species with avascular retinas, however, the only available source is the choriocapillaris, via the subretinal space. It is assumed that in these species the apical microvilli of Müller cells, which extend into the subretinal space, provide a greatly expanded surface area for the uptake of glucose and other metabolites. The microvilli membranes have features that are indicative of active exchange, such as abundant Na^+, K^+-ATPase sites (section 16.7.1) and a high concentration of lectin-binding glycoproteins (Reichenbach, 1989).

Like astrocytes in other parts of the brain, Müller cells are able to convert glucose into glycogen (Kuwabara and Cogan, 1961; Poitry-Yamate and Tsacopoulos, 1991) in a process that is stimulated by insulin (Reichenbach *et al.*, 1993). It is not surprising therefore, that Müller cells are the main repositories of retinal glycogen (Kuwabara and Cogan, 1961; Magelhães and Coimbra, 1970; Poitry-Yamate and Tsacopoulos, 1991), and they are the main source of glycogenolytic enzymes (Magelhães and Coimbra, 1970; Pfeiffer *et al.*, 1994). Since the entire metabolic pathway for glycogen is found within Müller cells, these cells are well equipped to function under anaerobic conditions. This idea is further supported by the observations that both retinal ischemia (Johnson, 1977) and reduced extracellular glucose (Magelhães and Coimbra, 1970) lead to a rapid depletion of glycogen stores in Müller cells. It has been proposed that these glycogen stores may also constitute an energy reserve that could be made available to neurons in times of need (Ripps and Witkovsky, 1985). This idea is supported by the observation that glycogenolysis in Müller cells increases in response to elevated levels of extracellular K^+ (Reichenbach *et al.*, 1993), which are a byproduct of neuronal activity (section 16.7). It is possible, however, that this increase is simply an attempt by Müller cells to meet their own increased metabolic demands, since these cells expend a considerable proportion of their energy budget in the uptake of K^+. If an energy source is provided to neurons by Müller cells, the substrate is unlikely to be glucose or glycogen because most neurons have weak glucose uptake systems and few glycolytic enzymes (reviewed by Ripps and Witkovsky, 1985). More likely candidates are pyruvate, lactate or alanine.

Since the retina is such a metabolically active tissue, considerable quantities of metabolic byproducts are released by the neurons into the extracellular space. One of these is carbon dioxide, which, if allowed to accumulate will cause lethal shifts in extracellular pH (see also Borgula *et al.*, 1989). Müller cells contain large stores of carbonic anhydrase, an enzyme that catalyzes the hydration of carbon dioxide (CO_2) to carbonic acid (reviewed by Moscona, 1983). The concentration of carbonic anhydrase is highest in those Müller cell processes that are located in the outer nuclear layer; this distribution presumably reflects the higher metabolic activity in this part of the retina. Much higher levels of carbonic anhydrase are found in chicken than in mouse, a difference that is probably due to the additional metabolic demands placed on Müller cells in avascular retinas (Moscona, 1983). A further indication of the role of Müller cells in pH regulation is the finding that their cell membrane contains an electrogenic Na^+/HCO_3^- co-transport system (Newman and Astion, 1991). As these co-transporters are concentrated in the basal endfoot membrane of Müller cells, they may be utilizing the great volume of the vitreous body as a means of stabilizing extracellular pH in the retina. Newman (1994) proposed a mechanism of 'CO_2 siphoning' by Müller cells (also section 16.7). CO_2 diffuses into the Müller cells where it is rapidly converted by the carbonic anhydrase into H^+ and HCO_3^-; the latter is transported into the vitreous body by the Na^+/HCO_3^- co-transport system of the endfoot membrane (Newman and Astion, 1991), and H^+ might be removed by an adjacent Na^+/H^+ exchanger which remains to be demonstrated.

16.5 SYNTHESIS AND RENEWAL OF VISUAL PIGMENTS

Cellular retinoid-binding proteins are involved in retinoid accumulation, synthesis and turnover. 11-*cis*-retinaldehyde-binding protein specifically binds 11-*cis*-retinol, and this protein has been shown by immunocytochemistry to be restricted to Müller cells in the retinas of several mammalian species (e.g. Bunt-Milam and Saari, 1983). In an elegant series of experiments, Das *et al.* (1992) have shown that Müller cells bind all-*trans* retinol (vitamin A), convert it into 11-*cis* retinol and then release it into the extracellular space for uptake by cone photoreceptors (see also Bridges *et al.*, 1984). The cones oxidize 11-*cis* retinol to 11-*cis*-retinaldehyde, which forms the basis of their photopigments. Rods however, are incapable of incorporating 11-*cis*-retinol and require 11-*cis*-retinaldehyde, which they obtain from the subretinal space after it has been released by pigment epithelial cells (Das *et al.*, 1990; Flannery *et al.*, 1990). Consequently, rhodopsin can only be regenerated in the presence of pigment epithelial cells, whereas cone opsins can be regenerated in retinas in which the pigment epithelium has been removed (reviewed by Das *et al.*, 1992). These observations indicate that Müller cells are intimately involved in the visual cycle through the synthesis and renewal of cone visual pigments, whereas pigment epithelial cells appear to be concerned principally with the renewal of rhodopsin.

16.6 NEUROTRANSMITTER RECYCLING

The inner and outer plexiform layers contain an extremely high density of synapses where leakage of neurotransmitters out of synaptic clefts and into the extracellular space may occur; this can potentially contaminate other synapses and degrade signal fidelity. By individually ensheathing synapses in both layers, Müller cells play a critical role in maintaining a high signal-to-noise ratio for retinal neurotransmission. However, the contribution made by Müller cells goes beyond mere containment of the transmitters released into synaptic clefts. Müller cells possess high affinity uptake sites for most neurotransmitters and by scavenging transmitters from synaptic clefts they shorten the time during which the transmitters are available to neuronal membrane receptors after their release from presynaptic terminals (e.g. Sarantis and Mobbs, 1992). Following uptake, Müller cells normally convert the transmitters into non-neuroactive compounds that are then recycled back to the presynaptic neurons. The importance of the contribution made by Müller cells in limiting the spread of neurotransmitters in the extracellular space is illustrated by the fact that isolated retinal neurons are 100-fold more sensitive to the toxic effects of glutamate than neurons in intact retinas (for review see Pow and Robinson, 1994).

The two principal neurotransmitters used in the OPL are the excitatory transmitter glutamate, and the inhibitory transmitter, γ-aminobutyric acid (GABA). A putative transmitter, taurine, is also abundant in the outer retina. The involvement of Müller cells in the recycling of these amino acids in the outer retina is as follows.

First, glutamate is released by photoreceptor cells onto second-order neurons (Chapter 6). It is also selectively taken up by Müller cells by means of an electrogenic Na^+ (and K^+) dependent uptake carrier (Ehinger, 1977; Brew and Attwell, 1987; Schwartz and Tachibana, 1990; Figure 16.2). Glutamate uptake sites are expressed at high density within those regions of the Müller cell membranes that face the plexiform layers (Brew and Attwell, 1987). Glutamine synthetase, an enzyme which transamidates glutamate to glutamine, has been localized exclusively in Müller cells (reviewed by Moscona, 1983). The glutamine synthesized by Müller cells is supplied to retinal neurons, as a precursor for their glutamate production (Figure 16.2). In rabbit retinas selective inhibition of glutamine synthetase in Müller cells with methionine sulfoximine not only leads to a total depletion of glutamine in Müller cells, but it also completely deprives both bipolar cells and ganglion cells of glutamate, in spite of the presence of both glutamate and glutamine in the superfusion medium (Pow and Robinson, 1994). These results indicate that bipolar and ganglion cells obtain all their glutamate in the form of glutamine from Müller cells. In contrast, photoreceptors exhibit an incomplete diminution of their glutamate content under these circumstances, indicating that they are able to obtain small quantities of glutamate or glutamine from alternative sources, such as the subretinal space, or in the form of other precursors, such as α-ketoglutarate.

Second GABA is thought to be the principal inhibitory transmitter used by horizontal cells, even in those mammals where adult horizontal cells lack a GABA uptake system (Chapter 10). In such mammals, the presence of an active glial GABA uptake carrier is essential if the accumulation and spread of the transmitter in the extracellular space are to be avoided. GABA is taken up by Müller cells by means of high-affinity, selective uptake carriers (Neal and Iversen, 1972; Ehinger, 1977; Sarthy, 1983). Uptake of GABA by Müller cells is Na^+ (and Cl^-) dependent and electrogenic (Quian *et al.*, 1993; Biedermann *et al.*, 1994; Figure 16.2). GABA uptake is competitively inhibited by nipecotic acid and β-alanine which both seem to be transported by the same mechanism

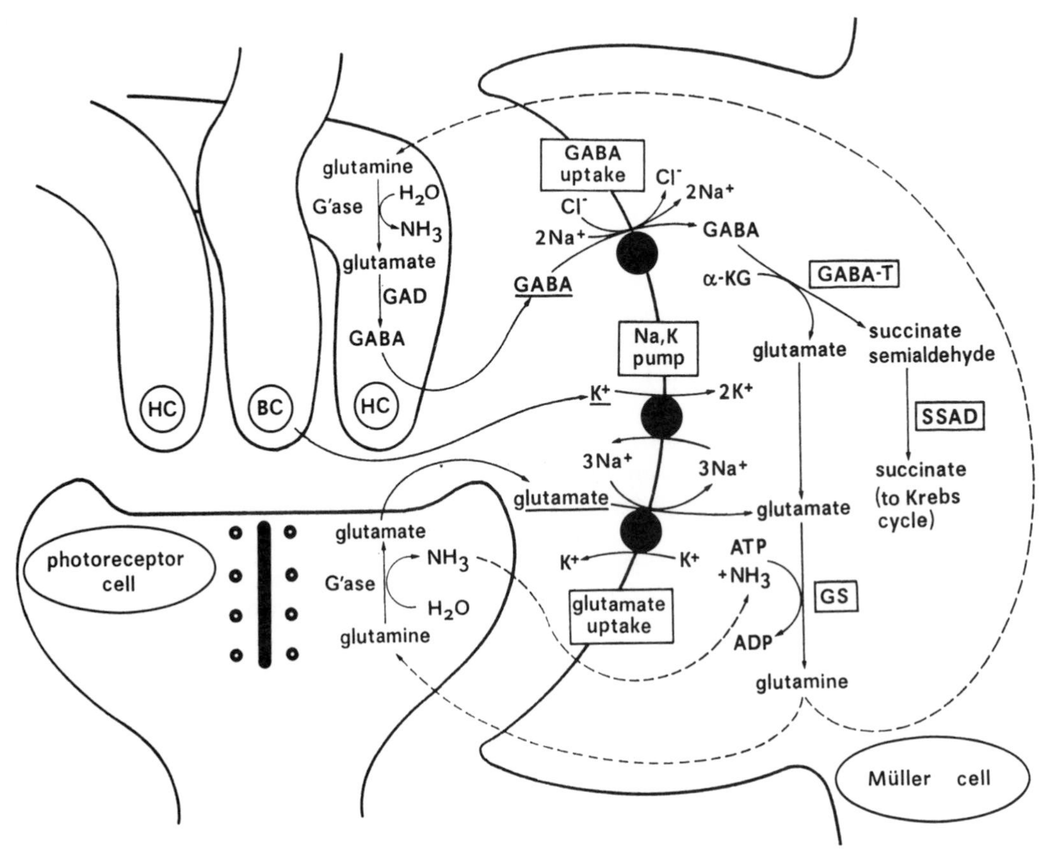

Figure 16.2 'Neurotransmitter recycling' in the outer plexiform layer. An overview of the mechanisms by which Müller cells (ensheathing a typical synaptic triade) contribute to clearance of the synaptic cleft. Photoreceptor terminals release glutamate, horizontal cell processes release GABA, and depolarized bipolar cells release K^+ ions. Müller cells possess high-affinity GABA- and glutamate uptake sites, as well as active Na,K pumps. After uptake, Müller cell-specific enzymes convert the amino acids into metabolites that are not neuroactive. One of these, glutamine, is released from Müller cells to be taken up and used by neurons as a precursor in the synthesis of glutamate and GABA. ADP, adenosine diphosphate; ATP, adenosine triphosphate; G'ase, glutaminase; GABA-T, GABA transaminase; GAD, glutamic acid decarboxylase; GS, glutamine synthetase; SSAD, succinic acid semialdehyde dehydrogenase; α-KG, α-ketoglutarate; HC, horizontal cell process; BC, bipolar cell dendrite.

(Biedermann *et al.*, 1994). Müller cells inactivate GABA by converting it into succinate semialdehyde via GABA-transaminase, an enzyme that in the retina is largely restricted to Müller cells (Sarthy and Lam, 1978). It is further degraded to succinate by the Müller cell enzyme succinate semialdehyde dehydrogenase, thereby returning GABA to the tricarboxylic acid cycle (Moore and Gruberg, 1974; Figure 16.2).

Third the retina contains extremely high levels of the sulfonic amino acid, taurine (for review see Lombardini, 1991). ^{3}H-labeled taurine is avidly taken up by Müller cells (Ehinger, 1973; Faff-Michalak *et al.*, 1994), and in normal retina, taurine is concentrated within photoreceptors and Müller cells (Pow and Robinson, 1994). It is reasonable to suppose therefore, that taurine is involved in the function of the outer retina, perhaps in a process that involves interaction between Müller cells and photoreceptors. Although the roles of taurine have not yet been determined, several possibilities, as follows have been proposed by Lombardini (1991).

1. In other regions of the brain, taurine has been suggested to be an inhibitory neurotransmitter, so in the retina it may also contribute to the inhibitory modulation of neuronal activity in the OPL.
2. It may regulate Ca^{2+} transport in cellular membranes.
3. Taurine is able to inactivate oxygen free radicals so it may help to protect photoreceptor membranes from peroxidation-induced damage arising from exposure to light.
4. It may assist in the regulation of signal transduction by second messengers.
5. Taurine may be involved in cell volume regulation.

The last possibility is supported by observations of an osmoregulatory release of taurine from cultured Müller cells in response to iso-osmotic high extracellular K^+ as well as to hypo-osmotic solutions (Faff-Michalak *et al.*, 1994). In view of the extremely narrow extracellular spaces within the retina, a means of minimizing cell swelling is needed so that tissue damage and interference to neural transmission can be avoided.

16.7 CLEARANCE OF EXTRACELLULAR POTASSIUM IONS

During signal transduction, neurons release K^+ into the extracellular space. In the retina, such release is maximal during and after illumination (e.g. Steinberg *et al.*, 1980). In order to limit fluctuations in neuronal excitability, extracellular K^+ must be cleared rapidly. This task is performed by glia, using two strategies: (1) by space-independent K^+ uptake, such as through the activity of Na^+, K^+-pumps, and (2) by space-dependent redistribution of K^+ ions via spatial buffering currents. The following two sections describe the way that these strategies are employed by Müller cells.

16.7.1 Na^+, K^+ PUMPS

The Na^+, K^+-ATPase of Müller cells consists mainly of α1 and α2 subunits whereas in most neurons it consists exclusively of the α3 isozyme (McGrail and Sweadner, 1989). This structural difference might account for the functional differences between neuronal and glial forms of the enzyme (reviewed by Reichenbach *et al.*, 1992). The Müller cell's enzyme activity mainly depends on $[K^+]$ over a wide range (maximum activity is seen at about 10–15 mM K^+ compared with 3 mM K^+ in rod photoreceptors) rather than $[Na^+]$ (maximum activation occurs at less than 10 mM Na^+ compared with 40 mM Na^+ in neurons). Thus, increased extracellular K^+ concentration ($[K^+]_e$) in the physiological range stimulates K^+ uptake by Müller cells in a manner that is independent of intracellular Na^+ concentration ($[Na^+]_i$). It follows that $[Na^+_i]$ must decrease during prolonged activity of the pump, a process that would inhibit the pump activity if the Müller cell's enzyme did not have special properties.

In rabbits, each Müller cell is estimated to have about 5×10^6 pump sites (Reichenbach *et al.*, 1988). In the plexiform layers, pump site density in Müller cell membranes is estimated to be almost 1000 mm^{-2}, which is about double the density in the nuclear layers. Maximum densities of about 3000 pump sites mm^{-2} were found in their microvilli (Reichenbach *et al.*, 1988). In turtles also, maximum pump site density was found in the microvillous membrane (Stirling and Sarthy, 1985). This distribution of Na^+, K^+-

pump sites seems to be well adapted to illumination-dependent changes of $[K^+]_e$ within the retina. Karwoski and Proenza (1987) analyzed the distribution of K^+ sinks and sources during light stimulation (i.e. light 'ON') across retinal layers. The authors found sources within the two plexiform layers; these correspond to sites of high pump density. After the cessation of illumination (i.e. light 'OFF'), there are large increases in $[K^+]_e$ reaching a peak in the subretinal space (Steinberg *et al*, 1980); this location corresponds to the peak density of pump sites within the microvillous membranes of Müller cells. Thus, Müller cells express high densities of Na^+, K^+-pump sites in those parts of their membrane that are exposed to increases in $[K^+]_e$ during and/or after physiological stimulation of the retina. It can be concluded that Müller cell Na^+, K^+-pumps contribute considerably to retinal K^+ clearance; this conclusion is supported by mathematical estimations of the pump currents (Reichenbach *et al.*, 1992).

16.7.2 K^+ CHANNELS AND SPATIAL BUFFERING

The ionic conductance of Müller cells at physiological membrane potentials is almost exclusively provided by K^+ channels. Although various types of K^+ channels have been described in Müller cells (Newman, 1985; Nilius and Reichenbach, 1988; Chao *et al.*, 1994), inwardly rectifying K^+ channels are the type that are responsible for most currents at physiological membrane potentials (Newman, 1985, 1993; Brew *et al.*, 1986; Chao *et al.*, 1994), particularly in the outer retina (Nilius and Reichenbach, 1988). The presence of inwardly rectifying K^+ channels at sites of transiently elevated $[K^+]_e$ facilitates the local influx of excess K^+ (see below).

The distribution of local membrane (inward) conductance has been studied across the surface of Müller cells in various species (Newman, 1984, 1987, 1988; Brew *et al.*, 1986; Reichenbach and Eberhardt, 1988). These experiments demonstrated that in avascular retinas, the endfoot membrane has a much higher K^+ conductivity than other parts of the membrane. Indeed, more than 80% of the total K^+ conductance of a Müller cell may be localized within its end foot membrane. In vascularized retinas, regions of high membrane conductance also occur in the outer retina where intraretinal blood vessels may be covered by 'en-passant' endfeet (Newman, 1987). These results have been interpreted as evidence that Müller cell membrane conductance is particularly high wherever the membrane faces a potential 'sink' for K^+ currents.

The term 'spatial buffer' is used to describe the redistribution by glial cells of excess extracellular K^+ that have been released by electrically active neurons. The principle underlying this process is that the membrane potential (E_m) of the Müller cells remains 'clamped' near the resting level because most of the cell's membrane is exposed to normal $[K^+]_e$ despite the fact that the value of the 'local' K^+ equilibrium potential (E_K) is decreased by local increases in $[K^+]_e$. This localized increase in $[K^+]_e$ triggers a local influx of K^+ into Müller cells which then flows through the Müller cell trunks toward their endfeet (Figure 16.3a). The high endfoot conductance, combined with the enormous volume of the vitreous body causes K^+ to be siphoned from the endfeet into the vitreous (hence 'K^+ siphoning'; Newman *et al.*, 1984; Karwoski *et al.*, 1989). This process is further supported by the presence of inwardly rectifying K^+ channels in that part of the Müller cell membrane that faces the inner plexiform layer (Nilius and Reichenbach, 1988). Excess K^+ within this layer may enter Müller cells, whereas the K^+ currents which flow from the OPL to the endfoot are prevented from extensive outflow. When compared with extracellular diffusion, K^+ siphoning is estim-

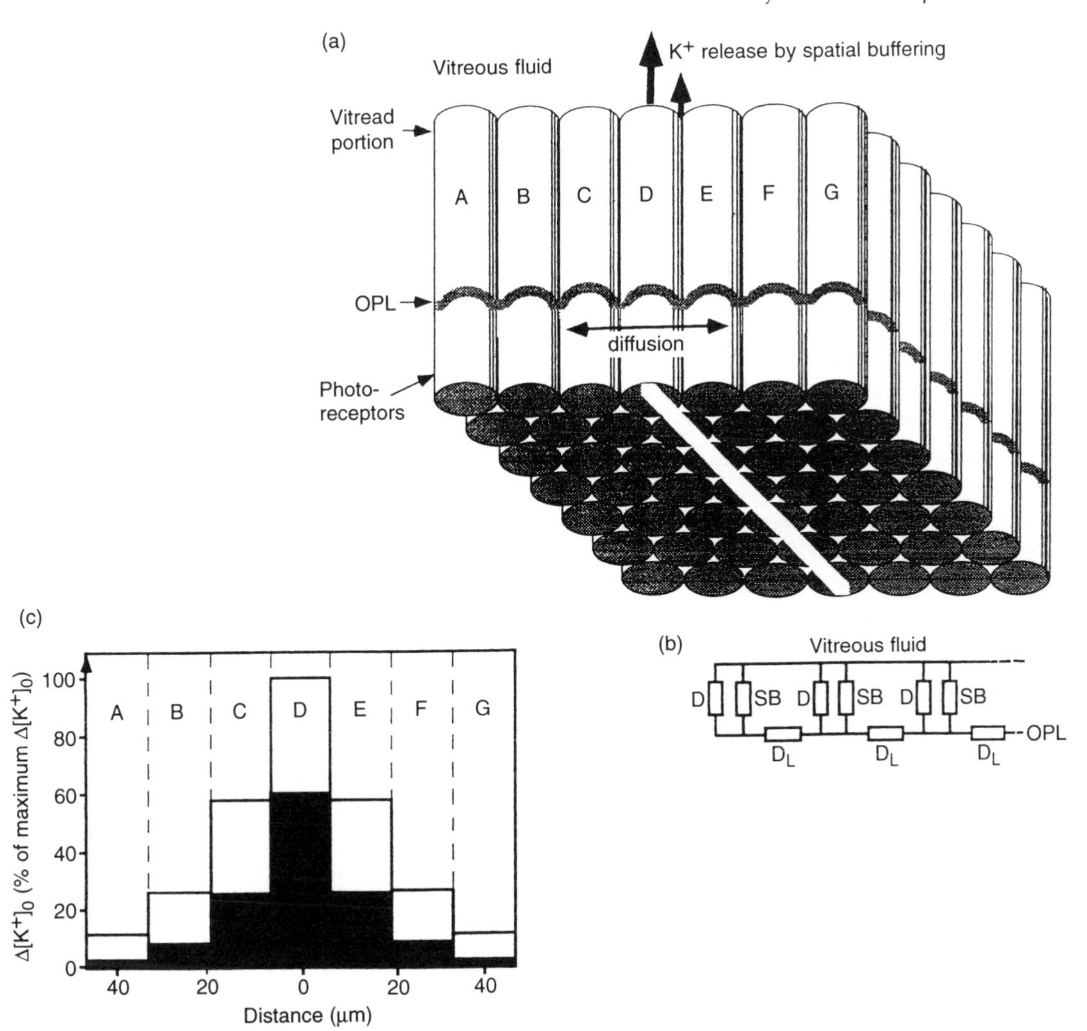

Figure 16.3 (a) Schematic drawing of columnar retinal units during illumination with a rod-shaped light stimulus (white line). The light-induced $[K^+]_e$ increases in the outer plexiform layer (OPL) occur primarily in the row of columns marked D, but lateral K^+ diffusion (D_L in (b)) may take place (towards columns A and G) and compromise visual acuity by 'unwarranted' stimulation of neighboring columns. Müller cells within each unit transport K^+ ions into the large sink provided by the vitreous fluid, by spatial buffering (SB in (b)); simultaneously, extracellular diffusion (D in (b)) contributes to 'K^+ drainage'. The effect of spatial buffering on lateral K^+ spread within the OPL was calculated as a cable problem, with the cable extending laterally within the OPL and the leaky mantle of the cable extending towards the vitreous body. The more 'leaky' this mantle (that is, the better SB is working), the less the lateral spread (b; see Reichenbach *et al.*, 1993, for details). The result is shown in (c) on the basis of either diffusion alone (open bars) or diffusion plus spatial buffering (closed bars). In the case of diffusion alone, the high $[K^+]_e$ in the OPL spreads much further (length constant λ of the 'cable': 18 μm) than in the case of additional spatial buffering (length constant λ of the 'cable': 11 μm). Thus, spatial buffering by Müller cells helps much in preventing lateral spread of excitation. (Reproduced from Reichenbach *et al.* (1993), with permission.)

ated to reduce the half-time of K^+ clearance from the OPL by 50–70% (e.g. Newman and Odette, 1984; Reichenbach *et al.*, 1992). Spatial buffering through Müller cells might reduce $[K^+]_e$ rapidly enough to prevent the spread of K^+ to neighboring (unstimulated) columns of retinal neurons (e.g. from D to E to F in Figure 16.3), thereby minimizing disturbances to visually driven acuity (Reichenbach *et al.*, 1993; Figure 16.3). Recently, it has been shown that K^+ currents in isolated Müller cells can be modulated by L-glutamate (Schwartz, 1993). If the same modulation occurs *in situ*, the rate of K^+ clearance will be influenced by the availability of glutamate in the extracellular space. Such modulation is of particular relevance to the OPL, since both K^+ and glutamate are released within this layer during physiological stimulation.

16.8 PHAGOCYTOSIS

An important function served by glial cells is the phagocytosis of neuronal debris. Müller cells are well equipped for this task since they contain a variety of hydrolases that can be used to break down phagocytosed membranes. They have for example, a high concentration of the hydrolytic enzyme acid phosphatase (Reichenbach, 1989), and they are also the primary source in the neural retina of cathepsin D, which is a main intracellular proteinase (Reichenbach, 1989; Yamada *et al.*, 1990). During retinal development Müller cells phagocytose the debris of dying neurons (Young, 1984), whereas in mature retinas, Müller cells assist pigment epithelium in the phagocytosis of the cast-off outer segments of photoreceptors (Long *et al.*, 1986), and assist microglia in the phagocytosis of debris associated with retinal degeneration (Caley *et al.*, 1972). Müller cells may also play an important, yet rarely recognized role in helping to maintain the clarity of the ocular media. This role is hinted at by the fact that they avidly phagocytose foreign particulate matter that has been injected into the vitreous humour, including red blood cells (Miller *et al.*, 1986). *In vitro*, Müller cells rapidly phagocytose latex beads (Burke, 1989; Mano and Puro, 1990; Stolzenburg *et al.*, 1992), a process that depends on the presence of extracellular Ca^{2+} (and active Ca^{2+} channels); the availability of which is modified by cyclic AMP and is stimulated by vitamin D_3 (Mano and Puro, 1990). Uptake of latex beads is so rapid that exposure for less than 20 min is sufficient to entirely label freshly isolated Müller cells (Stolzenburg *et al.*, 1992). The engulfed microspheres become distributed within the Müller cells at a rate that is equivalent to transport over a distance of one millimeter per day. This rate is within the same order of magnitude as slow axoplasmic transport. By contrast, the release of engulfed particles from cultured Müller cells occurs extremely slowly, if at all (Burke, 1989).

16.9 SIGNALING TO NEURONS

Müller cells release neuroactive substances when they are depolarized or are exposed to other neurotransmitters. This phenomenon has been demonstrated for GABA (Sarthy, 1983), and it has the potential to modify information processing in the OPL. Likewise, depolarization-induced release of taurine by Müller cells (Faff-Michalak *et al.*, 1994) may exert an inhibitory effect on the surrounding neurons. Thus, these (and other) substances may act as 'gliotransmitters'.

Another potential instance of neurotransmitter release by Müller cells involves dopamine. In all vertebrate retinas, horizontal cells are connected to each other by gap junctions which become uncoupled during light adaptation due to release of dopamine (Chapters 7 and 8). Since there are no dopaminergic neurons in the outer retina of most species, the source of this dopamine is not obvious. One hypothesis (e.g. Witkovsky and Schütte, 1991; Hampson *et al.*, 1994) proposes that dopamine diffuses from the inner plexiform layer to the OPL via the extracellular space. However, the extracellular clefts are very narrow (section 16.1) and ensheathment of neurons and synapses by Müller cells limits the spread of neuroactive substances (sections 16.4 and 16.5), so diffusion may not be a very effective means of delivery. However, in the teleost retina, interplexiform cell neurites may deliver dopamine directly to the outer retina and thus mediate the uncoupling of horizontal cells. In most vertebrates, however, interplexiform cells are few in number and their dendritic arbours in the OPL are sparse (Witkovsky and Schütte, 1991), so these cells may fall far short of providing the retinal coverage needed.

An alternative hypothesis, we would propose is that Müller cells mediate the uncoupling by transporting dopamine from dopaminergic amacrine cells to horizontal cells. It has not yet been shown that Müller cells take up and release dopamine, but they do possess large numbers of D2-like dopamine receptors (Muresan and Besharse, 1993; Reichenbach, unpublished data). A further indication that Müller cells may be involved in this process is the demonstration that the uncoupling of gap junctions between A-type horizontal cells in rabbit retinas is a pH-dependent process (Hampson *et al.*, 1994); as outlined in section 16.4, Müller cells play a central role in the control of retinal pH. Therefore, even if Müller cells are not directly responsible for providing horizontal cells with dopamine, they are nevertheless, likely to modulate horizontal cell coupling via pH-dependent processes. Similarly, Müller cells may provide photoreceptors with dopamine. The microvilli of the Müller cells project into the subretinal space where they lie adjacent to the inner segments of the rods and cones. The Müller cells may release dopamine into the subretinal space to act on the inner and outer segments of photoreceptors, via D2-dopamine receptors that are expressed on their surface membranes (Muresan and Besharse, 1993; Wagner *et al.* 1993). Although photoreceptors do not use dopamine as a neurotransmitter, dopamine is thought to regulate several aspects of photoreceptor function such as motility of photoreceptors in fish synaptic activity, and melatonin synthesis (in all vertebrate photoreceptors) (Chapters 2 and 7).

16.10 CONTRIBUTION TO THE ELECTRORETINOGRAM (ERG)

As described in section 16.7.2, light-induced changes of extracellular K^+ lead to local currents across Müller cell membranes. These currents are thought to contribute to the generation of retinal light-evoked potentials by causing extracellular voltage drops (e.g. Miller and Dowling, 1970: Newman and Odette 1984). Since Müller cell trunks lie orthogonal to the retinal surface, they are able to contribute to the generation of transretinal potentials such as the ERG waves. Direct evidence that Müller cells contribute to the generation of some components of the ERG comes from experiments in which the gliotoxin α-aminoadipic acid (α-AAA) has been injected into the vitreous chamber. This compound selectively lesions Müller cells (e.g. Olney *et al.*, 1971, 1980; Szamier *et al.*, 1981; Reichenbach and Wohlrab, 1985), thereby impairing the generation of the b-wave (Szamier *et al.*, 1981) and the slow P_{III} potential (Reichenbach and Wohlrab, 1985). The following ERG waves arising at light 'ON' in the outer retina have been related to Müller cells:

1. The (cornea-positive) b-wave has been attributed to Müller cell-mediated currents that flow from the OPL into the vitreous body as a result of K^+ siphoning (Figure 16.4a; Miller and Dowling, 1970; Newman

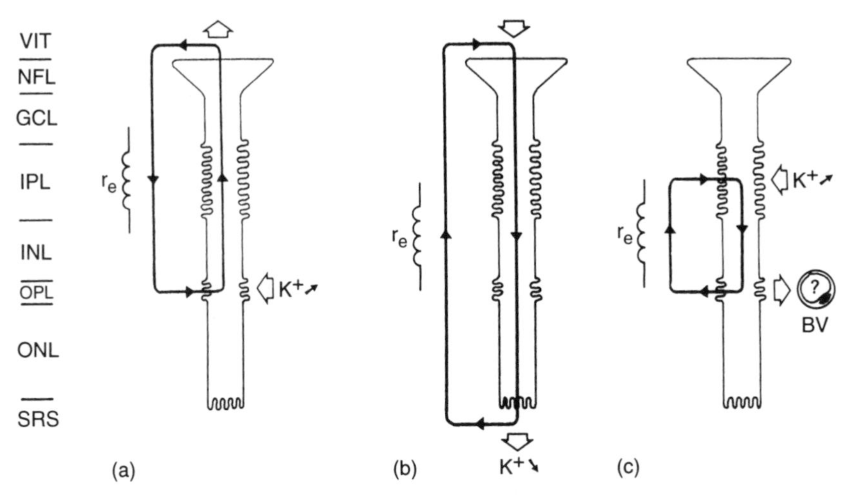

Figure 16.4 Schematic drawings of putative K^+ currents through Müller cells, contributing to ERG-wave light responses. (a) A light-induced distal $[K^+]_e$ increase in the outer plexiform layer is buffered by currents entering the vitreous body; the extracellular voltage drop, caused by current loops through the extracellular resistance, may generate the ERG b-wave. (b) During illumination, $[K^+]_e$ around the photoreceptor inner segments (in the subretinal space) decreases; this is buffered by K^+ currents from the vitreous body. The current loops through the extracellular resistance may generate the slow P_{III} component of the ERG which is of opposite polarity to the b-wave, and reaches its maximum later than the b-wave. (c) In dark-adapted, well-vascularized retinas such as from cat and man, a 'scotopic threshold response' (STR) has been observed. It seems to be caused by extracellular current loops due to $[K^+]_e$ increases in the inner plexiform layer being buffered by the outer retina. VIT, vitreous body; NFL, nerve fiber layer; GCL, ganglion cell layer; IPL, inner plexiform layer; INL, inner nuclear layer; OPL, outer plexiform layer; ONL, outer nuclear layer; SRS, subretinal space; BV, blood vessel; r_e, extracellular resistance across which the potential drop is generated. Intracellular currents through Müller cells are carried by K^+, while extracellular current loops are carried by Na^+ and Cl^-. See text for further details and sources.

and Odette, 1984). It should be noted, however, that depolarizing bipolar cells may also contribute to this current.

2. The (cornea-negative) slow P_{III} component of the ERG is thought to be due to spatial buffering currents that flow through Müller cells in an outward direction. This current arises because Müller cells transport K^+ ions from the vitreous body into the subretinal space where $[K^+]_e$ decreases during illumination (Karwoski and Proenza, 1977; Fujimoto and Tomita, 1979; Newman and Odette, 1984; Figure 16.4b).
3. In the cat retina, a (cornea-positive) 'scotopic threshold response' (STR) has been described at low stimulus intensities (Frishman and Steinberg, 1989; see Figure 16.4c). This wave has also been attributed to spatial buffering currents that pass through Müller cells. In this instance, the currents are thought to be due to Müller cells removing K^+ from the inner plexiform layer and releasing it into the OPL (Frishman and Steinberg, 1989). In vascularized retinas, Müller cell membranes in the OPL have a high conductivity (cat: Newman, 1987), and they exhibit morphological features that are similar to those in the endfoot membranes (mouse: Wolburg and Berg, 1988), suggesting that in such

retinas, excess K^+ are also ejected into capillaries within the OPL (Figure 16.4c).

16.11 INVOLVEMENT OF MÜLLER CELLS IN RETINAL PATHOLOGY

16.11.1 HYPERAMMONEMIA

As mentioned in section 16.6, Müller cells contain high concentrations of glutamine synthetase. This enzyme inactivates glutamate by converting it into glutamine, and by so doing it also binds ammonia. Consequently, Müller cells play an important role in detoxifying the extracellular space of ammonia. This property causes Müller cells to be particularly susceptible to hyperammonemia arising from liver failure. Thus the cells undergo structural changes that resemble those seen in Alzheimer-type II astrocytes, which are characteristic of hepatic encephalopathy. Affected Müller cells have enlarged, pale nuclei, swollen processes and an enhanced expression of GFAP and cathepsin D. This similarity between the Müller cell response and the response of astrocytes in the brain led Reichenbach *et al.* (1995), to refer to the retinal changes as 'hepatic retinopathy'. In patients with hepatic retinopathy, the amplitudes of the ERG waves (particularly the b-wave) are reduced, and their latency is increased. In this context it is noteworthy that the first ultrastructural changes in diabetic retinopathy are also detected in Müller cells (Hori and Mukai, 1980).

16.11.2 RETINAL DYSTROPHY

In other types of retinal pathology, Müller cells commonly respond to neuronal damage or degeneration by up-regulation of GFAP and vimentin expression (e.g. Bignami and Dahl, 1979; Guèrin *et al.*, 1990). In the RCS rat model of retinal dystrophy (as well as in other cases of retinal pathology), Müller cells up-regulate their expression of the bcl-2 protooncogene protein and of the β-amyloid precursor protein (A. Reichenbach, unpublished observations). Bcl-2 has been shown to function in an antioxidative pathway that prevents apoptotic cell death (Hockenbery *et al.*, 1993). The β-amyloid precursor protein supports neuroregenerative processes such as outgrowth of neurites, and has protective effects against excitotoxic amino acids by regulating intracellular Ca^{2+} levels (e.g. Ashall and Goate, 1994). It remains to be seen whether these metabolic responses support the surrounding neurons or the Müller cells themselves.

16.11.3 ISCHEMIA

Interruption of the oxygen and glucose supply to the retina by occlusion of the retinal vasculature (ischemia) leads to a rapid reduction in the responsiveness of the retina to light. After occlusions of short duration (e.g. 10 min), the light-evoked $[K^+]_e$ decrease (photoreceptors), the ERG b-waves (Müller cells) and the slow P_{III} waves (Müller cells) return to normal within about two hours (Hiroi *et al.*, 1994). Interruptions of blood supply lasting longer than 30–45 min often result in the death of photoreceptors, which are then phagocytosed by Müller cells (Buchi, 1992). The addition of taurine to the reperfusion medium following a 20 min period of hypoxic-ischemia significantly improves the recovery of the ERG b-wave and ameliorates damage to the mitochondria and inner segments of photoreceptors (Tseng *et al.*, 1990). High concentrations of taurine are normally present in photoreceptors and Müller cells (Pow and Robinson, 1994), so that it is surprising that taurine supplementation improves photoreceptor cell survival. The reason may be that taurine levels in Müller cells and photoreceptors are gradually reduced during ischemia (Pow and Crook, 1994), so the need for supplementation increases accordingly. The basis of the ability of taurine to protect photoreceptors from ischemic damage is not known, but as taurine can act as an antioxidant (section 16.6), it seems likely that taurine protects photoreceptor membranes and mitochondria from oxidation by free radicals, which are abundant during ischemia (for review see Phillis, 1994).

Apart from exposure to free radicals, the loss of neurons after prolonged ischemia seems to be related to the fact that the high energy demands of the retina result in the rapid breakdown of active membrane transport mechanisms, including the Na^+, K^+ pumps. This situation leads to neuronal depolarization and to the toxic accumulation of extracellular ammonia, K^+, glutamate and aspartate. There is disagreement concerning which of these components makes the greatest contribution to the lethal effects of ischemia. The view that extracellular glutamate and aspartate reach excitotoxic levels during ischemia is supported by the finding (Abu El-Asrar *et al.*, 1992) that prior exposure to NMDA receptor antagonists protects cultured retinal neurons from cell death under hypoxic conditions. However, Watanabe *et al.* (1994) claimed that extracellular glutamate is not significantly elevated in retinal ischemia because there is a compensatory reduction in the activity of phosphate-activated glutaminase (which converts glutamine to glutamate). They noted that the most significant change is seen for ammonia, which increases by 214%.

Ammonia, K^+, glutamate and aspartate accumulate in the extracellular space during ischemia because Müller cells fail to take up and inactivate these substances. Thus, while Müller cells are less susceptible than neurons to the lethal effects of ischemia (Hiroi *et al.*, 1994), their metabolism is clearly impaired. The profound effect exerted by ischemia is indicated by the fact that a significant upregulation of GFAP immunoreactivity persists in the Müller cells, six days after a 24-minute occlusion of the retinal blood supply (Osborne *et al.*, 1991).

16.11.4 SPREADING DEPRESSION

The accumulation of extracellular K^+ and glutamate in the retina triggers spreading depression, which is characterized by large field potentials, cessation of information processing, and changes in the translucency of the retina. Retinal spreading depression has been documented *in vivo* (Ripps *et al.*, 1981; De Jong *et al.*, 1991), which indicates that it is not simply an artifact of *in vitro* experimental conditions.

Müller cells are central to the phenomenon of retinal spreading depression. Intracellular recordings have shown that the intracellular K^+ concentration in Müller cells increases rapidly, probably due to highly active K^+ clearance mechanisms (Mori *et al.*, 1976). Simultaneously, the membrane potential of Müller cells is depolarized by as much as 30 mV; which may contribute to the generation of the large field potentials (Mori *et al.*, 1976). This depolarization is accompanied by a drastic decrease of the membrane resistance (Mori *et al.*, 1976). Furthermore, Müller cells swell during spreading depression (van Harreveld, 1982). This swelling, which may be excessive within the vitread endfeet (De Oliveira Castro and Martins-Ferreira, 1970), seems to contribute to the changes in translucency of the retina. The swelling of Müller cell endfeet has also been implicated as the cause of retinal surface deformations that are seen during spreading depression (De Oliveira Castro and Martins-Ferreira, 1970); such mechanical events might even facilitate the development of retinal detachment (see below).

In retinoschisis, the inner retinal blade splits apart from the outer retina, and the vitread endfeet are pulled off from the Müller cell trunks. In such instances it is inevitable that the spatial buffering of excess extracellular K^+ will be severely impaired. Such impairment has been suggested as a cause of spreading depression-like events such as the Mizuo phenomenon (De Jong *et al.*, 1991).

16.11.5 RETINAL DETACHMENT

Following retinal detachment, Müller cells undergo dramatic changes in their shape (Anderson *et al.*, 1986; Roque and Caldwell, 1990) and ultrastructure (Korte *et al.*, 1992). There are many reports of Müller cell proliferation following retinal damage or detachment, and this is often accompanied by a migration of some Müller cells into the vitreous cavity or subretinal space (e.g. Rentsch, 1979; Erickson *et al.*, 1983; Anderson *et al.*, 1986). These aberrant Müller cells contribute to the formation of pre- or subretinal membranes which can lead to further retinal detachment or degeneration (Laqua and Machemer, 1975). The factors that trigger this deleterious proliferation are not yet known, but there are likely to be several because Müller cells *in vitro* can be stimulated to divide by a variety of growth factors, as well as by glutamate and enhanced $[K^+]_e$.

16.12 CONCLUSIONS AND FUTURE PERSPECTIVES

Clearly, Müller cells are responsible for an extraordinary range of functions that vary according to whether the retina is mature, immature or undergoing pathological changes. Part of the uniqueness of Müller cells can be attributed to the fact that they are the principal glia of the retina and are the only cells to span the retinal thickness. It is evident from this review, however, that Müller cells should not be viewed in isolation; rather, Müller cells and retinal neurons should be seen as partners in an elaborate symbiosis. Müller cells supply neurons with endproducts of their own anaerobic metabolism (lactate/pyruvate) to fuel aerobic metabolism in neurons. In return, Müller cells receive byproducts of the neuronal tricarboxylic acid cycle, such as carbon dioxide (which can be used for lipid synthesis) and ammonia (which is needed for the generation of glutamine). Other examples of metabolic interactions between Müller cells and neurons include the recycling of amino acid transmitters and visual pigments.

The dependence of neurons on Müller cells extends well beyond metabolic recycling. Neurons and their synapses are cocooned within Müller cell processes to protect them from exposure to transmitters or other neuroactive substances that could degrade the fidelity of signal transduction. This situation, however, leaves neurons dependent on Müller cells for access to nutrients, and for the control of homeostasis by their uptake of transmitters, CO_2 and K^+ from the extracellular space. Fluctuations in the rate or efficiency of uptake of these substances by Müller cells are bound to influence neuronal function because neurons are extremely sensitive to perturbations in the composition of the extracellular fluid.

Since Müller cells are sensitive to changes in the extracellular fluid, the release of neuronal 'waste products' into the extracellular space can serve as an effective means of neuron-to-glia signaling. Such release not only stimulates glial uptake mechanisms and phagocytosis of debris, but also triggers a series of glial reactions such as the breakdown of glycogen for neuronal support, upregulation of beneficial gene products (e.g. bcl-2, amyloid precursor protein), and even Müller cell proliferation.

Finally, Müller cells are of clinical interest because their activity contributes to the generation of ERG waves (b-wave, slow P_{III}, STR) which are of diagnostic value in retinal pathology. Müller cells seem to be the primary targets of some pathogenic conditions, such as hepatic and diabetic retinopathy. Furthermore, in cases of retinal injury, Müller cells are a cause for concern because they can proliferate to form pre- and subretinal membranes that degrade vision and can lead to retinal detachment.

Future research in this field is likely to concentrate on theoretical and clinical issues. Much remains to be done to evaluate the electrophysiology, molecular biology, localization, and function(s) of ion channels and other membrane transport proteins present in Müller cells. There are still many unanswered questions concerning the contributions made by Müller cells to ERG potentials. The possibility that glial dysfunction forms the basis of many 'neuro' pathologies is a notion that is yet to gain credence. Nevertheless, it seems evident that Müller cell dysfunction is often a key feature. As future studies investigate other systemic and metabolic diseases affecting the retina, we expect attitudes to change considerably. Finally, the rapidly evolving field of transplantations of retinal (and pigment epithelial) cells will certainly require – and provide – more detailed knowledge of how Müller cells contribute to structural as well as functional integrity, including plasticity of retinal tissue.

REFERENCES

Abu El-Asrar, A.M., Morse, P.H., Maimone, D. *et al.* (1992) MK-801 protects retinal neurons from hypoxia and the toxicity of glutamate and aspartate. *Investigative Ophthalmology and Visual Science*, **33**, 3463–8.

Anderson, D.H., Guèrin, C.J., Erickson, P.A. *et al.* (1986) Morphological recovery in the reattached retina. *Investigative Ophthalmology and Visual Science*, **27**, 168–83.

Ashall, F. and Goate, A.M. (1994) Role of the β-amyloid precursor protein in Alzheimer's disease. *Trends in Biochemical Sciences*, **19**, 42–6.

Attwell, D., Brew, H. and Mobbs, P. (1986) Electrophysiology of the Müller cell network in the isolated axolotl retina. *Journal of Physiology (London)*, **369**, 33P.

Biedermann, B., Eberhardt, W. and Reichelt, W. (1994) GABA uptake into isolated retinal Müller glial cells of the guinea pig detected electrophysiologically. *Neuroreport*, **5**, 438–40.

Bignami, A. and Dahl, D. (1979) The radial glia of Müller in the rat retina and their response to injury. An immunofluorescence study with antibodies to the glial fibrillary acidic (GFA) protein. *Experimental Eye Research*, **28**, 63–9.

Borgula, G.A., Karwoski, C.J. and Steinberg, R.H. (1989) Light-evoked changes in extracellular pH in frog retina. *Vision Research*, **29**, 1069–77.

Brew, H. and Attwell, D. (1987) Electrogenic glutamate uptake is a major current carrier in the membrane of axolotl retinal glial cells. *Nature*, **327**, 707–9.

Brew, H., Gray, P.T.A., Mobbs, P. *et al.* (1986) Endfeet of retinal glial cells have higher densities of ion channels that mediate K^+ buffering. *Nature*, **324**, 466–8.

Bridges, C.D.B., Alvarez, R.A., Fong, S.-L. *et al.* (1984) Visual cycle in the mammalian eye. *Vision Research*, **24**, 1581–94.

Buchi, E.R. (1992) Cell death in rat retina after pressure-induced ischaemia-reperfusion insult: electron microscopic study. II. Outer nuclear layer. *Japanese Journal of Ophthalmology*, **36**, 62–8.

Bunt–Milam, A.H. and Saari, C.J. (1983) Immunocytochemical localization of two retinoid-binding proteins in vertebrate retina. *Journal of Cell Biology*, **97**, 703–12.

Burke, J.M. (1989) Growth in retinal glial cells *in vitro* is affected differentially by two types of cell contact-mediated interactions. *Experimental Cell Research*, **180**, 13–9.

Caley, D.W., Johnson, C. and Liebelt, R.A. (1972) The postnatal development of the retina in the normal and rodless CBA mouse: a light and electron microscopic study. *American Journal of Anatomy*, **133**, 179–212.

Chao, T.I., Henke, A., Reichelt, W., *et al.* (1994) Characterization and possible functional role(s) of K^+ channels in rabbit retinal Müller (glial) cells. *Pflüger's Archiv*, **426**, 51–60.

Das, S.R., Bhardwaj, N. and Gouras, P. (1990) Synthesis of retinoids by human retinal pigment epithelium and transfer to rod outer segments. *Biochemical Journal*, **268**, 201–6.

Das, S.R., Bhardwaj, N., Kjedbye, H. and Gouras, P. (1992) Müller cells of chicken retina synthesize 11-*cis*-retinol. *Biochemical Journal*, **285**, 907–13.

De Jong, P.T.V.M., Zrenner, E., Van Meel, G.J. *et al.* (1991) Mizuo phenomenon in X-linked retinoschisis. Pathogenesis of the Mizuo phenomenon. *Archives of Ophthalmology*, **109**, 1104–8.

De Oliveira Castro, G. and Martins-Ferreira, H. (1970) Deformations and thickness variations accompanying spreading depression in the retina. *Journal of Neurophysiology*, **33**, 891–900.

Dreher, Z., Robinson, S.R. and Distler, C. (1992) Müller cells in vascular and avascular retinae: a survey of seven mammals. *Journal of Comparative Neurology*, **323**, 59–80.

Ehinger, B. (1973) Glial uptake of taurine in the rabbit retina. *Brain Research*, **60**, 512–6.

Ehinger, B. (1977) Glial and neuronal uptake of GABA, glutamic acid, glutamine, and glutathione in the rabbit retina. *Experimental Eye Research*, **25**, 221–34.

Erickson, P.A., Fisher, S.K., Anderson, D.H. *et al.* (1983) Retinal detachment in the cat: the outer nuclear and outer plexiform layers. *Investigative Ophthalmology and Visual Science*, **24**, 927–42.

Fadool, J.M. and Linser, P.J. (1993) 5A11 antigen is a cell recognition molecule which is involved in neuronal-glial interactions in avian neural retina. *Developmental Dynamics*, **196**, 252–62.

Faff-Michalak, L., Reichenbach, A., Dettmer, D. *et al.* (1994) K^+-, hypoosmolarity-, and NH_4^+-induced taurine release from cultured rabbit Müller cells: role of Na^+ and Cl^- ions and relation to cell volume changes. *Glia*, **10**, 114–20.

Fain, G.L., Gold, G.H. and Dowling, J.E. (1976) Receptor coupling in the toad retina. *Cold Spring Harbor Symposia on Quantitative Biology*, **40**, 547–61.

Flannery, J.G., O'Day, W., Pfeffer, B.A. *et al.* (1990) Uptake, processing and release of retinoids by cultured human retinal pigment epithelium. *Experimental Eye Research*, **51**, 717–28.

Frishman, L.J. and Steinberg, R.H. (1989) Light-evoked increases in $[K^+]_o$ in proximal portion of the dark-adapted cat retina. *Journal of Neurophysiology*, **61**, 1233–43.

Fujimoto, M. and Tomita, T. (1979) Reconstruction of the slow PIII from the rod potential. *Investigative Ophthalmology and Visual Science*, **18**, 1090–3.

Guèrin, C.J., Anderson, D.H. and Fisher, S.K. (1990) Changes in intermediate filament immunolabelling occur in response to retinal detachment and reattachment in primates. *Investigative Ophthalmology and Visual Science*, **31**, 1474–82.

Hampson, E.C.G.M., Weiler, R. and Vaney, D.I. (1994) pH-gated dopaminergic modulation of horizontal cell gap junctions in mammalian retina. *Proceedings of the Royal Society, London, B*, **255**, 67–72.

Hiroi, K., Yamamoto, F. and Honda, Y. (1994) Intraretinal study of cat electroretinogram during retinal ischemia–reperfusion with extracellular K^+ concentration microelectrodes. *Investigative Ophthalmology and Visual Science*, **35**, 656–63.

Hockenbery, D.M., Oltvai, Z.N., Yin, X.M. *et al.* (1993) Bc1–2 functions in an antioxidant pathway to prevent apoptosis. *Cell*, **75**, 241–51.

Hori, S. and Mukai, N. (1980) Ultrastructural lesions of retinal arteries in streptozotocin-induced diabetic rats. *Japanese Journal of Ophthalmology*, **24**, 267–77.

Johnson, N.F. (1977) Retinal glycogen content during ischaemia. *Albrecht von Graefes Archiv für klinische und experimentelle Ophthalmologie*, **203**, 271–82.

Karwoski, C.J. and Proenza, L.M. (1977) Relationship between Müller cell responses, a local transretinal potential, and potassium flux. *Journal of Neurophysiology*, **40**, 244–59.

Karwoski, C.J. and Proenza, L.M. (1987) Sources and sinks of light-evoked $\Delta[K^+]_o$ in the vertebrate retina. *Canadian Journal of Physiology and Pharmacology*, **65**, 1009–17.

Karwoski, C.J., Frambach, D.A. and Proenza, L.M. (1985) Laminar profile of resistivity in frog retina. *Journal of Neurophysiology*, **54**, 1607–19.

Karwoski, C.J., Lu, H.-K. and Newman, E.A. (1989) Spatial buffering of light-evoked potassium increases by retinal Müller (glial) cells. *Science*, **244**, 578–80.

Kljavin, I.J. and Reh, T.A. (1991) Müller cells are a preferred substrate for in vitro neurite extension by rod photoreceptor cells. *Journal of Neuroscience*, **11**, 2985–94.

Korte, G.E., Hageman, G.S., Pratt, D.V. *et al.* (1992) Changes in Müller cell plasma membrane specializations during subretinal scar formation in the rabbit. *Experimental Eye Research*, **55**, 155–62.

Kuwabara, T. and Cogan, D.G. (1961) Retinal glycogen. *Archives of Ophthalmology*, **66**, 680–8.

Laqua, H. and Machemer, R. (1975) Glial cell proliferation in retinal detachment (massive preretinal proliferation). *Americal Journal of Ophthalmology*, **80**, 602–18.

Lombardini, J.B. (1991) Taurine: retinal function. *Brain Research Reviews*, **16**, 151–69.

Long, K.O., Fisher, S.K., Fariss, R.N. and Anderson, D.H. (1986) Disc shedding and autophagy in the cone-dominant ground squirrel retina. *Experimental Eye Research*, **43**, 193–205.

Magelhães, M.M. and Coimbra, A. (1970) Electron microscope radioautographic study of glycogen synthesis in the rabbit retina. *Journal of Cell Biology*, **47**, 263–75.

Mano, T. and Puro, D.G. (1990) Phagocytosis by human retinal glial cells in culture. *Investigative Ophthalmology and Visual Science*, **31**, 1047–55.

Mantych, G.J., Hageman, G.S. and Devascar, S.D. (1993) Characterization of glucose transporter isoforms in the adult and developing human eye. *Endocrinology*, **133**, 600–7.

McGrail, K.M. and Sweadner, K.J. (1989) Complex expression patterns for Na^+,K^+-ATPase isoforms in retina and optic nerve. *European Journal of Neuroscience*, **2**, 170–6.

Miller, B., Miller, H. and Ryan, S.J. (1986) Experimental epiretinal proliferation induced by intravitreal red blood cells. *American Journal of Ophthalmology*, **79**, 613–21.

Miller, R.F. and Dowling, J.E. (1970) Intracellular responses of the Müller (glial) cells of mudpuppy retina: their relation to b-wave of the electroretinogram. *Journal of Neurophysiology*, **33**, 323–41.

Moore, C.L. and Gruberg, E.R. (1974) The distribution of succinic semialdehyde dehydrogenase in the brain and retina of the tiger salamander. *Brain Research*, **67**, 467–78.

Mori, S., Miller, W.H. and Tomita, T. (1976) Müller cell function during spreading depression in frog retina. *Proceedings of the National Academy of Sciences of the USA*, **73**, 1351–4.

Moscona, A.A. (1983) On glutamine synthetase, carbonic anhydrase and Müller glia in the retina. *Progress in Retinal Research*, **2**, 111–35.

Müller, H. (1851) Zur Histologie der Netzhaut. *Zeitschrift für Wissenschaftliche Zoologie*, **3**, 234–7.

Muresan, Z. and Besharse, J.C. (1993) D2-like dopamine receptors in amphibian retina: localization with fluorescent ligands. *Journal of Comparative Neurology*, **331**, 149–60.

Neal, M.J. and Iversen, L.L. (1972) Autoradiographic localization of 3H-GABA in rat retina. *Nature*, **235**, 217–8.

Newman, E.A. (1984) Regional specialization of retinal glial cell membrane. *Nature*, **309**, 155–7.

Newman, E.A. (1985) Voltage-dependent calcium and potassium channels in retinal glial cells. *Nature*, **317**, 809–11.

Newman, E.A. (1987) Distributtion of potassium conductance in mammalian Müller (glial) cells. A comparative study. *Journal of Neuroscience*, **7**, 2423–32.

Newman, E.A. (1988) Potassium conductance in Müller cells of fish. *Glia*, **1**, 275–81.

Newman, E.A. (1993) Inward-rectifying potassium channels in retinal glial (Müller) cells. *Journal of Neuroscience*, **13**, 3333–45.

Newman, E.A. (1994) A physiological measure of carbonic anhydrase in Müller cells. *Glia*, **11**, 291–9.

Newman, E.A. and Astion, M.L. (1991) Localization and stoichiometry of electrogenic sodium bicarbonate cotransport in retinal glial cells. *Glia*, **4**, 424–8.

Newman, E.A. and Odette, L.L. (1984) Model of electroretinogram b-wave generation: a test of the K^+ hypothesis. *Journal of Neurophysiology*, **51**, 164–82.

Newman, E.A., Frambach, D.A. and Odette, L.L. (1984) Control of extracellular potassium levels by retinal glial cell K^+ siphoning. *Science*, **225**, 1174–5.

Nilius, B. and Reichenbach, A. (1988) Efficient K^+ buffering by mammalian retinal glial cells is due to cooperation of specialized ion channels. *Pflügers Archiv*, **411**, 654–60.

Olney, J.W., Ho, O.L. and Rhee, V. (1971) Cytotoxic effects of acidic and sulphur containing amino acids on the infant mouse central nervous system. *Experimental Brain Research*, **14**, 61–76.

Olney, J.W., DeGubareff, T. and Collins, J.F. (1980) Stereospecifity of the gliotoxic and anti-neurotoxic actions of alpha-aminoadipate. *Neuroscience Letters*, **19**, 277–82.

Osborne, N.N., Block, F. and Sontag, K.-H. (1991) Reduction of ocular blood flow results in glial fibrillary acidic protein (GFAP) expression in rat retinal Müller cells. *Visual Neuroscience*, **7**, 637–9.

Pfeiffer, B., Grosche, J., Reichenbach, A. and Hamprecht, B. (1994) Immunocytochemical demonstration of glycogen phosphorylase in Müller (glial) cells of the mammalian retina. *Glia*, **12**, 62–7.

Phillis, J.W. (1994) A 'radial' view of cerebral ischemic injury. *Progress in Neurobiology*, **42**, 441–8.

Poitry-Yamate, C. and Tsacopoulos, M. (1991) Glial (Müller) cells take up and phosphorylate [^{3}H]2-deoxy-F5D-glucose in a mammalian retina. *Neuroscience Letters*, **122**, 241–4.

Pow, D.H. and Crook, D.K. (1994) Rapid post-mortem changes in the cellular distribution of amino acid transmitters in retina. *Proceedings of the Australian Neuroscience Society*, **5**, 159.

Pow, D.W. and Robinson, S.R. (1994) Glutamate in some retinal neurons is derived solely from glia. *Neuroscience*, **60**, 355–66.

Quian, H., Malchow, R.P. and Ripps, H. (1993) The effects of lowered extracellular sodium on γ-aminobutyric acid (GABA-) induced currents of Müller (glial) cells of the skate retina. *Cellular and Molecular Neurobiology*, **13**, 147–58.

Reichenbach, A. (1989) Attempt to classify glial cells by means of their process specialization using the rabbit retinal Müller cell as an example of cytotopographic specialization of glial cells. *Glia*, **2**, 250–9.

Reichenbach, A. and Eberhardt, W. (1988) Cytotopographical specialization of enzymatically isolated rabbit retinal Müller (glial) cells: K^+ conductivity of the cell membrane. *Glia*, **1**, 191–7.

Reichenbach, A. and Robinson, S.R. (1995) Ependymoglia and ependymoglia-like cells, in *Neuroglia Cells* (eds B. Ransom and H. Kettenmann) (in press).

Reichenbach, A. and Wohlrab, F. (1985) Effects of α-aminoadipic acid on the glutamate-isolated PIII of the rabbit electroretinogram. *Documenta Ophthalmologica*, **59**, 359–64.

Reichenbach, A., Hagen, E., Schippel, K. and Eberhardt, W. (1988) Cytotopographical specialization of enzymatically isolated rabbit retinal Müller (glial) cells. Structure, ultrastructure, and (^{3}H)-ouabain binding sites. *Zeitschrift für mikroskopisch-anatomische Forschung*, **102**, 897–12.

Reichenbach, A., Schneider, H., Leibnitz, L. *et al.* (1989) The structure of rabbit retinal Müller (glial) cells is adapted to the surrounding retinal layers. *Anatomy and Embryology*, **180**, 71–9.

Reichenbach, A., Henke, A., Eberhardt, W. *et al.* (1992) K^+ regulation in retina. *Canadian Journal of Physiology and Pharmacology*, **70** (Suppl.), S239–47.

Reichenbach, A., Stolzenburg, J.-U., Eberhardt, W. *et al.* (1993) What do retinal Müller (glial) cells do for their neuronal 'small siblings'? *Journal of Chemical Neuroanatomy*, **6**, 201–13.

Reichenbach, A., Stolzenburg, J.-U., Wolburg, H. *et al.* (1995) Hepatic retinopathia. I. Morphological findings. *Acta Neuropathol.* (in press).

Reichenbach, A., Ziegert, M., Schnitzer, J. and Pritz-Hohmeir, S. (1994) Development of the rabbit retina. V. Columnar units. *Developmental Brain Research*, **79**, 72–84.

Rentsch, F.J. (1979) Preretinal proliferation of glial cells after mechanical injury of the rabbit retina. *Albrecht von Graefe's Archiv für klinische und experimentelle Ophthalmologie*, **188**, 79–87.

Ripps, H. and Witkovsky, P. (1985) Neuron–glia interaction in the brain and retina, in *Progress in Retinal Research*, Vol. 4 (eds N.N. Osborne and G.J. Chader), Pergamon Press, Oxford, New York, pp.181–219.

Ripps, H., Mehaffey. L. III and Siegel, I.M. (1981) 'Rapid regeneration' in the cat retina. A case for spreading depression. *Journal of General Physiology*, **77**, 335–46.

Robinson, S.R. (1991) Development of the mammalian retina, in *Neuroanatomy of the Visual Pathways and their Development* (eds B. Dreher and S.R. Robinson), Vol. 3 of *Vision and Visual Dysfunction* (series ed. J.R. Cronly-Dillon), Macmillan, London, pp.69–128.

Robinson, S.R., Hampson. E.C.G.M., Munro, M.N. and Vaney, D.I. (1993) Unidirectional coupling of gap junctions between neuroglia. *Proceedings of the Australian Neuroscience Society*, **3**, 167.

Roque, R.S. and Caldwell, R.B. (1990) Müller cell changes precede vascularization of the pigment epithelium in the dystrophic rat retina. *Glia*, **3**, 464–75.

Sarantis, M. and Mobbs, P. (1992) The spatial relationship between Müller cell processes and the photoreceptor output synapse. *Brain Research*, **584**, 299–304.

Sarthy, P.V. (1983) Release of (^{3}H) γ-aminobutyric acid from glial (Müller) cells of the rat retina: effects of K^+, veratridine, and ethylenediamine. *Journal of Neuroscience*, **3**, 2494–503.

Sarthy, P.V. and Lam, D.M.K. (1978) Biochemical studies of isolated glial (Müller) cells from the turtle retina. *Journal of Cell Biology*, **78**, 675–84.

Schwartz, E.A. (1993) L-Glutamate conditionally modulates the K^+ current of Müller glial cells. *Neuron*, **10**, 1141–9.

Schwartz, E.A. and Tachibana, M. (1990) Electrophysiology of glutamate and sodium co-transport in a glial cell of the salamander retina. *Journal of Physiology* (*London*), **426**, 43–80.

Steinberg, R.H., Oakley, B., II and Niemeyer, G. (1980) Light-evoked changes in $[K^+]_0$ in retina of intact cat eye. *Journal of Neurophysiology*, **44**, 897–921.

Stirling, C.E. and Sarthy, P.V. (1985) Localization of the Na–K pump in turtle retina. *Journal of Neurocytology*, **14**, 33–47.

Stolzenburg, J.-U., Haas, J., Härtig, W. *et al.* (1992) Phagocytosis of latex beads by rabbit retinal Müller (glial) cells *in vitro*. *Journal für Hirnforschung*, **33**, 557–64.

Szamier, R.B., Ripps, H. and Chappell, R.L. (1981) Changes in ERG b-wave and Müller cell structure induced by α-aminoadipic acid. *Neuroscience Letters*, **21**, 307–12.

Tseng, M.T., Liu, K.N. and Radtke, N.R. (1990) Facilitated ERG recovery in taurine-treated bovine eyes, an *ex vivo* study. *Brain Research*, **509**, 153–5.

Tout, S., Chan-Ling, T., Holländer, H. and Stone, J. (1993) The role of Müller cells in the formation of the blood retinal barrier. *Neuroscience*, **55**, 291–301.

Uga, S. and Smelser, G.K. (1973) Comparative study of the fine structure of retinal Müller cells in various vertebrates. *Investigative Ophthalmology*, **12**, 434–48.

Van Harreveld, A. (1982) Swelling of Müller fibers in the chicken retina. *Journal of Neurobiology*, **13**, 519–36.

Wagner, H.-J., Luo, B.-G., Ariano, M.A. *et al.* (1993) Localization of D2 dopamine receptors in vertebrate retinae with anti-peptide antibodies. *Journal of Comparative Neurology*, **331**, 469–81.

Watanabe, H., Tomita, H., Ishiguro, S.-I. and Tamai, M. (1994) Suppression of phosphate-activated glutaminase activity in ischemic rat retinas. *Investigative Ophthalmology and Visual Science*, **35**, 1866.

Witkovsky, P. and Schüte, M. (1991) The organization of dopaminergic neurons in vertebrate retinas. *Visual Neuroscience*, **7**, 113–24.

Wolburg, H. and Berg, K. (1988) Distribution of orthogonal arrays of particles in the Müller cell membrane of the mouse retina. *Glia*, **1**, 246–52.

Yamada, T., Hara, S. and Tamai, M. (1990) Immunohistochemical localization of cathepsin D in ocular tissues. *Investigative Ophthalmology and Visual Science*, **31**, 1217–23.

Young, R.W. (1984) Cell death during differentiation of the retina in the mouse. *Journal of Comparative Neurology*, **229**, 362–73.

17

Clinical aspects: outer retinal dystrophies

ALAN C. BIRD

17.1 INTRODUCTION

Dystrophies of the outer retina comprise a variety of disparate genetically determined conditions that differ from one to another in their mode of inheritance, their pattern of visual loss and their ophthalmoscopic appearances. Over the past few years, as a result of research by clinicians, biochemists, cell biologists and molecular biologists, an increasing number of distinct disorders have been recognized within this heterogeneous group, and some clues as to their pathogeneses have emerged. A number of nosological entities have been identified by detection of point mutations causing both autosomal dominant and recessive disease. In addition, in other conditions the locus of the abnormal gene has been defined. These discoveries are already having an impact on clinical management and the effect is likely to increase greatly in the near future.

It is possible to subdivide photoreceptor dystrophies into groups depending on their clinical features. Some cause loss of function early in disease associated with the scotopic system. Examination reveals defective vision in the mid zone of the visual field initially and morphological changes in the postequatorial fundus; most of the diseases in this category are known collectively as retinitis pigmentosa (RP). Other disorders causing loss of photopic function and morphological changes in the central fundus, are known as macular degenerations or cone dystrophies. This subdivision into 'peripheral degenerations' in which the rods may be the primary target of disease and 'central degenerations' in which the cones may be the cells initially affected is superficially attractive. However, the function of both rod and cone systems are compromised in most if not all progressive disorders, even in the early stages of their disease. There are also conditions in which there is involvement of the central and mid-peripheral fundus with relative sparing in the intermediate region. The last are often referred to as cone/rod dystrophies. The receptor dystrophies, therefore, comprise a spectrum of diseases ranging from predominant rod dystrophies to predominant cone dystrophies with disorders intermediate between the two in which there is varying involvement of the rod and cone systems. Despite these reservations, receptor dystrophies can be considered in two broad categories: peripheral receptor dystrophies typified

Neurobiology and Clinical Aspects of the Outer Retina
Edited by M.B.A. Djamgoz, S.N. Archer and S. Vallerga
Published in 1995 by Chapman & Hall, London
ISBN 0 412 60080 3

by RP, and central dystrophies. This chapter deals with central dystrophies; RP is covered in Chapter 18.

A number of genetically determined disorders cause progressive loss of visual functions associated with cones, namely loss of visual acuity, color vision and central visual field, and if diffuse, poor vision in bright light. In some the electrophysiological responses imply that the disorder is confined to the central retina whereas in others the whole cone system is affected. Unlike retinitis pigmentosa, the appearance of the ocular fundus varies greatly in macular dystrophies, which has led to subdivision of this group of disorders into purer samples of disease. In some disorders both clinical and histopathological observations imply that the major changes appear to occur at the level of the retinal pigment epithelium with good visual function at least for a period, whereas in others the photoreceptor cells appear to be affected initially. In a third group Bruch's membrane is affected initially, and visual loss occurs as a consequence of invasion of the outer retina by choroidal blood vessels or of deficient metabolic exchange between the outer retina and choroid.

Specific disorders, such as Best's disease and Sorsby's fundus dystrophy, can be identified as single nosological entities but the remainder include groups of diseases that cannot be clearly distinguished one from another. They have been subdivided with respect to inheritance and the appearance of the fundus, and it is likely that each subdivision contains more than one condition. Most disorders fall into two broad subdivisions, fundus flavimaculatus and bull's eye maculopathy (or cone dystrophy) which can be identified on the basis of fundus appearance although the distinction may not be absolute. The clinical entities are described according to the tissue which appears to be primarily affected although this may not necessarily indicate the site of the initiating metabolic defect as shown by identification of the responsible mutation.

17.2 CENTRAL PHOTORECEPTOR DISEASES

17.2.1 BULL'S EYE DYSTROPHIES (INCLUDING CONE AND CONE–ROD DYSTROPHIES)

The typical fundus changes in this group of cone and cone–rod dystrophies consist of one or more concentric rings of pigment epithelial change around the fovea, that gives rise to a characteristic appearance on fluorescein angiography. Visual loss may occur at any time during the second to the fifth decades of life and once started progresses slowly (Grey *et al.*, 1977). There may or may not be white deposits at the level of the pigment epithelium. The phenomenon of dark choroid is also seen in some families with bull's eye dystrophy (Bonin, 1971; Fish *et al.*, 1981; Uliss *et al.*, 1987). This phenomenon is due to absorption of short wavelength light by lipofuscin in the retinal pigment epithelium such that the choroidal fluorescence is not seen on fluorescein angiography.

These disorders can be transmitted as autosomal dominant (Goodman *et al.*, 1963, Berson *et al.*, 1968; Krill and Deutman, 1972; Pearlman *et al.*, 1974; Gouras *et al.*, 1983; Bresnick *et al.*, 1989; van Schooneveld *et al.*, 1991, Went *et al.*, 1992), autosomal recessive (Gouras *et al.*, 1983; Yagasaki and Jacobson, 1989) or X-linked (Heckenlively and Weleber, 1986; Jacobson *et al.*, 1989; Reichel *et al.*, 1989; van Everdingen *et al.*, 1992) conditions. Even within these different genetic forms of cone dystrophy there is heterogeneity. Such lesions may occur with disease restricted to the central region, or generalized cone loss in which case reduced visual acuity is associated with photophobia, and defective of color vision. In dominant cone dystrophies there is one recently described form with early and near complete absence of blue cone function, but with a blue-sensitive visual pigment locus indistinguishable from normal (van Schooneveld *et al.*, 1991; Went *et al.*, 1992; Bresnick *et al.*, 1989). In a family with X-linked cone dystrophy accompanied by loss of red cone function, a 6.5 kb deletion within the red cone pigment gene was found (Reichel *et al.*, 1989). Heterogeneity is further indicated by the presence of peripheral retinal degeneration in some conditions for which the term cone–rod dystrophy is used, and there is evidence that this group also includes several conditions (Pruett, 1983; Szlyk *et al.*, 1993). The different categories can be distinguished one from another on the basis of testing the electro-retingram (ERG). Drug-induced phenocopies occur (Kearns and Hollenhorst, 1966; Krill *et al.*, 1971) and comparable dystrophies have also been described in the Pierre–Marie type of hereditary ataxia (Bjork *et al.*, 1956), in fucosidosis (Snodgrass, 1976) and in amelanogenesis imperfecta (Jalili *et al.*, 1988).

There is considerable doubt as to whether the fundus changes of fundus flavimaculatus and bull's eye dystrophy indicate that the disorders are clearly separated into two broad groups. Studies of fundus appearance show that in some families the changes are constant, whereas in others bull's eye dystrophy may be seen in early disease and flavimaculatus lesions seen later.

17.2.2 CENTRAL AREOLAR CHOROIDAL SCLEROSIS

A dominantly inherited dystrophy with well-defined atrophy of the outer retina, retinal pigment epithelium and inner choroid was described by Sorsby (1939). Initially, the choriocapillaris was thought to be primarily affected (Sorsby, 1939; Sorsby and Crick, 1953; Noble, 1977) because atrophy of choroid at the macula was apparently the initial or at least the most prominent ophthalmoscopic change when the patient was first seen. However, Ashton (1953) demonstrated that the major choroidal blood vessels were normal in a case of central areolar choroidal sclerosis by histopathological examination such that the target cell of disease was uncertain.

Two mutations in the peripherin/RDS gene have recently been reported causing a macular dystrophy most closely resembling this condition (Wells *et al.*, 1992). The mutations are in codon 172. In two families arginine is substituted by tryptophan and in one other by glutamine. In each there are symptoms of difficulty in passing from light to dark in the third decade of life and retinal pigment epithelial changes are identifiable by this time which are centered at the fovea and extend outside the posterior pole. Profound atrophy of the outer retinal and inner choroid occurs during the next three decades at a variable rate. Peripheral rod function is normal. The diseases are qualitatively similar with the mutations but is more severe with the Arg-172-Trp substitution. Given that peripherin/rds is found only in photoreceptor cells, it is evident that the primary defect is in the outer neuro-retina despite the prominent changes seen in other structures.

17.2.3 PIGMENT EPITHELIAL DISEASES

In a series of disorders the abnormal phenotype is manifest initially at the level of the retinal pigment epithelium with accumulation of material appearing as pale deposits.

(a) Best's disease

Autosomal dominant inheritance of this disorder was first suggested by Best (1905) and was substantiated later by further observations on the same family (Vossius, 1921; Weisel and Bott, 1922; Jung, 1936).

The typical lesion of Best's disease is a round central yellow deposit at the level of the retinal pigment epithelium (Best, 1905) that may be identified within a short time of birth (Barkman, 1961), or that may develop later in a previously normal fundus (Deutman, 1971; Barricks, 1977; Mohler and Fine, 1981). Rarely, the lesion may be extramacular, the disease may be asymmetrical, or the fundus may be entirely normal (Deutman, 1971; Barricks, 1977; Krill, 1977; Mohler and Fine 1981; Godel *et al.*, 1986). Godel *et al.* (1986) stated that about half those with the abnormal gene have a normal or near normal fundus appearance and normal vision. The rise in ocular potential induced by light is always reduced in subjects with the abnormal gene for Best's disease, whatever their clinical status (Deutman, 1969, 1971). The universal reduction of the light-induced rise in ocular potential even in those patients with normal fundi, indicates a widespread dysfunction of the photoreceptor/pigment epithelial complex. Thus the distribution of the abnormal gene in a family cannot be identified by clinical examination alone, and can only be achieved by electro-oculography.

Histopathological studies imply that the abnormal material accumulates as lipofiscin in the pigment epithelium, and that the deposits seen clinically are probably between Bruch's membrane and the pigment epithelium (O'Gorman *et al.*, 1988; Frangieh *et al.*, 1982; Weingeist *et al.*, 1982). In patients with advanced disease, histopathology showed extensive atrophy of the receptors in the macular area (McFarland, 1955; Anderson, 1977). The abnormal material appears to be derived from degenerating pigment epithelial cells. Despite the fact that the abnormalities of structure and the electrical response to light were documented many years ago the pathogenetic relationship between the two is unknown. The lack of light-induced change in standing potential presumably implies a defective response of the apical cell membrane of the retinal pigment epithelium to changes in potassium concentration in the outer retina consequent upon transduction. Carbonic anhydrase induces a reduction of the standing potential demonstrating that the basolateral cell membrane domain functions normally.

The locus of the abnormal gene has been identified in chromosome 11 but the gene is as yet unknown (Stone *et al.*, 1992).

(b) Adult viteliform macular dystrophy

Adult viteliform macular dystrophy was first described by Gass (1974) as a peculiar foveomacular dystrophy. Although series have been reported where there was no evidence of familial involvement, this condition is now generally regarded as being transmitted as an autosomal dominant trait (Gass, 1974; Vine and Schatz, 1980; Kingham and Lochen, 1977; Brecher and Bird, 1990).

The disorder is characterized by a focal, round or oval shaped, subretinal yellowish foveal lesion, often with hyperpigmentation on the anterior surface of the lesion at the level of the pigment epithelium. The lesions may vary in size but are typically one-third to one-half a disc diameter, and are usually bilateral and symmetrical. Patients usually present in the fourth or fifth decade of life and tend to have minimal visual symptoms (Patrinely *et al.*, 1985). The disease differs from Best's disease in that the foveal lesions are smaller, it presents at a later age, it does not demonstrate evolutionary changes of the foveal lesion, and the light-induced rise in ocular potential is usually normal. Because of the variation in age of onset and variable severity, the distribution of the abnormal gene within a family cannot be established with certainty by clinical survey of the family.

The histopathological reports of this condition show loss of retinal pigment epithelium and disruption of photoreceptors at the fovea (Gass, 1974; Patrinely *et al.*, 1985). Perhaps of greater significance is the very intense autofluorescence seen in the intact retinal pigment epithelium outside the fovea implying that accumulation of lipofuscin at this site is central to the pathogenesis of the disorder.

A mutation in the peripherin gene, a stop codon at position 258, has been found in one patient with a retinal degeneration similar in appearance to adult viteliform macular dystrophy (Wells *et al.*, 1992) but there is no evidence that all families have such a mutation.

(c) Pattern dystrophies

In 1970, five members of a family from Holland were described with a unique fundus appearance consisting of a linear pattern of hyperpigmentation at the level of the retinal pigment epithelium in the macula (Deutman *et al.*, 1970). The disorder caused few symptoms since the worst visual acuity was 6/9. The most striking feature was the universal reduction of the light-induced rise in ocular potential to 130% or lower in the presence of good visual function. It was concluded that there was diffuse dysfunction of the retinal pigment epithelium with little associated visual deficit since there was no indication of rod dysfunction as gauged by the electroretinogram and no abnormality of color vision. In this respect butterfly-shaped dystrophy is similar to Best's disease. The original pedigree is suggestive of autosomal dominant inheritance; in the original communication the disease was identified in only two generations but large pedigrees have been published subsequently (Prensky *et al.*, 1983).

A number of families have been described with autosomal dominant disease characterized by linear or irregular changes of pigment at the level of the retinal pigment epithelium with considerable variability with in each family (Watzke *et al.*, 1981; de Jong *et al.*, 1991). Some members have been described as having macroreticular change and others have fundus pulverilentus, and an appearance of butterfly dystrophy as described by Deutmann or macular lesions (de Jong and Delleman, 1982; Gutman *et al.*, 1982; Guiffre, 1988; Guiffre and Lodato, 1988). The deposits at the level of the pigment epithelium may be pale or dark and may or may not be linear (Cortin *et al.*, 1980). In all families the light-induced rise in ocular potential may be depressed, and the condition is compatible with retention of good acuity throughout life. It is now believed that pattern dystrophy represents one or more than one nosological entity with variable expressivity. The problem may be resolved by comparison between families or by genomic analysis. In these families with variable expression the response of the standing potential to light is usually normal implying the possibility of at least two fundamentally different groups of disorder.

In some families with a pattern dystrophy a mutation has been detected in the peripherin/RDS gene (Nichols *et al.*, 1993; Kim *et al.*, 1994).

(d) Fundus flavimaculatus

This broad category comprises diseases in which there is deposition of white material at the level of the pigment epithelium; Stargardt's disease (Stargardt, 1909, 1913) and fundus flavimaculatus (Franceschetti, 1963; Franceschetti and François, 1965) have been used to denote these disorders, but there is no evidence that these two terms describe separate groups of conditions. Most cases show autosomal recessive inheritance as first described by Stargardt, although autosomal dominant forms are occasionally seen (Cibis *et al.*, 1980; Uliss *et al.*, 1987). These conditions usually cause rapid loss of central vision during a 6 month period in the first 15 years of life, although in some cases good visual acuity is maintained until the age of 50 years.

At the time of visual loss confluent atrophy of the outer retina pigment epithelium and choriocapillaris occurs at the fovea, and this area grows slowly during the rest of the patient's life. The white 'fish tail' flecks occupy the remaining part of the posterior pole with characteristic sparing of the peripapillary region. These lesions can be identified at the time of the initial visual loss and resolve as additional ones appear elsewhere (Hadden and Gass, 1976).

On fluorescein angiography the choroid appears normal in some patients but it is not seen in the majority (Bonin, 1971; Fish *et al.*, 1981; Fishman *et al.*, 1987; Uliss *et al.*, 1987). It seems likely that the lack of choroidal fluorescence signifies an even deposition of abnormal material at the level of the pigment epithelium that absorbs blue–green light (Eagle *et al.*, 1980). This conclusion has been supported by histopathological studies in which it has been demonstrated that the retinal pigment epithelial cells were packed with lipofuscin and melanolipofuscin (Klein and Krill, 1967; Eagle *et al.*, 1980; Lopez *et al.*, 1990); these changes start to develop in childhood (Steinmetz *et al.*, 1991). At the site of previous white lesions, atrophy of the pigment epithelium occurs and fluorescein angiography shows multifocal hyperfluorescence, corresponding to the areas of pigment epithelial atrophy (Hadden and Gass, 1976).

A study of various families with autosomal recessive fundus flavimaculatus found that all were compatible with the abnormal gene mapping to the short arm of chromosome 1 (Kaplan *et al.*, 1993).

(e) North Carolina dystrophy

This condition was well characterized in a large family from North Carolina (Lefler *et al.*, 1971), and several reports have followed (Frank *et al.*, 1974; Small, 1989; Small *et al.*, 1991). It is likely that previous reports from Europe described the same condition (Davenport, 1927; Clausen, 1928). It is dominantly inherited and appears to be fully penetrant although variable in its expressivity. It may cause profound atrophy of the outer retina and choroid at the macula in early life although visual acuity may be remarkably good. However, there may be drusen-like deposits only in the posterior pole. The disorder is widely believed not to progressive (Small *et al.*, 1991). In some families there is the consistent presence of skeletal abnormalities of the extremities (Sorsby, 1935; Turut *et al.*, 1991) implying that more than one disorder may cause the fundus lesion. The abnormal gene has been localized to the short arm of chromosome 6 in at least one family (6p) (Small *et al.*, 1992) but the gene has yet to be identified.

17.2.4 BRUCH'S MEMBRANE DISEASES

In some forms of central retinal disease the changes occur initially at the level of Bruch's membrane.

(a) Sorsby's fundus dystrophy (pseudoinflammatory macular dystrophy)

Sorsby *et al.* (1949) first reported an autosomal dominant disorder in which there is bilateral central visual loss in the fifth decade of life from subretinal neovascularization and progressive atrophy of the peripheral retina and choroid leading to loss of ambulatory vision by the seventh decade in most cases. Subsequent reports have shown that the disease is fully penetrant, and that the abnormal phenotype becomes evident in the third decade of life (Carr *et al.*, 1975; Hoskin *et al.*, 1981; Capon *et al.*, 1988; Polkinghorne *et al.*, 1989; Hamilton *et al.*, 1989; Steinmetz *et al.*, 1992). Patients report difficulty passing from light to dark and have slow recovery of sensitivity and rhodopsin following bleach. The fundi show drusen, yellow deposits and slow filling of the choroid on fluorescein angiography.

A light and electron microscopic study of the eyes of one patient showed a 30 μm thick deposit within Bruch's membrane that stained positive for lipids. In addition, there was gross loss of the outer retina, a discontinuous retinal pigment epithelium and atrophy of the choriocapillaris (Capon *et al.*, 1989).

In one family the locus of the gene has been identified at the telomeric region of 22q.

(b) Autosomal dominant drusen

Drusen of Bruch's membrane were first demonstrated microscopically as focal deposits at the level of Bruch's membrane (Wedl, 1854). They were first described clinically by Donders (1855). Historically, the diagnosis of dominantly inherited drusen has often been made in those patients in whom the drusen appear at a relatively early age, and various terms have been used to describe these conditions such as Hutchinson–Tay choroiditis (Hutchinson and Tay, 1875), guttata choroiditis (Juler, 1893; Clarke, 1932), Holthouse–Batten superficial chorioretinitis (Holthouse and Batten, 1897), Doyne's honeycomb retinal degeneration (Doyne, 1899; Tree, 1937; Pearce, 1968), family choroiditis (Doyne, 1910), Malattia levantinese (Klainguti, 1932), and crystalline retinal degeneration (Evans, 1950). However, neither Hutchinson and Tay, nor Holthouse and Batten reported sufficient evidence to indicate that they described a dominantly inherited disorder. By contrast in Doyne's honeycomb dystrophy, and Mallatia levantinese, there is good evidence of dominant inheritance.

The view that the various manifestations of drusen of the posterior pole represent the same disorder has been a popular one in recent decades (François and Deweer, 1952; Forni and Babel, 1962; Deutman and Jansen, 1970; Gass, 1973). However, the pattern of drusen in Doyne's honeycomb dystrophy, and Mallatia levantinese are constant within families and differ from each other. In the latter, there are small drusen in the peripheral macula distributed in a typical radial pattern with larger confluent drusen centrally. The small drusen have the clinical and angiographic appearance (Klainguti, 1932; Scarpatetti, *et al.*, 1978) of 'basal laminar drusen' (Gass, 1985), and it is claimed that they represent focal thickening of the basement membrane of the retinal pigment epithelium. None of Doyne's original families or their descendants were noted to have radially oriented basal laminar drusen (Doyne, 1899; Tree, 1937; Pearce, 1968). Based on the presence of radial basal laminar drusen, several families reported in the literature to have a Doyne's type 'colloid degeneration' might be more properly described as having Mallatia lavantinese (Pajtas, 1950; Alper and Alfano, 1953). In Mallatia levantinese the small discrete drusen represent focal thickening of the basement membrane of the retinal pigment epithelium (Forni and Babel, 1962), whereas in Doyne's honeycomb dystrophy the abnormal material accumulates between the basement membrane of the pigment epithelium and the inner collagenous layer of Bruch's membrane. In this respect the latter simulates age-related macular disease.

In both disorders visual acuity remains good until middle life. In Doyne's honeycomb dystrophy geographic atrophy may supervene at any time thereafter. In Mallatia levantinese, either geographic atrophy or occult choroidal neovascularization may supervene.

17.3 CONGENITAL RECEPTOR DEFECTS

A series of conditions has been recognized in which there is a genetically determined non-progressive visual defect that may be related to either the cone or the rod system.

17.3.1 CONE DEFECTS

(a) Defective color vision

Defects in color vision have been classified on the basis of the concept that color vision is determined by three classes of cones: red-sensitive, green-sensitive and blue-sensitive. The basis for this classification has been confirmed by the sequencing of the three cone pigment genes (Nathans *et al.*, 1986a). If color matching tests demonstrate that one of the three systems is defective but present (trichromats), the terms protanomaly (red), deuteranomaly (green) and tritanomaly (blue) are used, and if one is functionally absent (dichromats), the suffix -anopia replaces the suffix -anomaly.

Deuteranomaly, deuteranopia, protanopia and protanomaly are X-linked conditions: deuteranomaly is found in about 5% of the population of Western Europe and North America and deuteranopia, protanomaly and protanopia in about 1% each (Waardenburg *et al.*, 1963). Tritanomaly and tritanopia, that are inherited as autosomal dominant disorders, are much less common affecting between 0.002 and 0.007% of the population (Verriest, 1974).

Individuals with these abnormalities of color vision have normal visual acuity and anomalous trichromats are often unaware of this condition. Apart from the color defect, the eye is normal.

The molecular genetic basis of the red and green color defects has now been demonstrated. Abnormalities of the red and green pigment genes to differing degrees occur in deuteranopia, deuteranomaly, protanopia and protanomaly. People with deuteranopia were found to have a normal red pigment gene but no green pigment gene(s). Those with protanopia lacked a normal red pigment gene, that was replaced by a hybrid red–green pigment gene. Deuteranomalous and protanomalous trichromats had hybrid red–green genes (Nathans *et al.*, 1986b; Vollrath *et al.*, 1988).

(b) Monochromatism

Patients suffering from monochromatism have absent or markedly impaired color vision. In complete rod monochromatism there is little evidence of cone function, so that visual acuity is poor, usually the patient has nystagmus and extreme photophobia is characteristic. The fundi show no gross abnormality, although, the foveola may appear abnormal, and changes in the central pigment epithelium are sometimes identified (Waardeburg *et al.*, 1963). The dark adaptation curve is typically monophasic, fusion frequency is very low and the photopic (cone) ERG is absent but the scotopic (rod) ERG is normal. Rod monochromatism is an autosomal recessive trait.

Blue cone monochromatism presents in a similar fashion to complete rod monochromatism, except that there appears to be an intact blue cone system and the inheritance is X-linked (Alpern *et al.*, 1965). The gene for this form of monochromatism maps to Xq28, and it is possible that this rare disorder results from alterations of the red and green pigment genes, or from their deletion (Lewis *et al.*, 1987).

An incomplete form of rod monochromatism has been described where the symptoms are less severe, visual acuity is better and photophobia and nystagmus may be absent (Sloan and Newhall, 1942; Siegel *et al.*, 1966; Krill, 1977). In another form of monochromatism there is absence of central cones (Turut *et al.*, 1991).

Four eyes from individuals with monochromatism have been examined histopathologically (Larsen, 1921; Harrison *et al.*, 1960; Falls *et al.*, 1965; Glickstein and Heath, 1975). These showed a change at the fovea with a reduced number of cones and an abnormality of those that remained.

17.3.2 ROD DEFECTS

(a) Congenital stationary night blindness

Congenital stationary night blindness (CSNB) is characterized by night blindness, normal fundus appearance, the absence of rod dark adaptation (monophasic dark adaptation curve) and lack of progression. It may be inherited as an autosomal dominant, an autosomal recessive or an X-linked trait.

The most widely reported form of CSNB is an autosomal dominant disorder (Dejean and Gassenc, 1949). These patients appear to have normal cone function but little rod function. Visual acuity, color vision and photopic visual fields are normal or are at most mildly abnormal and dark adaptation shows only a cone segment that may be abnormally prolonged (Krill and Martin, 1971); there is no shift from cone to rod characteristics in the dark adaptation curve, and the ERG shows no prolongation of the b-wave implicit time between the photopic and scotopic records. However, fundus reflectometry indicates a normal concentration of bleachable rhodopsin in one case (Carr *et al.*, 1966) and histological studies have shown no structural abnormalities either in the retina as a whole or in the rods in particular (Vaghefi *et al.*, 1978).

The majority of patients with autosomal recessive and X-linked CSNB, and occasional patients with autosomal dominant CSNB, have an ERG with a near normal a-wave and a substantially reduced b-wave on testing under scotopic conditions (negative ERG) (Noble *et al.*, 1990; Sharp *et al.*, 1990). With increasing intensity of the test stimulus the amplitude of the a-wave increases but that of the b-wave is unchanged (Schubert and Bornschein, 1952). These patients can be further divided into two groups: one group lacking rod function (complete type), the other with some rod function (incomplete type) (Miyake *et al.*, 1986). Patients with the complete type are myopic; those with the incomplete type may be hyperopic or myopic. Although complete and incomplete CSNB did not coexist in any of the families in the original study (Miyake *et al.*, 1986), others have found patients with both complete and incomplete CSNB within the same X-linked pedigrees (Khouri *et al.*, 1988; Pearce *et al.*, 1990). This led Pearce *et al.* (1990) to propose that X-linked CSNB is a single clinical entity manifesting a wide variation in clinical expression. It has also been suggested that incomplete CSNB and Åland Island eye disease may be the same condition (Weleber *et al.*, 1989; Alitalo *et al.*, 1991).

Myopia is almost always associated with the X-linked form of CSNB when visual acuity will be reduced and nystagmus may be present. Carriers of X-linked CSNB are not night blind but may show abnormal oscillatory potentials on ERG (Miyake and Kawase, 1984; Young *et al.*, 1989). Myopia also occurs in those cases of autosomal recessive CSNB that have abnormal vision and, again, these patients may have nystagmus (Krill, 1977). These forms of CSNB may present in infancy with blindness (Weleber and Tongue, 1987).

The complete form of X-linked CSNB has been assigned to Xp11.3 (Musarella *et al.*, 1989; Gal *et al.*, 1989) or Xp11.22 (Bech-Hansen *et al.*, 1990). A mutation in the rhodopsin gene has now been identified in autosomal recessive CSNB (Dryja *et al.*, 1993). It is believed that the mutant rhodopsin may create noise by activating transducin in the absence of light, causing profound reduction in sensitivity of the scotopic system.

Light and electron microscopic studies of one eye with CSNB and a negative ERG showed a normal arrangement of discs of rod outer segments, and normal synaptic ends of the photoreceptors. It was suggested that the cause in this case of CSNB may be related to mechanisms inhibitory to cells of the bipolar layer (Watanabe *et al.*, 1986). The absence of rod–cone interaction, together with an absent scotopic b-wave, also implies that the defect is in the midretinal layers (Siegel *et al.*, 1987).

(b) Prolonged dark adaptation

Three distinct conditions have been described in which the final threshold of dark adaptation is normal but in which the rod phase is abnormally prolonged; the defect in each appears to be static.

Oguchi's disease

This autosomal recessively inherited disorder with prolonged dark adaptation was first reported in Japan (Oguchi, 1907) and most subsequent reported cases have come from the same country, although non-Japanese patients have been described (Klein, 1939; Winn *et al.*, 1969). In most patients defective night vision is the only complaint, visual acuity being normal. The characteristic feature of this condition is the abnormal coloration of the light-adapted fundus, the abnormal white or cream-colored appearance being derived from the inner limiting membrane and presumably from the foot plates of Müller cells. The abnormal color resolves over a period of 0.5–8 hours in darkness and has been termed the Mizuo–Nakamura phenomenon after the authors who provided the first description (Mizuo, 1913; Nakamura, 1920). Dark adaptation is characteristically slow and a final rod threshold may be attained only after several hours, and even then may be slightly elevated (François and Verriest, 1954).

Histopathological examination has been undertaken on three eyes. An excess of cones compared with rods together with an abnormal layer of material between the receptors and pigment epithelium was reported by Oguchi (1925). Parallel histological studies on the other half of the same eye by Yamanaka (1924) failed to confirm this additional layer. It was then considered that an abundance of round lipofuscin granules confined to the apical portion of the pigment cells was the characteristic feature. Histopathological and electron microscopic study of another eye supported the view that there was an abnormal layer between the outer segments of the photoreceptors and the pigment epithelial cells (Kuwabara *et al.*, 1963). However, the constituents were normal components of the retina consisting of lipofuscin granules and protrusions of the pigment epithelium with complex interdigitations of the outer segments. There was no abnormal cone distribution. A third histological study (Yamanaka, 1969) is open to question as the patient had reduced vision with retinal pigmentary changes and both parents had RP. On electrophysiological testing, both the a-wave of the ERG and the light rise in the standing potential of the eye are normal, but the scotopic b-wave of the ERG is severely depressed even in the fully dark-adapted eye (Sharp *et al.*, 1990; Carr *et al.*, 1967).

The pathogenesis of Oguchi's disease is not understood. Rhodopsin regeneration is normal (Carr and Ripps, 1967). These observations imply that the primary abnormality is unrelated to light catch and rhodopsin bleaching, but is related to other systems of transduction. From the electrophysiological results the region of bipolar cells appears to be the earliest stage in the visual pathway exhibiting signs of defective function. Given that the abnormal reflex arises from the inner limiting membrane at which site the major cellular component is the foot plate of Müller cells, and the Müller cells are the source of the b-wave, it is likely that these cells are primarily at fault as suggested by de Jong *et al.* (1991). Unfortunately the histopathological studies did not report on the state of Müller cells.

Fundus albipunctatus

Fundus albipunctatus is a stationary autosomal recessive condition in which the only symptoms are related to defective dark adaptation (Lauber, 1924). The condition should not be confused with albipunctate dystrophy (retinitis punctata albescens), that is progressive and that represents a variant of RP. The fundus shows widespread distribution of uniform-sized almost white dots at the level of the pigment epithelium, that are most dense in the postequatorial fundus; the macula may or may not be involved. Changes in the distribution of the white dots have been described (Marmor, 1977), as have their change from flecks in childhood to relatively permanent punctate dots that increase in number over the years (Marmor, 1990). Diffuse pigmentary changes in the pigment epithelium are unusual. Fluorescein angiography shows punctate hyperfluorescence that does not, however, correspond to the white dots (Marmor, 1990). Although fundus albipunctatus usually involves both eyes, unilateral disease has been reported (Henkes, 1963).

Typically, the visual acuity and visual fields are normal, but minor loss of field has been described (Marmor, 1977). In a majority of cases the dark adaptation of both cones and rods is markedly prolonged (Henkes, 1963; Marmor, 1977, 1990; Margolis *et al.*, 1987; Sharp *et al.*, 1990), and there is delay in the acquisition of scotopic ERG thresholds (François *et al.*, 1956; Smith *et al.*, 1959; Margolis *et al.*, 1987; Sharp *et al.*, 1990). Variation from this pattern has been described in which the dark adaptation and ERG are normal (François *et al.*, 1956; Franceschetti *et al.*, 1963) or dark adaptation shows a cone segment only (Mandelbaum, 1941). It is not clear whether this variation implies various degrees of severity of a single disease or that several disorders have been described that share this fundus abnormality.

Studies of rhodopsin kinetics showed slow rhodopsin regeneration that parallels dark adaptation (Carr *et al.*, 1974). This implies that by contrast with Oguchi's disease the sensory defect is due to abnormal photopigment kinetics.

Fleck retina of Kandori

This rare condition in which there is prolonged dark adaptation giving rise to difficulty with night vision but no other symptoms, has been described only in Japan (Kandori *et al.*, 1972). Dark adaptation shows a prolonged rod phase reaching normal thresholds within 40 min. The fundus presents large irregular white lesions at the level of the pigment epithelium that are most concentrated in the equatorial region. The photopic ERG is normal and a prolonged interval of dark adaptation is needed to reach scotopic potentials.

17.4 PATHOGENESIS OF PHOTORECEPTOR DYSTROPHIES

17.4 GENERAL CONSIDERATIONS

There are many questions which are amenable to investigation concerning the disease phenotype, and those factors which determine the form and severity dysfunction. Hereditary disorders are caused by defects in the genetic code, which, if expressed, result in an abnormal amino acid composition of specific proteins. If a gene which is confined to a single cell type is expressed, the primary effect will be localized in that cell, even though secondary effects may occur in other cells. For example, the mutation may either reside in the photoreceptor cell giving rise to degeneration of rods or cones or both, or it may be expressed in a support tissue leading to the same consequences. The loss of photoreceptors due to retinal pigment epithelial disease in the RCS rat would be an example of the latter (Mullen and Lavail, 1976). Alternatively a systemic metabolic abnormality may result in the degeneration of a specific cell type, such as visual cells, by depriving them of vital metabolites.

In autosomal dominant disease the abnormality is produced in the heterozygous state. The consequent disease may be due to the influence of the abnormal product of the mutant gene or shortage of the normal product from the single normal gene. The first is illustrated by the observation that mice transfected with a mutant rhodopsin gene develop retinal degeneration (Olsson *et al.*, 1992). Evidence of the second mechanism is shown in the RDS-mouse in which transgenic rescue has been achieved by insertion of a normal peripherin/*rds* gene on a rhodopsin promoter (Travis *et al.*, 1992). Retinal changes in the heterozygous RDS-mouse are compatible with this concept (Hawkins *et al.*, 1985). The relevance of this model to human disease is illustrated by the finding of mutations in the peripherin/RDS gene in autosomal dominant retinal dystrophies (Farrar *et al.*, 1991a, b; Kajiwara *et al.*, 1991, 1993; Wells *et al.*, 1992; Nichols *et al.*, 1993).

The identification of mutations in the rhodopsin and peripherin/RDS genes allows investigation of pathogenetic mechanisms involved. It is fortunate that much was known concerning the structure and function of both proteins prior to the discovery of the mutations.

17.4.2 RHODOPSIN

In families with autosomal dominant RP due to mutations in the rhodopsin gene on chromosome 3q, different patterns of retinal dysfunction have been demonstrated with different mutations (Fishman *et al.*, 1991, 1992a, b; Heckenlively *et al.*, 1991; Jacobson *et al.*, 1991; Richards *et al.*, 1991; Stone *et al.*, 1991; Kemp *et al.*, 1992; Moore *et al.*, 1992; Kim *et al.*, 1994; Owens *et al.*, 1994). In three mutations, proline to leucine at codon 347, lysine to glutamic acid at codon 296 (that is the binding site for retinal), and an isoleucine deletion at codon 255 or 256, functional loss compatible with diffuse (type I) autosomal dominant RP was identified. With all three mutations poor night vision was consistently identified in early life and rod ERGs were severely reduced or unmeasurable. In those with measurable visual function, psychophysical testing showed rod function to be severely affected throughout the retina (with threshold elevations of more than 3 log units), even in younger individuals, whereas loss of cone function varied widely between families, and was less widespread and less severe than loss of rod function. In the most severely affected family (Lys-296-Glu) there was little visual function after the age of 30 years in most members. With the isoleucine deletion at codon 256, cone function was limited to the central 10° by the age of 25 years but little further loss occurred thereafter. Cone function was retained over most of the visual field until middle life in the 347 mutation despite widespread and severe early loss of rod function.

The functional abnormality in families with mutations Thr-17-Arg, Pro-23-His, Thr-58-Arg, Gly-106-Arg and Gua-182-Ade is qualitatively similar with altitudinal distribution of disease and this attribute appears to be constant within the families (Fishman *et al.*, 1991, 1992; Heckenlively *et al.*, 1991; Jacobson *et al.*, 1991; Richards *et al.*, 1991; Stone *et al.*, 1991; Kemp *et al.*, 1992; Moore *et al.*, 1992). This characteristic of disease is in marked contrast with that seen in patients with other rhodopsin mutations examined to date. Rod sensitivity is severely depressed in the superior field but is nearly normal in the inferior field. Loss of cone sensitivity closely follows that of the rods. In this respect the pattern of disease resembles regional (type II) RP. An additional striking finding in these families is a characteristic abnormality in one component of the kinetics of dark adaptation following exposure to a bright light. Prior to bleaching, measurements made in the relatively intact portion of the visual field show mild threshold elevations of approximately one log unit. Following light adaptation there is a marked delay in the recovery of sensitivity. The initial portion of recovery mediated by cones and rods is normal. However by one hour, when in the normal subjects recovery of sensitivity is complete, these patients showed residual threshold elevations of 1–2 log units from the prebleach values. Even after nearly two hours thresholds are still elevated by more than 1.0 log unit. In two subjects it was found that the time course of this slow recovery of sensitivity was of the order of 80–120 hours (Moore, 1992). Using a model based on primate data of rod outer segment length and turnover, is has been calculated that the delayed phase of the recovery of rod sensitivity following strong light adaptation could be due in part to the formation of new disc membranes with a normal concentration of rhodopsin rather than *in situ* regeneration of photopigment (Moore *et al.*, 1992). The model requires that the outer segments are short as a result of RP, and that a major portion of the outer segment is shed following strong light adaptation.

Some observations on the behavior of the abnormal protein made in the laboratory allow predictions to be made concerning possible effects of the mutation on photoreceptor function. It is uncertain as to whether or not the mutant rhodopsin is expressed in the outer segment. It has been shown that abnormal proteins may or may not pass from the rough endoplasmic reticulum to the Golgi apparatus, and that this depends on the influence of the mutation upon molecular folding (Lodish, 1988). Those proteins that do not pass on may be destroyed or may accumulate in the rough endoplasmic reticulum which in turn may interfere with cell function (Carlson *et al.*, 1988; Lippincott–Schwartz *et al.*, 1988; Cheng *et al.*, 1990; Klausner and Sitia, 1990).

The fate of protein produced by rhodopsin genes with mutations found in human RP has been investigated (Doi *et al.*, 1990; Sung *et al.*, 1991). COS and 293S cells were transfected with various mutant rhodopsin genes. The abnormal proteins could be divided into distinct classes according to their behavior when compared to normal (Sung *et al.*, 1991). In class I mutations rhodopsin was expressed normally on the plasma membrane and bound 11-*cis*-retinal creating a chromophore with an absorbance spectrum similar to that of rhodopsin. Class II showed little if any ability to form a pigment when exposed to 11-*cis*-retinal, inefficient transport to the plasma membrane, and low levels of cell surface localization. However, in no instance was expression on the plasma membrane absent. In a similar experiment, mutant rhodopsin was prepared by site-directed mutagenesis, COS cells transfected and incubated in the presence of 11-*cis* retinal (Min *et al.*, 1993). There was no binding of 11-*cis* retinal with the 68–72 deletion. Binding did occur with Thr-58-Arg, Arg-137-Leu and Arg-135-Trp mutations giving rise to a pigment with maximum absorption (λ_{max}) at 500 nm. Although light caused conversion to metarhodopsin II, when incubated with GDP and transducin there was defective activation. Several of the mutations were identical to those found in forms of RP in which the functional deficit has been characterized. Pro-347-Leu was designated as biochemical class I and the same mutation in man causes diffuse RP. The mutations Pro-23-His, Thr-58-Arg and Gly-106-Trp were designated as biochemical class II, and in man are associated with regional RP with altitudinal distribution of disease and slow recovery from strong light adaptation. It cannot be assumed that the behavior in the photoreceptor cell would necessarily mirror that in COS cells but it is possible that the magnitude of expression may differ between different mutations.

It is possible that expression of a single gene only may cause disease. Such a situation may exist in the family described by Berson *et al.* (1968), in which a stop codon caused reduction of the ERG in the heterozygous state but no symptoms (see also Rosenfeld *et al.*, 1992). In the remainder of the mutations known to date it is more likely that some of the mutant protein expresses in the outer segment.

If mutant protein is incorporated into the outer segment it is possible to speculate on its influence on function. Rhodopsin is an intrinsic protein of rod disk membranes and its activation by light to the photoexcited form R*, is the first step in the enzymic amplification cascade of vertebrate phototransduction. Thus specific functional and structural domains enable (a) effective quantum capture, (b) appropriate energy transfer kinetics and pathways for conformational alterations thereby leading to (c) the unmasking of receptor sites for information transfer to cytoplasmic proteins. The basic membrane topography of the rhodopsin molecule is known but details of structural reorganization during photostimulation and consequent alterations in receptor unmasking are incompletely understood. The rhodopsin molecule consists of 348 amino acid residues with two asparagine linked oligosaccharide side chains. Peptide folding results in seven predominantly α-helices being embedded in the disk bilayer with interconnecting hydrophilic segments protruding from the membranous surface (Albert and Litman, 1978; Michel-Villaz *et al.*, 1979; Hargrave *et al.*, 1980; Rothschild *et al.*, 1980; Applebury and Hargrave, 1982; Hargrave, 1982; Davison and Findlay, 1986). A lysine residue (no. 296) situated at the mid-point of the seventh helix is the attachment site for the 11-*cis* retinal chromophore. 11-*cis* retinal also interacts with the other helices and since all seven are aligned in a plane perpendicular to the disk surface, they effectively form a cage around the chromophore (Honig *et al.*, 1979; Rothschild *et al.*, 1980; Kakitani *et al.*, 1985). Photon absorption by 11-*cis* retinal causes isomerization of the chromophore to the all-*trans* form with simultaneous rearrangement of the helices. This conformational change in rhodopsin exposes a binding site (residues 231–252) for the G-protein, transducin, on the cytoplasmic surface (Kühn and Hargrave, 1981). This is the first step in signal amplification and activation of the enzymic cascade leading to visual transduction (McNaughton, 1990). On the cytoplasmic surface, the carboxy terminal contains many serines and threonines, which, on formation of R*, are phosphorylated by a rhodopsin kinase. The ensuing binding of arrestin to phosphorylated rhodopsin blocks the further activation of transducin; this is therefore one of the termination steps in visual transduction. Of paramount importance to the structural stability of rhodopsin are the two highly conserved cysteines in the intradiskal space (Karnik *et al.*, 1988).

Thus, physical instability of the membrane may occur if glycosylation on the *N*-terminal or the disulfide bond between the cysteine residues 110 and 187 were abnormal (Kühn and Hargrave, 1981; Karnik and Khorana, 1990). Transduction efficiency may be reduced by mutations near the *C*-terminal causing a reduction of sensitivity due to a reduced signal or an increase of noise. It has been shown that rhodopsin with a mutation at the 296 amino acid, which is the binding site for retinal, fails to form a complex with 11-*cis* retinal and reacts constantly with transducin (Robinson *et al.*, 1992). It would be predicted from this observation that the retina may act as if it were in constant lighting and would not dark adapt. These possibilities may be distinguishable one from another by clinical testing. The interaction of the abnormal with the normal protein may also influence function.

Perhaps of most importance is that it has been shown for the first time that some forms of RP are due to defects of metabolic systems that are limited to rod photoreceptors, and yet it is evident that loss of photopic function is consistently found in these disorders. The question arises as to why there is cone cell death. It is possible that cone loss occurs as a result of either a release of endotoxins by dying rods or the cones being metabolically dependent upon the presence of rods. It follows that loss of cone function, although important to the patient, is a secondary effect.

17.4.3 PERIPHERIN

Even more striking is the identification that RP and macular dystrophy may be due to mutations in the peripherin/RDS gene (Farrar *et al.*, 1991b; Kajiwara *et al.*, 1991, 1993; Wells *et al.*, 1992; Nichols *et al.*, 1993). The RP is of early onset and is not obviously different from that seen with rhodopsin mutations (Wroblewski *et al.*, 1994a) with one exception in which the disease simulated retinitis punctata albescens (Kajiwara *et al.*, 1993). Two families with a form of macular dystrophy have been shown to have a tryptophan substitution for arginine at codon 172, and another family with almost identical disease has a glutamine substitution at the same codon (Wells *et al.*, 1992). A mutation resulting in a stop codon at position 258 has been identified in a family with a retinal degeneration similar in appearance to adult viteliform macular dystrophy (Wells *et al.* 1992, Wroblewski *et al.*, 1994b), and mutations at codons 140 and 168 cause pattern dystrophy (Nichols *et al.*, 1993; Kim *et al.*, 1994). These findings demonstrate that some forms of autosomal dominant RP and dominantly inherited macular dystrophies are caused by mutations in the same (peripherin/RDS) gene. Furthermore, the macular diseases cause either primary demise of the photoreceptor cells or changes in the retinal pigment epithelium. The variability of phenotype caused by mutations at different codons suggests that the functional significance of certain amino acids to cones and rods may be different.

The current knowledge of the putative function of peripherin/RDS provides a potential explanation for diseases being different with different mutations in this gene. Peripherin/RDS, has an amino acid sequence of 346 amino acids with four transmembrane hydrophobic domains and two putative *N*-linked glycosylation sites (Connell and Molday, 1990). One of these is conserved across four species and is thought to be important to the protein's function in stabilizing photoreceptor outer segment membranes (Travis *et al.*, 1991a; Arokawa *et al.*, 1992). Immunohistochemical studies have shown that the protein is limited to the membranes of outer segments of both rods and cones although there has been disagreement over its precise localization. One study using a polyvalent antibody to a short peptide sequence near the carboxyl terminus implied that the protein was distributed over the entire length of the outer segment disk membrane (Travis *et al.*,) 1991). It was suggested that covalent bonding between peripherin molecules was responsible for the maintenance of the parallel arrangement of the outer segment membranes. By contrast, a separate study using a monoclonal antibody to purified photoreceptor disk showed that labeling was confined to the rims of the outer segment implying that the primary function of the protein was to stabilize the unfavorable thermodynamic bend at the disk rim (Arokawa *et al.*, 1992). To explain these different findings it was hypothesized that the epitopes available for antibody binding may differ at the disk rim when compared with other parts of the disk membrane (Travis *et al.*, 1991b). Regardless of the precise mechanism, the general belief is that peripherin is important to the structural stability of the outer segment disk membrane.

Recently it has been shown that peripherin/RDS may bind non-covalently to ROM1, a protein structurally related to peripherin. ROM1 has been localized to the disk rims of rod outer segments but has not been identified in cones (Bascom *et al.*, 1992). It has been proposed that the formation and stability of the bond in the disk membrane in rods is dependent upon the association between peripherin and ROM1. However, the absence of ROM1 in cones implies that differences exist between the precise mechanisms by which peripherin stabilizes outer segment membranes in the two classes of photoreceptor. In rods the association between peripherin and ROM1 may be important, whereas in cones peripherin may bind to a different membrane protein or act alone. If the binding sites are different in rods and cones, constancy of one amino acid of the peripherin molecule may be important to rods only, and a mutation causing an abnormality at this site would cause a dystrophy falling within the category of RP in which rods were the target cell of disease with relative preservation of cones. Conversely a different mutation in the peripherin/*RDS* gene may disrupt the metabolism or structure of either cones alone causing macular dystrophy of both rods and cones. It appears that the presence of arginine at 172 is important to the structure and function of cones but not rods (Wells *et al.*, 1992). There are at present no data to explain the apparent preferential loss of central cones as opposed to peripheral cones, or peripheral rods as opposed to paracentral rods (Wroblewski *et al.*, 1994a, b).

The disorders associated with the 4 pb insert at codon 140, a stop sequence at codon 258 and the mutation at codon 167 are different from others with mutations in the peripherin/RDS gene in that there is little evidence of functional loss, and the changes in the ocular fundus are apparently at the level of the retinal pigment epithelium. It is possible that these mutations produce metabolic changes similar to those seen in the mouse heterozygous for the abnormal peripherin/RDS gene in which there is a 10 kb insert in the gene at codon 238 (Travis *et al.*, 1991a). In the homozygous rds-mouse (RDS/RDS) a relatively high molecular weight mRNA is produced demonstrating that the whole insert is transcribed (Travis *et al.*, 1991a). However, it is unlikely that the mutant protein is expressed since the mRNA does not appear to leave the nucleus. As a consequence outer segment disks are not formed, the disk membrane being discharged as small vesicles into the subretinal space (Cohen, 1983; Jansen and Sanyal, 1984, Usukura and Bok, 1987; Sanyal and Jansen, 1992). Predictably, only half the normal amount of protein would be available in the heterozygous state (RDS/+) as a result of expression of the normal gene. The photoreceptor outer segments develop but contain long lengths of disk membrane (Hawkins *et al.*, 1985) which is compatible with there being less than the normal quantity of peripherin rds. However, the ERG is well preserved, and 50% of the photoreceptors survive after 18 months of life (Hawkins *et al.*, 1985; Sanyal and Hawkins, 1987) which is close to the life expectancy of the mouse. The outer segments appear to be unstable, and the retinal pigment epithelium contains large and abnormal phagosomes (Sanyal and Hawkins, 1987). Such a situation may exist in some patients with adult vitelliform macular degeneration since the mutations are close. As in the heterozygous rds mouse it is likely that the photoreceptors receive only half the normal quantity of peripherin rds, and that the abnormal protein does not pass into the outer segment. If the homology is close, it would be understandable that excessive shedding of the photoreceptor outer segments over many years would cause change in the retinal pigment epithelium, but little photoreceptor dysfunction. A similar situation may pertain in pattern dystrophies. If this reasoning is correct, a primary photoreceptor disease would cause changes which are recognizable clinically only at the level of the retinal pigment epithelium.

That a mutation in a gene which expresses in the photoreceptor cell causes recognizable changes in the retinal pigment epithelium in the absence of major dysfunction of the photoreceptor cells is relevant to previous concepts concerning the classification of macular dystrophies.

17.4.4 CAUSE OF CELL DEATH

There has been increasing evidence that the metabolic defect caused by the mutation does not cause cell death directly at least in some disorders in both man and animals. This is evident with respect to cone loss in patients with RP due to mutations in the rhodopsin gene. A similar situation exists in mice transfected with a mutant rhodopsin gene (Naash *et al.*, 1993). This is also illustrated by loss of photoreceptors in the RCS rat in which the primary defect is in the retinal pigment epithelium (Mullen and Lavail, 1976). In the setter with progressive atrophy of both the rod and cone photoreceptors, PDE activity is defective in rods but is normal in cones. (Suber *et al.*, 1993). A chimera was created of an albino mouse transfected with a mutant rhodopsin gene and of a wild-type pigmented mouse. Although there was regional distribution of pigmented and non-pigmented cells, the photoreceptor cell death was diffuse rather than regional implying that the cells containing the mutant and wild-type rhodopsin genes degenerated simultaneously (Huang *et al.*, 1993). The only recorded variation between animals was that the proportions of mutant to wild-type populations determined the speed of degeneration. Atrophy was faster in animals in which the chimera comprised of more than 50% of mutant cell as judged by skin pigmentation when compared with animals with less than 50%.

This dilemma has been explained by the observation that cell loss is due to apoptosis. Apoptosis (as opposed to necrosis) is a genetically encoded potential of all cells and is an essential part of embryonic development. It has been well recorded in many tissues of both vertebrates and invertebrates as a cause of cell loss (Wyllie, 1980; Wyllie *et al.*, 1980, Truman and Schwartz 1984, Oppenheim, 1991; Raff, 1992). It may also represent the mechanism of cell turnover, and of removal of cells infected by virus and harboring mutations. It is characterized morphologically by disintegration of the nucleolus and generalized condensation of the chromatin. This is associated with incision of most of the nuclear DNA into short but well-organized chains of nucleosomes in multiples of 200 base pairs, by an endogenous non-lysosomal nuclease (Wyllie *et al.*, 1981). These DNA fragments can be identified by gel electrophoresis of a pooled DNA or *in situ* (Gravielli *et al.*, 1992). After cell death, a cell corpse is engulfed and quickly degraded by a neighboring cell. In contrast to necrosis the process affects individual cells within a tissue, its neighbors remaining healthy, and takes place in the absence of inflammation.

There is increasing evidence from work on a nematode, *Caenorhabditis elegans* that apoptosis is regulated by a group of genes some of which induce (*ced 1, 3, 4*), and others which inhibit cell death (*ced 2, 9*) (Ellis *et al.*, 1991a. Hengartner *et al.*, 1992). It is the expression of these genes that governs the production of endonucleases. A homolog to *ced 9, bcl 2* has been identified in vertebrates (Vaux and Weissan, 1993a; Zhong *et al.*, 1993). It had been shown that *bcl 2* inhibited cell death in vertebrates including man (Vaux *et al.*, 1988), and subsequently that *bcl 2* and *ced 9* had considerable structural and functional similarities (Vaux, 1993). The engulfment of the apoptotic cells is also dependent on a further series of genes (Ellis *et al.*, 1991b). For degradation to occur, engulfing cells must recognize, phagocytose and digest the corpses of dying cells. The factors which induce expression of the genes which induce or suppress apoptosis are as yet unknown.

Apoptosis has been shown to be the cause of cell death in all animals with genetically determined retinal degeneration examined to date, whether naturally occurring or induced by transfection by mutant genes (Chang *et al.*, 1993, Lolley *et al.*, 1994; Portera-Cailliau *et al.*, 1994; Tso *et al.*, 1994).

17.5 SIGNIFICANCE TO MANAGEMENT OF ADVANCES IN MOLECULAR BIOLOGY

The recent advances in molecular biology whereby specific genomic defects have been identified in retinal dystrophies has already had an impact on clinical management, as well as advancing our understanding of the pathogenesis of disease. Hopefully this work will eventually lead to therapy for patients with retinal dystrophies.

17.5.1 DISEASE MECHANISMS

Identification of the genomic defect, and demonstration of the influence of abnormal proteins on cell function render the findings of clinical investigation much more significant. In the past the functional deficits in retinal dystrophies were of value in distinguishing one disorder from another or at least in identifying categories of disease. It is now possible to relate these characteristics to putative disease mechanisms, and to generate hypotheses that are amenable to testing in the laboratory. It is also the case that observations are possible in man, particularly with respect to detailed recording of functional loss in those with retinal dystrophies, that cannot be made in animals. Thus the results of clinical and laboratory studies are relevant one to the other.

Although there are both quantitative and qualitative differences in functional loss between families the severity of disease varies considerably within families. Analysis of the variation within families would allow assessment of the potential influence of mechanisms influencing phenotypic expression of abnormal genes by genomic imprinting and allelic competition (Moore and Haig, 1991, Willison, 1991).

The initial subdivision of retinal dystrophies was achieved on the basis of its inheritance; morphological and functional studies added to our understanding of this complex group of disorders. The recent identification of a number of different mutations in the rhodopsin and peripherin genes are at present being correlated with clinical and functional abnormalities, and we can expect, in the near future, to find other genes responsible for further members of this group of disorders. The mechanism by which alteration of the amino acid sequence in rhodopsin or peripherin molecules may influence photoreceptor function is unclear, but much may depend on the site of expression of the abnormal protein. The recognition that a disease affecting the retinal pigment epithelium is due to a gene which is expressed in photoreceptor cells has already induced a change in our appreciation of potential disease processes. More sophisticated visual testing is required to correlate the genotype with the phenotypic expression of the disease. The generation of transgenic mice carrying a specific human mutation would accelerate our understanding of functional aberrations and more importantly the mechanisms leading to eventual death of the photoreceptor. Although there has been a very rapid advance in the understanding of disease mechanisms in RP, there is every prospect that the expansion of knowledge will accelerate in the next few years, particularly if there is continued and, if possible, increasing collaboration between clinicians and basic scientists.

17.5.2 IMPACT ON CLINICAL MANAGEMENT

Knowledge of the mutation allows the distribution of the abnormal gene in a family to be documented with almost 100% certainty, and the genetic status of a subject can be established at any age and at any stage of evolution of disease. Furthermore, advice on visual prognosis may be made on a firmer basis if single nosological entities are identified by genomic studies.

It is now possible to reduce the risk of affected individuals being born as a consequence of these discoveries. Selective termination of pregnancy may be requested by patients in a small number of conditions in which the disease is severe. Selective implantation of normal embryos following *in vitro* fertilization is also now feasible. These management techniques have been initiated in severe diseases such as cystic fibrosis. Population screening has also been initiated for cystic fibrosis by which it is proposed to identify the distribution of the mutant genes within the population (Willison, 1991; Williamson, 1991, 1993; Watson *et al.*, 1991). It is not clear if this would ever be feasible or desirable for retinal dystrophies; to some extent this would depend on the number of genes involved.

Various approaches to treatment have been investigated over the last few years. These include replacement of the defective cells such as retinal pigment epithelium and photoreceptor cells (Lazar and del Cerro, 1992; Schuschereba and Silverman, 1992; Yamaguchi *et al.*, 1992; Banerjee and Lund, 1992). It has been shown that retinal pigment epithelium survives as a monolayer when inserted in the subretinal space, and that the photoreceptor cells in RCS rat overlying the donor cells survive better than cells elsewhere. Fetal photoreceptor cells have also been inserted into the subretinal space in dystrophic animals and animals with photoreceptor loss due to light exposure. The cells survive and have limited functional capability. They express photopigment, and neural connections are formed. However, the cells do not have normal morphology and the visual potential of such cells has not been proven. The techniques of grafting RPE and photoreceptor cells are still in their infancy, but these forms of management may one day be feasible.

In both RD- and RDS- mice, transfection of the fertilized ovum with the appropriate wild-type gene causes photoreceptor rescue for weeks in the case of the RD mouse and months in the RDS mouse (Travis *et al.*, 1992; Flannery *et al.*, 1992). However, there are major differences between insertion of genes into the fertilized ovum of animals and into non-dividing photoreceptor cells in man. Attempts to transfect mature photoreceptor cells has had limited success using adenovirus to introduce reporter genes with rhodopsin promoters (Bennett *et al.*, 1994). However, the expression occurred only in a very limited number of cells and the expression was not long lasting.

An alternative approach to treatment involves the injection of growth factors into the eye which has been shown to result in long-term survival of photoreceptors in the RCS rat (Faktorovich *et al.*, 1990). The mechanism by which this occurs is not certain but it is possible that this influences apoptosis (Barres *et al.*, 1992a, b). If this were shown to be the case, the application may become widespread such that at least some cells may be induced to survive.

If any treatment becomes available by which functional loss may be modified in retinal dystrophies, it is likely that it will be specific to a disease or diseases identified by the target cell involved, the causative gene or the mutation within that gene. Therefore, foreknowledge of the genetic defect in a subject would be necessary.

17.6 CONCLUSIONS AND FUTURE PERSPECTIVES

It is evident that great advances have been made in the understanding of retinal dystrophies. This has been achieved as a result of the combined research of workers in many scientific disciplines. It has been shown that the findings in each discipline are relevant to all workers in the field. The most encouraging aspect of the work is that therapy for subjects with retinal dystrophies is now seen as a realistic long-term goal by many. If therapy becomes available, the management of retinal dystrophies will require a team of workers with a variety of disciplines. The inheritance can only be established if the family is well known to the patient. The pedigree may be constructed by the patient on the basis of memory or enquiries within the family. Failing this, the help of a genealogist may be needed to search other sources of information such as registers of births, marriages and deaths, and census or parish records. Examination of relatives at risk of having the abnormal gene but who are asymptomatic is often helpful. Once the pedigree is established the relevant mutation would be identified by the molecular biologist. The availability of treatment would depend on the nature of the resultant disease which would be determined by the biochemists and cell biologists. Delivery of treatment and monitoring of its effects would be undertaken by clinicians, psychophysicists and electrophysiologists. Therefore, it is crucial that multidisciplinary teams be created to take full clinical advantage of the advances taking place.

REFERENCES

Albert, A.D. and Litman, B.J. (1978) Independent structural domains in the membrane protein bovine rhodopsin. *Biochemistry*, **17**, 3893–900.

Alitalo, T., Kruse, T.A., Forsius, H. *et al.* (1991) Localization of the Åland Island eye disease locus to the pericentromeric region of the X chromosome by linkage analysis. *American Journal of Human Genetics*, **48**, 31–8.

Alper, M.G. and Alfano, J.A. (1953) Honeycomb colloid degeneration of the retina. *Archives of Ophthalmology*, **49**, 392–9.

Alpern, M., Lee, G.B. and Spivey, B.E. (1965) Pi cone monochromatism. *Archives of Ophthalmology*, **74**, 334–7.

Anderson, S. (1977) Quoted in *Hereditary and Choroidal Diseases* (A.E. Krill), Harper & Row, New York, pp.697.

Applebury, M.L. and Hargrave, P.A. (1982) Molecular biology of the visual pigments. *Vision Research*, **26**, 1881–95.

Arokawa, K., Molday, M.M., Molday, R.S. *et al.* (1992) Localization of peripherin/rds in the disk membranes of cone and rod photoreceptors; relationship to disk membrane morphogenesis and retinal degeneration. *Journal of Cell Biology*, **116**, 659–67.

Ashton, N. (1953) Central areolar choroidal sclerosis. A histopathological study. *British Journal of Ophthalmology*, **37**, 140–7.

Banerjee, R. and Lund, R.D. (1992) A role for microglia in the maintenance of photoreceptors in retinal transplants lacking pigment epithelium. *Journal of Neurocytology*, **21**, 235–43.

Barkman, Y. (1961) A clinical study of a central tapetoretinal degeneration. *Acta Ophthalmologica*, **39**, 663–71.

Barres, B.A., Hart, I.K., Coles. H.S. *et al.* (1992a) Cell death in the oligodendrocyte lineage. *Journal of Neurobiology*, **23**, 1221–30.

Barres, B.A., Hart, I.K., Coles, H.S. *et al.* (1992b) Cell death and control of cell survival in the oligodendrocyte lineage. *Cell*, **70**, 31–46.

Barricks, M.E. (1977) Vitelliform lesions developing in normal fundi. *American Journal of Ophthalmology*, **83**, 324–7.

Bascom, R.A., Manara, S., Collins, L. *et al.* (1992) Cloning of the cDNA for a novel photoreceptor membrane (rom-1) identifies a disk rim protein family implicated in human retinopathies. *Neuron*, **8**, 1171–84.

Bech-Hansen, N.T., Field, L.L., Schramm, A.M. *et al.* (1990) A locus for X-linked congenital stationary night blindness is located on the proximal portion of the short arm of the X chromosome. *Human Genetics*, **84**, 406–8.

Bennett, J., Wilson, J., Sun, D. *et al.* (1994) Adenovirus vector-mediated in vivo gene transfer into adult murine retina. *Investigative Ophthalmology and Visual Science*, **35**, 2535–42.

Berson, E.L., Gouras, P.G. and Gunkel, R.D. (1968) Progressive cone degeneration, dominantly inherited. *Archives of Ophthalmology*, **80**, 77–83.

Best, F. (1905) Über eine hereditäre Maculaaffektion: Beitrag zur Vererbungslehre. *Zeitschrift fur Augenheilkunde*, **13**, 199–212.

Bjork, A., Lindbalm, V. and Wadanstein, L. (1956) Retinal degeneration in hereditary ataxia. *Journal of Neurology, Neurosurgery and Psychiatry*, **19**, 186–193.

Bonin, P. (1971) Le signe du silence choroidien dans les dégénérescences tapéto-rétiniennes centrales éxaminées sous fluorescence. *Bulletin des Societés d'Ophtalmologie de France*, **71**, 348–51.

Brecher, R. and Bird, A.C. (1990) Adult vitelliform macular dystrophy. *Eye*, **4**, 210–215.

Bresnick, G.H., Smith, V.C. and Pokorny, J. (1989) Autosomal dominantly inherited macular dystrophy with preferential short wavelength sensitive cone involvement. *American Journal of Ophthalmology*, **108**, 265–76.

Capon, M.R.C., Polkinghorne, P.J., Fitzke, F.W. *et al.* (1988) Sorsby's pseudoinflammatory macula dystrophy – Sorsby's fundus dystrophies. *Eye*, **2**, 114–22.

Capon, M.R.C., Marshall, J., Krafft, J.I. *et al.* (1989) Sorsby's fundus dystrophy: a light and electron microscopic study. *Ophthalmology*, **96**, 1769–77.

Carlson, J.A., Rogers, B.B., Sifers, R.N. *et al.* (1988) Multiple tissues express alpha 1- antitrypsin in transgenic mice and man. *Investigative Ophthalmology*, **5**, 497–507.

Carr, R.E. and Ripps, H. (1967) Rhodopsin kinetics and rod adaptation in Oguchi's disease. *Investigative Ophthalmology and Visual Science*, **6**, 426–36.

Carr, R.E. and Ripps, H., Siegel, I.M. and Weale, R.A. (1966) Rhodopsin and the electrical activity of the retina in congenital night blindness. *Investigative Ophthalmology*, **5**, 497–507.

Carr, R.E., Ripps, H. and Siegel, I.M. (1974) Visual pigment kinetics and adaptation in fundus albipunctatus. *Documenta Ophthalmologica*, **4**, 193–204.

Carr, R.E., Mittl, R.N. and Noble, K.G. (1975) Choroidal abiotrophies. *Transactions of the American Academy of Ophthalmology and Otolaryngology*, **79**, 796–816.

Chang, G.Q., Hao, Y. and Wong, F. (1993) Apoptosis: final common pathway of photoreceptor death in rd, rds, and rhodopsin mutant mice. *Neuron*, **11**, 595–605.

Cheng, S.H., Gregory, R.J., Marshall, J. *et al.* (1990) Defective intracellular transport and processing of CFT is the molecular basis of most cystic fibrosis. Cell, **63**, 827–34.

Cibis, G.N., Morey, M. and Harris, D.J. (1980) Dominantly inherited macular dystrophy with flecks (Stargardt). *Archives of Ophthalmology*, **98**, 1785–9.

Clarke, E. (1932) Tay's 'guttata choroiditis'. *Proceedings of the Royal Society of Medicine*, **25**(12), 59–60.

Clausen, W. (1928) Zur Frage der Verebung der Makula-Kolobome. *Klinioche Monatsblatter für Augenheilkunde*, **81**, 385.

Cohen, A.I. (1983) Some cytological and initial biochemical observations on photoreceptors in retinas of rds mice. *Investigative Ophthalmology and Visual Science* **24**, 832–43.

Connell, G. and Molday, R.S. (1990) Molecular cloning, primary structure and orientation of he vertebrate photoreceptor cell protein peripherin in the rod disc membrane. *Biochemistry*, **29**, 4691–8.

Cortin, P., Archer, D. and Maumennee, I.H. (1980) A patterned macular dystrophy with yellow plaques and atrophic changes. *British Journal of Ophthalmology*, **64**, 127–34.

Davenport, R.C. (1927) Bilateral macular coloboma in mother and son. *Proceeding of the Royal Society of Medicine*, **21**, 109–10.

Davison, M.D. and Findlay, J.B.C. (1986) Modification of ovine opsin with the photosensitive hydrophobic probe 1-azido-4[^{125}I]iodobenzene. *Biochemical Journal*, **224**, 413–20.

Dejean, C. and Gassenc, R. (1949) Note sur la généalogie de la famille Nougaret, Vendémian. *Bulletin des Societés d'Ophtalmologie de France*, **1**, 96–9.

de Jong, P.T.V.M. and Delleman, J.W. (1982) Pigment epithelial pattern dystrophy. *Archives of Ophthalmology*, **3**, 1416–21.

de Jong, P.T., Zrenner, E., van Meel, G.J. *et al.* (1991). Mizuo phenomenon in X-linked retinoschisis. Pathogenesis of the Mizuo phenomenon. *Archives of Ophthalmology*, **109**, 1104–8.

Deutman, A.F. (1969) Electro-oculography in families with vitelliform dystrophy of the fovea. *Archives of Ophthalmology*, **81**, 305–16.

Deutman, A.F. (1971) *The Hereditary Dystrophies of the Posterior Pole of the Eye*. Assen, Van Gorcum.

Deutman, A.F., van Blommestein, J.D.A., Henkes, H.E. *et al.* (1970) Butterfly shaped pigment dystrophy of the fovea. *Archives of Ophthalmology*, **83**, 558–69.

Deutman, A.F. and Jansen, L.M. (1970) Dominantly inherited drusen of Bruch's membrane. *British Journal of Ophthalmology*, **54**, 373–82.

Doi, T., Molday, R.S. and Khorana, H.G. (1990) Role of the intradiscal domain in rhodopsin assembly and function. *Proceedings of the National Academy of Sciences USA*, **87**, 4991–5.

Donders, F.C. (1855) Beitrage zür pathologischen Anatomie des Auges. *Graefes Archive of Ophthalmology*, **1**, 106–18.

Doyne, R.W. (1899) Peculiar condition of choroiditis occurring in several members of the same family. *Transactions of the Ophthalmological Society, UK*, **19**, 71.

Doyne, R.W. (1910) A note on family choroiditis. *Transactions of the Ophthalmological Society, UK*, **30**, 93–5.

Dryja, T.P., Berson, E.L. Rao, V. and Oprian, D.D. (1993) Heterozygous missence mutation in the rhodopsin gene as a cause of stationary night bindness. *Investigative Ophthalmology and Visual Science*, **34** (Suppl), 1150.

Eagle, R.C., Lucier, A.C., Bernardino, J.R. *et al.* (1980) Retinal pigment epithelial abnormalities in fundus flavimaculatus; a light and electron microscopic study. *Ophthalmology*, **87**, 1189–200.

Ellis, R.E., Horvitz, H.R. and Two, C. (1991a) Elegans genes control the programmed deaths of specific cells in the pharynx. *Development*, **112**, 591–603.

Ellis, R.E., Jacobson, D.M. and Horvitz, H.R. (1991b) Genes required for the engulfment of cell corpses during programmed cell death in *Caenorhabditis elegans*. *Genetics*, **129**, 79–94.

Evans, P.J. (1950) Five cases of familial retinal abiotrophy. *Transactions of the Ophthalmological Society UK*, **70**, 96.

Faktorovich, E.G., Steinberg, R.H., Yasumura, D. *et al.* (1990) Photoreceptor degeneration in inherited retinal dystrophy delayed by basic fibroblast growth factor. *Nature*, **347**, 83–6.

Falls H.F., Wolter, J.R. and Alpern, M. (1965) Typical total monochromacy. *Archives of Ophthalmology*, **74**, 610–16.

Farrar, G.J., Jordan, S.A. and Kenna, P. (1991a) Autosomal dominant retinitis pigmentosa; localization of a disease gene (RP6) to the short arm of chromosome 6. *Genomics*, **11**, 870–4.

Farrar, G.J., Kenna, P., Jordan, S.A. *et al.* (1991b). A three-base-pair deletion in the peripherin-RDS gene in one form of retinitis pigmentosa. *Nature*, **354**, 478–80.

Fish, G., Grey, R.H.B., Sehmi, K.S. *et al.* (1981) The dark choroid in posterior retinal dystrophies. *British Journal of Ophthalmology*, **65**, 359–63.

Fishman, G.A., Farber, M, Patel, B.S. *et al.* (1987) Visual acuity loss in patients with Stargardt's macular dystrophy. *Ophthalmology*, **94**, 809–14.

Fishman, G.A., Stone, E.M. Gilbert, L.D. *et al.* (1991) Ocular findings associated with a rhodopsin gene codon 58 transversion mutation in autosomal dominant retinitis pigmentosa. *Archives of Ophthalmology*, **109**, 1387–93.

Fishman, G.A., Stone, E.M., Sheffield, V.C. *et al.* (1992a) Ocular findings associated with rhodopsin gene codon 17 and codon 182 transition mutations in dominant retinitis pigmentosa. *Archives of Ophthalmology*, **110**, 54–62.

Fishman, G.A., Vandenberg, K., Stone, E.M. *et al.* (1992b) Ocular findings associated with rhodopsin gene codon 267 and codon 190 mutations in dominant retinitis pigmentosa. *Archives of Ophthalmology*, **110**, 1582–8.

Flannery, J., Lem, J., Simon, M. *et al.* (1992) Transgenic rescue of the rd/rd mouse. *Investiga-*

tive Ophthalmology and Visual Science, **33**(Suppl), 945.

Forni, S. and Babel, J. (1962) Étude clinique et histologique de la Malattia levantinese. Affection appartenant au groupe des dégénerescences hyalines du pole posterieur. *Ophthalmologica*, 144, 213–322.

Franceschetti, A. and François, J. (1965) Fundus flavimaculatus. *Archives d'Ophtalmologie (Paris)*, **25**, 505–30.

Franceschetti, A. (1963) Über tapeto-retinale Degeneration in Kindesalter, in *Entwicklung and Fortschritte in der Augenheilk*, Enke Verlag, Stuttgart, pp. 107.

Franceschetti, A., Dieterle, P., Amman, P. and Marty, F. (1963) Une nouvelle forme de fundus albipunctatus cum hemeralopia. *Ophthalmologica*, **145**, 403–10.

François J. and Deweer, J.P. (1952) Dégénérescence maculaire sénile et hérédité. *Annals Occulist*, **185**, 136–54.

François, J. and Verriest, G. (1954) La Maladie d'Oguchi. *Bulletin de la Societé Belge d'Ophtalmologie*, **108**, 465–506.

François, J., Verriest, G. and De Rouck, A. (1956) Les fonctions visuelles dans les dégénérescences tapéto-rétiniennes. *Ophthalmologica*, **131** (Suppl. 43), 1–40.

Frangieh, G.T., Green. R. and Fine, S.L. (1982) A histopathological study of Best's macular dystrophy. *Archives of Ophthalmology* **100**, 1115–21.

Frank, H.R., Landers, M.B., Williams, R.J. *et al.* (1974) A new dominant progressive foveal dystrophy. *American Journal of Ophthalmology*, **78**, 903–16.

Gal, A., Schinel, A., Orth, U. *et al.* (1989) Gene of X-chromosomal congenital stationary night blindness is closely linked to DXS7 on Xp. *Human Genetics*, **81**, 315–18.

Gass, J.D.M. (1973) Drusen and disciform macular detachment and degeneration. *Archives of Ophthalmology*, **90**, 206–17.

Gass, J.D.M. (1974) A clinicopathologic study of a peculiar foveomacular dystrophy. *Transactions of the American Ophthalmological Society*, **72**, 139–56.

Gass, J.D.M. (1973) Adult vitelliform macular detachment occurring in patients with basal laminar drusen. *American Journal of Ophthalmology* **99**, 445–59.

Glickstein, M. and Heath, G.G. (1975) Receptors in the monochromat eye. *Vision Research*, **15**, 633–6.

Godel, V., Chaine, G., Regenbogen, L. *et al.* (1986) Best's vitelliform macular dystrophy. *Acta Ophthalmologica Suppl (Copenhagen)*, **175**, 1–31.

Goodman, G., Ripps, H. and Siegel, I.M. (1963) Cone dysfunction syndromes. *Archives of Ophthalmology*, **70**, 214–31.

Gouras, P., Eggars, H.M. and MacKay, C.J. (1983) Cone dystrophy, nyctalopia and supernormal rod responses. A new retinal degeneration. *Archives of Ophthalmology*, **101**, 718–24.

Gravielli, Y., Sherman, Y. and Ben-Sasson, S.A. (1992) Identification of programmed cell death in situ via specific labeling of nuclear DNA fragmentation. *Journal of Cell Biology*, **119**, 493–501.

Grey, R.H.B., Blach, R.K. and Barnard, W.M. (1977) Bull's eye maculopathy with early cone degeneration. *British Journal of Ophthalmology*, **61**, 702–18.

Guiffre, G. (1988) Autosomal dominant pattern dystrophy of the retinal pigment epithelium. *Retina*, **8**, 169–73.

Guiffre, G. and Lodato, G. (1988) Viteliform dystrophy and pattern dystrophy of the retinal pigment epithelium: concomitant presence in a family. *British Journal of Ophthalmology* **70**, 526–32.

Gutman, I., Walsh, J.B. and Henkind, P. (1982) Viteliform macular dystrophy and butterfly-shaped epithelial dystrophy. *British Journal of Ophthalmology*, **66**, 170–3.

Hadden, O.B. and Gass, J.D.M. (1976) Fundus flavimaculatus and Stargardt's disease. *American Journal Ophthalmology*, **82**, 527–39.

Hamilton, W.K., Ewing, C.C., Ives, E.J. *et al.* (1989) Sorsby's Fundus dystrophy. *Ophthalmology*, **96**, 1755–62.

Hargrave, P.A., Fong, S.L., McDowell, J.H. *et al.* (1980) The partial primary structure of bovine rhodopsin and its topography in the retinal rod cell disc membrane. *Neurochemistry International*, **1**, 231–44.

Hargrave, P.A. (1982) Rhodopsin chemistry, structure and topography. *Progress in Retinal Research*, **1**, 1–51.

Harrison, R., Hoeffnagel, D. and Hayward, J.N. (1960) Congenital total color blindness: a clinicopathological report. *Archives of Ophthalmology*, **64**, 685–92.

Hawkins, R.K., Jansen, H.G. and Sanyal, S. (1985) Development and degeneration of retina in rds mutant mice: photoreceptor abnormalities in the heterozygotes. *Experimental Eye Research*, **41**, 701–20.

Heckenlively, J.R., Weleber, R.G. (1986) X-linked recessive cone dystrophy with tapetal like sheen. A newly recognised entity with Mizuo–Nakamura phenomenon. *Archives of Ophthalmology*, **104**, 1322–8.

Heckenlively, J.R., Rodriguez, J.A. and Daiger, S.P. (1991) Autosomal dominant sectoral retinitis pigmentosa. Two families with transversion mutation in codon 23 of rhodopsin. *Archives of Ophthalmology*, **109**, 84–91.

Hengartner, M.O., Ellis, R.E. and Horvitz, H.R. (1992) *Caenorhabditis elegans* gene ced-9 protects cells from programmed cell death. *Nature*, **356**, 494–9.

Henkes, H.E. (1963) Unilateral fundus albipunctatus. *Ophthalmologica*, **145**, 470–80.

Holthouse, E.H. and Batten, R.D. (1897) A case of superficial chorioretinitis of peculiar form and doubtful causation. *Transactions of the Ophthalmological Society*, **17**, 62–3.

Honig, B., Dinur, U., Nakanishi, K. *et al.* (1979) An external point-charge model for wavelength regulation in visual pigments. *Journal of the American Chemical Society*, **101**, 7084–6.

Hoskin, A., Bird, A.C. and Sehmi, K. (1981) Sorsby's pseudoinflammatory macular dystrophy. *British Journal of Ophthalmology*, **65**, 859–65.

Huang, P.C., Gaitan, A.E., Hao, Y. *et al.* (1993) Cellular interactions implicated in the mechanism of photoreceptor degeneration in transgenic mice expressing a mutant rhodopsin gene. *Proceedings of the National Academy of Sciences USA*, **90**, 8484–8.

Hutchinson, J. and Tay, W. (1875) Symmetrical central chorioretinal disease occurring in senile persons. *Royal London Ophthalmology Hospital Report*, **8**, 231–44.

Jacobson, D.M., Thompson, H.S. and Bartley, J.A. (1989) X-linked progressive cone dystrophy. Clinical characteristics of affected males and female carriers. *Ophthalmology*, **96**, 885–95.

Jacobson, S.G., Kemp, C.M., Sung, C.H. *et al.* (1991) Retinal function and rhodopsin levels in autosomal dominant retinitis pigmentosa with rhodopsin mutations. *American Journal of Ophthalmology*, **112**, 256–71.

Jansen, H.G. and Sanyal, S. (1984) Development and degeneration of retina in rds mutant mice: electron microscopy. *Journal of Comparative Neurology*, **224**, 71–84.

Juler, H. (1893) Guttata choroiditis. *Transactions of the Ophthalmological Society, UK*, **13**, 143.

Jung, E.E. (1936) Über eine Sippe mit angeborener Maculadegeneration. *Ber. Dtsch. Ophthalmol. Ges.* **51**, 81 (Quoted in Duke–Elder, S. *System of Ophthalmology*, Volume X, Henry Kimpton, London, pp. 632)

Kajiwara, K., Hahn, L.B., Mukai, S. *et al.* (1991) Mutations in the human retinal degeneration slow gene in autosomal dominant retinitis pigmentosa. *Nature*, **354**, 480–3.

Kajiwara, K., Sandberg, M.A., Berson, E.L. *et al.* (1993) A null mutation in the human peripherin/RDS gene in a family with autosomal dominant retinitis punctata albescens. *Nature Genet.*, **3**, 208–12.

Kakitani, H., Kakitani, T. Rodman, H. and Honig, B. (1985) On the mechanism of wavelength regulation in visual pigments. *Photochemistry and Photobiology*, **41**, 471–9.

Kandori, F., Tamai, A., Kurimoto, S. and Fukunaga, K. (1972) Fleck retina. *American Journal of Ophthalmology*, **73**, 673–85.

Kaplan, J., Gerber, S., Larget–Piet, D. *et al.* (1993) A gene for Stargardt's disease (fundus flavimaculatus) maps to the short arm of chromosome 1. *Nature Genet.*, **5**, 308–11.

Karnik, S.S. and Khorana, H.G. (1990) Assembly and functional rhodopsin requires a disulphide bond between cysteine residues 110 and 187. *Journal of Biological Chemistry*, **265**, 17520–4.

Karnik, S.S., Sakmar, T.P., Chen, H.B. and Khorana, H.G. (1988) Cysteine residues 110 and 187 are essential for the formation of correct structure in bovine rhodopsin. *Proceeding of the National Academy of Sciences, USA*, **85**, 8459–63.

Kearns, T.P. and Hollenhorst, R.W. (1966) Chloroquine retinopathy. Evaluation by fluorescein angiography. *Archives of Ophthalmology*, **76**, 378–84.

Kemp, C.M., Jacobson, S.G., Roman, A.J. *et al.* (1992) Abnormal rod adaptation in autosomal dominant retinitis pigmentosa with Pro-23-His rhodopsin mutation. *American Journal of Ophthalmology*, **113**, 165–74.

Khouri, G., Mets, M.B., Smith, V.C. *et al.* (1988) X-linked congenital stationary night blindness. Review and report of a family with hyperopia. *Archives of Ophthalmology*, **106**, 1417–22.

Kim, R.Y., Dollfus, H., Keen, T.J. *et al.* (1994) Autosomal dominant pattern dystrophy of the retina associated with a 4 bp insertion at codon 140 in the *RDS/peripherin* gene. *Archives of Ophthalmology*, in press.

Kingham, J.D. and Lochen, G.P. (1977) Vitelliform macular degeneration. *American Journal of Ophthalmology*, **84**, 526–31.

Klainguti, R. (1932) Die tapeto-retinal Degeneration im Kanton Tessin. *Klinische Monatsblatter fur Augenheilkunde*, **89**, 253–4.

Klausner, R.D. and Sitia, R. (1990) Protein degradation in the endoplamic reticulum. *Cell*, **62**, 611–14.

Klein, B.A. (1939) A case of so-called Oguchi's disease in the USA. *American Journal of Ophthalmology*, **22**, 953–5.

Klein, B.A. and Krill, A.E. (1967) Fundus flavimaculatus: clinical, functional and histologic observations. *American Journal of Ophthalmology*, **64**, 2–23.

Krill, A.E. (1977) Incomplete rod–cone degenerations, in *Hereditary Retinal and Choroidal Diseases* (eds A.E. Krill and D. Archer), Harper & Row, Hagerstown, pp. 625–36.

Krill, A.E. and Deutman, A.F. (1972) Dominant macular degenerations. The cone dystrophies. *American Journal of Ophthalmology*, **73**, 352–69.

Krill, A.E. and Martin, D. (1971) Photopic abnormalities in congenital stationary night blindness. *Investigative Ophthalmology*, **10**, 625–36.

Krill, A.E., Potts, A.M. and Johanson, C.E. (1971) Chloroquine retinopathy. Investigation of discrepancy between dark adaptation and electroretinographic findings in advanced stages. *American Journal of Ophthalmology*, **71**, 530–43.

Kühn, H. and Hargrave, P.A. (1981) Light-induced binding of gunosinetriphosphate to bovine photoreceptor membranes: effect of limited proteolysis of the membranes. *Biochemistry*, **20**, 2410–17.

Kuwabara, Y., Ishikara, K. and Akiyas, S. (1963) Histologic and electron microscopic studies of the retina in Oguchi's disease. *Acta Societatis Ophthalmological Japanical*, **67**, 1323–51.

Larsen, H. (1921) Demonstration mikroskopischer Präparate von einem monochromatischen Auge. *Klinische Monatsblatter Augenheilkunde*, **67**, 301–2.

Lauber, H. (1924) The origin of hyalin formations within the eye. *Berl. D. Ophthalmol. Gesund.*, **44**, 216–20.

Lazar, E. and del Cerro, M. (1992) A new procedure for multiple intraretinal transplantation into mammalian eyes. *Journal of Neuroscience Methods*, **43**, 157–69.

Lefler, W.H., Wadsworth, J.A.C. and Sidbury, J.B. (1971) Hereditary macular degeneration and amino-acid urea. *American Journal of Ophthalmology*, **71**, 224–30.

Lewis, R.A., Holcomb, J.D., Bromley, W.C. *et al.* (1987) Mapping X-linked ophthalmic diseases. III. Provisional assignment of the locus for blue cone monochromacy to Xq28. *Archives of Ophthalmology*, **105**, 1055–9.

Lippincott–Schwartz, J.L., Bonifacio, J.S., Yuan, L.C. and Klausner, R.D. (1988) Degradation from the endoplasmic reticulum: disposing of newly synthesised protein. *Cell*, **54**, 209–20.

Lodish, H.F. (1988) Transport of secretory and membrane glycoproteins form the rough endoplasmic reticulum to the Golgi. *Journal of Biological Chemistry*, **263**, 2107–10.

Lolley, R.N., Rong, H. and Craft, C.M. (1994) Linkage of photoreceptor degeneration by apoptosis with inherited defect in phototransduction. *Investigative Ophthalmology and Visual Science*, **35**, 358–62.

Lopez, P.F., Maumenee, I.H., de la Cruz, Z. *et al.* (1990) Autosomal-dominant fundus flavimaculatus. Clinicopathologic correlation. *Ophthalmology*, **97**, 798–809.

Mandelbaum, J. (1941) Dark adaptation: some physiologic and clinical observations. *Archives of Ophthalmology*, **26**, 203–239.

Margolis, S., Siegel, I.M. and Ripps, H. (1987) Variable expressivity in fundus albipunctatus. *Ophthalmology*, **94**, 1416–22.

Marmor, M.F. (1977) Defining fundus albipunctatus. *Documenta Ophthalmologica*, **13**, 227–34.

Marmor, M.F. (1990) Long-term follow-up of the physiologic abnormalities and fundus changes in fundus albipunctatus. *Ophthalmology*, **97**, 380–4.

McFarland, C.B. (1955) Heredodegeneration of macula lutea; study of clinical and pathological aspects. *Archives of Ophthalmology*, **53**, 224–228.

McNaughton, P.A. (1990) Light response of vertebrate photoreceptors. *Physiological Reviews*, **70**, 847–83.

Michel–Villaz, M., Saibil, H.R. and Chabre, M. (1979) Orientation of rhodopsin α-helices in retinal rod outer segment membranes studied by infrared linear dichroism. *Proceedings of the National Academy of Sciences, USA*, **76**, 4405–8.

Min, K.C., Zvyaga, T.A., Cypess, A.M. and Sakmar, T.P. (1993) Characterization of mutant rhodopsins responsible for autosomal dominant retinitis pigmentosa. *Journal of Biological Chemistry*, **13**, 9400–4.

Miyake, Y. and Kawase, Y. (1984) Reduced amplitude of oscilatory potentials in female carriers of X-linked recessive congenital stationary night blindness. *American Journal of Ophthalmology*, **98**, 208–15.

Miyake, Y., Yagasaki, K., Horiguchi, M. *et al.* (1986) Congenital stationary night blindness with negative electroretinogram. A new classification. *Archives of Ophthalmology*, **104**, 1013–20.

Mizuo, A. (1913) On new discovery in dark adaptation in Oguchi's disease. *Acta Societatis Ophthalmological Japonicae*, **17**, 1148–50.

Mohler, C.W. and Fine, S.L. (1981) Long-term evaluation of patients with Best's vitelliform dystrophy. *Ophthalmology*, **88**, 688–92.

Moore, T. and Haig, D. (1991) Genomic imprinting in mammalian development: a parental tug of war. *Trends in Genetics*, **7**, 45–9.

Moore, A.T., Fitzke, F.W., Kemp, C.M. *et al.* (1992) Abnormal dark adaptation kinetics in autosomal dominant sector retinitis pigmentosa due to rod opsin mutation. *British Journal of Ophthalmology*, **76**, 465–9.

Moore, A.T., Fitzke, F., Jay, M. *et al.* (1993) Autosomal dominant retinitis pigmentosa with apparent incomplete penetrance: a clinical, electrophysiological, psychophysical and molecular biological study. *British Journal of Ophthalmology*, **77**, 473–9.

Mullen, R.J. and Lavail, M.M. (1976) Inherited retinal dystrophy: primary defect in pigment epithelium determined with experimental rat chimeras. *Science*, **192**, 799–801.

Musarella, M.A., Weleber, R.G., Murphey, W.H. *et al.* (1989) Assignment of the gene for complete X-linked congenital stationary night blindness (CSNB1) to Xp11.3. *Genomics*, **5**, 727–37.

Naash, M.I., Hollyfield, J.G., al Ubaidi, M.R. and Baehr, W. (1993) Simulation of human autosomal dominant retinitis pigmentosa in transgenic mice expressing a mutated murine opsin gene. *Proceedings of the National Academy of Sciences USA*, **90**, 5499–503.

Nakamura, B. (1920) Über ein neues Phänomen der Farberveränderung des menschlichen Augenhintergrundes im Zusammenhang mit der fortschreitenden Dunkeladaptation. *Klinische Monatsblatter Augenheilkunde*, **65**, 83–5.

Nathans, J., Thomas, D. and Hogness, D.S. (1986a) Molecular genetics of human color vision: the genes encoding blue, green, and red pigments. *Science*, **232**, 193–202.

Nathans, J., Piantandida, T.P. Eddy, R.L. *et al.* (1986b) Molecular genetics of inherited variation in human color vision. *Science*, **232**, 203–10.

Nichols, B.E., Sheffield, V.C., Vandenburgh, K. *et al.* (1993) Butterfly-shaped pigment dystrophy of the fovea caused by a point mutation in codon 167 of the RDS gene. *Nature Genet*, **3**, 202–7.

Noble, K.G. (1977) Central areolar choroidal dystrophy. *American Journal of Ophthalmology*, **84**, 310–18.

Noble, K.G., Carr, R.E. and Siegel, I.M. (1990) Autosomal dominant congenital stationary night blindness and normal fundus with an electronegative electroretinogram. *American Journal of Ophthalmology*, **109**, 44–8.

O'Gorman, S., Flaherty, W.A., Fishman, G.A. *et al.* (1988) Histopathologic findings in Best's vitelliform macular dystrophy. *Archives of Ophthalmology*, **106**, 1261–8.

Oguchi, C. (1907) Über einen Fall von eigenartiger Hemeralopie. *Nippon Ganka Gakkai Zasshi*, **11**, 123.

Oguchi, C. (1925) Zur Anatomie der sogenannten Oguchi'schen Krankheit. *Graefes Archive for Clinical and Experimental Ophthalmology* **115**, 234–4.

Olsson, J.E., Gordon, J.W., Pawlyk, B.S. *et al.*

(1992) Transgenic mice with a rhodopsin mutation (Pro23His): a mouse model of autosomal dominant retinitis pigmentosa. *Neuron*, **9**, 815–30.

Oppenheim, R.W. (1991) Cell death during development of the nervous system. *Annual Review of Neuroscience*, **14**, 453–501.

Owens, S.L., Fitzke, F.W., Inglehearn, C.F. *et al.* (1994) Ocular manifestations in autosomal dominant retinitis pigmentosa with a Lys-296-Glu rhodopsin mutation at the retinal binding site. *British Journal of Ophthalmology*, **78**, 153–8.

Pajtas, J. (1950) A case of Doyne's honeycomb choroiditis. *Ceskoslov Oftal*, **6**, 282–6.

Patrinely, J.R., Lewis, R.A. and Font, R.L. (1985) Foveomacular vitelliform macular dystrophy, adult type. A clinicopathological study including electron microscopic observations. *Ophthalmology*, **92**, 1712–18.

Pearce, W.G. (1968) Doyne's honeycomb retinal degeneration. Clinical and genetic features. *British Journal of Ophthalmology*, **52**, 73–78.

Pearce, W.G., Reedyk, M. and Coupland, S.G. (1990) Variable expressivity in X-linked congenital stationary night blindness. *Canadian Journal of Ophthalmology*, **25**, 3–10.

Pearlman, J.T., Owen, G.W., Brounley, D.W. *et al.* (1974) Cone dystrophy with dominant inheritance. *American Journal of Ophthalmology*, **77**, 293–303.

Polkinghorne, P.J., Capon, M.R.C., Berninger, T. *et al.* (1989) Sorsby's fundus dystrophy: a clinical study. *Ophthalmology* **96**, 1763–8.

Portera-Cailliau, C., Sung, C-H., Nathans, J. *et al.* (1994) Apoptotic photoreceptor cell death in mouse models of retintitis pigmentosa. *Proceedings of the National Academy of Sciences USA*, **91**, 974–8.

Prensky, J.G., Bresnic, G.H. *et al* (1983) Butterfly-shaped macular dystrophy in four generations. *Archives of Ophthalmology*, **101**, 1198–203.

Raff, M.C. (1992) Social controls on cell survival and cell death. *Nature*, **356**, 397–400.

Ragnetti, E. (1962) An atypical form of retinitis pigmentosa. *Bollettino di Oculistica*, **41**, 617–25.

Reichel, E., Bruce, A.M., Sandberg, M.A. *et al.* (1989) An electroretinographic and molecular genetic study of X-linked cone degeneration. *American Journal of Ophthalmology*, **108**, 540–7.

Richards, J.E., Kuo, C.Y. Boehnke, M. *et al.* (1991) Rhodopsin Thr58 Arg mutation in a family with autosomal dominant retinitis pigmentosa. *Ophthalmology*, **98**, 1797–805.

Robinson, P.R., Cohen, G.B. Zhukovsky, E.A. and Oprian, D.D. (1992) Constitutively active mutants of rhodopsin. *Neuron*, **9**, 719–25.

Rosenfeld, P., Cowley, G.S., McGee, T.L. *et al.* (1992) A null mutation in the rhodopsin gene causes rod photoreceptor dysfunction and autosomal recessive retinitis pigmentosa. *Nature Genet.*, **1**, 209–13.

Rothschild, K.J., Sanches, R., Hsiao, T.L. and Clark, N.A. (1980) A spectroscopic study of rhodopsin alpha-helix orientation. *Biophysical Journal*, **31**, 53–64.

Sanyal, S. and Hawkins, R.K. (1987) Development and degeneration of retina in rds mutant mice. Altered disc shedding pattern in the albino heterozygotes and its relation to light exposure. *Vision Research*, **28**, 1171–8.

Sanyal, S. and Jansen, H. (1992) Absence of receptor outer segments in the retina of rds mutant mice. *Neurosciences Letters*, **21**, 23–26.

Scarpatetti, A., Forni, S. and Niemeyer, G. (1978) Die Netzhautfunktion bei Malattia leventinese (dominant drusen). *Klinische Monatsblatter Augenheilkunde*, **172**, 590–7.

Schubert, G. and Bornschein, H. (1952) Beitrag zur Analyse des menschlichen Elektroretinogramms. *Ophthalmologica*, **123**, 396–412.

Schuschereba, S.T. and Silverman, M.S. (1992) Retinal cell and photoreceptor transplantation between adult New Zealand red rabbit retinas. *Experimental Neurology*, **115**, 95–9.

Sharp, D.M., Arden, G.B., Kemp, C.R. *et al.* (1990) Mechanisms and sites of loss of scotopic sensitivity: a clinical analysis of congenital night blindness. *Clinics in Visual Science*, **5**, 217–30.

Siegel, I.M., Graham, C.H., Ripps, H. *et al.* (1966) Analysis of photopic and scotopic function in an incomplete achromat. *Journal of the Optical Society of America*, **56**, 699–704.

Siegel, I.M., Greenstein, V.C., Seiple, W.H. and Carr, R.E. (1987) Cone function in congenital nyctalopia. *Documenta Ophthalmologica*, **65**, 307–18.

Sloan, L.L. and Newhall, S.M. (1942) Comparison of cases of atypical and typical achromatopsia. *American Journal of Ophthalmology*, **25**, 945.

Small, K.W. (1989) North Carolina macula dystrophy revisited. *Ophthalmology*, **96**, 1747–54.

Small, K.W., Killian, J. and McLean, W.C. (1991) North Carolina's dominant progressive foveal dystrophy: how progressive is it? *British Journal of Ophthalmology*, **75**, 401–6.

Small, K.W., Weber, J.L., Roses, A. *et al.* (1992) North Carolina macular dystrophy is assigned to chromosome 6. *Genomics*, **13**, 681–5.

Smith, B.F., Ripps, H.A. and Goodman, G. (1959) Retinitis punctata albescens. A functional and diagnostic evaluation. *Archives of Ophthalmology*, **61**, 93–101.

Snodgrass, N.B. (1976) Ocular findings in fucosidosis. *British Journal of Ophthalmology*, **60**, 508–11.

Sorsby, A. (1935) Congenital coloboma of the macula, together with an account of the familial occurrence of bilateral coloboma in association with apical dystrophy of the hands and feet. *British Journal Ophthalmology*, **19**, 65–90.

Sorsby, A. (1939) Choroidal angiosclerosis with special reference to its hereditary character. *British Journal of Ophthalmology*, **23**, 433–44.

Sorsby, A. and Crick, R.P. (1953) Central areolar choroidal sclerosis. *British Journal of Ophthalmology*, **37**, 129–39.

Sorsby, A., Mason, M.E.J. and Gardener, N. (1949) A fundus dystrophy with unusual features. *British Journal of Ophthalmology* **33**, 67–97.

Stargardt, K. (1909) Über familiäre, progressive Degeneration in der Makulagegend des Auges. *Graefes Archiv fur Klinical and Experimental Ophthalmology*, **71**, 534–50.

Stargardt, K. (1913) Über familiäre, progressive Degeneration in der Makulagegend des Auges. *Zeitschrift Augenheilk*, **30**, 95–116.

Steinmetz, R.L., Garner, A., Maguire, J.I. *et al.* (1991) Histopathology of incipient fundus flavimaculatus. *Ophthalmology*, **98**, 953–6.

Steinmetz, R.L., Polkinghorne, P.C., Fitzke, F.W. *et al.* (1992) Abnormal dark adaptation and rhodopsin kinetics in Sorsby's fundus dystrophy. *Investigative Ophthalmology and Visual Science*, **33**, 1633–6.

Stone, E.M., Kimura, A.E. Nichols, B.E. *et al.* (1991) Regional distribution of retinal degeneration in patients with the proline to histidine mutation in codon 23 of the rhodopsin gene. *Ophthalmology*, **98**, 1806–13.

Stone, E.M., Nichols, B.E. Streb, L.M. *et al.* (1992) Genetic linkage of vitelliform macular degeneration (Best's disease) to chromosome 11q13. *Nature Genet*, **1**, 246–250.

Suber, M.L., Pittler, S.J., Qin, N. *et al.* (1993) Irish setter dogs affected with rod/cone dysplasia contain a nonsense mutation in the rod cGMP phosphodiesterase beta-subunit gene. *Proceedings of the National Academy of Sciences USA*, **90**, 3968–72.

Sung, C.H., Schneider, B.G., Agerwal, N. *et al.* (1991) Functional heterogeneity on mutant rhodopsins responsible for autosomal retinitis pigmentosa. *Proceedings of the National Academy of Sciences USA*, **88**, 8840–4.

Szlyk, J.P., Fishman, G.A., Alexander, K.R. *et al.* (1993) Clincial subtypes of cone–rod dystrophies. *Archives of Ophthalmology*, **111**, 781–8.

Travis, G.H., Christerson, L., Danielson, P.E. *et al.* (1991a) The human retinal degeneration slow (RDS) gene: chromosome assignment and structure of the mRNA. *Genomics*, **10**, 733–9.

Travis, G., Sutcliffe, J.G. and Bok, D. (1991b) The retinal degeneration slow (rds) gene product is a photoreceptor disc membrane associated glycoprotein. *Neuron*, **6**, 61–70.

Travis, G., Lloyd, M. and Bok, D. (1992) Complete reversal of photoreceptor dysplasia in transgenic retinal denegeration slow (rds) mice. *Neuron*, **9**, 113–20.

Tree, M. (1937) Familial hyaline dystrophy in the fundus oculi or Doyne's family honeycomb choroiditis. *British Journal of Ophthalmology*, **21**, 65–91.

Truman, J.W. and Schwartz, L.M. (1984) Steroid regulation of neuronal death in the moth nervous system. *Journal of Neuroscience*, **4**, 274–80.

Tso, M.O.M., Zhang, C., Abler, A.S. *et al.* (1994) Apoptosis leads to photoreceptor degeneration in inherited retinal dystrophy of RCS rat. *Investigative Ophthalmology and Visual Science*, **35**, 2693–9.

Turut, P., Chaine, G., Puech, B. *et al.* (1991) Les dystrophies héréditaires de la macula. *Bulletin des Societés d'Ophtalmologie de France*, (numéro spécial); 237–44.

Uliss, A.E., Moore, A.T. and Bird, A.C. (1987) The

dark choroid in posterior retinal dystrophies. *Ophthalmology*, **95**, 1423–7.

Usukura, J. and Bok, D. (1987) Changes in the localization and content of opsin during retinal development in the rds mutant mouse: immunocytochemistry and immunoassay. *Experimental Eye Research*, **45**, 501–15.

Vaghefi, H.A., Green, R., Kelly, J.S. *et al.* (1978) Correlation of clinicopathological findings in a patient: congenital night blindness, branch retinal vein occlusion, cilioretinal artery, drusen of the nerve head and intraretinal pigmented lesion. *Archives of Ophthalmology*, **96**, 2079–104.

van Everdingen, J.A.M., Went, L.N., Keunen, J.E.E. *et al.* (1992) X-linked progressive cone dystrophy with specific attention to carrier detection. *Journal of Medical Genetics*, **29**, 291–4.

van Schooneveld, M.J., Went, L.N. and Oosterhuis, J.A. (1991) Dominant cone dystrophy starting with blue cone involvement. *British Journal of Ophthalmology*, **75**, 332–6.

Vaux, D.L. (1993) Toward an understanding of the molecular mechanisms of physiological cell death. *Proceedings of the National Academy of Sciences USA*, **90**, 786–9.

Vaux, D.L. and Weissman, I.L. (1993) Neither macromolecular synthesis nor myc is required for cell death via the mechanism that can be controlled by Bcl-2. *Molecular and Cellular Biology*, **13**, 7000–5.

Vaux, D.L. Cory, S. and Adams, J.M. (1988) *Bcl-2* gene promotes haemopoetic cell survival and promotes cooperates with *c-myc* to immortanize pre-B cells. *Nature*, **335**, 440–2.

Verriest, M.G. (1974) Recent progress in the study of acquired deficiencies of colour vision. *Bulletin des Societes d'Ophtalmologie de France*, **74**, 595–620.

Vine, A.K. and Schatz, H. (1980) Adult-onset foveomacular pigment epithelial dystrophy. *American Journal of Ophthalmology*, **89**, 680–91.

Vollrath, D., Nathans, J. and Davis, R.W. (1988) Tandem array of human visual pigment genes at Xq28. *Science*, **240**, 1669–72.

Vossius, A. (1921) *Graefe's Archive for Clinical and Experimental Ophthalmology*, **105**, 1050.

Waardenburg, P.J., Franceschetti, A. and Klein, D. (1963) *Genetics and Ophthalmology*, Vol. 2. Charles C. Thomas, Springfield, IL, pp.1736.

Watanabe, I., Taniguchi, Y. Morioka, K and Kato, M. (1986) Congenital stationary night blindness with myopia: a clinico-pathologic study. *Documenta Ophthalmologica*, **63**, 55–62.

Watson, E.K., Williamson, R. and Chapple, J. (1991) Attitudes to carrier screening for cystic fibrosis: a survey of health care professionals, relatives of sufferers and other members of the public. *British Journal of General Practice*, **41**, 237–40.

Watzke, R.C., Folk, J.C. and Lang, R.M. (1981) Pattern dystrophy of the retinal pigment epithelium. *Ophthalmology*, **66**, 1400–6.

Wedl, C. (1854) *Rudiments of Pathological History*, George Busk, London, pp.282.

Weingeist, T.A., Kobrin, J.L. and Watzke, R.C. (1982) Histopathology of Best's macular dystrophy. *Archives of Ophthalmology*, **100**, 1108–14.

Weisel, G. and Beitr, Z. (1922) Bestschen hereditären Maculaerkrankung (Diss.), Geissen. (Quoted in Duke-Elder, S. *System of Ophthalmology*, Volume X, Henry Kimpton, London, pp.632.)

Weleber, R.G. and Tongue, A.C. (1987) Congenital stationary night blindness presenting as Leber's congenital amaurosis. *Archives of Ophthalmology*, **105**, 360–5.

Weleber, R.G., Pillers, D.A. and Powell, B.R. (1989) Åland Island eye disease (Forsius–Eriksson syndrome) associated with contiguous deletion syndrome at Xp21. Similarity to incomplete congenital stationary night blindness. *Archives of Ophthalmology*, **107**, 1170–9.

Wells, J., Wroblewski, J., Keen, J. *et al.* (1992) Mutations in the human retinal degeneration slow (rds) gene can cause either retinitis pigmentosa or macular dystrophy. *Nature Genet*, **3**, 213–17.

Went, L.N., van Schooneveld, M.J. and Oosterhuis, J.A. (1992) Late onset dominant cone dystrophy with early blue cone involvement. *Journal of Medical Genetics*, **29**, 295–8.

Williamson, R. (1991) Cystic fibrosis – a strategy for the future. *Advances in Experimental Medicine and Biology*, **290**, 1–7.

Williamson, R. (1993) Universal community carrier screening for cystic fibrosis? *Nature Genet*, **3**, 195–201.

Willison, K. (1991) Opposite imprinting of the mouse Igf2 and Igf2r genes. *Trends in Genetics*, **7**, 107–9.

Winn, S., Tasman, W., Spaeth, G. *et al.* (1969) Ouguchi's disease in Negroes. *Archives of Ophthalmology*, **1**, 501–7.

Wroblewski, J.J., Wells, J.A., Eckstein, A. *et al.* (1994a) Ocular findings associated with a three-base pair deletion in the peripherin-RDS gene in autosomal dominant retinitis pigmentosa. *British Journal of Ophthalmology*, **78**, 831–6.

Wroblewski, J.J., Wells, J.A., Eckstein, A. *et al.* (1994b) Macular dystrophy associated with mutations at codon 172 in the human retinal degeneration slow (RDS) gene. *Ophthalmology*, **101**, 12–22.

Wyllie, A.H. (1980) Glucocorticoid-induced thymocyte apoptosis is associated with endogenous endonuclease activation. *Nature*, **284**, 555–6.

Wyllie, A.H., Kerr, J.F. and Currie, A.R. (1980) Cell death: the significance of apoptosis. *International Review of Cytology*, **68**, 251–306.

Wyllie, A.H., Beattie, G.J. and Hargreaves, A.D. (1981) Chromatin changes in apoptosis. *Histochemical Journal* **13**, 681–92.

Yagasaki, Y. and Jacobson, S.G. (1989) Cone–rod dystrophy. Phenotypic diversity by retinal function testing. *Archives of Ophthalmology*, **107**, 701–8.

Yamaguchi, K., Yamaguchi, K., Young, R.W. *et al.* (1992) Vitreoretinal surgical technique for transplanting retinal pigment epithelium in rabbit retina. *Japanese Journal of Ophthalmology* **36**, 142–50.

Yamanaka, J. (1924) Existiert die Pigmentverschiebung im Retinalepithel im menschlichen Auge? Der erste Sektionsfall von sogenannter Oguchischer Krankheit. *Klinische Monatsblatter Augenheilkunde*, **73**, 742–52.

Yamanaka, M. (1969) Histologic study of Oguchi's disease: its relationship to pigmentary degeneration of the retina. *American Journal of Ophthalmology*, **68**, 19–26.

Young, R.S., Chaparro, A., Price, J. *et al.* (1989) Oscillatory potentials of X-linked carriers of congenital stationary night blindness. *Investigative Ophthalmology and Visual Science*, **30**, 806–12.

Zhong, L.T., Sarafian, T., Kane, D.J. *et al.* (1993) bcl–2 inhibits death of central neural cells induced by multiple agents. *Proceedings of the National Academy of Sciences, USA*, **90**, 4533–7.

18

Clinical aspects: retinitis pigmentosa

EBERHART ZRENNER, ECKART APFELSTEDT-SYLLA and KLAUS RÜTHER

18.1 INTRODUCTION

The term 'retinitis pigmentosa' (RP) is used for a group of progressive retinal diseases representing one of the most frequent retinal hereditary dystrophies with a prevalence of 1:3000 to 1:5000. Detailed studies over the past years have shown that this condition is made up of genetically and clinically heterogeneous subtypes with different modes of genetic transmission and different types of progression (Merin and Auerbach, 1976; Heckenlively, 1988; Pagon, 1988). In their final stages, however, the disease conditions are identical showing diffusely affected photoreceptors and retinal pigment epithelial cells. Consequently, they are ophthalmologically difficult to distinguish from one another.

Over the past few years, molecular-genetic research, biochemical studies and refined methods of functional examinations have made it possible to distinguish particular delineated subgroups of RP. The elucidation of physiological processes such as the various functions of the pigment epithelium, the precise mechanisms of the phototransduction process, the morphology of the photoreceptor outer segments and their renewal have yielded a tremendous increase in our knowledge of physiological and possible pathophysiological mechanisms. It is the purpose of this chapter to give a survey and a summary of the latest findings in RP research.

18.2 CLINICAL PICTURE OF RETINITIS PIGMENTOSA

18.2.1 DIAGNOSIS AND MAIN SYMPTOMS

The diagnosis of RP should only be established when the following findings are present (Marmor *et al.*, 1983):

- bilateral involvement;
- loss of peripheral vision;
- an elevated final dark adaptation (DA) threshold and/or abnormal rod responses in the electroretinogram (ERG) reflecting rod dysfunction; and
- progressive loss in photoreceptor function.

The corresponding visual symptoms comprise nightblindness or delayed adaptation to dim light, impaired orientation caused by the

Neurobiology and Clinical Aspects of the Outer Retina
Edited by M.B.A. Djamgoz, S.N. Archer and S. Vallerga
Published in 1995 by Chapman & Hall, London
ISBN 0 412 60080 3

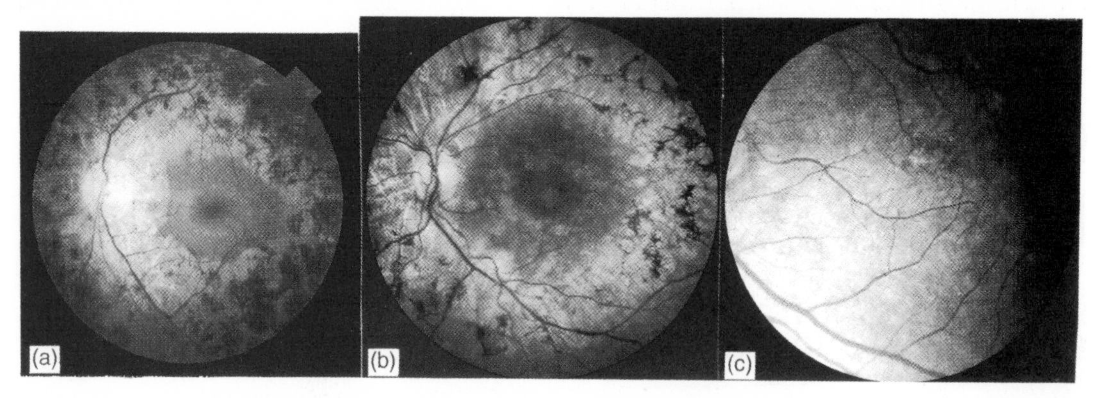

Figure 18.1 (a) A typical fundus in RP with waxy disk, narrowed vessels and bone-spicule like pigmentation. (b) Phenocopy of RP in a patient suffering from syphilis. (c) Retinopathy associated with congenital rubella can mimic RP.

loss of the peripheral visual field, abnormal sensitivity to glare, and, eventually, difficulties with color discrimination and a decrease of visual acuity. Most patients become legally blind in the late stages of the disease.

18.2.2 MORPHOLOGICAL FINDINGS

The following signs are characteristic of RP: waxy yellow-colored optic disks, narrowed blood vessels, almost always macular changes such as wrinkling of the inner limiting membrane, 'bull's eye' or diffuse atrophy or cystoid macular edema. However, the first and the foremost changes are in the midperipheral retinal pigment epithelium beginning with a depigmentation, followed by dark pigments in a typical 'bone-spicule' configuration, reflecting an intraretinal pigment migration (Figure 18.1a). Almost always degenerative vitreous changes are to be found and myopia is very frequent. In middle-aged (older than 40) patients, subcapsular posterior cataract is very common. The diagnosis is made by carefully recording the case history with the onset of symptoms, biomicroscopy of the anterior and posterior segments of the eye, kinetic perimetry, measurement of the final DA threshold, and a standard ERG procedure (Marmor *et al.*, 1989; Heckenlively and Arden, 1991).

18.2.3 GENETICS

Retinitis pigmentosa occurs with local variations (Jay, 1982) in about 25% of patients in an autosomal-dominantly inherited manner, frequently with a relatively mild course; three consecutively affected generations and male-to-male transmission are observed in such families. The risk of transmitting the disease in this case is about 50%. The autosomal-recessive form is almost as frequent (20%), proven by affected siblings, called 'multiplex-cases', with about 25% risk of the disease manifesting itself within this generation; consanguinity of the parents is not uncommon in such patients. As a rule, patients from families with autosomal-recessive inheritance of RP, must count on a risk of less than 1% chance of having affected children, provided the partner is healthy and from a non-affected family.

By contrast, X-chromosomal recessive inherited RP is rare (about 8% of all cases), but onset is earlier and a severe course of the disease is common; on average 50% of all male descendants of female carriers are affected. The majority of female carriers of X-linked RP are asymptomatic, but most of them can be detected by fundoscopy and ERG testing. Fundus changes include a 'tapetal reflex', a dust-like golden sheen of the posterior pole, and/or varying areas of pigment epithelium atrophy with or without intraretinal pigment migration (Bird, 1975). In ERG recordings, amplitudes of rod and cone activities are reduced, cone implicit times are prolonged, or both (Berson *et al.*, 1979; Fishman *et al.*, 1986). In 47% of all affected patients, no heredity is provable: these isolated cases are called 'simplex patients', some of which are autosomal recessive cases.

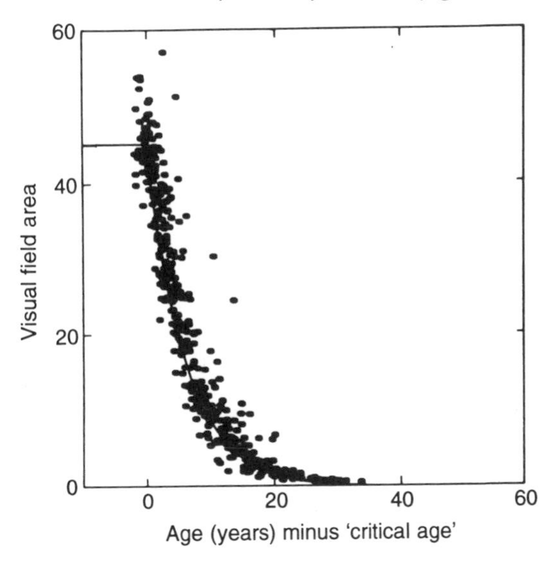

Figure 18.2 The two-stage course of visual field loss in retinitis pigmentosa in 400 patients. The visual field initially is only slowly deteriorating. From a critical age on (zero on the abscissa), the area of the visual field decreases exponentially. The critical age lies between 19 and 33 years, depending on the mode of inheritance. (From Massof and Finkelstein, (1987).)

18.2.4 NATURAL COURSE OF THE DISEASE

Based on a large study, Massof and Finkelstein (1987) suggested that the loss of visual field in RP occurs in two stages (Figure 18.2). Stage 1 is characterized by a very slow progression. Beyond a certain critical age, the rate of visual field loss follows an e-shaped function (stage 2), so that every year about 20% of the remaining visual field are lost. On average, the critical age is 32 years in the autosomal dominant form and about 20 years in the X-linked type. The autosomal recessive form lies somewhere in between. As disease progression can significantly deviate from these values in individual cases, in another study over a three-year period, Berson *et al.* (1985) noted that RP patients lost on average 16–18.5% of the remaining full-field ERG amplitude per year.

18.2.5 CLINICALLY DEFINED SUBTYPES OF RP

Various studies point to two clinical subtypes of autosomal dominant RP (Lyness *et al.*, 1985; Massof and Finkelstein, 1979), each of which is obviously genetically heterogeneous (Jacobson *et al.*, 1991). Type 1 is characterized by a relatively early diffuse loss of rod sensitivity with increasing concentric visual field impairment, while cone function is preserved for a very long time. Type 2 is characterized by a more regional simultaneous loss of rod and cone function; the onset of night blindness occurs late and the loss of peripheral visual field progresses relatively slowly. The distinction between the two types can be made by means of two-color dark-adapted threshold perimetry with blue–green and deep red test targets. In the normal retina, blue–green stimuli are detected by the rods, deep red targets are detected by cones. There is, however, a great number of subgroups of autosomal dominant RP that can only be differentiated by molecular genetic techniques, as described below.

18.2.6 RETINAL DEGENERATIONS RELATED TO RP

There are other genetic disorders that cause retinal degeneration associated with symptoms identical or similar to RP, but can be clinically and/or genetically differentiated.

(a) Choroideremia

Choroideremia (TCD), is a X-linked disease which shows characteristic fundus findings of diffuse choroidal and retinal atrophy. Its pattern of visual function loss is identical to that of typical RP.

(b) Gyrate atrophy

Gyrate atrophy of the choroid and retina, a rare autosomal recessive disorder with clinical signs similar to those seen in RP, is characterized by sharply defined garland-shaped areas of chorioretinal atrophy. An elevation of plasma ornithine levels, caused by a deficiency of ornithine-aminotransferase, is involved in the disease. Therefore, therapy has been attempted by lowering plasma ornithine with pyridoxine and/or low arginine diet (Kaiser-Kupfer and Valle, 1987.)

(c) Cone–rod dystrophies

These differ clinically from RP in that photoaversion and decrease of color vision and visual acuity occur before the onset of peripheral visual field loss and impaired dark adaptation. In the ERG a preponderant loss of cone function or equally affected rod and cone activities are found. The fundus changes, however, may be confused with RP. Cone–rod dystrophies may be transmitted in any of the Mendelian modes of inheritance with predominance of the autosomal dominant mode.

(d) Leber's congenital amaurosis

Leber's congenital amaurosis (LCA) denotes a heterogenous group of inherited retinal dystrophies causing severe visual loss with non-recordable or largely reduced ERG responses, either congenitally or in very early childhood. It may be accompanied by systemic disorders. The fundus findings may be indistinguishable from those seen in typical RP.

(e) Syndromes

Typical or atypical RP may be part of a systemic disorder appearing as a syndromic disease. Frequent symptoms associated with RP are sensorineural hearing loss, as in different types of Usher syndrome, or various kinds of neurologic involvement, such as mental retardation, ataxia, or other signs of degeneration of the central or peripheral nervous system. Various metabolic disorders can also present with retinal degeneration. Examples are the group of mitochondriopathies, peroxisomal disorders, mucopolysaccharidoses, or abetalipoproteinemia (Bassen–Kornzweig). For a review in detail see Pagon (1988).

18.2.7 PHENOCOPIES

The fundus appearance of several non-genetic, acquired diseases can mimic that seen in retinitis pigmentosa. It is important to differentiate these forms as some of them can be treated or are non-progressive. Drug exposure with high doses of chloroquine or thioridazine, a neurolepticum, may result in a pigmentary retinopathy and atrophy of the choriocapillaris. Visual function loss may be reversible after withdrawal of the drugs. Vitamin A deficiency following longstanding malnutrition or malabsorption can produce visual symptoms of retinitis pigmentosa. The fundus findings of this condition numbers among the 'flecked retina' diseases (Zrenner, 1991). Substitution of Vitamin A will generally restore normal visual function.

Infectious chorioretinal diseases may cause fundus changes resembling RP. The fundus shown in Figure 18.1(b) stems from a female patient with healed lues, treated 30 years ago. It looks very much like RP. The patient had mildly constricted visual fields, normal visual acuity, slightly elevated threshold dark adaptation but a normal rod-and cone-ERG, which precludes the diagnosis of RP. In cases of suspected syphilis, specific antitreponemal antibody tests, such as FTA-ABS, should be performed.

Other forms of chorioretinitis can elicit similar pictures. In particular, a retinopathy associated with congenital rubella can look confusingly like an incipient RP, as illustrated in the fundus in Figure 18.1(c), which shows massive changes in the retinal pigment epithelium of a 7-year-old boy.

18.3 RECENT ADVANCES

Over the past 10 years, RP research has increased significantly, both with regard to clinical and to basic science research. Various animal models have been established. Molecular genetic research of RP has also made great progress. In the following, some selected examples are given.

18.3.1 ANIMAL MODELS

Why do the photoreceptors degenerate in one person out of 3000 to 5000? Why does the outer segment of the photoreceptor increasingly shorten, why does the outer segment suffocate in the surrounding cellular debris? We do not yet have a fully satisfying answer to these questions. Results from animal experiments of the past years have led us to assume with great certainty that the underlying pathophysiological mechanisms are not one but rather many (Voaden, 1988).

In a certain animal model, the RCS rat, there is unambiguous proof that the phagocytosing function of the pigment epithelium is disturbed (Voaden, 1988). As shown in Figure 18.3, the uppermost part of our photoreceptor disks are being phagocytosed by the overlying pigment epithelium night after night. In this form of RP, it is very likely that the pigment epithelium cells of the degenerating retina do recognize the photoreceptors, that they also try to 'embrace' the outer segment and try to separate the uppermost part, but that phagocytosis does not succeed. It is as yet unknown whether there is a human counterpart to this condition.

In the *rds* (retinal degeneration slow) mouse strain, a protein exclusively located in the photoreceptor outer segment disks with a molecular weight of 38 000 is defective. This protein, called 'rds/peripherin', appears to be situated near the rim, i.e. the hinge of the membrane-folds of the photoreceptor outer segments (Molday *et al.*, 1987). When this rim protein is modified by a genetic defect, apparently the proper folding of new disk membranes and/or disk shedding, is disturbed.

The responsible RDS/peripherin gene defect has been characterized by Travis *et al.* (1989). Mice homozygous and heterozygous for the mutation both show a progressive photoreceptor degeneration. Unlike in the RCS rat, a number of human phenotypes corresponding to the heterozygous *rds* mouse have been identified since the discovery of mutations in the human RDS/peripherin gene associated with autosomal dominant retinal disease (Farrar *et al.*, 1993, see also below).

In mice homozygous for the autosomal recessive *rd* (retinal degeneration) gene there is a defect of the cGMP-phosphodiesterase (PDE), which breaks down cGMP into 5′-GMP in the phototransduction process (Chapter 5). The enzyme defect leads to a toxic accumulation of cGMP in the photoreceptors. In affected animals, arrested development of photoreceptors and rapid degeneration of rods followed by slow cone

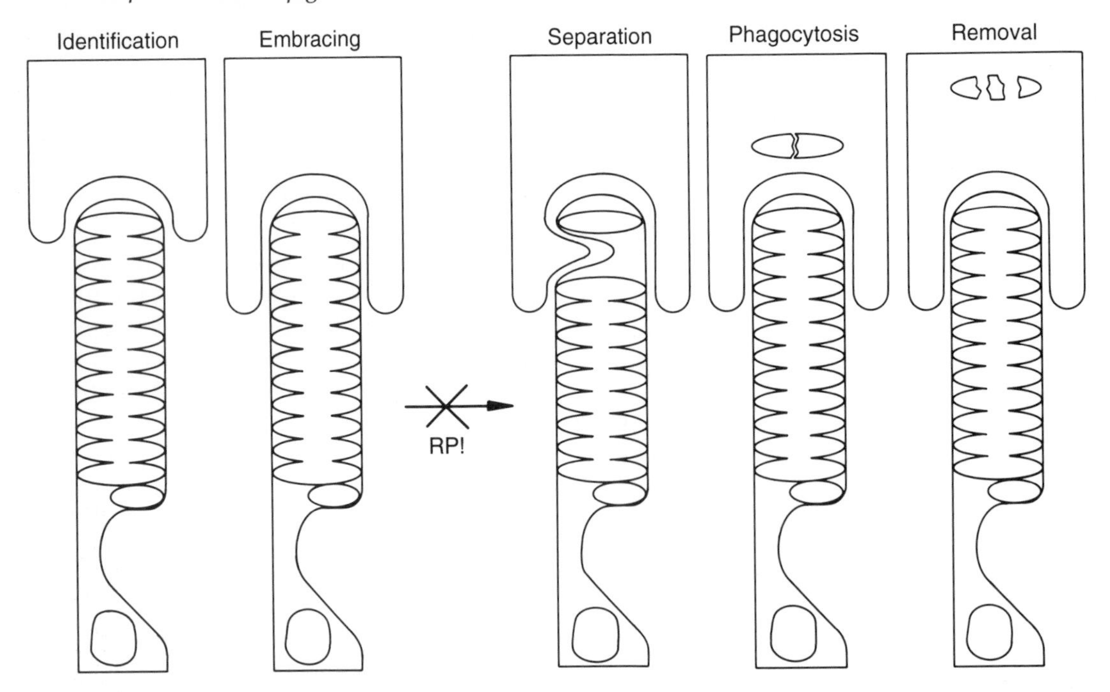

Figure 18.3 Schematic drawing of the phagocytosis of the tip of the outer segment by the pigment epithelial cell. In some forms of RP the mechanisms of separation and phagocytosis of the tip of the rod outer segment are affected, leading to accumulation of cell debris, finally visible as pigmentation.

degeneration is observed. The responsible gene defect is a mutation in the gene encoding the beta subunit of cGMP-PDE (Bowes *et al.*, 1990).

It has subsequently been shown that Irish setter dogs affected with rod–cone dysplasia (RCD) also carry a mutation in the rod cGMP-PDE beta subunit gene (Suber *et al.*, 1993). From a collie model of RP, it is also known that the activity of cGMP-PDE within the photoreceptors is diminished (Chader, 1987). However, it is still unclear whether this reduction in enzyme activity constitutes a primary cause rather than a consequence of another primary defect. Again, it has been shown that there is a human counterpart to the retinal disease in rd mice and Irish setters with RCD (McLaughlin *et al.*, 1993).

Mutations in the human rhodopsin gene have been proved to be one primary cause of autosomal dominant retinitis pigmentosa (see below). A natural corresponding animal phenotype is not known, but progress in molecular biological techniques has made it feasible to establish transgenic mouse models carrying various rhodopsin gene mutations. In transgenic mice carrying a rhodopsin Pro23His mutation, Peachey *et al.* (1994) found photoreceptor function characteristics similar to those detected in patients with the corresponding human rhodopsin mutation. Therefore, these new models might become very important in studying underlying pathophysiological events, exogenous factors and therapeutic approaches in photoreceptor degeneration (Sandberg *et al.*, 1993; Naash *et al.*, 1994; Roof *et al.*, 1994).

Why defects in various photoreceptor proteins not only lead to retinal dysfunction but also to photoreceptor cell death remains to be

Table 18.1 List of deletions and point mutations in the rhodopsin gene in RP according to Farrar *et al.* (1993), as illustrated in Figure 18.4

Codon	Rhodopsin Mutations	Disorders	Codon	Rhodopsin Mutations	Disorders
4	Thr → Lys	adRP	181	Glu → Lys	adRP
15	Asn → Ser	adRP	182	Gly → Ser	adRP
17	Thr → Met	adRP	186	Ser → Pro	adRP
23	Pro → His	adRP	188	Gly → Arg	adRP
23	Pro → Leu	adRP	190	Asp → Gly	adRP
28	Gln → His	adRP	190	Asp → Asn	adRP
45	Phe → Leu	adRP	190	Asp → Tyr	adRP
46	Leu → Arg	adRP	207	Met → Arg	adRP
51	Gly → Val	adRP	211	His → Pro	adRP
51	Gly → Arg	adRP	220	Phe → Cys	adRP
53	Pro → Arg	adRP	222	Cys → Arg	adRP
58	Thr → Arg	adRP	249	Glu → Stop	adRP
68–71	12bp → Del	adRP	255, 256	He → Del	adRP
87	Val → Asp	adRP	267	Pro → Leu	adRP
89	Gly → Asp	adRP	296	Lys → Glu	adRP
90	Gly → Asp	DCCN	296	Lys → Met	adRP
106	Gly → Try	adRP	340	lbp → Del	adRP
106	Gly → Arg	adRP	340–348	42bp → Del	adRP
110	Cys → Tyr	adRP	341–343	8bp → Del	adRP
125	Leu → Arg	adRP	341	Glu → Lys	adRP
135	Arg → Leu	adRP	342	Thr → Met	adRP
135	Arg → Trp	adRP	344	Gln → Stop	adRP
135	Arg → Gly	adRP	345	Val → Met	adRP
167	Cys → Arg	adRP	347	Pro → Leu	adRP
171	Pro → Leu	adRP	347	Pro → Ser	adRP
171	Pro → Ser	adRP	347	Pro → Arg	adRP
178	Tyr → Cys	adRP	347	Pro → Ala	adRP

clarified. One recent hypothesis is that this decay of photoreceptors occurs by a mechanism that links a defect of a photoreceptor protein with a program for cell death, called apoptosis (Lolley *et al.*, 1994).

18.3.2 MOLECULAR GENETIC RESEARCH

A point mutation in codon 23 of the human rhodopsin gene has been detected in one family with autosomal-dominant RP which should lead to an exchange of the amino acid proline for histidine (Dryja *et al.*, 1990). Since then, more than 50 different mutations in the human rhodopsin gene have been shown to cosegregate with autosomal dominant RP. These deletions and point mutations are represented in Table 18.1, according to Farrar *et al.* (1993). Figure 18.4 represents schematically an opsin molecule with its amino acid chain in a disk of the photoreceptor outer segment; each disk carries about 10 000 such molecules.

In large patient groups, about 25% of all patients with dominant RP have a mutation in the rhodopsin gene (Dryja *et al.*, 1991). Up to now, the association of a 'rhodopsin' mutation with autosomal recessive RP has

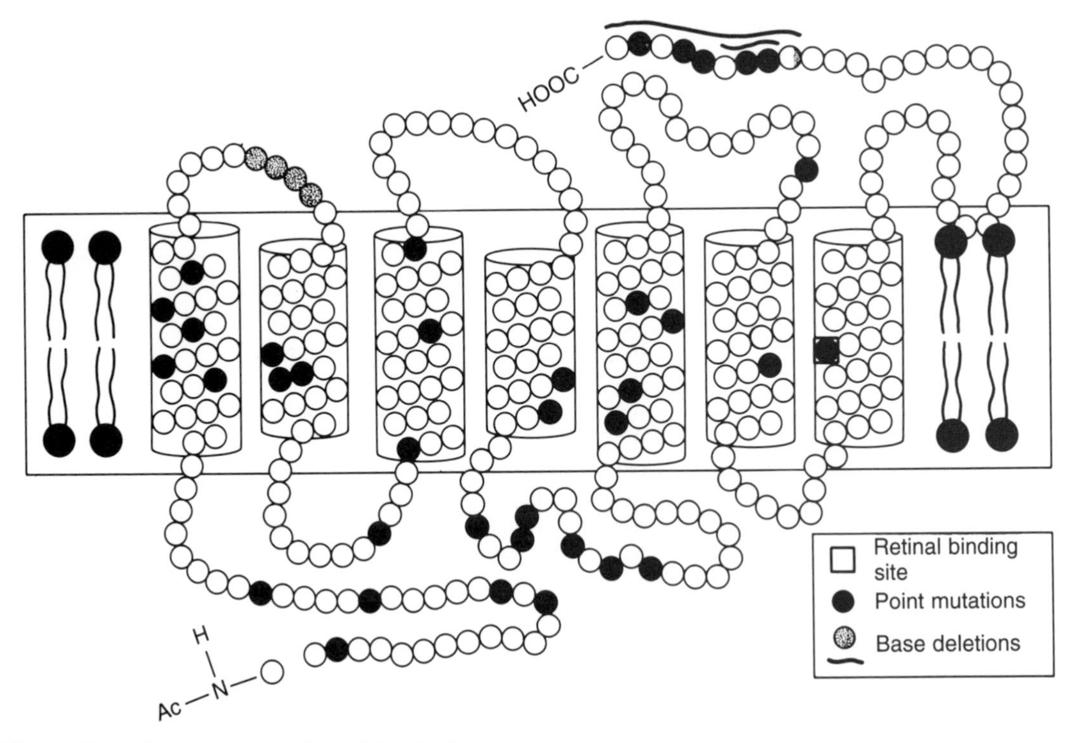

Figure 18.4 Schematic drawing of the rhodopsin molecule, a sequence of 348 amino acids that form the opsin. Retinol is bound at lysine 296. Point mutations found in autosomal dominant RP are marked by dark gray dots, deletions of base pairs by light gray dots or lines.

only been described in one single family (Rosenfeld *et al.*, 1992).

Why these mutations finally lead to photoreceptor death is still unclear; it is conceivable that changes in the rhodopsin molecule may cause a disturbance in the phototransduction process, as explained above. On the other hand, there are hints from cell culture experiments that some rhodopsin mutants might lead to disturbed transport of synthesized molecules to the outer segment plasma membrane and accumulation in the inner segment, thereby possibly compromising normal cell metabolism (Sung *et al.*, 1991). It is hoped that future studies of experimental animals with specific rhodopsin mutations (e.g. transgenic mice) will clarify these mechanisms.

Another protein involved in the phototransduction cascade has also been shown to be defective in RP. First, the human gene encoding the beta subunit of the cGMP-PDE, which is located on chromosome 3 was sequenced by Riess *et al.* (1992). Soon thereafter mutations in the human gene were found in rare families with autosomal recessive RP (McLaughlin *et al.*, 1993).

After the human RDS/peripherin gene had been characterized, it was found to be defective in families with autosomal dominant RP (Farrar *et al.*, 1993). Subsequently, RDS/peripherin mutations were also found to be associated with surprisingly different phenotypes such as macular dystrophies, retinitis punctata albescens and fundus flavimaculatus (Weleber *et al.*, 1993). The mechanism

Table 18.2 Localization of genes associated with various forms of RP (from the Concerted Action News, 1994)

Retinitis pigmentosa	Chromosomal location	Protein implicated
X-linked	Xp11.4-p11.23	?
X-linked	Xp21	?
Autosomal dominant	3q21-q24	rhodopsin
Autosomal dominant	6p21-cen	rds/peripherin
Autosomal dominant	7p13-p15	?
Autosomal dominant	7q31-q35	?
Autosomal dominant	8p11-q21	?
Autosomal dominant	19qter	?
Autosomal recessive	3q21-q24	rhodopsin
Autosomal recessive	4p16	beta cGMP-PDE
Usher syndrome II	1q32-q41	?
Usher syndrome I	11p15-p14	?
Usher syndrome I	11q13	?
Usher syndrome I	14q31-qter	?
Other related retinal degenerations		
Bardet Biedl syndrome	11q13	?
Bardet Biedl syndrome	16q21	?
Leber's congenital amaurosis	17q ?	
Dominant cone–rod dystrophy	19q13.1–13.2	?

leading from the defective protein to retinal disease remains to be clarified. As mentioned above, one hypothesis is that RDS/peripherin may act as a structural protein stabilizing the outer segment disk region.

Another protein, ROM 1, is also thought to play a role in the maintenance of proper outer segment disk rim structure and possibly to interact with RDS/peripherin (Bascom *et al.*, 1992). It therefore also seems to be a strong candidate gene for retinitis pigmentosa.

In choroideremia, a chorioretinal degeneration closely related to RP, the responsible gene has been characterized and mutations have been found in a number of patients (van den Hurk *et al.*, 1992). The mechanisms leading to photoreceptor death, however, remain unclear in this entity.

In a number of other forms of RP with different modes of inheritance, the precise gene defects have not yet been found, but the genes have been localized to certain chromosomal regions (Table 18.2; Concerted Action News, 1994). In the USA, the genes for 30–40% of all RP cases have been mapped so far. Hopefully, the localization of these various genes will soon be followed by their identification.

18.3.3 GENOTYPE–PHENOTYPE CORRELATION

With increasing knowledge of the molecular genetic foundations of RP, it becomes interesting to try to correlate defined genotypes, that is, precisely known gene mutations, with clinical patterns of visual function loss. This aim has been rendered difficult by the fact that there is some inter- and even intrafamilial phenotypic variation between individuals carrying the same mutation, which should be due to genetic and exogenous cofactors. For example, in patients with identical mutations of the RDS/peripherin gene, clinically distinct entities such as RP, pattern dystrophy and fundus flavimaculatus occur even within one family (Weleber *et al.*, 1993). This points to the fact that additional endogenous and exogenous factors influence the clinical picture.

In autosomal dominant RP with rhodopsin mutations, however, tendencies are detectable that certain mutations or classes of mutations frequently show up with special clinical pictures or lead to a certain degree of severity of visual function loss. Some rhodopsin mutations (e.g. at codons 17,23,58,106) tend to present with regional patterns of photoreceptor function loss whereas others (e.g. located at codon 347) have been consistently reported to be associated with diffuse and relatively severe function loss (for review see Apfelstedt–Sylla *et al.*, 1993). Sandberg *et al.* (1994) has shown that rhodopsin mutations affecting the cytoplasmatic domain of the rhodopsin molecule as a group result in a significantly more severe disease course than those located in the intradiskal region of the molecule. In the future, when genotype–phenotype correlation will have been performed in large groups of patients with identical gene mutations, it should also be possible to identify risk factors aggravating the course of the disease.

18.3.4 REFRACTION

In RP patients, the incidence of myopia is higher than in the normal population, particularly in the X-chromosomal inherited form (Fishman *et al.*, 1988). Until now, no reliable explanation for this fact exists. One may hypothesize a disturbed feedback loop during eye growth, when development of proper refraction might depend on a precise object imaging on the retina (Schaeffel and Howland, 1991). However, in RP functional impairment typically takes place only at a more advanced age. It is also striking that there are subgroups of RP that are accompanied by hyperopia. A familiar example would be Leber's congenital amaurosis (Foxman *et al.*, 1985) and 'preserved para-arteriolar retinal pigment epithelium retinitis pigmentosa' (Heckenlively, 1988). As underlying cause for such correlations one might conceive of a genetic cosegregation.

18.3.5 LIPID METABOLISM

Changes in plasma lipids have been found in RP (Converse *et al.*, 1987). Deviations from normal were found among others in LDL-cholesterine, in the apolipoprotein E_2 and above all in the polyunsaturated fatty acids. In an animal model that closely resembles human RP (a miniature poodle with progressive rod–cone degeneration), a decreased level of docosahexa-arachidonic acid was described (22:6w3) that could be caused by a defect in delta-desaturase (Anderson *et al.*, 1991). Furthermore, it is known that docosahexa-arachidonic acid constitutes the most important fatty acid in the rod outer segments. Such results may have a pathophysiological importance in connection with disturbances in the structure of the cell membrane.

18.3.6 IMMUNOLOGICAL ASPECTS

Immunological phenomena in RP have elicited great interest (Newsome and Fishman, 1986). Although it is considered improbable that RP might result from autoimmune processes, immunological epiphenomena could negatively influence the progression of the disease. This is plausible as important molecules may be exposed to the immunological system in an unphysiological manner (e.g. in the destruction of the photoreceptors). It is known that the retinal antigens rhodopsin, the 'interphotoreceptor retinoid-binding protein' as well as the retinal S-antigen in animal models can cause experimental uveitis, and therefore are potentially pathogenic to the retina. Thirkill *et al.* (1991) reported that 11 of 52 patients examined were immunoreactive towards these antigens, where reactions towards the retinal S-antigen dominated. However, various attempts of immunomodulatory treatment, which were meant to influence this suspected aggravation factor, have so far been unsuccessful.

18.4 THERAPEUTIC APPROACHES TO RETINAL DEGENERATIONS

Only two rare forms of RP associated with the Bassen–Kornzweig syndrome and Refsum disease can currently be treated as their underlying biochemical defects are known (Pagon, 1988).

Other therapeutic approaches are limited to isolated problems: alleviating discomfort in glare-sensitive patients of any RP type is tried by the prescription of cut-off filters (Krastel and Moreland, 1987); in some patients with cystoid macular edema treatment with acetazolamide (Diamox) has improved central visual function (Fishman *et al.*, 1989). Therefore, patients continue to stress the desirability of research concentrating on possible therapeutic measures. Indeed, attempts have been made to transplant retinal tissue or tissue from the pigment epithelium. The goal of this type of research is to replace degenerate host cells and thus to retain or to reestablish visual function. Various approaches are being followed. In animal experiments, it was possible to transplant cultured pigment epithelium cells (Gouras *et al.*, 1985). In RCS rats, such RPE cell transplants are capable of not only rescuing photoreceptor cells but also of maintaining synaptic components in the plexiform layers of the retina (Sheedlo *et al.*, 1993).

Treatment with growth factors seems to be a non-specific way of transiently saving photoreceptors in various animal models of RP (Faktorovich *et al.*, 1990; Wiedemann, 1992). Currently, it can not be said whether this approach will also be promising in human retinal degeneration.

Another approach to therapy would be to rescue photoreceptor function by transfering wild-type genes into the retinas of individuals who carry a mutation leading to retinal degeneration (somatic gene therapy). This measure is still far from being applied to humans, but gene transfer *in vivo* has been successfully performed in normal and rd mice using adenoviral vectors (Li *et al.*, 1994). Further study will show whether this approach can reliably protect photoreceptor function in animals with mutations causing retinal disease.

A large drug treatment trial over a 6-year period has been performed by Berson *et al.* (1993). They compared the effects of various doses of vitamin A and/or vitamin E on the natural course of human RP. They found that vitamin A palmitate at a daily dosage of 15 000 IU significantly lowered the decline of ERG amplitudes over that period of time when compared to untreated persons, whereas vitamin E had a maligning effect. They concluded that most patients affected with RP should take vitamin A palmitate in the specified dosage. These results are still discussed controversially among ophthalmologists and they do not promise a real therapy for RP. Nevertheless, there is some hope for many patients in that there might be a way of slowing down the natural course of the disease. Hopefully, the current and future therapeutic efforts trial will result in new ways of relieving the symptoms of RP, which is still not curable.

18.5 CONCLUSIONS AND FUTURE PERSPECTIVES

New clinical pathophysiological, cell biological and molecular genetic findings have helped in recent years to understand better the disease mechanisms associated with RP.

There is a surprising heterogeneity of causes with probably hundreds of different genes involved. However, the links to pathophysiological mechanisms, some of which are quite well understood, are still missing. Extensive studies in suitable animal models that will include techniques of genetic engineering, combined with careful clinical studies involving genotype–phenotype correlations, ultrastructural, biochemical and physiological investigation, seem necessary to further elucidate the still unknown causes of this disease. The enormous progress made in recent years provides justified hope that causes and therapies will be found – at least for some of the subgroups of RP – during the coming decades.

REFERENCES

Anderson, R.E., Maude, M.B., Alvarez, R.A. *et al.* (1991) Plasma lipid abnormalities in the miniature poodle with progressive rod–cone degeneration. *Experimental Eye Research*, **52**, 349–55.

Apfelstedt-Sylla, E., Bunge, S., David, D. *et al.* (1993) Phenotypes of carboxyl-terminal rhodopsin mutations in autosomal dominant retinitis pigmentosa, in: *Retinal Degeneration* (eds J.G. Hollyfield *et al.*), Plenum Press, New York.

Bascom, R.A., Manara, S., Collins, L. *et al.* (1992) Cloning of the cDNA for a novel photoreceptor membrane protein (rom-1) identifies a disk rim protein family indicated in human retinopathies. *Neuron*, **8**, 1171–84.

Berson, E.L., Rosen, J.B. and Simonoff, E.A. (1979) Electroretinographic testing as an aid in detection of carrier of X-chromosome linked retinitis pigmentosa. *American Journal of Ophthalmology*, **87**, 460–8.

Berson, E.L., Sandberg, M.A., Rosner, B. *et al.* (1985) Natural course of retinitis pigmentosa. *American Journal of Ophthalmology*, **99**, 240–51.

Berson, E.L., Rosner, B., Sandberg, M.A. *et al.* (1993) A randomized trial of vitamin A and vitamin E supplementation for retinitis pigmentosa. *Archives of Ophthalmology*, **111**, 761–72.

Bird, A.C. (1975) X-linked retinitis pigmentosa. *British Journal of Ophthalmology*, **59**, 177–99.

Bowes, C., Li, T., Dancinger, M. *et al.* (1990) Retinal degeneration in the rd mouse is caused by a defect in the beta subunit of rod cGMP-phosphodiesterase. *Nature*, **347**, 677–80.

Chader, G.J. (1987) Biochemical studies of retinal degeneration in animal models and in the human, in *Research in Retinitis Pigmentosa. Advances in the Biosciences*, vol. 62 (eds E. Zrenner, H. Krastel and H.H. Goebel), Pergamon, Oxford–New York.

Concerted Action of the European Communities (1994) *Concerted Action News*, March.

Converse, C.A., McLachlan, T., Packard, C.J. and Shepherd, J. (1987) Lipid abormalities in retinitis pigmentosa, in Zrenner E., Krastel H., Goebel H.H. (eds) *Research in Retinitis Pigmentosa. Advances in the Biosciences*, vol. 62 (eds E. Zrenner, H. Krastel and H.H. Goebel), Pergamon, Oxford–New York.

Dryja, T.P., McGee, T., Reichel, E. *et al.* (1990) A point mutation of the rhodopsin gene in one form of retinitis pigmentosa. *Nature*, **343**, 364–6.

Dryja, T.P., Hahn, L.B., Cowley, G.S. *et al.* (1991) Mutation spectrum among patients with autosomal dominant retinitis pigmentosa of the rhodopsin gene. *Proceedings of the National Academy of Sciences USA*, **88**, 9370–4.

Faktorovich, E.G., Steinberg, R.H., Yasamura, D. *et al.* (1990) Photoreceptor degeneration in inherited retinal dystrophy delayed by basic fibroblast growth factor. *Nature*, **347**, 83–6.

Farrar, G.J., Jordan, S.A., Kumar-Singh, R. *et al.* (1993) Extensive genetic heterogeneity in autosomal dominant retinitis pigmentosa, in *Retinal Degeneration* (eds J.G. Hollyfield *et al.*), Plenum Press, New York, pp.63–77.

Fishman, G.A., Weinberg, A.B. and McMahon, T.T. (1986) X-linked recessive retinitis pigmentosa: clinical characteristics of carriers. *Archives of Ophthalmology*, **104**, 1329–35.

Fishman, G.A., Farber, M.D. and Derlacki, D.J. (1988) X-linked retinitis pigmentosa. Profile of clinial findings. *Archives of Ophthalmology*, **106**, 369–75.

Fishman, G.A., Gilbert, G.D., Fiscella, R.G. *et al.* (1989) Acetazolamide for treatment of chronic macular edema in retinitis pigmentosa. *Archives of Ophthalmology*, **107**, 1445–53.

Foxman, S.G., Heckenlively, J.R., Bateman, J.B. and Wirtschafter, J.D. (1985) Classification of congenital and early onset retinitis pigmentosa. *Archives of Ophthalmology*, **103**, 1502–6.

Gouras, P.G., Flood, M.T., Kjeldbye, H. *et al.* (1985) Tranplantation of cultured human retinal epithelium to Bruch's membrane of the owl monkey's eye. *Current Eye Research*, **43**, 253.

Heckenlively, J.R. (1988) *Retinitis Pigmentosa*. Lippincott, Philadelphia.

Heckenlively, J.R. and Arden, G.B. (1991) *Principles and Practice of Visual Electrophysiology*, Mosby, St Louis.

Jacobson, S.G., Kemp, C.M., Sung, C.H. and Nathans, J. (1991) Retinal function and rhodopsin levels in autosomal dominant retinitis pigmentosa with rhodopsin mutations. *American Journal of Ophthalmology*, **112**, 256–71.

Jay, M. (1982) On the heredity of retinitis pigmentosa. *British Journal of Ophthalmology*, **66**, 405–16.

Kaiser-Kupfer, M.I. and Valle, D.L. (1987) Clinical, biochemical and therapeutic aspects of gyrate atrophy. *Progress in Retinal research*, **6**, 179–206.

Krastel, H. and Moreland, J. (1987) Cut-off filters and the acuity–luminance function in retinitis pigmentosa, in *Advances in the Biosciences*, vol. 62 (eds E. Zrenner, H. Krastel and H.H. Goebel), Pergamon, Oxford, New York.

Li, T., Berson, E.L., Dryja, T.P. and Roessler, B.J. (1994) *In vivo* gene transfer to normal and rd mouse retinas mediated by adenoviral vectors. *Investigative Ophthalmology and Visual Science Suppl.*, **35**, 1716.

Lolley, R.N., Rong, H. and Craft, C.M. (1994) Linkage of photoreceptor degeneration by apoptosis with inherited defect in phototransduction. *Investigative Ophthalmology and Visual Science*, **35**, 358–62.

Lyness, A.L., Ernst, W., Quinlan, M.P. *et al.* (1985) A clinical, psychophysical and electroretinographic survey of patients with autosomal dominant retinitis pigmentosa. *British Journal of Ophthalmology*, **69**, 326–39.

Marmor, M.F., Aguirre, G., Arden, G.B. *et al.* (1983) Retinitis pigmentosa, a symposium on terminoloy and methods of examination. *Ophthalmology*, **90**, 126–31.

Marmor, M.F., Arden, G.B., Nilsson, S.E. and Zrenner, E. (1989) Standard for clinical electrophysiology. *Archives of Ophthalmology*, **107**, 816–19.

Massof, R.W. and Finkelstein, D. (1979) Rod sensitivity relative to cone sensitivity in retinitis pigmentosa. *Investigative Ophthalmology and Visual Science*, **18**, 263–72.

Massof, R.W. and Finkelstein, D. (1987) A two-stage hypothesis for the natural course of retinitis pigmentosa, in *Advances in the Biosciences*, vol 62 (eds E. Zrenner, H. Krastel and H.H. Goebel), Pergamon, Oxford, pp.29–58.

McLaughlin, M.E., Sandberg, M.A., Berson, E.L. and Dryja, T.P. (1993) Recessive mutations in the gene encoding the beta subunit or rod phosphodiesterase in patients with retinitis pigmentosa. *Nature Genet.*, **4**, 130–4.

Merin, S. and Auerbach, E. (1976) Retinitis pigmentosa. *Survey of Ophthalmology*, **20**, 303–46.

Molday, S., Hicks, D. and Molday, L. (1987) Peripherin: a rim-specific protein of rod outer segment discs. *Investigative Ophthalmology and Visual Science*, **78**, 50–61.

Naash, M.I., Tong, C., Goto, Y. *et al.* (1994) Studies on the mechanisms of the photoreceptor cell degeneration in a transgenic mouse model for human adRP. *Investigative Ophthalmology and Visual Science*, Suppl. **35**, 1834.

Newsome, D.A. and Fishman, G.A. (1986) Research update. Report on retinitis pigmentosa and the immune system workshop 7–8 November 1985, *Experimental Eye Research*, **43**, 1.

Pagon, R. (1988) Retinitis pigmentosa. *Survey of Ophthalmology*, **33**, 137–77.

Peachey, N.S., Goto, Y., Al-Ubaidi, M.R., and Naash, M.I. (1994) Rod and cone dysfunction in a transgenic, mouse model of human adRP. *Investigative Ophthalmology and Visual Science Suppl.*, **35**, 2044.

Riess, O., Noeremoelle, A., Collins, C. *et al.* (1992) Exclusion of DNA changes in the beta-subunit of the cGMP phosphodiesterase gene as the cause for Huntington's disease. *Nature Genet.*, **1**, 104–8.

Roof, D.J., Adamian, M. and Hayes, A. (1994) Analysis of gene function by the use of transgenic animals. *Investigative Ophthalmology and Visual Science Suppl.*, **35**, 1480.

Rosenfeld, P.J., Cowley, G.S., McGee, T.L., *et al.*

(1992) A null mutation in the rhodopsin gene causes rod photoreceptor dysfunction and autosomal recessive retinitis pigmentosa. *Nature Genet.*, **1**, 209–13.

Sandberg, M.A., Pawlyk, B.S., Franson, W.K. *et al.* (1993) Effect of light on the rate of ERG decline in transgenic mice with the human rhodopsin mutation pro2shis *Investigative Ophthalmology and Visual Science Suppl.*, **34**, 1078.

Sandberg, M.A., Weigel-Di Franco, C., Dryja, T.P. and Berson, E.L. (1994) ERG amplitude and visual field area correlate with categories rhodopsin mutations causing dominant retinitis pigmentosa. *Investigative Ophthalmology and Visual Science Suppl.*, **35**, 1478.

Schaeffel, F. and Howland, C.(1991) Properties of visual feedback loops controlling eye growth and refraction state in chickens. *Vision Research*, **31**, 717–734.

Sheedlo, H.J., Li, L., Barnstable, C.J, and Turner, J.E. (1993) Synaptic and photoreceptor components in retinal pigment epithelial cell transplanted retinas of the Royal College of Surgeons dystrophic rats. Journal of Neuroscience Research, **36**, 424–31.

Suber, M.L., Pittler, S.J., Qin, N. *et al.* (1993) Irish setter dogs affected with rod/cone dysplasia contain a nonsense mutation in the rod cGMP phosphodiesterase beta-subunit gene. *Proceedings of the National Academy of Sciences USA*, **90**, 3968–72.

Sung, C.H., Schneider, B.G., Agarwal, N. *et al.* (1991) Functional heterogeneity of mutant rhodopsins responsible for autosomal dominant retinitis pigmentosa. *Proceedings of the National Academy of Sciences USA*, **88**, 8840–4.

Thirkill, C.E., Roth, A.M., Takemoto, D.J. *et al.* (1991) Antibody indications of secondary and superimposed retinal hypersensitivity in retinitis pigmentosa. *American Journal of Ophthalmology*, **112**, 132–7.

Travis, G.H., Brennan, M.B, Danielson, P.E. *et al.* (1989) Indentification of a photoreceptor-specific mRNA encoded by the gene responsible for retinal degeneration slow (rds). *Nature*, **388**, 70–3.

van den Hurk, J.A., van de Pol, T.J., Molloy, C.M. *et al.* (1992) Detection and characterization of point mutations in the chorioderemia candidate gene by PCR-SSCP analysis and direct DNA sequencing. *American Journal of Human Genetics* **50**, 1195–202.

Voaden, M.J. (1988) Retinitis pigmentosa and its models. *Progress in Retinal Research*, **7**, 293–331.

Weleber, R.G., Carr, R.E., Murphey, W.H. *et al.* (1993) Phenotypic variation including retinitis pigmentosa, pattern dystrophy, and fundus flavimaculatus in a single family with a deletion of codon 153 and 154 of the peripherine/RDS gene. *Archives of Opthalmology*, **111**, 1531–42.

Wiedemann, P. (1992) Growth factors in retinal diseases: proliferative a vitreoretinopathy, proliferative diabetic retinopathy, and retinal degeneration. *Survey of Ophthalmology*, **36**, 373–84.

Zrenner, E. (1991) Differential diagnosis of 'Flecked retina' disease, in *La Retinite Pigmentosa Movimenti Oculari e Ambliopia Valutazione del Danno e Monitoraggio nel Glaucoma Fondamenti per le Indagini di Base* (eds M. Cordella, G. Baratta and C. Macaluso), Casa Editrice Mattioli, Parma.

19

Clinical aspects: paraneoplastic retinopathy

ANN H. MILAM

19.1 INTRODUCTION

Some cancer patients develop neurologic deficits not directly attributable to their primary tumor or metastases. These so-called remote effects of cancer are termed paraneoplastic and can result from secretion of toxins or hormones by the tumor cells, or as side effects of chemotherapy. Another cause of paraneoplastic disease is development of tumor-specific antibodies, which may play an important role in limiting tumor cell growth but can also cross react with the patient's own tissues, including neurons of the central and peripheral nervous system (Anderson, 1989). More specifically, certain patients with carcinoma or melanoma and visual symptoms have circulating antibodies that recognize specific proteins and subsets of neurons in the human retina. This chapter will review the two best-studied forms of paraneoplastic retinopathy, cancer-associated retinopathy (CAR) and melanoma-associated retinopathy (MAR).

19.2 CANCER-ASSOCIATED RETINOPATHY

CAR occurs in a small percentage of patients with carcinoma, most commonly small cell carcinoma of the lung, but also cancers of the cervix, ovary, colon, prostate, breast and others. Typically, the CAR patient's visual symptoms precede or occur at the time of tumor diagnosis and include progressive dysfunction of both rods and cones over a period of weeks to months. Some CAR patients complain of bizarre visual sensations or photopsias, described as 'swarms of bees', 'spaghetti-like', or 'shimmering curtains' (Sawyer *et al.*, 1976; Rizzo and Gittinger, 1992). The CAR patients have diminished visual fields, including ring or central scotomata, decreased visual acuity, abnormal or absent color vision, nyctalopia, and variable retinal inflammation (Thirkill *et al.*, 1989, 1993a; Jacobson *et al.*, 1990; Cogan *et al.*, 1990; Matsui *et al.*, 1992; Keltner *et al.*, 1992). The light- and dark-adapted electroretinographic (ERG) responses are reduced or extinguished, consistent with loss of both rod and cone function (Rizzo and Gittinger, 1992; Thirkill *et al.*, 1993a). Histopathologic studies of post mortem retinas from CAR patients have revealed loss of photoreceptors (Sawyer *et al.*, 1976; Keltner *et al.*, 1983; Buchanan *et al.*, 1984; Rizzo and Gittinger, 1992; Adamus *et al.*, 1993b) or ganglion cells (Grunwald *et al.*, 1987).

Neurobiology and Clinical Aspects of the Outer Retina
Edited by M.B.A. Djamgoz, S.N. Archer and S. Vallerga
Published in 1995 by Chapman & Hall, London
ISBN 0 412 60080 3

The first indication that CAR is an autoimmune disease came from observations that serum samples from CAR patients produced specific immunolabeling of certain retinal cells, including rods and cones (Sawyer *et al.*, 1976; Keltner *et al.*, 1983; Buchanan *et al.*, 1984; Rizzo and Gittinger, 1992) and ganglion cells (Kornguth *et al.*, 1982a; Grunwald *et al.*, 1985, 1987). Next it was found that some CAR patients' sera that labeled retinal photoreceptors by immunocytochemistry were reactive with a 23 kDa retinal protein (Thirkill *et al.*, 1987, 1992; Jacobson *et al.*, 1990). This protein corresponded in amino acid sequence and immunoreactivity to recoverin (Polans *et al.*, 1991, 1993; Thirkill *et al.*, 1992, 1993b; Adamus *et al.*, 1993b), a Ca^{2+}-binding protein present in both rods and cones in a number of mammalian retinas (Dizhoor *et al.*, 1991; McGinnis *et al.*, 1992; Stepanik *et al.*, 1993; Wiechmann and Hammarback, 1993), and in OFF-center/midget (flat) cone bipolar cells in human and monkey retinas (Milam *et al.*, 1993b; Wässle *et al.*, 1994) (Figure 19.1).

It has been hypothesized that antibodies generated initially against a tumor antigen can cross-react with a retina-specific protein. This was supported by the observation that CAR antibodies reactive with 23 kDa retinal recoverin also recognized a 65 kDa protein in small cell carcinoma of the lung (Thirkill *et al.*, 1989; Adamus *et al.*, 1993b). A cell line of this tumor was also found to express authentic recoverin when cultured *in vivo* (Thirkill *et al.*, 1993b). Sera from other CAR patients labeled retinal rods and some cones, but had low reactivity with recoverin (Rizzo and Gittinger, 1992, 1993), suggesting that the autoantibodies may have recognized another photoreceptor protein. Antibodies from some CAR patients recognized retinal enolase (Adamus *et al.*, 1993a) or some other, as yet unidentified, 50 kDa and 48 kDa retinal antigens (Crofts *et al.*, 1988; Jacobson *et al.*, 1990; Rizzo and Gittinger, 1992). These findings emphasize that CAR is actually a family

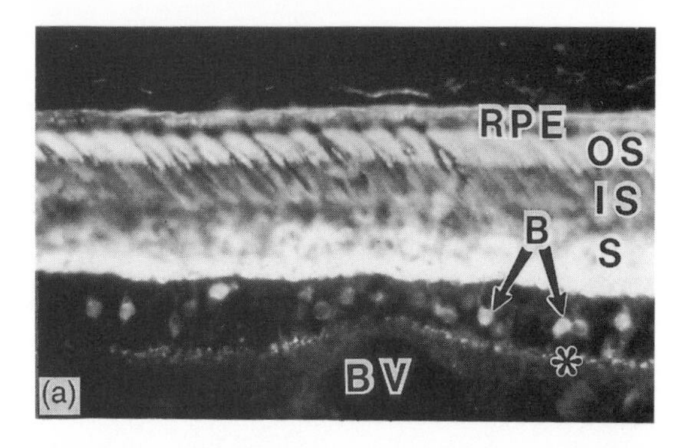

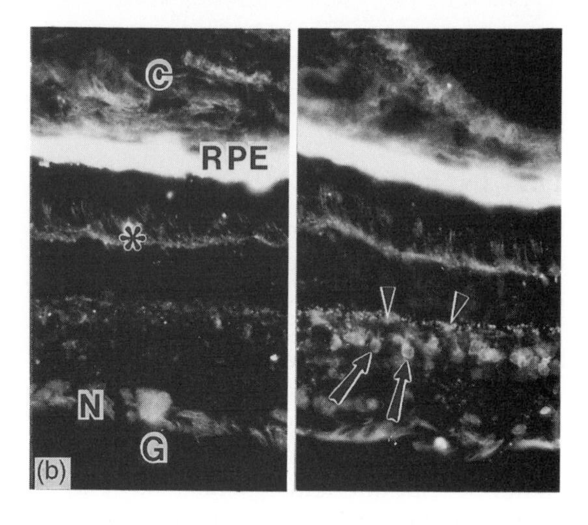

Figure 19.1 (a) Cryostat section of fixed monkey retina processed for indirect immunofluorescence using anti-recoverin. Note labeling of all photoreceptors, including outer segments (OS), inner segments (IS) and somata (S). There is also labeling of a subpopulation of cone bipolar cells (B), whose axons stratify (*) in the outermost portion of the inner plexiform layer. RPE, retinal pigment epithelium; BV, retinal blood vessel. ×120; (b) Sections of unfixed human retina processed for indirect immunofluorescence using IgG from a normal subject (left panel) and a MAR patient (right panel). The retinal pigment epithelium (RPE) contains autofluorescent lipofuscin granules. The choroid (C), external limiting membrane (*), ganglion cell (G), and nerve fiber layer (N) are also autofluorescent. Note specific labeling of bipolar somata (arrows) and dendrites (arrowheads) with the MAR IgG. ×120.

of autoimmune retinopathies, involving a number of different retinal proteins. It is likely that additional retinal autoantigens will be identified as sera of further CAR patients are subjected to laboratory study.

Autoantibodies that recognize retinal neurofilaments and several antigens in small cell carcinoma of the lung have been found in certain CAR patients with this tumor (Grunwald *et al.*, 1985; Kornguth *et al.*, 1986). However, this pattern of reactivity is not unique to CAR, as antibodies against neurofilaments have also been demonstrated in sera of some normal subjects with no visual or neurologic symptoms (Stefansson *et al.*, 1985) and in non-cancer patients with retinitis pigmentosa (Galbraith *et al.*, 1986) or age-related macular degeneration (Gurne *et al.*, 1991). Finally, non-cancer patients with retinal inflammation, due to various causes, can develop leakiness of the blood–retinal barrier, accompanied by release of soluble retinal proteins into the circulation. Certain photoreceptor proteins, including rhodopsin, arrestin (S-antigen) and interphotoreceptor retinoid binding protein, are highly immunogenic (Adamus *et al.*, 1992; Chader, 1994). Elevated titers against these proteins are found in patients with various retinal diseases, including vasculitis and retinitis pigmentosa, and commonly follow retinal inflammation (Nussenblatt, 1991; Yamamoto *et al.*, 1992). This type of autoimmune response may also occur in CAR patients and exacerbate their retinal disease (Rizzo and Gittinger, 1992).

An autoimmune etiology is strongly indicated for CAR because many cancer patients with progressive loss of retinal function have circulating antibodies against retina-specific proteins. However, details of the disease process are poorly understood. It has generally been considered that the autoantibodies cross the blood–retinal barrier, bind to, and interfere with function of the neurons containing the reactive protein. This was supported by the demonstration of antibody deposits in a CAR patient's retina that had lost ganglion cells (Grunwald *et al.*, 1987). Functional effects of antibody exposure were evaluated by Stanford *et al.* (1992), who found that perfusion of rat retina with rat antibodies against S-antigen produced ERG changes not found with an ovalbumin antibody or with antibodies against S-antigen raised in other species. This study indicated that antibodies against a retinal protein are able to cross the blood–retinal barrier and interfere with retinal function.

Based on the histopathology of CAR retinas (see above), it is likely that the antibodies ultimately cause degeneration of the targeted neurons. Few studies have tested this hypothesis directly, although in one study intravitreal injection in cats of antibodies specific for large retinal ganglion cells caused selective loss of this population of neurons (Kornguth *et al.*, 1982b), and abolished the Y-cell retinogeniculate pathway (Spear *et al.*, 1982). The role of circulating antibodies in producing retinal cell death has recently been addressed by immunizing rats with bovine recoverin (Adamus *et al.*, 1994; Gery *et al.*, 1994). The rats developed antibodies against recoverin and inflammatory retinal infiltrates, and the photoreceptors ultimately degenerated. Uveitis could adoptively be transferred to naive rats using lymphocytes from the immunized rats. The effects on the recoverin-containing retinal bipolar cells in these animals were not reported.

9.3 MELANOMA-ASSOCIATED RETINOPATHY

The clinical characteristics of MAR patients differ from those with CAR in several features. MAR develops in patients with cutaneous malignant melanoma and visual symptoms typically occur months to years after diagnosis of metastatic disease. These patients experience relatively acute (over hours to several days) onset of night blindness and persistent photopsias, which have been variously described as 'pulsing or shimmering lights', 'exploding points', 'looking through water', or 'amoebae' (Milam *et al.*, 1993a,c; Andreasson *et al.*, 1993; Rush, 1993; Weinstein *et al.*, 1994). The photopsias are annoying to the patients, and interfere with testing of acuity and visual fields. MAR patients characteristically have elevated dark-adapted thresholds and ERGs that resemble those of patients with congenital stationary night blindness, having normal rod and cone a-waves but a reduced or absent scotopic b-wave (Ripps *et al.*, 1984; Berson and Lessell, 1988; Alexander *et al.*, 1992; Pollack *et al.*, 1992; Milam *et al.*, 1993a,c; Andreasson *et al.*, 1993; Weinstein *et al.*, 1994). There is marked loss of rod-mediated sensitivity across the visual field by dark-adapted perimetry (Milam *et al.*, 1993c). Photoreceptor function is intact, including normal rhodopsin levels determined by fundus reflectometry (Ripps *et al.*, 1984), but signal transmission between photoreceptors and second-order neurons is defective (Berson and Lessell, 1988; Alexander *et al.*, 1992), with a specific defect in ON-center bipolar cell function (Alexander *et al.*, 1992). Several reports indicate that the cone b-wave can be abnormal (Alexander *et al.*, 1992; Andreasson *et al.*, 1993) and that blue-sensitive cone sensitivity can be reduced or absent (Alexander *et al.*, 1992; Milam *et al.*, 1993a,c).

We recently evaluated a series of twelve MAR patients (Milam and Saari, 1994), ten of whom developed visual symptoms within a year after diagnosis of metastatic disease. The mean time between diagnosis of the primary skin tumor and development of visual symptoms was 3.45 years. The onset of nyctalopia and photopsias was characteristically abrupt, occurring over a period of hours to several days, and often one eye preceded the other by several days or weeks. The visual acuity remained relatively normal in ten of the patients, but nine of them had some loss of visual field. Nine had a normal fundus, with three showing mild inflammation. One had severe retinal inflammation and marked loss of visual acuity (Kellner *et al.*, 1994). The ERG was abnormal in all patients, including the diminished rod b-wave that is characteristic of MAR and thought to reflect defective ON-center bipolar cell function. Two of the MAR patients, in addition to defects at the level of the rod bipolar cells, had visual test results attributed to selective loss of magnocellular elements proximal to the retinal ganglion cells (Wolf and Arden, 1994).

Although MAR was initially thought to be caused by vincristine toxicity (Ripps *et al.*, 1984), an autoimmune etiology was later suggested (Berson and Lessell, 1988; Alexander *et al.*, 1992). Based on visual test results, it was hypothesized that autoantibodies directed initially against a melanoma-specific antigen might cross react with post-receptoral retinal neurons, and specifically with the ON-bipolar cells (Alexander *et al.*, 1992). In support of this hypothesis, recent studies indicate that a number of MAR patients have circulating antibodies that recognize retinal bipolar cells (Milam *et al.*, 1993a,c; Milam and Saari, 1994; Weinstein *et al.*, 1994), specifically the population of rod bipolars (Milam *et al.*, 1993a,c; Milam and Saari, 1994). In our series of twelve MAR patients, all of whom had similar visual symptoms and test results, all twelve were found to have circulating antibodies (IgG but not IgM class) that recognized bipolar cells in sections of human, bovine and rat retinas (Milam and Saari, 1994). Bipolar labeling was not found with sera from six normal subjects or 28 patients with metastatic cutaneous malignant melanoma but no visual symptoms (Milam *et al.*, 1993a,c).

Labeling of bipolar cells with MAR sera was fixation sensitive, and most intense labeling was found in sections of unfixed retina (Milam *et al.*, 1993a,c; Weinstein *et al.*, 1994). Bipolar labeling was found over the somata but not the nuclei (Figures 19.1b and 19.2a), and the labeled bipolar dendrities formed a punctate pattern in the outer plexiform layer (Figure 19.2a). The bipolar axons appeared to be unlabeled, but this was difficult to discern because the inner plexiform layer was poorly preserved in the unfixed sections.

The bipolar cells labeled with MAR sera were further characterized by double labeling immunofluorescence using a mouse antibody against the α-isoenzyme of protein kinase C, a marker for rod bipolars (Grünert and Martin, 1991; Wässle *et al.*, 1991; Kolb *et al.*, 1993). Virtually all the bipolar cells labeled by the MAR IgG were also labeled with anti-α-protein kinase C (Milam *et al.*, 1993a,c) (Figure 19.2b), We used other markers for cone bipolar cells, including anticholecystokinin for the blue-sensitive cone and ON-midget (invaginating) bipolars (Wässle *et al.*, 1994) and antirecoverin for the OFF-midget (flat) bipolars (Milam *et al.*, 1993b; Wässle *et al.*, 1994). However, the soluble bipolar antigens reactive with these antibodies were lost from the unfixed sections during serial incubations, precluding identification of any cone bipolar cell types as being reactive with MAR IgG. As virtually all of the bipolar cells labeled by MAR IgG were also reactive for α-protein kinase C, we conclude that the rod bipolars are selectively recognized by the MAR patients' autoantibodies.

Attempts were made to demonstrate the MAR bipolar antigen in human retinal homogenates and extracts using ELISA and SDS-PAGE followed by Western blotting. No differences were found using sera from MAR patients, other melanoma patients with normal vision, and normal subjects, suggesting that the MAR retinal antigen is not a protein (Milam *et al.*, 1993a,c). Bipolar labeling was absent in unfixed retinal sections that had been pretreated with chloroform/methanol, methanol or ethanol, but persisted after pretreatment with chloroform, acetone, or acetonitrile. These extraction characteristics suggest that the bipolar antigen is a polar lipid, and attempts are underway to determine if the MAR retinal antigen is a ganglioside (Milam and Saari, 1994). There is a large literature on melanoma-specific antigens, including gangliosides, which are polar glycolipids present in the surface membrane (Lloyd, 1993). Many melanoma patients develop tumor ganglioside-specific antibodies, which are thought to be protective against tumor cell growth and to prolong survival. Antibodies against melanoma-specific gangliosides are cytotoxic to melanoma cells in culture (Kawashima *et al.*, 1990), and the robust immune reponse to melanoma gangliosides in some patients forms the basis for several experimental vaccines (Portoukalian *et al.*, 1991; Morton, *et al.*, 1992; Houghton *et al.*, 1993). However, if the circulating antibodies to melanoma antigens cross-react with a ganglioside, or another epitope on retinal bipolar cells, the incidence of MAR may be expected to increase in melanoma patients undergoing treatment with these new vaccines.

19.4 CONCLUSIONS AND FUTURE PERSPECTIVES

A small subset of patients with certain forms of cancer develop paraneoplastic retinopathy, which is thought to result from generation of antibodies against tumor cell antigens that cross-react with specific neuron types in the retina. CAR patients typically show gradual loss of both rod and cone function over a period of weeks to months, and some have circulating antibodies reactive with the Ca^{2+}-binding protein, recoverin, found in rods, cones, and certain cone bipolar cells. The presence of the autoantigen in both rods and cones correlates with the patients' loss of photoreceptor function, but the mechanism by which the autoantibodies

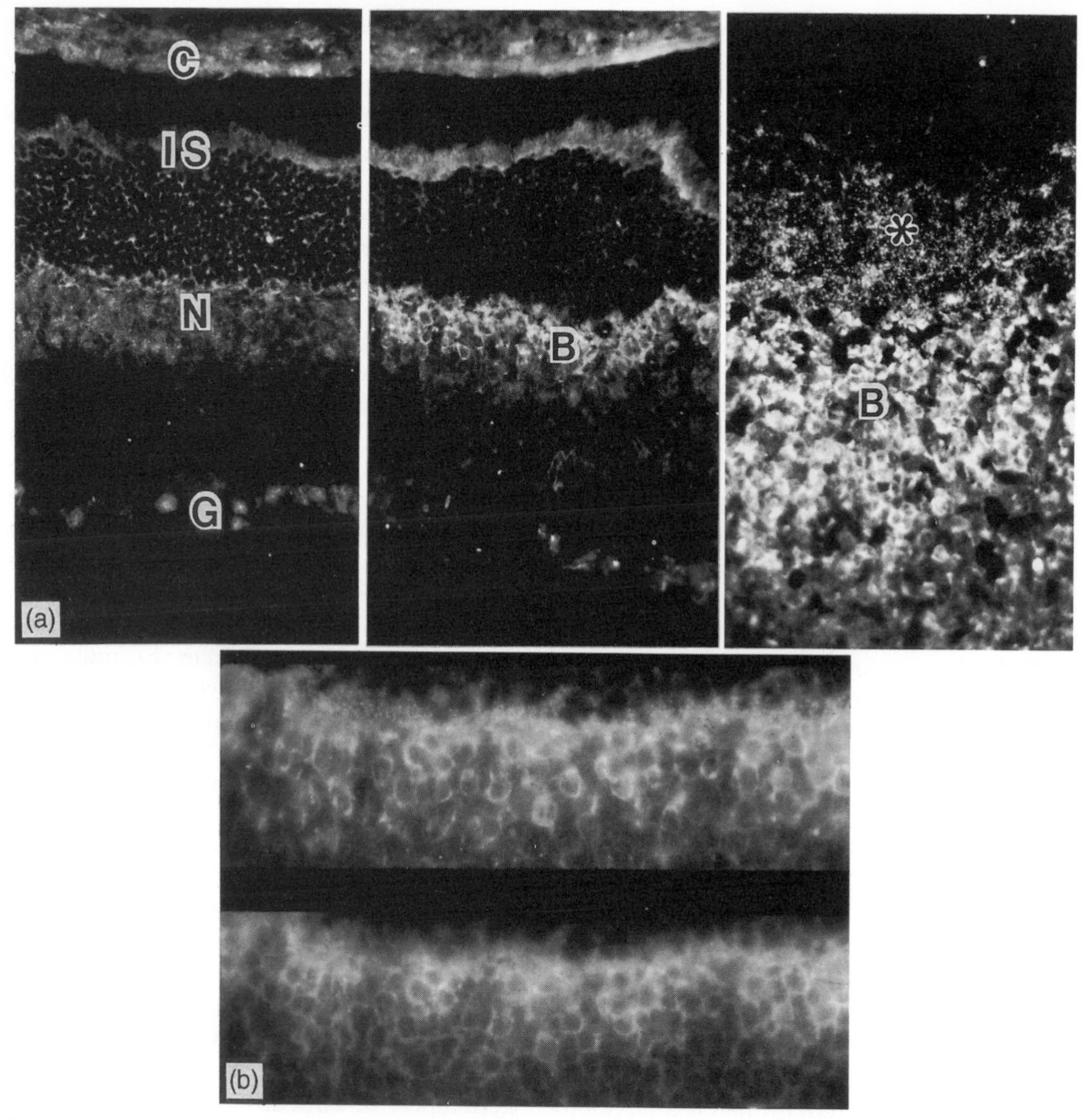

Figure 19.2 (a) Sections of unfixed rat retina processed for indirect immunofluorescence using IgG from a normal subject (left panel) and a MAR patient (middle panel). The choroid (C), inner segments (IS), inner nuclear layer (N) and ganglion cells (G) are autofluorescent. Note specific labeling of bipolar cells (B) with MAR IgG. ×100. The right panel is an oblique section of unfixed rat retina processed in the same way using MAR IgG. Note labeling of the bipolar cell somata (B) and the punctate pattern formed by their labeled dendrites (*) in the outer plexiform layer. ×250; (b) This section of unfixed rat retina was processed using a mixture of MAR IgG and an antibody against α-protein kinase C made in mouse, followed by a mixture of FITC-labeled anti-human IgG and rhodamine-labeled anti-mouse IgG. The upper panel was photographed with a filter for detection of FITC fluorescence and shows the numerous bipolar cells labeled with MAR IgG. The lower panel was photographed with a filter for detection of rhodamine fluorescence, and demonstrates that the cells labeled with MAR IgG are also labeled with anti-α-protein kinase C, a marker for rod bipolar cells. ×360.

cause dysfunction and death of these cells is incompletely understood. Sera of some CAR patients contain autoantibodies that recognize neurofilaments, enolase, or other retinal proteins, suggesting that CAR is a family of autoimmune retinopathies involving a number of retinal antigens. Exposure of animal retinas to antibodies reactive with certain neuron-specific proteins has been shown to alter retinal function and selectively destroy the reactive cells. However, less is known of the disease mechanism in CAR patients.

A few patients with cutaneous malignant melanoma develop a different clinical syndrome known as MAR, which is characterized by abrupt onset over hours to days of nyctalopia and annoying photopsias, elevated dark-adapted thresholds, and loss of the scotopic ERG b-wave. Photoreceptor function is intact, but transmission is defective between photoreceptors and second order neurons, specifically the ON-center bipolar cell system. A number of MAR patients have circulating antibodies that recognize rod bipolar cells in sections of unfixed retina. Localization of the MAR retinal antigen to rod bipolar cells correlates well with the patients' visual defect, but details of the disease mechanism are incomplete.

Damage to the retina in CAR and MAR has been suggested to result as a side effect of chemotherapy (Ripps *et al.*, 1984); from a circulating toxic tumor cell product (Sawyer *et al.*, 1976; Buchanan *et al.*, 1984; Pollack *et al.*, 1992); or from autoantibodies directed initially against a tumor-specific epitope that cross-react with a retinal antigen (Keltner *et al.*, 1983; Berson and Lessell, 1988; Alexander *et al.*, 1992). Specific labeling of retinal photoreceptors with sera from CAR patients, and of rod bipolar cells with MAR sera, adds support to the latter hypothesis. However, no studies have yet demonstrated cross-reactivity of these patients' sera with their own tumor cells, principally because these cells have not been available for analysis.

It should be emphasized that whereas retinal function tests and immunologic staining implicate components of the same pathways in both CAR (rods and cones) and MAR (rod bipolars), there is still no direct evidence that the autoantibodies cause retinal dysfunction. It is not known how circulating antibodies can bypass the blood–retinal barrier, recognize antigens that are sequestered intracellularly in specific neuron types, interfere with retinal function, and ultimately cause neuronal cell death. The studies to date on CAR and MAR have concentrated on humoral aspects of the immune reponse. The role of cell-mediated immunity is probably also important because uveitis, induced by immunization of rats with recoverin, can been adoptively transferred to other rats using cells isolated from lymph nodes and spleen (Adamus *et al.*, 1994; Gery *et al.*, 1994). No information is yet available on the role of lymphocytes in CAR or MAR patients.

Why do certain cancer patients develop autoimmune retinopathy? Do their tumor cells present novel antigens, or does the defect lie in the patients immune response? Virtually nothing is known about the tumor antigens that elicit an immune response in CAR and MAR. As more tumor samples become available, it will be important to characterize the tumor antigens recognized by the antibodies that cross-react with retinal antigens. It will also be instructive to screen these patients for HLA antigens known to be associated with other autoimmune diseases.

The sera of some but not all CAR patients label retinal proteins identified as recoverin and enolase (Polans *et al.*, 1991; Thirkill *et al.*, 1992; Adamus *et al.*, 1993a,b), but a retinal protein has not been found that is recognized uniquely by MAR sera. Extraction characteristics of the rod bipolar antigen, coupled with the high level of expression of certain gangliosides by melanoma cells, suggest that the

MAR bipolar antigen may be a lipid rather than a protein. Additional biochemical studies are required to resolve this question.

The main therapy for paraneoplastic retinopathy has been immunosuppression with steroids, which reduced retinal inflammation and improved acuity and visual fields in some cases of CAR (Jacobson *et al.*, 1990; Keltner *et al.*, 1992; Rizzo and Gittinger, 1992). However, some CAR and MAR patients did not recover normal vision, possibly because the affected retinal neurons had already been destroyed (Keltner *et al.*, 1992; Milam *et al.*, 1993a,c). Furthermore, the patients' circulating autoantibodies may play an important role in holding their tumor cells in check, and attempts to ameliorate the retinal disease are contraindicated if they also compromise immune surveillance of the malignancy. Clinical trials are underway to assess the efficacy of administering autoantigens orally as therapy for a number of autoimmune diseases, including uveitis (Weiner *et al.*, 1994). This new therapy may also be of benefit to patients with CAR and MAR.

In order to develop rational therapies, additional information is required on the epitopes shared by tumor and retinal antigens, as well as the mechanisms of retinal disease. New knowledge of photoreceptor- and bipolar cell-specific molecules recognized by the patients' antibodies is important not only for understanding CAR and MAR, but also for elucidating the normal function of these retinal neurons. Finally, additional histopathologic studies are needed of post mortem retinas from CAR and MAR patients to determine if their photoreceptors and bipolar cells are abnormal or absent, and if immune complexes are present in the tissue.

ACKNOWLEDGEMENTS

This work was supported by the National Retinitis Pigmentosa Foundation, Inc., NIH grants EY0-1311 and -1730, a departmental award from Research to Prevent Blindness, Inc., and by the Chatlos Foundation. AHM is a Senior Scholar of Research to Prevent Blindness, Inc. The author thanks Ms Jean Chang, Mr Greg Garwin, Ms Jing Huang, Ms Ingrid Klock and Mr Chuck Stephens for technical assistance, and Dr John Saari for critical review of the manuscript.

REFERENCES

Adamus, G., Schmied, J.L., Hargrave, P.A. *et al.* (1992) Induction of experimental autoimmune uveitis with rhodopsin synthetic peptides in Lewis rats. *Current Eye Research*, **11**, 657–67.

Adamus, G., Aptisiauri, N., Guy, J. *et al.* (1993a) Anti-enolase antibodies in cancer associated retinopathy. *Investigative Ophthalmology and Visual Science*, **34**, 1485.

Adamus, G., Guy, J., Schmied, J.L. *et al.* (1993b) Role of anti-recoverin autoantibodies in cancer-associated retinopathy. *Investigative Ophthalmology and Visual Science*, **34**, 2626–33.

Adamus, G., Ortega, H., Witkowska, D. and Polans, A. (1994) Recoverin: a potent uveitogen for the induction of photorecepter degeneration in Lewis rats. *Experimental Eye Research*, **59**, 447–55.

Alexander, K.R., Fishman, G.A., Peachey, N.S. *et al.* (1992) 'On' response defect in paraneoplastic night blindness with cutaneous malignant melanoma. *Investigative Ophthalmology and Visual Science*, **33**, 477–83.

Anderson, N.E. (1989) Anti-neuronal autoantibodies and neurological paraneoplastic syndromes. *Australian and New Zealand Journal of Medicine*, **19**, 379–87.

Andreasson, S., Ponjavic, V. and Ehinger, B. (1993) Full-field electroretinogram in a patient with cutaneous melanoma-associated retinopathy. *Acta Ophthalmologica*, **71**, 487–90.

Berson, E.L. and Lessell, S. (1988) Paraneoplastic night blindness with malignant melanoma. *American Journal of Ophthalmology*, **106**, 307–11.

Buchanan, T.A.S., Gardiner, T.A. and Archer, D.B. (1984) An ultrastructural study of retinal photoreceptor degeneration associated with bronchial carcinoma. *American Journal of Ophthalmology*, **97**, 277–87.

Chader, G.J. (1994) Retinal degenerations of hereditary, viral and autoimmune origins: studies on opsin and IRBP. *Progress in Retinal and Eye Research*, **13**, 65–99.

Cogan D.G., Kuwabara, T., Currie, J. and Kattah, J. (1990) Paraneoplastic retinopathy simulating cone dystrophy with achromatopsia. *Klinische Monatsblalter Augenheilkunde*, **197**, 156–8.

Crofts, J.W., Bachynski, B.N. and Odel, J.G. (1988) Visual paraneoplastic syndrome associated with undifferentiated endometrial carcinoma. *Canadian Journal of Ophthalmology*, **23**, 128–32.

Dizhoor, A.M., Ray, S., Kumar, S. *et al.* (1991) Recoverin: a calcium sensitive activator of retinal rod guanylate cyclase. *Science*, **251**, 915–18.

Galbraith, G.M.P., Emerson, D., Fudenberg, H.H. *et al.* (1986) Antibodies to neurofilament protein in retinitis pigmentosa. *Journal of Clinical Investigation*, **78**, 865–9.

Gery, I., Chanaud, N.P., 3rd and Anglade, E. (1994) Recoverin is highly uveitogenic in rats. *Investigative Ophthalmology and Visual Science*, **35**, 3342–5.

Grünert, U. and Martin, P.R. (1991) Rod bipolar cells in the macaque monkey retina: immunoreactivity and connectivity. *Journal of Neuroscience*, **11**, 2742–58.

Grunwald, G.B., Klein, R., Simmonds, M.A. and Kornguth, S.E. (1985) Autoimmune basis for visual paraneoplastic syndrome in patients with small-cell lung carcinoma. *Lancet*, **i**, 658–61.

Grunwald, G.B., Kornguth, S.E., Towfighi, J. *et al* (1987) Autoimmune basis for visual paraneoplastic syndrome in patients with small cell lung carcinoma. Retinal immune deposits and ablation of retinal ganglion cells. *Cancer*, **60**, 780–6.

Gurne, D.H., Tso, M.O.M., Edward, D.P. and Ripps, H. (1991) Antiretinal antibodies in serum of patients with age-related macular degeneration. *Ophthalmology*, **98**, 602–7.

Houghton, A.N., Asaradhi, S.V., Bouchard, B. *et al.* (1993) Recognition of autoantigens by patients with melanoma. *Annals of the New York Academy of Sciences*, **690**, 59–68.

Jacobson D.M., Thirkill, C.E. and Tipping, S.J. (1990) A clinical triad to diagnose paraneoplastic retinopathy. *Annals of Neurology*, **28**, 162–7.

Kawashima I., Tada, N, Fujimori, T. and Tai, T. (1990) Monoclonal antibodies to disialogangliosides: characterization of antibody-mediated cytotoxicity against human melanoma and neuroblastoma cells *in vitro*. *Journal of Biochemistry*, **108**, 109–15.

Kellner U., Bornfield, N. and Foerster, M.H. (1994) Severe course of cutaneous melanoma associated retinopathy. *Investigative Ophthalmology and Visual Science*, **35**, 2117.

Keltner, J.L., Roth, A.M. and Chang, R.S. (1983) Photoreceptor degeneration. Possible autoimmune disorder. *Archives of Ophthalmology*, **101**, 564–9.

Keltner, J.L., Thirkill, C.E. Tyler, N.K. and Roth, A.E. (1992) Management and monitoring of cancer-associated retinopathy. *Archives of Ophthalmology*, **110**, 48–53.

Kolb, H., Zhang, L. and Dekorver, L. (1993) Differential staining of neurons in the human retina with antibodies to protein kinase C isozymes. *Visual Neuroscience* **10**, 341–51.

Kornguth, S.E., Klein, R., Appen R. and Choate, J. (1982a) Occurrence of antiretinal ganglion cell antibodies in patients with small cell carcinoma of the lung. *Cancer*, **50**, 1289–93.

Kornguth, S.E., Spear, P.D. and Langer, E. (1982b) Reduction of large ganglion cells in cat retina following intravitreous injection of antibodies. *Brain Research*, **245**, 35–45.

Kornguth, S.E., Kalinke, T., Grunwald, G.B. *et al.* (1986) Anti-neurofilament antibodies in the sera of patients with small cell carcinoma of the lung and with visual paraneoplastic syndrome. *Cancer Research*, **46**, 2588–95.

Lloyd, K.O. (1993) Tumor antigens known to be immunogenic in man. *Annals of the New York Academy of Sciences*, **690**, 50–8.

Matsui, Y., Mehta, M.C., Katsumi, O. *et al.* (1992) Electrophysiological findings in paraneoplastic retinopathy. *Graefes Archives of Clinical and Experimental Ophthalmology*, **230**, 324–28.

McGinnis, J.F., Stepanik, P.L., Baehr, W. *et al.* (1992) Cloning and sequencing of the 23 kDa mouse photoreceptor cell-specific protein. *Federation of European Biochemical Societies Letters*, **302**, 172–6.

Milam, A.H. and Saari, J.C. (1994) Autoantibodies against retinal rod bipolar cells in melanoma-associated retinopathy. *Investigative Ophthalmology and Visual Science*, **35**, 2116.

Milam, A.H., Alexander, K.R., Jacobson, S.G. *et al.* (1993a) Cutaneous melanoma-associated retinopathy, in *Retinal Degeneration* (eds J.G. Hollyfield *et al.*), Plenum Press, New York, pp.335–42.

Milam, A.H., Dacey, D.M. and Dizhoor, A.M. (1993b) Recoverin immunoreactivity in mammalian cone bipolar cells. *Visual Neuroscience*, **10**, 1–12.

Milam, A.H., Saari, J.C., Jacobson, S.G *et al.* (1993c) Autoantibodies against retinal bipolar cells in cutaneous melanoma-associated retinopathy. *Investigative Ophthalmology and Visual Science*, **34**, 91–100.

Morton, D.L., Foshag, L.J., Hoon, D.S.B. (1992). Prolongation of survival in metastatic melanoma after active specific immunotherapy with a new polyvalent melanoma vaccine. *Annals of Surgery*, **216**, 463–82.

Nussenblatt, R.B. (1991) Proctor lecture. Experimental autoimmune uveitis: mechanisms of disease and clinical therapeutic indications. *Investigative Ophthalmology and Visual Science*, **32**, 3131–41.

Polans, A.S., Buczylko, J., Crabb, J. and Palczewski, K. (1991) A photoreceptor calcium-binding protein is recognized by autoantibodies obtained from patients with cancer-associated retinopathy. *Journal of Cell Biology*, **112**, 981–9.

Polans, A.S., Burton, M.D., Halet, T.L. *et al* (1993) Recoverin, but not visinin, is an autoantigen in the human retina identified with a cancer-associated retinopathy. *Investigative Ophthalmology and Visual Science*, **34**, 81–90.

Pollack, S.C., Kuwashima, L.H., Kelman, S.E. and Tiedeman, J.S. (1992) Melanoma-associated retinopathy (MAR). Presented at the 1992 meeting of the North American Neuro-Ophthalmology Society; 23–27 February, San Diego, CA.

Portoukalian, J., Carrel, S., Dore, J.-F. and Rümke, P. (1991) Humoral immune response in disease-free advanced melanoma patients after vaccination with melanoma-asociated gangliosides. *International Journal of Cancer*, **49**, 893–9.

Ripps, H., Carr, R.E., Siegel, I.M. and Greenstein, V.C. (1984) Functional abnormalities in vincristine-induced night blindness. *Investigative Ophthalmology and Visual Science*, **25**; 787–94.

Rizzo III, J.F. and Gittinger, Jr., J.W. (1992) Selective immunohistochemical staining in the paraneoplastic retinopathy syndrome. *Ophthalmology*, **99**, 1286–95.

Rizzo III, J.F. and Gittinger, J.W. (1993) Paraneoplastic retinopathy syndrome. Authors' reply. *Ophthalmology*, **100**, 147.

Rush, J.A. (1993) Paraneoplastic retinopathy in malignant melanoma. *American Journal of Ophthalmology*, **115**, 390–1.

Sawyer, R.A., Selhorst, J.B., Zimmerman, L.E. and Hoyt, W.F. (1976) Blindness caused by photoreceptor degeneration as a remote effect of cancer. *American Journal of Ophthalmology*, **81**, 606–13.

Spear, P.D., Jones, K.R., Zetlan, S.R. *et al* (1982) Effect of antibodies to large ganglion cells on the cat's retinogeniculate pathway. *Journal of Neurophysiology*, **47**, 1174–95.

Stanford, M.R., Robbins, J., Kasp, E. and Dumonde, D.C. (1992) Passive administration of antibody against retinal S-antigen induces electroretinographic supernormality. *Investigative Ophthalmology and Visual Science*, **33**, 30–5.

Stefansson, K., Marton, L.S., Dieperink, M.E. *et al.* (1985) Circulating autoantibodies to the 200,000-dalton protein of neurofilaments in the serum of healthy individuals. *Science*, **228**, 1117–19.

Stepanik, P.L., Lerious, V. and McGinnis, J.F. (1993) Developmental appearance, species and tissue specificity of mouse 23-kDa, a retinal calcium-binding protein (recoverin). *Experimental Eye Research*, **57**, 189–97.

Thirkill, C.E., Roth, A.M. and Keltner, J.L (1987). Cancer-associated retinopathy. *Archives of Ophthalmology*, **105**, 372–5.

Thirkill, C.E., FitzGerald, P., Sergott, R.C. *et al.* (1989) Cancer-associated retinopathy (CAR syndrome) with antibodies reacting with retinal, optic-nerve, and cancer cells. *New England Journal of Medicine*, **321**, 1589–94.

Thirkill, C.E., Tait, R.C. Tyler, N.K. *et al.* (1992) The cancer-associated retinopathy antigen is a recoverin-like protein. *Investigative Ophthalmology and Visual Science*, **33**, 2768–72.

Thirkill, C.E., Keltner, J.L., Tyler, N.K. and Roth, A.M. (1993a) Antibody reactions with retina and cancer-associated antigens in ten patients with cancer-associated retinopathy. *Archives of Ophthalmology*, **111**, 931–7.

Thirkill, C.E., Tait, R.C., Tyler, N.K. *et al.* (1993b) Intraperitoneal cultivation of small-cell carcinoma induces expression of the retinal cancer-associated retinopathy antigen. *Archives of Ophthalmology*, **111**, 974–8.

Wässle, H., Yamashita, M., Greferath, U. *et al.* (1991) The rod bipolar cell of the mammalian retina. *Visual Neuroscience*, **7**, 99–112.

Wassle, H., Grünert, U., Martin, P.R. and Boycott, B.B. (1994) Immunocytochemical characterization and spatial distribution of midget bipolar cells in the macaque monkey retina. *Vision Research*, **34**, 561–79.

Weinstein, J., Kelman, S., Bresnick, G. and Kornguth, S.E. (1994) Paraneoplastic retinopathy associated with anti-retinal bipolar cell antibodies in cutaneous malignant melanoma. *Ophthalmology*, **101**, 1236–43.

Wiechmann, A.F and Hammarback, J.A. (1993) Expression of recoverin mRNA in the human retina: localization by *in situ* hybridization. *Experimental Eye Research*, **57**, 763–9.

Wolf, J.E. and Arden, G.B. (1994) Selective loss of magnocellular function in melanoma associated retinopathy. *Investigative Ophthalmology and Visual Science*, **35**, 2117.

Yamamoto, J.H., Okajima, O., Mochizuki, M. *et al* (1992) Cellular immune responses to retinal antigens in retinis pigmentosa. *Graefes Archives of Clinical and Experimental Ophthalmology* **230**, 119–23.

Weiner, H.L., Friedman, A., Miller, A. *et al.* (1994) Oral tolerance: immunologic mechanisms and treatement of animal and human organ-specific autoimmune diseases by oral administration of autoantigens. *Annual Review of Immunology*, **12**, 809–37.

20

Clinical aspects: Parkinson's disease

IVAN BODIS-WOLLNER and ANDREA ANTAL

20.1 INTRODUCTION

A significant number of patients suffering from the more common neurodegenerative disorders such as Parkinson's (PD) or Alzheimer's diseases (AD) complain of some form of visual impairment, blurred vision or double vision even though their visual acuity score is within the normal range (for a summary see Bodis-Wollner *et al.*, 1993). Neurodegenerative diseases are characterized by progressive loss of discrete groups of nerve cell populations resulting in a spectrum of motor and behavioral abnormalities, and, as we now know, sensory changes. Although the neuroanatomico-pathology of visual dysfunction in most neurodegenerative diseases is unknown, in PD it is certain that the retina is involved.

Two main approaches have been used to study visual losses in neurodegenerative diseases. One of them is to apply tests which have the power to quantify select as opposed to general visual losses in clinical populations. A second approach has been to create chronic or acute animal models of a neurodegenerative disorder by the applications of selective neurotoxins and receptor blocker substances. The tools of visual measurements, both in humans and animals, include visual psychophysical and electrophysiological measurements. Particularly, visual-evoked potential (VEP), and electroretinogram (ERG) measurements have found applications in this field. The ERG directly reflects retinal processing and in PD a number of studies have shown abnormal flash (FERG) and pattern (PERG) electroretinograms (Krejkova *et al.*, 1985; Nightingale *et al.*, 1986; Bodis-Wollner *et al.*, 1991). Dopaminergic retinal neurons and dopamine (DA) release mechanisms have been recognized in all vertebrates, including humans (Nguyen-Legros *et al.*, 1982, 1984; Zarbin *et al.*, 1986; Ikeda *et al.*, 1987; Dowling, 1991; Hankins and Ikeda, 1991; Djamgoz and Wagner, 1992). Drugs like haloperidol, levodopa and sulpiride which affect the dopaminergic system, influence the ERG (Bartel *et al.*, 1990; Stanzione *et al.*, 1991). In addition to electrophysiology, neurochemical findings also point to dopaminergic deficiency in the retina. Finally, a primate model of the disease exists and has been studied including retinal morphology, chemistry and electrophysiology.

Although it is true that all vertebrate retinas have dopaminergic neuronal circuits,

Neurobiology and Clinical Aspects of the Outer Retina
Edited by M.B.A. Djamgoz, S.N. Archer and S. Vallerga
Published in 1995 by Chapman & Hall, London
ISBN 0 412 60080 3

there are substantial species differences. The retinas of lower species, including commonly researched rabbits, cats and less commonly used dogs, differ from the human and monkey retina in essential details such as aqueous dynamics, retinal architecture, spatiotemporal frequency segregation and center-surround properties of magno- and parvocellular neurons. Whereas in the cat DA was reported only in retinal ganglion cells, in Cebus monkeys it was found in interplexiform cells (IPCs) and in cynomolgous monkeys in amacrine cells (ACs) and IPCs. Cynomolgous and rhesus retinal processing are sufficiently similar to humans as the available morphological and physiological data suggest (Ehinger and Floren, 1979). DA receptors exist in both the inner and outer retinal layers in man and monkey (Zarbin *et al.*, 1986; Schorderet and Nowak, 1990; Stormann *et al.*, 1990). They can be subdivided in the retina into major subtypes of D1 and D2 receptors as in the striatonigral system (although recently further subclasses have been added to the major division of D1 and D2 receptors). For these reasons, the non-human primate model does provide a reasonable tool to study normal dopaminergic and abnormal (DA-deficient) visual processing. An important purpose of studying visual processing in the monkey, in reference to human PD, is to provide better understanding of the neurophysiopathology and behavioral aspects of this disorder, to develop new therapeutic avenues and last but not least, to understand the functional role of dopaminergic circuitry of the human retina.

20.2 RETINAL DA DEFICIENCY IN THE MONKEY

The pattern electroretinogram (PERG) is a mass electrical response of the cells of the inner central neuronal retina (Maffei and Fiorentini, 1986) (See Appendix for defini-

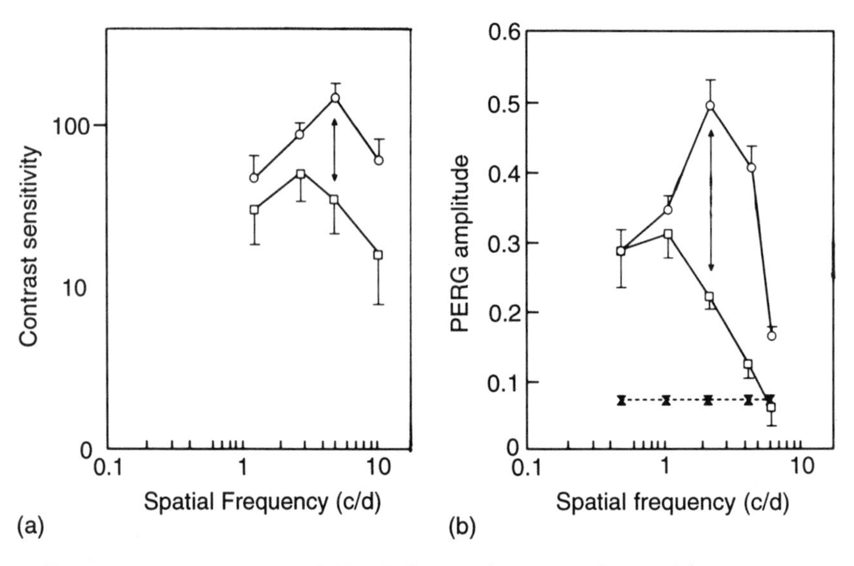

Figure 20.1 Contrast sensitivity (a) and PERG (b) as a function of spatial frequency in normal (circle) and PD patients (square). Note attenuation and shift to the left of the peak response both for CS and PERG measurements. Although there are similar changes in both the CS and PERG transfer functions, they do not peak at the same spatial frequencies. Dotted line denotes noise level.

tions of key terms). The response depends on the existence of neurons with center-surround receptive field organization. These include, to some degree, bipolar neurons but predominantly retinal ganglion cells. A further requirement for a PERG generation is the presence of essential preganglionic non-linearities (Brannan *et al.*, 1992). The PERG varies as a function of spatial and temporal parameters of the stimulation (Hess and Baker, 1984). In humans and in primates, the PERG shows spatial tuning, peaking at medium and decreasing at both low and high spatial frequencies (Hess and Baker, 1984; Hess *et al.*, 1986) (Figure 20.1). The PERG is affected by pathological changes of the ganglion cells or by processes that alter the input to ganglion cells (Bobak *et al.*, 1983; Maffei and Fiorentini, 1990; Nesher and Trick, 1991).

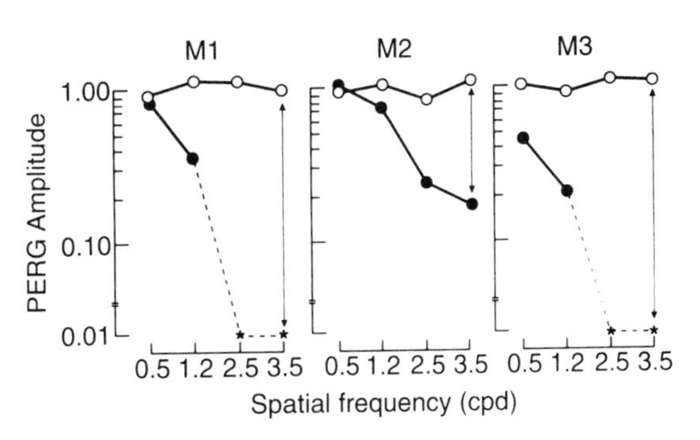

Figure 20.2 PERG amplitude ratios before (open circle) and after (filled circle) the intravitreal injection of 6-OHDA are illustrated as a function of spatial frequency for three monkeys (M1, M2 and M3). The responses from the 6-OHDA-treated eyes were reduced or absent for the higher spatial frequencies. The data were collected on day 46 (M1), on day 2 (M2) and on day 10 (M3) after the drug administration. (*) shows noise level. (After Ghilardi *et al.* (1989).)

20.2.1 THE 6-HYDROXY-DOPAMINE MODEL

Several approaches have been pursued to develop animal models for PD. These included systemic or local neurotoxins, which either destroy dopaminergic neurons or deplete or decrease dopaminergic neurotransmission. In lower species one of the most successfully employed drug was 6-hydroxy-dopamine (6-OHDA), which was injected into the nigrostriatal system of rats in order to mimic symptoms of PD (Ungerstedt and Arbuthnott, 1970). After a unilateral 6-OHDA lesion of the nigrostriatal system a lack of movement initiation is observed and the animals turn toward the injected side even in the absence of any stimulation. Bilateral injections of 6-OHDA lead to, in addition to akinesia, limb dysfunction and rigidity and also to an impairment in orientation to sensory stimuli (Marshall, 1979; Kozlowski and Marshall, 1981). 6-OHDA was also used to destroy DA retinal neurons in turtle and rat (Piccolino *et al.*, 1987; Hadjiconstantinou *et al.*, 1991).

In cynomolgous monkeys, monocular DA depletion using 6-OHDA affected the VEPs and PERGs evoked by vertical sinusoidal grating stimuli (Ghilardi *et al.*, 1989). PERG attenuation was detected for medium spatial frequency stimuli (2.5 and 3.5 cycles/degree), whereas responses to stimuli of lower frequencies (0.5 and 1.2 cpd) remained unaffected (Figure 20.2). DA retinal content was measured in the eyes of one monkey following treatment with 750 μg 6-OHDA. The levels of DA and its metabolite, dihydroxyphenylacetic acid (DOPAC) in the treated eye were 14.7% and 29%, respectively, of the values measured in the saline-treated eye (Ghilardi *et al.*, 1989).

20.2.2 THE MPTP PRIMATE MODEL

MPTP (1-methyl-4-phenyl-1,2,5,6-tetrahydropyridine) and particularly its oxidation product, MPP^+ (1-methyl-4-phenylpyridinium) have proven to be effective neurotoxins in several animal species. This neurotoxin helped to establish *in vitro* and *in vivo* experimental models for studying the behavior, pharmacological, electrophysiological, metabolic and toxic mechanisms involved in PD. It has been used extensively in lower animal species, particularly in mice where retinal studies were also performed (Hadjiconstantinou *et al.*, 1985). MPTP not only provokes the pathological and biochemical changes well known in patients with PD but characteristic behavioral defects following MPTP induced parkinsonism also closely resemble those of PD (Burns *et al.*, 1983). In monkeys, 1–2 weeks after MPTP administration, neuropathologic examination demonstrates selective destruction of dopaminergic cells in the substantia nigra pars compacta and their projection to the striatum. At about the same time, the animals show the cardinal symptoms of advanced PD such as tremor, bradykinesia, rigidity, motor freezing, eating problems, delay in initiating movements (Burns *et al.*, 1983; Langston *et al.*, 1984). These abnormalities are observable for a long period in primates following MPTP administration (Burns *et al.*, 1983; Redmond *et al.*, 1986). The cardinal symptoms of the MPTP-treated monkeys improve under the effects of dopaminergic drugs which are commonly used in human PD.

Retinal impairment following MPTP was first described in the rabbit (Wong *et al.*, 1985). In cynomolgous monkeys, using flash stimulation, a decrease of the amplitude of the so-called 'oscillatory potential' of the ERG was observed after MPTP (Harnois *et al.*, 1987). Profoundly abnormal pattern VEPs and PERGs, elicited by grating stimuli, were recorded in cynomolgous monkeys following systematic MPTP administration (Onofrj *et al.*, 1986a; Ghilardi *et al.*, 1988a). However, the electrophysiological abnormalities were evident only for spatial frequencies above 2 cpd. Both the PVEP and PERG to 0.5 cpd stimuli returned to normal 30–40 days after MPTP administration but PVEP latency and amplitude and PERG amplitude to 2.5 and 3.5 cpd stimuli remained abnormal. Post mortem data showed that the level of DA and DOPAC were 6.5 and 21.5 times lower in the retina of the MPTP-treated animals than in normal animals (Ghilardi *et al.*, 1988b). The effects of MPTP on the VEP and PERG were very similar to those induced by intraocular 6-OHDA, suggesting a specific role of dopaminergic neurons in spatial processing in the monkey retina. The visual abnormalities (PERG and VEP) could also be improved by L-Dopa therapy, similarly to the main motor deficits. It is interesting to consider some possible mechanisms which enable levodopa to exert a beneficial effect following MPTP-induced destruction of dopaminergic retinal neurons.

In the retina, as elsewhere in the central nervous system (CNS), DA receptors have been classified as either D1 or D2 types on the basis of different sensitivities to various ligands and for the differences in their respective second messenger systems (Makman *et al.*, 1982, Makman and Dvorkin, 1986). For D1 receptors, cyclic adenosine monophosphatase (cAMP) has been identified as a second messenger whereas D2 receptors are probably negatively linked to cAMP or not linked to it at all. Five types of DA receptors have been revealed by molecular genetic and ligand studies in various vertebrate tissues (Sibley and Monsma, 1992). Second messenger involvement may be similar in groups D1 and D5, whereas D2, D3 and D4 belong to an other major subdivision. Within each subdivision, affinities to the same ligand can be orders of magnitude different (Van Tol *et al.*, 1991). For instance the DA receptor blocker clozapine, more commonly used in psychiatry than in neurology, has an order of magnitude higher affinity for D4 than for D2 receptors. Specific mRNA probes for D2 and D4 receptors revealed non-identical distribution in retinas

(Cohen *et al.*, 1992). At present, little is known about the details of the specific signal transduction pathways for DA receptor subtypes.

There is also only limited information concerning the differential effects of pathology on various subtypes of retinal DA receptors. MPP^+ leads to depletion of retinal DA and induces D1 receptor supersensitivity in mice (Qu *et al.*, 1988). After MPTP administration, an increase of the density of D2 receptors in the striatum and retina was reported in the cynomolgous monkey (Joyce *et al.*, 1985). It is thought that a good response to dopaminergic therapy following either selective dopaminergic neurotoxins or DA depletion (using reserpine or alpha-methylparathyrosine) indicates receptor up-regulation. At present we do not know whether or not D1 and D2 receptors are differentially up-regulated following MPTP treatment. In the monkey, in any case, the beneficial visual effects of levodopa therapy following MPTP, which destroys DA neurons, suggest that post- rather than presynaptic receptors mediate improved spatiotemporal tuning under levodopa therapy. This is important in reference to the visual effects of D2 blocker substances since theoretically they could act via pre- or postsynaptic receptors.

20.3 RETINAL EFFECTS OF DA RECEPTOR LIGANDS IN THE MONKEY

There is clinical/therapeutic interest in pharmaceutical research for developing more selective DA receptor ligands to various subtypes of DA receptors. A dysfunction of D2 receptors plays an important role in the motor impairment of PD; however, some studies suggest that the number of postsynaptic receptors are only mildly affected if at all in idiopathic PD. Activation of D1 and D2 receptors in the CNS results in antagonistic membrane changes at the cellular level, but their parallel activation achieves functional synergism at the behavioral level. In this respect analyzing the role of selective DA ligands and their interactions with other neurotransmitter systems in the retina could provide further insights.

Selective DA receptor ligand studies in lower vertebrates have yielded fundamental information concerning horizontal cell (HC) interactions under the action of dopaminergic agents (for a review see Witkovsky and Schutte, 1991). Stimulation of HCs of the turtle retina in the presence of the D2 agonist LY 171555, produced an opposite effect to that induced by DA (Piccolino *et al.*, 1987) which suggests the opposite functional role of D1 and D2 receptors. In mouse retina, D4 receptor activation leads to a modulation of the light-sensitive cAMP pool in photoreceptors (Cohen *et al.*, 1992). In primates and in humans, the function of the different DA receptors in retinal processing is not known with any precision due to the lack of detailed intracellular studies. However, recent PERG studies (Tagliati *et al.*, 1994) suggest a possible resolution of the paradox that there is antagonism of D1 and D2 receptor-mediated cellular responses whereas D1 and D2 receptor-mediated functions are synergistic at the behavioral level.

20.3.1 HALOPERIDOL

The mixed DA receptor antagonist haloperidol has a concentration-dependent affinity to both D1 and D2 receptors: nanomolar potency for D2 and micromolar potency for D1 receptors (Kebabian and Calne, 1979). In monkeys, it produces akinesia, restlessness and tremor (Bedard *et al.*, 1982). DA receptor blockade of the primate (cynomolgous monkey) retina by low dose of haloperidol (0.1 mg/kg) enhanced the amplitude of both the light and dark-adapted FERGs, but mostly for very low spatial frequencies (Bodis-Wollner *et al.*, 1989).

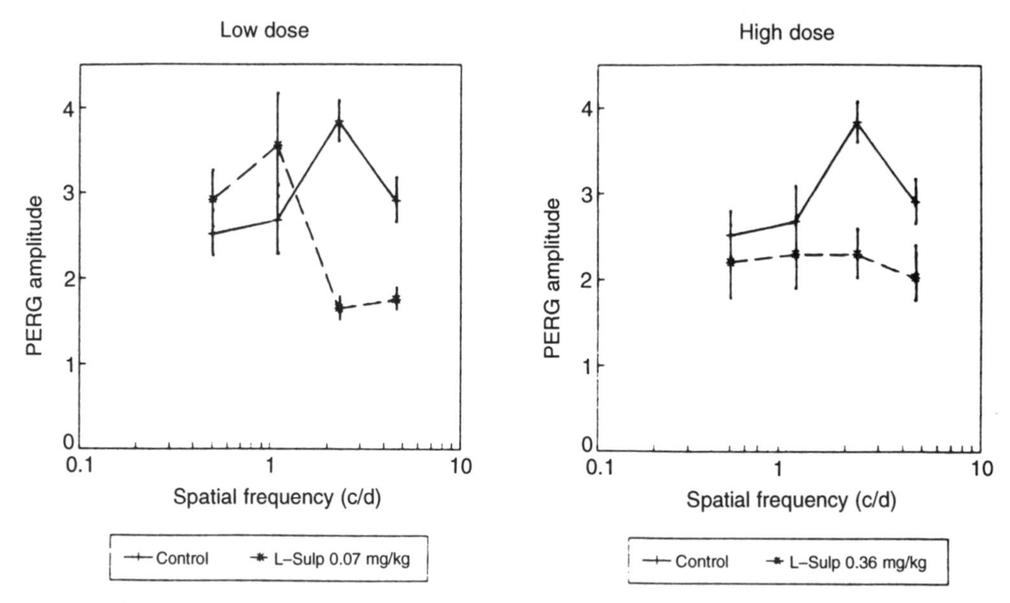

Figure 20.3 The effect of L-sulpiride on the primate ERG. Mean PERG amplitude (μV ± SE) in control conditions (seven experiments) and after the i.m. administration of two different doses (0.07 and 0.35 mg/kg) of L-sulpiride in a monkey (four experiments). (After Tagliati *et al.* (1994).)

20.3.2 SULPIRIDE

The specific D2-receptor blocker, L-sulpiride (0.07–0.35 mg/kg), caused a spatial frequency-dependent decrease of the PERG amplitude in monkeys (Tagliati *et al.*, 1994). Its effect was evident mainly for medium and higher and not for lower frequencies. Tagliati *et al.* (1994) calculated a spatial 'tuning ratio' (amplitudes of the 4.6 cpd and 0.5 cpd responses divided) and report reversal at the tuning ratio following sulpiride administration (Figure 20.3).

20.3.3 CY 208–243.

Preliminary results using a D1 agonist, CY 208–243 (Markstein *et al.*, 1988) in cynomolgous and rhesus monkeys partially mimicked a D2 antagonist action, i.e. the PERG amplitude was significantly attenuated for 2.3 cpd stimuli. This suggested that D1 and D2 receptor-mediated functions are antagonistic and that this D1 agonist had the same effect as the D2 blocker sulpiride (see above). However, CY 208–243 also depressed the PERG amplitude to low spatial frequency SF (0.5 cpd) stimulation in two monkeys, although at the concentration used (1 mg/kg) the change was statistically insignificant for the group of four (Peppe *et al.*, unpublished. Conversely, sulpiride raised low spatial frequency response amplitude in two monkeys, although the group effect was insignificant.

In these experiments, the PERG reflected an antagonistic role for D1 and D2 receptors for middle spatial frequency. For low frequencies the D1 agonist may have an opposite effect. A possible explanation may involve the location of D1 and D2 receptors with different weighting for preganglionic center and surround mechanisms or their location on altogether different neurons.

20.4 VISUAL DYSFUNCTION IN PD

Although, upon questioning, PD patients may complain of some form of visual impairment, the relationship of visual symptoms to the main manifestation of the disease are not known with precision. Visual dysfunction in PD has received increasing attention since it was demonstrated that a high percentage of the patients have abnormal VEPs and impairment of retinal dopaminergic transmission was suggested as one possibility (Bodis-Wollner and Yahr, 1978). This suggestion was consistent with VEP results obtained in DA-depleted rats (Dyer *et al.*, 1981; Onofrj and Bodis-Wollner, 1982). Nevertheless, VEP changes in PD remained controversial until studies began to emerge which critically evaluated the importance of spatial and temporal parameters of stimulation (Bodis-Wollner, 1985; Onofrj *et al.*, 1986b; Tartaglione *et al.*, 1987). The abnormality of VEP is highest for stimuli of medium and high spatial frequency (above 2 cpd) where normal human observers are most sensitive to the visual stimuli (Bodis-Wollner, 1985; Tartaglione *et al.*, 1984; Onofrj *et al.*, 1986b).

20.4.1 CONTRAST SENSITIVITY AND VISUAL ACUITY

Several studies have reported static spatial contrast sensitivity (CS) impairment in patients affected by PD (Kupersmith *et al.*, 1982; Regan and Neima, 1984; Bulens *et al.*, 1986). In addition to static CS, dynamic CS measurements have also been extensively used to detect visual deficits. These studies have established that visual dysfunction in PD is dependent on the spatial and temporal parameters of stimuli. The normal observer needs a minimum contrast for seeing a target of 4 c/d modulated at 8 Hz (Robson, 1966); in PD, contrast detection is degraded for visual stimuli to which the normal observer is most sensitive (Bodis-Wollner *et al.*, 1987; Bodis-Wollner, 1990) (Figure 20.4). However, it has also to be noted that not all CS results fit into a theory of pure retinal dysfunction in PD. It has also been found that in some patients, the loss of contrast sensitivity depends on the orientation of the grating pattern (Bulens *et al.*, 1988, Regan and Maxner, 1987).

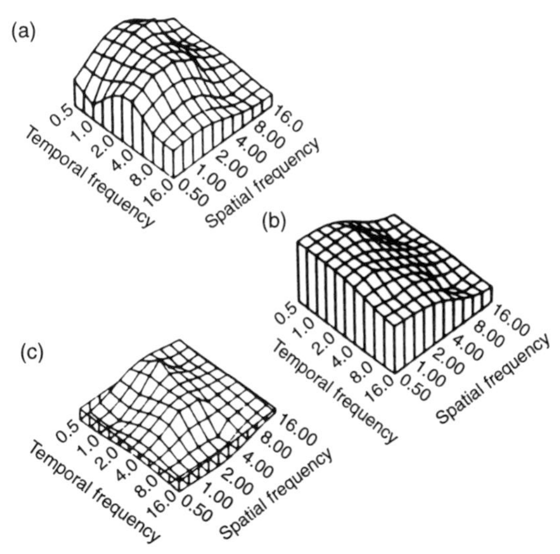

Figure 20.4 Spatiotemporal CS to a 9 degree field is shown for six spatial frequencies and six temporal frequencies in a three-dimensional surface plot for a 46-year-old PD patient who shows the on-off syndrome. Snellen visual acuity was 20/20 and 20/30. (a) Representation of the three-dimensional surface measured when this patient was in the on phase. (b) The surface measured when this patient was in the off phase. This patient serves as his own control by providing data in both on and off periods. (c) The three-dimensional surface resulting when the off-phase responses are subtracted from the on-phase responses. The shape of this surface suggests that DA enhances sensitivity at the peak of the spatiotemporal surface and attenuates sensitivity at the lower spatial frequencies. (From Bodis–Wollner *et al.* (1987).)

Most studies agree that the CS and the PERG are affected in PD without visual acuity changes (Nightingale *et al.*, 1986; Ellis *et al.*, 1987, Gottlob *et al.*, 1987; Jaffe *et al.*, 1987;

Stanzione *et al.*, 1989). However, in one study it was found that visual acuity to high-contrast stimuli is impaired in PD and dopaminergic therapy does not improve the acuity impairment (Jones *et al.*, 1992). This result should argue that visual acuity changes in this group of patients may have been due to other (optical or retinal) factors, rather than retinal DA deficiency. On the whole acuity is only mildly affected in most PD patients without evident optical or ocular problems.

20.4.2 COLOR VISION LOSSES

Studies concerning color vision in PD patients are contradictory. Kupersmith *et al.* (1982) reported normal color vision in PD patients using pseudoisochromatic plates. On the other hand, the utilization of the Farnsworth-Munsell 100 Hue Test (Smith *et al.*, 1985) revealed significant differences between PD patients and age-matched controls (Price *et al.*, 1992). Color vision loss appeared to correlate with the duration of the disease. It has not been established whether or not deficiency in color vision in PD patients reflects abnormalities of dopaminergic neurotransmission in the retina.

20.4.3 THE ELECTRORETINOGRAM IN PD

In the last decade several attempts have been made to assess retinal function in PD using PERG or FERG. The PERG is a spatially (relatively) narrow-band stimulus whereas the FERG is both a spatially and a temporally broad-band. The results of FERG measurement appeared contradictory: normal b-wave amplitudes were found in patients without any kind of therapy, but in patients overdosed with L-Dopa (1000 mg + 100 mg carbidopa) a reduction of FERG amplitude was found (Iudice *et al.*, 1980). Kupersmith *et al.* (1982) studied the oscillatory potentials of the FERG in PD patients maintained on L-Dopa, bromocriptine, and occasionally, anticholinergic drugs. No difference was found between PD patients and normal controls. In another study, significantly larger b-wave amplitudes were found in PD patients in comparison with normal subjects (Krejkova *et al.*, 1985). By contrast, a diminished amplitude of the b-wave was recorded for FERG in PD patients maintained on levodopa, bromociptine or anticholinergics (Nightingale *et al.*, 1986). Gottlob *et al.* (1987) also found reduced a- and b-wave amplitudes in PD patients, but they were independent of the DA medication. Similarly, using diffuse light flash stimulation, either a decrease in amplitude (Jaffe *et al.*, 1987), or an increase in b-wave latency (Ellis *et al.*, 1987; Jaffe *et al.*, 1987; Ellis and Ikeda, 1988) have been reported in PD patients. Ellis *et al.* (1987) recorded a significant b-wave latency decrease which was also found by Jaffe *et al.* (1987) who reported a- and b-wave amplitude increase after levodopa administration with the dosage predetermined individually for each patient. In contrast, Murayama *et al.* (1989) found that the FERG was not significantly altered during or after L-Dopa therapy (100 mg L-Dopa + 25 mg benserazide), although PVEP and PERG latencies were prolonged when the treatment was withheld.

It would thus appear that FERG results are not consistent across diverse groups of PD patients. The FERG is a mass retinal response (reflecting mostly the photoreceptor, Müller cell and possibly weak ON bipolar cell activity) to which DA-containing cells would only contribute in a modulatory manner. The FERG may not, therefore, be a suitable method for assessing possible retinal abnormalities in PD patients. It must be admitted however, that if DA neurons of the primate retina provide input into HCs which in turn adjust receptor sensitivity by feedback, then the 'a' -wave of the FERG could also reflect DA deficiency. One reason why this may not occur is because the contribution to the FERG of foveal/parafoveal receptors affected by DA deficiency may be small compared to the massive contributions of peripheral receptors. Peripheral receptors, having little to do with contrast vision, may not be much affected by dopaminergic mechanisms of the retina. On the other hand the input of all foveal/parafoveal, preganglionic retinal neurons could theoretically be evaluated by registering the PERG, since ganglion cell responses would be subject to all possible preganglionic changes. In order to reveal dopaminergic visual abnormalities, therefore, it was suggested to use patterned stimuli, which would evoke a response reflecting the activity of the cells of the inner retina, predominantly ganglion cells (Maffei and Fiorentini, 1981, 1986). In PD patients, the amplitude of the PERG b-wave was found to be decreased by about 25% for reversal stimulation with checkerboard pattern (with 25′ and 50′ checks), whereas no effect on PERG latency was noticed (Nightingale *et al.*, 1986; Gottlob *et al.*, 1987). In these studies, the patients were under levodopa therapy (200 mg) and the contrast of the stimuli was high (97%). Gottlob *et al.* (1989) also tested the effect of L-Dopa on PERG and PVEP of normal subjects using different check sizes and luminance levels. After L-Dopa administration (200 mg) the mean PERG latencies decreased but only for the lower luminance levels (0.05–2 cd/m^2). Check size had no effect in this study. Hess and Baker (1984) found that the amplitude of the steady-state PERG increased linearly with contrast in normal subjects. In another study, PERG latency increased as the contrast decreased both in normal subjects and PD patients, but it was more pronounced in patients without therapy (Peppe *et al.*, 1992). Dopaminergic therapy (275–750 mg/day) induced PERG latency recovery in PD patients. These results suggested that in PD patient visual abnormalities may become more evident with low contrast stimuli.

The results of different PERG studies are less contradictory than the results of FERG studies in PD, although not entirely consistent. One reason may be due to stimulus differences. As shown in Figure 20.1, both contrast sensitivity and PERG measures reveal spatial frequency-dependent visual effects of DA deficiency. A checkerboard is a complex stimulus and activates neurons with various receptive field sizes (Tartaglione *et al.*, 1984). Hence it is difficult to demonstrate a spatial frequency tuned defect using checkerboard stimuli. Using grating pattern reversal stimulation, an increase of the latency of PERG b-wave was reported in PD patients coming off therapy (Stanzione *et al.*, 1989). The increase was larger for the 2 cpd stimulus than for the 0.75 cpd stimulus. Similar results were obtained for the pattern-evoked VEP. These spatial frequency selective abnormalities were reversed by L-Dopa therapy (375–750 mg/day). Furthermore, CS and VEP studies in monkey and man (e.g. Marx *et al.*, 1986) suggest that stimulation with a temporal frequency range of 4–8 Hz is optimal to elicit any abnormal responses in the DA-deficient visual system.

20.4.4 MORPHOLOGY AND HISTOLOGY OF THE RETINA IN PD

Dopaminergic systems of vertebrate retinas generally have already been described in detail (Chapter 15). As regards the human retina, two types of dopaminergic cells have been found: ACs and IPCs. The ACs project mainly to the innerplexiform layer, whereas the IPCs transmit signals from the inner to outer plexiform layer. Dopaminergic ACs connect with bipolar cells and with other ACs with different neurotransmitters (GABA, acetylcholine, etc.). The perikarya of the IPCs are located among the amacrines and they have two parallel dendritic trees: one in the inner plexiform layer, where they receive input from amacrines, the other in the outer plexiform layer, where the dendritic tree forms a presynaptic feedback pathway, mainly onto HCs (Nguyen-Legros, 1988). The detailed function of DA in preganglionic circuits is known mainly for lower vertebrates. In lower vertebrates DA may enhance excitatory amino acid gated conductance in HCs (Knapp and Dowling, 1987) and synaptic plasticity (Weiler *et al.*, 1988). In mammalian retina, AII amacrine cells are connected to each other and to ON-center bipolar cells by gap junctions. Input to AII cells from DA amacrines uncouples these junctions via D1 receptor activation (Hampson *et al.*, 1992). Although it is tempting to speculate how such a pathway could contribute to visual changes observed in PD patients (section 20.6), knowledge about functional details of D1 receptor involvement in regulation of amacrine cell gap junctions in primates remains to be elucidated. It is still a matter of conjecture, therefore, how dopaminergic deficiency in the primate affects the preganglionic organization, the interaction with other neurotransmitter systems and hence the spatiotemporal profile of the final common output (ganglion cell responses).

DA and its metabolites in the retina can be demonstrated with autoradiographic and histochemical methods. However, evaluation of neurotransmitter levels in post mortem human tissue is difficult because profound changes occur after death and there are also historical differences among patients (treatment, etc.). Using tyrosine hydroxylase (TH) immunohistochemistry the decrease of DA innervation was observed in the central retina in post mortem of PD patients, although the cell density did not change (Nguyen-Legros *et al.*, 1988). In a later study, a decrease of retinal DA content as well as morphological alterations of the dendritic pattern of the dopaminergic neurons was reported (Nguyen-Legros *et al.*, 1993). The number of heavily labeled TH immunoreacture cells was significantly decreased when it was counted in the total retina (mainly in the superior and peripheral retina), but this decrease was not dramatic. In 50% of the cases the decreased DA level was related to a significant loss of cells with TH immunoreactivity. In another post mortem study, using high-performance liquid chromatography methods patients without levodopa therapy showed lower DA content compared to normal (Harnois and Di Paolo, 1990). The retinal DA level in patients who were on levodopa therapy, was similar to that of controls. However, a wide variety of the retinal levels of DOPAC and homovallinic acid (HVA) was observed in both groups (treated patients and controls).

20.5 THE EFFECTS OF DA RECEPTOR LIGANDS IN THE NORMAL HUMAN RETINA

Although new subclasses of DA receptors have been identified, the distinction of D1 and D2 receptors remains the major one. Presynaptic D2 receptors in the striatum have been found to regulate the release of DA from dopaminergic neurons (Starke *et al.*, 1989). In human retinas, mRNA for a cloned D2 receptor was localized to the inner nuclear layer (Stormann *et al.*, 1990) and it was suggested that these D2 receptors may be autoreceptors inhibiting retinal DA synthesis and release.

Haloperidol is a mixed D1 and D2 receptor antagonist. As such, in some circumstances it imitates the effects of dopaminergic impairment. Acute administration of haloperidol (10 mg) prolonged the duration of the b-wave component of the FERG, whereas it had no effect for the PERG and VEPs (Bartel *et al.*, 1990). In another study, a significant increase of PERG b-wave latency was found following haloperidol (0.05 mg/kg) administration (Stanzione *et al.*, 1991). The simultaneously recorded VEP was not significantly modified by the drug, suggesting that retinal visual mechanisms are more sensitive to the pharmacological impairment of dopaminergic transmission, due to mixed DA receptor blockade, than cortical ones. However, data obtained with haloperidol must be interpreted cautiously, because of its possible coupler effects on pre- and postsynaptic D2 receptors in addition to possible effects on D1 and other DA receptors.

To determine the functional role of D2 receptors in the human retina, sulpiride (a D2 receptor antagonist) was administered to normal subjects and its effect was contrasted to haloperidol. Sulpiride (100 mg) reduced the steady state (7.5 Hz) and delayed transient (1 Hz) PERG using 1 cpd square-wave pattern stimulation (Stanzione *et al.*, 1992), suggesting that D2 receptors may also be involved in neurons with larger receptive fields in the human retina.

20.6 PHARMACO-FUNCTIONAL CONSIDERATIONS

It is thought that modifications of the receptive fields of HCs predominantly influence the diameter of the surround mechanism in bipolar cells (Nguyen-Legros, 1988). If the receptive field is narrower, the contrast signal of the cell becomes higher and vice versa. Two major neurotransmitters act antagonistically on HCs: DA narrows and GABA expands the receptive field in the turtle retina (Piccolino *et al.*, 1985). Dopaminergic IPCs act directly on HCs, whereas GABA-ergic interplexiforms can act directly or indirectly by inhibiting dopaminergic IPCs. In goldfish, DA blocks the release of GABA (Yazulla and Kleinschmidt, 1982). In this species the receptive field of HCs are larger in the absence of DA (Chapter 8).

There is strong evidence that light-induced stimulation of DA turnover occurs in the retina of several species (Bauer *et al.*, 1980, Reading, 1983; Hamasaki *et al.*, 1986; Lin and Yazulla, 1994). Whereas the onset of darkness decreases the overflow of DA, it remains at a low level if darkness is prolonged (Weiler *et al.*, 1989). Light onset can increase the metabolism or it can increase the release of DA in the retina (Ehinger, 1983). Retinas of humans who died during the night were found to have lower DA concentration than those who died in the daytime (Di Paolo *et al.*, 1987). Exposure of rabbits to constant light causes a decrease of D2 DA receptor sites (Dubocovich, 1983; Dubocovich and Weiner, 1985). This synapse regulation appears to be associated with the lack of the activation of inhibitory melatonin receptors on DA receptors because these effects can be reversed by melatonin administration or darkness. Dubocovich (1983) suggested that if the melatonin level increases, the decrease of the sensitivity of presynaptic melatonin receptors leads to an increase in DA release and to D2 autoreceptor down-regulation. Not only light and melatonin, but other neurotransmitters, such as acetycholine (Gerschenfeld *et al.*, 1982), and neuromodulators, such as vaso-intestinal peptide (VIP) or enkephalin (Weiler and Ball, 1989), have an effect on retinal DA content and release. Unfortunately, little is known about the action of these substances in higher vertebrates. It may be relevant to note, however, that the reduction in contrast sensitivity in PD exhibits a diurnal rhythm (Struck *et al.*, 1990). PD patients performed as well as controls early in the morning, but not later in the day. Enhanced DA release could augment stimulus contrast by narrowing the surround receptive fields. However, the more intense the light the less the contrast, and DA release depends on the intensity of the light (Ehinger, 1983). Hence, one would expect minimal effects of exogenous levodopa in normal volunteers.

20.7 DO ALL VISUAL PERCEPTUAL CHANGES IN PD RESULT FROM RETINAL IMPAIRMENT?

One technique of evaluating visual impairment beyond retinal processing uses VEP measurements. In fact, the earliest description of visual deficits in PD were provided by VEP studies (Bodis-Wollner and Yahr, 1978). However, although increased VEP latency and contrast sensitivity losses may be caused by DA depletion in the brain, abnormal ERGs evoked by either flash or pattern visual stimuli must have a retinal cause. There is similar spatial frequency dependent CS and PERG change in PD (Figure 20.1). Furthermore, it is a fact that in some patients the VEP is asymmetrically affected in the two eyes (Bodis-Wollner and Yahr, 1978). These considerations suggest that simple visual changes in PD and DA deficiency are predominantly caused by retinal pathology. Are VEP changes in the patients the direct consequences of ERG changes? In other words, do VEP abnormalities passively reflect retinal abnormality? Nightingale *et al.* (1986) found a significant relationship between the VEP latency and PERG amplitude in PD patients. For those subjects who demonstrated an interocular difference in simultaneous VEP and PERG recordings, the abnormality was commonly found in the same eye. In another study using vertical grating pattern stimuli, an increase of both VEP and PERG latencies were documented suggesting a retinal contribution to the VEP (Stanzione *et al.*, 1989). On the other hand, Calzetti *et al.* (1990) did not find significant interdependence between VEP abnormalities and PERG changes in PD patients with a mean duration of the disease of only 10 months. However, for proper evaluation of these results the duration of the disease and the age of the patients must be taken into account. The latency of VEP increases with age (Celesia and Daly, 1977) and it is conceivable that aging differently affects dopaminergic circuits of the retina and the cortex (Porciatti *et al.*, 1992). Some studies reported orientation-dependent VEP changes in PD. Another argument concerning the independence of PERG and VEP changes in PD emanates from studies of patients with surgical implants. Improved VEP was reported in PD patients after implantation of human fetal ventral mesencephalon to the right caudate nucleus of the brain (Henderson *et al.*, 1992), whereas PERG amplitude and latency improvements were (as expected) not significant. All these data suggest that, whereas retinal visual impairment affects the cortex in PD, additional postretinal mechanisms must be taken into account.

20.8 CONCLUSIONS AND FUTURE PERSPECTIVES

One significant conclusion is that PD affects retinal DA and by demonstrating retinal involvement, current concepts of PD as an exclusive motor system disorder have to be revised. A general relevance of finding retinal pathology in this disease is that PD itself may be regarded as a model of a distributed neurodegenerative disease. In the case of PD, the vulnerable subsystems are distinguished by being dopaminergic. However, there is another, more abstract interpretation of these data, which concerns the logic function that retinal DA circuits with antagonistic and diverse DA receptors may perform. To discuss this function, it may be useful to summarize some of the results as they pertain to the role of DA in spatiotemporal processing in the retina.

In DA deficiency, as well as in D2 receptor blockade, the response to peak spatial frequency patterns is affected most. This could be the result of attenuation of retinal preganglionic signals involved in the 'center' or the 'surround' receptive field mechanisms, or it may represent a degradation of the interactive processes of center and surround signals. Conversely, and by either mechanism, the action of DA may result in enhanced center dominated responses. This could be achieved by either lateral and/or by feedback inhibition so that the peak of the center response and the trough of the inhibitory surround response (or vice versa) are coincident in time. Pathology distorting the optimal temporal relationship could result in attenuated peak response. In *Limulus* for instance, elegant calculations (Ratliff *et al.*, 1970) have shown how a given artificial delay alters the spatial transfer function. Using the concepts of Ratliff *et al.* (1970), one would argue that in PD there is a change of the time constant of a feedback inhibitory network. Depending on the delay, in addition to attenuated peak responses, secondary peak responses may occur at the low frequency end. Our data, concerning the similarity of D_2 receptor blockade and MPTP-induced cell destruction, argue that D_2 postsynaptic receptors mediate the effect of DA on spatial tuning. If we accept results of studies from lower vertebrates as applicable to the primate retina, the following scheme could be considered. HCs, provide dominant, if not exclusive, input to the surround organization. HCs may be coupled into larger fields in the absence of DA, or in the presence of D_2 receptor blockade. For peak response (tuning) the timing of center–surround interactions is crucial. There may be two reasons for a low-pass function resulting from insufficiency of retinal DA. One is due to delayed or dispersed DA release (especially in the paracrine mode of operation; Bodis-Wollner, 1990) due to pathology of proximal DA neurons. A second reason may be the result of insufficient DA release. This would result in consequent signal attenuation and dispersion in the surround (HC) organization. For both of these reasons, phase shifts may occur in the dopaminergic feedback loop. These delays may result in attenuated peak response and perhaps secondary low spatial frequency maxima. Contrast signals to lower spatial frequency stimuli will be less attenuated due to signal spread in large HC receptive fields. Whether or not this skeleton scheme (derived from analytical data in *Limulus* and experimental data in lower vertebrates) constructed to explain the loss of spatiotemporal tuning in DA deficiency, applies to the primate retina, is unknown at present.

The role of D1 receptors is less well explored in the primate retina, but may be less complicated. To fit the available scarce data into the above scheme, we would propose that D1 receptor activation results in attenuating the contrast signal to low spatial frequency stimuli by the uncoupling action of postsynaptic D1 receptors on HCs. At the same time, the sensitivity of the center mechanism may also be reduced (if only sightly) due to the spatially less dispersed signal of HC feedback onto photo receptors, crucial for the center mechanism of foveal ganglion cell receptive fields. Hence, the net simultaneous effect of D1 and D2 receptor activation will be the attenuation of low- and augmentation of peak-spatial frequency responses and augmented tuning (Figure 20.5).

Although the proposed scheme is undoubtedly highly speculative, it could account for the behavioral observation that D1 and D2 receptors act antagonistically at the cell membrane level whereas DA itself (acting upon D1 and D2 receptors) has a synergistic therapeutic effect in PD and in various monkey models. Selective DA receptor ligand studies in humans suggested that simultaneous activation of the two types of receptors is needed for a beneficial effect on the motor impairment of PD (Barone *et al.*, 1986). In this sense, our retinal scheme suggesting the importance of the dynamics of lateral and feedback inhibition of local neuronal circuits may be considered as a 'formal' model for evaluating complex interactions in other CNS structures, in particular in the striatonigral–striatal loop, known to be involved in PD. Further advances in retinal pharmacotherapy may well take advantage of a retinal model where signal transduction and physiological significance of various DA

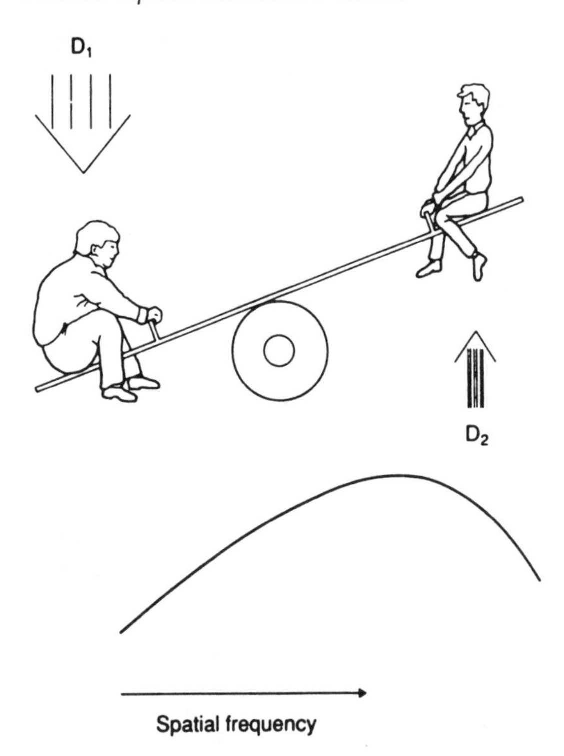

Figure 20.5 A sketch symbolizing the antagonistic effects of D1 and D2 receptor activation acting on two different arms of the see-saw. As a consequence, the doubly opposite effects produce an overall synergistic action. The space underneath the curve represents the overall spatial frequency transfer function of the retina: low frequency decline occurs where D1 receptors are active. The peak of the curve is created by the see-saw pointing to the right, where D2 activity is high. (From Bodis-Wollner *et al.* (1993).)

receptor subtypes can be studied in a realistic network *in vivo*. We could envisage the use of appropriately timed combination therapy with agents selective for specific functions of the neural circuits involving DA. Lastly, we should mention the importance of vision to coordinated movements and mobility. Future studies may address the question of how a retinally disturbed spatiotemporal transfer function affects motor programming in the basal ganglia.

APPENDIX: GLOSSARY OF TERMS

CONTRAST SENSITIVITY (CS)

Rather than their absolute luminance value, the detectability of visual stimuli is determined by their visibility against their background. This is quantified by their contrast value. One commonly used contrast measure is based on the Michelson formula:

$$C = L_{max} - L_{min}/L_{max} + L_{min}$$

where L is luminance, L_{max} is the highest luminance value and L_{min} is the minimum. It can be easily shown that when the surround of a target is absolutely black (hence its luminance value is 0) then, irrespective of the luminance of the target, C equals 1 (100% contrast). On the other hand, when the target and background are equiluminant, the contrast value is 0. The contrast threshold is the contrast needed to detect a given pattern. For example, a contrast threshold of 1 means that the observer needs 100% contrast to detect a pattern. On the other hand, a more sensitive observer may have a contrast threshold of 0.1, needing only 10% contrast to detect the pattern. In order to express contrast sensitivity with an integer number, the inverse of the experimentally established contrast threshold value is used. Hence, a contrast threshold of 1 becomes 1/1 or 1. In several neurodegenerative diseases, contrast sensitivity is affected without a change in visual acuity.

ELECTRORETINOGRAM (ERG)

Recorded with a corneal electrode, the ERG reflects the light-evoked mass electrical response, and consists of a number of components. The two major ones are labelled the 'a' and 'b' waves. Using 'patterned' (e.g. grating) stimuli, the resulting ERG (known as a PERG) mostly reflects electrical activity of the inner (ganglion cell) layer of the retina, whereas the FERG ('flicker' ERG) predominantly reflects receptor and bipolar cell responses.

RECEPTIVE FIELD (RF)

This is the restricted area of the complete visual field subserving a visual neuron. Like other sensory neurons, most visual neurons have spatially antagonistic RFs; in other words, the response to two adjacent visual stimuli may subtract from each other rather than add. Some neurons have more complicated RF organization with the property of responding well to motion. Neurons responding to the direction of movement do not exist in the primate precortical visual pathway.

SPATIAL FREQUENCY (SF)

The visual stimulus commonly used in physiological studies is an alternating bar pattern, where 'white' and adjacent 'black' bars are of equal width. They are together considered a 'cycle'. The number of cycles subtended in 1 degree of visual angle is called the spatial frequency. The higher the spatial frequency number, the finer the pattern. Primates are most sensitive (see 'contrast sensitivity' above) to spatial frequencies between 4 and 6 cycles/degree (c/d). In angular subtense, a bar of a 6 c/d pattern is 5′ wide on the retina.

VISUAL ACUITY

All different acuity measures (Snellen, Jaeger, Landolt, etc.) rely on the concept of visual resolution. Usually visual acuity is quantified as the transformed measure of the smallest detectable image at 100% contrast.

REFERENCES

Barone, B., Davis, T.A., Braun, A.R. and Chase, T.N. (1986) Dopaminergic mechanisms and motor functions: characterization of D1 and D2 dopamine receptor interactions. *European Journal of Pharmacology*, **123**, 109–14.

Bartel, P., Blom, M., Robinson, E. *et al.* (1990) The effect of levodopa and haloperidol on flash and pattern ERGs and VEPs in normal humans. *Documenta Ophthalmologica*, **76**, 55–64.

Bauer, B., Ehinger, B. and Aberg, L. (1980) ^{3}H-dopamine release from the rabbit retina. *Albrecht von Graefes Archives Klinische und Experimentelle Ophthalmolologie*, **215**, 71–8.

Bedard, P.J., Boucher, R. and Larochelle, L. (1982) Experimental tardive dyskinesia. *Progress in Neuropsychopharmacology and Biological Psychiatry*, **6**, 551–4.

Bobak, P., Bodis-Wollner, I., Harnois, C. *et al.* (1983) Pattern electroretinograms and visual evoked potentials in glaucoma and multiple sclerosis. *American Journal of Ophthalmology*, **96**, 72–83.

Bodis-Wollner, I. (1985) Pattern evoked potential changes in Parkinson's disease are stimulus-dependent. *Neurology*, **35**, 1675–6.

Bodis-Wollner, I. (1990) Visual deficits related to dopamine deficiency in experimental animals and Parkinson's disease patients. *Trends in Neuroscience*, **13**, 296–302.

Bodis-Wollner, I. and Yahr, M.D. (1978) Measurements of visual evoked potentials in Parkinson's disease. *Brain*, **101**, 661–71.

Bodis-Wollner, I, Marx, M.S., Mitra, S. *et al.* (1987) Visual dysfunction in Parkinson's disease – loss in spatiotemporal contrast sensitivity. *Brain*, **110**, 1675–98.

Bodis-Wollner, I., Marx, M. and Ghilardi, M.F. (1989) Systemic haloperidol administration increases the amplitude of the light-and dark-adapted flash ERG in the monkey. *Clinical Vision Science*, **4**, 19–26.

Bodis-Wollner, I., Pang, S., Falk, A. *et al.* (1991) Vision and visual discrimination in Parkinson's disease, in *The Basal Ganglia III* (eds G. Bernardi, M.B. Carpenter, G. Di Chiara *et al.*), Plenum, New York, pp. 743–50.

Bodis-Wollner, I., Tagliati, M., Peppe A. and Antal, A. (1993) Visual and visual perceptual disorders in neurodegenerative diseases. *Bailliere's Clinical Neurology*, **2**, 461–91.

Brannan, J.R., Bodis-Wollner, I. and Storch, R.L. (1992) Evidence for two distinct nonlinear components in the human pattern ERG. *Vision Research*, **32**, 11–17.

Bulens C., Meerwaldt, J.D., van der Wildt, G.J. and Keemink, C. (1986) Contrast sensitivity in Parkinson's Disease. *Neurology*, **36**, 1121–5.

Bulens, C., Meerwaldt, J.D. and van der Wildt, G.J. (1988) Effect of stimulus orientation on contrast sensitivity in Parkinson's disease. *Neurology*, **38**, 76–81.

Burns, R.S., Chiuech, C.C., Markey, S. *et al.* (1983) A primate model of Parkinson's disease: selective destruction of substantia nigra pars compacta dopaminergic neurons by *N*-methyl-4-phenyl-1, 2, 3, 6-tetrahydropyridine. *Proceeding of the National Academy of Sciences in the USA*, **80**, 4546–50.

Calzetti, S., Franchi, A., Taratufolo, G. and Groppi, E. (1990) Simultaneous VEP and PERG investigations in early Parkinson's disease. *Journal of Neurology, Neurosurgery and Psychiatry*, **53**, 114–17.

Celesia, G.A. and Daly, R.F. (1977) Effects of aging on visual evoked response. *Archives of Neurology*, **34**, 403–7.

Cohen, A.I., Todd, R.D., Harmon, S. and O'Malley, K.L. (1992) Photoreceptors of mouse retinas possess D4 receptors coupled to adenylate cyclase. *Proceedings of the National Academy of Sciences in the USA*, **89**, 12093–7.

Di Paolo, T., Harnois, C. and Daigle, M. (1987) Assay of dopamine and its metabolites in human and rat retina. *Neuroscience Letters*, **74**, 250–4.

Djamgoz, M.B. and Wagner, H.J. (1992) Localization and function of dopamine in the adult vertebrate retina. *Neurochemistry International*, **20**, 139–91.

Dowling, J.E. (1991) Retinal neuromodulation: the role of dopamine. *Vision Neuroscience*, **7**, 87–97.

Dubocovich, M.L. (1983) Melatonin is a potential modulator of dopamine release in the retina. *Nature*, **306**, 782–4.

Dubocovich, M. and Weiner, N. (1985) Pharmacological differences between D2 autoreceptor and D1 dopamine receptor in the rabbit retina. *Journal of Pharmacology and Experimental Therapy*, **233**, 747–54.

Dyer, R.S., Howell, W.E. and MacPhail, R.C. (1981) Dopamine depletion slows retinal transmission. *Experimental Neurology*, **71**, 326–40.

Ehinger, B. (1983) Functional role of dopamine in the retina. *Progress in Retinal Research*, **2**, 213–32.

Ehinger, B. and Floren, I. (1979) Absence of indoleamine-accumulating neurons in the retina of humans and cynomologous monkeys. *Albrecht von Graefes Archives Klinische und Experimentelle Ophthalmolologie*, **209**, 145.

Ellis, C.J.K. and Ikeda, H. (1988) Evidence for retinal dopamine deficiency in Parkinson's disease, in *Dopaminergic Mechanism in Vision* (eds I. Bodis-Wollner and M. Piccolino), Alan R. Liss, New York, pp.239–51.

Ellis, C.J.K., Allen, T.G.L., Marsden, C.D. and Ikeda, H. (1987) Electroretinographic abnormalities in idiopathic Parkinson's disease and the effect of levodopa administration. *Clinical Vision Science*, **1**, 347–55.

Gerschenfeld, H.M., Neyton, J., Piccolino, M. and Witkowsky, P. (1982) L-horizontal cells of the turtle: network organization and coupling modulation, In *Biomedical Research* (*Suppl*) (eds A. Kaneko, N. Tsukahara and K. Uchnizono), Biomedical Research Federation, Tokyo, pp.21–32.

Ghilardi, M.F., Bodis-Wollner, I., Onofrj, M.C. *et al.* (1988a) Spatial frequency-dependent abnormalities of pattern electroretinogram and visual evoked potentials in a parkinsonian monkey model. *Brain*, **111**, 131–49.

Ghilardi, M.F., Chung, E., Bodis-Wollner, I. *et al.* (1988b) Systemic 1-methyl, 4-phenyl, 1–2–3–6-tetrahydropyridine (MPTP) administration decreases retinal dopamine content in primates. *Life Sciences*, **43**, 255–62.

Ghilardi, M.F., Marx, M.S., Bodis-Wollner, I. *et al.* (1989) The effect of intraocular 6-hydroxydopamine on retinal processing of primates. *Annals of Neurology*, **25**, 359–64.

Gottlob, I., Schneider, E., Heider, W. and Skrandies, W. (1987) Alteration of visual evoked potentials and electroretinograms in Parkinson's disease. *Electroencephalography and Clinical Neurophysiology*, **66**, 349–57.

Gottlob, I., Ewghaut, H., Vass, C. and Auff, E. (1989) Effect of levodopa on the human pattern electroretinogram and pattern visual evoked potentials. *Albrecht von Graefes Archives Klinische und Experimentelle Ophthalmolologie*, **227**, 421–7.

Hadjiconstantinou, M., Cavalla, D., Anthoupoulou, E. *et al.* (1985) *N*-methyl-4-phenyl-1,2,3,6-

tetrahydropyridine increases acetylcholine and decreases dopamine in mouse striatum: both responses are blocked by anticholinergic drugs. *Journal of Neurochemistry*, **45**, 1957–9.

Hadjiconstantinou, M., Qu, Z.X. and Neff, N.H. (1991) Differential changes of retina dopamine binding sites and adenylyl cyclase responses following 6-hydroxydopamine treatment. *Brain Research*, **538**, 193–5.

Hamasaki, D.I., Trattler, W.B. and Hajek, A.S. (1986) Light ON depresses and light OFF enhances the release of dopamine from the cat's retina. *Neuroscience Letters*, **68**, 112–16.

Hampson, E.C.G.M., Vaney, D.I., Weiler, R. (1992) Dopaminergic modulation of gap junction permeability between amacrine cells in mammalian retina. *Journal of Neuroscience*, **12**, 4911–22.

Hankins, M.W. and Ikeda, H. (1991) The role of dopaminergic pathways at the outer plexiform layer of the mammalian retina. *Clinical Vision Sciences*, **6**, 87–94.

Harnois, C. and Di Paolo, T. (1990) Decreased dopamine in the retinas of patients with Parkinson's disease. *Investigative Ophthalmology and Vision Science*, **31**, 2473–5.

Harnois, C., Marcotte, G. and Bedard, P.J. (1987) Alteration of monkey retinal oscillatory potentials after MPTP injection. *Documenta Ophthalmologica*, **67**, 363–9.

Henderson, B., Good, P.A., Hitchcock, E.R. *et al.* (1992) Visual evoked cortical responses and electroretinograms following implantation of human fetal mesencephalon to the right caudate nucleus in Parkinson's disease. *Journal of Neurological Science*, **107**, 183–90.

Hess, R.F. and Baker, C.L.Jr (1984) Human pattern-evoked electroretinogram. *Journal of Neurophysiology*, **51**, 939–51.

Hess, R.F. and Baker, C.L., Zrenner, E. and Schwarzer, J. (1986) Differences between electroretinograms of cat and primate. *Journal of Neurophysiology*, **56**, 747–768.

Ikeda, H., Robbins, J. and Wakakauwa, K. (1987) Evidence for dopaminergic innervation on kitten retinal ganglion cells. *Brain Research*, **432**, 83–9.

Iudice, A., Virgili, P. and Muratorio A. (1980) The electroretinogram in Parkinson's disease. *Research Communications in Psychology, Psychiatry and Behavior*, **5**, 283–9.

Jaffe, M.J., Bruno, G., Campbell, G. *et al.* (1987) Ganzfeld electroretinographic findings in Parkinsonism: untreated patients and the effect of levodopa intravenous infusion. *Journal of Neurology, Neurosurgery and Psychiatry*, **50**, 847–52.

Jones, R.D., Donaldson, I.M. and Timmings, P.L. (1992) Impairment of high-contrast visual acuity in Parkinson's disease. *Movement Disorders*, **7**, 232–8.

Joyce, J.N., Marshall, J.F., Bankievicz K.S. *et al.* (1985) Hemiparkinsonism in a monkey after unilateral internal carotid artery infusion of 1-methyl-4-phenyl-1, 2, 3, 6-tetrahydropyridine (MPTP) is associated with regional ipsilateral changes in striatal dopamine D2 receptor density. *Brain Research*, **382**, 360–73.

Kebabian, J.W. and Clane, D.B. (1979) Multiple receptors for dopamine. *Nature*, **277**, 93–6.

Kozlowski, M.R. and Marshall, J.F. (1981) Plasticity of neostriatal metabolic activity and behavioral recovery from nigrostrial injury. *Experimental Neurology*, **74**, 318–23.

Knapp, A.G. and Dowling, J.E. (1987) Dopamine enhance excitatory amino acid-gated conductances in cultured retinal horizontal cells. *Nature*, **325**, 437–8.

Krejkova, H., Jerabek, J., Filipova, M. *et al.* (1985) Vestibulo-ocular and ERG changes in Parkinson patients. *Journal of Neurology Supplement*, **232**, 130.

Kupersmith, M.J., Shakin, E., Siegel, I.M. and Lieberman, A. (1982) Visual system abnormalities in patients with Parkinson's disease. *Archives of Neurology*, **39**, 284–6.

Langston, J.W., Forno, L.S., Robert, C.S. and Irwin, I. (1984) Selective nigral toxicity after systematic administration of 1-methyl-4-phenyl-1, 2, 3, 6-tetrahydropyridine (MPTP) in the squirrel monkey. *Brain Research*, **292**, 390–4.

Lin, Z. and Yazulla, S. (1994) Depletion of retinal dopamine increases brightness perception in the goldfish. *Visual Neuroscience*, **11**, 683–93.

Maffei, L. and Fiorentini, A. (1981) Electroretinographic responses to alternating gratings before and after section of the optic nerve. *Science*, **211**, 953–5.

Maffei, L. and Fiorentini, A. (1986) Generator sources of the pattern ERG in man and animals, in *Evoked Potentials* (eds I. Bodis-Wollner and J. Cracco), Alan R. Liss, New York, pp.101–16.

Maffei, L. and Fiorentini, A. (1990) Pattern visual evoked potentials and electroretinograms in man and animals, in *Visual Evoked Potentials* (ed. J.E. Desmedt), Elsevier, Amsterdam, pp.25–33.

Makman, M.H. and Dvorkin, B. (1986) Binding sites for [^{3}H] SCH 23390 in retina: properties and possible relationship to dopamine d1-receptors mediating stimulation of adenylate cyclase. *Brain Research*, **387**, 261–70.

Makman, M.H., Dvorkin, B. and Klein, P. (1982) Sodium ion modulates D2 receptor characteristics of dopamine agonist and antagonist binding sites in striatum and retina. *Proceedings of the National Academy of Sciences in the USA*, **79**, 4212–16.

Markstein, R., Seiler, M.P., Vigouret, J.M. *et al.* (1988) Pharmacological properties of CY 208–243, a novel D1 agonist, in *Progress in Catecholamine Research*, Part B: Central Aspects. Alan R. Liss, New York, pp.59–64.

Marshall, J.F. (1979) Somatosensory inattention after dopamine-depleting intracerebral 6-OHDA injections: spontaneous recovery and pharmacological control. *Brain Research*, **177**, 311–24.

Marx, M., Bodis-Wollner, I., Bobak, P. *et al.* (1986) Temporal frequency dependent VEP changes in Parkinson's disease. *Vision Research*, **26**, 185–93.

Murayama, K., Adachi–Usami, E. and Yamada, Y. (1989) Effects of L-Dopa on electroretinographic responses and pattern visual evoked potentials in juvenile Parkinsonism. *Clinical Vision Science*, **4**, 265–8.

Nesher, R. and Trick, G.L. (1991) The pattern electroretinogram in retinal and optic nerve disease. *Documenta Ophthalmologica*, **77**, 225–35.

Nightingale, S., Mitchel, K.W. and Howe, J.W. (1986) Visual and cortical potentials and pattern electroretinograms in Parkinson's disease and control subjects. *Journal of Neurology, Neurosurgery and Psychiatry*, **49**, 1280–7.

Nguyen-Legros, J. (1988) Functional neuroarchitecture of the retina: hypothesis of the retinal dysfunction of dopaminergic circuity in Parkinson's disease. *Surgical Radiological Anatomy*, **10**, 137–144.

Nguyen-Legros, J., Berger, B., Vigny, A. and Alvarez, C. (1982) Presence of interplexiform dopaminergic neurons in the rat retina. *Brain Research Bulletin*, **9**, 379–81.

Nguyen-Legros, J., Botteri, C., Phuc, L.H. *et al.* (1984) Morphology of primate's dopaminergic amacrine cells as revealed by TH-like immunoreactivity on retinal flat-mounts. *Brain Research*, **295**, 145–53.

Nguyen-Legros, J., Harnois, C., Di Paolo, H. and Simon, A. (1993) The retinal dopamine system in Parkinson's disease. *Clinical Vision Science*, **8**, 1–12.

Onofrj, M. and Bodis-Wollner, I. (1982) Dopaminergic deficiency causes delayed visual evoked potentials in rats. *Annals of Neurology*, **11**, 484–90.

Onofrj, M., Bodis-Wollner, I., Ghilardi, M.F. *et al.* (1986a) Pattern vision in monkeys with parkinsonism: a stimulus-specific effect of MPTP on retinal and cortical processing, in *A Neurotoxin Producing a Parkinsonian Syndrome* (eds S.P. Markey, N. Castagnoli, Jr, A.J. Trevor and I.J. Kopin), pp.683–688.

Onofrj, M., Ghilardi, M.F., Basciani, M. and Gambi, D. (1986b) Visual evoked potentials in parkinsonism and dopamine blockade reveal a stimulus-dependent dopamine function in humans. *Journal of Neurology, Neurosurgery and Psychiatry*, **49**, 1150–9.

Peppe, A., Stanzione, P., Pierelli, F. *et al.* (1992) Low contrast stimuli enhance PERG sensitivity to the visual dysfunction in Parkinson's disease. *Electroencephalography and Clinical Neurophysiology*, **82**, 453–7.

Piccolino, M., Witkowsky, P., Neyton, J. *et al.* (1985) Modulation of gap junction permeability by dopamine and GABA in the network of horizontal cells of the turtle retina, in *Neurocircuitry of the Retina* (eds A. Gallego and P. Gouras), Elsevier, Amsterdam and New York, pp.66–76.

Piccolino, M., De Montis, G., Witkowsky, P. *et al.* (1987) D1 and D2 dopamine receptors involved in the control of electrical transmission between retinal horizontal cells, in *Symposium of Neuroscience* (eds G. Biggio, P.F. Spano, G. Toffano and G.L. Gessa), Living Press–Springer Verlag, Berlin, pp.1–12.

Porciatti, V., Burr, D.C., Morrone, M.C. and Fiorentini, A. (1992) The effects of ageing on the pattern electroretinogram and visual evoked potential in humans. *Vision Research*, **32**, 1199–209.

Price, M.J., Feldman, R.G., Adelberg, D. and Kayne, H. (1992) Abnormalities in color vision and contrast sensitivity in Parkinson's disease. *Neurology*, **42**, 887–90.

Qu, Z., Neff, N.H. and Hadjiconstantinou, M. (1988) MPP+ depletes retinal dopamine and induces D1 receptor supersensitivity. *European Journal of Pharmacology*, **148**, 453–5.

Ratliff, F., Knight, B.W. and Milkman, N. (1970) Superposition of excitatory and inhibitory influences in the retina of Limulus: effect of delayed inhibition. *Proceedings of the National Academy of Sciences of the USA*, **67**, 1558–64.

Reading, H.W. (1983) Dopaminergic receptors in bovine retina and their interaction with thyrotropin-releasing hormone. *Journal of Neurochemistry*, **41**, 1587–95.

Redmond, D.E., Roth, R.H., Elsworth, J.D. *et al.* (1986) Fetal neuronal grafts in monkeys given methylphenyltetrahydropyridine. *Lancet*, **ii**, 1125–7.

Regan, D. and Maxner, C. (1987) Orientation-selective visual loss in patients with Parkinson's disease. *Brain*, **110**, 415–32.

Regan, D. and Neima, D. (1984) Low-contrast letter charts in early diabetic retinopathy, ocular hypertension, glaucoma, and Parkinson's disease. *British Journal of Ophthalmology*, **68**, 885–9.

Robson, J.G. (1966) Spatial and temporal contrast-sensitivity functions of the visual system. *Journal of Ophthalmological Society of America*, **56**, 1141–2.

Schorderet, M. and Nowak, J.Z. (1990) Retinal dopamine D1 and D2 receptors: characterization by binding or pharmacological studies and physiological functions. *Cellular and Molecular Neurobiology*, **10**, 303–25.

Sibley, D.R. and Monsma, F.J. (1992) Molecular biology of dopamine receptors. *Trends in Pharmacological Science*, **13**, 61–9.

Smith, V.C., Pokorny, J. and Pass, A.S. (1985) Color-axis determination on the Farnsworth–Munsell 100-hue test. *American Journal of Ophthalmology*, **100**, 176–82.

Stanzione, P., Pierelli, F., Peppe, A. *et al.* (1989) Pattern visual evoked potentials and electroretinogram in Parkinson's disease: effects of L-dopa therapy. *Vision Research*, **4**, 115–27.

Stanzione, P., Tagliati, M., Silvestrini, M. *et al.* (1991) Haloperidol delays pattern electroretinogram more than VEPs in normal humans. *Clinical Vision Science*, **6**, 137–47.

Stanzione, P., Traversa, R., Pierantozzi, M. *et al.* (1992) An electrophysiological study of D2 dopaminergic actions in human retina: a tool in Parkinson's disease. *Neuroscience Letters*, **136**, 125–8.

Starke, K., Gothert, M. and Kilbinger, H. (1989) Modulation of neurotransmitter release by presynaptic autoreceptors. *Physiological Reviews*, **69**, 864–9.

Stormann, T., Gdula, D., Weiner, D. and Brann, M. (1990) Molecular cloning and expression of a dopamine D2 receptor from human retina. *Molecular Pharmacology*, **37**, 1–6.

Struck, L., Rodnitzky, R. and Dobson, J. (1990). Circadian fluctuations of contrast sensitivity in Parkinson's disease. *Neurology*, **40**, 467–70.

Tagliati, M., Bodis-Wollner, I., Kovanecz, I. and Stanzione, P. (1994) Spatial frequency tuning of the monkey pattern ERG depends on D2 receptor-linked action of dopamine. *Vision Research*, **34**, 2051–7.

Tartaglione, A., Pizio, N., Spadavecchia, L. and Favale, E. (1984) VEP changes in Parkinson's disease are stimulus dependent. *Journal of Neurology, Neurosurgery and Psychiatry*, **47**, 305–7.

Tartaglione, A., Oneto, A., Bandini, F. and Favale, E. (1987) Visual evoked potentials and pattern electroretinograms in Parkinson's disease and control subjects. *Journal of Neurology, Neurosurgery and Psychiatry*, **50**, 1243–4.

Ungerstedt, U. and Arbuthnott, G.W. (1970) Quantitative recording of rotational behavior in rats after 6-hydroxy-dopamine lesions of the nigrostriatal dopamine system. *Brain Research*, **24**, 485–93.

Van Tol, H.H.M., Bunzow, J.R., Guan, H.C. *et al.* (1991) Cloning of the gene for a human dopamine D4 receptor with high affinity for the antipsychotic clozapine. *Nature*, **350**, 610–14.

Weiler, R. and Ball, A.K. (1989) Enkephalinergic

modulation of the dopamine system in the turtle retina. *Visual Neuroscience*, **3**, 455–61.

Weiler, R., Kower, R., Kiesch, M. and Wagner, H. (1988) Glutamate and dopamine modulate synaptic plasticity in horizontal cell dendrites of fish retina. *Neuroscience Letters*, **87**, 205–9.

Weiler, R., Kolbinger, W. and Kohler, K. (1989) Reduced light responsiveness of the cone pathway during prolonged darkness does not result from an increase of dopaminergic activity in the fish retina. *Neuroscience Letters*, **99**, 214–18.

Witkovsky, P. and Schutte, M. (1991) The organization of dopaminergic neurons in vertebrate retinas. *Vision Neuroscience*, **7**, 113–24.

Witkovsky, P., Alones, V. and Piccolino, M. (1987) Morphological changes induced in turtle retinal neurons by exposure to 6-hydroxydopamine and 5,6-dihydroxytryptamine. *Journal of Neurocytology*, **16**, 55–67.

Wong, C.G., Ishibashi, T., Tucker, G. and Hamasaki, D. (1985) Responses of the pigmented rabbit retina to NMPTP, a chemical inducer of Parkinsonism. *Experimental Eye Research*, **40**, 509–19.

Yazulla, S. and Kleinschmidt, J. (1982) Dopamine blocks carrier-mediated release of GABA from retinal horizontal cells. *Brain Research*, **233**, 211–15.

Zarbin, M.A., Wamsley, J.K., Palacios, J.M. and Kuhar, M.J. (1986) Autoradiographic localization of high affinity GABA, benzodiazepine, dopaminergic, adrenergic and muscarinic cholinergic receptors in the rat monkey and human retina. *Brain Research*, **373**, 75–92.

Index